Mathematics
for the Trades

Mathematics for the Trades

A Guided Approach

Ninth Edition

Robert A. Carman

Hal M. Saunders

Prentice Hall

Boston   Columbus   Indianapolis   New York   San Francisco   Upper Saddle River
Amsterdam   Cape Town   Dubai   London   Madrid   Milan   Munich   Paris   Montreal   Toronto
Delhi   Mexico City   Sao Paulo   Sydney   Hong Kong   Seoul   Singapore   Taipei   Tokyo

Editor in Chief: Vernon Anthony
Senior Acquisitions Editor: Gary Bauer
Development Editor: Linda Cupp
Editorial Assistant: Megan Heintz
Director of Marketing: David Gesell
Marketing Manager: Leigh Ann Sims
Marketing Assistant: Les Roberts
Senior Managing Editor: JoEllen Gohr
Project Manager: Christina Taylor
Operations Specialist: Laura Weaver
Senior Art Director: Diane Ernsberger

Text Designer: Mike Fruhbeis
Cover Designer: Delgado and Company
Manager, Rights and Permissions, Image Resource Center: Zina Arabia
Image Permission Coordinator: Vicki Menanteaux
Cover Art: ERproductions Ltd./Blend Images/Jupiter Images
Full-Service Project Management: Kelly Ricci
Composition: Aptara®, Inc.
Printer/Binder: Edwards Brothers Malloy
Cover Printer: Lehigh-Phoenix Color/Hagerstown
Text Font: Times

Credits and acknowledgments borrowed from other sources and reproduced, with permission, in this textbook appear on appropriate page within text.

Library of Congress Cataloging-in-Publication Data

Carman, Robert A.
Mathematics for the trades: a guided approach/Robert A. Carman, Hal M. Saunders.—9th ed.
p. cm.
Includes index.
ISBN-13: 978-0-13-609708-2
ISBN-10: 0-13-609708-1
1. Mathematics.   I. Saunders, Hal M.   II. Title.
QA39.3.C37 2011
513'.14—dc22

2009039455

10 9 8 7 6 5
Prentice Hall
is an imprint of

www.pearsonhighered.com

ISBN 13: 978-0-13-609708-2
ISBN 10:   0-13-609708-1

*Dedicated to Laurie and Eric
with much love
and appreciation.*

Brief Contents

Contents

Preface

Mathematics for the Trades: A Guided Approach provides the practical mathematics skills needed in a wide variety of trade, technical, and other occupational areas, including plumbing, automotive, electrical and construction trades, machine technology, landscaping, HVAC, allied health, and many more. It is especially intended for students who find math challenging and for adults who have been out of school for a time. This text assists students by providing a direct, practical approach that emphasizes careful, complete explanations and actual on-the-job applications. It is intended to provide practical help with real math, beginning at each student's individual level of ability. Careful attention has been given to readability, and reading specialists have helped plan both the written text and the visual organization.

Several special features are designed with the math-challenged student in mind. Each chapter begins with a preview quiz keyed to the textbook which shows concepts that will be covered in that chapter. A summary is provided for each chapter, and each chapter ends with a set of problems to check the student's progress. The book can be used for a traditional course, a course of self-study, or a tutor-assisted study program.

The format is clear and easy to follow. It respects the individual needs of each reader, providing immediate feedback at each step to ensure understanding and continued attention. The emphasis is on *explaining* concepts rather than simply *presenting* them. This is a practical presentation rather than a theoretical one.

A calculator is a necessary tool for workers in trade and technical areas. We have integrated calculators extensively into the text—in finding numerical solutions to problems (including specific keystroke sequences) and in determining the values of transcendental functions. We have taken care to first explain all concepts and problem solving without the use of the calculator and describe how to estimate and check answers. Many realistic problems in the exercise sets involve large numbers, repeated calculations, and large quantities of information and are thus ideally suited to calculator use. They are representative of actual job situations in which a calculator is needed. Detailed instruction on the use of calculators is included in special sections at the end of Chapters 1, 2, and 3 as well as being integrated into the text.

Special attention has been given to on-the-job math skills, by using a wide variety of real problems and situations. Many problems parallel those that appear on professional and apprenticeship exams. The answers to the odd-numbered exercises are given in the back of the book.

New to This Edition

- Approximately 300 new exercises and examples have been added, over 180 of which are applications. Based on research and interviews with people in the trades, the new applications are authentic, topical, and unique.
- Over 150 existing applications have been revised to reflect current practices and updated information.
- The calculator instruction and key sequences have been revised as needed to reflect the way that calculators with multi-line displays operate. With these models, calculations are entered in the same order that they are written and spoken. The majority of students using this text now own these models.
- A Calculator Appendix has been added to provide alternate key sequences for those students using calculators with single-line displays. These models sometimes operate differently from the newer, multi-line models. Students using the single-line models will be alerted to look in this appendix for alternate instructions when a reference number appears to the left of the calculator symbol.
- The step-by-step solutions to worked examples have been revised as needed so that all step explanations are to the left of the mathematics that they describe.
- Two sections of the text, section 2-4 and section 10-3, have been extensively revised. In section 2-4, additional instruction and many new examples have been added to enhance the treatment of addition and subtraction of mixed numbers. Section 10-3 has been totally reorganized so that students will be better able to recognize and distinguish among the different types of right triangle problems. Additionally, generic triangles have been relabeled throughout Chapter 10 to reflect the standard labeling practice of uppercase for vertices and lowercase for sides.
- The following new categories have been added to better describe and distinguish among the applied problems: Flooring and Carpeting, General Interest, General Trades, Life Skills, and Trades Management.
- All examples have been numbered on a section-by-section basis for easy reference.
- Hundreds of miscellaneous improvements have been made, including revised and expanded explanations, more worked solutions, new examples, new and revised boxes such as the Units of Measure box at the end of section 1-1, and the introduction of inequality symbols in section 6-1.
- The answers to the odd-numbered exercises are provided at the back of the text.

Instructor Resources

An extensive package of supplementary materials is available with adoption of this text, including:

- An **Instructor's Solution Manual**
- **TestGen** (computerized test generation software) The test bank for this edition has been expanded. Use TestGen to create tests with alternate versions and print them out or post them to the Web. You can also easily add your own questions to the test bank.

The Instructor's Solutions Manual and TestGen can be downloaded from our **Instructor Resource Center**. To access supplementary materials online, instructors need to request an instructor access code. Go to **www.pearsonhighered.com/irc**, where you can register for an instructor access code. Within 48 hours of registering, you will receive a confirming e-mail including an instructor access code. Once you have received your code, locate your text in the online catalog and click on the Instructor Resources button on the left side of the catalog product page. Select a supplement and a log-in page will appear. Once you have logged in, you can access instructor material for all Pearson textbooks.

NEW!  and **MyMathLab**

MyMathLab® Online Course (access code required) MyMathLab® is a text-specific, easily customizable online course that integrates interactive multimedia instruction with textbook content. MyMathLab gives you the tools you need to deliver all or a portion of your course online, whether your students are in a lab setting or working from home.

- **Interactive homework exercises,** correlated to your textbook at the objective level, are algorithmically generated for unlimited practice and mastery. Most exercises are free-response and provide guided solutions, sample problems, and learning aids for extra help.
- **Personalized Study Plan,** generated when students complete a test or quiz, indicates which topics have been mastered and links to tutorial exercises for topics students have not mastered. Instructors can customize the available topics in the study plan to match their course concepts.
- **Multimedia learning aids,** such as video lectures, animations, and a complete multimedia textbook, help students independently improve their understanding and performance.
- **Assessment Manager** lets you assign media resources (such as a video segment or a textbook passage), homework, quizzes, and tests. If you prefer, you can create your own online homework, quizzes, and tests that are automatically graded. Select just the right mix of questions from the MyMathLab exercise bank, instructor-created custom exercises, and/or TestGen® test items.
- **Gradebook,** designed specifically for mathematics and statistics, automatically tracks students' results, lets you stay on top of student performance, and gives you control over how to calculate final grades. You can also add offline (paper-and-pencil) grades to the gradebook.
- **MathXL Exercise Builder** allows you to create static and algorithmic exercises for your online assignments. You can use the library of sample exercises as an easy starting point, or you can edit any course-related exercise.
- **Pearson Tutor Center** (www.pearsontutorservices.com) access is automatically included with MyMathLab. The Tutor Center is staffed by qualified math instructors who provide textbook-specific tutoring for students via toll-free phone, fax, email, and interactive Web sessions.

The new, Flash®-based MathXL Player is compatible with almost any browser (Firefox®, Safari™, or Internet Explorer®) on almost any platform (Macintosh® or Windows®). MyMathLab is powered by CourseCompass™, Pearson Education's online teaching and learning environment, and by MathXL®, our online homework, tutorial, and assessment system. MyMathLab is available to qualified adopters. For more information, visit www.mymathlab.com or contact your Pearson representative.

MathXL® Online Course (access code required)

MathXL® is a powerful online homework, tutorial, and assessment system that accompanies Pearson Education's textbooks in mathematics or statistics. With MathXL, instructors can:

- Create, edit, and assign online homework and tests using algorithmically generated exercises correlated at the objective level to the textbook.
- Create and assign their own online exercises and import TestGen tests for added flexibility.
- Maintain records of all student work tracked in MathXL's online gradebook.

With MathXL, students can:

- Take chapter tests in MathXL and receive personalized study plans based on their test results.
- Use the study plan to link directly to tutorial exercises for the objectives they need to study and retest.
- Access supplemental animations and video clips directly from selected exercises.

MathXL is available to qualified adopters. For more information, visit our website at www.mathxl.com, or contact your Pearson representative.

MathXL® Tutorials on CD

This interactive tutorial CD-ROM provides algorithmically generated practice exercises that are correlated at the objective level to the exercises in the textbook. Every practice exercise is accompanied by an example and a guided solution designed to involve students in the solution process. Selected exercises may also include a video clip to help students visualize concepts. The software provides helpful feedback for incorrect answers and can generate printed summaries of students' progress.

Acknowledgments

It is a pleasure to acknowledge the help of the many people who have contributed to the development of this book. Lyn Carman spent many hours assisting us, and we greatly appreciate her contributions. The staff of Prentice Hall provided outstanding assistance at every step of the development and production of this ninth edition. We are especially grateful to our executive editor Gary Bauer and development editor Linda Cupp for their guidance and assistance, and to project manager Christina Taylor, copy editor Susan Nodine, and production supervisor Kelly Ricci for their help in moving this edition through the production process.

We would like to give special thanks to Laurie von Melchner for her invaluable contributions to this new edition. Laurie provided numerous applications, and spent countless hours reviewing the text and proofreading copy throughout all stages of production. We would also like to thank Eric Carman for lending his expertise in crafting many of the new applied problems, and for his excellent proofreading.

We would also like to thank the reviewers of this edition: Marylynne Abbott, Ozarks Technical Community College, Missouri; Joseph Bonee, Spencerian College, Kentucky; Elizabeth C. Cunningham, Bellingham Technical College, Washington; Tami Kinkaid, Louisville Technical Institute, Louisville, Kentucky; Francois Nguyen, Saint Paul College, Minnesota; and Rose Martinez, Bellingham Technical College.

We are indebted to the following teachers who read prior versions of the text and offered many helpful suggestions:

Robert Ahntholz, Coordinator—Learning Center, Central City Occupational Center, Los Angeles, California

Peter Arvanites, Rockland Community College, Suffern, New York

Dean P. Athans, East Los Angeles College, Monterey Park, California

Frances L. Brewer, Vance-Granville Community College, Henderson, North Carolina

Gerald Barkley, Glen Oaks Community College, Centreville, Michigan

James W. Brennan, Boise State University, Boise, Idaho

Robert S. Clark, Spokane Community College, Spokane, Washington

James W. Cox, Merced College, Merced, California

Harry Craft, Spencerian College, Lexington, Kentucky

Liz Cunningham, Bellingham Technical College, Bellingham, Washington

Gordon A. DeSpain, San Juan College, Farmington, New Mexico

B. H. Dwiggins, Systems Development Engineer, Technovate, Inc., Tacoma, Washington

Kenneth R. Ebernard, Chabot College, Hayward, California

Hal Ehrenreich, North Central Technical College, Wausau, Wisconsin

Donald Fama, Chairperson, Mathematics-Engineering Science Department, Cayuga Community College, Auburn, New York

James Graham, Industrial Engineering College of Chicago, Chicago, Illinois

Mary Glenn Grimes, Bainbridge College, Bainbridge, Georgia

Ronald J. Gryglas, Tool and Die Institute, Chicago, Illinois

Vincent J. Hawkins, Chairperson, Department of Mathematics, Warwick Public Schools, Warwick, Rhode Island

Bernard Jenkins, Lansing Community College, Lansing, Michigan

Chris Johnson, Spokane Community College, Spokane, Washington

Judy Ann Jones, Madison Area Technical College, Madison, Wisconsin

Robert Kimball, Wake Technical College, Raleigh, North Carolina

J. Tad Martin, Technical College of Alamance, Haw River, North Carolina

Gregory B. McDaniel, Texas State Technical College, Texas

Sharon K. Miller, North Central Technical College, Mansfield, Ohio

David C. Mitchell, Seattle Central Community College, Seattle, Washington

Voya S. Moon, Western Iowa Technical Community College, Sioux City, Iowa

Robert E. Mullaney, Santa Barbara School District, Santa Barbara, California

Joe Mulvey, Salt Lake Community College, Salt Lake City, Utah

Jack D. Murphy, Penn College, Williamsport, Pennsylvania

R. O'Brien, SUNY–Canton, Canton, New York

Steven B. Ottmann, Southeast Community College, Lincoln, Nebraska

Emma M. Owens, Tri-County Technical College, Pendleton, South Carolina

David A. Palkovich, Spokane Community College, Spokane, Washington

William Poehler, Santa Barbara School District, Santa Barbara, California

Richard Powell, Fullerton College, Fullerton, California

Martin Prolo, San Jose City College, San Jose, California

Ilona Ridgeway, Fox Valley Technical College, Appleton, Wisconsin

Kurt Schrampfer, Fox Valley Technical College, Appleton, Wisconsin

Richard C. Spangler, Developmental Instruction Coordinator, Tacoma Community College, Tacoma, Washington

Arthur Theobald, Bergen County Vocational School, Wayne, New Jersey

Curt Vander Vere, Pennsylvania College of Technology, Williamsport, Pennsylvania

Cathy Vollstedt, North Central Technical College, Wausau, Wisconsin

Joseph Weaver, Associate Professor, State University of New York, Delhi, New York

Kay White, Walla Walla Community College, Walla Walla, Washington

Raymond E. Wilhite, Solano Community College, Vacaville, California

Edward Graper, LeBow Co., Goleta, California

This book has benefited greatly from their excellence as teachers.

Finally, through every step of the seemingly endless sequence of researching, interviewing, testing, writing, and rewriting that makes a textbook, we have benefited from the patience, understanding, and concern of our wives, 'Lyn and Chris. They have made it a better book and a more pleasant experience than we would have otherwise had.

Robert A. Carman
Hal M. Saunders

Santa Barbara, California

How to Use
This Book

In this book you will find many questions—not only at the end of each chapter or section, but on every page. This textbook is designed for those who need to learn the practical math used in the trades, and who want it explained carefully and completely at each step. The questions and explanations are designed so that you can:

- Start either at the beginning or where you need to start.
- Work on only what you need to know.
- Move as fast or as slowly as you wish.
- Receive the guidance and explanation you need.
- Skip material that you already understand.
- Do as many practice problems as you need.
- Test yourself often to measure your progress.

In other words, if you find mathematics difficult and you want to be guided carefully through it, this book is designed for you.

This is no ordinary book. You cannot browse through it; you don't read it. You *work* your way through it.

This textbook has been designed for students who will work through it to achieve understanding and practical skills. The alert student will look for and use the helpful features described below.

You will learn to do mathematics problems because you will follow our examples. The signal that a worked example is coming up looks like this:

Example 1

You should respond by following each step and questioning each step. If necessary, seek specific help from your instructor or tutor.

When you are confident that you understand the process, move on to the section labeled

Your Turn

This is your chance to show that you are ready to try a similar problem or two on your own. Use the step-by-step procedure you learned in *Example* to solve the problems.

The *Your Turn* is followed by the answers or worked solutions to the problems in the *Your Turn*.

Solution

Further drill problems are often provided in another set, which is labeled as follows:

More Practice

These problems may be followed by worked solutions or by a list of answers.

Answers

Keep on the lookout for the following helpers:

▶ Note Every experienced teacher knows that certain mathematical concepts and procedures will present special difficulties for students. To help you with these, special notes are included in the text. A banner and a warning word appear in the left margin to indicate the start of the comment, and another triangle ◀ shows when it is completed. The word **Note**, as used at the start of this paragraph, calls your attention to conclusions or consequences that might be overlooked, common mistakes students make, or alternative explanations. ◀

▶ Careful A **Careful** comment points out a common mistake that you might make and shows you how to avoid it. ◀

▶ Learning Help A **Learning Help** gives you an alternative explanation or slightly different way of thinking about and working with the concepts being presented. ◀

▶ A Closer Look This phrase signals a follow-up to an answer or a worked solution. It may provide a more detailed or an alternative look at the whole process. ◀

Examples often include step-by-step explanations that look like this:

Step 1 The solution of each worked example is usually organized in a step-by-step format, similar to this paragraph.

Step 2 In each worked example, explanations for each step are provided alongside the corresponding mathematical operations.

Step 3 Color, ◁ boxed comments , and other graphical aids are used to highlight the important or tricky aspects of a solution, if needed.

 A **check icon** appears in a problem solution to remind you to check your work.

The calculator is an important tool for the modern trades worker or technician. We assume in this textbook that once you have learned the basic operations of arithmetic you will use a calculator. Problems in the exercise sets or examples in the text that involve the use of a scientific calculator are preceded by the calculator symbol shown here.

Solutions often include a display of the proper calculator key sequences. For example, the calculation

$$\frac{85.7 + (12.9)^2}{71.6}$$

would be shown as

85.7 $\boxed{+}$ **12.9** $\boxed{x^2}$ $\boxed{=}$ $\boxed{\div}$ **71.6** $\boxed{=}$ → `3.521089385`

Not every student owns the same brand or model of calculator, and there are often variations in the way different calculators operate. There are four main types of calculators used by students at this level: (1) scientific calculators with a one-line display window; (2) scientific calculators with a two-line display window; (3) scientific calculators with a four-line display window; and (4) graphing calculators, normally with eight lines of display. Types 2, 3, and 4 are very similar in the way they work, and we will often refer to them all as "multi-line" calculators. The display windows on these models show the entire calculation, as well as the answer. In addition, all multi-line calculators allow you to enter calculations exactly as they are written in the textbook or spoken aloud. Single-line scientific calculators display the last entry, the intermediate results, and the final answer, but do not have room to display the entire entry. Furthermore, on most single-line calculators, certain operations are entered in the reverse order of the way they are written or spoken.

Our survey showed that the multi-line scientific calculators, particularly the two-line models, are the most popular at this level. Therefore, all calculator instruction and key sequences in the text are written for the two-line scientific calculators. Those students using four-line scientific calculators or graphing calculators will find that—in most instances—their key sequences will be the same, and that any differences will be noted in the text. For those students using the single-line scientific calculators, we have provided a Calculator Appendix that contains their key sequences whenever they differ from the ones shown in the text. Those using the single-line models will be alerted to look in the appendix by the letter **C** followed by reference number, and this will appear just to the left of the calculator symbol. For example, on page 365 in Chapter 6, there is **C6-4** to the left of the calculator symbol. To find the alternate sequence for single-line models, look in the Calculator Appendix at key sequence 6-4. If there is no such code next to the calculator symbol, then the key sequence shown in the text is valid for the single-line models.

Finally, to save space, our calculator answer displays show only the line with the final result and do not show the line listing the entries.

Exercises At the conclusion of each section of each chapter you will find a set of problems covering the work of that section. These will include a number of routine or drill problems as well as applications or word problems. Each applied problem begins with an indication of the occupational area from which it has been taken.

Summary A chapter summary is included at the end of each chapter. It contains a list of objectives with corresponding reviews and worked examples.

Problem Set Following each chapter summary is a set of problems reviewing all of the material covered in the chapter.

> Important rules, definitions, equations, or helpful hints are often placed in a box like this so that they will be easy to find.

If your approach to learning mathematics is to skim the text lightly on the way to puzzling through a homework assignment, you will have difficulty with this or any other textbook. If you are motivated to study mathematics so that you understand it and can use it correctly, this textbook is designed for you.

According to an old Spanish proverb, the world is an ocean and one who cannot swim will sink to the bottom. A study published by the U.S. Department of Education revealed that two-thirds of the skilled and semiskilled job opportunities in today's labor market are available only to those who have an understanding of the basic principles of arithmetic, algebra, and geometry. If the modern world of work is an ocean, the skill needed to keep afloat or even swim to the top is clearly mathematics. It is the purpose of this book to help you learn these basic skills.

Now, turn to page 1 and let's begin.

R. A. C.
H. M. S.

Mathematics for the Trades

Arithmetic of Whole Numbers

Objective	Sample Problems	For help, go to
When you finish this chapter you will be able to:		
1. Work with whole numbers.	(a) Write 250,374 in words _____	Page 4
	(b) Write in numerical form: "One million, sixty-five thousand, eight" _____	Page 6
	(c) Round 214,659	
	(1) to the nearest ten-thousand _____	Page 7
	(2) to the nearest hundred _____	
2. Add and subtract whole numbers.	(a) 67 + 58 _____	Page 6
	(b) 7009 + 1598 _____	
	(c) 82 − 45 _____	Page 20
	(d) 4035 − 1967 _____	
	(e) 14 + 31 + 59 − 67 + 22 + 37 − 19 _____	
3. Multiply and divide whole numbers.	(a) 64 × 37 _____	Page 28
	(b) 305 × 243 _____	
	(c) 908 × 705 _____	
	(d) 2006 ÷ 6 _____	Page 38
	(e) 7511 ÷ 37 _____	

Name _____

Date _____

Course/Section _____

Objective	Sample Problems	For help, go to
4. Solve word problems with whole numbers.	**Machine Trades** A metal casting weighs 680 lb; 235 lb of metal is removed during shaping. What is its finished weight? _____	
5. Determine factors and prime factors.	(a) List all the factors of 12. _____	Page 43
	(b) Write 12 as a product of its prime factors. _____	Page 44
6. Use the correct order of operations with addition, subtraction, multiplication, and division.	(a) $6 + 9 \times 3$ _____	Page 48
	(b) $35 - 14 \div 7$ _____	
	(c) $56 \div 4 \times 2 + 9 - 4$ _____	
	(d) $(23 - 7) \times 24 \div (12 - 4)$ _____	

(Answers to these preview problems are given in the Appendix. Also, worked solutions to many of these problems appear in the chapter Summary. Don't peek.)

If you are certain that you can work *all* these problems correctly, turn to page 57 for a set of practice problems. If you cannot work one or more of the preview problems, turn to the page indicated to the right of the problem. For those who wish to master this material with the greatest success, turn to Section 1-1 and begin to work there.

Chapter 1

Arithmetic of Whole Numbers

Reprinted with permission of Universal Press Syndicate.

The average person a century ago used numbers to tell time, count, and keep track of money. Today, most people need to develop technical skills based on their ability to read, write, and work with numbers in order to earn a living. Although we live in an age of computers and calculators, much of the simple arithmetic used in industry, business, and the skilled trades is still done mentally or by hand. In fact, most trade and technical areas require you to *prove* that you can do the calculations by hand before you can get a job.

In the first part of this book we take a practical, how-to-do-it look at basic arithmetic: addition, subtraction, multiplication, and division, including fractions, decimal numbers, negative numbers, powers, and roots. Once we are past the basics, we will show you how to use a calculator to do such calculations. There are no quick and easy formulas here, but we do provide a lot of help for people who need to use mathematics in their daily work.

The simplest numbers are the whole numbers—the numbers we use for counting the number of objects in a group. The whole numbers are 0, 1, 2, 3, . . ., and so on.

Example 1

How many letters are in the collection shown in the margin?

We counted 23. Notice that we can count the letters by grouping them into sets of ten:

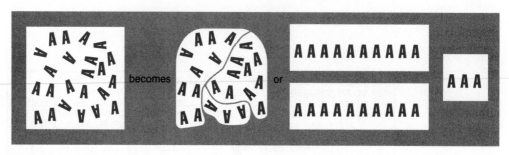

2 tens + 3 ones

20 + 3   or 23

Expanded Form

Mathematicians call this the *expanded form* of a number. For example,

$$46 = \ 40 + 6 \qquad = 4 \text{ tens} + 6 \text{ ones}$$
$$274 = 200 + 70 + 4 = 2 \text{ hundreds} + 7 \text{ tens} + 4 \text{ ones}$$
$$305 = 300 + 5 \qquad = 3 \text{ hundreds} + 0 \text{ tens} + 5 \text{ ones}$$

Only ten numerals or number symbols—0, 1, 2, 3, 4, 5, 6, 7, 8, and 9—are needed to write any number. These ten basic numerals are called the *digits* of the number. The digits 4 and 6 are used to write 46, the number 274 is a three-digit number, and so on.

→ Your Turn

Write out the following three-digit numbers in expanded form:

(a) 362 = ____ + ____ + ____ = ____ hundreds + ____ tens + ____ ones

(b) 425 = ____ + ____ + ____ = ____ hundreds + ____ tens + ____ ones

(c) 208 = ____ + ____ + ____ = ____ hundreds + ____ tens + ____ ones

→ Solutions

(a) *362 = 300 + 60 + 2 = 3* hundreds + *6* tens + *2* ones

(b) *425 = 400 + 20 + 5 = 4* hundreds + *2* tens + *5* ones

(c) *208 = 200 + 0 + 8 = 2* hundreds + *0* tens + *8* ones

A Closer Look

Notice that the 2 in 362 means something different from the 2 in 425 or 208. In 362 the 2 signifies two ones. In 425 the 2 signifies two tens. In 208 the 2 signifies two hundreds. Ours is a *place-value* system of naming numbers: the value of any digit depends on the place where it is located.

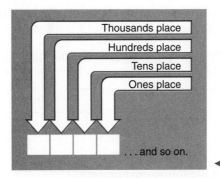

Being able to write a number in expanded form will help you to understand and remember the basic operations of arithmetic—even though you'll never find it on a blueprint or in a technical handbook.

This expanded-form idea is useful especially in naming very large numbers. Any large number given in numerical form may be translated to words by using the following diagram:

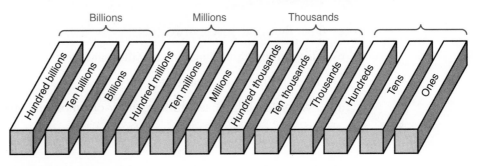

Example 2

The number 14,237 can be placed in the diagram like this:

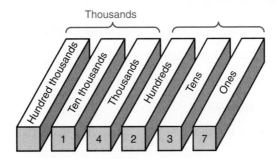

and read "fourteen thousand, two hundred thirty-seven."

Example 3

The number 47,653,290,866 becomes

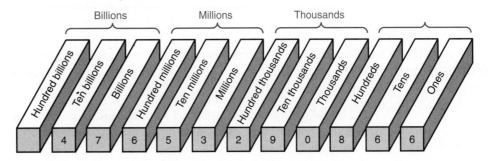

and is read "forty-seven billion, six hundred fifty-three million, two hundred ninety thousand, eight hundred sixty-six."

In each block of three digits read the digits in the normal way ("forty-seven," "six hundred fifty-three") and add the name of the block ("billion," "million"). Notice that the word "and" is not used in naming these numbers.

→ **Your Turn**

Use the diagram to name the following numbers.

(a) 4072

(b) 1,360,105

(c) 3,000,210

(d) 21,010,031,001

→ **Answers**

(a) Four thousand, seventy-two

(b) One million, three hundred sixty thousand, one hundred five

(c) Three million, two hundred ten

(d) Twenty-one billion, ten million, thirty-one thousand, one

---

It is also important to be able to write numbers correctly when you hear them spoken or when they are written in words.

→ **More Practice**

Read each of the following aloud and then write them in correct numerical form.

(a) Fifty-eight thousand, four hundred six

(b) Two hundred seventy-three million, five hundred forty thousand

(c) Seven thousand, sixty

(d) Nine billion, six million, two hundred twenty-three thousand, fifty-eight

→ **Answers**

(a) 58,406

(b) 273,540,000

(c) 7060

(d) 9,006,223,058

Addition of Whole Numbers

Adding whole numbers is fairly easy provided that you have stored in your memory a few simple addition facts. It is most important that you be able to add simple one-digit numbers mentally.

The following sets of problems in one-digit addition are designed to give you some practice. Work quickly. You should be able to answer all problems in a set in the time shown.

Problems   One-Digit Addition

A. Add.

7	5	2	5	8	2	3	8	9	7
3	6	9	7	8	5	6	7	3	6

6	8	9	3	7	2	9	9	7	4
4	5	6	5	7	7	4	9	2	7

9	2	5	8	4	9	6	4	8	8
7	6	5	9	5	5	6	3	2	3

5	6	7	7	5	6	2	3	6	9
8	7	5	9	4	5	8	7	8	8

7	5	9	4	3	8	4	8	5	7
4	9	2	6	8	6	9	4	8	8

Average time = 90 seconds
Record = 35 seconds

B. Add. Try to do all addition mentally.

2	7	3	4	2	6	3	5	9	5
5	3	6	5	7	7	4	7	6	2
4	2	5	8	9	8	4	8	3	8

6	5	4	8	6	9	7	4	8	1
2	4	2	1	8	3	1	9	4	8
7	5	9	9	8	5	6	1	6	7

1	9	3	1	7	2	9	9	8	5
9	9	1	6	9	9	8	5	3	4
2	1	4	3	6	1	2	1	3	7

Average time = 90 seconds
Record = 41 seconds

The answers are given in the Appendix.

Rounding Whole Numbers

In many situations a simplified approximation of a number is more useful than its exact value. For example, the accountant for a business may calculate its total monthly revenue as $247,563, but the owner of the business may find it easier to talk about the revenue as "about $250,000." The process of approximating a number is called *rounding*. Rounding numbers comes in handy when we need to make estimates or do "mental mathematics."

A number can be rounded to any desired place. For example, $247,563 is approximately

$247,560 rounded to the nearest ten,
$247,600 rounded to the nearest hundred,
$248,000 rounded to the nearest thousand,
$250,000 rounded to the nearest ten thousand, and
$200,000 rounded to the nearest hundred thousand.

To round a whole number, follow this **step-by-step** process:

Example 4

Round 247,563	to the nearest hundred thousand	to the nearest ten thousand
Step 1 Determine the place to which the number is to be rounded. Mark it on the right with a $\wedge$.	$2\underset{\wedge}{\,}47,563	$24\underset{\wedge}{\,}7,563
Step 2 If the digit to the right of the mark is less than 5, replace all digits to the right of the mark with zeros.	$200,000	
Step 3 If the digit to the right of the mark is equal to or larger than 5, increase the digit to the left by 1 and replace all digits to the right with zeros.		$250,000

→ **Your Turn**

Try these for practice. Round

(a) 73,856 to the nearest thousand

(b) 64 to the nearest ten

(c) 4852 to the nearest hundred

(d) 350,000 to the nearest hundred thousand

(e) 726 to the nearest hundred

→ **Solutions**

(a) **Step 1** Place a mark to the right of the thousands place. The digit 3 is in the thousands place. $\qquad$ 73$\underset{\wedge}{\,}$856

   Step 2 Does not apply.

   Step 3 The digit to the right of the mark, 8, is larger than 5. Increase the 3 to a 4 and replace all digits to the right with zeros. $\qquad$ 74,000

(b) **Step 1** Place a mark to the right of the tens place. The digit 6 is in the tens place. $\qquad$ 6$\underset{\wedge}{\,}$4

   Step 2 The digit to the right of the mark, 4, is less than 5. Replace it with a zero. $\qquad$ 60

(c) 4900   (d) 400,000   (e) 700

Doing arithmetic with one-digit numbers is very important. It is the key to any mathematical computation—even if you do the work on a calculator. Suppose that you need to find the total time spent on a job by two workers. You need to find the sum

31 hours + 48 hours = _____

Estimating

What is the first step? Start adding digits? Punch in some numbers on your trusty calculator? Rattle your abacus? None of these. The first step is to *estimate* your answer. The most important rule in any mathematical calculation is:

> Know the approximate answer to any calculation before you calculate it.

Never do an arithmetic calculation until you know roughly what the answer is going to be. Always know where you are going.

Rounding to the nearest 10 hours, the preceding sum can be estimated as 31 hours + 48 hours or approximately 30 hours + 50 hours or 80 hours, not 8 or 8000 or 800 hours. Having the estimate will keep you from making any major (and embarrassing) mistakes. Once you have a rough estimate of the answer, you are ready to do the arithmetic work.

Calculate 31 + 48 = _____

You don't really need an air-conditioned, solar-powered, talking calculator for that, do you?

You should set it up like this:

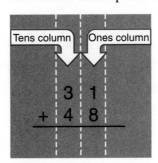

1. The numbers to be added are arranged vertically (up and down) in columns.

2. The right end or ones digits are placed in the ones column, the tens digits are placed in the tens column, and so on.

Avoid the confusion of
$$\begin{array}{r} 31 \\ + 48 \end{array} \quad \text{or} \quad \begin{array}{r} 31 \\ + 48 \end{array}$$

Careful Most often the cause of errors in arithmetic is carelessness, especially in simple tasks such as lining up the digits correctly. ◄

Once the digits are lined up the problem is easy.

$$\begin{array}{r} 31 \\ + 48 \\ \hline 79 \end{array}$$

Does the answer agree with your original estimate? Yes. The estimate, 80, is roughly equal to the actual sum, 79.

What we have just shown you is called the *guess n' check method* of doing mathematics.

1. **Estimate** the answer using rounded numbers.

2. **Work** the problem carefully.

 3. **Check** your answer against the estimate. If they disagree, repeat both steps 1 and 2. The check icon reminds you to check your work.

Most students worry about estimating, either because they won't take the time to do it or because they are afraid they might do it incorrectly. Relax. You are the only one who will know your estimate. Do it in your head, do it quickly, and make it reasonably accurate. Step 3 helps you to find incorrect answers before you finish the problem. The guess n' check method means that you never work in the dark; you always know where you are going.

Note Estimating is especially important in practical math, where a wrong answer is not just a mark on a piece of paper. An error may mean time and money lost. ◄

→ **Your Turn**

Here is a slightly more difficult problem:

27 lb + 58 lb = _____

Try it, then check your answer.

→ **Solution**

First, estimate the answer. 27 + 58 is roughly
$$30 + 60 \text{ or about } 90.$$
The answer is about 90 lb.

Second, line up the digits in columns.

$$\begin{array}{r} 27 \\ + 58 \\ \hline \end{array}$$

The numbers to be added, 27 and 58 in this case, are called *addends*.

Third, add carefully.

$$\begin{array}{r} {}^{1} \\ 27 \\ + 58 \\ \hline 85 \end{array}$$

 Finally, check your answer by comparing it with the estimate. The estimate 90 lb is roughly equal to the answer 85 lb—at least you have an answer in the right ballpark.

What does that little 1 above the tens digit mean? What really happens when you "carry" a digit? Let's look at it in detail. In expanded notation,

$$27 \rightarrow 2 \text{ tens } + 7 \text{ ones}$$
$$+ 58 \rightarrow \underline{5 \text{ tens } + 8 \text{ ones}}$$
$$= 7 \text{ tens } + 15 \text{ ones}$$
$$= 7 \text{ tens } + 1 \text{ ten } + 5 \text{ ones}$$
$$= 8 \text{ tens } \qquad + 5 \text{ ones}$$
$$= 85$$

The 1 that is carried over to the tens column is really a ten.

▶ **Learning Help** Trades people often must calculate exact answers mentally. To do a problem such as 27 + 58, a trick called "balancing" works nicely. Simply add 2 to the 58 to get a "round" number, 60, and subtract 2 from 27 to balance, keeping the total the same. Therefore, 27 + 58 is the same as 25 + 60, which is easy to add mentally to get 85. ◀

We use the same procedure to add three or more numbers. Estimating and checking become even more important when the problem gets more complicated.

Example 5

To add 536 + 1473 + 875 + 88

Estimate: Rounding each number to the nearest hundred,

$$500 + 1500 + 900 + 100 = 3000$$

Step 1 To help avoid careless errors, put the number with the most digits, 1473, on top. Put the number with the fewest digits, 88, on the bottom.

$$
\begin{array}{r}
1473 \\
536 \\
875 \\
+\ \ 88 \\
\hline
\end{array}
$$

Step 2

$$
\begin{array}{r}
2 \\
1473 \\
536 \\
875 \\
+\ \ 88 \\
\hline
2
\end{array}
$$
  $3 + 6 + 5 + 8 = 22$ Write 2; carry 2 tens

Step 3

$$
\begin{array}{r}
22 \\
1473 \\
536 \\
875 \\
+\ \ 88 \\
\hline
72
\end{array}
$$
  $2 + 7 + 3 + 7 + 8 = 27$ Write 7; carry 2 tens

Step 4

$$
\begin{array}{r}
122 \\
1473 \\
536 \\
875 \\
+\ \ 88 \\
\hline
972
\end{array}
$$
  $2 + 4 + 5 + 8 = 19$ Write 9; carry 1 ten

Step 5

$$
\begin{array}{r}
122 \\
1473 \\
536 \\
875 \\
+\ \ 88 \\
\hline
2972
\end{array}
$$
  $1 + 1 = 2$ Write 2.

 Check: The estimate 3000 and the answer 2972 are very close.

→ **More Practice**

The following is a short set of problems. Add, and be sure to estimate your answers first. Check your answers with your estimates.

(a) 429 + 738 = _____   (b) 446 + 867 = _____

(c) 2368 + 744 = _____   (d) 409 + 2572 + 3685 + 94 = _____

(e) **Masonry**   Three bricklayers working together on a job each laid the following number of bricks in a day: 927, 1143, and 1065. How many bricks did all three lay that day?

→ **Solutions**

(a) **Estimate:** Rounding each number to the nearest hundred, $400 + 700 = 1100$
     Line up the digits: 429
                          + 738

Calculate:

Step 1
$$\begin{array}{r} \overset{1}{4}29 \\ +738 \\ \hline 7 \end{array}$$
$9 + 8 = 17$ Write 7; carry 1 ten.

Step 2
$$\begin{array}{r} \overset{1}{4}29 \\ +738 \\ \hline 67 \end{array}$$
$1 + 2 + 3 = 6$ Write 6.

Step 3
$$\begin{array}{r} \overset{1}{4}29 \\ +738 \\ \hline 1167 \end{array}$$
$4 + 7 = 11$ Write 11.

Check: The estimate 1100 and the answer 1167 are roughly equal.

(b) **Estimate:** $400 + 900 = 1300$

Calculate:

Step 1
$$\begin{array}{r} \overset{1}{4}46 \\ +867 \\ \hline 3 \end{array}$$
$6 + 7 = 13$ Write 3; carry 1 ten.

Step 2
$$\begin{array}{r} \overset{1\,1}{4}46 \\ +867 \\ \hline 13 \end{array}$$
$1 + 4 + 6 = 11$ Write 1; carry 1 hundred.

Step 3
$$\begin{array}{r} \overset{1\,1}{4}46 \\ +867 \\ \hline 1313 \end{array}$$
$1 + 4 + 8 = 13$ Write 13.

Check: The estimate 1300 and the answer 1313 are roughly equal.

(c) **Estimate:** Rounding each number to the nearest hundred, $2400 + 700 = 3100$

Calculate:
$$\begin{array}{r} \overset{1\,1\,1}{2}368 \\ +744 \\ \hline 3112 \end{array}$$

Check: The estimate 3100 and the answer 3112 are roughly equal.

(d) **Estimate:** $400 + 2600 + 3700 + 100 = 6800$

Calculate:
$$\begin{array}{r} \overset{1\,2\,2}{2}572 \\ 3685 \\ 409 \\ +\ \ 94 \\ \hline 6760 \end{array}$$

Check: The estimate 6800 and the answer 6760 are roughly equal.

(e) In word problems or applications, addition is often used when there are two or more individual quantities and a total must be found. In this case, we have the amounts for the individual bricklayers, and we must find the total number of bricks.

Estimate: $900 + 1100 + 1100 = 3100$

Calculate:
$$\begin{array}{r} \overset{1\,1\,1}{9}27 \\ 1143 \\ +1065 \\ \hline 3135 \end{array}$$

Check: The estimate 3100 is roughly equal to the answer 3135.

Estimating answers is a very important part of any mathematics calculation, especially for the practical mathematics used in engineering, technology, and the trades. A successful builder, painter, or repairperson must make accurate estimates of job costs—business success depends on it. If you work in a technical trade, getting and keeping your job may depend on your ability to get the correct answer *every* time.

If you use a calculator to do the actual arithmetic, it is even more important to get a careful estimate of the answer first. If you plug a wrong number into the calculator, accidentally hit a wrong key, or unknowingly use a failing battery, the calculator may give you a wrong answer—lightning fast, but wrong. The estimate is your best insurance that a wrong answer will be caught immediately. Convinced?

UNITS OF MEASURE

Units of measure are important in every trade. Just as you want to make sure that a customer is not thinking in centimeters while you are thinking yards, you will want to make sure that that the measurements required for a project have consistent units.

An in-depth study of measurement units is presented in Chapter 5, and we will require only a few basic conversions between units until then. However, because it is nearly impossible to discuss numbers without discussing their units, many of the practical word problems in Chapters 1 to 4 contain units of measure. Therefore, we are providing you with the following table listing the units of measure used in Chapters 1 to 4, along with their most common abbreviations.

Type of Measurement	English Units	Metric Units
Length or distance	inch (in.* or ″) foot (ft, or ′) yard (yd) mile (mi)	millimeter (mm) centimeter (cm) meter (m) kilometer (km)
Weight	ounce (oz) pound (lb) ton (t)	microgram (μg) milligram (mg) gram (g) kilogram (kg)
Area	square inch (sq in.) square foot (sq ft) square yard (sq yd) acre (a)	square centimeter (sq cm) square meter (sq m) square kilometer (sq km) hectare (ha)
Capacity or volume	pint (pt) quart (qt) gallon (gal) bushel (bu) cubic inch (cu in.) cubic feet (cu ft) cubic yard (cu yd)	cubic centimeter (cu cm, cc) milliliter (mL) liter (L) cubic meter (cu m)
Velocity or speed	miles per hour (mph or mi/hr) beats per minute (bpm) cycles per second (hertz) revolutions per minute (rpm or rev/min)	meters per second (m/sec) kilometers per hour (km/hr)

(continued)

Type of Measurement	English Units	Metric Units
Temperature	degrees Fahrenheit (°F)	degrees Celsius (°C)
Power and energy	ohm (Ω) watt (W) volt (V) ampere (A) horsepower (hp)	cubic foot per meter (cfm) kilohertz (kHz) picofarad (pF) kilowatt (kW)
Pressure	pounds per square inch (psi or lb/in.^{2})	pascal (Pa)

*For abbreviations that might be mistaken for a word (e.g., "in" for inches), a period is included at the end of the abbreviation. For abbreviations that would not be mistaken for a word (e.g, "ft"), no period is added.

Now, try the following problems for practice in working with whole numbers.

Exercises 1-1 **Working with Whole Numbers**

A. Add.

1. 47 <u>23</u>	2. 27 <u>38</u>	3. 45 <u>35</u>	4. 38 <u>65</u>	5. 75 <u>48</u>
6. 26 <u>98</u>	7. 48 <u>84</u>	8. 67 <u>69</u>	9. 189 <u>204</u>	10. 508 <u>495</u>
11. 684 <u>706</u>	12. 432 <u>399</u>	13. 621 <u>388</u>	14. 747 <u>59</u>	15. 375 <u>486</u>
16. 4237 <u>1288</u>	17. 5076 <u>4385</u>	18. 7907 <u>1395</u>	19. 3785 <u>7643</u>	20. 6709 <u>9006</u>
21. 18745 <u>6972</u>	22. 40026 <u>7085</u>	23. 10674 <u>397</u>	24. 9876 <u>4835</u>	25. 78044 <u>97684</u>
26. 83754 66283 <u>5984</u>	27. 498321 65466 95873 <u>3604</u>	28. 843592 710662 497381 <u>25738</u>		

B. Arrange vertically and add.

1. 487 + 29 + 526 = _____

2. 715 + 4293 + 184 + 19 = _____

3. 1706 + 387 + 42 + 307 = _____

4. 456 + 978 + 1423 + 3584 = _____

5. 6284 + 28 + 674 + 97 = _____

6. 6842 + 9008 + 57 + 368 = _____

7. 322 + 46 + 5984 = _____

8. 7268 + 209 + 178 = _____

9. 5016 + 423 + 1075 = _____

10. 8764 + 85 + 983 + 19 = _____

11. 4 + 6 + 11 + 7 + 14 + 3 + 9 + 6 + 4 = _____

12. 12 + 7 + 15 + 16 + 21 + 8 + 10 + 5 + 30 + 17 = _____

13. 1 + 2 + 3 + 4 + 5 + 6 + 7 + 8 + 9 + 10 = _____

14. 22 + 31 + 43 + 11 + 9 + 1 + 19 + 12 = _____

15. 75 + 4 + 81 + 12 + 14 + 65 + 47 + 22 + 37 = _____

16. 89,652 + 57,388 + 6506 = _____

17. 443,700 + 629,735 + 85,962 + 6643 = _____

18. 784,396 + 858,390 + 662,043 + 965,831 + 62,654 = _____

C. Writing and Rounding Whole Numbers

Write in words.

1. 357   2. 2304   3. 17,092   4. 207,630   5. 2,000,034

6. 10,007   7. 740,106   8. 5,055,550   9. 118,180,018   10. 6709

Write as numbers.

11. Three thousand, six

12. Seventeen thousand, twenty-four

13. Eleven thousand, one hundred

14. Three million, two thousand, seventeen

15. Four million, forty thousand, six

16. Seven hundred twenty million, ten

Round as indicated.

17. 357 to the nearest ten

18. 4386 to the nearest hundred

19. 4386 to the nearest thousand

20. 5386 to the nearest thousand

21. 225,799 to the nearest ten thousand

22. 225,799 to the nearest thousand

D. Applied Problems

1. **Electrical Trades**   In setting up his latest wiring job, an electrician cut the following lengths of wire: 387, 913, 76, 2640, and 845 ft. Find the total length of wire used.

2. **Construction**   The Acme Lumber Co. made four deliveries of 1-in. by 6-in. flooring: 3280, 2650, 2465, and 2970 fbm. What was the total number of board feet of flooring delivered? (The abbreviation for "board feet" is fbm, which is short for "feet board measure.")

3. **Machine Trades**   The stockroom has eight boxes of No. 10 hexhead cap screws. How many screws of this type are in stock if the boxes contain 346, 275, 84, 128, 325, 98, 260, and 120 screws, respectively?

4. **Trades Management**   In calculating her weekly expenses, a contractor found that she had spent the following amounts: materials, $13,860; labor, $3854; salaried help, $942; overhead expense, $832. What was her total expense for the week?

5. **Trades Management** The head machinist at Tiger Tool Co. is responsible for totaling time cards to determine job costs. She found that five different jobs this week took 78, 428, 143, 96, and 384 minutes each. What was the total time in minutes for the five jobs?

6. **Roofing** On a home construction job, a roofer laid 1480 wood shingles the first day, 1240 the second, 1560 the third, 1320 the fourth, and 1070 the fifth day. How many shingles did he lay in five days?

7. **Industrial Technology** Eight individually powered machines in a small production shop have motors using 420, 260, 875, 340, 558, 564, 280, and 310 watts each. What is the total wattage used when (a) the total shop is in operation? (b) the three largest motors are running? (c) the three smallest motors are running?

8. **Automotive Trades** A mechanic is taking inventory of oil in stock. He has 24 quarts of 10W-30, 8 quarts of 30W, 42 quarts of 20W-50, 16 quarts of 10W-40, and 21 quarts of 20W-40. How many total quarts of oil does he have in stock?

9. **Construction** The Happy Helper building materials supplier has four piles of bricks containing 1250, 865, 742, and 257 bricks. What is the total number of bricks they have on hand?

10. **Machine Trades** A machinist needs the following lengths of 1-in. diameter rod: 8 in., 14 in., 6 in., 27 in., and 42 in. How long a rod is required to supply all five pieces? (Ignore cutting waste.)

11. **Landscaping** A new landscape maintenance business requires the following equipment:

    1 rototiller for $499
    1 gas trimmer for $249
    1 mower for $369
    1 hedge trimmer for $79

    What is the total cost of this equipment?

12. **Electrical Trades** The Radius Electronics Company orders 325 resistors, 162 capacitors, 25 integrated circuit boards, and 68 transistors. Calculate the total number of parts ordered.

13. **Electronics** When resistors are connected in series, the total resistance is the sum of the individual resistors. If resistances of 520, 1160, 49, and 1200 ohms are connected in series, calculate the total resistance.

14. **Electronics** Kirchhoff's law states that the sum of the voltage drops around a closed circuit is equal to the source voltage. $V_s = E_1 + E_2 + E_3 + E_4 + E_5$. Calculate the source voltage V_s for the circuit shown.

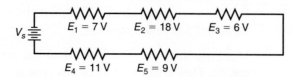

15. **Automotive Trades** To balance an engine, a mechanic must know the total weight of a piston assembly, also known as the reciprocating weight of the assembly. A certain piston assembly consisted of a 485-gram piston, a 74-gram wrist-pin, 51 grams of compression and oil rings, and the small end of the connecting rod weighing 146 grams. What is the reciprocating weight of this assembly?

16. **Allied Health** The following standard amounts of food were recorded in one patient's Intake/Output record: one carton of milk, 8 oz; one carton of juice, 8 oz; one bowl of soup, 150 oz; one cup of coffee, 6 oz; one serving of sherbet, 120 oz; one glass of water, 8 oz. Determine the total fluid intake for this patient.

17. **Construction** For working on remote construction sites, GVM Construction needs a generator that can supply a 1400-W (watt) circular saw, an 1800-W table saw, a 600-W hand drill, and a 100-W radio simultaneously. What total wattage does the generator need to supply?

18. **Automotive Trades** The base price for a 2008 BMW 135i Coupe, including destination charge, is $35,675. The cost of "extras" is given as follows:

Item	Cost	Item	Cost
Premium package	$3300	Premium sound	$875
Satellite radio	$ 595	Heated seats	$500
Comfort access system	$ 500	Metallic paint	$475
iPod and USB adaptor	$ 400	HD radio	$350

What would be the total cost if a customer added premium sound, satellite radio, heated seats, metallic paint, and the iPod and USB adaptor?

19. **Sports and Leisure** In the 1988 Olympics in Seoul, South Korea, Jackie Joyner-Kersee (USA) won the gold medal in the heptathlon with a world-record performance. The heptathlon consists of seven separate events performed over two days, and the points for each event are added to determine the total score. Jackie's points (and performances) for each event were:

Day	Event	Performance	Points earned
Day 1	100-meter hurdles	12.69 sec	1172
	High jump	1.86 m	1054
	Shot-put	15.80 m	915
	200-meter dash	22.56 sec	1123
Day 2	Long jump	7.27 m	1264
	Javelin	45.66 m	776
	800-meter run	2 min 8.51 sec	987

(a) How many points did she earn on Day 1?

(b) How many points did she earn on Day 2?

(c) What was Jackie's world-record-setting point total?

E. Calculator Problems

You probably own a calculator and, of course, you are eager to put it to work doing practical math calculations. In this book we include problem sets for calculator users. These problems are taken from real-life situations and, unlike most textbook problems, involve big numbers and lots of calculations. If you think that having an electronic brain-in-a-box means that you do not need to know basic arithmetic, you will be disappointed. The calculator helps you to work faster, but it will not tell you *what* to do or *how* to do it.

Detailed instruction on using a calculator with whole numbers appears on page 60.

Here are a few helpful hints for calculator users:

1. Always *estimate* your answer before doing a calculation.

2. *Check* your answer by comparing it with the estimate or by the other methods shown in this book. Be certain that your answer makes sense.

3. If you doubt the calculator (they do break down, you know), put a problem in it whose answer you know, preferably a problem like the one you are solving.

1. **Electronics**   An electronics mixing circuit adds two given input frequencies to produce an output signal. If the input frequencies are 35,244 kHz and 61,757 kHz, calculate the frequency of the output signal.

2. **Manufacturing**   The following table lists the number of widget fasteners made by each of the five machines at the Ace Widget Co. during the last ten working days.

| Day | Machine | | | | | Daily Totals |
	A	B	C	D	E	
1	347	402	406	527	237	
2	451	483	312	563	316	
3	406	511	171	581	289	
4	378	413	0	512	291	
5	399	395	452	604	342	
6	421	367	322	535	308	
7	467	409	256	578	264	
8	512	514	117	588	257	
9	302	478	37	581	269	
10	391	490	112	596	310	
Machine Total						

(a) Complete the table by finding the number of fasteners produced each day. Enter these totals under the column "Daily Totals" on the right.

(b) Find the number of fasteners produced by each machine during the ten-day period and enter these totals along the bottom row marked "Machine Totals."

(c) Does the sum of the daily totals equal the sum of the machine totals?

3. Add the following as shown.

(a) $ 67429   (b) $216847   (c) $693884
      6070   9757   675489
      4894   86492   47039
   137427   4875   276921
    91006   386738   44682
      399   28104   560487

(d) $4299 + $137 + $20 + $177 + $63 + $781 + $1008 + $671 = ?

4. **Trades Management** Joe's Air Conditioning Installation Co. has not been successful, and he is wondering if he should sell it and move to a better location. During the first three months of the year his expenses were:

Rent $4260   Utilities $815
Supplies $2540   Advertising $750
Part-time helper $2100 Miscellaneous $187
Transportation $948

His monthly income was:

January $1760
February $2650
March $3325

(a) What was his total expense for the three-month period?

(b) What was his total income for the three-month period?

(c) Now turn your calculator around to learn what Joe should do about this unhappy situation.

5. **Electrical Trades** A mapper is a person employed by an electrical utility company who has the job of reading diagrams of utility installations and listing the materials to be installed or removed by engineers. Part of a typical job list might look like this:

INSTALLATION (in feet of conductor)

Location Code	No. 12 BHD (bare, hard-drawn copper wire)	#TX (triplex)	410 AAC (all-aluminum conductor)	110 ACSR (aluminum-core steel-reinforced conductor)	6B (No. 6, bare conductor)
A3	1740	40	1400		350
A4	1132		5090		2190
B1	500			3794	
B5		87	3995		1400
B6	4132	96	845		
C4		35		3258	2780
C5	3949		1385	1740	705

(a) How many total feet of each kind of conductor must the installer have to complete the job?

(b) How many feet of conductor are to be installed at each of the seven locations?

When you have completed these exercises, check your answers to the odd-numbered problems in the Appendix and then continue with Section 1-2.

1-2 Subtraction of Whole Numbers

Subtraction is the reverse of addition.

Addition: $3 + 4 = \square$

Subtraction: $3 + \square = 7$

Written this way, a subtraction problem asks the question: How much must be added to a given number to produce a required amount?

Most often, however, the numbers in a subtraction problem are written using a minus sign (−):

$17 - 8 = \square$ means that there is a number $\square$ such that $8 + \square = 17$

But we should remember that

$8 + 9 = 17$ or $17 - 8 = 9$ ⇐ Difference

The *difference* is the name given to the answer in a subtraction problem.

Solving simple subtraction problems depends on your knowledge of the addition of one-digit numbers.

For example, to solve the problem

$9 - 4 =$ _____

you probably go through a chain of thoughts something like this:

> Nine minus four. Four added to what number gives nine? Five? Try it: four plus five equals nine. Right.

Subtraction problems with small whole numbers will be easy for you if you know your addition tables.

Example 1

Here is a more difficult subtraction problem:

$47 - 23 =$ _____

The **first** step is to estimate the answer—remember?

$47 - 23$ is roughly $50 - 20$ or $30.$

The difference, your answer, will be about 30—not 3 or 10 or 300.

The **second** step is to write the numbers vertically as you did with addition. Be careful to keep the ones digits in line in one column, the tens digits in a second column, and so on.

$$\begin{array}{r} 4\ 7 \\ -2\ 3 \end{array}$$ Notice that the larger number is written above the smaller number.

Once the numbers have been arranged in this way, the difference may be written by performing the following two steps:

Step 1

$$\begin{array}{r} 4\ \boxed{7} \\ -2\ \boxed{3} \\ \hline \boxed{4} \end{array}$$ ones digits: $7 - 3 = 4$

Step 2

$$\begin{array}{r} \boxed{4}\ 7 \\ -\boxed{2}\ 3 \\ \hline \boxed{2}\ 4 \end{array}$$ tens digits: $4 - 2 = 2$

 The difference is 24, which agrees roughly with our estimate.

Example 2

With some problems it is necessary to rewrite the larger number before the problem can be solved. Let's try this one:

$64 - 37 =$ _____

First, estimate the answer. Rounding to the nearest ten, $64 - 37$ is roughly $60 - 40$ or 20.

Second, arrange the numbers vertically in columns.
$$\begin{array}{r} 64 \\ -37 \\ \hline \end{array}$$

Because 7 is larger than 4 we must "borrow" one ten from the 6 tens in 64. We are actually rewriting 64 (6 tens + 4 ones) as 5 tens + 14 ones. In actual practice our work would look like this:

Step 1
$$\begin{array}{r} 6\ 4 \\ -3\ 7 \\ \hline \end{array}$$

Step 2
$$\begin{array}{r} ^{5}\!\!\not6\ ^{14}\!\!\not4 \\ -3\ 7 \\ \hline 7 \end{array}$$ ⟵ Borrow one ten, change the 6 in the tens place to 5, change 4 to 14, subtract $14 - 7 = 7$.

Step 3
$$\begin{array}{r} ^{5}\!\!\not6\ ^{14}\!\!\not4 \\ -3\ 7 \\ \hline 2\ 7 \end{array}$$

$\boxed{5 - 3 = 2}$ $\boxed{14 - 7}$

 Double-check subtraction problems by adding the answer and the smaller number; their sum should equal the larger number.

Step 4   Check:
$$\begin{array}{r} 37 \\ +27 \\ \hline 64 \end{array}$$

> **Learning Help**   If you need to get an exact answer to a problem such as $64 - 37$ mentally, add or subtract to make the smaller number, 37, a "round" number. In this case, add 3 to make it 40. Because we're subtracting, we want the *difference,* not the *total,* to be the same or balance. Therefore, we also add 3 to the 64 to get 67. The problem becomes $64 - 37 = (64 + 3) - (37 + 3) = 67 - 40$. Subtracting a round number is easy mentally: $67 - 40 = 27$. ◄

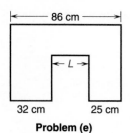

Problem (e)

→ **Your Turn**

Try these problems for practice.

(a)   71   (b)   263   (c)   426   (d)   902
   -39 -127 -128 -465

(e)   Find the missing dimension L in the drawing in the margin.

→ **Solutions**

(a)   **Estimate:** $70 - 40 = 30$

Step 1	**Step 2**
	6 11
7 1	$\not7$ $\not1$
-3 9	-3 9
	3 2

Borrow one ten from 70,
change the 7 in the tens place to 6,
change the 1 in the ones place to 11.

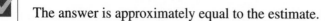

$11 - 9 = 2$ Write 2

$6 - 3 = 3$ Write 3

✓ The answer 32 is approximately equal to the estimate 30. As a shortcut, mentally add 1 to each number,

$71 - 39 = 72 - 40$

then subtract.

$72 - 40 = 32$

(b)   **Estimate:** $260 - 130 = 130$.

Step 1	**Step 2**
	5 13
2 6 3	2 $\not6$ $\not3$
-1 2 7	-1 2 7
	1 3 6

Borrow one ten from 60,
change the 6 in the tens place to 5,
change the 3 in the ones place to 13.

$13 - 7 = 6$ Write 6

$5 - 2 = 3$ Write 3

$2 - 1 = 1$ Write 1

✓ The answer is approximately equal to the estimate.

(c)   **Estimate:** $400 - 100 = 300$

Step 1	**Step 2**	**Step 3**
	1 16	3 11 16
4 2 6	4 $\not2$ $\not6$	$\not4$ $\not2$ $\not6$
-1 2 8	-1 2 8	-1 2 8
	8	2 9 8

In this case we borrow twice. Borrow one ten from the 20 in 426 and make 16. Then borrow one hundred from the 400 in 426 to make 11 in the tens place.

$16 - 8 = 8$ Write 8

$11 - 2 = 9$ Write 9

$3 - 1 = 2$ Write 2

The answer 298 is approximately equal to the estimate 300.

(d) **Estimate:** $900 - 500 = 400$

Step 1	Step 2	Step 3
	8 10	8 9 12
9 0 2	$\cancel{9}$ $\cancel{0}$ 2	$\cancel{9}$ $\cancel{0}$ $\cancel{2}$
−4 6 5	−4 6 5	−4 6 5
		4 3 7 ⟵ $12 - 5 = 7$ Write 7

$9 - 6 = 3$ Write 3

$8 - 4 = 4$ Write 4

 The answer 437 is roughly equal to the estimate 400.

In problem (d) we first borrow one hundred from 900 to get a 10 in the tens place. Then we borrow one 10 from the tens place to get a 12 in the ones place.

(e) In word problems or applications, we use subtraction whenever we have a total and one of the quantities adding up to that total is missing.

In this drawing, we see that the three smaller horizontal dimensions must add up to the 86-cm length along the top of the drawing. Therefore, the missing dimension is

$$L = 86 \text{ cm} - 32 \text{ cm} - 25 \text{ cm}$$

Estimate: $80 - 30 - 20 = 30$

Step 1	Step 2
	4 14
8 6	$\cancel{8}$ $\cancel{6}$
−3 2	−2 5
5 4	2 9

The missing dimension L is 29 cm. This is very close to our estimate of 30 cm.

Example 3

Let's work through a few examples of subtraction problems involving zero digits.

(a) $400 - 167 = ?$

Step 1	Step 2	Step 3	Check
	3 10	3 9 10	
4 0 0	$\cancel{4}$ $\cancel{0}$ $\cancel{0}$	$\cancel{4}$ $\cancel{0}$ $\cancel{0}$	1 6 7
−1 6 7	−1 6 7	−1 6 7	+2 3 3
		2 3 3	4 0 0

Do you see in steps 2 and 3 that we have rewritten
400 as $300 + 90 + 10$?

(b) $5006 - 2487 = ?$

Step 1	Step 2	Step 3	Step 4	Check
	4 10	4 9 10	4 9 9 16	
5 0 0 6	$\cancel{5}$ $\cancel{0}$ 0 6	$\cancel{5}$ $\cancel{0}$ $\cancel{0}$ 6	$\cancel{5}$ $\cancel{0}$ $\cancel{0}$ $\cancel{6}$	2 4 8 7
−2 4 8 7	−2 4 8 7	−2 4 8 7	−2 4 8 7	+2 5 1 9
			2 5 1 9	5 0 0 6

Here is an example involving repeated borrowing.

(c) $24632 - 5718 = ?$

Step 1	Step 2	Step 3	Step 4	Check

Step 1
```
  2 4 6 3 2
−     5 7 1 8
```

Step 2
```
          2 12
  2 4 6 3̶ 2̶
−     5 7 1 8
            1 4
```

Step 3
```
      3 16 2 12
  2 4̶ 6̶ 3̶ 2̶
−     5 7 1 8
        9 1 4
```

Step 4
```
    1 13 16 2 12
  2̶ 4̶ 6̶ 3̶ 2̶
−     5 7 1 8
    1 8 9 1 4
```

Check
```
      5 7 1 8
+ 1 8 9 1 4
  2 4 6 3 2
```

Any subtraction problem that involves borrowing should always be checked in this way. It is very easy to make a mistake in this process.

→ **More Practice**

Subtract.

(a) 85
     28

(b) 500
     234

(c) 7008
     3605

(d) 37206
     4738

→ **Answers**

(a) 57   (b) 266   (c) 3403   (d) 32,468

Now check your progress on subtraction in Exercises 1-2.

Exercises 1-2 Subtraction of Whole Numbers

A. Subtract.

1. 13 7	2. 12 5	3. 8 6	4. 8 0	5. 11 7	6. 16 7
7. 10 7	8. 5 5	9. 12 9	10. 11 8	11. 10 2	12. 14 6
13. 12 3	14. 15 6	15. 9 0	16. 14 5	17. 9 6	18. 11 5
19. 15 7	20. 12 7	21. 13 6	22. 18 0	23. 16 9	24. 12 4
25. 0 0	26. 13 8	27. 17 9	28. 18 9	29. 14 8	30. 16 8
31. 15 9	32. 17 8	33. 14 9	34. 15 8	35. 13 9	36. 12 8

B. Subtract.

1. 40 27	2. 78 49	3. 51 39	4. 36 17	5. 42 27	6. 52 16
7. 65 27	8. 46 17	9. 84 38	10. 70 48	11. 34 9	12. 56 18
13. 546 357	14. 409 324	15. 476 195	16. 330 76	17. 504 96	18. 747 593

19. 400	20. 803	21. 632	22. 438	23. 6218	24. 6084
127	88	58	409	3409	386

25. 13042	26. 57022	27. 5007	28. 10000	29. 48093	30. 27004
524	980	266	386	500	4582

C. Applied Problems

1. **Painting**   In planning for a particular job, a painter buys $486 worth of materials. When the job is completed, she returns some unused rollers and brushes for a credit of $27. What was the net amount of her bill?

2. **Construction**   How many square feet (sq ft) of plywood remain from an original supply of 8000 sq ft after 5647 sq ft is used?

3. **Welding**   A storage rack at the Tiger Tool Company contains 3540 ft of 1-in. stock. On a certain job 1782 ft is used. How much is left?

4. **Welding**   Five pieces measuring 26, 47, 38, 27, and 32 cm are cut from a steel bar that was 200 cm long. Allowing for a total of 1 cm for waste in cutting, what is the length of the piece remaining?

5. **Trades Management**   Taxes on a group of factory buildings owned by the Ace Manufacturing Company amounted to $875,977 eight years ago. Taxes on the same buildings last year amounted to $1,206,512. Find the increase in taxes.

6. **Trades Management**   To pay their bills, the owners of Edwards Plumbing Company made the following withdrawals from their bank account: $72, $375, $84, $617, and $18. If the original balance was $5820, what was the amount of the new balance?

7. **Manufacturing**   Which total volume is greater, four drums containing 72, 45, 39, and 86 liters, or three drums containing 97, 115, and 74 liters? By how much is it greater?

8. **Machine Trades**   Determine the missing dimension (L) in the following drawings. (In the figures, feet are abbreviated with the ($'$) symbol, and inches are abbreviated with the ($''$) symbol.)

(a)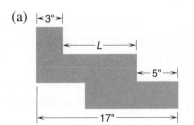

(b)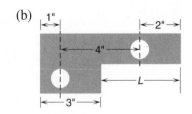

(c)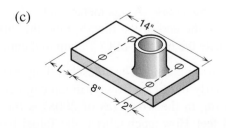

(d)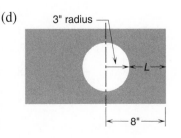

9. **Automotive Trades** A service department began the day with 238 gallons of coolant. During the day 64 gallons were used. How many gallons remained at the end of the day?

10. **Construction** A truck loaded with rocks weighs 14,260 lb. If the truck weighs 8420 lb, how much do the rocks weigh?

11. **Printing** A press operator has a total of 22,000 impressions to run for a job. If the operator runs 14,250 the first day, how many are left to run?

12. **Plumbing** In the following plumbing diagram, find pipe lengths A and B.

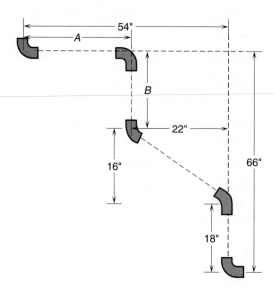

Problem 12

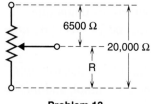

Problem 13

13. **Electronics** A *potentiometer* is a device that acts as a variable resistor, allowing a resistance to change from 0 ohms to some maximum value. If a 20,000-ohm potentiometer is set to 6500 ohms (Ω) as shown, calculate the resistance R.

14. **Allied Health** The white blood cell (WBC) count is an important indicator of health. Before surgery, the WBC count of a patient was 9472. After surgery, his WBC count had dropped to 5786. Calculate the difference in WBC count for this patient.

15. **Electronics** An electronic mixer can produce a signal with frequency equal to the difference between two input signals. If the input signals have frequency 1,350,000 and 850,000 hertz, calculate the difference frequency.

16. **Automotive Trades** A set of tires was rated at 40,000 miles. A car's odometer read 53,216 when the tires were installed and 91,625 when they needed replacing.

    (a) How long did they actually last?

    (b) By how many miles did they miss the advertised rating?

17. **Office Services** A gas meter read 8701 at the beginning of the month and 8823 at the end of the month. Find the difference between these readings to calculate the number of CCF (hundred cubic feet) used.

18. **General Interest** The tallest building in the United States is the Willis Tower (formerly the Sears Tower) in Chicago at 1451 feet. The tallest completed building in the world (as of 2008) is the Taipei 101 in Taipei, Taiwan, at 1671 feet. How much taller is the Taipei 101 than the Willis Tower?

D. Calculator Problems

1. **Plumbing**   The Karroll Plumbing Co. has 10 trucks and, for the month of April, the following mileage was recorded on each.

Truck No.	Mileage at Start	Mileage at End
1	58352	60027
2	42135	43302
3	76270	78007
4	40006	41322
5	08642	10002
6	35401	35700
7	79002	80101
8	39987	40122
9	10210	11671
10	71040	73121

Find the mileage traveled by each truck during the month of April and the total mileage of all vehicles.

2. Which sum is greater?

987654321		123456789
87654321		123456780
7654321		123456700
654321		123456000
54321	or	123450000
4321		123400000
321		123000000
21		120000000
1		100000000

3. **Trades Management**   If an electrician's helper earns $28,245 per year and she pays $3814 in withholding taxes, what is her take-home pay?

4. **Trades Management**   The revenue of the Smith Construction Company for the year is $3,837,672 and the total expenses are $3,420,867. Find the difference, Smith's profit, for that year.

5. **Life Skills**   Balance the following checking account record.

Date	Deposits	Withdrawals	Balance
7/1			$6375
7/3		$ 379	
7/4	$1683		
7/7	$ 474		
7/10	$ 487		
7/11		$2373	
7/15		$1990	
7/18		$ 308	
7/22		$1090	
7/26		$ 814	
8/1			A

(a) Find the new balance A.

(b) Keep a running balance by filling each blank in the balance column.

6. **Construction** A water meter installed by the BetterBilt Construction Company at a work site read 9357 cubic feet on June 1 and 17,824 cubic feet on July 1. How much water did they use during the month of June?

When you have completed these exercises, check your answers to the odd-numbered problems in the Appendix, then turn to Section 1-3 to study the multiplication of whole numbers.

1-3 Multiplication of Whole Numbers

In a certain football game, the West Newton Waterbugs scored five touchdowns at six points each. How many total points did they score through touchdowns? We can answer the question several ways:

1. Count points, ...

2. Add touchdowns, $6 + 6 + 6 + 6 + 6 = ?$

or

3. Multiply $5 \times 6 = ?$

We're not sure about the mathematical ability of the West Newton scorekeeper, but most people would multiply. Multiplication is a shortcut method of performing repeated addition.

How many points did they score?

Product

$$5 \times 6 = 30$$

Factors

In a multiplication problem the *product* is the name given to the result of the multiplication. The numbers being multiplied are the *factors* of the product.

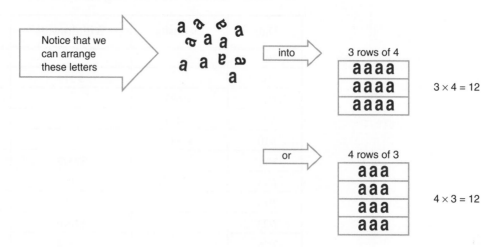

Changing the order of the factors does not change their product. This is called the *commutative* property of multiplication.

To become skillful at multiplication, you must know the one-digit multiplication table from memory. Even if you use a calculator for your work, you need to know the one-digit multiplication table to make estimates and to check your work.

→ **Your Turn**

Complete the table below by multiplying the number at the top by the number at the side and placing their product in the proper square. We have multiplied $3 \times 4 = 12$ and $2 \times 5 = 10$ for you.

Multiply	2	5	8	1	3	6	9	7	4
1									
7									
5	10								
4					12				
9									
2									
6									
3									
8									

→ **Answers**

Here is the completed multiplication table:

Multiplication Table

×	2	5	8	1	3	6	9	7	4
1	2	5	8	1	3	6	9	7	4
7	14	35	56	7	21	42	63	49	28
5	10	25	40	5	15	30	45	35	20
4	8	20	32	4	12	24	36	28	16
9	18	45	72	9	27	54	81	63	36
2	4	10	16	2	6	12	18	14	8
6	12	30	48	6	18	36	54	42	24
3	6	15	24	3	9	18	27	21	12
8	16	40	64	8	24	48	72	56	32

Multiplying by Zero and One

Notice that the product of any number and 1 is that same number. For example,

$1 \times 2 = 2$

$1 \times 6 = 6$

or even

$1 \times 753 = 753$

Zero has been omitted from the table because the product of any number and zero is zero. For example,

$$0 \times 2 = 0$$

$$0 \times 7 = 0$$

$$395 \times 0 = 0$$

Problems   One-Digit Multiplication

Multiply as shown. Work quickly; you should be able to answer all problems in a set correctly in the time indicated. (These times are for community college students enrolled in a developmental math course.)

A. Multiply.

6	4	9	6	3	9	7	8	2	8
2	8	7	6	4	2	0	3	7	1

6	8	5	5	2	3	9	7	3	1
8	2	9	6	5	3	8	5	6	4

7	5	4	7	4	8	6	9	8	6
4	3	9	7	2	5	7	6	8	4

5	3	5	9	9	6	1	8	4	7
4	0	5	3	9	1	1	6	4	9

Average time = 100 sec
Record = 37 sec

B. Multiply.

2	6	3	5	6	4	4	8	2	7
8	5	3	7	3	5	7	6	6	9

8	0	2	3	1	5	6	9	5	8
4	6	9	8	9	5	4	5	2	9

3	7	5	6	9	2	7	8	9	2
5	7	8	9	4	4	6	8	0	2

5	9	1	8	6	4	9	0	2	7
5	3	7	7	6	3	9	4	1	8

Average time = 100 sec
Record = 36 sec

Check your answers in the Appendix.

If you are not able to perform these one-digit multiplications quickly from memory, you should practice until you can do so. A multiplication table is given in the Appendix on Page 805. Use it if you need it.

Multiplying by Larger Numbers

The multiplication of larger numbers is based on the one-digit number multiplication table.

Example 1

Consider the problem:

$$34 \times 2 = \underline{\hspace{2cm}}$$

First, estimate the answer: $30 \times 2 = 60$. The actual product of the multiplication will be about 60.

Second, arrange the factors to be multiplied vertically, with ones digits in a single column, tens digits in a second column, and so on.

Finally, to make the process clear, let's write it in expanded form.

$$\begin{array}{r} 34 \\ \times 2 \end{array}$$ $$\begin{array}{r} 3 \text{ tens } + 4 \text{ ones} \\ \times 2 \\ \hline 6 \text{ tens } + 8 \text{ ones } = 60 + 8 = 68 \end{array}$$

✓ The guess 60 is roughly equal to the answer 68.

→ Your Turn

Now write the following multiplication in expanded form.

$$\begin{array}{r} 28 \\ \times 3 \end{array}$$

→ Solution

Estimate: $30 \times 3 = 90$ The answer is about 90.

$$\begin{array}{r} 28 \\ \times 3 \end{array}$$ ⇨ $$\begin{array}{r} 2 \text{ tens } + 8 \text{ ones} \\ \times 3 \\ \hline 6 \text{ tens } + 24 \text{ ones} \end{array}$$
$$\begin{aligned} &= 6 \text{ tens } + 2 \text{ tens } + 4 \text{ ones} \\ &= 8 \text{ tens } + 4 \text{ ones} \\ &= 80 + 4 \\ &= 84 \end{aligned}$$

 90 is roughly equal to 84.

Of course, we do not normally use the expanded form; instead we simplify the work like this:

$$\begin{array}{r} \overset{2}{2} \ 8 \\ \times \quad 3 \\ \hline 8 \ 4 \end{array}$$
$3 \times 8 = 24$ Write 4 and carry 2 tens.
$3 \times 2 \text{ tens} = 6 \text{ tens}$ $6 \text{ tens} + 2 \text{ tens} = 8 \text{ tens}$
Write 8.

→ More Practice

Now try these problems to be certain you understand the process. Multiply as shown.

(a) $\begin{array}{r} 43 \\ \times 5 \end{array}$ (b) $\begin{array}{r} 73 \\ \times 4 \end{array}$ (c) $\begin{array}{r} 29 \\ \times 6 \end{array}$ (d) $\begin{array}{r} 258 \\ \times 7 \end{array}$

(a) **Estimate:** $40 \times 5 = 200$ The answer is roughly 200.

$$\begin{array}{r} \overset{1}{4}\ 3 \\ \times\ \ \ \ \ 5 \\ \hline 2\ 1\ 5 \end{array}$$

$5 \times 3 = 15$ Write 5; carry 1 ten.
5×4 tens $= 20$ tens

20 tens $+ 1$ ten $= 21$ tens

✔ The answer 215 is roughly equal to the estimate 200.

(b) **Estimate:** $70 \times 4 = 280$

$$\begin{array}{r} \overset{1}{7}\ 3 \\ \times\ \ \ \ \ 4 \\ \hline 2\ 9\ 2 \end{array}$$

$4 \times 3 = 12$ Write 2; carry 1 ten.
4×7 tens $= 28$ tens

28 tens $+ 1$ ten $= 29$ tens

✔ The answer 292 is roughly equal to the estimate 280.

(c) **Estimate:** $30 \times 6 = 180$

$$\begin{array}{r} \overset{5}{2}\ 9 \\ \times\ \ \ \ \ 6 \\ \hline 1\ 7\ 4 \end{array}$$

$6 \times 9 = 54$ Write 4; carry 5 tens.
6×2 tens $= 12$ tens

12 tens $+ 5$ tens $= 17$ tens

✔ The answer 174 is roughly equal to the estimate 180.

(d) **Estimate:** $300 \times 7 = 2100$

$$\begin{array}{r} \overset{4}{2}\ \overset{5}{5}\ 8 \\ \times\ \ \ \ \ \ \ \ 7 \\ \hline 1\ 8\ 0\ 6 \end{array}$$

$7 \times 8 = 56$ Write 6; carry 5 tens.
7×5 tens $= 35$ tens
35 tens $+ 5$ tens $= 40$ tens   Write 0; carry 4 hundreds.
7×2 hundreds $= 14$ hundreds
14 hundreds $+ 4$ hundreds $= 18$ hundreds

✔ The answer and estimate are roughly equal.

> **Learning Help** Multiplying a two- or three-digit number by a one-digit number can be done mentally. For example, think of 43×5 as 40×5 plus 3×5, or $200 + 15 = 215$. For a problem such as 29×6, think 30×6 minus 1×6 or $180 - 6 = 174$. Tricks like this are very useful on the job when neither paper and pencil nor a calculator is at hand. ◀

MULTIPLICATION SHORTCUTS

There are hundreds of quick ways to multiply various numbers. Most of them are quick only if you are already a math whiz. If you are not, the shortcuts will confuse more than help you. Here are a few that are easy to do and easy to remember.

(continued)

1. To multiply by 10, attach a zero to the right end of the first factor. For example,

    $34 \times 10 = 340$

    $256 \times 10 = 2560$

    Multiplying by 100 or 1000 is similar.

    $34 \times 100 = 3400$

    $256 \times 1000 = 256000$

2. To multiply by a number ending in zeros, carry the zeros forward to the answer. For example,

    $$\begin{array}{r} 26 \\ \times\,20 \\ \hline \end{array} \Rightarrow \begin{array}{r} 26 \\ \times\,20 \\ \hline 520 \end{array}$$
    Multiply 26×2 and attach the zero on the right. The product is 520.

    $$\begin{array}{r} 34 \\ \times\,2100 \\ \hline \end{array} \Rightarrow \begin{array}{r} 34 \\ \times\,2100 \\ \hline 34 \\ 68 \\ \hline 71400 \end{array}$$

3. If both factors end in zeros, bring all zeros forward to the answer.

    $$\begin{array}{r} 230 \\ \times\,200 \\ \hline \end{array} \Rightarrow \begin{array}{r} 230 \\ 200 \\ \hline 46000 \end{array}$$
    Attach three zeros to the product of 23×2.

    $$\begin{array}{r} 1000 \\ \times\ \ 100 \\ \hline 100{,}000 \end{array}$$
    This kind of multiplication is mostly a matter of counting zeros.

Two-Digit Multiplications

Calculations involving two-digit multipliers are done in a similar way.

Example 2

To multiply

$$\begin{array}{r} 89 \\ \times\,24 \\ \hline \end{array}$$

First, estimate the answer. Rounding each number to the nearest ten, $90 \times 20 = 1800$.

Second, multiply by the ones digit 4.

$$\begin{array}{r} \overset{3}{} \\ 89 \\ \times\,24 \\ \hline 356 \end{array}$$

$4 \times 9 = 36$ Write 6; carry 3 tens.
4×8 tens $= 32$ tens
32 tens $+ 3$ tens $= 35$ tens

Third, multiply by the tens digit 2.

$$\begin{array}{r} \overset{1}{} \\ \overset{3}{} \\ 89 \\ \times\,24 \\ \hline 356 \\ 178 \leftarrow \end{array}$$

$2 \times 9 = 18$ Write 8; carry 1.
$2 \times 8 = 16$
$16 + 1 = 17$
Leave a blank space here, because we are actually multiplying $89 \times 20 = 1780$.

Fourth, add the products obtained.

$$\begin{array}{r} 356 \\ \underline{178} \\ 2136 \end{array}$$

 Finally, check it. The estimate and the answer are roughly the same, or at least in the same ballpark.

Notice that the product in the third step, 178, is written one digit space over from the product from the second step. When we multiplied 2×9 to get 18 in Step 3, we were actually multiplying $20 \times 9 = 180$, but the zero in 180 is usually omitted to save time.

→ **Your Turn**

Try these:

(a) $\quad\begin{array}{r} 64 \\ \times\ 37 \\ \hline \end{array}$ (b) $\quad\begin{array}{r} 327 \\ \times\ 145 \\ \hline \end{array}$ (c) $\quad\begin{array}{r} 342 \\ \times\ 102 \\ \hline \end{array}$

(d) **Office Services** The Good Value hardware store sold 168 toolboxes at $47 each. What was their total income on this item?

→ **Solutions**

(a) **Estimate:** $60 \times 40 = 2400$

$$\begin{array}{r} 64 \\ \times\ 37 \\ \hline 448 \\ 192 \\ \hline 2368 \end{array}$$

$\begin{cases} 7 \times 4 = 28 \\ 7 \times 6 = 42 \\ 3 \times 4 = 12 \\ 3 \times 6 = 18 \end{cases}$ $\quad$ Write 8; carry 2.
Add carry 2 to get 44; write 44.
Write 2; carry 1.
Add carry 1 to get 19; write 19.

Add to obtain the answer.

(b) **Estimate:** $300 \times 150 = 45{,}000$

$$\begin{array}{r} 327 \\ \times\ 145 \\ \hline 1635 \\ 1308 \\ 327 \\ \hline 47415 \end{array}$$

$\begin{cases} 5 \times 7 = 35 \\ 5 \times 2 = 10 \\ 5 \times 3 = 15 \\ 4 \times 7 = 28 \\ 4 \times 2 = 8 \\ 4 \times 3 = 12 \\ 1 \times 327 = 327 \end{cases}$ $\quad$ Write 5; carry 3.
Add carry 3 to get 13; write 3, carry 1.
Add carry 1 to get 16; write 16.
Write 8, carry 2.
Add carry 2 to get 10; write 0, carry 1.
Add carry 1 to get 13; write 13.

The product is 47,415.

(c) **Estimate:** $300 \times 100 = 30{,}000$

$$\begin{array}{r} 342 \\ \times\ 102 \\ \hline 684 \\ 000 \\ 342 \\ \hline 34884 \end{array}$$

$2 \times 342 = 684$
$0 \times 342 = 000$
$1 \times 342 = 342$

The product is 34,884.

(d) When working with word problems or applications, we use multiplication whenever we have a given number of items all sharing in some characteristic. It may be possible to multiply to find the total. In this problem, the number of items is 168 toolboxes. They all share a common cost of $47. To find the total cost, multiply 168 by 47.

$$
\begin{array}{r}
168 \\
\times\ 47 \\
\hline
1176 \\
672 \\
\hline
7896
\end{array}
$$

The total income is $7896.

Note Be very careful when there are zeros in the multiplier; it is very easy to misplace one of those zeros. Do not skip any steps, and be sure to estimate your answer first. ◄

THE SEXY SIX

Here are the six most often missed one-digit multiplications:

Inside
digits

$9 \times 8 = 72$
$9 \times 7 = 63$
$9 \times 6 = 54$
$8 \times 7 = 56$
$8 \times 6 = 48$
$7 \times 6 = 42$

It may help you to notice that in these multiplications the "inside" digits, such as 8 and 7, are consecutive and the digits of the answer add to nine: $7 + 2 = 9$. This is true for *all* one-digit numbers multiplied by 9.

Be certain that you have these memorized.

(There is nothing very sexy about them, but we did get your attention, didn't we?)

Go to Exercises 1-3 for a set of practice problems on the multiplication of whole numbers.

Exercises 1-3 Multiplication of Whole Numbers

A. Multiply.

1. 7 6	2. 7 8	3. 6 8	4. 8 9	5. 9 7	6. 29 3
7. 9 6	8. 4 7	9. 5 9	10. 12 7	11. 37 8	12. 24 7
13. 72 8	14. 47 9	15. 64 5	16. 39 4	17. 58 5	18. 94 6
19. 32 13	20. 46 14	21. 72 11	22. 68 16	23. 54 26	24. 17 9
25. 47 6	26. 77 4	27. 48 15	28. 64 27	29. 90 56	30. 86 83
31. 34 57	32. 66 25	33. 59 76	34. 29 32	35. 78 49	36. 94 95

B. Multiply.

1. 305   2. 2006   3. 8043   4. 809   5. 3706
    123   125   37   47   102

6. 708   7. 684   8. 2043   9. 2008   10. 563
     58   45   670   198   107

11. 809   12. 609   13. 500   14. 542   15. 7009
      9   7   50   600   504

16. 407   17. 316   18. 514   19. 807   20. 560
     22   32   62   111   203

C. Practical Problems

1. **Plumbing**   A plumber receives $75 per hour. How much is she paid for 40 hours of work?

2. **Electrical Trades**   What is the total length of wire on 14 spools if each spool contains 150 ft?

3. **Construction**   How many total linear feet of redwood are there in 65 2-in. by 4-in. boards each 20 ft long?

4. **Automotive Trades**   An auto body shop does 17 paint jobs at a special price of $859 each. How much money does the shop receive from these jobs?

5. **Office Services**   Three different-sized boxes of envelopes contain 50, 100, and 500 envelopes, respectively. How many envelopes total are there in 18 boxes of the first size, 16 of the second size, and 11 of the third size?

6. **Machine Trades**   A machinist needs 25 lengths of steel each 9 in. long. What is the total length of steel that he needs? No allowance is required for cutting.

7. **Machine Trades**   The Ace Machine Company advertises that one of its machinists can produce 2 parts per hour. How many such parts can 27 machinists produce if they work 45 hours each?

8. **Construction**   What is the horizontal distance in inches covered by 12 stair steps if each is 11 in. wide? (See the figure.)

9. **Automotive Trades**   Five bolts are needed to mount a tire. How many bolts are needed to mount all four tires on 60 cars?

10. **Electrical Trades**   A voltmeter has a 4000 ohms per volt rating. This means that, for example, on the 3-volt scale, the meter reads 3 $\times$ 4000 or 12,000 ohms resistance. Find the resistance on the 50-volt range.

11. **Printing**   A printer finds that nine 6-in. by 8-in. cards can be cut from a 19-in. by 25-in. sheet. How many such cards can be cut from 850 of the sheets?

12. **Industrial Technology**   The Hold Tite fastener plant produces 738,000 cotter pins per shift. If it operates two shifts per day, five days each week, how many cotter pins will the plant produce in six weeks?

13. **Machine Trades**   If a machine produces 16 screws per minute, how many screws will it produce in 24 hours?

14. **Electronics**   For a transistor amplifier circuit, the collector current is calculated by multiplying the gain by the base current. For a circuit with a gain of 72, and base current 210 microamps, calculate the collector current.

15. **Roofing**   In the Happy Homes development cedar shingles are laid to expose 5 in. per course. If 23 courses of shingles are needed to cover a surface, how long is the surface?

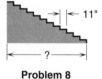

11"

?

Problem 8

16. **Electronics**   A 78-ohm resistor R draws 3 amps of current i in a given circuit. Calculate the voltage drop E across the resistor by multiplying resistance by current.

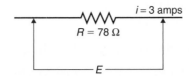

17. **Automotive Trades**   The maximum resistance for a new spark plug wire is 850 ohms per inch. A 25-in. wire had a resistance of 21,500 ohms. Does this fall within the acceptable limit?

18. **Automotive Trades**   According to the EPA rating, the average gas mileage for a certain car is expected to be 24 miles per gallon. How many total miles can be expected from a 16-gallon tank?

19. **Agriculture**   At KneeHi Farms, they expect a yield of 170 bushels of corn per acre. How much corn should they plan for if they plant 220 acres?

20. **HVAC**   Home ventilation standards require a controlled ventilation rate of 15 cubic feet per minute per person. If a house is occupied by four people, what total ventilation rate is required?

21. **Electrical Trades**   If a solar energy system consists of 96 solar panels, and each panel puts out 5 A (amperes) of current, how many total amperes does the system generate?

22. **Allied Health**   Pulse rate is generally expressed as beats per minute (bpm). To save time, health practitioners often make a quick check of pulse rate by counting the number of beats in 15 seconds, and then multiplying this number by 4 to determine beats per minute. If a nurse counts 23 pulse beats in 15 seconds, what is the patient's pulse rate in bpm?

23. **Allied Health**   During a blood donation drive in Goleta, California, 176 people donated blood. If each person donated the maximum recommended amount of 500 milliliters (mL), how much blood was collected?

D. Calculator Problems

1. **Business and Finance**   If an investment company started in business with $1,000,000 of capital on January 1 and, because of mismanagement, lost an average of $873 every day for a year, what would be its financial situation on the last day of the year?

2. **Life Skills**   Which of the following pay schemes gives you the most money over a one-year period?

   (a) $100 per day

   (b) $700 per week

   (c) $400 for the first month and a $400 raise each month

   (d) 1 cent for the first two-week pay period, 2 cents for the second two-week period, 4 cents for the third two-week period, and so on, the pay doubling each two weeks.

3. Multiply.

   (a) $12{,}345{,}679 \times 9 =$ (b) $15{,}873 \times 7 =$
        $12{,}345{,}679 \times 18 =$ $15{,}873 \times 14 =$
        $12{,}345{,}679 \times 27 =$ $15{,}873 \times 21 =$

(c) $1 \times 1 =$
$11 \times 11 =$
$111 \times 111 =$
$1111 \times 1111 =$
$11111 \times 11111 =$

(d) $6 \times 7 =$
$66 \times 67 =$
$666 \times 667 =$
$6666 \times 6667 =$
$66666 \times 66667 =$

Can you see the pattern in each of these?

4. **Machine Trades**   In the Aztec Machine Shop there are nine lathes each weighing 2285 lb, five milling machines each weighing 2570 lb, and three drill presses each weighing 395 lb. What is the total weight of these machines?

5. **Manufacturing**   The Omega Calculator Company makes five models of electronic calculators. The following table gives the hourly production output. Find the weekly (five eight-hour days) production costs for each model.

Model	Alpha	Beta	Gamma	Delta	Tau
Cost of production of each model	$6	$17	$32	$49	$78
Number produced during typical hour	117	67	29	37	18

Check your answers to the odd-numbered problems in the Appendix.

1-4   Division of Whole Numbers

Division is the reverse of multiplication. It enables us to separate a given quantity into equal parts. The mathematical phrase $12 \div 3$ is read "twelve divided by three," and it asks us to separate a collection of 12 objects into 3 equal parts. The mathematical phrases

$$12 \div 3 \qquad 3\overline{)12} \qquad \frac{12}{3} \qquad \text{and} \qquad 12/3$$

all represent division, and they are all read "twelve divided by three."

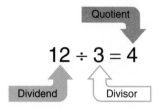

In this division problem, 12, the number being divided, is called the *dividend;* 3, the number used to divide, is called the *divisor;* and 4, the result of the division, is called the *quotient,* from a Latin word meaning "how many times."

One way to perform division is to reverse the multiplication process.

$24 \div 4 = \square$ means that $4 \times \square = 24$

If the one-digit multiplication tables are firmly in your memory, you will recognize immediately that $\square = 6$.

Try these.

35 ÷ 7 = _____	30 ÷ 5 = _____	18 ÷ 3 = _____
28 ÷ 4 = _____	42 ÷ 6 = _____	63 ÷ 9 = _____
45 ÷ 5 = _____	56 ÷ 7 = _____	72 ÷ 8 = _____
70 ÷ 10 = _____		

→ **Answers**

35 ÷ 7 = 5	30 ÷ 5 = 6	18 ÷ 3 = 6
28 ÷ 4 = 7	42 ÷ 6 = 7	63 ÷ 9 = 7
45 ÷ 5 = 9	56 ÷ 7 = 8	72 ÷ 8 = 9
70 ÷ 10 = 7		

You should be able to do all these quickly by working backward from the one-digit multiplication tables.

Careful

Division by zero has no meaning in arithmetic. (Any number) $\div$ 0 or $\dfrac{\text{(any number)}}{0}$ is not defined. But $0 \div \text{(any number)} = \dfrac{0}{\text{(any number)}} = 0$ ◄

How do we divide numbers that are larger than 9×9 and therefore not in the multiplication table? Obviously, we need a better procedure.

Example 1

Here is a step-by-step explanation of the division of whole numbers:

Divide 96 ÷ 8 = _____.

First, *estimate* the answer: 100 ÷ 10 is about 10. The quotient or answer will be about 10.

Second, arrange the numbers in this way:

$$8\overline{)96}$$

Notice that the order in which the numbers are written is reversed. With the $\div$ symbol, the divisor (8) is on the right. With the $\overline{)}$ symbol, the divisor is on the left.

Third, divide using the following step-by-step procedure.

Step 1 $8\overline{)\,96}^{\,1}$ 8 into 9? Once. Write 1 in the answer space above the 9.

Step 2 $8\overline{)\,96}^{\,1}$ Multiply $8 \times 1 = 8$ and write the product 8 under the 9.
$\phantom{8\overline{)\,9}}8$

Step 3 $8\overline{)\,96}^{\,1}$ Subtract $9 - 8 = 1$ and write 1. Bring down the next digit, 6.
$\phantom{8\overline{)\,9}}\underline{-8\downarrow}$
$\phantom{8\overline{)\,9}}16$

Step 4 $\overset{12}{8)\,96}$ 8 into 16? Twice. Write 2 in the answer space above the 6.
$\quad\underline{-8}$
$\quad\ 16$

Step 5 $\overset{12}{8)\,96}$ Multiply $8 \times 2 = 16$ and write the product 16 under the 16.
$\quad\underline{-8}$
$\quad\ 16$
$\quad\underline{-16}$

Step 6 $\overset{12}{8)\,96}$ Subtract $16 - 16 = 0$. Write 0.
$\quad\underline{-8}$
$\quad\ 16$
$\quad\underline{-16}$
$\quad\quad\ 0$ ← The remainder is zero. 8 divides into 96 exactly 12 times.

 Finally, *check* your answer. The answer 12 is roughly equal to the original estimate 10. As a second check, multiply the divisor 8 and the quotient 12. Their product should be the original dividend number, $8 \times 12 = 96$, which is correct.

→ **Your Turn**

Practice this step-by-step division process by finding $112 \div 7$.

→ **Solution**

Estimate: $112 \div 7$ is roughly $100 \div 5$ or 20. The answer will be roughly 20.

$\overset{016}{7)\,112}$
$\ \underline{-\ 7\downarrow}$
$\quad\ 42$
$\ \underline{-\ 42}$
$\quad\quad 0$

Step 1 7 into 1. Won't go, so put a zero in the answer place.

Step 2 7 into 11. Once. Write 1 in the answer space as shown.

Step 3 $7 \times 1 = 7$. Write 7 below 11.

Step 4 $11 - 7 = 4$. Write 4. Bring down the 2.

Step 5 7 into 42. Six. Write 6 in the answer space.

Step 6 $7 \times 6 = 42$. Subtract 42. The remainder is zero.

 The answer 16 is roughly equal to our original estimate of 20. Check again by multiplying. $7 \times 16 = 112$. Finding a quick estimate of the answer will help you avoid making mistakes.

Careful Notice that once the first digit of the answer has been obtained, there will be an answer digit for every digit of the dividend.

$$16$$
$$\Uparrow\Downarrow\Downarrow$$
$$7)\overline{112}\ ◄$$

Example 2

$203 \div 7 = ?$

Estimate: $203 \div 7$ is roughly $210 \div 7$ or 30. The answer will be roughly 30.

$$\begin{array}{r} 029 \\ 7{\overline{\smash{)}\,203}} \\ -14\downarrow \\ \hline 63 \\ -\ 63 \\ \hline 0 \end{array}$$

Step 1 7 into 2. Won't go, so put a zero in the answer space.

Step 2 7 into 20. Twice. Write 2 in the answer space as shown.

Step 3 $7 \times 2 = 14$. Write 14 below 20.

Step 4 $20 - 14 = 6$. Write 6. Bring down the 3.

Step 5 7 into 63. Nine times. Subtract 63. The remainder is zero.

 The answer is 29, roughly equal to the original estimate of 30. Double-check by multiplying: $7 \times 29 = 203$.

The remainder is not always zero, of course. This is shown in the next example.

Example 3

To calculate $153 \div 4$,

$$\begin{array}{r} 38 \qquad \text{Quotient} \\ 4{\overline{\smash{)}\,153}} \qquad \text{Dividend} \\ -12 \\ \hline 33 \\ -32 \\ \hline 1 \qquad \text{Remainder} \end{array}$$

Write the answer as $38r1$ to indicate that the quotient is 38 and the remainder is 1.

 To check an answer with a remainder, first multiply the quotient by the divisor, then add the remainder. In this example,

$38 \times 4 = 152$

$152 + 1 = 153$ which checks.

→ Your Turn

Ready for some guided practice? Try these problems.

(a) $3174 \div 6$ (b) $206 \div 6$ (c) $59 \div 8$

(d) $5084 \div 31$ (e) $341 \div 43$

(f) **Plumbing**   Pete the Plumber billed \$12,852 for the labor on a large project. If his regular rate is \$63 per hour, how many hours did he work?

→ Solutions

(a) $3174 \div 6 = $ _____

Estimate: $6 \times 500 = 3000$ The answer will be roughly 500.

$$\begin{array}{r} 0529 \\ 6{\overline{\smash{)}\,3174}} \\ -30 \\ \hline 17 \\ -\ 12 \\ \hline 54 \\ -\ 54 \\ \hline 0 \end{array}$$

Step 1   6 into 3? No. Write a zero. 6 into 31? 5 times. Write 5 above the 1.

Step 2 $6 \times 5 = 30$. Subtract $31 - 30 = 1$.

Step 3   Bring down 7. 6 into 17? Twice. Write 2 in the answer.

Step 4 $6 \times 2 = 12$. Subtract $17 - 12 = 5$.

Step 5   Bring down 4. 6 into 54? 9 times. Write 9 in the answer.

Step 6 $6 \times 9 = 54$. Subtract $54 - 54 = 0$.

Quotient $= 529$

The answer is 529.

☑ The estimate, 500, and the answer, 529, are roughly equal.

Double-check: $6 \times 529 = 3174$.

(b) $206 \div 6 = $ _____

   Estimate: $6 \times 30 = 180$ The answer is about 30.

$$\begin{array}{r} 034 \\ 6\overline{)\ 206} \\ -18 \\ \hline 26 \\ -\ 24 \\ \hline 2 \end{array}$$

Step 1   6 into 2? No. Write a zero. 6 into 20? Three times. Write 3 above the zero in the answer.

Step 2 $6 \times 3 = 18$. Subtract $20 - 18 = 2$.

Step 3   Bring down 6. 6 into 26? 4 times. Write 4 in the answer.

Step 4 $6 \times 4 = 24$. Subtract $26 - 24 = 2$. The remainder is 2.

Quotient $= 34$

The answer is 34 with a remainder of 2.

☑ The estimate 30 and the answer are roughly equal.

Double-check: $6 \times 34 = 204$ $204 + 2 = 206$.

(c) $59 \div 8 = $ _____

   Estimate: $7 \times 8 = 56$. The answer will be about 7.

$$\begin{array}{r} 7 \\ 8\overline{)\ 59} \\ -56 \\ \hline 3 \end{array}$$

8 into 5? No. There is no need to write the zero.

The quotient is 7 with a remainder of 3.

☑ Check it, and double-check by multiplying and adding the remainder.

(d) $5084 \div 31 = $ _____

   Estimate: This is roughly about the same as $5000 \div 30$ or $500 \div 3$ or about 200. The quotient will be about 200.

$$\begin{array}{r} 164 \\ 31\overline{)\ 5084} \\ -31 \\ \hline 198 \\ -186 \\ \hline 124 \\ -\ 124 \\ \hline 0 \end{array}$$

Step 1   31 into 5: No. 3 i to 50? Yes, once. Write 1 above the zero.

Step 2 $31 \times 1 = 31$. Subtract $50 - 31 = 19$.

Step 3   Bring down 8. 31 into 198? (That is, about the same as 3 into 19.) Yes, 6 times; write 6 in the answer.

Step 4 $31 \times 6 = 186$. Subtract $198 - 186 = 12$.

Step 5   Bring down 4. 31 into 124? (That is, about the same as 3 into 12.) Yes, 4 times. Write 4 in the answer.

Step 6 $31 \times 4 = 124$. Subtract $124 - 124 = 0$.

The quotient is 164.

☑ The estimate is reasonably close to the answer.

Double-check: $31 \times 164 = 5084$.

Notice that in step 3 it is not at all obvious how many times 31 will go into 198. Again, you must make an educated guess and check your guess as you go along.

(e) $341 \div 43 =$ _____

Estimate: This is roughly $300 \div 40$ or $30 \div 4$ or about 7. The answer will be about 7 or 8.

$$\begin{array}{r} 7 \\ 43\overline{)\,341} \\ -301 \\ \hline 40 \end{array}$$

Your first guess would probably be that 43 goes into 341 8 times (try 4 into 34), but $43 \times 8 = 344$, which is larger than 341. The quotient is 7 with a remainder of 40.

 Check it.

(f) To determine the total bill, Pete multiplied his hourly rate by the number of hours. Therefore, if we know the total and the hourly rate, we reverse this process and divide the total by the hourly rate to find the number of hours.

Estimate: $13,000 \div 60$ or $1300 \div 6$ is about 200.

$$\begin{array}{r} 204 \\ 63\overline{)\,12852} \\ 126 \\ \hline 25 \\ 0 \\ \hline 252 \\ 252 \\ \hline 0 \end{array}$$

On the second step, notice that 63 does not go into 25. So we must write a zero in the tens place of the answer, above the 5.

Then multiply 0 by 63 to get 0, which we write below the 25.

Subtract and bring down the 2. 63 into 252 is 4.

The quotient is 204. Pete worked a total of 204 hours on the project.

 The answer 204 and the estimate 200 are roughly equal.

Double-check: $63 \times 204 = 12,852$.

Factors

It is sometimes useful in mathematics to be able to write any whole number as a product of other numbers. If we write

$6 = 2 \times 3$ 2 and 3 are called **factors** of 6.

Of course, we could also write

$6 = 1 \times 6$ and see that 1 and 6 are also factors of 6.

The factors of 6 are 1, 2, 3, and 6.

The factors of 12 are 1, 2, 3, 4, 6, and 12.

The factors of 30 are 1, 2, 3, 5, 6, 10, 15, and 30.

Any number is exactly divisible by its factors; that is, every factor divides the number with zero remainder.

→ **Your Turn**

List all the factors of each of the following whole numbers.

| 1. 4 | 2. 10 | 3. 24 | 4. 18 | 5. 14 |
| 6. 32 | 7. 20 | 8. 26 | 9. 44 | 10. 90 |

Primes

For some numbers the only factors are 1 and the number itself. For example, the factors of 7 are 1 and 7 because $7 = 1 \times 7$. There are no other numbers that divide 7 with remainder zero. Such numbers are known as prime numbers. A **prime number** is one for which there are no factors other than 1 and the prime itself.

Here is a list of the first few prime numbers.

2, 3, 5, 7, 11, 13, 17, 19, 23, 29, 31, 37, 41, 43, 47

(Notice that 1 is not listed. All prime numbers have two different factors: 1 and the number itself. The number 1 has only 1 factor—itself.)

Prime Factors

The **prime factors** of a number are those factors that are prime numbers. For example, the prime factors of 6 are 2 and 3. The prime factors of 30 are 2, 3, and 5. You will see later that the concept of prime factors is useful when working with fractions.

To find the prime factors of a number a **factor-tree** is often helpful.

Example 4

Find the prime factors of 132.

First, write the number 132 and draw two branches below it.

Second, beginning with the smallest prime in our list, 2, test for divisibility. $132 = 2 \times 66$. Write the factors below the branches.

Next, repeat this procedure on the nonprime factor, 66, so that $66 = 2 \times 33$.

Finally, continue dividing until all branches end with a prime.

Therefore, $132 = 2 \times 2 \times 3 \times 11$.

Check by multiplying.

→ **Your Turn**

Write each whole number as a product of its prime factors. If the given number is a prime, label it with a letter P.

1. 9
2. 11
3. 15
4. 3
5. 17
6. 21
7. 23
8. 29
9. 52
10. 350

Exercises 1-4 provide some practice in the division of whole numbers.

SOME DIVISION "TRICKS OF THE TRADE"

1. If a number is exactly divisible by 2, that is, divisible with zero remainder, it will end in an even digit, 0, 2, 4, 6, or 8.

   Example: We know that 374 is exactly divisible by 2 because it ends in the even digit 4.

   $$\begin{array}{r} 187 \\ 2\overline{)374} \end{array}$$

2. If a number is exactly divisible by 3, the sum of its digits is exactly divisible by 3.

   Example: The number 2784 has the sum of digits $2 + 7 + 8 + 4 = 21$. Because 21 is exactly divisible by 3, we know that 2784 is also exactly divisible by 3.

   $$\begin{array}{r} 928 \\ 3\overline{)2784} \end{array}$$

3. A number is exactly divisible by 4 if its last two digits are exactly divisible by 4 or are both zero.

   Example: We know that the number 3716 is exactly divisible by 4 because 16 is exactly divisible by 4.

   $$\begin{array}{r} 929 \\ 4\overline{)3716} \end{array}$$

4. A number is exactly divisible by 5 if its last digit is either 5 or 0.

   Example: The numbers 875, 310, and 33,195 are all exactly divisible by 5.

5. A number is exactly divisible by 8 if its last three digits are divisible by 8.

   Example: The number 35,120 is exactly divisible by 8 because the number 120 is exactly divisible by 8.

6. A number is exactly divisible by 9 if the sum of its digits is exactly divisible by 9.

   Example: The number 434,673 has the sum of digits 27. Because 27 is exactly divisible by 9, we know that 434,673 is also exactly divisible by 9.

   $$\begin{array}{r} 48297 \\ 9\overline{)434673} \end{array}$$

7. A number is exactly divisible by 10 if it ends with a zero.

   Example: The numbers 230, 7380, and 100,200 are all exactly divisible by 10.

Can you combine the tests for divisibility by 2 and 3 to get a rule for divisibility by 6?

A. Divide.

1. $63 \div 7$ 2. $92 \div 8$ 3. $72 \div 0$

4. $37 \div 5$ 5. $71 \div 7$ 6. $6 \div 6$

7. $\dfrac{32}{4}$ 8. $\dfrac{28}{7}$ 9. $\dfrac{54}{9}$

10. $245 \div 7$ 11. $167 \div 7$ 12. $228 \div 4$

13. $310 \div 6$ 14. $3310 \div 3$ 15. $\dfrac{147}{7}$

16. $7\overline{)364}$ 17. $6\overline{)222}$ 18. $4\overline{)201}$

19. $322 \div 14$ 20. $382 \div 19$ 21. $936 \div 24$

22. $700 \div 28$ 23. $730 \div 81$ 24. $\dfrac{901}{17}$

25. $31\overline{)682}$ 26. $27\overline{)1724}$ 27. $42\overline{)371}$

B. Divide.

1. $61\overline{)7320}$ 2. $33\overline{)303}$ 3. $16\overline{)904}$

4. $2001 \div 21$ 5. $2016 \div 21$ 6. $1000 \div 7$

7. $2000 \div 9$ 8. $2400 \div 75$ 9. $14\overline{)4275}$

10. $71\overline{)6005}$ 11. $53\overline{)6307}$ 12. $3\overline{)9003}$

13. $7\overline{)3507}$ 14. $6\overline{)48009}$ 15. $6\overline{)3624}$

16. $3\overline{)62160}$ 17. $15\overline{)3000}$ 18. $67\overline{)3354}$

19. $24\overline{)2596}$ 20. $47\overline{)94425}$ 21. $38\overline{)22800}$

22. $231\overline{)14091}$ 23. $411\overline{)42020}$ 24. $603\overline{)48843}$

25. $111\overline{)11111}$ 26. $102\overline{)2004}$ 27. $405\overline{)7008}$

C. For each of the following whole numbers, (a) list all its factors, and (b) write it as a product of its prime factors.

1. 6   2. 16   3. 19   4. 27   5. 40   6. 48

D. Applied Problems

1. **Machine Trades** A machinist has a piece of bar stock 243 in. long. If she must cut nine equal pieces, how long will each piece be? (Assume no waste to get a first approximation.)

2. **Carpentry** How many braces 32 in. long can be cut from a piece of lumber 192 in. long?

3. **Construction** If subflooring is laid at the rate of 85 sq ft per hour, how many hours will be needed to lay 1105 sq ft?

4. **Construction** The illustration shows a stringer for a short flight of stairs. Determine the missing dimensions H and W.

5. **Construction** How many joists spaced 16 in. o.c. (on center) are required for a floor 432 in. long? (Add one joist for a starter.)

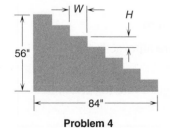

Problem 4

6. **Roofing** If rafters are placed 24 in. o.c., how many are required for both sides of a gable roof that is 576 in. long? Add one rafter on each side for a starter.

7. **Carpentry** A stairway in a house is to be built with 18 risers. If the distance from the top of the first floor to the top of the second floor is 126 in., what is the height of each step?

8. **Manufacturing** What is the average horsepower of six engines having the following horsepower ratings: 385, 426, 278, 434, 323, and 392? (*Hint:* To find the average, add the six numbers and divide their sum by 6.)

9. **Trades Management** Mr. Martinez, owner of the Maya Restaurant, hired a plumber who worked for 18 hours installing some new equipment. The total bill for wages and supplies used was $4696. If the cost for materials was $3400, calculate the plumber's hourly wage.

10. **Transportation** A delivery truck traveling at an average rate of 54 miles per hour must cover a distance of 486 miles in one run. Allowing 2 hours for refueling and meals, how long will it take to complete the trip?

11. **Automotive Trades** A mechanic needs to order 480 spark plugs. If there are ten plugs in a box, how many boxes does he need?

12. **Metalworking** One cubic foot of a certain alloy weighs 375 lb. How many cubic feet does a 4875-lb block occupy?

13. **Printing** A printer needs 13,500 sheets of paper for a job. If there are 500 sheets in a ream, how many reams does the printer need to get from the stockroom?

14. **Wastewater Technology** A water tank contains 18,000 gallons. If 500 gallons per day are used from this tank, how long will it last?

15. **HVAC** A geothermal heat pump uses loops of piping under a lawn to heat or cool a house. If a house requires 18 kW (kilowatts) of heating capacity and each loop has a capacity of 3 kW, how many loops of piping must be installed?

16. **Allied Health** A home health care worker distributes one packet of multivitamins each week to each of his elderly patients, for a total of 52 packets per patient over an entire year. Last year he distributed a total of 1404 multivitamin packets. How many patients did he have?

E. Calculator Problems

1. Calculate the following.

(a) $\dfrac{1347 \times 46819}{3}$

(b) $\dfrac{76459 + 93008 + 255}{378}$

(c) $\dfrac{4008 + 408 + 48}{48}$

(d) $\dfrac{9909 + 9090}{3303}$

2. **Masonry** If 6-in. common red brick is on sale at $41 per hundred, calculate the cost of the 14,000 bricks needed on a construction project.

3. **Manufacturing** The specified shear strength of a $\frac{1}{8}$-in. 2117-T3 (AD) rivet is 344 lb. If it is necessary to assure a strength of 6587 lb to a riveted joint in an aircraft structure, how many rivets must be used?

4. **Transportation**   A train of 74 railway cars weighed a total of 9,108,734 lb. What was the average weight of a car?

5. **Manufacturing**   A high-speed stamping machine can produce small flat parts at the rate of 96 per minute. If an order calls for 297,600 parts to be stamped, how many hours will it take to complete the job? (60 minutes = 1 hour.)

6. **Painting**   Following are four problems from the national apprentice examination for painting and decorating contractors.

   (a) $11,877,372 \div 738$ (b) $87,445,005 \div 435$

   (c) $1,735,080 \div 760$ (d) $206,703 \div 579$

7. **Construction**   Max, the master mason, is constructing a wall. From similar jobs he estimates that he can lay an average of 115 bricks an hour. How long will it take him to lay 4830 bricks?

Check your answers to the odd-numbered problems in the Appendix, then turn to Section 1-5 to study another important topic in the arithmetic of whole numbers.

1-5   Order of Operations

What is the value of the arithmetic expression

$3 \times 4 + 2 = ?$

If we multiply first, $(3 \times 4) + 2 = 12 + 2 = 14$

but if we add first, $3 \times (4 + 2) = 3 \times 6 = 18$

To avoid any possible confusion in situations like this, mathematicians have adopted a standard order of operations for arithmetic calculations. When two or more of the four basic arithmetic operations (addition, subtraction, multiplication, and division) are combined in the same calculation, follow these rules.

Rule 1	Perform any calculations shown inside parentheses first.
Rule 2	Perform all multiplications and divisions next, working from left to right.
Rule 3	Perform additions and subtractions last, working from left to right.

Example 1

To calculate $3 \times 4 + 2$,

multiply first (Rule 2): $3 \times 4 = 12$. $3 \times 4 + 2 = 12 + 2$

Add last (Rule 3). $= 14$

→ Your Turn

Perform the following calculations, using the order of operation rules.

(a) $5 + 8 \times 7$

(b) $12 \div 6 - 2$

(c) $(26 - 14) \times 2$

(d) $12 \times (6 + 2) \div 6 - 7$

→ Solutions

(a) Do the multiplication first (Rule 2): $8 \times 7 = 56$

Then, add (Rule 3):

$$5 + 8 \times 7 = 5 + 56$$
$$= 61$$

(b) First, divide (Rule 2): $12 \div 6 = 2$

Then, subtract (Rule 3).

$$12 \div 6 - 2 = 2 - 2$$
$$= 0$$

(c) Perform the operation in parentheses first (Rule 1):
$$26 - 14 = 12$$

Then, multiply (Rule 2).

$$(26 - 14) \times 2 = 12 \times 2$$
$$= 24$$

(d) Work inside parentheses first (Rule 1):
$$6 + 2 = 8$$

Then, multiply (Rule 2).

Next, divide (Rule 2).

Finally, subtract (Rule 3).

$$12 \times (6 + 2) \div 6 - 7$$
$$= 12 \times 8 \div 6 - 7$$
$$= 96 \div 6 - 7$$
$$= 16 - 7$$
$$= 9$$

> **A Closer Look**

Notice in Exercise (d) that the multiplication and division without parentheses are performed working left to right.

Using parentheses helps guard against confusion. For example, the calculation $12 - 4 \div 2$ should be written $12 - (4 \div 2)$ to avoid making a mistake. ◄

When division is written with a fraction bar or *vinculum,* all calculations above the bar or below the bar should be done before the division.

> **Example 2**

$$\frac{24 + 12}{6 - 2} \qquad \text{should be thought of as} \qquad \frac{(24 + 12)}{(6 - 2)}$$

First, simplify the top and bottom to get $\dfrac{36}{4}$.

Then, divide to obtain the answer $\dfrac{36}{4} = 9$.

> **Note**

This is very different from the calculation

$$\frac{24}{6} + \frac{12}{2} = 4 + 6 = 10$$

where we use Rules 2 and 3, doing the divisions first and the additions last. ◄

The order of operations is involved in a great many practical problems.

Example 3

Office Services   To encourage water conservation, a water district began charging its customers according to the following monthly block rate system:

Block	Quantity Used	Rate per HCF*
1	The first 20 HCF	$3
2	The next 30 HCF (from 21–50 HCF)	$4
3	The next 50 HCF (from 51–100 HCF)	$5
4	Usage in excess of 100 HCF	$6

*Hundred cubic feet

Suppose that a customer used 72 HCF in a month. The first 20 HCF will be charged at a rate of $3 per HCF. The next 30 HCF will be charged at a rate of $4 per HCF, and the remaining amount (72 HCF–50 HCF) will be charged at a rate of $5 per HCF. We can calculate their bill as follows:

$$\underbrace{20 \times \$3}_{\text{Cost for Block 1}} + \underbrace{30 \times \$4}_{\text{Cost for Block 2}} + \underbrace{(72 - 50) \times \$5}_{\text{Cost for part of Block 3}}$$

Using the order of operations,

First, simplify within parentheses. $= 20 \times \$3 + 30 \times \$4 + 22 \times \$5$

Next, multiply from left to right. $= \$60 + \$120 + \$110$

Finally, add from left to right. $= \$290$

→ More Practice

Perform the following calculations, using the three rules of order of operations.

(a) $240 \div (18 + 6 \times 2) - 2$ (b) $6 \times 5 + 14 \div (6 + 8) - 3$

(c) $6 + 2 \times 3 \div (7 - 4) + 5$

(d) $\dfrac{39 - 5 \times 3}{11 - 5}$

(e)   **Construction**   A remodeling job requires 128 sq ft of countertops. Two options are being considered. The more expensive option is to use all Corian at $64 per sq ft. The less expensive option is to use 66 sq ft of granite at $89 per sq ft and 62 sq ft of laminate at $31 per sq ft. Write out a mathematical statement to calculate the difference in cost between these two options. Then calculate this difference.

(f)   **Office Services**   Use the table in Example 3 to calculate the charge for using 116 HCF in a month.

→ Solutions

(a)   Perform the multiplication inside parentheses $240 \div (18 + 6 \times 2) - 2$

first (Rules 1 and 2): $6 \times 2 = 12$ $= 240 \div (18 + 12) - 2$

Add inside parentheses next (Rule 1): $18 + 12 = 30$ $= 240 \div 30 - 2$

Then, divide (Rule 2): $240 \div 30 = 8$ $= 8 - 2$

Finally, subtract (Rule 1). $= 6$

(b) First, work inside parentheses: $6 \times 5 + 14 \div (6 + 8) - 3$

$6 + 8 = 14$ $= 6 \times 5 + 14 \div 14 - 3$

Next, multiply and divide from left to right: $= 30 + 14 \div 14 - 3$

$6 \times 5 = 30 \qquad 14 \div 14 = 1$ $= 30 + 1 - 3$

Finally, add and subtract from left to right. $= 31 - 3 = 28$

(c) First, work inside parentheses: $6 + 2 \times 3 \div (7 - 4) + 5$

$7 - 4 = 3$ $= 6 + 2 \times 3 \div 3 + 5$

Next, multiply and divide from left to right: $= 6 + 6 \div 3 + 5$

$2 \times 3 = 6 \qquad 6 \div 3 = 2$ $= 6 + 2 + 5$

Finally, add from left to right. $= 8 + 5 = 13$

(d) Simplify the top first by multiplying (Rule 2), $\dfrac{39 - 5 \times 3}{11 - 5} = \dfrac{39 - 15}{11 - 5}$

and then subtracting (Rule 3). $= \dfrac{24}{11 - 5}$

Then, simplify the bottom. $= \dfrac{24}{6}$

Finally, divide. $= 4$

(e) The more expensive option is: $128 \times \$64 = \$8192$

The less expensive option is: $\$89 \times 66 + \$31 \times 62$

$= \$5874 + \$1922$

$= \$7796$

The difference is: $\$8192 - \$7796 = \$396$

(f) From the table, we see that 116 HCF includes all of Blocks 1, 2, and 3. The remainder $(116 - 100)$ HCF is charged to Block 4. The total monthly bill is:

$$20 \times \$3 + 30 \times \$4 + 50 \times \$5 + (116 - 100) \times \$6$$

To calculate, simplify parentheses first. $= 20 \times \$3 + 30 \times \$4 + 50 \times \$5 + 16 \times \$6$

Then multiply. $= \$60 + \$120 + \$250 + \$96$

Finally add. $= \$526$

Now go to Exercises 1-5 for more practice on the order of operations.

Exercises 1-5 Order of Operations

A. Perform all operations in the correct order.

1. $2 + 8 \times 6$ 2. $20 - 3 \times 2$

3. $40 - 20 \div 5$ 4. $16 + 32 \div 4$

5. $16 \times 3 + 9$

6. $2 \times 9 - 4$

7. $48 \div 8 - 2$

8. $64 \div 16 + 8$

9. $(5 + 9) \times 3$

10. $(18 - 12) \div 6$

11. $24 \div (6 - 2)$

12. $9 \times (8 + 3)$

13. $16 + 5 \times (3 + 6)$

14. $8 + 3 \times (9 - 4)$

15. $(23 + 5) \times (12 - 8)$

16. $(17 - 9) \div (6 - 2)$

17. $6 + 4 \times 7 - 3$

18. $24 - 8 \div 2 + 6$

19. $5 \times 8 + 6 \div 6 - 12 \times 2$

20. $24 \div 8 - 14 \div 7 + 8 \times 6$

21. $2 \times (6 + 4 \times 9)$

22. $54 \div (8 - 3 \times 2)$

23. $(4 \times 3 + 8) \div 5$

24. $(26 \div 2 - 5) \times 4$

25. $8 - 4 + 2$

26. $24 \div 6 \times 2$

27. $18 \times 10 \div 5$

28. $22 + 11 - 7$

29. $12 - 7 - 3$

30. $48 \div 6 \div 2$

31. $12 - (7 - 3)$

32. $18 \div (3 \times 2)$

33. $\dfrac{36}{9} + \dfrac{27}{3}$

34. $\dfrac{36 - 27}{9 - 6}$

35. $\dfrac{44 + 12}{11 - 3}$

36. $\dfrac{44}{11} + \dfrac{12}{3}$

37. $\dfrac{6 + 12 \times 4}{15 - 3 \times 2}$

38. $\dfrac{36 - (7 - 4)}{5 + 3 \times 2}$

39. $\dfrac{12 + 6}{3 + 6} + \dfrac{24}{6} - 6 \div 6$

40. $8 \times 5 - \dfrac{2 + 4 \times 12}{18 - 4 \times 2} + 72 \div 9$

B. Applied Problems

1. **Painting**   A painter ordered 3 gallons of acrylic vinyl paint for $34 a gallon and 5 gallons of acrylic eggshell enamel for $39 a gallon. Write out a mathematical statement giving the total cost of the paint. Calculate the cost.

2. **Landscaping**   On a certain landscaping job, Steve charged a customer $468 for labor and $90 each for eight flats of plants. Write a single mathematical statement giving the total cost of this job, then calculate the cost.

3. **Electrical Trades**   An electrician purchased 12 dimmer switches at $25 each and received a $6 credit for each of the three duplex receptacles she returned. Write out a mathematical statement that gives the amount of money she spent, then calculate this total.

4. **Automotive Trades**   At the beginning of the day on Monday, the parts department has on hand 520 spark plugs. Mechanics in the service department estimate they will need about 48 plugs per day. A new shipment of 300 will arrive on Thursday. Write out a mathematical statement that gives the number of spark plugs on hand at the end of the day on Friday. Calculate this total.

5. **Trades Management**   A masonry contractor is preparing an estimate for building a stone wall and gate. He estimates that the job will take a 40-hour

work week. He plans to have two laborers at $12 per hour and three masons at $20 per hour. He'll need $3240 worth of materials and wishes to make a profit of $500. Write a mathematical statement that will give the estimated cost of the job; then calculate this total.

6. **Allied Health**   A pregnant woman gained 4 pounds per month during her first trimester (3 months) of pregnancy, 3 pounds per month during her second trimester, and 2 pounds per month during her last trimester. What was her total weight gain during the pregnancy?

7. **Graphic Design**   Girilla Graphics charges $80 per hour for the work of its most senior graphic artists, $40 per hour for the work of a production designer, and $18 per hour for the work of a trainee. A recent project required 33 hours of work from a senior artist, 12 hours from a production designer, and 45 hours from a trainee. Write out a mathematical statement for calculating the total labor cost, and then compute this cost.

8. **Culinary Arts**   A caterer must estimate the amount and cost of strawberries required for an anniversary dinner involving 200 guests. She estimates that she will need about 6 strawberries per person for decorations and cocktails, and about 10 strawberries per person for the dessert. A local farmer sells strawberries in flats containing 80 strawberries each for $7.00 per flat. How many flats will she need, and how much should she expect to pay for the strawberries?

9. **Sports and Leisure**   In the 2008 Olympics, the United States ended up with the most total medals, but China won the most gold medals. This created a controversy over which country had the best overall performance. One way to objectively settle this issue is to score the medals much like a high school or college track meet—that is, award 5 points for first place (gold), 3 points for second place (silver), and 1 point for third place (bronze). Use the following final medal counts to determine a score for each country based on this point system. According to these scores, which country "won" the 2008 Olympics?

	Gold	Silver	Bronze
China	51	21	28
United States	36	38	36

10. **Office Services**   Use the table in Example 3 to calculate the monthly water bill for the following amounts of water usage: (a) 32 HCF   (b) 94 HCF   (c) 121 HCF

C. Calculator Problems

1. $462 + 83 \times 95$

2. $425 \div 25 + 386$

3. $7482 - 1152 \div 12$

4. $1496 - 18 \times 13$

5. $(268 + 527) \div 159$

6. $2472 \times (1169 - 763)$

7. $612 + 86 \times 9 - 1026 \div 38$

8. $12 \times 38 + 46 \times 19 - 1560 \div 24$

9. $3579 - 16 \times (72 + 46)$

10. $273 + 25 \times (362 + 147)$

11. $864 \div 16 \times 27$

12. $973 - (481 + 327)$

13. $(296 + 18 \times 48) \times 12$

14. $(27 \times 18 - 66) \div 14$

15. $\dfrac{3297 + 1858 - 493}{48 \times 16 - 694}$ 16. $\dfrac{391}{17} + \dfrac{4984}{89} - \dfrac{1645}{47}$

Check your answers to the odd-numbered problems in the Appendix, then turn to Problem Set 1 on page 57 for practice on the arithmetic of whole numbers, with many practical applications of this mathematics. If you need a quick review of the topics in this chapter, visit the chapter Summary first.

Summary

Arithmetic of Whole Numbers

Objective

Review

Write whole numbers in words then translate words to numbers.
(pp. 4–6)

Use place value and expanded form to help translate in both directions.

Example: The number 250,374 can be placed in a diagram;

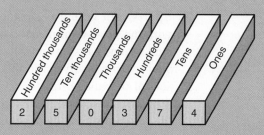

and read "two hundred fifty thousand, three hundred seventy-four".

The number "one million, sixty-five thousand, eight" can be written as 1,065,008.

Round whole numbers.
(p. 7)

Place a ∧ mark to the right of the place to which the number must be rounded. If the digit to the right of the mark is less than 5, replace all digits to the right of the mark with zeros. If the digit to the right of the mark is equal to or larger than 5, increase the digit to the left of the mark by 1 and replace all digits to the right with zeros.

Example: Round 214,659

(a) to the nearest ten thousand $21\underset{\wedge}{\,}4659 = 210,000$

(b) to the nearest hundred $2146\underset{\wedge}{\,}59 = 214,700$

Add whole numbers.
(p. 6)

First, estimate your answer. Then, line up the digits and add.

Example: Add: 7009 + 1598

Estimate: 7000 + 1600 = 8600

Line up and add:
$$\begin{array}{r} \scriptstyle 1\ 1 \\ 7009 \\ +1598 \\ \hline 8607 \end{array}$$

The answer is approximately equal to the estimate.

Objective	Review

Subtract whole numbers. (p. 20)

First, estimate your answer. Then, line up the digits and subtract column by column from right to left. "Borrow" when necessary.

Example: Subtract: $4035 - 1967$

Estimate: $4000 - 2000 = 2000$

Line up and subtract:

$$
\begin{array}{r}
{\scriptstyle 3\ 10}\\
4\cancel{0}35\\
-1967\\
\hline
\end{array}
\quad
\begin{array}{r}
{\scriptstyle 3\ 9\ 13}\\
4\cancel{0}\cancel{3}5\\
-1967\\
\hline
\end{array}
\quad
\begin{array}{r}
{\scriptstyle 3\ 9\ 12\ 15}\\
4\cancel{0}\cancel{3}\cancel{5}\\
-1967\\
\hline
2068
\end{array}
$$

Multiply whole numbers. (p. 28)

First, estimate your answer. Then, multiply the top number by each digit in the bottom number. Leave a blank space on the right end of each row of the calculation. Finally, add the rows of products.

Example: Multiply: 305×243

Estimate: $300 \times 250 = 75,000$

$$
\begin{array}{r}
305\\
\times 243\\
\hline
915 \quad \leftarrow \boxed{305 \times 3}\\
1220 \quad\ \leftarrow \boxed{305 \times 4}\\
610 \quad\ \ \leftarrow \boxed{305 \times 2}\\
\hline
74115
\end{array}
$$

The product is 74,115 and this agrees with the estimate.

Divide whole numbers. (p. 38)

First, estimate your answer. Then divide, step by step, as shown in the following example.

Example: Divide $7511 \div 37$

$$
\begin{array}{r}
203\\
37\overline{)7511}\\
74\\
\hline
11\\
0\\
\hline
111\\
111\\
\hline
0
\end{array}
$$

Step 1 37 into 7? No. 37 into 75? Yes, twice. Write 2 above the 5.

Step 2 $37 \times 2 = 74$. Subtract $75 - 74 = 1$

Step 3 Bring down 1. 37 into 11? No. Write 0 in the answer.

Step 4 $37 \times 0 = 0$. Subtract $11 - 0 = 11$

Step 5 Bring down 1. 37 into 111? Yes, three times. Write 3 in the answer.

Step 6 $37 \times 3 = 111$. Subtract $111 - 111 = 0$
The quotient is 203. Check it.

Determine factors and prime factors. (p. 43)

The factors of a whole number are all the numbers that divide it with zero remainder. The prime factors of a whole number are those factors that are prime; that is, they are evenly divisible only by themselves and 1.

Example: The factors of 12 are 1, 2, 3, 4, 5, 6, and 12.

To write 12 as a product of its prime factors, use the factor-tree shown.

$12 = 2 \times 2 \times 3$

$$
\begin{array}{c}
12\\
\diagup\ \diagdown\\
2 \qquad 6\\
\diagup\ \diagdown\\
2 \qquad 3
\end{array}
$$

Objective

Use the correct order of operations with addition, subtraction, multiplication, and division.
(p. 48)

Review

Follow these steps in this order:

Step 1   Perform any calculations shown inside parentheses.

Step 2   Perform all multiplications and divisions, working from left to right.

Step 3   Perform all additions and subtractions, working from left to right.

> **Example:** $8 + 56 \div 4 \times 2 + 8 - 3$
> $= 8 + 14 \times 2 + 8 - 3$ Divide: $56 \div 4 = 14$
> $= 8 + 28 + 8 - 3$ Multiply next: $14 \times 2 = 28$
> $= 44 - 3$ Add next: $8 + 28 + 8 = 44$
> $= 41$ Finally, subtract: $44 - 3 = 41$

Solve word problems with whole numbers.

Read the problem carefully, looking for key words or phrases that indicate which operation to use.

> **Example:**   A metal casting weighs 680 lb. What is the finished weight after 235 lb of metal is removed during shaping?
>
> Solution: The word "removed" indicates that subtraction is involved.
>
> $680 - 235 = 445$ lb

Arithmetic of Whole Numbers

Answers to odd-numbered problems are given in the Appendix.

A. Writing and Rounding Whole Numbers

Write in words.

1. 593
2. 6710
3. 45,206

4. 137,589
5. 2,403,560
6. 970,001

7. 10,020
8. 1,528,643

Write as numbers.

9. Four hundred eight
10. Six thousand, three hundred twenty-seven

11. Two hundred thirty thousand, fifty-six
12. Five million, ninety-eight thousand, one hundred seven

13. Sixty-four thousand, seven hundred
14. Eight hundred fifty-two million

Round as indicated.

15. 692 to the nearest ten
16. 5476 to the nearest hundred

17. 17,528 to the nearest thousand
18. 94,746 to the nearest hundred

19. 652,738 to the nearest hundred thousand
20. 705,618 to the nearest ten thousand

B. Perform the arithmetic as shown.

1. $24 + 69$
2. $38 + 45$
3. $456 + 72$

4. $43 + 817$
5. $396 + 538$
6. $2074 + 906$

7. $43 - 28$
8. $93 - 67$
9. $734 - 85$

10. $315 - 119$
11. $543 - 348$
12. $3401 - 786$

13. 376×4
14. 489×7
15. 67×21

16. 45×82
17. 207×63
18. 314×926

19. 5236×44
20. 4018×392
21. $259 \div 7$

22. $1704 \div 8$
23. $42\overline{)2394}$
24. $34\overline{)2108}$

25. $1440 \div 160$
26. $11309 \div 263$
27. $\dfrac{1314}{73}$

28. $\dfrac{23 \times 51}{17}$
29. $\dfrac{36 \times 91}{13 \times 42}$
30. $(18 + 5 \times 9) \div 7$

Name

Date

Course/Section

31. $120 - 40 \div 8$ **32.** $32 \div 4 + 16 \div 2 \times 4$ **33.** $3 \times 4 - 15 \div 3$

34.
$$\begin{array}{r} 139 \\ 407 \\ + \ 81 \\ \hline \end{array}$$

35.
$$\begin{array}{r} 308 \\ 793 \\ + \ 144 \\ \hline \end{array}$$

36.
$$\begin{array}{r} 194 \\ 271 \\ + \ 368 \\ \hline \end{array}$$

C. For each of the following whole numbers, (a) list all its factors, and (b) write each as a product of its prime factors.

1. 8 **2.** 28 **3.** 31

4. 35 **5.** 36 **6.** 42

D. Practical Problems

1. **Electrical Trades**   From a roll of No. 12 wire, an electrician cut the following lengths: 6, 8, 20, and 9 ft. How many feet of wire did he use?
2. **Machine Trades**   A machine shop bought 14 steel rods of $\frac{7}{8}$-in.-diameter steel, 23 rods of $\frac{1}{2}$-in. diameter, 8 rods of $\frac{1}{4}$-in. diameter, and 19 rods of 1-in. diameter. How many rods were purchased?
3. **Interior Design**   In estimating the floor area of a house, Jean listed the rooms as follows: living room, 346 sq ft; dining room, 210 sq ft; four bedrooms, 164 sq ft each; two bathrooms, 96 sq ft each; kitchen, 208 sq ft; and family room, 280 sq ft. What is the total floor area of the house?
4. **Allied Health**   A patient was instructed to take one tablet of blood pressure medication twice daily (one tablet in the morning and one in the evening). Before the patient leaves for a 12-week summer holiday, the medical technician must determine how many tablets the patient will need over her entire holiday. How many tablets will the patient require?
5. **Electrical Trades**   Roberto has 210 ft of No. 14 wire left on a roll. If he cuts it into pieces 35 ft long, how many pieces will he get?
6. **Flooring and Carpeting**   How many hours will be required to install the flooring of a 2160 sq ft house if 90 sq ft can be installed in 1 hour?
7. **Construction**   The weights of seven cement platforms are 210, 215, 245, 217, 220, 227, and 115 lb. What is the average weight per platform? (Add up and divide by 7.)
8. **Flooring and Carpeting**   A room has 234 square feet of floor space. If carpeting costs $5 per sq ft, what will it cost to carpet the room?
9. A plasma TV set can be bought for $500 down and 12 payments of $110 each. What is its total cost?
10. **Electronics**   The output voltage of an amplifier is calculated by multiplying the input voltage by the voltage gain of the amplifier. Calculate the output voltage, in millivolts (mV), for a circuit if the input voltage is 45 mV and the gain is 30.
11. **Automotive Trades**   During a compression check on an engine, the highest compression in any cylinder was 136 pounds per square inch (psi), and the lowest was 107 psi. The maximum allowable difference between highest and lowest is 30 psi. Did the engine fall within the maximum allowed?
12. **Office Services**   During the first three months of the year, the Print Rite Company reported the following sales of printers:

January	$55,724
February	$47,162
March	$62,473

    What is their sales total for this quarter of the year?
13. **Fire Protection**   A fire control pumping truck can move 156,000 gallons of water in 4 hours of continuous pumping. What is the flow rate (gallons per minute) for this truck?

14. **Construction** A pile of lumber contains 170 boards 10 ft long, 118 boards 12 ft long, 206 boards 8 ft long, and 19 boards 16 ft long.
    (a) How many boards are in the pile?
    (b) Calculate the total length, or linear feet, of lumber in the pile.

15. **Marine Technology** If the liquid pressure on a surface is 17 psi (pounds per square inch), what is the total force on a surface area of 167 square inches?

16. **Automotive Trades** A rule of thumb useful to auto mechanics is that a car with a pressure radiator cap will boil at a temperature of (3 × cap rating) + 212 degrees. What temperature will a Dino V6 reach with a cap rating of 17?

17. **Machine Trades** If it takes 45 minutes to cut the teeth on a gear blank, how many hours will be needed for a job that requires cutting 32 such gear blanks?

18. **Electrical Trades** A 4-ft lighting track equipped with three fixtures costs $45. Fluorescent bulbs for the fixtures cost $15 each. Write out a mathematical statement that gives the total cost for three of these tracks, with bulbs, plus $9 shipping. Then calculate this total.

19. **Manufacturing** A gallon is a volume of 231 cubic inches. If a storage tank holds 380 gallons of oil, what is the volume of the tank in cubic inches?

20. **Office Services** The Ace Machine Tool Co. received an order for 15,500 flanges. If two dozen flanges are packed in a box, how many boxes are needed to ship the order?

21. **Manufacturing** At 8 A.M. the revolution counter on a diesel engine reads 460089. At noon it reads 506409. What is the average rate, in revolutions per minute, for the machine?

22. **Fire Protection** The pressure reading coming out of a pump is 164 lb. It is estimated that every 50-ft section of hose reduces the pressure by 7 lb. What will the estimated pressure be at the nozzle end of nine 50-ft sections?

23. **Carpentry** A fireplace mantel 6 ft long is to be centered along a wall 18 ft long. How far from each end of the wall should the ends of the mantel be?

24. **Electronics** When capacitors are connected in parallel in a circuit as shown, the total capacitance can be calculated by adding the individual capacitor values C_1, C_2, C_3, C_4, and C_5. Find the total capacitance of this circuit, in picofarads (pF).

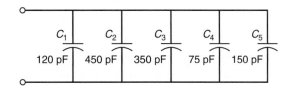

25. **Trades Management** RAD Electric submitted a bid of $2400 for a certain job. Using billing for T&M (time and materials), RAD normally charges $60 per hour. If materials for the job come to $350, and the electrician actually worked for 36 hours, by how much money did RAD underbid the job?

26. **Life Skills** An electrician is deciding between two job offers. One is with Company A, paying him $24 per hour plus health benefits. The other is with Company B, paying him $28 per hour with no health benefits. Health insurance would cost him $500 per month. Assume he works an 8-hour day and an average of 22 work days per month, then answer the following questions:
    (a) How much would he earn per month with Company A?
    (b) How much would he earn per month with Company B?
    (c) What are his net earnings with Company B after paying for his health benefits?
    (d) Which company provides the best overall compensation?

27. **Office Services**   Use the table in Example 3 on page 50 to determine the monthly cost for 87 HCF of water usage.

28. **Life Skills**   Gorilla glue comes in two sizes: an 8-ounce container for $4 and an 18-ounce container for $6.
   (a) How many of the smaller containers would you need in order to get 72 ounces of glue? How much would this cost?
   (b) How many of the larger containers would you need in order to get 72 ounces? How much would this cost?
   (c) How much do you save by purchasing 72 ounces in the larger-sized container instead of in the smaller-sized container?

29. **Automotive Trades**   A 2008 Volvo V70 is EPA-rated at 16 mi/gal for city driving and 24 mi/gal for highway driving.
   (a) How many gallons of gas would be used on a trip that included 80 miles of city driving and 288 miles of highway driving?
   (b) At approximately $2 per gallon, how much would the gas cost for this trip?

30. **Landscaping**   A landscape company charges $55 per hour for the work of a designer, $40 per hour for the work of a foreman, and $25 per hour for the work of a laborer. A particular installation required 6 hours of design work, 11 hours from the foreman, and 33 hours of labor. Write out a mathematical statement for calculating the total labor cost and then compute this cost.

USING A CALCULATOR, I: WHOLE-NUMBER CALCULATIONS

Since its introduction in the early 1970s, the handheld electronic calculator has quickly become an indispensable tool in modern society. From clerks, carpenters, and shoppers to technicians, engineers, and scientists, people in virtually every occupation now use calculators to perform mathematical tasks. With this in mind, we have included in this book special instructions, examples, and exercises demonstrating the use of a calculator wherever it is appropriate.

To use a calculator intelligently and effectively in your work, you should remember the following:

1. Whenever possible, make an estimate of your answer before entering a calculation. Then check this estimate against your calculator result. You may get an incorrect answer by accidentally pressing the wrong keys or by entering numbers in an incorrect sequence.

2. Always try to check your answers to equations and word problems by substituting them back into the original problem statement.

3. Organize your work on paper before using a calculator, and record any intermediate results that you may need later.

4. Round your answer whenever necessary.

This is the first of several sections, spread throughout the text, designed to teach you how to use a calculator with the mathematics taught in this course. In addition to these special sections, the solutions to many worked examples include the appropriate calculator key sequences and displays. Most exercise sets contain a special section giving calculator problems.

Selecting a Calculator   As noted in the "How to Use This Book" section (see page xiv), there are four main types of calculators used by students at this level: (1) scientific calculators with one line of display, (2) scientific calculators with two lines of display, (3) scientific calculators with four lines of display, and (4) graphing calculators with eight lines of display. A reasonably priced scientific calculator is ideal for this textbook, although for students planning to continue with more advanced mathematics, a graphing calculator may eventually be needed. The calculator instructions and key sequences shown in this text reflect the way that the two-line scientific calculators operate. In most cases, these instructions will also be appropriate for the four-line scientific calculators as well as for graphing calculators. We will often refer to all three of these types as "multi-line calculators." If you have selected a single-line scientific calculator, your key sequences are shown in the calculator appendix whenever they differ from the ones in the text.

Becoming Familiar with Your Calculator

First check out the machine. Note whether it has a one-line or a multi-line display window. Look for the "On" switch, usually located in the upper-right- or lower-left-hand corner of the keyboard. There may be a separate "Off" switch, or it may be a second function of the "On" switch. When the calculator is first turned on or cleared, it will either display a zero followed by a decimal point (*0.*), or a blinking cursor on the left end of the display window, or both. The lower portion of the keyboard will have both numerical keys (0, 1, 2, 3, . . . , 9) and basic function keys ($+$, $-$, $\times$, $\div$), as well as a decimal point key ($\cdot$), and either an equals ($=$) or ENTER key.

There is no standard layout for the remaining function keys, but some numerical keys and almost all non-numerical keys have a second function written above them. To activate these second functions, you must first press the 2nd key and then press the appropriate function key. The second function key is located in the upper-left-hand corner of the keyboard and may also be labeled 2ndF or SHIFT. Some calculators even have keys containing a third function, and a 3rd key is used to activate them. Unfortunately, there is no standard agreement among different models as to which operations are main functions and which are second functions. Therefore, we will not generally include the 2nd key in our key sequences, and we will leave it to the student to determine when to use this key.

If you have a calculator with two or more lines of display, you will find four arrow keys located in the upper-right portion of the keyboard and arranged like this:

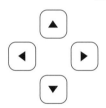

These multi-line calculators will have a CLEAR key, also labeled AC or simply C, and it may be combined with the "On" key. They will also have a delete (DEL) key with insert (INS) printed above it. The CLEAR key clears the entire calculation and the answer, but it is not generally necessary to clear the previous calculation in order to start a new one. To clear only part of a calculation, use the arrow keys to place the cursor at the appropriate spot, and then use DEL or INS as needed. The up-arrow key ($\blacktriangle$) can also be used to retrieve a prior entry, even after the display has been cleared. Calculators with a one-line display have a "clear all" key (usually marked AC), as well as a "clear entry" key (CE). The AC key clears the entire calculation, whereas the CE key removes the last entry only.

All calculators have two basic memory keys, one for storing information, labeled STO or Min, and one for retrieving information, labeled RCL or MR. These are often part of the same key, one being a second function of the other. Some calculators also have M+ and M– keys for adding and subtracting numbers to an existing memory total. A final pair of keys useful for basic whole number calculations are the parentheses keys. There is one for opening parentheses, labeled (, and one for closing parentheses, labeled).

Basic Operations

Every number is entered into the calculator digit by digit, left to right. For example, to enter the number 438, simply press the 4, 3, and 8 keys followed by the $=$ key, and the display will read ▨▨▨▨▨ **438.** ▨. If you make an error in entering the number, press either the clear Ⓒ or clear entry Ⓒᴇ key and begin again. Note that the calculator does not know if the number you are entering is 4, 43, 438, or something even larger, until you stop entering numerical digits and press a function key such as ⊞, ⊟, ⊠, ⊘ or ⊜. For the four basic functions, the key sequences are the same for both single-line and multi-line calculators.

Example 1

To add $438 + 266$

first, estimate the answer: $400 + 300 = 700$

then, enter this sequence of keys,

4 3 8 ⊞ **2 6 6** ⊜ → ▨▨▨▨ **704.**

Notice that we only show the final result, **704** in this case, on our answer display. However, on a multi-line calculator, the line above the answer will show all of your entries except for the equals sign, or in this case **438 + 266.** On a single-line calculator, the **438** is replaced by the **266** after the ⊞ key is pressed, and the sum of **704** appears after the ⊜ key is pressed.

All four basic operations work the same way.

Try the following problems for practice.

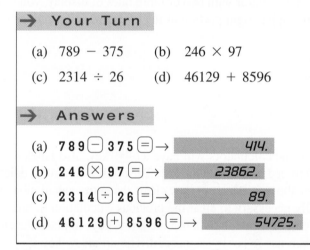

→ **Your Turn**

(a) $789 - 375$ (b) 246×97

(c) $2314 \div 26$ (d) $46129 + 8596$

→ **Answers**

(a) **7 8 9** ⊟ **3 7 5** ⊜ → ▨▨ **414.**

(b) **2 4 6** ⊠ **9 7** ⊜ → ▨▨ **23862.**

(c) **2 3 1 4** ⊘ **2 6** ⊜ → ▨▨ **89.**

(d) **4 6 1 2 9** ⊞ **8 5 9 6** ⊜ → ▨▨ **54725.**

Don't forget to estimate the answer before using the calculator.

Combined Operations

Scientific calculators are programmed to follow the correct order of operations. This means that you can enter a long string of calculations without having to find an intermediate result.

Example 2

To calculate $2 + 3 \times 4$, we know from Section 1-5 that the multiplication ($3 \times 4 = 12$) must be performed first, and the addition ($2 + 12 = 14$) must be performed last.

With a scientific calculator, simply enter the calculation exactly as it is written and then press the $\boxed{=}$ key:

2 $\boxed{+}$ 3 $\boxed{\times}$ 4 $\boxed{=}$ → ▨ 14.

→ **Your Turn**

Enter each of the following calculations as written.

(a) 480 − 1431 ÷ 53   (b) 72 × 38 + 86526 ÷ 69

(c) 2478 − 726 + 598 × 12   (d) 271440 ÷ 48 × 65

→ **Answers**

(a) 480 $\boxed{-}$ 1431 $\boxed{÷}$ 53 $\boxed{=}$ → ▨ 453.

(b) 72 $\boxed{\times}$ 38 $\boxed{+}$ 86526 $\boxed{÷}$ 69 $\boxed{=}$ → ▨ 3990.

(c) 2478 $\boxed{-}$ 726 $\boxed{+}$ 598 $\boxed{\times}$ 12 $\boxed{=}$ → ▨ 8928.

(d) 271440 $\boxed{÷}$ 48 $\boxed{\times}$ 65 $\boxed{=}$ → ▨ 367575.

A Closer Look   When the top line of the display of a two-line scientific calculator fills up, every additional entry causes characters to disappear from the left end of the display. You can make these entries reappear by pressing the left-arrow key ($\boxed{◄}$). Once you press $\boxed{=}$, the top line shifts back to the start of the calculation, and overflow on the right end will disappear. If you then press the up-arrow key ($\boxed{▲}$), you can edit the calculation using the right- and left- arrow keys in combination with delete ($\boxed{DEL}$) and insert ($\boxed{INS}$). You can then press $\boxed{=}$ at any time to get the new answer. ◄

Parentheses and Memory   As you have already learned, parentheses, brackets, and other grouping symbols are used in a written calculation to signal a departure from the standard order of operations. Fortunately, every scientific calculator has parentheses keys. To perform a calculation with grouping symbols, either enter the calculation inside the parentheses first and press $\boxed{=}$, or enter the calculation as it is written but use the parentheses keys.

Example 3

To enter 12 × (28 + 15) use either the sequence

28 $\boxed{+}$ 15 $\boxed{=}$ $\boxed{\times}$ 12 $\boxed{=}$ → ▨ 516.

or

12 $\boxed{\times}$ $\boxed{(}$ 28 $\boxed{+}$ 15 $\boxed{)}$ $\boxed{=}$ → ▨ 516.

On multi-line calculators, the $\boxed{\times}$ key is optional when multiplying an expression in parentheses. Therefore, the following sequence will also work on these models:

12 $\boxed{(}$ 28 $\boxed{+}$ 15 $\boxed{)}$ $\boxed{=}$ → ▨ 516.

When division problems are given in terms of fractions, parentheses are implied but not shown.

Example 4

The problem

$\dfrac{48 + 704}{117 - 23}$ is equal to $(48 + 704) \div (117 - 23)$

The fraction bar acts as a grouping symbol. The addition and subtraction in the top and bottom of the fraction must be performed before the division. Again you may use either parentheses or the $=$ key to simplify the top half of the fraction, but after pressing $=$ you must use parentheses to enter the bottom half. Here are the two possible sequences:

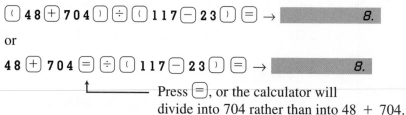

or

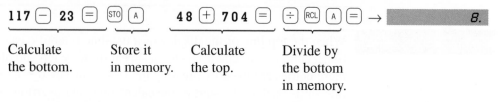

Press $=$, or the calculator will divide into 704 rather than into $48 + 704$.

The memory keys provide a third option for this problem. Memory is used to store a result for later use. In this case, we can calculate the value of the bottom first, store it in memory, then calculate the top and divide by the contents of memory. The entire sequence looks like this:

117 $-$ **23** $=$ (STO) (A)   **48** $+$ **704** $=$ $\div$ (RCL) (A) $=$ → ▓▓▓▓ *8.*

Calculate	Store it	Calculate	Divide by
the bottom.	in memory.	the top.	the bottom
			in memory.

Although all memory storing begins with the (STO) key, and all memory retrieval begins with the (RCL) key, most calculators require you to then select a memory location, and the procedures for selecting these vary greatly from model to model. You may need to use your right-arrow key to select a location such as A, B, C, etc., from the display and then press $=$. Or you may need to press a number key or a key that has a letter of the alphabet written above it to select a location. As shown in the previous key sequence, we shall always use the generic sequences "(STO) (A)" and "(RCL) (A)" to indicate storing and recalling results using a particular memory location. Be sure to consult your calculator's instruction manual to learn precisely how your memory works.

A Closer Look   Most multi-line calculators have a last answer, or (ANS) key, usually the second function of the negative ((−)) key. This key provides still another way to do the previous example without using memory. The sequence would look like this:

117 $-$ **23** $=$ (**48** $+$ **704**) $\div$ (ANS) $=$ → ▓▓▓▓ *8.*

After you press the first $=$, the calculator automatically stores the result of $117 - 23$ as the "last answer," which acts as another memory location. It will stay there until you complete the next calculation, and it can be retrieved with the (ANS) function. Notice that we must use parentheses around $48 + 704$ to avoid the need for $=$ after 704. Pressing $=$ at that point would replace the original "last answer." ◀

→ Your Turn

Here are a few practice problems involving grouping symbols. Use the method indicated.

(a) $(4961 - 437) \div 52$ Do not use the parentheses keys.

(b) $56 \times (38 + 12 \times 17)$ Use the parentheses keys.

(c) $2873 - (56 + 83) \times 16$ Use the parentheses keys.

(d) $\dfrac{263 \times 18 - 41 \times 12}{18 \times 16 - 17 \times 11}$ Use memory.

(e) $12 \times 16 - \dfrac{7 + 9 \times 17}{81 - 49} + 12 \times 19$ Use parentheses.

(f) Repeat problem (e) using the memory to store the value of the fraction, then doing the calculation left to right.

(g) Repeat problem (d) using the last answer key.

→ **Answers**

(a) $4\,9\,6\,1\;\boxed{-}\;4\,3\,7\;\boxed{=}\;\boxed{\div}\;5\,2\;\boxed{=}\;\rightarrow$ 87.

(b) $5\,6\;\boxed{\times}\;\boxed{(}\;3\,8\;\boxed{+}\;1\,2\;\boxed{\times}\;1\,7\;\boxed{)}\;\boxed{=}\;\rightarrow$ 13552.

(c) $2\,8\,7\,3\;\boxed{-}\;\boxed{(}\;5\,6\;\boxed{+}\;8\,3\;\boxed{)}\;\boxed{\times}\;1\,6\;\boxed{=}\;\rightarrow$ 649.

(d) $1\,8\;\boxed{\times}\;1\,6\;\boxed{-}\;1\,7\;\boxed{\times}\;1\,1\;\boxed{=}\;\boxed{\text{STO}}\;\boxed{\text{A}}\;2\,6\,3\;\boxed{\times}\;1\,8\;\boxed{-}\;4\,1\;\boxed{\times}\;1\,2$

  $\boxed{=}\;\boxed{\div}\;\boxed{\text{RCL}}\;\boxed{\text{A}}\;\boxed{=}\;\rightarrow$ 42.

(e) $1\,2\;\boxed{\times}\;1\,6\;\boxed{-}\;\boxed{(}\;7\;\boxed{+}\;9\;\boxed{\times}\;1\,7\;\boxed{)}\;\boxed{\div}\;\boxed{(}\;8\,1\;\boxed{-}\;4\,9\;\boxed{)}\;\boxed{+}\;1\,2$

  $\boxed{\times}\;1\,9\;\boxed{=}\;\rightarrow$ 415.

(f) $7\;\boxed{+}\;9\;\boxed{\times}\;1\,7\;\boxed{=}\;\boxed{\div}\;\boxed{(}\;8\,1\;\boxed{-}\;4\,9\;\boxed{)}\;\boxed{=}\;\boxed{\text{STO}}\;\boxed{\text{A}}\;1\,2\;\boxed{\times}\;1\,6\;\boxed{-}\;\boxed{\text{RCL}}\;\boxed{\text{A}}$

  $\boxed{+}\;1\,2\;\boxed{\times}\;1\,9\;\boxed{=}\;\rightarrow$ 415.

(g) $1\,8\;\boxed{\times}\;1\,6\;\boxed{-}\;1\,7\;\boxed{\times}\;1\,1\;\boxed{=}\;\boxed{(}\;2\,6\,3\;\boxed{\times}\;1\,8\;\boxed{-}\;4\,1\;\boxed{\times}\;1\,2\;\boxed{)}$

  $\boxed{\div}\;\boxed{\text{ANS}}\;\boxed{=}\;\rightarrow$ 42.

Objective	Sample Problems	For help, go to
When you finish this chapter you will be able to:		
1. Work with fractions.	(a) Write as a mixed number $\dfrac{31}{4}$. _____	Page 72
	(b) Write as an improper fraction $3\dfrac{7}{8}$. _____	Page 73
	Write as an equivalent fraction.	
	(c) $\dfrac{5}{16} = \dfrac{?}{64}$ _____	Page 74
	(d) $1\dfrac{3}{4} = \dfrac{?}{32}$ _____	
	(e) Write in lowest terms $\dfrac{10}{64}$. _____	Page 75
	(f) Which is larger, $1\dfrac{7}{8}$ or $\dfrac{5}{3}$? _____	Page 78
2. Multiply and divide fractions.	(a) $\dfrac{7}{8} \times \dfrac{5}{32}$ _____	Page 81
	(b) $4\dfrac{1}{2} \times \dfrac{2}{3}$ _____	
	(c) $\dfrac{3}{5}$ of $1\dfrac{1}{2}$ _____	
	(d) $\dfrac{3}{4} \div \dfrac{1}{2}$ _____	Page 87

_____
Name

_____
Date

_____
Course/Section

Objective	Sample Problems	For help, go to
	(e) $2\frac{7}{8} \div 1\frac{1}{4}$	_____
	(f) $4 \div \frac{1}{2}$	_____
3. Add and subtract fractions.	(a) $\frac{7}{16} + \frac{3}{16}$	_____ Page 94
	(b) $1\frac{3}{16} + \frac{3}{4}$	_____
	(c) $\frac{3}{4} - \frac{1}{5}$	_____ Page 103
	(d) $4 - 1\frac{5}{16}$	_____
4. Solve practical problems involving fractions.	**Carpentry** A tabletop is constructed using $\frac{3}{4}$-in. plywood with $\frac{5}{16}$-in. wood veneer on both sides. What is the total thickness of the table top?	_____

(Answers to these preview problems are given in the Appendix. Also, worked solutions to many of these problems appear in the chapter Summary.)

If you are certain that you can work *all* these problems correctly, turn to page 115 for a set of practice problems. If you cannot work one or more of the preview problems, turn to the page indicated to the right of the problem. For those who wish to master this material with the greatest success, turn to Section 2-1 and begin work there.

Chapter 2

Fractions

THE FAR SIDE® BY GARY LARSON

© 1983 FarWorks, Inc. All Rights Reserved/Dist. by Creators Syndicate

The Far Side® by Gary Larson © 1983 FarWorks, Inc. All Rights Reserved. The Far Side® and the Larson® signature are registered trademarks of FarWorks, Inc. Used with permission.

"Fool! This is an eleven-sixteenths. ... I asked for a five-eighths!"

2-1   Working with Fractions

The word *fraction* comes from a Latin word meaning "to break," and fraction numbers are used when we need to break down standard measuring units into smaller parts. Anyone doing practical mathematics problems will find that fractions are used in many different trade and technical areas. Unfortunately, fraction arithmetic usually *must* be done with pencil and paper, because not all calculators handle fractions directly.

Look at the following piece of lumber. What happens when we break it into equal parts?

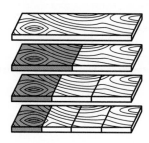

2 parts, each one-half of the whole

3 parts, each one-third of the whole

4 parts, each one-fourth of the whole

Suppose we divide this area into fifths by drawing vertical lines.

Notice that the five parts or "fifths" are equal in area. ⬛⬛⬛⬛⬛

A fraction is normally written as the division of two whole numbers: $\frac{2}{3}$, $\frac{3}{4}$, or $\frac{9}{16}$. One of the five equal areas below would be "one-fifth" or $\frac{1}{5}$ of the entire area.

⬛⬛⬛⬛⬛

Example 1

How would you label this portion of the area? ⬛⬛⬛⬛⬛ = ?

⬛⬛⬛⬛⬛ $\frac{3}{5} = \frac{3 \text{ shaded parts}}{5 \text{ total parts}}$

The fraction $\frac{3}{5}$ implies an area equal to three of the original parts.

$$\frac{3}{5} = 3 \times \left(\frac{1}{5}\right)$$

There are three equal parts, and the name of each part is $\frac{1}{5}$ or one-fifth.

→ **Your Turn**

In this collection of letters, HHHHPPT, what fraction are Hs?

→ **Solution**

Fraction of Hs $= \dfrac{\text{number of Hs}}{\text{total number of letters}} = \dfrac{4}{7}$ (read it "four sevenths")

The fraction of Ps is $\frac{2}{7}$, and the fraction of Ts is $\frac{1}{7}$.

Numerator   The two numbers that form a fraction are given special names to simplify talking about them. In the fraction $\frac{3}{5}$ the upper number 3 is called the *numerator* from the Latin *numero* meaning "number." It is a count of the number of parts.

Denominator   The lower number 5 is called the *denominator* from the Latin *nomen* or "name." It tells us the name of the part being counted. The numerator and denominator are called the *terms* of the fraction.

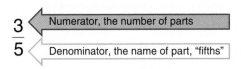

$\frac{3}{5}$ ← Numerator, the number of parts

← Denominator, the name of part, "fifths"

> **Learning Help**   A handy memory aid is to remember that the denominator is the "down part"—D for down. ◄

→ **Your Turn**

A paperback book costs $6 and I have $5. What fraction of its cost do I have? Write the answer as a fraction.

numerator = _____ , denominator = _____

→ **Answer**

$ $ $ $ $ $5 is $\frac{5}{6}$ of the total cost.
$\underbrace{}_{5}$ └─numerator = 5, denominator = 6

→ **More Practice**

Complete the following sentences by writing in the correct fraction.

(a) If we divide a length into eight equal parts, each part will be _____ of the total length.

(b) Then three of these parts will represent _____ of the total length.

(c) Eight of these parts will be _____ of the total length.

(d) Ten of these parts will be _____ of the total length.

→ **Answers**

(a) $\frac{1}{8}$ (b) $\frac{3}{8}$ (c) $\frac{8}{8}$ (d) $\frac{10}{8}$

Proper Fraction   The original length is used as a standard for comparison, and any other length—smaller or larger—can be expressed as a fraction of it. A *proper fraction* is a number less than 1, as you would suppose a fraction should be. It represents a quantity less than the standard. For example, $\frac{1}{2}$, $\frac{2}{3}$, and $\frac{17}{20}$ are all proper fractions. Notice that for a proper fraction, the numerator is less than the denominator—the top number is less than the bottom number in the fraction.

Improper Fraction   An *improper fraction* is a number greater than 1 and represents a quantity greater than the standard. If a standard length is 8 in., a length of 11 in. will be $\frac{11}{8}$ of the standard. Notice that for an improper fraction the numerator is greater than the denominator—top number greater than the bottom number in the fraction.

Circle the proper fractions in the following list.

$$\frac{3}{2} \quad \frac{3}{4} \quad \frac{7}{8} \quad \frac{5}{4} \quad \frac{15}{12} \quad \frac{1}{16} \quad \frac{35}{32} \quad \frac{7}{50} \quad \frac{65}{64} \quad \frac{105}{100}$$

→ **Answers**

You should have circled the following proper fractions: $\frac{3}{4}$, $\frac{7}{8}$, $\frac{1}{16}$, $\frac{7}{50}$. In each fraction the numerator (top number) is less than the denominator (bottom number). Each of these fractions represents a number less than 1.

Mixed Numbers

The improper fraction $\frac{7}{3}$ can be shown graphically as follows:

Unit standard = ▢▢▢ is equal to 1

$\frac{1}{3}$ = ▢

then, $\frac{7}{3}$ = ▨▨▨▨▨▨▨ (seven; count 'em!)

We can rename this number by regrouping.

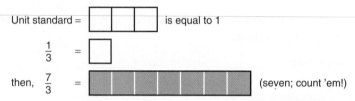

$\boxed{\frac{1}{3} \mid \frac{1}{3} \mid \frac{1}{3}}$ equals 1

$\boxed{\frac{1}{3} \mid \frac{1}{3} \mid \frac{1}{3}}$ equals 1

$\boxed{\frac{1}{3}}$ $\frac{7}{3} = 2 + \frac{1}{3}$ or $2\frac{1}{3}$

A *mixed number* is an improper fraction written as the sum of a whole number and a proper fraction.

$$\frac{7}{3} = 2 + \frac{1}{3} \text{ or } 2\frac{1}{3}$$

We usually omit the + sign and write $2 + \frac{1}{3}$ as $2\frac{1}{3}$, and read it as "two and one-third." The numbers $1\frac{1}{2}$, $2\frac{2}{5}$, and $16\frac{2}{3}$ are all written as mixed numbers.

Example 2

To write the improper fraction $\frac{13}{5}$ as a mixed number, divide numerator by denominator and form a new fraction as shown:

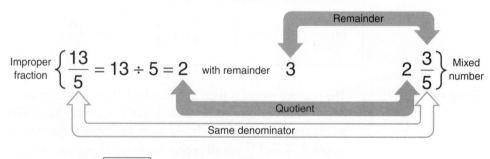

→ **Your Turn**

Now you try it. Rename $\frac{23}{4}$ as a mixed number. $\frac{23}{4}$ = _____

→ **Solution**

$\frac{23}{4} = 23 \div 4 = 5$ with remainder $3 \longrightarrow 5\frac{3}{4}$

If in doubt, check your work with a diagram like this:

→ **More Practice**

Now try these for practice. Write each improper fraction as a mixed number.

(a) $\frac{9}{5}$ (b) $\frac{13}{4}$ (c) $\frac{27}{8}$ (d) $\frac{31}{4}$ (e) $\frac{41}{12}$ (f) $\frac{17}{2}$

→ **Answers**

(a) $\frac{9}{5} = 1\frac{4}{5}$ (b) $\frac{13}{4} = 3\frac{1}{4}$ (c) $\frac{27}{8} = 3\frac{3}{8}$

(d) $\frac{31}{4} = 7\frac{3}{4}$ (e) $\frac{41}{12} = 3\frac{5}{12}$ (f) $\frac{17}{2} = 8\frac{1}{2}$

The reverse process, rewriting a mixed number as an improper fraction, is equally simple.

Example 3

We rewrite the mixed number $2\frac{3}{5}$ in improper fraction form as follows:

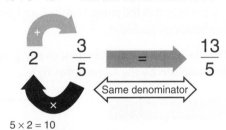

Graphically, $2\frac{3}{5} = 1 + 1 + \frac{3}{5}$

$1 = $ ⬚⬚⬚⬚⬚ $= \frac{5}{5}$

$1 = $ ⬚⬚⬚⬚⬚ $= \frac{5}{5}$

$\frac{3}{5} = $ ⬚⬚⬚ $= \frac{3}{5}$

or $\frac{13}{5}$; (count them)

→ **Your Turn**

Now you try it. Rewrite these mixed numbers as improper fractions.

(a) $3\frac{1}{5}$ (b) $4\frac{3}{8}$ (c) $1\frac{1}{16}$ (d) $5\frac{1}{2}$ (e) $15\frac{3}{8}$ (f) $9\frac{3}{4}$

→ **Answers**

(a) **Step 1** $3 \times 5 = 15$

　　 Step 2 $15 + 1 = 16 \leftarrow$ The new numerator

　　 Step 3 $3\frac{1}{5} = \dfrac{16}{5} \swarrow$ The original denominator

(b) $4\frac{3}{8} = \dfrac{35}{8}$　　　　　(c) $1\frac{1}{16} = \dfrac{17}{16}$　　　　　(d) $5\frac{1}{2} = \dfrac{11}{2}$

(e) $15\frac{3}{8} = \dfrac{123}{8}$　　　　　(f) $9\frac{3}{4} = \dfrac{39}{4}$

Equivalent Fractions

$\frac{1}{2}$

$\frac{2}{4}$

Two fractions are said to be *equivalent* if they represent the same number. For example, $\frac{1}{2} = \frac{2}{4}$ since both fractions represent the same portion of some standard amount.

There is a very large set of fractions equivalent to $\frac{1}{2}$.

$$\frac{1}{2} = \frac{2}{4} = \frac{3}{6} = \frac{4}{8} = \frac{5}{10} = \cdots = \frac{46}{92} = \frac{61}{122} = \frac{1437}{2874} \quad \text{and so on}$$

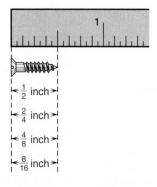

$\frac{1}{2}$ inch

$\frac{2}{4}$ inch

$\frac{4}{8}$ inch

$\frac{8}{16}$ inch

Each fraction is the same number, and we can use these fractions interchangeably in any mathematics problem.

To obtain a fraction equivalent to any given fraction, multiply the original numerator and denominator by the same nonzero number.

Example 4

Rewrite the fraction $\frac{2}{3}$ as an equivalent fraction with denominator 15.

$$\frac{2}{3} = \frac{2 \times \boxed{5}}{3 \times \boxed{5}} = \frac{10}{15}$$

⬆

| Multiply top and bottom by 5 |

Rewrite the fraction $\frac{3}{4}$ as an equivalent fraction with denominator equal to 20.

$$\frac{3}{4} = \frac{?}{20}$$

→ **Solution**

$$\frac{3}{4} = \frac{3 \times \square}{4 \times \square} = \frac{?}{20} \qquad 4 \times \square = 20, \text{ so } \square \text{ must be 5.}$$

$$= \frac{3 \times 5}{4 \times 5} = \frac{15}{20}$$

The number value of the fraction has not changed; we have simply renamed it.

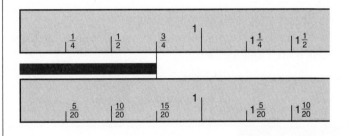

→ **More Practice**

Practice with these.

(a) $\dfrac{5}{8} = \dfrac{?}{32}$ (b) $\dfrac{7}{16} = \dfrac{?}{48}$ (c) $1\dfrac{2}{3} = \dfrac{?}{12}$

→ **Solutions**

(a) $\dfrac{5}{8} = \dfrac{5 \times 4}{8 \times 4} = \dfrac{20}{32}$ (b) $\dfrac{7}{16} = \dfrac{7 \times 3}{16 \times 3} = \dfrac{21}{48}$

(c) $1\dfrac{2}{3} = \dfrac{5}{3} = \dfrac{5 \times 4}{3 \times 4} = \dfrac{20}{12}$

Writing in Lowest Terms

Very often in working with fractions you will be asked to *write a fraction in lowest terms*. This means to replace it with the most simple fraction in its set of equivalent fractions. This is sometimes called "reducing" the fraction to lowest terms. To write $\frac{15}{30}$ in its lowest terms means to replace it by $\frac{1}{2}$. They are equivalent.

$$\frac{15}{30} = \frac{15 \div 15}{30 \div 15} = \frac{1}{2}$$

We have divided both the top and bottom of the fraction by 15.

In general, you would write a fraction in lowest terms as follows:

Example 5

Write $\dfrac{30}{48}$ in lowest terms.

First, find the largest number that divides both top and bottom of the fraction exactly.

$30 = 5 \times 6$

$48 = 8 \times 6$

In this case, the factor 6 is the largest number that divides both parts of the fraction exactly.

Second, eliminate this common factor by dividing.

$$\frac{30}{48} = \frac{30 \div 6}{48 \div 6} = \frac{5}{8}$$

The fraction $\frac{5}{8}$ is the simplest fraction equivalent to $\frac{30}{48}$. No whole number greater than 1 divides both 5 and 8 exactly.

Example 6

Write $\dfrac{90}{105}$ in lowest terms.

$$\frac{90}{105} = \frac{90 \div 15}{105 \div 15} = \frac{6}{7}$$

This process of eliminating a common factor is usually called *canceling*. When you cancel a factor, you divide both top and bottom of the fraction by that number. We would write $\dfrac{90}{105}$ as $\dfrac{6 \times \cancel{15}}{7 \times \cancel{15}} = \dfrac{6}{7}$. When you "cancel," you must divide by a *pair* of common factors, one factor in the numerator and the same factor in the denominator.

A Closer Look Sometimes, it may be difficult to find the *largest* number that divides both the numerator and denominator. You can still reduce the fraction to lowest terms by using two or more steps. In Example 6, for instance, you might have just divided by 5 at first:

$$\frac{90}{105} = \frac{90 \div \boxed{5}}{105 \div \boxed{5}} = \frac{18}{21}$$

Checking this result, you can see that 3 is also a factor. Now divide the numerator and denominator by 3.

$$\frac{18}{21} = \frac{18 \div \boxed{3}}{21 \div \boxed{3}} = \frac{6}{7}$$

Even though it took two steps to do it, the fraction is now written in lowest terms. ◄

Example 7

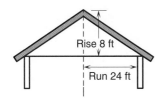

Rise 8 ft

Run 24 ft

Roofing In roof construction, the *pitch* or steepness of a roof is defined as the fraction

$$\text{Pitch} = \frac{\text{rise*}}{\text{run}}$$

The **rise** is the increase in height and the **run** is the corresponding horizontal distance. Typically, roofers express the pitch as the amount of rise per foot, or per 12 inches, of run. Therefore, rather than write the pitch as a fraction in lowest terms, they express it as an equivalent fraction with a denominator of 12. For the roof shown,

$$\text{Pitch} = \frac{\text{Rise}}{\text{Run}} = \frac{8\text{ ft}}{24\text{ ft}} = \frac{8 \div 2}{24 \div 2} = \frac{4}{12}$$

Divide by 2 . . . to get a denominator of 12

→ **Your Turn**

Write the following fractions in lowest terms.

(a) $\dfrac{6}{8}$ (b) $\dfrac{12}{16}$ (c) $\dfrac{2}{4}$ (d) $\dfrac{4}{12}$

(e) $\dfrac{15}{84}$ (f) $\dfrac{21}{35}$ (g) $\dfrac{12}{32}$

(h) **Roofing** Calculate the pitch of a storage shed roof having a rise of 6 ft over a run of 36 ft.

(i) **Allied Health** A technician needs to dilute antiseptic for use in hand-washing before and after surgery. The instructions say that 250 milliliters (mL) of antiseptic should be added to 750 mL water. What fraction of the final solution is water?

→ **Solutions**

(a) $\dfrac{6}{8} = \dfrac{6 \div 2}{8 \div 2} = \dfrac{3}{4}$ (b) $\dfrac{12}{16} = \dfrac{12 \div 4}{16 \div 4} = \dfrac{3}{4}$

(c) $\dfrac{2}{4} = \dfrac{2 \div 2}{4 \div 2} = \dfrac{1}{2}$ (d) $\dfrac{4}{12} = \dfrac{4 \div 4}{12 \div 4} = \dfrac{1}{3}$

(e) $\dfrac{15}{84} = \dfrac{15 \div 3}{84 \div 3} = \dfrac{5}{28}$ (f) $\dfrac{21}{35} = \dfrac{21 \div 7}{35 \div 7} = \dfrac{3}{5}$

(g) $\dfrac{12}{32} = \dfrac{12 \div 4}{32 \div 4} = \dfrac{3}{8}$ (h) $\text{Pitch} = \dfrac{6\text{ ft}}{36\text{ ft}} = \dfrac{6 \div 3}{36 \div 3} = \dfrac{2}{12}$

(i) We wish to know what fraction of the final solution is water, so we write the amount of water, 750 mL, in the numerator and the *total* amount of solution, 750 mL + 250 mL, in the denominator.

$$\frac{750\text{ mL}}{750\text{ mL} + 250\text{ mL}} = \frac{750\text{ mL}}{1000\text{ mL}} = \frac{3}{4}$$

---

* This is the most commonly used definition of pitch today. However, in the past, pitch was often defined as $\dfrac{\text{rise}}{\text{span}}$, where the span is twice the run.

Example 8

This one is a little tricky. Write $\frac{6}{3}$ in lowest terms.

$$\frac{6}{3} = \frac{6 \div 3}{3 \div 3} = \frac{2}{1} \qquad \text{or simply 2.}$$

Any whole number may be written as a fraction by using a denominator equal to 1.

$$3 = \frac{3}{1} \qquad 4 = \frac{4}{1} \qquad \text{and so on.}$$

Writing whole numbers in this way will be helpful when you learn to do arithmetic with fractions.

Comparing Fractions

If you were offered your choice between $\frac{2}{3}$ of a certain amount of money and $\frac{5}{8}$ of it, which would you choose?

Example 9

Which is the larger fraction, $\frac{2}{3}$ or $\frac{5}{8}$?

Can you decide? Rewriting the fractions as equivalent fractions will help.

To compare two fractions, rename each by changing them to equivalent fractions with the same denominator.

$$\frac{2}{3} = \frac{2 \times 8}{3 \times 8} = \frac{16}{24} \qquad \text{and} \qquad \frac{5}{8} = \frac{5 \times 3}{8 \times 3} = \frac{15}{24}$$

Now compare the new fractions: $\frac{16}{24}$ is greater than $\frac{15}{24}$ and therefore $\frac{2}{3}$ is larger than $\frac{5}{8}$.

Learning Help

1. The new denominator is the product of the original ones ($24 = 8 \times 3$).

2. Once both fractions are written with the same denominator, the one with the larger numerator is the larger fraction. (16 of the fractional parts is more than 15 of them.) ◄

Example 10

Construction A $\frac{5}{8}$-in. drill bit is too small for a job and a $\frac{3}{4}$-in. bit is too large. What size drill bit should be tried next?

To solve this problem, we must keep changing the given sizes to equivalent fractions until we can find a size between the two.

If we express them both in eighths, we have:

$$\frac{5}{8} \qquad \text{and} \qquad \frac{3}{4} = \frac{3 \times 2}{4 \times 2} = \frac{6}{8}$$

But there is no "in-between" size drill bit using these denominators. However, if we change them both to sixteenths, we have:

$$\frac{5}{8} = \frac{5 \times 2}{8 \times 2} = \frac{10}{16} \quad \text{and} \quad \frac{6}{8} = \frac{6 \times 2}{8 \times 2} = \frac{12}{16}$$

Now we can clearly see that an $\frac{11}{16}$-in. bit is larger than the $\frac{5}{8}$-in. bit and smaller than the $\frac{3}{4}$-in. bit. This "in-between" size bit should be tried next.

→ **Your Turn**

Which of the following quantities is the larger?

(a) $\frac{3}{4}$ in. or $\frac{5}{7}$ in.   (b) $\frac{7}{8}$ or $\frac{19}{21}$ (c) 3 or $\frac{40}{13}$

(d) $1\frac{7}{8}$ lb or $\frac{5}{3}$ lb   (e) $2\frac{1}{4}$ ft or $\frac{11}{6}$ ft

(f) **Construction** If a $\frac{3}{8}$-in. drill bit is too small for a job, and a $\frac{1}{2}$-in. bit is too large, what size bit should be tried next?

→ **Solutions**

(a) $\frac{3}{4} = \frac{21}{28}, \frac{5}{7} = \frac{20}{28}; \frac{21}{28}$ is larger than $\frac{20}{28}$, so $\frac{3}{4}$ in. is larger than $\frac{5}{7}$ in.

(b) $\frac{7}{8} = \frac{147}{168}, \frac{19}{21} = \frac{152}{168}; \frac{152}{168}$ is larger than $\frac{147}{168}$, so $\frac{19}{21}$ is larger than $\frac{7}{8}$.

(c) $3 = \frac{39}{13}; \frac{40}{13}$ is larger than $\frac{39}{13}$, so $\frac{40}{13}$ is larger than 3.

(d) $1\frac{7}{8} = \frac{15}{8} = \frac{45}{24}, \frac{5}{3} = \frac{40}{24}; \frac{45}{24}$ is larger than $\frac{40}{24}$, so $1\frac{7}{8}$ lb is larger than $\frac{5}{3}$ lb.

(e) $2\frac{1}{4} = \frac{9}{4} = \frac{54}{24}, \frac{11}{6} = \frac{44}{24}; \frac{54}{24}$ is larger than $\frac{44}{24}$, so $2\frac{1}{4}$ ft is larger than $\frac{11}{6}$ ft.

(f) To find a size between the two, we must change them both to sixteenths.

$$\frac{3}{8} = \frac{6}{16} \quad \text{and} \quad \frac{1}{2} = \frac{4}{8} = \frac{8}{16}$$

A $\frac{7}{16}$-in. bit is between the two in size and therefore should be tried next.

Now turn to Exercises 2-1 for some practice in working with fractions.

Exercises 2-1 Working with Fractions

A. Write as an improper fraction.

1. $2\frac{1}{3}$ 2. $7\frac{1}{2}$ 3. $8\frac{3}{8}$ 4. $1\frac{1}{16}$ 5. $2\frac{7}{8}$

6. 2   7. $2\frac{2}{3}$ 8. $4\frac{3}{64}$ 9. $4\frac{5}{6}$ 10. $1\frac{13}{16}$

B. Write as a mixed number.

1. $\dfrac{17}{2}$
2. $\dfrac{8}{5}$
3. $\dfrac{11}{8}$
4. $\dfrac{40}{16}$
5. $\dfrac{3}{2}$

6. $\dfrac{11}{3}$
7. $\dfrac{100}{6}$
8. $\dfrac{4}{3}$
9. $\dfrac{80}{32}$
10. $\dfrac{5}{2}$

C. Write in lowest terms.

1. $\dfrac{12}{16}$
2. $\dfrac{4}{6}$
3. $\dfrac{6}{16}$
4. $\dfrac{18}{4}$
5. $\dfrac{4}{10}$

6. $\dfrac{35}{30}$
7. $\dfrac{24}{30}$
8. $\dfrac{10}{4}$
9. $4\dfrac{3}{12}$
10. $\dfrac{34}{32}$

11. $\dfrac{42}{64}$
12. $\dfrac{10}{35}$
13. $\dfrac{15}{36}$
14. $\dfrac{45}{18}$
15. $\dfrac{38}{24}$

D. Complete.

1. $\dfrac{7}{8} = \dfrac{?}{16}$
2. $\dfrac{3}{4} = \dfrac{?}{16}$
3. $\dfrac{1}{8} = \dfrac{?}{64}$

4. $\dfrac{3}{8} = \dfrac{?}{64}$
5. $1\dfrac{1}{4} = \dfrac{?}{16}$
6. $2\dfrac{7}{8} = \dfrac{?}{32}$

7. $3\dfrac{3}{5} = \dfrac{?}{10}$
8. $1\dfrac{1}{16} = \dfrac{?}{32}$
9. $1\dfrac{40}{60} = \dfrac{?}{3}$

10. $4 = \dfrac{?}{6}$
11. $2\dfrac{5}{8} = \dfrac{?}{16}$
12. $2\dfrac{5}{6} = \dfrac{?}{12}$

E. Which is larger?

1. $\dfrac{3}{5}$ or $\dfrac{4}{7}$
2. $\dfrac{3}{2}$ or $\dfrac{13}{8}$
3. $1\dfrac{1}{2}$ or $1\dfrac{3}{7}$

4. $\dfrac{3}{4}$ or $\dfrac{13}{16}$
5. $\dfrac{7}{8}$ or $\dfrac{5}{6}$
6. $2\dfrac{1}{2}$ or $1\dfrac{11}{8}$

7. $1\dfrac{2}{5}$ or $\dfrac{6}{4}$
8. $\dfrac{3}{16}$ or $\dfrac{25}{60}$
9. $\dfrac{13}{5}$ or $\dfrac{5}{2}$

10. $3\dfrac{1}{2}$ or $2\dfrac{7}{4}$
11. $\dfrac{3}{8}$ or $\dfrac{5}{12}$
12. $1\dfrac{1}{5}$ or $\dfrac{8}{7}$

F. Practical Problems

1. **Carpentry** An apprentice carpenter measured the length of a 2-by-4 as $15\dfrac{6}{8}$ in. Express this measurement in lowest terms.

2. **Electrical Trades** An electrical light circuit in John's welding shop had a load of 2800 watts. He changed the circuit to ten 150-watt bulbs and six 100-watt bulbs. What fraction represents a comparison of the new load with the old load?

3. The numbers $\dfrac{22}{7}$, $\dfrac{19}{6}$, $\dfrac{47}{15}$, $\dfrac{25}{8}$, and $\dfrac{41}{13}$ are all reasonable approximations to the number π. Which is the largest approximation? Which is the smallest approximation?

4. **Sheet Metal Trades** Which is thicker, a $\dfrac{3}{16}$-in. sheet of metal or a $\dfrac{13}{64}$-in. fastener?

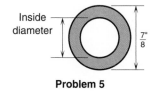

Problem 5

5. **Plumbing** Is it possible to have a $\frac{7}{8}$-in. pipe with an inside diameter of $\frac{29}{32}$ in.?

6. **Sheet Metal Trades** Fasteners are equally spaced on a metal vent cover, with nine spaces between fasteners covering 24 in. Write the distance between spaces as a mixed number.

$$\frac{24}{9} = ?$$

7. **Printing** A printer has 15 rolls of newsprint in the warehouse. What fraction of this total will remain if six rolls are used?

8. **Machine Trades** A machinist who had been producing 40 parts per day increased the output to 60 parts per day by going to a faster machine. How many times faster is the new machine? Express your answer as a mixed number.

9. **Landscaping** Before it can be used, a 12-ounce container of liquid fertilizer must be mixed with 48 ounces of water. What fraction of fertilizer is in the final mixture?

10. **Roofing** A ridge beam rises 18 in. over a horizontal run of 72 in. Calculate the pitch of the roof and express it as a fraction with a denominator of 12.

11. **Construction** A $\frac{3}{4}$-in. drill bit is too large for a job, and an $\frac{11}{16}$-in. bit is too small. What size should be tried next?

12. **Allied Health** During a visit to an elementary school, a health care worker determined that 15 of the 48 girls were overweight and 12 of 40 boys were overweight.

   (a) Express in lowest terms the fraction of girls who were overweight.

   (b) Express in lowest terms the fraction of boys who were overweight.

   (c) Is the fraction of overweight children higher among the girls or the boys? [*Hint:* Convert your answers to parts (a) and (b) to a common denominator.]

When you have had the practice you need, check your answers to the odd-numbered problems in the Appendix.

2-2 Multiplication of Fractions

The simplest arithmetic operation with fractions is multiplication and, happily, it is easy to show graphically. The multiplication of a whole number and a fraction may be illustrated this way.

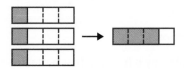

$$3 \times \frac{1}{4} = \frac{1}{4} + \frac{1}{4} + \frac{1}{4} = \frac{3}{4} \qquad \text{three segments each } \frac{1}{4} \text{ unit long.}$$

Any fraction such as $\frac{3}{4}$ can be thought of as a product: $\qquad 3 \times \frac{1}{4}$

The product of two fractions can also be shown graphically.

$$\frac{1}{2} \times \frac{1}{3} \quad \text{means} \quad \frac{1}{2} \text{ of } \frac{1}{3}$$

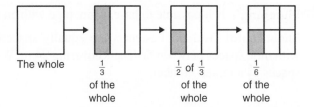

The whole $\frac{1}{3}$ $\frac{1}{2}$ of $\frac{1}{3}$ $\frac{1}{6}$
of the   of the   of the
whole   whole   whole

The product $\frac{1}{2} \times \frac{1}{3}$ is

$$\frac{1}{2} \times \frac{1}{3} = \frac{1}{6} = \frac{1 \text{ shaded area}}{6 \text{ equal areas in the square}}$$

In general, we calculate this product as

$$\frac{1}{2} \times \frac{1}{3} = \frac{1 \times 1}{2 \times 3}$$

Multiply the numerators (top)

Multiply the denominators (bottom)

$$\frac{1}{2} \times \frac{1}{3} = \frac{1}{6}$$

The product of two fractions is a fraction whose numerator is the product of their numerators and whose denominator is the product of their denominators.

Example 1

Fraction Times a Fraction

Multiply $\frac{5}{6} \times \frac{2}{3}$.

$$\frac{5}{6} \times \frac{2}{3} = \frac{5 \times 2}{6 \times 3} = \frac{10}{18} = \frac{5 \times \cancel{2}^{1}}{9 \times \cancel{2}_{1}} = \frac{5}{9}$$

> **Note**   Always write your answer in lowest terms. In Example 1 you probably recognized that both 10 and 18 were evenly divisible by 2, so you canceled out that common factor. It will save you time and effort if you cancel common factors, such as the 2 above, *before* you multiply this way:
>
> $$\frac{5 \times \cancel{2}^{1}}{\cancel{6}_{3} \times 3} = \frac{5}{9} \blacktriangleleft$$

Example 2

Whole Number Times a Fraction

When multiplying a whole number by a fraction, first write the whole number as a fraction.

Multiply $6 \times \frac{1}{8}$.

Step 1   Rewrite 6 as a fraction $6 \times \frac{1}{8} = \frac{6}{1} \times \frac{1}{8}$
with denominator 1.

Step 2   Divide by common $= \frac{\cancel{6}^{3} \times 1}{1 \times \cancel{8}_{4}} = \frac{3}{4}$
factors and multiply.

Example 3

Mixed Number Times a Fraction

Multiply $2\frac{1}{2} \times \frac{3}{4}$.

Step 1 Write the mixed number $2\frac{1}{2}$

as the improper fraction $\frac{5}{2}$.

$$2\frac{1}{2} \times \frac{3}{4} = \frac{5}{2} \times \frac{3}{4}$$

Step 2 Multiply.

$$= \frac{5 \times 3}{2 \times 4} = \frac{15}{8}$$

Step 3 Write the product as a mixed number.

$$= 1\frac{7}{8}$$

Example 4

Mixed Number Times a Mixed Number

Multiply $3\frac{1}{3} \times 2\frac{1}{4}$.

Step 1 Write as improper fractions.

$$3\frac{1}{3} \times 2\frac{1}{4} = \frac{10}{3} \times \frac{9}{4}$$

Step 2 Multiply.

$$= \frac{\overset{5}{\cancel{10}} \times \overset{3}{\cancel{9}}}{\underset{1}{\cancel{3}} \times \underset{2}{\cancel{4}}}$$

Step 3 Write the answer as a mixed number.

$$= \frac{15}{2} = 7\frac{1}{2}$$

Note In many word problems the words "of" or "product of" appear as signals that you are to multiply. For example, the phrase "one-half of 16" means

$$\frac{1}{2} \times 16 = \frac{1}{2} \times \frac{16}{1} = \frac{16}{2} = 8,$$ and the phrase "the product of $\frac{2}{3}$ and $\frac{1}{4}$"

should be translated as $\frac{2}{3} \times \frac{1}{4} = \frac{2}{12} = \frac{1}{6}$. ◀

→ Your Turn

Now test your understanding with these problems. Multiply as shown. Change any mixed numbers to improper fractions *before* you multiply.

(a) $\frac{7}{8}$ of $\frac{2}{3} = $ _____

(b) $\frac{8}{12} \times \frac{3}{16} = $ _____

(c) $\frac{3}{32}$ of $\frac{4}{15} = $ _____

(d) $\frac{15}{4} \times \frac{9}{10} = $ _____

(e) $\frac{3}{2}$ of $\frac{2}{3} = $ _____

(f) $1\frac{1}{2} \times \frac{2}{5} = $ _____

(g) $4 \times \frac{7}{8} = $ _____

(h) $3\frac{5}{6} \times \frac{3}{10} = $ _____

(i) $1\frac{4}{5} \times 1\frac{3}{4} = $ _____

(j) **Plumbing** Polly the Plumber needs six lengths of PVC pipe each $26\frac{3}{4}$ in. long. What total length of pipe will she need?

Remember to write your answer in lowest terms.

→ **Solutions**

(a) (*Hint:* "Of" means multiply.)

$$\frac{7}{8} \times \frac{2}{3} = \frac{7 \times 2}{(4 \times 2) \times 3} = \frac{7}{4 \times 3} = \frac{7}{12}$$

Eliminate common factors before you multiply.

Your work will look like this when you learn to do these operations mentally:

$$\frac{7}{\overset{8}{\underset{4}{8}}} \times \frac{\overset{1}{2}}{3} = \frac{7}{12}$$

(b) $$\frac{8}{12} \times \frac{3}{16} = \frac{\overset{1}{8} \times \overset{1}{3}}{(4 \times \underset{1}{3}) \times (\underset{1}{8} \times 2)} = \frac{1}{4 \times 2} = \frac{1}{8}$$

$$\text{or } \frac{\overset{1}{8}}{\underset{4}{12}} \times \frac{\overset{1}{3}}{\underset{2}{16}} = \frac{1}{8}$$

(c) $$\frac{\overset{1}{3}}{\underset{8}{32}} \times \frac{\overset{1}{4}}{\underset{5}{15}} = \frac{1}{40}$$

(d) $$\frac{\overset{3}{15}}{4} \times \frac{9}{\underset{2}{10}} = \frac{27}{8} = 3\frac{3}{8}$$

(e) $$\frac{\overset{1}{3}}{\underset{1}{2}} \times \frac{2}{\underset{1}{3}} = 1$$

(f) $$1\frac{1}{2} \times \frac{2}{5} = \frac{3}{2} \times \frac{\overset{1}{2}}{5} = \frac{3}{5}$$

If you don't remember how to change a mixed number to an improper fraction see page 73.

(g) $$4 \times \frac{7}{8} = \frac{\overset{1}{4}}{1} \times \frac{7}{\underset{2}{8}} = \frac{7}{2} = 3\frac{1}{2}$$

(h) $$3\frac{5}{6} \times \frac{3}{10} = \frac{23}{\underset{2}{6}} \times \frac{\overset{1}{3}}{10} = \frac{23}{20} = 1\frac{3}{20}$$

(i) $$1\frac{4}{5} \times 1\frac{3}{4} = \frac{9}{5} \times \frac{7}{4} = \frac{63}{20} = 3\frac{3}{20}$$

(j) Total length of pipe = number of pieces $\times$ length of each piece

$$= 6 \times 26\frac{3}{4}$$

$$= 6 \times \frac{107}{4}$$

$$= \frac{\overset{3}{6}}{1} \times \frac{107}{\underset{2}{4}}$$

$$= \frac{321}{2} = 160\frac{1}{2} \text{ in.}$$

Exercises 2-2 Multiplication of Fractions

A. Multiply and write the answer in lowest terms.

1. $\dfrac{1}{2} \times \dfrac{1}{4}$
2. $\dfrac{2}{5} \times \dfrac{2}{3}$
3. $\dfrac{4}{5} \times \dfrac{1}{6}$
4. $6 \times \dfrac{1}{2}$

5. $\dfrac{8}{9} \times 3$
6. $\dfrac{11}{12} \times \dfrac{4}{15}$
7. $\dfrac{8}{3} \times \dfrac{5}{12}$
8. $\dfrac{7}{8} \times \dfrac{13}{14}$

9. $\dfrac{12}{8} \times \dfrac{15}{9}$
10. $\dfrac{4}{7} \times \dfrac{49}{2}$
11. $4\dfrac{1}{2} \times \dfrac{2}{3}$
12. $6 \times 1\dfrac{1}{3}$

13. $2\dfrac{1}{6} \times 1\dfrac{1}{2}$
14. $\dfrac{5}{7} \times 1\dfrac{7}{15}$
15. $4\dfrac{3}{5} \times 15$
16. $10\dfrac{5}{6} \times 3\dfrac{3}{10}$

17. $34 \times 2\dfrac{3}{17}$
18. $7\dfrac{9}{10} \times 1\dfrac{1}{4}$
19. $11\dfrac{6}{7} \times \dfrac{7}{8}$

20. $18 \times 1\dfrac{5}{27}$
21. $\dfrac{1}{2} \times \dfrac{1}{2} \times \dfrac{1}{2}$
22. $1\dfrac{4}{5} \times \dfrac{2}{3} \times \dfrac{1}{4}$

23. $\dfrac{1}{4} \times \dfrac{2}{3} \times \dfrac{2}{5}$
24. $2\dfrac{1}{2} \times \dfrac{3}{5} \times \dfrac{8}{9}$
25. $\dfrac{2}{3} \times \dfrac{3}{2} \times 2$

B. Find.

1. $\dfrac{1}{2}$ of $\dfrac{1}{3}$
2. $\dfrac{1}{4}$ of $\dfrac{3}{8}$
3. $\dfrac{2}{3}$ of $\dfrac{3}{4}$
4. $\dfrac{7}{8}$ of $\dfrac{1}{2}$

5. $\dfrac{1}{2}$ of $1\dfrac{1}{2}$
6. $\dfrac{3}{4}$ of $1\dfrac{1}{4}$
7. $\dfrac{5}{8}$ of $2\dfrac{1}{10}$
8. $\dfrac{5}{3}$ of $1\dfrac{2}{3}$

9. $\dfrac{4}{3}$ of $\dfrac{3}{4}$
10. $\dfrac{3}{5}$ of $1\dfrac{1}{6}$
11. $\dfrac{7}{8}$ of $1\dfrac{1}{5}$
12. $\dfrac{3}{5}$ of 4

13. $\dfrac{7}{16}$ of 6
14. $\dfrac{5}{16}$ of $1\dfrac{1}{7}$
15. $\dfrac{3}{8}$ of $2\dfrac{2}{3}$
16. $\dfrac{15}{16}$ of $1\dfrac{3}{5}$

C. Practical Problems

1. **Flooring and Carpeting**   Find the width of floor space covered by 38 boards with $3\dfrac{5}{8}$-in. exposed surface each.

2. **Construction**   There are 14 risers in the stairs from the basement to the first floor of a house. Find the total height of the stairs if the risers are $7\dfrac{1}{8}$ in. high.

3. **Roofing**   Shingles are laid so that 5 in. or $\dfrac{5}{12}$ ft is exposed in each layer. How many feet of roof will be covered by 28 courses?

4. **Carpentry**   A board $5\dfrac{3}{4}$ in. wide is cut to three-fourths of its original width. Find the new width.

5. **Carpentry**   What length of 2-in. by 4-in. material will be required to make six bench legs each $28\dfrac{1}{4}$ in. long?

6. **Electrical Trades**   Find the total length of 12 pieces of wire each $9\dfrac{3}{16}$ in. long.

7. **Automotive Trades**   If a car averages $22\dfrac{3}{4}$ miles to a gallon of gas, how many miles can it travel on 14 gallons of gas?

8. **Machine Trades**   What is the shortest bar that can be used for making six chisels each $6\dfrac{1}{8}$ in. in length?

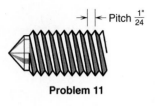

→| |← Pitch $\frac{1"}{24}$

Problem 11

9. **Manufacturing** How many pounds of grease are contained in a barrel if a barrel holds $46\frac{1}{2}$ gallons, and a gallon of grease weighs $7\frac{2}{3}$ lb?

10. **Plumbing** A plumber cut eight pieces of copper tubing from a coil. Each piece is $14\frac{3}{4}$ in. long. What is the total length used?

11. **Machine Trades** How far will a nut advance if it is given 18 turns on a $\frac{1}{4}$-in. 20-NF (National Fine thread) bolt? (*Hint:* The designation 20-NF means that the nut advances $\frac{1}{20}$ in. for each complete turn.)

12. **Flooring and Carpeting** What width of floor space can be covered by 48 boards each with $4\frac{3}{8}$ in. of exposed surface?

13. **Drafting** If $\frac{3}{8}$ in. on a drawing represents 1 ft, how many inches on the drawing will represent 26 ft?

14. **Manufacturing** What is the volume of a rectangular box with interior dimensions $12\frac{1}{2}$ in. long, $8\frac{1}{8}$ in. wide, and $4\frac{1}{4}$ in. deep? (*Hint:* Volume = length × width × height.)

15. **Machine Trades** How long will it take to machine 45 pins if each pin requires $6\frac{3}{4}$ minutes? Allow 1 minute per pin for placing stock in the lathe.

16. **Manufacturing** There are 231 cu in. in a gallon. How many cubic inches are needed to fill a container with a rated capacity of $4\frac{1}{3}$ gallons?

17. **Printing** In a print shop, 1 unit of labor is equal to $\frac{1}{6}$ hour. How many hours are involved in 64 units of work?

18. **Carpentry** Find the total width of 36 2-by-4s if the finished width of each board is actually $3\frac{1}{2}$ in.

19. **Photography** A photograph must be reduced to four-fifths of its original size to fit the space available in a newspaper. Find the length of the reduced photograph if the original was $6\frac{3}{4}$ in. long.

20. **Printing** A bound book weighs $1\frac{5}{8}$ lb. How many pounds will 20 cartons of 12 books each weigh?

21. **Printing** There are 6 picas in 1 in. If a line of type is $3\frac{3}{4}$ in. long, what is this length in picas?

22. **Wastewater Technology** The normal daily flow of raw sewage into a treatment plant is 32 MGD (million gallons per day). Because of technical problems one day, the plant had to cut back to three-fourths of its normal intake. What was this reduced flow?

23. **Masonry** What is the height of 12 courses of $2\frac{1}{4}$-in. bricks with $\frac{3}{8}$-in. mortar joints? (12 rows of brick and 11 mortar joints)

24. **Plumbing** To find the degree measure of the bend of a pipe fitting, multiply the fraction of bend by 360 degrees. What is the degree measure of a $\frac{1}{8}$ bend? A $\frac{1}{5}$ bend? A $\frac{1}{6}$ bend?

25. **Plumbing** A drain must be installed with a grade of $\frac{1}{8}$ in. of vertical drop per foot of horizontal run. How much drop will there be for 26 ft of run?

26. **Machine Trades** The center-to-center distance between consecutive holes in a strip of metal is $\frac{5}{16}$ in. What is the total distance x between the first and last centers as shown in the figure?

27. **Sheet Metal Trades** The allowance for a wired edge on fabricated metal is $2\frac{1}{2}$ times the diameter of the wire. Calculate the allowance for a wired edge if the diameter of the wire is $\frac{3}{16}$ in.

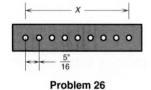

|← x →|

$\frac{5"}{16}$

Problem 26

Chapter 2 Fractions

28. **Allied Health**   One caplet of a certain cold medicine contains 240 mg of medication. How much medication is contained in $3\frac{1}{2}$ caplets?

29. **Allied Health**   From the age of 1 year to 3 years, a child's weight increases at an average rate of $4\frac{5}{8}$ pounds per year. How much weight would an average child be expected to gain in these two years?

30. **Graphic Design**   A graphic artist must produce 12 different illustrations for a report. Each illustration takes about $\frac{3}{4}$ of an hour to prepare, and she will need an additional 2 hours to insert the illustrations into the report. How many hours should she set aside for this project?

31. **Construction**   A patio $15\frac{1}{2}$ feet wide is being constructed next to a house. To drain water off the patio and away from the house, a slope of $\frac{1}{4}$ inch per foot is necessary. What total difference in height is required from one edge of the patio to the other?

32. **Carpentry**   A carpenter is drilling a pilot hole for a $\frac{1}{8}$-inch wood screw. Pilot holes must be slightly smaller than the actual size of the screw. Which of the following sized pilot holes should the carpenter drill: $\frac{9}{64}$ in., $\frac{5}{64}$ in., $\frac{5}{32}$ in., or $\frac{3}{16}$ in.?

Check your answers to the odd-numbered problems in the Appendix, then continue in Section 2-3.

2-3   Division of Fractions

Addition and multiplication are both reversible arithmetic operations. For example,

2 × 3 and 3 × 2 both equal 6

4 + 5 and 5 + 4 both equal 9

The order in which you add or multiply is not important. This reversibility is called the *commutative* property of addition and multiplication.

In division this type of exchange is not allowed, and because it is not allowed many people find division very troublesome. In the division of fractions it is very important that you set up the problem correctly.

The phrase "8 divided by 4" can be written as

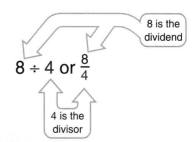

In the previous problem you were being asked to divide a set of eight objects into sets of four objects.

Example 1

In the division $5 \div \frac{1}{2}$, which number is the divisor?

The divisor is $\frac{1}{2}$.

The division $5 \div \frac{1}{2}$, read "5 divided by $\frac{1}{2}$," asks how many $\frac{1}{2}$-unit lengths are included in a length of 5 units.

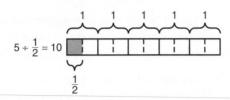

5 units $\qquad$ $\frac{1}{2}$ unit

Division answers the question: How many of the divisor are in the dividend?

$8 \div 4 = \square$ asks you to find how many 4s are in 8.
It is easy to see that $\square = 2$.

$5 \div \frac{1}{2} = \square$ asks you to find how many $\frac{1}{2}$s are in 5. Do you see that $\square = 10$?

$$5 \div \frac{1}{2} = 10$$

There are ten $\frac{1}{2}$-unit lengths contained in the 5-unit length.

Using a drawing like this to solve a division problem is difficult and clumsy. We need a simple rule.

Reciprocal   We can simplify the division of fractions with the help of a new concept called the **reciprocal.** The reciprocal of a fraction is obtained by switching its numerator and denominator. (This is often called *inverting* the fraction.)

Example 2

The reciprocal of $\frac{5}{6}$ is $\frac{6}{5}$.

The reciprocal of $\frac{1}{4}$ is $\frac{4}{1}$ or 4.

The reciprocal of 8 or $\frac{8}{1}$ is $\frac{1}{8}$.

To find the reciprocal of a mixed number, first convert it to an improper fraction.

The reciprocal of $2\frac{3}{5}$ or $\frac{13}{5}$ is $\frac{5}{13}$.

→ Your Turn

Find the reciprocal of each of the following numbers.

(a) $\frac{2}{5}$ (b) $\frac{7}{8}$ (c) $\frac{1}{6}$ (d) 7   (e) $2\frac{1}{5}$

→ Answers

(a) $\frac{5}{2}$ (b) $\frac{8}{7}$ (c) $\frac{6}{1}$ or 6   (d) $7 = \frac{7}{1}$ so its reciprocal is $\frac{1}{7}$.

(e) $2\frac{1}{5}$ is $\frac{11}{5}$ and its reciprocal is $\frac{5}{11}$.

We can now use the following simple rule to divide fractions.

> To divide by a fraction, multiply by its reciprocal.

Example 3

$5 \div \frac{1}{2} = ?$

$$5 \div \frac{1}{2} = 5 \times \frac{2}{1} = \frac{10}{1} = 10$$ as shown graphically on the previous page

(Reciprocal: $\frac{1}{2} \to \frac{2}{1}$; Multiply: $\div \to \times$)

Example 4

$\frac{2}{5} \div \frac{4}{3} = ?$

$$\frac{2}{5} \div \frac{4}{3} = \frac{2}{5} \times \frac{3}{4} = \frac{\overset{1}{\cancel{2}} \times 3}{5 \times \underset{2}{\cancel{4}}} = \frac{3}{10}$$

(Multiply: $\div \to \times$; The reciprocal: $\frac{4}{3} \to \frac{3}{4}$)

Careful When dividing fractions, never attempt to cancel common factors until *after* converting the division to multiplication by the reciprocal. ◄

Example 5

$\frac{5}{8} \div 3\frac{3}{4} = ?$

$$\frac{5}{8} \div 3\frac{3}{4} = \frac{5}{8} \div \frac{15}{4} = \frac{\overset{1}{\cancel{5}}}{\underset{2}{\cancel{8}}} \times \frac{\overset{1}{\cancel{4}}}{\underset{3}{\cancel{15}}} = \frac{1}{6}$$

We have converted division problems that are difficult to picture into simple multiplication.

✔ The final, and very important, step in every division is checking the answer. To check, multiply the divisor by the quotient and compare this answer with the original fraction or dividend.

If $\frac{2}{5} \div \frac{4}{3} = \frac{3}{10}$ then $\frac{4}{3} \times \frac{3}{10}$ should equal $\frac{2}{5}$.

$$\frac{\overset{2}{\cancel{4}}}{\underset{1}{\cancel{3}}} \times \frac{\overset{1}{\cancel{3}}}{\underset{5}{\cancel{10}}} = \frac{2}{5}$$

→ **Your Turn**

Divide:

$$\frac{7}{8} \div \frac{3}{2} = \underline{\hspace{2cm}}$$

→ **Solution**

Reciprocal

$$\frac{7}{8} \div \frac{3}{2} = \frac{7}{8} \times \frac{2}{3} = \frac{7 \times \overset{1}{\cancel{2}}}{\underset{4}{\cancel{8}} \times 3} = \frac{7}{12}$$

☑ $\dfrac{\cancel{3}}{2} \times \dfrac{7}{\underset{4}{\cancel{12}}} = \dfrac{7}{8}$

Multiply

Learning Help The chief source of confusion for many people in dividing fractions is deciding which fraction to invert. It will help if you:

1. Put every division problem in the form

    (dividend) ÷ (divisor)

    then find the reciprocal of the divisor, and finally, multiply to obtain the quotient.

2. Check your answer by multiplying. The product

    (divisor) × (quotient or answer)

    should equal the dividend. ◄

→ **More Practice**

Here are a few problems to test your understanding.

(a) $\dfrac{2}{5} \div \dfrac{3}{8}$ (b) $\dfrac{7}{40} \div \dfrac{21}{25}$ (c) $3\dfrac{3}{4} \div \dfrac{5}{2}$

(d) $4\dfrac{1}{5} \div 1\dfrac{4}{10}$ (e) $3\dfrac{2}{3} \div 3$ (f) Divide $\dfrac{3}{4}$ by $2\dfrac{5}{8}$.

(g) Divide $1\dfrac{1}{4}$ by $1\dfrac{7}{8}$.   (h) **Carpentry** How many sheets of plywood, each $\dfrac{3}{4}$ in. thick, are in a stack 18 in. high?

Work carefully and check each answer.

→ **Solutions**

(a) $\dfrac{2}{5} \div \dfrac{3}{8} = \dfrac{2}{5} \times \dfrac{8}{3} = \dfrac{16}{15} = 1\dfrac{1}{15}$

The answer is $1\dfrac{1}{15}$.   ☑ $\dfrac{\cancel{3}}{8} \times \dfrac{\overset{2}{\cancel{16}}}{\underset{5}{\cancel{15}}} = \dfrac{2}{5}$

(b) $\dfrac{7}{40} \div \dfrac{21}{25} = \dfrac{\overset{1}{\cancel{7}}}{\underset{8}{\cancel{40}}} \times \dfrac{\overset{5}{\cancel{25}}}{\underset{3}{\cancel{21}}} = \dfrac{5}{24}$ ✔ $\dfrac{\overset{7}{\cancel{21}}}{\underset{5}{\cancel{25}}} \times \dfrac{\overset{1}{\cancel{8}}}{\underset{8}{\cancel{24}}} = \dfrac{7}{40}$

(c) $3\dfrac{3}{4} \div \dfrac{5}{2} = \dfrac{15}{4} \div \dfrac{5}{2} = \dfrac{\overset{3}{\cancel{15}}}{\underset{2}{\cancel{4}}} \times \dfrac{\overset{1}{\cancel{2}}}{\underset{1}{\cancel{8}}} = \dfrac{3}{2} = 1\dfrac{1}{2}$ ✔ $\dfrac{5}{2} \times \dfrac{3}{2} = \dfrac{15}{4} = 3\dfrac{3}{4}$

(d) $4\dfrac{1}{5} \div 1\dfrac{4}{10} = \dfrac{21}{5} \div \dfrac{14}{10} = \dfrac{\overset{3}{\cancel{21}}}{\underset{1}{\cancel{5}}} \times \dfrac{\overset{2}{\cancel{10}}}{\underset{2}{\cancel{14}}} = \dfrac{6}{2} = 3$

✔ $1\dfrac{4}{10} \times 3 = \dfrac{\overset{7}{\cancel{14}}}{\underset{5}{\cancel{10}}} \times \dfrac{3}{1} = \dfrac{21}{5} = 4\dfrac{1}{5}$

(e) $3\dfrac{2}{3} \div 3 = \dfrac{11}{3} \div \dfrac{3}{1} = \dfrac{11}{3} \times \dfrac{1}{3} = \dfrac{11}{9} = 1\dfrac{2}{9}$

✔ $3 \times \dfrac{11}{9} = \dfrac{\overset{1}{\cancel{3}}}{1} \times \dfrac{11}{\underset{3}{\cancel{9}}} = \dfrac{11}{3} = 3\dfrac{2}{3}$

(f) $\dfrac{3}{4} \div 2\dfrac{5}{8} = \dfrac{3}{4} \div \dfrac{21}{8} = \dfrac{\overset{1}{\cancel{3}}}{\underset{1}{\cancel{4}}} \times \dfrac{\overset{2}{\cancel{8}}}{\underset{7}{\cancel{21}}} = \dfrac{2}{7}$ ✔ $2\dfrac{5}{8} \times \dfrac{2}{7} = \dfrac{\overset{3}{\cancel{21}}}{\underset{4}{\cancel{8}}} \times \dfrac{\overset{1}{\cancel{2}}}{\underset{1}{\cancel{7}}} = \dfrac{3}{4}$

(g) $1\dfrac{1}{4} \div 1\dfrac{7}{8} = \dfrac{5}{4} \div \dfrac{15}{8} = \dfrac{\overset{1}{\cancel{5}}}{\underset{1}{\cancel{4}}} \times \dfrac{\overset{2}{\cancel{8}}}{\underset{3}{\cancel{15}}} = \dfrac{2}{3}$

✔ $1\dfrac{7}{8} \times \dfrac{2}{3} = \dfrac{\overset{5}{\cancel{15}}}{\underset{4}{\cancel{8}}} \times \dfrac{\overset{1}{\cancel{2}}}{\underset{1}{\cancel{3}}} = \dfrac{5}{4} = 1\dfrac{1}{4}$

(h) A question of the form "How many X are in Y?" tells us we must divide Y by X. In this problem, divide 18 in., the total height, by $\frac{3}{4}$ in., the thickness of each sheet.

$18 \div \dfrac{3}{4} = \dfrac{\overset{6}{\cancel{18}}}{1} \times \dfrac{4}{\underset{1}{\cancel{8}}} = 24$ There are 24 sheets of plywood in the stack.

WHY DO WE USE THE RECIPROCAL WHEN WE DIVIDE FRACTIONS?

Notice that $8 \div 4 = 2$ and $8 \times \dfrac{1}{4} = \dfrac{\overset{2}{\cancel{8}}}{1} \times \dfrac{1}{\underset{1}{\cancel{4}}} = 2.$

Dividing by a number gives the same result as multiplying by its reciprocal. The following shows why this is so.

(*continued*)

The division $8 \div 4$ can be written $\dfrac{8}{4}$. Similarly, $\dfrac{1}{2} \div \dfrac{2}{3}$ can be written $\dfrac{\frac{1}{2}}{\frac{2}{3}}$.

To simplify this fraction, multiply by $\dfrac{\frac{3}{2}}{\frac{3}{2}}$ (which is equal to 1).

$$\frac{\frac{1}{2}}{\frac{2}{3}} = \frac{\frac{1}{2} \times \frac{3}{2}}{\frac{2}{3} \times \frac{3}{2}} = \frac{\frac{1}{2} \times \frac{3}{2}}{1} = \frac{1}{2} \times \frac{3}{2}$$

$$\frac{2}{3} \times \frac{3}{2} = \frac{2 \times 3}{3 \times 2} = \frac{6}{6} = 1$$

Therefore, $\dfrac{1}{2} \div \dfrac{2}{3} = \dfrac{1}{2} \times \dfrac{3}{2}$, the reciprocal of the divisor $\dfrac{2}{3}$.

Turn to Exercises 2-3 for a set of practice problems on dividing fractions.

Exercises 2-3 Division of Fractions

A. Divide and write the answer in lowest terms.

1. $\dfrac{5}{6} \div \dfrac{1}{2}$
2. $6 \div \dfrac{2}{3}$
3. $\dfrac{5}{12} \div \dfrac{4}{3}$
4. $8 \div \dfrac{1}{4}$

5. $\dfrac{6}{16} \div \dfrac{3}{4}$
6. $\dfrac{1}{2} \div \dfrac{1}{2}$
7. $\dfrac{3}{16} \div \dfrac{6}{8}$
8. $\dfrac{3}{4} \div \dfrac{5}{16}$

9. $1\dfrac{1}{2} \div \dfrac{1}{6}$
10. $6 \div 1\dfrac{1}{2}$
11. $3\dfrac{1}{7} \div 2\dfrac{5}{14}$
12. $3\dfrac{1}{2} \div 2$

13. $6\dfrac{2}{5} \div 5\dfrac{1}{3}$
14. $10 \div 1\dfrac{1}{5}$
15. $8 \div \dfrac{1}{2}$
16. $\dfrac{2}{3} \div 6$

17. $\dfrac{12}{\frac{2}{3}}$
18. $\dfrac{\frac{3}{4}}{\frac{7}{8}}$
19. $\dfrac{\frac{5}{2}}{3}$
20. $\dfrac{1\frac{1}{2}}{2\frac{1}{2}}$

21. $\dfrac{5}{16} \div \dfrac{3}{8}$
22. $\dfrac{7}{12} \div \dfrac{2}{3}$
23. $\dfrac{7}{32} \div 1\dfrac{3}{4}$
24. $1\dfrac{2}{3} \div 1\dfrac{1}{4}$

B. Practical Problems

1. **Drafting** How many feet are represented by a 4-in. line if it is drawn to a scale of $\frac{1}{2}$ in. = 1 ft?

2. **Drafting** If $\frac{1}{4}$ in. on a drawing represents 1 ft 0 in., then $3\frac{1}{2}$ in. on the drawing will represent how many feet?

3. **Flooring and Carpeting** How many boards $4\frac{5}{8}$ in. wide will it take to cover a floor 222 in. wide?

92   Chapter 2 Fractions

4. **Construction** How many supporting columns $88\frac{1}{2}$ in. long can be cut from six pieces each 22 ft long? (*Hint:* Be careful of units.)

5. **Construction** How many pieces of $\frac{1}{2}$-in. plywood are there in a stack 42 in. high?

6. **Drafting** If $\frac{1}{4}$ in. represents 1 ft 0 in. on a drawing, how many feet will be represented by a line $10\frac{1}{8}$ in. long?

7. **Masonry** If we allow $2\frac{5}{8}$ in. for the thickness of a course of brick, including mortar joints, how many courses of brick will there be in a wall $47\frac{1}{4}$ in. high?

8. **Plumbing** How many lengths of pipe $2\frac{5}{8}$ ft long can be cut from a pipe 21 ft long?

9. **Machine Trades** How many pieces $6\frac{1}{4}$ in. long can be cut from 35 metal rods each 40 in. long? Disregard waste.

10. **Architecture** The architectural drawing for a room measures $3\frac{5}{8}$ in. by $4\frac{1}{4}$ in. If $\frac{1}{8}$ in. is equal to 1 ft on the drawing, what are the actual dimensions of the room?

11. **Printing** How many full $3\frac{1}{2}$-in. sheets can be cut from $24\frac{3}{4}$-in. stock?

12. **Machine Trades** The feed on a boring mill is set for $\frac{1}{32}$ in. How many revolutions are needed to advance the tool $3\frac{3}{8}$ in.?

13. **Machine Trades** If the pitch of a thread is $\frac{1}{18}$ in., how many threads are needed for the threaded section of a pipe to be $2\frac{1}{2}$ in. long?

14. **Construction** The floor area of a room on a house plan measures $3\frac{1}{2}$ in. by $4\frac{5}{8}$ in. If the drawing scale is $\frac{1}{4}$ in. represents 1 ft, what is the actual size of the room?

15. **Construction** The family room ceiling of a new Happy Home is to be taped and mudded. A $106\frac{1}{2}$ sq ft area has already been done, and this represents $\frac{3}{8}$ of the job. How large is the area of the finished ceiling?

16. **Plumbing** A pipe fitter needs to divide a pipe $32\frac{5}{8}$ inches long into three pieces of equal length. Calculate the length of each piece.

17. **Automotive Trades** Over a period of $3\frac{1}{2}$ years, $\frac{7}{16}$ in. of tread has worn off a tire. What was the average tread wear per year?

18. **Automotive Trades** On a certain vehicle, each turn of the tie-rod sleeve changes the toe-in setting by $\frac{1}{8}$ in. How many turns of the sleeve are needed to change the toe-in setting by $\frac{5}{32}$ in.?

19. **Carpentry** A carpenter must mount an electrical panel on an uneven surface. She must raise one corner of the panel $1\frac{1}{8}$ in. off the surface. If $\frac{3}{16}$-in. washers are used as spacers, how many washers are needed to mount the panel?

20. **Allied Health** The average height of a 2-year old girl is $36\frac{1}{2}$ inches, and the average height of an 11-year old girl is 57 inches. What is the average annual growth in inches during this 9-year period? (*Hint:* $57 - 36\frac{1}{2} = 20\frac{1}{2}$)

Check your answers to the odd-numbered problems in the Appendix, then continue in Section 2-4.

Addition   At heart, adding fractions is a matter of counting:

$$\frac{1}{5} + \frac{3}{5} = \frac{1+3}{5} = \frac{4}{5}$$

$\frac{1}{5}$ [ ][ ][ ][ ][ ]   1 fifth
            +
$\frac{3}{5}$ [ ][ ][ ][ ][ ]   + 3 fifths
            =
$\frac{4}{5}$ [ ][ ][ ][ ][ ]   = 4 fifths, count them.

Example 1

Add $\dfrac{1}{8} + \dfrac{3}{8} =$ _____

This one is easy to see with measurements:

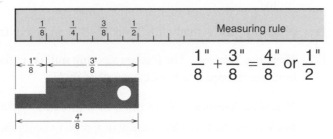

$$\frac{1}{8}" + \frac{3}{8}" = \frac{4}{8}" \text{ or } \frac{1}{2}"$$

→ **Your Turn**

Add $\dfrac{2}{7} + \dfrac{3}{7} =$ _____

→ **Solution**

$$\frac{2}{7} + \frac{3}{7} = \frac{2+3}{7} = \frac{5}{7}$$

$\frac{2}{7}$ [ ][ ][ ][ ][ ][ ][ ]   2 sevenths
            +
$\frac{3}{7}$ [ ][ ][ ][ ][ ][ ][ ]   + 3 sevenths
            =
$\frac{5}{7}$ [ ][ ][ ][ ][ ][ ][ ]   5 sevenths or $\frac{5}{7}$

Like Fractions   Fractions having the same denominator are called *like* fractions. In the preceding problem, $\frac{2}{7}$ and $\frac{3}{7}$ both have denominator 7 and are like fractions. Adding like fractions is easy: *first,* add the numerators to find the numerator of the sum and *second,* use the denominator the fractions have in common as the denominator of the sum.

$$\frac{2}{9} + \frac{5}{9} = \frac{2+5}{9} \longleftarrow \boxed{\text{Add numerators}}$$

$$= \frac{7}{9} \longleftarrow \boxed{\text{Same denominator}}$$

We can use this same technique to add three or more like fractions.

Example 2

Add $\dfrac{3}{12} + \dfrac{1}{12} + \dfrac{5}{12}$ like this:

$$\frac{3}{12} + \frac{1}{12} + \frac{5}{12} = \frac{3 + 1 + 5}{12} = \frac{9}{12} = \frac{3}{4}$$

Notice that we write the sum in lowest terms.

→ **Your Turn**

Try these problems for exercise.

(a) $\dfrac{1}{8} + \dfrac{3}{8}$ (b) $\dfrac{7}{9} + \dfrac{5}{9} + \dfrac{4}{9} + \dfrac{8}{9}$

→ **Solutions**

(a) $\dfrac{1}{8} + \dfrac{3}{8} = \dfrac{1 + 3}{8} = \dfrac{4}{8} = \dfrac{1}{2}$

(b) $\dfrac{7}{9} + \dfrac{5}{9} + \dfrac{4}{9} + \dfrac{8}{9} = \dfrac{7 + 5 + 4 + 8}{9} = \dfrac{24}{9} = \dfrac{8}{3} = 2\dfrac{2}{3}$

When the addition problem involves mixed numbers or whole numbers, it may be easier to arrange the sum vertically, as shown in the next example.

Example 3

To add the mixed numbers $2\dfrac{1}{5} + 3\dfrac{3}{5}$

First, arrange the numbers vertically, with
whole numbers in one column and fractions
in another column.

$$\begin{array}{r} 2\dfrac{1}{5} \\ + \; 3\dfrac{3}{5} \\ \hline \end{array}$$

Then, add fractions and whole numbers
separately.

$$\begin{array}{r} 2\dfrac{1}{5} \\ + \; 3\dfrac{3}{5} \\ \hline 5\dfrac{4}{5} \end{array} \longleftarrow \boxed{\text{Answer}}$$

If the sum of the fraction parts is greater than 1, an extra step will be required to simplify the answer. This is illustrated in Example 4.

Example 4

Add $1\frac{5}{8} + 3\frac{7}{8}$.

If we arrange the mixed numbers vertically and find the sum of the whole numbers and fractions separately, we have:

$$\begin{array}{r} 1\frac{5}{8} \\ + 3\frac{7}{8} \\ \hline 4\frac{12}{8} \end{array}$$

Notice that the sum of the fraction parts is greater than 1. Therefore, we must simplify the answer as follows:

$$4\frac{12}{8} = 4 + \frac{12}{8} = 4 + \frac{3}{2} = 4 + 1\frac{1}{2} = 5\frac{1}{2}$$

Learning Help   If you notice in advance that the sum of the fraction parts of the mixed numbers will be greater than or equal to 1, you may find it easier to first rewrite the mixed numbers as improper fractions and then add. For Example 4, this alternate solution would go as follows:

First, rewrite each mixed number as an improper fraction.
$$1\frac{5}{8} + 3\frac{7}{8} = \frac{13}{8} + \frac{31}{8}$$

Then, add.
$$= \frac{44}{8}$$

Finally, rewrite in lowest terms and convert back to a mixed number.
$$= \frac{11}{2} = 5\frac{1}{2} \blacktriangleleft$$

→ **Your Turn**

Find each sum:

(a) $6 + 3\frac{2}{3}$ (b) $1\frac{5}{8} + 2\frac{1}{8}$ (c) $4\frac{1}{6} + 2\frac{5}{6} + 3\frac{5}{6}$ (d) $3\frac{1}{4} + 2\frac{3}{4}$

(e) **Construction** A stud $3\frac{4}{8}$-in. thick is covered on both sides with $\frac{3}{8}$-in. sheetrock. What is the total thickness of the resulting wall?

→ **Solutions**

(a) $\begin{array}{r} 6 \\ + 3\frac{2}{3} \\ \hline 9\frac{2}{3} \end{array}$

(b) $\begin{array}{r} 1\frac{5}{8} \\ + 2\frac{1}{8} \\ \hline 3\frac{6}{8} \end{array} = 3\frac{3}{4}$

Always express the answer in lowest terms.

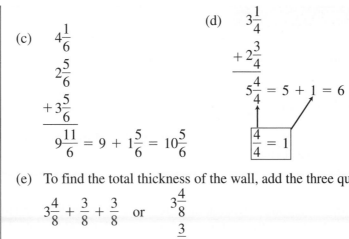

(c) $4\frac{1}{6}$

$2\frac{5}{6}$

$+3\frac{5}{6}$

$9\frac{11}{6} = 9 + 1\frac{5}{6} = 10\frac{5}{6}$

(d) $3\frac{1}{4}$

$+2\frac{3}{4}$

$5\frac{4}{4} = 5 + 1 = 6$

$\boxed{\frac{4}{4} = 1}$

(e) To find the total thickness of the wall, add the three quantities:

$3\frac{4}{8} + \frac{3}{8} + \frac{3}{8}$ or $3\frac{4}{8}$

$\frac{3}{8}$

$+\frac{3}{8}$

$3\frac{10}{8} = 3 + \frac{5}{4} = 3 + 1\frac{1}{4} = 4\frac{1}{4}$

The total thickness of the wall is $4\frac{1}{4}$ in.

▶ **A Closer Look**   In problems (c), (d), and (e), the sum of the fraction parts is greater than or equal to 1. If we use improper fractions to add, the solutions look like this:

(c) $4\frac{1}{6} + 2\frac{5}{6} + 3\frac{5}{6} = \frac{25}{6} + \frac{17}{6} + \frac{23}{6} = \frac{65}{6} = 10\frac{5}{6}$

(d) $3\frac{1}{4} + 2\frac{3}{4} = \frac{13}{4} + \frac{11}{4} = \frac{24}{4} = 6$

(e) $3\frac{4}{8} + \frac{3}{8} + \frac{3}{8} = \frac{28}{8} + \frac{3}{8} + \frac{3}{8} = \frac{34}{8} = \frac{17}{4} = 4\frac{1}{4}$ ◀

▶ **Note**   If the addition is done using improper fractions, large and unwieldy numerators may result. Be careful. ◀

Unlike Fractions   How do we add fractions whose denominators are not the same? For example, how do we add $\frac{2}{3} + \frac{3}{4}$?

The problem is to find a name for this new number. One way to find it is to change these fractions to equivalent fractions with the same denominator.

$\frac{2}{3} = \frac{2 \times \boxed{4}}{3 \times \boxed{4}} = \frac{8}{12}$ We chose $\frac{4}{4}$ as a multiplier because it changes $\frac{2}{3}$ to an equivalent fraction with a denominator of 12.

$\frac{3}{4} = \frac{3 \times \boxed{3}}{4 \times \boxed{3}} = \frac{9}{12}$ We chose $\frac{3}{3}$ as a multiplier because it changes $\frac{3}{4}$ to an equivalent fraction with a denominator of 12.

The two fractions now have the same denominator.

▶ **Note**   You should remember that we discussed equivalent fractions on page 74. Return for a quick review if you need it. ◀

Example 5

Add $\frac{2}{3} + \frac{3}{4}$ using the equivalent fractions.

$$\frac{2}{3} + \frac{3}{4} = \frac{8}{12} + \frac{9}{12} = \frac{17}{12} = 1\frac{5}{12}$$

We change the original fractions to equivalent fractions with the same denominator and then add as before.

Least Common Denominator

How do you know what number to use as the new denominator? In general, you cannot simply guess at the best new denominator. We need a method for finding it from the denominators of the fractions to be added. The new denominator is called the *least common denominator,* abbreviated LCD.

Example 6

Suppose that we want to add the fractions $\frac{1}{8} + \frac{5}{12}$.

The first step is to find the LCD of the denominators 8 and 12. To find the LCD follow this procedure.

Step 1 Write each denominator as a product of its prime factors. If you need to review this concept, see page 44.

$$8 = 2 \cdot 2 \cdot 2 \qquad\qquad 12 = 2 \cdot 2 \cdot 3$$

Step 2 To form the LCD, write each factor that appears in either denominator, then repeat it for the most number of times it appears in any one denominator.

$$\text{LCD} = 2 \cdot 2 \cdot 2 \cdot 3 \qquad = \qquad 24$$

There are 3 factors of 2 in 8.   There is 1 factor of 3 in 12.

The LCD of 8 and 12 is 24. This means that 24 is the smallest number that is exactly divisible by both 8 and 12.

Example 7

Find the LCD of 12 and 45.

Step 1 Find the prime factors of each number.

$$12 = 2 \cdot 2 \cdot 3 \qquad\qquad 45 = 3 \cdot 3 \cdot 5$$

Step 2 The LCD must contain the factors 2, 3, and 5. The factor 2 occurs twice in 12, the factor 3 occurs twice in 45, and the factor 5 occurs just once in 45.

$$\text{LCD} = 2 \cdot 2 \cdot 3 \cdot 3 \cdot 5 = 180$$

The number 180 is the smallest number that is exactly divisible by both 12 and 45.

→ **Your Turn**

Use the method described to find the LCD of the numbers 28 and 42.

→ **Solution**

Step 1 Find the prime factors of each number.

$$28 = 2 \cdot 2 \cdot 7 \qquad 42 = 2 \cdot 3 \cdot 7$$

Step 2 The factors 2, 3, and 7 all appear. The factor 2 appears at most twice (in 28) and both 3 and 7 appear at most just once in any single number. Therefore the LCD is

$$2 \cdot 2 \cdot 3 \cdot 7 = 84$$

A CALCULATOR METHOD FOR FINDING THE LCD

Here is an alternative method for finding the LCD that you might find easier than the method just described. Follow these two steps.

Step 1 Choose the larger denominator and write down a few multiples of it.

Example: To find the LCD of 12 and 15, first write down a few of the multiples of the larger number, 15. The multiples are 15, 30, 45, 60, 75, and so on.

Step 2 Test each multiple until you find one that is exactly divisible by the smaller denominator.

Example: 15 is not exactly divisible by 12. 30 is not exactly divisible by 12. 45 is not exactly divisible by 12. 60 *is* exactly divisible by 12. The LCD is 60.

This method of finding the LCD has the advantage that you can use it with an electronic calculator. For this example the calculator steps would look like this:

15 ÷ **12** = → *1.25* *Not* a whole number; therefore, *not* exactly divisible by 12.

15 × **2** ÷ **12** = → *2.5* Second multiple: answer is *not* a whole number.

15 × **3** ÷ **12** = → *3.75* Third multiple: answer is *not* a whole number.

15 × **4** ÷ **12** = → *5.* Fourth multiple: answer *is* a whole number; therefore, the LCD is 4 × 15 or 60.

With smaller numbers, this process may be done mentally.

Ready for more practice in finding LCDs?

→ **Your Turn**

Find the LCD of 4, 10, and 15.

Step 1 Write the prime factors of all three numbers.

$$4 = 2 \cdot 2 \qquad 10 = 2 \cdot 5 \qquad 15 = 3 \cdot 5$$

Step 2 The factors 2, 3, and 5 all appear and therefore must be included in the LCD. The factor 2 appears at most twice (in 4), and the factors 3 and 5 both appear at most just once in any single number. Therefore the LCD is

$$2 \cdot 2 \cdot 3 \cdot 5 = 60$$

This means that 60 is the smallest number that is exactly divisible by 4, 10, and 15.

→ **More Practice**

Practice by finding the LCD for each of the following sets of numbers.

(a) 2 and 4 (b) 8 and 4 (c) 6 and 3

(d) 5 and 4 (e) 9 and 15 (f) 15 and 24

(g) 4, 5, and 6 (h) 4, 8, and 12 (i) 12, 15, and 21

→ **Answers**

(a) 4 (b) 8 (c) 6 (d) 20 (e) 45

(f) 120 (g) 60 (h) 24 (i) 420

To use the LCD to add fractions, rewrite the fractions with the LCD as the new denominator, then add the new equivalent fractions.

Example 8

Add $\dfrac{1}{6} + \dfrac{5}{8}$.

First, find the LCD. The LCD of 6 and 8 is 24.

Next, rewrite the two fractions with denominator 24.

$$\frac{1}{6} = \frac{1 \times 4}{6 \times 4} = \frac{4}{24}$$

$$\frac{5}{8} = \frac{5 \times 3}{8 \times 3} = \frac{15}{24}$$

Finally, add the new equivalent fractions.

$$\frac{1}{6} + \frac{5}{8} = \frac{4}{24} + \frac{15}{24} = \frac{19}{24}$$

→ **Your Turn**

Add $\dfrac{3}{8} + \dfrac{1}{10}$.

The LCD of 8 and 10 is 40.

$$\frac{3}{8} = \frac{3 \times 5}{8 \times 5} = \frac{15}{40}$$

$$\frac{1}{10} = \frac{1 \times 4}{10 \times 4} = \frac{4}{40}$$

$$\frac{3}{8} + \frac{1}{10} = \frac{15}{40} + \frac{4}{40} = \frac{19}{40}$$

When mixed numbers contain unlike denominators, find the LCD of the fractions, rewrite the fractions with the LCD, and then add.

Example 9

To add $2\frac{3}{8} + 3\frac{1}{6}$,

Step 1 Find the LCD of the fractions.   The LCD of 8 and 6 is 24.

Step 2 Rewrite the two fractions with the LCD and arrange the sum vertically.

$$2\frac{9}{24} \quad \Leftarrow \quad \frac{3}{8} = \frac{3 \times 3}{8 \times 3} = \frac{9}{24}$$

$$+3\frac{4}{24} \quad \Leftarrow \quad \frac{1}{6} = \frac{1 \times 4}{6 \times 4} = \frac{4}{24}$$

Step 3 Add the whole numbers $5\frac{13}{24} \quad \Leftarrow \quad$ Answer
and fractions separately.

If the sum of the fraction parts of the mixed numbers is greater than or equal to 1, an additional step will be required to simplify the answer. This is illustrated in Example 10.

Example 10

To add $5\frac{1}{2} + 3\frac{2}{3}$,

Step 1 The LCD of 2 and 3 is 6.

Step 2 Rewrite the two fractions with the LCD and arrange the sum vertically.

$$5\frac{3}{6} \quad \Leftarrow \quad \frac{1}{2} = \frac{1 \times 3}{2 \times 3} = \frac{3}{6}$$

$$+3\frac{4}{6} \quad \Leftarrow \quad \frac{2}{3} = \frac{2 \times 2}{3 \times 2} = \frac{4}{6}$$

Step 3 Add. $8\frac{7}{6}$

Step 4 Simplify the answer. $8\frac{7}{6} = 8 + \frac{7}{6} = 8 + 1\frac{1}{6} = 9\frac{1}{6}$

Using improper fractions, the solution to Example 10 looks like this:

Step 1 Convert each mixed number to an improper fraction.

$$5\frac{1}{2} + 3\frac{2}{3} = \frac{11}{2} + \frac{11}{3}$$

Step 2 Rewrite each fraction with the LCD of 6.

$$= \frac{33}{6} + \frac{22}{6}$$

Step 3 Add.

$$= \frac{55}{6}$$

Step 4 Convert the result back to a mixed number.

$$= 9\frac{1}{6} \blacktriangleleft$$

→ More Practice

Find the LCD, rewrite the fractions, and add.

(a) $\dfrac{1}{2} + \dfrac{1}{4}$ (b) $\dfrac{3}{8} + \dfrac{1}{4}$ (c) $\dfrac{1}{6} + \dfrac{2}{3} + \dfrac{5}{9}$

(d) $4\dfrac{3}{4} + \dfrac{1}{6}$ (e) $1\dfrac{4}{5} + 2\dfrac{3}{8}$

(f) **Manufacturing** The Monterey Canning Co. packs $8\frac{1}{2}$ ounces of tuna into each can. If the can itself weighs $1\frac{7}{8}$ ounces, what is the total weight of a can of Monterey tuna?

→ Solutions

(a) The LCD of 2 and 4 is 4.

$$\frac{1}{2} = \frac{?}{4} = \frac{1 \times \boxed{2}}{2 \times \boxed{2}} = \frac{2}{4}$$

$$\frac{1}{2} + \frac{1}{4} = \frac{2}{4} + \frac{1}{4} = \frac{3}{4}$$

(b) The LCD of 8 and 4 is 8.

$$\frac{1}{4} = \frac{?}{8} = \frac{1 \times \boxed{2}}{4 \times \boxed{2}} = \frac{2}{8}$$

$$\frac{3}{8} + \frac{1}{4} = \frac{3}{8} + \frac{2}{8} = \frac{5}{8}$$

(c) The LCD of 6, 3, and 9 is 18.

$$\frac{1}{6} = \frac{?}{18} = \frac{1 \times \boxed{3}}{6 \times \boxed{3}} = \frac{3}{18} \qquad\qquad \frac{2}{3} = \frac{?}{18} = \frac{2 \times \boxed{6}}{3 \times \boxed{6}} = \frac{12}{18}$$

$$\frac{5}{9} = \frac{?}{18} = \frac{5 \times \boxed{2}}{9 \times \boxed{2}} = \frac{10}{18}$$

$$\frac{1}{6} + \frac{2}{3} + \frac{5}{9} = \frac{3}{18} + \frac{12}{18} + \frac{10}{18} = \frac{25}{18} = 1\frac{7}{18}$$

(d) The LCD of 4 and 6 is 12.

$$4\frac{9}{12} \Leftarrow \boxed{\frac{3\times3}{4\times3}=\frac{9}{12}}$$

$$+\frac{2}{12} \Leftarrow \boxed{\frac{1\times2}{6\times2}=\frac{2}{12}}$$

$$4\frac{11}{12} \Leftarrow \boxed{\text{Answer}}$$

(e) The LCD of 5 and 8 is 20.

Using mixed numbers:

$$1\frac{32}{40} \Leftarrow \boxed{\frac{4\times8}{5\times8}=\frac{32}{40}}$$

$$+2\frac{15}{40} \Leftarrow \boxed{\frac{3\times5}{8\times5}=\frac{15}{40}}$$

$$3\frac{47}{40} = 3 + 1\frac{7}{40} = 4\frac{7}{40}$$

Using improper fractions:

$$1\frac{4}{5} + 2\frac{3}{8} = \frac{9}{5} + \frac{19}{8} = \frac{9\times8}{5\times8} + \frac{19\times5}{8\times5}$$

$$= \frac{72}{40} + \frac{95}{40} = \frac{167}{40} = 4\frac{7}{40}$$

(f) To find the total weight, add the weight of the tuna to the weight of the can. The LCD of 2 and 8 is 8.

Using mixed numbers:

$$8\frac{4}{8} \Leftarrow \boxed{\frac{1\times4}{2\times4}=\frac{4}{8}}$$

$$+1\frac{7}{8}$$

$$9\frac{11}{8} = 9 + 1\frac{3}{8} = 10\frac{3}{8}$$

Using improper fractions:

$$8\frac{1}{2} + 1\frac{7}{8} = \frac{17}{2} + \frac{15}{8}$$

$$= \frac{17\times4}{2\times4} + \frac{15}{8}$$

$$= \frac{68}{8} + \frac{15}{8}$$

$$= \frac{83}{8} = 10\frac{3}{8}$$

The total weight is $10\frac{3}{8}$ ounces.

Note Remember, it is not necessary to find the LCD when you *multiply* fractions. ◀

Subtraction Once you have mastered the process of adding fractions, subtraction is very simple indeed. To find $\frac{3}{8} - \frac{1}{8}$, notice that the denominators are the same. We can subtract the numerators and write this difference over the common denominator.

$$\frac{3}{8} - \frac{1}{8} = \frac{3-1}{8} \Leftarrow \boxed{\text{Subtract numerators}}$$

$$\boxed{\text{Same denominator}}$$

$$= \frac{2}{8} \quad \text{or} \quad \frac{1}{4}$$

To subtract fractions with unlike denominators, first find the LCD and then subtract the equivalent fractions with like denominators.

Example 11

To find $\dfrac{3}{4} - \dfrac{1}{5}$,

first, determine that the LCD of 4 and 5 is 20.

Then, find equivalent fractions with a denominator of 20.

$$\frac{3}{4} = \frac{?}{20} = \frac{3 \times \boxed{5}}{4 \times \boxed{5}} = \frac{15}{20}$$

$$\frac{1}{5} = \frac{?}{20} = \frac{1 \times \boxed{4}}{5 \times \boxed{4}} = \frac{4}{20}$$

Finally, subtract the equivalent fractions:

$$\frac{3}{4} - \frac{1}{5} = \frac{15}{20} - \frac{4}{20} = \frac{15 - 4}{20} = \frac{11}{20}$$

The procedure is exactly the same as for addition.

→ **Your Turn**

Subtract as indicated.

(a) $\dfrac{9}{20} - \dfrac{3}{20}$

(b) $\dfrac{5}{6} - \dfrac{3}{10}$

→ **Solutions**

(a) $\dfrac{9}{20} - \dfrac{3}{20} = \dfrac{9 - 3}{20} = \dfrac{6}{20} = \dfrac{3}{10}$ Be sure to write your final answer in lowest terms.

(b) The LCD of 6 and 10 is 30.

$$\frac{5}{6} = \frac{5 \times \boxed{5}}{6 \times \boxed{5}} = \frac{25}{30} \qquad \frac{3}{10} = \frac{3 \times \boxed{3}}{10 \times \boxed{3}} = \frac{9}{30}$$

$$\frac{5}{6} - \frac{3}{10} = \frac{25}{30} - \frac{9}{30} = \frac{25 - 9}{30} = \frac{16}{30} = \frac{8}{15}$$

If the fractions to be subtracted are given as mixed numbers, it is sometimes easier to work with them as improper fractions. The next two examples illustrate when to leave them as mixed numbers, and when to convert them to improper fractions.

Example 12

Consider the problem $5\dfrac{5}{6} - 3\dfrac{1}{6}$.

Because $\dfrac{5}{6}$ is larger than $\dfrac{1}{6}$, it will be easy to subtract using mixed numbers. Simply arrange the problem vertically and subtract the whole numbers and fractions separately.

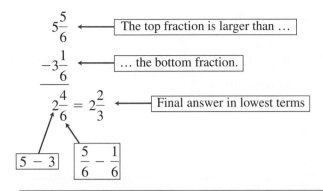

$5\dfrac{5}{6}$ ← The top fraction is larger than …

$-3\dfrac{1}{6}$ ← … the bottom fraction.

$2\dfrac{4}{6} = 2\dfrac{2}{3}$ ← Final answer in lowest terms

$\boxed{5 - 3}$ $\boxed{\dfrac{5}{6} - \dfrac{1}{6}}$

Example 13

Consider the problem $6\dfrac{1}{4} - 1\dfrac{3}{4}$.

Because $\dfrac{1}{4}$ is smaller than $\dfrac{3}{4}$, leaving them as mixed numbers will require borrowing.

(See "A Closer Look" following this example.) To avoid this complication, convert the mixed numbers to improper fractions and then subtract.

$$6\dfrac{1}{4} - 1\dfrac{3}{4} = \dfrac{25}{4} - \dfrac{7}{4} = \dfrac{18}{4} = \dfrac{9}{2} = 4\dfrac{1}{2}$$

Answer in lowest terms   Answer converted to a mixed number

A Closer Look   In the previous example, if we try to subtract using mixed numbers, we must borrow before subtracting. Here is how the problem would be done:

$6\dfrac{1}{4}$

$-1\dfrac{3}{4}$

$\boxed{\dfrac{3}{4} \text{ cannot be subtracted from } \dfrac{1}{4}}$

$\boxed{\begin{array}{l}\text{We must borrow 1}\\ \text{from the 6 so that}\\ \text{it becomes 5.}\end{array}}$ → $5\dfrac{5}{4}$ ← $\boxed{\begin{array}{l}\text{The 1 we borrow is added to the fraction part:}\\[4pt] 1 + \dfrac{1}{4} = \dfrac{4}{4} + \dfrac{1}{4} = \dfrac{5}{4}\end{array}}$

$-1\dfrac{3}{4}$

$4\dfrac{2}{4} = 4\dfrac{1}{2}$ ← Answer in lowest terms

If you wish to avoid having to borrow, always use improper fractions in this situation. ◀

→ **Your Turn**

Subtract as indicated.

(a) $9\dfrac{7}{8} - 3\dfrac{5}{8}$ (b) $4\dfrac{1}{5} - 2\dfrac{4}{5}$

(a) Because $\dfrac{7}{8}$ is larger than $\dfrac{5}{8}$, we can proceed with the mixed numbers. Arranging the problem vertically, we have

$$
\begin{array}{r}
9\dfrac{7}{8} \\[2ex]
-3\dfrac{5}{8} \\[1ex]
\hline
6\dfrac{2}{8} = 6\dfrac{1}{4}
\end{array}
$$

⟵ Answer in lowest terms

(b) Because $\dfrac{1}{5}$ is smaller than $\dfrac{4}{5}$, we will use improper fractions to avoid having to borrow.

$$4\dfrac{1}{5} - 2\dfrac{4}{5} = \dfrac{21}{5} - \dfrac{14}{5} = \dfrac{7}{5} = 1\dfrac{2}{5}$$

When subtracting mixed numbers with unlike denominators, find the LCD of the fraction parts first. You can then decide which method is best.

Example 14

(a) To subtract $3\dfrac{1}{2} - 1\dfrac{1}{3}$

Step 1 Find the LCD. The LCD of 2 and 3 is 6.

Step 2 Find equivalent fractions with a denominator of 6.

$$\dfrac{1}{2} = \dfrac{1 \times 3}{2 \times 3} = \dfrac{3}{6} \qquad\qquad \dfrac{1}{3} = \dfrac{1 \times 2}{3 \times 2} = \dfrac{2}{6}$$

so that $3\dfrac{1}{2} - 1\dfrac{1}{3} = 3\dfrac{3}{6} - 1\dfrac{2}{6}$

Step 3 Because $\dfrac{3}{6}$ is larger than $\dfrac{2}{6}$, we can proceed with the mixed numbers.

$$
\begin{array}{r}
3\dfrac{3}{6} \\[2ex]
-1\dfrac{2}{6} \\[1ex]
\hline
2\dfrac{1}{6}
\end{array}
$$

⟵ Answer

(b) To subtract $5\dfrac{1}{4} - 2\dfrac{3}{8}$,

Step 1 Find the LCD. The LCD of 4 and 8 is 8.

Step 2 Only the first fraction must be rewritten.

$$\dfrac{1}{4} = \dfrac{1 \times 2}{4 \times 2} = \dfrac{2}{8}$$

so that $5\dfrac{1}{4} - 2\dfrac{3}{8} = 5\dfrac{2}{8} - 2\dfrac{3}{8}$

Step 3 Because $\dfrac{2}{8}$ is smaller than $\dfrac{3}{8}$, it will be easier to use improper fractions to perform the subtraction.

$$5\frac{2}{8} - 2\frac{3}{8} = \frac{42}{8} - \frac{19}{8} = \frac{23}{8} = 2\frac{7}{8}$$

The following two-part example shows you what to do if one of the numbers is a whole number.

Example 15

(a) If you are subtracting a whole number from a mixed number, simply subtract the numbers in their original form. To subtract $4\dfrac{5}{16} - 2$, arrange vertically and subtract as usual:

$$\begin{array}{r} 4\dfrac{5}{16} \\ -2\phantom{\dfrac{5}{16}} \\ \hline 2\dfrac{5}{16} \end{array} \quad \longleftarrow \boxed{\text{Answer}}$$

(b) If you are subtracting a mixed number from a whole number, use improper fractions to avoid borrowing. To subtract $8 - 3\dfrac{2}{3}$, recall that 8 in fraction form is $\dfrac{8}{1}$, so that

$$\boxed{\frac{8}{1} = \frac{8 \times 3}{1 \times 3} = \frac{24}{3}}$$

$$8 - 3\frac{2}{3} = \frac{8}{1} - \frac{11}{3} = \frac{24}{3} - \frac{11}{3}$$

$$= \frac{13}{3}$$

$$= 4\frac{1}{3}$$

→ More Practice

Try these problems for practice in subtracting fractions.

(a) $\dfrac{7}{8} - \dfrac{5}{8}$ (b) $\dfrac{4}{5} - \dfrac{1}{6}$ (c) $9\dfrac{13}{16} - 3\dfrac{1}{4}$

(d) $6 - 2\dfrac{3}{4}$ (e) $7\dfrac{1}{6} - 2\dfrac{5}{8}$

(f) **Plumbing** If a pipe has a $2\frac{1}{2}$-in. O.D. (outside diameter) and a wall thickness of $\frac{1}{8}$ in., what is its I.D. (inside diameter)? (See the figure.)

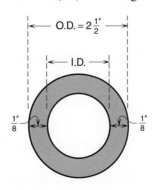

(g) **Machine Trades** A bar $6\frac{3}{16}$ in. long is cut from a piece $24\frac{3}{8}$ in. long. If $\frac{5}{32}$ in. is wasted in cutting, what length remains?

(a) $\dfrac{7}{8} - \dfrac{5}{8} = \dfrac{7-5}{8} = \dfrac{2}{8} = \dfrac{1}{4}$

(b) The LCD of 5 and 6 is 30.

$$\dfrac{4}{5} = \dfrac{?}{30} = \dfrac{4 \times \boxed{6}}{5 \times \boxed{6}} = \dfrac{24}{30}$$

$$\dfrac{1}{6} = \dfrac{?}{30} = \dfrac{1 \times \boxed{5}}{6 \times \boxed{5}} = \dfrac{5}{30}$$

so $\dfrac{4}{5} - \dfrac{1}{6} = \dfrac{24}{30} - \dfrac{5}{30}$

$$= \dfrac{19}{30}$$

(c) The LCD of 4 and 16 is 16. Only the second fraction must be rewritten:

$$9\dfrac{13}{16} - 3\dfrac{1}{4} = 9\dfrac{13}{16} - 3\dfrac{4}{16}$$

$$9\dfrac{13}{16}$$
$$-3\dfrac{4}{16}$$
$$\overline{6\dfrac{9}{16}} \longleftarrow \boxed{\text{Answer}}$$

(d)

$\overset{\displaystyle \overbrace{}^{(2 \times 4) + 3}}{}$

$$6 - 2\dfrac{3}{4} = \dfrac{24}{4} - \dfrac{11}{4} = \dfrac{24 - 11}{4} = \dfrac{13}{4} = 3\dfrac{1}{4}$$

$$6 = \dfrac{6}{1} = \dfrac{6 \times \boxed{4}}{1 \times \boxed{4}}$$

(e) The LCD of 6 and 8 is 24. $\quad \dfrac{1}{6} = \dfrac{4}{24} \qquad \dfrac{5}{8} = \dfrac{15}{24}$

so $7\dfrac{1}{6} = 7\dfrac{4}{24} = \dfrac{172}{24}$ and $2\dfrac{5}{8} = 2\dfrac{15}{24} = \dfrac{63}{24}$

$$7\dfrac{1}{6} - 2\dfrac{5}{8} = \dfrac{172}{24} - \dfrac{63}{24} = \dfrac{109}{24} = 4\dfrac{13}{24}$$

(f) From the figure, notice that the I.D. plus *twice* the wall thickness equals the O.D. Therefore, to find the I.D., subtract twice the wall thickness from the O.D.

$$\text{I.D.} = 2\dfrac{1}{2} - 2\left(\dfrac{1}{8}\right) = 2\dfrac{1}{2} - \dfrac{2}{1} \times \dfrac{1}{8}$$

Step 1   According to the order of operations, we must first multiply.
$$= 2\dfrac{1}{2} - \dfrac{\overset{1}{\cancel{2}}}{1} \times \dfrac{1}{\underset{4}{\cancel{8}}}$$

$$= 2\dfrac{1}{2} - \dfrac{1}{4}$$

Step 2   Rewrite with the LCD of 4. $= 2\dfrac{2}{4} - \dfrac{1}{4}$

Step 3   Subtract. $= 2\dfrac{1}{4}$

The inside diameter is $2\dfrac{1}{4}$ inches.

(g) "Cutting" and "wasting" both imply subtraction. So the problem becomes

$$24\dfrac{3}{8} - 6\dfrac{3}{16} - \dfrac{5}{32}$$

Rewritten with a common denominator of 32, we have

$$24\frac{12}{32} - 6\frac{6}{32} - \frac{5}{32}$$

According to the order of operations, we must subtract from left to right. Therefore, we subtract $24\frac{12}{32} - 6\frac{6}{32}$ first, and then subtract $\frac{5}{32}$ from the result.

$$\begin{array}{ccc} 24\frac{12}{32} & & 18\frac{6}{32} \\ -\ 6\frac{6}{32} & \longrightarrow & -\ \frac{5}{32} \\ \hline 18\frac{6}{32} & & 18\frac{1}{32} \end{array}$$

The remaining length is $18\frac{1}{32}$ in.

Now turn to Exercises 2-4 for a set of problems on adding and subtracting fractions.

Exercises 2-4 **Addition and Subtraction of Fractions**

A. Add or subtract as shown.

1. $\dfrac{1}{16} + \dfrac{3}{16}$ 2. $\dfrac{5}{12} + \dfrac{11}{12}$ 3. $\dfrac{5}{16} + \dfrac{7}{16}$

4. $\dfrac{2}{6} + \dfrac{3}{6}$ 5. $\dfrac{3}{4} - \dfrac{1}{4}$ 6. $\dfrac{13}{16} - \dfrac{3}{16}$

7. $\dfrac{3}{5} - \dfrac{1}{5}$ 8. $\dfrac{5}{12} - \dfrac{2}{12}$ 9. $\dfrac{5}{16} + \dfrac{3}{16} + \dfrac{7}{16}$

10. $\dfrac{1}{8} + \dfrac{3}{8} + \dfrac{7}{8}$ 11. $1\dfrac{7}{8} - \dfrac{3}{8}$ 12. $3\dfrac{9}{16} - 1\dfrac{5}{16}$

13. $\dfrac{1}{4} + \dfrac{1}{2}$ 14. $\dfrac{7}{16} + \dfrac{3}{8}$ 15. $\dfrac{5}{8} + \dfrac{1}{12}$

16. $\dfrac{5}{12} + \dfrac{3}{16}$ 17. $\dfrac{1}{2} - \dfrac{3}{8}$ 18. $\dfrac{5}{16} - \dfrac{3}{32}$

19. $\dfrac{15}{16} - \dfrac{1}{2}$ 20. $\dfrac{7}{16} - \dfrac{1}{32}$ 21. $\dfrac{3}{5} + \dfrac{1}{8}$

22. $\dfrac{2}{3} + \dfrac{4}{5}$ 23. $\dfrac{7}{8} - \dfrac{2}{5}$ 24. $\dfrac{4}{9} - \dfrac{1}{4}$

25. $\dfrac{1}{2} + \dfrac{1}{4} - \dfrac{1}{8}$ 26. $\dfrac{11}{16} - \dfrac{1}{8} - \dfrac{1}{3}$ 27. $1\dfrac{1}{2} + \dfrac{1}{4}$

28. $2\dfrac{7}{16} + \dfrac{3}{4}$ 29. $2\dfrac{1}{2} + 1\dfrac{5}{8}$ 30. $2\dfrac{8}{32} + 1\dfrac{1}{10}$

31. $2\frac{1}{3} + 1\frac{1}{5}$ 　　　32. $1\frac{7}{8} + \frac{1}{4}$ 　　　33. $4\frac{1}{8} - 1\frac{3}{4}$

34. $5\frac{3}{4} - 2\frac{1}{12}$ 　　　35. $3\frac{1}{5} - 2\frac{1}{12}$ 　　　36. $5\frac{1}{3} - 2\frac{2}{5}$

B. Add or subtract as shown.

1. $8 - 2\frac{7}{8}$ 　　　　　2. $3 - 1\frac{3}{16}$ 　　　　3. $3\frac{5}{8} - \frac{13}{16}$

4. $\frac{1}{2} + \frac{1}{3} + \frac{1}{4} + \frac{1}{5}$ 　　　5. $\frac{1}{2} + \frac{1}{4} + \frac{1}{8}$ 　　6. $6\frac{1}{2} + 5\frac{3}{4} + 8\frac{1}{8}$

7. $\frac{7}{8} + 2\frac{1}{2} - 1\frac{1}{4}$ 　　　8. $1\frac{3}{8}$ subtracted from $4\frac{3}{4}$

9. $2\frac{3}{16}$ less than $4\frac{7}{8}$ 　　　10. $6\frac{2}{3}$ reduced by $1\frac{1}{4}$

11. $2\frac{3}{5}$ less than $6\frac{1}{2}$ 　　　12. By how much is $1\frac{8}{7}$ larger than $1\frac{7}{8}$?

C. Practical Problems

1. **Construction** The exterior wall of a small office building under construction is constructed of $\frac{1}{4}$-in. paneling, $\frac{5}{8}$-in. firecode sheetrock, $5\frac{3}{4}$-in. studs, $\frac{1}{2}$-in. CDX plywood sheathing, $1\frac{1}{4}$-in. insulation board, and $\frac{5}{8}$-in. exterior surfacing. Calculate the total thickness of the wall.

2. **Carpentry** A countertop is made of $\frac{5}{8}$-in. particleboard and is covered with $\frac{3}{16}$-in. laminated plastic. What width of metal edging is needed to finish off the edge?

3. **Welding** A welder needs a piece of half-inch pipe $34\frac{3}{4}$ in. long. She has a piece that is $46\frac{3}{8}$ in. long. How much must she cut off from the longer piece?

4. **Plumbing** If a piece of $\frac{3}{8}$-in.-I.D. (inside diameter) copper tubing measures $\frac{9}{16}$ in. O.D. (outside diameter), what is the wall thickness?

5. **Manufacturing** What is the outside diameter of tubing whose inside diameter is $1\frac{5}{16}$ in. and whose wall thickness is $\frac{1}{8}$ in.? (See the figure.)

6. **Machine Trades** How long a bolt is needed to go through a piece of tubing $\frac{5}{8}$ in. long, a washer $\frac{1}{16}$ in. thick, and a nut $\frac{1}{4}$ in. thick? (See the figure.)

7. **Office Services** Newspaper ads are sold by the column inch (c.i.). What is the total number of column inches for a month in which a plumbing contractor has had ads of $6\frac{1}{2}$, $5\frac{3}{4}$, $3\frac{1}{4}$, $4\frac{3}{4}$, and 5 c.i.?

8. **Plumbing** While installing water pipes, a plumber used pieces of pipe measuring $2\frac{3}{4}$, $4\frac{1}{8}$, $3\frac{1}{2}$, and $1\frac{1}{4}$ ft. How much pipe would remain if these pieces were cut from a 14-ft length of pipe? (Ignore waste in cutting.)

9. **Electrical Trades** A piece of electrical pipe conduit has a diameter of $1\frac{1}{2}$ in. and a wall thickness of $\frac{3}{16}$ in. What is its inside diameter? (See the figure.)

10. **Machine Trades** What is the total length of a certain machine part that is made by joining four pieces that measure $3\frac{1}{8}$, $1\frac{5}{32}$, $2\frac{7}{16}$, and $1\frac{1}{4}$ in.?

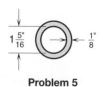

$1\frac{5}{16}"$ 　 $\frac{1}{8}"$

Problem 5

$\frac{1}{16}"$ 　 $\frac{5}{8}"$ 　 $\frac{1}{4}"$

?

Problem 6

$\frac{3}{16}"$

$1\frac{1}{2}"$

Problem 9

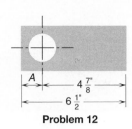

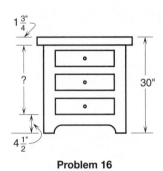

Problem 12

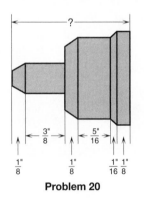

Problem 16

Problem 20

11. **Carpentry** A blueprint requires four separate pieces of wood measuring $5\frac{3}{8}$, $8\frac{1}{4}$, $6\frac{9}{16}$, and $2\frac{5}{8}$ in. How long a piece of wood is needed to cut these pieces if we allow $\frac{1}{2}$ in. for waste?

12. **Drafting** Find the missing dimension A in the drawing shown.

13. **Machine Trades** Two splice plates are cut from a piece of sheet steel that has an overall length of $18\frac{5}{8}$ in. The plates are $9\frac{1}{4}$ in. and $6\frac{7}{16}$ in. long. How much material remains from the original piece if each saw cut removes $\frac{1}{16}$ in.?

14. **Printing** A printer has $2\frac{3}{4}$ rolls of a certain kind of paper in stock. He must do three jobs that require $\frac{5}{8}$, $1\frac{1}{2}$, and $\frac{3}{4}$ roll, respectively. Does he have enough?

15. **Automotive Trades** The wheel stagger of an automobile is the difference between the axle-to-axle lengths on the right and left sides. If this length is $101\frac{1}{4}$ in. on the right side of a particular car and is $100\frac{7}{8}$ in. on the left side, find the wheel stagger of this automobile.

16. **Carpentry** A cabinet 30 in. high must have a $4\frac{1}{2}$-in. base and a $1\frac{3}{4}$-in. top. How much space is left for drawers? (See the figure.)

17. **Printing** Before it was trimmed, a booklet measured $8\frac{1}{4}$ in. high by $6\frac{3}{4}$ in. wide. If each edge of the height and one edge of the width were trimmed $\frac{1}{4}$ in. what is the finished size?

18. **Carpentry** A wall has $\frac{1}{2}$-in. paneling covering $\frac{3}{4}$-in. drywall attached to a $3\frac{3}{4}$-in. stud. What is the total thickness of the three components?

19. **Machine Trades** The large end of a tapered pin is $2\frac{15}{16}$ in. in diameter, while the small end is $2\frac{3}{8}$ in. in diameter. Calculate the difference to get the amount of taper.

20. **Machine Trades** Find the total length of the metal casting shown.

21. **Carpentry** A joiner is set to remove $\frac{7}{64}$ in. from the width of an oak board. If the board was $4\frac{5}{8}$ in. wide, find its width after joining once.

22. **Carpentry** A rule of thumb used in constructing stairways is that the rise and the run should always add up to 17 inches. Applying this rule, what should be the run of a stairway if the rise is $7\frac{3}{4}$ in.?

23. **Automotive Trades** During an oil change, $5\frac{1}{4}$ quarts of oil were needed to fill up an engine. During the next oil change, only $4\frac{1}{2}$ quarts of oil drained from the engine. How much oil was consumed between oil changes?

24. **Carpentry** A 2-in. wood screw is used to join two pieces of a wooden workbench frame that are each $1\frac{1}{4}$ in. thick. How far into the second piece does the wood screw penetrate?

25. **Construction** The concrete slab for a patio requires $4\frac{1}{3}$ cubic yards (cu yd) of concrete. If the truck delivering the concrete has a capacity of 9 cu yd and is full, how many cubic yards will remain in the concrete truck after delivery?

When you have finished these exercises, check your answers to the odd-numbered problems in the Appendix, and turn to Problem Set 2 on page 115 for practice working with fractions. If you need a quick review of the topics in this chapter, visit the chapter Summary first.

Objective

Review

Write an improper fraction as a mixed number. (p. 72)

Divide the numerator by the denominator. Express the remainder in fraction form.

Example: To write $\frac{31}{4}$ as a mixed number divide as follows:

$$
\begin{array}{r}
7 \\
4\overline{)31} \\
28 \\
\hline
3
\end{array}
\quad = \quad 7\frac{3}{4}
$$

Quotient

Remainder

Write a mixed number as an improper fraction. (p. 73)

Multiply the whole number portion by the denominator and add the numerator. Place this total over the denominator.

Example:

$$3\frac{7}{8}$$

New Numerator $= 3 \times 8 + 7$

Denominator

$$\frac{31}{8}$$

Write equivalent fractions with larger denominators. (p. 74)

Multiply both numerator and denominator by the same nonzero number.

Example:

$$\frac{5}{16} = \frac{?}{64} \qquad \frac{5 \times \square}{16 \times \square} = \frac{?}{64} \qquad \frac{5 \times \boxed{4}}{16 \times \boxed{4}} = \frac{20}{64}$$

Rewrite a fraction so that it is in lowest terms. (p. 75)

Divide numerator and denominator by their largest common factor.

Example:

$$\frac{10}{64} = \frac{10 \div \boxed{2}}{64 \div \boxed{2}} = \frac{5}{32}$$

Multiply fractions. (p. 81)

First, change whole or mixed numbers to improper fractions. Then, eliminate common factors. Finally, multiply numerator by numerator and denominator by denominator.

Example:

$$4\frac{1}{2} \times \frac{2}{3} = \frac{\overset{3}{\cancel{9}}}{\underset{1}{\cancel{2}}} \times \frac{\overset{1}{\cancel{2}}}{\underset{1}{\cancel{3}}} = \frac{3}{1} = 3$$

Objective

Divide fractions.
(p. 87)

Add and subtract
fractions. (p. 94)

Solve practical problems
involving fractions.

Review

First, change whole or mixed numbers to improper fractions. Then, multiply by the reciprocal of the divisor.

Example:

$$2\frac{7}{8} \div 1\frac{1}{4} = \frac{23}{8} \div \frac{5}{4} = \frac{23}{\underset{2}{8}} \times \frac{\overset{1}{4}}{5} = \frac{23}{10} \quad \text{or} \quad 2\frac{3}{10}$$

If necessary, rewrite the fractions as equivalent fractions with the least common denominator (LCD). Then add or subtract the numerators and write the result as the numerator of the answer, with the LCD as denominator.

Example:

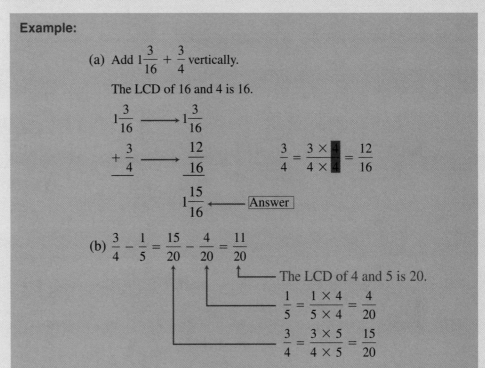

(a) Add $1\frac{3}{16} + \frac{3}{4}$ vertically.

The LCD of 16 and 4 is 16.

$$1\frac{3}{16} \longrightarrow 1\frac{3}{16}$$
$$+\frac{3}{4} \longrightarrow \frac{12}{16} \qquad \frac{3}{4} = \frac{3 \times 4}{4 \times 4} = \frac{12}{16}$$
$$1\frac{15}{16} \longleftarrow \boxed{\text{Answer}}$$

(b) $\frac{3}{4} - \frac{1}{5} = \frac{15}{20} - \frac{4}{20} = \frac{11}{20}$

The LCD of 4 and 5 is 20.

$$\frac{1}{5} = \frac{1 \times 4}{5 \times 4} = \frac{4}{20}$$
$$\frac{3}{4} = \frac{3 \times 5}{4 \times 5} = \frac{15}{20}$$

Read the problem carefully. Look for key words or phrases that indicate which operation to use, and then perform the calculation.

Example: A tabletop is constructed using $\frac{3}{4}$-in. plywood with $\frac{5}{16}$-in. veneer on both sides. Calculate the total thickness of the tabletop.

To solve, add two thicknesses of veneer to one thickness of plywood.

$$2\left(\frac{5}{16}\right) + \frac{3}{4} = \frac{5}{8} + \frac{3}{4}$$
$$= \frac{5}{8} + \frac{6}{8}$$
$$= \frac{11}{8} \quad \text{or} \quad 1\frac{3}{8} \text{ in.}$$

Fractions

Answers to odd-numbered problems are given in the Appendix.

A. Write as an improper fraction.

1. $1\dfrac{1}{8}$ 2. $4\dfrac{1}{5}$ 3. $1\dfrac{2}{3}$ 4. $2\dfrac{3}{16}$

5. $3\dfrac{3}{32}$ 6. $2\dfrac{1}{16}$ 7. $1\dfrac{5}{8}$ 8. $3\dfrac{7}{16}$

Write as a mixed number.

9. $\dfrac{10}{4}$ 10. $\dfrac{19}{2}$ 11. $\dfrac{25}{3}$ 12. $\dfrac{9}{8}$

13. $\dfrac{25}{16}$ 14. $\dfrac{21}{16}$ 15. $\dfrac{35}{4}$ 16. $\dfrac{7}{3}$

Write in lowest terms.

17. $\dfrac{6}{32}$ 18. $\dfrac{8}{32}$ 19. $\dfrac{12}{32}$ 20. $\dfrac{18}{24}$

21. $\dfrac{5}{30}$ 22. $1\dfrac{12}{21}$ 23. $1\dfrac{16}{20}$ 24. $3\dfrac{10}{25}$

Complete these.

25. $\dfrac{3}{4}=\dfrac{?}{12}$ 26. $\dfrac{7}{16}=\dfrac{?}{64}$ 27. $2\dfrac{3}{4}=\dfrac{?}{16}$ 28. $1\dfrac{3}{8}=\dfrac{?}{32}$

29. $5\dfrac{2}{3}=\dfrac{?}{12}$ 30. $1\dfrac{4}{5}=\dfrac{?}{10}$ 31. $1\dfrac{1}{4}=\dfrac{?}{12}$ 32. $2\dfrac{3}{5}=\dfrac{?}{10}$

Circle the larger number.

33. $\dfrac{7}{16}$ or $\dfrac{2}{15}$ 34. $\dfrac{2}{3}$ or $\dfrac{4}{7}$ 35. $\dfrac{13}{16}$ or $\dfrac{7}{8}$ 36. $1\dfrac{1}{4}$ or $\dfrac{7}{6}$

37. $\dfrac{13}{32}$ or $\dfrac{3}{5}$ 38. $\dfrac{2}{10}$ or $\dfrac{3}{16}$ 39. $1\dfrac{7}{16}$ or $\dfrac{7}{4}$ 40. $\dfrac{3}{32}$ or $\dfrac{1}{9}$

B. Multiply or divide as shown.

Name _____

1. $\dfrac{1}{2}\times\dfrac{3}{16}$ 2. $\dfrac{3}{4}\times\dfrac{2}{3}$ 3. $\dfrac{7}{16}\times\dfrac{4}{3}$ 4. $\dfrac{15}{64}\times\dfrac{1}{12}$

Date _____

5. $1\dfrac{1}{2}\times\dfrac{5}{6}$ 6. $3\dfrac{1}{16}\times\dfrac{1}{5}$ 7. $\dfrac{3}{16}\times\dfrac{5}{12}$ 8. $14\times\dfrac{3}{8}$

Course/Section _____

9. $\dfrac{3}{4}\times10$ 10. $\dfrac{1}{2}\times1\dfrac{1}{3}$ 11. $18\times1\dfrac{1}{2}$ 12. $16\times2\dfrac{1}{8}$

13. $2\frac{2}{3} \times 4\frac{3}{8}$ 14. $3\frac{1}{8} \times 2\frac{2}{5}$ 15. $\frac{1}{2} \div \frac{1}{4}$ 16. $\frac{2}{5} \div \frac{1}{2}$

17. $4 \div \frac{1}{8}$ 18. $8 \div \frac{3}{4}$ 19. $\frac{2}{3} \div 4$ 20. $1\frac{1}{2} \div 2$

21. $3\frac{1}{2} \div 5$ 22. $1\frac{1}{4} \div 1\frac{1}{2}$ 23. $2\frac{3}{4} \div 1\frac{1}{8}$ 24. $3\frac{1}{5} \div 1\frac{5}{7}$

C. Add or subtract as shown.

1. $\frac{3}{8} + \frac{7}{8}$ 2. $\frac{1}{2} + \frac{3}{4}$ 3. $\frac{3}{32} + \frac{1}{8}$ 4. $\frac{3}{8} + 1\frac{1}{4}$

5. $\frac{3}{5} + \frac{5}{6}$ 6. $\frac{5}{8} + \frac{1}{10}$ 7. $\frac{9}{16} - \frac{3}{16}$ 8. $\frac{7}{8} - \frac{1}{2}$

9. $\frac{11}{16} - \frac{1}{4}$ 10. $\frac{5}{6} - \frac{1}{5}$ 11. $\frac{7}{8} - \frac{3}{10}$ 12. $1\frac{1}{2} - \frac{3}{32}$

13. $2\frac{1}{8} + 1\frac{1}{4}$ 14. $1\frac{5}{8} + \frac{13}{16}$ 15. $6 - 1\frac{1}{2}$ 16. $3 - 1\frac{7}{8}$

17. $3\frac{2}{3} - 1\frac{7}{8}$ 18. $2\frac{1}{4} - \frac{5}{6}$ 19. $\frac{1}{2} + \frac{1}{3} + \frac{1}{5}$ 20. $1\frac{1}{2} + 1\frac{1}{4} + 1\frac{1}{5}$

21. $3\frac{1}{2} - 2\frac{1}{3}$ 22. $2\frac{3}{5} - 1\frac{4}{15}$ 23. $2 - 1\frac{3}{5}$ 24. $4\frac{5}{6} - 1\frac{1}{2}$

D. Practical Problems

1. **Welding**   In a welding job three pieces of 2-in. I-beam with lengths $5\frac{7}{8}$, $8\frac{1}{2}$, and $22\frac{3}{4}$ in. are needed. What is the total length of I-beam needed? (Do not worry about the waste in cutting.)

2. **Machine Trades**   How many pieces of $10\frac{5}{16}$-in. bar can be cut from a stock 20-ft bar? The metal is torch cut and allowance of $\frac{3}{16}$ in. kerf (waste) should be made for each piece. (*Hint:* 20 ft = 240 in.)

3. **Welding**   A piece of metal must be cut to a length of $22\frac{3}{8}$ in. $\pm \frac{1}{16}$ in. What are the longest and shortest acceptable lengths? (*Hint:* The symbol $\pm$ means to add $\frac{1}{16}$ in. to get the longest length and subtract $\frac{1}{16}$ in. to get the shortest length. Longest = $22\frac{3}{8}$ in. $+ \frac{1}{16}$ in. = ? Shortest = $22\frac{3}{8}$ in. $- \frac{1}{16}$ in. = ?)

4. **Automotive Trades**   A damaged car is said to have "sway" when two corresponding diagonal measurements under the hood are different. If these diagonals are found to be $64\frac{1}{4}$ in. and $62\frac{7}{8}$ in., calculate the magnitude of the sway, the difference between these measurements.

5. **Machine Trades**   A shaft $1\frac{7}{8}$ in. in diameter is turned down on a lathe to a diameter of $1\frac{3}{32}$ in. What is the difference in diameters?

6. **Machine Trades**   A bar $14\frac{5}{16}$ in. long is cut from a piece $25\frac{1}{4}$ in. long. If $\frac{3}{32}$ in. is wasted in cutting, will there be enough left to make another bar $10\frac{3}{8}$ in. long?

7. **Manufacturing**   A cubic foot contains roughly $7\frac{1}{2}$ gallons. How many cubic feet are there in a tank containing $34\frac{1}{2}$ gallons?

8. **Manufacturing**   Find the total width of the three pieces of steel plate shown.

9. **Machine Trades**   What would be the total length of the bar formed by welding together the five pieces of bar stock shown?

10. **Machine Trades**   The Ace Machine Shop has the job of producing 32 zinger bars. Each zinger bar must be turned on a lathe from a piece of stock $4\frac{7}{8}$ in. long. How many feet of stock will they need?

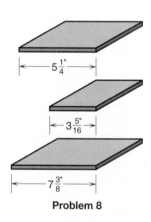

$5\frac{1}{4}''$

$3\frac{5}{16}''$

$7\frac{3}{8}''$

Problem 8

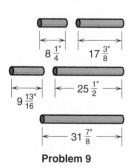

$8\frac{1}{4}''$ $17\frac{3}{8}''$

$9\frac{13}{16}''$ $25\frac{1}{2}''$

$31\frac{7}{8}''$

Problem 9

Chapter 2 Fractions

11. **Carpentry** What is the thickness of a tabletop made of $\frac{3}{4}$-in. plywood and covered with a $\frac{3}{16}$-in. sheet of glass?

12. **Construction** For the wooden form shown, find the lengths *A, B, C,* and *D.*

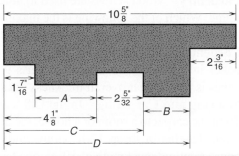

Problem 12

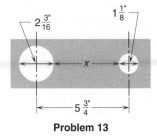

Problem 13

13. **Carpentry** Find the spacing *x* between the holes.

14. **Masonry** Find the height of the five-course (five-bricks-high) brick wall shown if each brick is $2\frac{1}{2}$ in. by $3\frac{7}{8}$ in. by $8\frac{1}{4}$ in. and all mortar joints are $\frac{1}{2}$ in.

15. **Masonry** If the wall in problem 14 has 28 stretchers (bricks laid lengthwise), what is its length?

16. **Electrical Trades** An electrical wiring job requires the following lengths of 14/2 BX cable: seven pieces each $6\frac{1}{2}$ ft long, four pieces each $34\frac{3}{4}$ in. long, and nine pieces each $19\frac{3}{8}$ in. long. What is the total length of cable needed?

17. **Printing** An invitation must be printed on card stock measuring $4\frac{1}{4}$ in. wide by $5\frac{1}{2}$ in. long. The printed material covers a space measuring $2\frac{1}{8}$ in. wide by $4\frac{1}{8}$ in. long. If the printed material is centered in both directions, what are the margins?

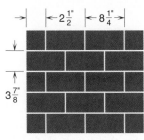

Problem 14

18. **Printing** As a rule of thumb, the top margin of a page of a book should be $\frac{2}{5}$ of the total margin, and the bottom margin should be $\frac{3}{5}$ of the total margin. If the print takes up $9\frac{1}{2}$ in. of an 11-in.-long page, what should the top and bottom margins be? (*Hint:* The total margin = 11 in. $- 9\frac{1}{2}$ in. = $1\frac{1}{2}$ in.)

19. **Welding** A 46-in. bar must have 9 equally spaced holes drilled through the centerline. If the centers of the two end holes are each $2\frac{1}{4}$ in. in from their respective ends, what should the center-to-center distance of the holes be? (*Hint:* There are 8 spaces between holes.)

20. **Construction** If an I-beam is to be $24\frac{3}{8}$ in. long with a tolerance of $\pm\frac{1}{4}$ in., find the longest and shortest acceptable lengths.

21. **Machine Trades** If a positioner shaft turns at 18 revolutions per minute, and the tool feed is $\frac{1}{16}$ in. per revolution, how long will it take to advance $7\frac{1}{2}$ in.?

22. **Sheet Metal Trades** The total allowance for both edges of a grooved seam is three times the width of the seam. Half of this total is added to each edge of the seam. Find the allowance for each edge of a grooved seam if the width of the seam is $\frac{5}{16}$ in.

23. **Machine Trades** Four pieces of steel plate are bolted together. Each piece is $\frac{5}{8}$ in. thick. What is the total thickness of the four combined pieces?

24. **Machine Trades** A machine tech must bolt together two steel plates that are $\frac{7}{8}$ in. thick and $1\frac{1}{4}$ in. thick. If the bolt must be at least $\frac{1}{2}$ in. longer than the combined thickness of the plates, how long a bolt is needed?

25. **Automotive Trades** On a certain vehicle, each turn of the tie-rod sleeve changes the toe-in setting by $\frac{3}{16}$ in. If the tie-rod sleeve makes $1\frac{1}{2}$ turns, how much will the toe-in setting change?

26. **Automotive Trades** A certain engine has a stroke of $3\frac{1}{4}$ in. If the throw of the crankshaft is half the distance of the stroke, what is the throw of the crankshaft for this engine?

27. **Allied Health** One bottle of NoGo pain medicine contains 60 tablets. How many tablets are contained in $3\frac{3}{4}$ bottles?

28. **Construction** The standard slope for a drainage pipe carrying water is $\frac{1}{4}$ in. of vertical drop per foot of horizontal distance. How many inches of vertical drop would an 18-ft drainage pipe required?

29. **Carpentry** Wood screws are used to attach a $\frac{3}{8}$-in. wood panel to a $\frac{3}{4}$-in. wood frame. Which of the following is the maximum screw length that will be less than the combined thickness of the panel and the frame: $\frac{7}{8}$ in., 1 in., $1\frac{1}{4}$ in., or $1\frac{1}{2}$ in.?

30. **Carpentry** A carpenter making a window frame will use dual-pane glass for better insulation. If each pane of glass is $\frac{3}{32}$ in. thick and the space between panes is $\frac{5}{16}$ in., what overall thickness does the frame need to enclose?

31. **Flooring and Carpeting** In the U.S. carpet comes in standard 12-ft widths. To fit the carpet to a room or a house, the installer may need to order several different lengths of these 12-ft widths and seam them together. One particular house required one length of $10\frac{1}{4}$ ft, one length of $8\frac{1}{2}$ ft, one length of $22\frac{2}{3}$ ft, and two lengths of $15\frac{1}{3}$ ft each.

    (a) What was the total length of carpet needed?

    (b) Multiply your answer to part (a) by 12 to find the total area of carpet needed in square feet.

    (c) Multiply your answer to part (b) by $6 per square foot to find the total cost of the carpet.

USING A CALCULATOR, II: FRACTIONS

Fractions can be entered directly on most scientific calculators, and the results of arithmetic calculations with fractions can be displayed as fractions or decimals. If your calculator has an $\boxed{A\frac{b}{c}}$ key, you may enter fractions or mixed numbers directly into your machine without using the division key. Alternate names for this key include $\boxed{a\frac{b}{c}}$, $\boxed{\frac{n}{d}}$, and $\boxed{\frac{\blacksquare}{\square}}$.

Note Most graphing calculators do not have a key such as $\boxed{A\frac{b}{c}}$ for entering fractions. Instead, you simply use the division key to enter them, and answers appear in decimal form. You can then convert them to fraction form using the function " ▶ Frac" normally found on the "MATH" menu. ◄

The calculator display will usually indicate a common or improper fraction with a ⌐, a Γ, or a / symbol.

Example 1

To enter the fraction $\frac{3}{4}$, we will show the following key sequence and answer display:

3 $\boxed{A\frac{b}{c}}$ 4 $\boxed{=}$ → ▓▓▓ 3 / 4 ▓▓▓

Depending on your model, you may also see one of the following displays:

▓▓ 3 ⌐ 4 ▓▓   or   ▓▓ 3 Γ 4 ▓▓ .

If your calculator has a four-line display, you can enter the fraction in vertical form using the arrow keys, and the result will be displayed as ▓▓ $\frac{3}{4}$. See your instruction manual for further details.

If you enter an unsimplified fraction, pressing $\boxed{=}$ will automatically reduce it to lowest terms.

Example 2

Enter the fraction $\frac{6}{16}$, we would have

6 $\boxed{A_c^b}$ **16** $\boxed{=}$ $\rightarrow$ | $3/8$

To enter a mixed number, use the $\boxed{A_c^b}$ key twice.

Example 3

To enter $2\frac{7}{8}$, we will show the following key sequence and display:

2 $\boxed{A_c^b}$ **7** $\boxed{A_c^b}$ **8** $\boxed{=}$ $\rightarrow$ | $2\llcorner7/8$

On your calculator, you might also see this mixed number displayed in one of the following ways:

| $2\_7\lrcorner8$ | or | $2\lrcorner7\lrcorner8$ | or | $2\ulcorner7\ulcorner8$

Calculators with a four-line display are able to show the mixed number as

$2\frac{7}{8}$

Note With a graphing calculator, you would enter $2\frac{7}{8}$ as **2** $\boxed{+}$ **7** $\boxed{\div}$ **8** $\boxed{=}$. If you convert this to a fraction, the graphing calculator will display it only as an improper fraction. ◄

When you enter an improper fraction followed by $\boxed{=}$, the calculator will automatically convert it to a mixed number.

Example 4

Entering the fraction $\frac{12}{7}$ followed by $\boxed{=}$ will result in a display of $1\frac{5}{7}$, like this:

12 $\boxed{A_c^b}$ **7** $\boxed{=}$ $\rightarrow$ | $1\llcorner5/7$

To convert this back to an improper fraction, look for a second function key labeled $\boxed{A_c^b\leftrightarrow\frac{d}{e}}$ or simply $\boxed{\frac{d}{c}}$.

Example 5

The conversion of the previous answer will look like one of the following:

$\boxed{2^{nd}}$ $\boxed{A_c^b\leftrightarrow\frac{d}{e}}$ $\boxed{=}$ $\rightarrow$ | $12/7$ or $\boxed{2^{nd}}$ $\boxed{\frac{d}{c}}$ $\rightarrow$ | $12\lrcorner7$

Notice that in the second sequence, pressing the $\boxed{=}$ key is not required. Repeating either sequence will convert the answer back to a mixed number. We shall use the first sequence in the text, but the second sequence will appear in the calculator appendix.

To perform arithmetic operations with fractions, enter the calculation in the usual way.

Example 6

Compute $\dfrac{26}{8} - 1\dfrac{2}{3}$ like this:

26 $\boxed{A\frac{b}{c}}$ **8** $\boxed{-}$ **1** $\boxed{A\frac{b}{c}}$ **2** $\boxed{A\frac{b}{c}}$ **3** $\boxed{=}$ → ▨ $1 \sqcup 7/12$

Note   Some models require you to press $\boxed{\text{ENTER}}$ or $\boxed{=}$ after keying in each fraction. Consult your instruction manual for details. ◄

→ Your Turn

Work the following problems using your calculator. Where possible, express your answer both as a mixed number and as an improper fraction.

(a) $\dfrac{2}{3} + \dfrac{7}{8}$ (b) $1\dfrac{3}{4} - \dfrac{2}{5}$ (c) $\dfrac{25}{32} + 1\dfrac{7}{8}$ (d) $8\dfrac{1}{5} \div 2\dfrac{1}{6}$

(e) $\dfrac{17}{20} \times \dfrac{1}{3}$

→ Solutions

(a) **2** $\boxed{A\frac{b}{c}}$ **3** $\boxed{+}$ **7** $\boxed{A\frac{b}{c}}$ **8** $\boxed{=}$ → ▨ $1 \sqcup 13/24$

            $\boxed{A\frac{b}{c} \leftrightarrow \frac{d}{e}}$ $\boxed{=}$ → ▨ $37/24$ $\left(1\dfrac{13}{24} \text{ or } \dfrac{37}{24}\right)$

(b) **1** $\boxed{A\frac{b}{c}}$ **3** $\boxed{A\frac{b}{c}}$ **4** $\boxed{-}$ **2** $\boxed{A\frac{b}{c}}$ **5** $\boxed{=}$ → ▨ $1 \sqcup 7/20$

            $\boxed{A\frac{b}{c} \leftrightarrow \frac{d}{e}}$ $\boxed{=}$ → ▨ $27/20$ $\left(1\dfrac{7}{20} \text{ or } \dfrac{27}{20}\right)$

(c) **25** $\boxed{A\frac{b}{c}}$ **32** $\boxed{+}$ **1** $\boxed{A\frac{b}{c}}$ **7** $\boxed{A\frac{b}{c}}$ **8** $\boxed{=}$ → ▨ $2 \sqcup 21/32$

            $\boxed{A\frac{b}{c} \leftrightarrow \frac{d}{e}}$ $\boxed{=}$ → ▨ $85/32$ $\left(2\dfrac{21}{32} \text{ or } \dfrac{85}{32}\right)$

(d) **8** $\boxed{A\frac{b}{c}}$ **1** $\boxed{A\frac{b}{c}}$ **5** $\boxed{\div}$ **2** $\boxed{A\frac{b}{c}}$ **1** $\boxed{A\frac{b}{c}}$ **6** $\boxed{=}$ → ▨ $3 \sqcup 51/65$

            $\boxed{A\frac{b}{c} \leftrightarrow \frac{d}{e}}$ $\boxed{=}$ → ▨ $246/65$ $\left(3\dfrac{51}{65} \text{ or } \dfrac{246}{65}\right)$

(e) **17** $\boxed{A\frac{b}{c}}$ **20** $\boxed{\times}$ **1** $\boxed{A\frac{b}{c}}$ **3** $\boxed{=}$ → ▨ $17/60$ $\left(\dfrac{17}{60}\right)$

Decimal Numbers

Objective	Sample Problems		For help, go to

When you finish this chapter you will be able to:

1. Add, subtract, multiply, and divide decimal numbers.

   (a) $5.82 + 0.096$ _____ Page 127

   (b) $3.78 - 0.989$ _____ Page 129

   (c) $27 - 4.03$ _____

   (d) 7.25×0.301 _____ Page 135

   (e) $104.2 \div 0.032$ _____ Page 139

   (f) $0.09 \div 0.0004$ _____

   (g) $20.4 \div 6.7$ (round to three decimal places)   _____

2. Find averages.

  Find the average of 4.2, 4.8, 5.7, 2.5, 3.6, 5.0   _____ Page 146

3. Work with decimal fractions.

   (a) Write as a decimal number $\frac{3}{16}$ _____ Page 153

   (b) Write as a decimal number: One hundred six and twenty-seven ten-thousandths   _____ Page 125

   (c) Write in words 26.035   _____ Page 123

   (d) $1\frac{2}{3} + 1.785$ _____

   (e) $4.1 \times 2\frac{1}{4}$ _____

   (f) $1\frac{5}{16} \div 4.3$ (round to three decimal places)   _____

Name

_____

Date

_____

Course/Section

Objective	Sample Problems	For help, go to

4. Solve practical problems involving decimal numbers.

(a) **Electrical Trades**   Six recessed lights must be installed in a strip of ceiling. The housings cost $16.45 each, the trims cost $19.86 each, the lamps cost $15.20 each, miscellaneous hardware and wiring cost $33.45, and labor is estimated at 3.25 hours at $65 per hour. What will be the total cost of the job?   _____

(b) **Machine Trades**   A container of 175 bolts weighs 61.3 lb. If the container itself weighs 1.8 lb, how much does each bolt weigh?   _____

(Answers to these preview problems are given in the Appendix. Also, worked solutions to many of these problems appear in the chapter Summary.)

If you are certain that you can work *all* these problems correctly, turn to page 165 for a set of practice problems. If you cannot work one or more of the preview problems, turn to the page indicated to the right of the problem. Those who wish to master this material with the greatest success should turn to Section 3-1 and begin work there.

Decimal Numbers

Reprinted with permission of Universal Press Syndicate.

3-1 Addition and Subtraction of Decimal Numbers

Place Value of Decimal Numbers

By now you know that whole numbers are written in a form based on powers of ten. A number such as

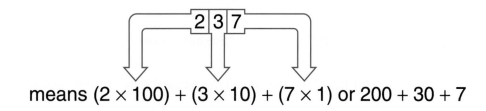

means $(2 \times 100) + (3 \times 10) + (7 \times 1)$ or $200 + 30 + 7$

This way of writing numbers can be extended to fractions. A *decimal* number is a fraction whose denominator is 10 or some power of 10.

A decimal number may have both a whole-number part and a fraction part. For example, the number 324.576 means

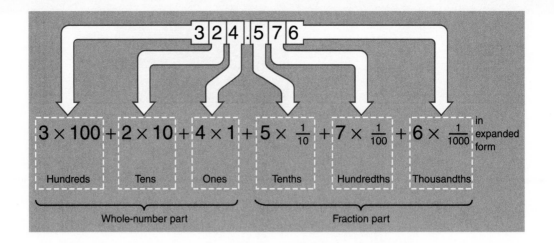

You are already familiar with this way of interpreting decimal numbers from working with money.

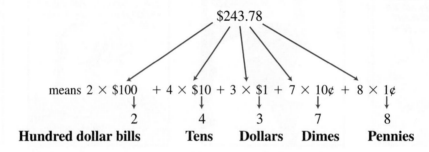

To write a decimal number in words, remember this diagram:

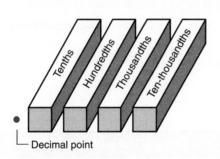

L— Decimal point

Example 1

(a)   .6 = Six tenths $= \dfrac{6}{10}$

(b)   7.05 = Seven and five hundredths $= 7\dfrac{5}{100}$

(c) 12.267 = Twelve and two hundred sixty-seven thousandths $= 12\dfrac{267}{1000}$

(d) .0065 = Sixty-five ten-thousandths $= \dfrac{65}{10,000}$

Decimal form

Fraction form

→ **Your Turn**

Write the following decimal numbers in words.

(a) 0.56   (b) 19.278   (c) 6.4   (d) 0.07   (e) 5.064

(f) **Automotive Trades** The clearance between a piston and its bore is measured at 0.0018 in. Express this in words.

→ **Answers**

(a) Fifty-six hundredths

(b) Nineteen and two hundred seventy-eight thousandths

(c) Six and four tenths

(d) Seven hundredths

(e) Five and sixty-four thousandths

(f) Eighteen ten-thousandths

Learning Help The word **and** represents the decimal point. Everything preceding **and** is the whole-number part, and everything after **and** is the decimal part of the number. ◄

It is also important to be able to perform the reverse process—that is, write a decimal in numerical form if it is given in words.

Example 2

To write "six and twenty-four thousandths" as a numeral,

First, write the whole-number part (six). This is the part before the "and." Follow this by a decimal point—the "and."   6.

Next, draw as many blanks as the decimal part indicates. In this case, the decimal part is "thousandths," so we allow for three decimal places. Draw three blanks.   6.__ __ __

Finally, write a number giving the decimal (twenty-four). Write it so it *ends* on the far right blank. Fill in any blank decimal places with zeros.   6.__ 2 4

6.024

→ **Your Turn**

Now you try it. Write each of the following as decimal numbers.

(a) Five thousandths

(b) One hundred and six tenths

(c) Two and twenty-eight hundredths

(d) Seventy-one and sixty-two thousandths

(e) Three and five hundred eighty-nine ten-thousandths

→ **Answers**

(a) 0.005   (b) 100.6   (c) 2.28   (d) 71.062   (e) 3.0589

Expanded Form

The decimal number .267 can be written in expanded form as

$$\frac{2}{10} + \frac{6}{100} + \frac{7}{1000}$$

Example 3

Write the decimal number .526 in expanded form.

$$.526 = \frac{5}{10} + \frac{2}{100} + \frac{6}{1000}$$

We usually write a decimal number less than 1 with a zero to the left of the decimal point.

.526 would be written 0.526

.4   would be written 0.4

.001 would be written 0.001

It is easy to mistake .4 for 4, but the decimal point in 0.4 cannot be overlooked. That zero out front will help you remember where the decimal point is located.

→ **Your Turn**

To help get these ideas clear in your mind, write the following in expanded form.

(a) 86.42   (b) 43.607   (c) 14.5060   (d) 235.22267

→ **Answers**

(a) 86.42 $= 8 \times 10 + 6 \times 1 + 4 \times \frac{1}{10} + 2 \times \frac{1}{100}$

$= \quad 80 \quad + \quad 6 \quad + \quad \frac{4}{10} \quad + \quad \frac{2}{100}$

(b) 43.607 $= 4 \times 10 + 3 \times 1 + 6 \times \frac{1}{10} + 0 \times \frac{1}{100} + 7 \times \frac{1}{1000}$

$= \quad 40 \quad + \quad 3 \quad + \quad \frac{6}{10} \quad + \quad \frac{0}{100} \quad + \quad \frac{7}{1000}$

(c) 14.5060 $= 10 + 4 + \frac{5}{10} + \frac{0}{100} + \frac{6}{1000} + \frac{0}{10000}$

(d) 235.22267 $= 200 + 30 + 5 + \frac{2}{10} + \frac{2}{100} + \frac{2}{1000} + \frac{6}{10000} + \frac{7}{100000}$

Learning Help Notice that the denominators in the decimal fractions change by a factor of 10. For example,

3247 . 8956

3 × 1000	thousands	3000 . 0006	ten-thousandths	6 × 0.0001
2 × 100	hundreds	200 . 005	thousandths	5 × 0.001
4 × 10	tens	40 . 09	hundredths	9 × 0.01
7 × 1	ones	7 . 8	tenths	8 × 0.1

Each row changes by a factor of ten

Decimal Digits In the decimal number 86.423 the digits 4, 2, and 3 are called *decimal digits*.

The number 43.6708 has four decimal digits: 6, 7, 0, and 8.

The number 5376.2 has one decimal digit: 2.

All digits to the right of the decimal point, those that name the fractional part of the number, are decimal digits.

→ **Your Turn**

How many decimal digits are included in each of these numbers?

(a)   1.4   (b)   315.7   (c)   0.425   (d)   324.0075

→ **Answers**

(a)   one   (b)   one   (c)   three   (d)   four

We will use the idea of decimal digits often in doing arithmetic with decimal numbers.

The decimal point is simply a way of separating the whole-number part from the fraction part. It is a place marker. In whole numbers the decimal point usually is not written, but it is understood to be there.

The whole number 2 is written 2. as a decimal.

$$2 = 2. \qquad \text{or} \qquad 324 = 324.$$

The decimal point   The decimal point

This is very important. Many people make big mistakes in arithmetic because they do not know where that decimal point should go.

Very often, additional zeros are attached to the decimal number without changing its value. For example,

8.5 = 8.50 = 8.5000   and so on

6 = 6. = 6.0 = 6.000   and so on

The value of the number is not changed, but the additional zeros may be useful, as we shall see.

Addition Because decimal numbers represent fractions with denominators equal to powers of ten, addition is very simple.

$$2.34 = 2 + \frac{3}{10} + \frac{4}{100}$$
$$+5.23 = \underline{5 + \frac{2}{10} + \frac{3}{100}}$$
$$7 + \frac{5}{10} + \frac{7}{100} = 7.57$$

Adding like fractions

Of course, we do not need this clumsy business in order to add decimal numbers. As with whole numbers, we may arrange the digits in vertical columns and add directly.

Example 4

Let's add 1.45 + 3.42.

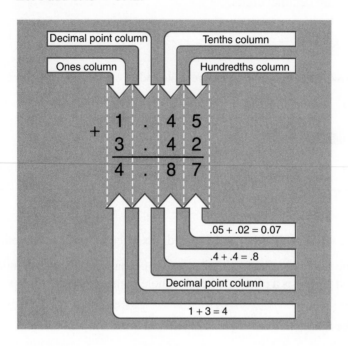

Digits of the same power of ten are placed in the same vertical column. Decimal points are always lined up vertically.

If one of the numbers is written with fewer decimal digits than the other, attach as many zeros as needed so that both have the same number of decimal digits.

```
 2.345              2.345
+1.5     becomes   +1.500
```

Except for the preliminary step of lining up decimal points, addition of decimal numbers is exactly the same process as addition of whole numbers.

→ **Your Turn**

Add the following decimal numbers.

(a) $4.02 + $3.67   = _____   (b) 13.2 + 1.57   = _____

(c) 23.007 + 1.12   = _____   (d) 14.6 + 1.2 + 3.15   = _____

(e) 5.7 + 3.4   = _____   (f) $9 + $0.72 + $6.09   = _____

(g) 0.07 + 6.79 + 0.3 + 3 = _____   (h) **Sheet Metal Trades**   A sheet metal worker uses 2.36, 7, 3.9, and 0.6 ounces (oz) of cleaning concentrate on successive jobs. What total amount of concentrate has he used during this time?

Arrange each sum vertically, placing the decimal points in the same column, then add as with whole numbers.

Decimal points are lined up vertically.

(a) $4.02
    $3.67
    ─────
    $7.69   0.02 + 0.07 = 0.09   Add cents.
            0.0 + 0.6 = 0.6   Add 10-cent units.
            4   + 3   = 7   Add dollars.

As a check, notice that the sum is roughly $4 + $4 or $8, which agrees with the actual answer. Always check your answer by first estimating it, then comparing your estimate or rough guess with the final answer.

Decimal points are in line.

(b) 13.20 ────── Attach a zero to provide the same number of
    1.57   decimal digits as in the other addend.
    ─────
    14.77

Place answer decimal point in the same vertical line.

✓ 13 + 2 = 15, which agrees roughly with the answer.

(c)   23.007
      +   1.120 ← Attach extra zero.
      ─────────
         24.127

(d)   14.60 ┐
          1.20 ┘ Attach extra
      +   3.15   zeros.
      ─────────
         18.95

(e)   ¹
       5.7
      +3.4
      ─────
       9.1   0.7 + 0.4 = 1.1 Write 0.1.
                             Carry 1.
                Add: 1 + 5 + 3 = 9

(f)   ¹
     $ 9.00
       0.72
       6.09
      ──────
     $15.81

(g)   ¹ ¹
       0.07
       6.79
       0.30 ┐
       3.00 ┘ Attach extra zeros.
      ──────
      10.16

(h)   ¹
       2.36
       7.00 ←
       3.90 ← Attach extra
       0.60 ←   zeros.
      ──────
      13.86

13.86 oz of concentrate was used.

Careful You must line up the decimal points vertically to be certain of getting a correct answer. ◄

Subtraction Subtraction is equally simple if you line up the decimal points carefully and attach any needed zeros before you begin work. As in the subtraction of whole numbers, you may need to "borrow" to complete the calculation.

Example 5

$437.56 − $41 = _____ is

      3   13
$ 4 $\cancel{3}$ 7 .5 6 ┐ Decimal points are in a vertical line.
−$ 4 1 .0 0   Attach zeros (remember that $41 is $41. or $41.00).
─────────────────
$ 3 9 6 .5 6

or again $\qquad$ 19.452 − 7.3617 = _____

```
            3  15  1  10  └─ Decimal points are in a vertical line.
   1   9  ⁰4   ⁵8  ⁵2  ⁰0      Attach zero.
 −     7  .3   6   1   7
   1   2  .0   9   0   3
          ↑└──────────────── Answer decimal point is in same vertical line.
```

→ **Your Turn**

Try these problems to test yourself on the subtraction of decimal numbers.

(a) $37.66 − $14.57 = _____   (b) 248.3 − 135.921 = _____

(c) 6.4701 − 3.2 = _____   (d) 7.304 − 2.59 = _____

(e) $20 − $7.74 = _____   (f) 36 − 11.132 = _____

(g) **Machine Trades** If 0.037 in. of metal is machined from a rod exactly 10 in. long, what is the new length of the rod?

Work carefully.

→ **Solutions**

(a)
```
            ⌐5  16 └─ Line up decimal points.
   $  3   7  ⁶6  ⁶6
 − $  1   4  .5   7
   $  2   3  .0   9
```
 $14.57 + $23.09 = $37.66

(b)
```
            7   12  9  10
   2   4   8  ⁹3  ⁰0  ⁰0 ⌐─ Line up decimal points.
 − 1   3   5  .9   2   1 └─ Attach zeros.
   1   1   2  .3   7   9
             ↑└──────────────
```
 135.921 + 112.379 = 248.300
Answer decimal point is in the same vertical line.

(c)
```
   6  .4   7   0   1
 − 3  .2   0   0   0 ◄── Attach zeros.
   3  .2   7   0   1
```
 3.2000 + 3.2701 = 6.4701

(d)
```
        6  12  10
   7   ⁷3  ⁰0   4
 − 2   .5   9   0 ◄── Attach a zero.
   4   .7   1   4
```
 2.590 + 4.714 = 7.304

(e)
```
            1   9   9  10
   $  2   0  ⁹0  ⁰0  ⁰0 ◄── Attach zeros.
 − $      7  .7   4
   $  1   2  .2   6
```
 $7.74 + $12.26 = $20.00

(f)
```
            5   9   9  10
   3   6  ⁵0  ⁰0  ⁰0 ◄── Attach zeros.
 − 1   1  .1   3   2
   2   4  .8   6   8
```
 11.132 + 24.868 = 36.000

(g)
```
            9   9   9  10
   1   0  ⁹0  ⁰0  ⁰0 ◄── Attach zeros.
 −     0  .0   3   7
   9      .9   6   3
```
 0.037 + 9.963 = 10.000

Notice that each problem is checked by comparing the sum of the answer and the number subtracted with the first number. You should also start by estimating the answer. Whether you use a calculator or work it out with pencil and paper, checking your answer is important if you are to avoid careless mistakes.

Learning Help The balancing method used to add and subtract whole numbers mentally can also be used with decimal numbers.

For example, to add 12.8 and 6.3 mentally, first increase 12.8 by 0.2 to get a whole number 13. To keep the total the same, balance by subtracting 0.2 from 6.3 to get 6.1. Now 12.8 + 6.3 = 13.0 + 6.1, which is easy to do mentally. The answer is 19.1.

To subtract 34.6 − 18.7, add 0.3 to 18.7 to get the whole number 19. Balance this by adding 0.3 to 34.6 to get 34.9. Now 34.6 − 18.7 = 34.9 − 19.0. If this is still too tough to do mentally, balance again to make it 35.9 − 20.0 or 15.9. ◀

Now, for a set of practice problems on addition and subtraction of decimal numbers, turn to Exercises 3-1.

Exercises 3-1 Addition and Subtraction of Decimal Numbers

A. Write in words.

1.	0.72	2.	8.7	3.	12.36	4.	0.05
5.	3.072	6.	14.091	7.	3.0024	8.	6.0083

Write as a number.

9. Four thousandths

10. Three and four tenths

11. Six and seven tenths

12. Five thousandths

13. Twelve and eight tenths

14. Three and twenty-one thousandths

15. Ten and thirty-two thousandths

16. Forty and seven tenths

17. One hundred sixteen ten-thousandths

18. Forty-seven ten-thousandths

19. Two and three hundred seventy-four ten-thousandths

20. Ten and two hundred twenty-two ten-thousandths

B. Add or subtract as shown.

1.	14.21 + 6.8	2.	75.6 + 2.57
3.	$2.83 + $12.19	4.	$52.37 + $98.74
5.	0.687 + 0.93	6.	0.096 + 5.82
7.	507.18 + 321.42	8.	212.7 + 25.46

9. 45.6725 + 18.058   10. 390 + 72.04

11. 19 − 12.03   12. 7.83 − 6.79

13. $33.40 − $18.04   14. $20.00 − $13.48

15. 75.08 − 32.75   16. 40 − 3.82

17. $30 − $7.98   18. $25 − $0.61

19. 130 − 16.04   20. 19 − 5.78

21. 37 + 0.09 + 3.5 + 4.605   22. 183 + 3.91 + 45 + 13.2

23. $14.75 + $9 + $3.76   24. 148.002 + 3.4

25. 68.708 + 27.18   26. 35.36 + 4.347

27. 47.04 − 31.88   28. 180.76 − 94.69

29. 26.45 − 17.832   30. 92.302 − 73.6647

31. 6.4 + 17.05 + 7.78   32. 212.4 + 76 + 3.79

33. 26.008 − 8.4   34. 36.4 − 7.005

35. 0.0046 + 0.073   36. 0.038 + 0.00462

37. 28.7 − 7.38 + 2.9   38. 0.932 + 0.08 − 0.4

39. 6.01 − 3.55 − 0.712   40. 2.92 − 1.007 − 0.08

C. Practical Problems

1. **Machine Trades** What is the combined thickness of these five shims: 0.008, 0.125, 0.150, 0.185, and 0.005 in.?

2. **Electrical Trades** The combined weight of a spool and the wire it carries is 13.6 lb. If the weight of the spool is 1.75 lb, what is the weight of the wire?

3. **Electrical Trades** The following are diameters of some common household wires: No. 10 is 0.102 in., No. 11 is 0.090 in., No. 12 is 0.081 in., No. 14 is 0.064 in., and No. 16 is 0.051 in.

   (a) The diameter of No. 16 wire is how much smaller than the diameter of No. 14 wire?

   (b) Is No. 12 wire larger or smaller than No. 10 wire? What is the difference in their diameters?

   (c) John measured the thickness of a wire with a micrometer as 0.059 in. Assuming that the manufacturer was slightly off, what wire size did John have?

4. **Trades Management** In estimating the cost of a job, a contractor included the following items:

Material	$ 877.85
Trucking	$ 62.80
Permits	$ 250.00
Labor	$1845.50
Profit	$ 450.00

   What was his total estimate for the job?

5. **Metalworking** Find A, B, and C.

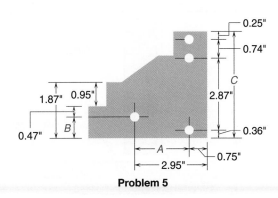

Problem 5

6. **Plumbing** A piece of pipe 8.4 in. long is cut from a piece 40.8 in. long. What is the length of the longer piece remaining if the width of the saw cut is 0.2 in.?

7. **Machine Trades** A certain machine part is 2.345 in. thick. What is its thickness after 0.078 in. is ground off?

8. **Masonry** Find the total cost of the materials for a certain masonry job if sand cost $16.63, mortar mix cost $99.80, and brick cost $1476.28.

9. **Masonry** The specifications for a reinforced masonry wall called for 1.5 sq in. of reinforcing steel per square foot of cross-sectional area. If the three pieces of steel being used had cross sections of 0.125, 0.200, and 1.017 sq in., did they meet the specifications for a 1-sq-ft area?

10. **Construction** A plot plan of a building site showed that the east side of the house was 46.35 ft from the east lot line, and the west side of the house was 41.65 ft from the west lot line. If the lot was 156.00 ft wide along the front, how wide is the house?

11. **Automotive Trades** A mechanic must estimate the total time for a particular servicing. She figures 0.3 hour for an oil change, 1.5 hours for a tune-up, 0.4 hour for a brake check, and 1.2 hours for air-conditioning service. What is the total number of hours of her estimate?

12. **Automotive Trades** A heated piston measures 8.586 cm in diameter. When cold it measures 8.573 cm in diameter. How much does it expand when heated?

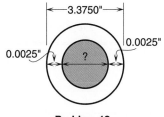

Problem 13

13. **Automotive Trades** A piston must fit in a bore with a diameter of 3.3750 in. What must be the diameter of the piston given the clearance shown in the diagram?

14. **Automotive Trades** In squaring a damaged car frame, an autobody worker measured the diagonals between the two front cross-members. One diagonal was 196.1 cm and the other was 176.8 cm. What is the difference that must be adjusted?

15. **Machine Trades** The diameter of a steel shaft is reduced 0.007 in. The original diameter of the shaft was 0.850 in. Calculate the reduced diameter of the shaft.

16. **Allied Health** A bottle of injectable vertigo medication contains 30 mL. If 5.65 mL are removed, what volume of medicine remains?

17. **Construction** A structural steel Lally column is mounted on a concrete footing as shown on the next page. Find the length of the Lally column.

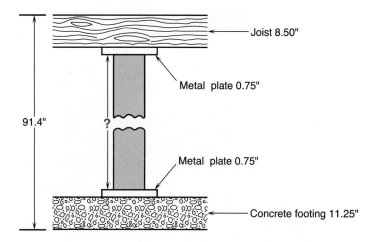

Joist 8.50"

Metal plate 0.75"

91.4"

?

Metal plate 0.75"

Concrete footing 11.25"

18. **Allied Health**   A patient has body mass index (BMI) of 28.3. If the optimum range of BMI is between 18.5 and 25, how much above the upper limit is the patient's BMI?

19. **Automotive Trades**   A technician renovating a truck is replacing the spark plugs. The recommended spark plug gap is 0.044 in., but the new plugs have a gap of 0.035 in. By what amount must the electrodes be bent in order to achieve the recommended gap?

20. **Electronics**   A satellite dish must be angled at an elevation of 43.1 degrees above horizontal to capture the signal from a particular satellite. When installing the dish, the technician will first approximate the proper position and then adjust the angle as needed. If the angle is initially set at 35.7 degrees, how many more degrees must the angle be increased so that the dish will capture the signal?

21. **Automotive Trades**   An automotive technician is paid by the job, or "flat rate," so that his hours will vary on a daily basis. During one particular day, he completed jobs that credited him with the following number of hours: 1.5, 0.6, 0.4, 1.0, 2.2, and 1.5. What was the total number of hours he accumulated that day?

D. Calculator Problems (If you need help using a calculator with decimals, turn to page 170.)

1. **Life Skills**   Balance this checkbook by finding the closing balance as of November 4.

Date	Balance	Withdrawals	Deposits
Oct. 1	$367.21		
Oct. 3		$167.05	
Oct. 4		$104.97	
Oct. 8			$357.41
Oct. 16		$ 87.50	
Oct. 18		$ 9.43	
Oct. 20		$ 30.09	
Oct. 22			$364.85
Oct. 27		$259.47	
Oct. 30		$100.84	
Nov. 2		$ 21.88	
Nov. 4	?		

2. **Automotive Trades**   What is the actual cost of the following car?

"Sticker price"	$21,325.00
Destination and delivery	635.00
Leather interior	875.40
5-speed automatic transmission	900.00
CD player	575.95
Moonroof	735.50
Tax and license	1602.34
Less trade-in	$ 1780.00

3. Add as shown.

(a) 0.0067
0.032
0.0012
0.0179
0.045
0.5
0.05
0.0831
0.004

(b) 1379.4
204.5
16.75
300.04
2070.08
167.99
43.255
38.81
19.95

(c) 14.07
67.81
132.99
225.04
38.02
4
16.899
7.007
4.6

(d) $0.002 + 17.1 + 4.806 + 9.9981 - 3.1 + 0.701 - 1.001 - 14 - 8.09 + 1.0101$

Check your answers to the odd-numbered problems in the Appendix, then continue in Section 3-2.

3-2 Multiplication and Division of Decimal Numbers

A decimal number is really a fraction with a power of 10 as denominator. For example,

$$0.5 = \frac{5}{10} \qquad 0.3 = \frac{3}{10} \qquad \text{and} \qquad 0.85 = \frac{85}{100}$$

Multiplication of decimals is easy to understand if we think of it in this way:

$$0.5 \times 0.3 = \frac{5}{10} \times \frac{3}{10} = \frac{15}{100} = 0.15$$

> **Learning Help**   To estimate the product of 0.5 and 0.3, remember that if two numbers are both less than 1, their product must be less than 1. ◄

Multiplication   Of course it would be very, very clumsy and time-consuming to calculate every decimal multiplication this way. We need a simpler method. Here is the procedure most often used:

Step 1   Multiply the two decimal numbers as if they were whole numbers. Pay no attention to the decimal points.

Step 2   The sum of the decimal digits in the two numbers being multiplied will give you the number of decimal digits in the answer.

Example 1

Multiply 3.2 by 0.41.

Step 1   Multiply, ignoring the decimal points.

$$\begin{array}{r} 32 \\ \times\ 41 \\ \hline 1312 \end{array}$$

Step 2 Count decimal digits in each number: 3.2 has *one* decimal digit (the 2), and 0.41 has *two* decimal digits (the 4 and the 1). The total number of decimal digits in the two factors is three. The answer will have *three* decimal digits. Count over *three* digits from right to left in the answer.

1.312　　three decimal digits

 3.2 × 0.41 is roughly $3 \times \frac{1}{2}$ or about $1\frac{1}{2}$. The answer 1.312 agrees with our rough guess. Remember, even if you use a calculator to do the actual work of arithmetic, you must *always* estimate your answer first and check it afterward.

→ **Your Turn**

Try these simple decimal multiplications.

(a) 2.5 × 0.5 = _____

(b) 0.1 × 0.1 = _____

(c) 10 × 0.6 = _____

(d) 2 × 0.4 = _____

(e) 2 × 0.003 = _____

(f) 0.01 × 0.02 = _____

(g) 0.04 × 0.005 = _____

(h) **Roofing** If asphalt tile weighs 1.1 lb/sq ft (pounds per square foot), what is the weight of tile covering 8 sq ft?

→ **Solutions**

(a) 2.5 × 0.5 = _____

First, multiply 25 × 5 = 125.

Second, count decimal digits.
2.5 × 0.5

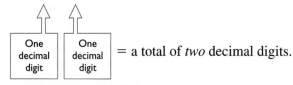 = a total of *two* decimal digits.

Count over *two* decimal digits from the right: 1.25 two decimal digits.

The product is 1.25.

 $2 \times \frac{1}{2}$ is about 1, so the answer seems reasonable.

(b) 0.1 × 0.1   1 × 1 = 1

Count over *two* decimal digits from the right. Because there are not two decimal digits in the product, attach a zero on the left: 0.01

So 0.1 × 0.1 = 0.01.

 Two decimal digits

$\frac{1}{10} \times \frac{1}{10} = \frac{1}{100}$.

(c) 10 × 0.6   10 × 6 = 60

Count over *one* decimal digit from the right: 6.0

So 10 × 0.6 = 6.0.

Notice that multiplication by 10 simply shifts the decimal place one digit to the right.

$$10 \times 6.2 \quad = 62$$
$$10 \times 0.075 = 0.75$$
$$10 \times 8.123 = 81.23 \qquad \text{and so on} \blacktriangleleft$$

(d) $2 \times 0.4 \qquad 2 \times 4 = 8$

Count over *one* decimal digit: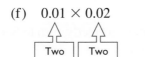

$$2 \times 0.4 = 0.8$$

(e) $2 \times 0.003 \qquad 2 \times 3 = 6$

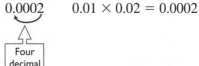

Count over three decimal digits. Attach two zeros as place holders.

$$0.006 \qquad 2 \times 0.003 = 0.006$$

(f) $0.01 \times 0.02 \qquad 1 \times 2 = 2$

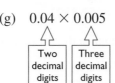

Total of four decimal digits

Count over four decimal digits. Attach three zeros.

$$0.0002 \qquad 0.01 \times 0.02 = 0.0002$$

Four decimal digits

(g) $0.04 \times 0.005 \qquad 4 \times 5 = 20$

Two decimal digits | Three decimal digits

Count over five decimal digits. Attach three zeros.

$$0.00020 \qquad 0.04 \times 0.005 = 0.00020 = 0.0002$$

Total of five decimal digits

Five decimal digits

(h) We are given the weight of 1 sq ft and we need to calculate the weight of 8 sq ft. Multiply

$$1.1 \times 8 \qquad 11 \times 8 = 88$$

One decimal digit

The answer will have *one* decimal digit: $1.1 \times 8 = 8.8$

The weight of 8 sq ft of tile is 8.8 lb.

1. Do not try to do this entire process mentally until you are certain you will not misplace zeros.

2. Always estimate before you begin the arithmetic, and check your answer against your estimate. ◀

Multiplication of larger decimal numbers is performed in exactly the same manner.

Try these:

(a) 4.302 × 12.05 = _____   (b) 6.715 × 2.002 = _____

(c) 3.144 × 0.00125 = _____

(d) **Office Services** An automotive tech earns a base pay of $21.46 per hour plus time-and-a-half for overtime (weekly time exceeding 40 hours). If the tech worked 46.5 hours one week, what was her gross pay?

→ **Solutions**

(a) **Estimate:** 4 × 12 = 48   The answer will be about 48.

Multiply:
```
      4302
    × 1205
    5183910
```

(If you cannot do this multiplication correctly, turn to Section 1-3 for help with the multiplication of whole numbers.)

The two factors being multiplied have a total of five decimal digits (three in 4.302 and two in 12.05). Count over five decimal digits from the right in the answer.

51.83910

So that 4.302 × 12.05 = 51.83910 = 51.8391

✔ The answer 51.8391 is approximately equal to the estimate of 48.

(b) **Estimate:** 6.7 × 2 is about 7 × 2 or 14.

Multiply:
```
       6.715          6.715 has three decimal digits
     × 2.002          2.002 has three decimal digits
     13.443430        a total of six decimal digits
```

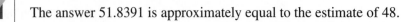

Six decimal digits

6.715 × 2.002 = 13.44343

✔ The answer and estimate are roughly equal.

(c) **Estimate:** 3 × 0.001 is about 0.003.

Multiply:
```
        3.144         3.144 has three decimal digits
     × 0.00125        0.00125 has five decimal digits
      .00393000       a total of eight decimal digits
```

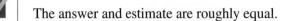

Eight decimal digits

To count over 8 digits we had to attach 2 zeros on the left.

3.144 × 0.00125 = 0.00393

✔ The answer and estimate are roughly equal.

(d) **Estimate:** For her first 40 hours, 40 × $21.46 is about 40 × $20 or $800. "Time-and-a-half" for $20 would be $30 per hour. Since we *under*estimated the pay for the first 40 hours, we will *over*estimate the overtime pay by rounding 6.5 hr to 7 hr. Her overtime pay is approximately 7 × \$30 or \$210. Her total gross pay should be about \$800 + \$210 or \$1010.

Calculation:

For her first 40 hours:

$21.46 ← *two* decimal digits
× 40 ← *no* decimal digits
$858.40 ← *two* decimal digits

For her overtime hours:

First, find her hourly rate. Multiply by 1.5 to find "time-and-a-half":

$21.46 ← *two* decimal digits
× 1.5 ← *one* decimal digit
$32.190 ← *three* decimal digits

Then, multiply this overtime rate by her overtime hours.
46.5 hr − 40 hr = 6.5 overtime hours

$32.19 ← *two* decimal digits
× 6.5 ← *one* decimal digit
$209.235 ← *three* decimal digits

For her gross pay, add the two amounts together:

$858.40
+ 209.235
$1067.635   or approximately \$1067.64

The answer and estimate are roughly equal.

> **Note**   When multiplying, you need not line up the decimal points. ◄

One nice thing about a calculator is that it counts decimal digits and automatically adjusts the answer in any calculation. But to get an estimate of the answer, you must still understand the process.

Division

Division of decimal numbers is very similar to the division of whole numbers. For example,

$6.8 \div 1.7$ can be written $\dfrac{6.8}{1.7}$

and if we multiply both top and bottom of the fraction by 10 we have

$$\frac{6.8}{1.7} = \frac{6.8 \times \boxed{10}}{1.7 \times \boxed{10}} = \frac{68}{17}$$

But you should know how to divide these whole numbers.

$68 \div 17 = 4$

Therefore, $6.8 \div 1.7 = 4$.

To divide decimal numbers, use the following procedure.

Step 1   Write the divisor and dividend in standard long-division form. $6.8 \div 1.7$
$$1.7\overline{)6.8}$$

Step 2   Shift the decimal point in the divisor to the right so as to make the divisor a whole number. $1.7.\overline{)}$

Step 3 Shift the decimal point in the dividend the same amount. (Attach zeros if necessary.)

$$1.7.\overline{)6.8.}$$

Step 4 Place the decimal point in the answer space directly above the new decimal position in the dividend.

$$17.\overline{)68.}$$

Step 5 Now divide exactly as you would with whole numbers. The decimal points in divisor and dividend may now be ignored.

$$6.8 \div 1.7 = 4$$

$$\begin{array}{r} 4. \\ 17.\overline{)68.} \\ -68 \\ \hline 0 \end{array}$$

Notice in Steps 2 and 3 that we have simply multiplied both divisor and dividend by 10.

If there is a remainder in Step 5, we must add additional zeros to complete the division. This will be necessary in the next example.

Example 2

Divide: $1.38 \div 2.4$

Let's do it step by step.

Write the problem in standard long-division form.

$$2.4\overline{)1.38}$$

Shift both decimal points one digit to the right to make the divisor (2.4) a whole number (24).

$$2.4.\overline{)13.8}$$

Place the decimal point in the answer space.

$$24.\overline{)13.8}$$

Divide as usual. 24 goes into 138 five times.

$$\begin{array}{r} .5 \\ 24.\overline{)13.8} \\ 120 \\ \hline 18 \end{array}$$

There is a remainder (18), so we must keep going.

Attach a zero to the dividend and bring it down.

$$\begin{array}{r} .57 \\ 24.\overline{)13.80} \\ 120 \\ \hline 180 \\ 168 \\ \hline 12 \end{array}$$

Divide 180 by 24. 24 goes into 180 seven times.

Now the remainder is 12.

Attach another zero and bring it down.

$$\begin{array}{r} .575 \\ 24.\overline{)13.800} \\ 120 \\ \hline 180 \\ 168 \\ \hline 120 \\ 120 \\ \hline 0 \end{array}$$

24 goes into 120 exactly five times.

The remainder is zero. $1.38 \div 2.4 = 0.575$

 $1.38 \div 2.4$ is roughly $1 \div 2$ or 0.5.

Double-Check: $2.4 \times 0.575 = 1.38$. Always multiply the answer by the divisor to double-check the work. Do this even if you are using a calculator.

Note In some problems, we may continue to attach zeros and divide yet never get a remainder of zero. We will examine problems like this later. ◄

How would you do this one?

$2.6 \div 0.052 = ?$

To make the divisor a whole number, shift the decimal point three digits to the right.

$$0.\underset{\smile}{052.})\overline{2.6}$$

To shift the decimal place three digits in the dividend, we must attach two zeros to its right.

$$0.\underset{\smile}{052.})\overline{2.\underset{\smile}{600.}}$$

Now place the decimal point in the answer space above that in the dividend and divide normally.

$$
\begin{array}{r}
50. \\
52.)\overline{2600.} \\
\underline{260} \longleftarrow 5 \times 52 = 260 \\
0 \\
\underline{0}
\end{array}
$$

$2.6 \div 0.052 = 50$ ✓ $0.052 \times 50 = 2.6$

Shifting the decimal point three digits and attaching zeros to the right of the decimal point in this way is equivalent to multiplying both divisor and dividend by 1000.

Note In this last example inexperienced students think the problem is finished after the first division because they get a remainder of zero. You must keep dividing at least until all the places up to the decimal point in the answer are filled. The answer here is 50, not 5. ◄

Try these problems.

(a) $3.5 \div 0.001$ = _____   (b) $9 \div 0.02$ = _____

(c) $0.365 \div 18.25$ = _____   (d) $8.8 \div 3.2$ = _____

(e) $7.230 \div 6$ = _____   (f) $3 \div 4$ = _____

(g) $30.24 \div 0.42$ = _____   (h) $273.6 \div 0.057$ = _____

(i) **Construction** How many pieces of plywood 0.375 in. thick are in a stack 30 in. high?

(a) $0.\underset{\smile}{001.})\overline{3.\underset{\smile}{500.}}$

$$
\begin{array}{r}
3500. \\
1.)\overline{3500.}
\end{array}
$$
$3.5 \div 0.001 = 3500$

✓ $0.001 \times 3500 = 3.5$

(b) Shift the decimal point two places to the right. Divide 900 by 2.

$$0.02\overline{)9.00}$$

$$\begin{array}{r} 450. \\ 2\overline{)900.} \end{array} \qquad 9 \div 0.02 = 450$$

 $0.02 \times 450 = 9$

Dividing a whole number by a decimal is very troublesome for most people.

(c) $18.25\overline{)0.365}$

$$\begin{array}{r} .02 \\ 1825\overline{)36.50} \\ \underline{36.50} \\ 0 \end{array}$$

1825 does not go into 365, so place a zero above the 5. Attach a zero after the 5.

1825 goes into 3650 twice. Place a 2 in the answer space above the zero.

$0.365 \div 18.25 = 0.02$

 $18.25 \times 0.02 = 0.365$

(d) $3.2\overline{)8.8}$

$$\begin{array}{r} 2.75 \\ 32\overline{)88.00} \\ \underline{64}\downarrow \\ 240 \\ \underline{224}\downarrow \\ 160 \\ \underline{160} \\ 0 \end{array}$$

$2 \times 32 = 64$
Subtract, attach a zero, and bring it down.
$7 \times 32 = 224$
Subtract, attach another zero, and bring it down.
$5 \times 32 = 160$
$8.8 \div 3.2 = 2.75$

 The estimated answer is $9 \div 3$ or 3.

Double-check: $3.2 \times 2.75 = 8.8$

(e) $6\overline{)7.230}$

$$\begin{array}{r} 1.205 \\ 6\overline{)7.230} \\ \underline{-6}\downarrow \\ 12 \\ \underline{-12}\downarrow \\ 03 \\ 0\downarrow \\ \overline{30} \\ \underline{-30} \\ 0 \end{array}$$

The divisor 6 is a whole number, so we can bring the decimal point in 7.23 up to the answer space.

 $1.205 \times 6 = 7.230$

$7.230 \div 6 = 1.205$

(f) $4\overline{)3.00}$

$$\begin{array}{r} .75 \\ 4\overline{)3.00} \\ 28 \\ \overline{20} \\ \underline{20} \\ 0 \end{array}$$

 $0.75 \times 4 = 3.00$

$3 \div 4 = 0.75$

Chapter 3 Decimal Numbers

(g)
$$
\begin{array}{r}
72. \\
0.42.\overline{)30.24.} \\
29\ 4 \\
\hline
84 \\
84 \\
\hline
0
\end{array}
$$

✓ $72 \times 0.42 = 30.24$

(h)
$$
\begin{array}{r}
4\,800. \\
0.057.\overline{)273.600.} \\
228 \\
\hline
45\ 6 \\
45\ 6 \\
\hline
0
\end{array}
$$

✓ $4800 \times 0.057 = 273.6$

(i)
$$
\begin{array}{r}
80. \\
0.375.\overline{)30.000.} \\
30\ 00 \\
\hline
0
\end{array}
$$

✓ $80 \times 0.375 = 30$

There are 80 pieces of plywood in the stack.

If the dividend is not exactly divisible by the divisor, either stop the process after some preset number of decimal places in the answer, or round the answer. We do not generally indicate a remainder in decimal division.

ARITHMETIC "TRICKS OF THE TRADE"

1. To divide a number by 5, use the fact that 5 is one-half of 10. First, multiply the number by 2, then divide by 10 by shifting the decimal point one place to the left.

   Example: $64 \div 5$ $\quad$ $64 \times 2 \div 10 = 128 \div 10 = 12.8$

2. To divide a number by 25, first multiply by 4, then divide by 100, shifting the decimal point two places left.

   Example: $304 \div 25 = 304 \times 4 \div 100 = 1216 \div 100 = 12.16$

3. To divide a number by 20, first divide by 2, then divide by 10.

   Example: $86 \div 20 = 86 \div 2 \div 10 = 43 \div 10 = 4.3$

4. To multiply a number by 20, first multiply by 2, then multiply by 10, shifting the decimal point one place to the right.

   Example: $73 \times 20 = 73 \times 2 \times 10 = 146 \times 10 = 1460$

When doing "mental arithmetic" like this, it is important that you start with a rough estimate of the answer.

Rounding Decimal Numbers

Calculations involving decimal numbers often result in answers having more decimal digits than are justified. When this occurs, we must round our answer. The process of rounding a decimal number is very similar to the procedure for rounding a whole number. The only difference is that after rounding to a given decimal place, all digits to the right of that place are dropped. The following examples illustrate this difference.

Example 3

Round 35782.462 to the

	nearest thousand	**nearest hundredth**
Step 1 Place a $\wedge$ mark to the right of the place to which the number is to be rounded.	35 $\wedge$ 782.462	35782.46 $\wedge$ 2

Step 2 If the digit to the right of the mark is less than 5, replace all digits to the right of the mark with zeros.

35782.460

Drop this decimal digit zero.

If these zeros are decimal digits, discard them.

The rounded number is 35782.46

Step 3 If the digit to the right of the mark is equal to or larger than 5, increase the digit to the left by 1 and replace all digits to the right with zeros.

36 000.000

Drop these right-end decimal zeros

Drop all right-end decimal zeros, but keep zero placeholders.

36000

Keep these zeros as placeholders.

The rounded number is 36000.

Careful Drop only the decimal zeros that are to the right of the $\wedge$ mark. For example, to round 6.4086 to three decimal digits, we write 6.408$\wedge$6, which becomes 6.4090 or 6.409. We dropped the end zero because it was a decimal zero to the *right* of the mark. But we retained the other zero because it is needed as a placeholder. ◄

→ Your Turn

Try rounding these numbers.

(a) Round 74.238   to two decimal places.

(b) Round 8.043   to two decimal places.

(c) Round 0.07354   to three decimal places.

(d) Round 7.98   to the nearest tenth.

→ Answers

(a) 74.238 is 74.24 to two decimal places.
(Write 74.23$\wedge$8 and note that 8 is larger than 5, so increase the 3 to 4 and drop the last digit because it is a decimal digit.)

(b) 8.043 is 8.04 to two decimal places.
(Write 8.04∧3 and note that 3 is less than 5 so drop it.)

(c) 0.07354 is 0.074 to three decimal places.
(Write 0.073∧54 and note that the digit to the right of the mark is 5; therefore, change the 3 to a 4 to get 0.074∧00. Finally, drop the digits on the right to get 0.074.)

(d) 7.98 is 8.0 to the nearest tenth.
(Write 7.9∧8 and note that 8 is greater than 5, so increase the 9 to 0 and the 7 to 8. Drop the digit in the hundredths place.)

There are a few very specialized situations where this rounding rule is not used:

1. Some engineers use a more complex rule when rounding a number that ends in 5.

2. In business, fractions of a cent are usually rounded up to determine selling price. Three items for 25 cents or $8\frac{1}{3}$ cents each is rounded to 9 cents each.

Our rule will be quite satisfactory for most of your work in arithmetic.

Rounding During Division   In some division problems, this process of dividing will never result in a zero remainder. At some point the answer must be rounded.

To round answers in a division problem, first continue the division so that your answer has one place more than the rounded answer will have, then round it.

Example 4

In the division problem $4.7 \div 1.8 = ?$

to get an answer rounded to one decimal place, first, divide to two decimal places:

$1.8.)\overline{4.7.}$

$$
\begin{array}{r}
2.61 \\
18)\overline{47.00} \quad \longleftarrow \text{ Attach two zeros to the dividend} \\
36 \quad \longleftarrow 2 \times 18 = 36 \\
\overline{110} \\
108 \quad \longleftarrow 6 \times 18 = 108 \\
\overline{20} \\
18 \quad \longleftarrow 1 \times 18 = 18 \\
\end{array}
$$

Therefore, $4.7 \div 1.8 = 2.61 \ldots$. We can now round back to one decimal place to get 2.6.

→ Your Turn

For the following problem, divide and round your answer to two decimal places.

$6.84 \div 32.7 = $ _____

Careful now.

$$32.\overset{\curvearrowright}{7.)}6.\overset{\curvearrowright}{8.}4$$

$$
\begin{array}{r}
.209 \\
327.\overline{)68.400} \\
65\,4 \downarrow\downarrow \\
\hline
3\,000 \\
2\,943 \\
\hline
\end{array}
$$

← Carry the answer to three decimal places.
← Notice that two zeros must be attached to the dividend.
← 2 × 327 = 654

← 9 × 327 = 2943

0.209 rounded to two decimal places is 0.21.

6.84 ÷ 32.7 = 0.21   rounded to two decimal places

 32.7 × 0.21 = 6.867, which is approximately equal to 6.84. (The check will not be exact because we have rounded.)

The ability to round numbers is especially important for people who work in the practical, trade, or technical areas. You will need to round answers to practical problems if they are obtained "by hand" or with a calculator. Rounding is discussed in more detail in Chapter 5.

Averages   Suppose that you needed to know the diameter of a steel connecting pin. As a careful and conscientious worker, you would probably measure its diameter several times with a micrometer, and you might come up with a sequence of numbers like this (in inches):

   1.3731, 1.3728, 1.3736, 1.3749, 1.3724, 1.3750

What is the actual diameter of the pin? The best answer is to find the **average** value or **arithmetic mean** of these measurements.

$$\text{Average} = \frac{\text{sum of the measurements}}{\text{number of measurements}}$$

For the preceding problem,

$$\text{Average} = \frac{1.3731 + 1.3728 + 1.3736 + 1.3749 + 1.3724 + 1.3750}{6}$$

$$= \frac{8.2418}{6} = 1.3736333\ldots$$

$$= 1.3736 \quad \text{rounded to four decimal places}$$

When you calculate the average of a set of numbers, the usual rule is to round the answer to the same number of decimal places as the least precise number in the set—that is, the one with the fewest decimal digits. If the numbers to be averaged are all whole numbers, the average will be a whole number.

Example 5

The average of 4, 6, 4, and 5 is $\dfrac{19}{4} = 4.75$ or 5 when rounded.

If one number of the set has fewer decimal places than the rest, round off to agree with the least precise number.

Example 6

The average of 2.41, 3.32, 5.23, 3.51, 4.1, and 4.12 is

$\dfrac{22.69}{6} = 3.78166\ldots$ and this answer should be rounded to 3.8 to agree in precision with the least precise number, 4.1. (We will learn more about precision in Chapter 5.)

To avoid confusion, we will usually give directions for rounding.

→ Your Turn

Try it. Find the average for each of these sets of numbers.

(a) 8, 9, 11, 7, 5   (b) 0.4, 0.5, 0.63, 0.2

(c) 2.35, 2.26, 2.74, 2.55, 2.6, 2.31

→ Solutions

(a) Average $= \dfrac{8 + 9 + 11 + 7 + 5}{5} = \dfrac{40}{5} = 8$

(b) Average $= \dfrac{0.4 + 0.5 + 0.63 + 0.2}{4} = \dfrac{1.73}{4} = 0.4325 \approx 0.4$ rounded

(c) Average $= \dfrac{2.35 + 2.26 + 2.74 + 2.55 + 2.6 + 2.31}{6}$

$= \dfrac{14.81}{6} = 2.468333\ldots$

≈ 2.5 rounded to agree in precision with 2.6, the least precise number in the set. (The symbol $\approx$ means "approximately equal to.")

Now turn to Exercises 3-2 for a set of problems on the multiplication and division of decimal numbers.

Exercises 3-2 Multiplication and Division of Decimals

A. Multiply or divide as shown.

1. 0.01×0.001	2. 10×2.15	3. 0.04×100
4. 0.3×0.3	5. 0.7×1.2	6. 0.005×0.012
7. 0.003×0.01	8. 7.25×0.301	9. 2×0.035
10. $0.2 \times 0.3 \times 0.5$	11. $0.6 \times 0.6 \times 6.0$	12. $2.3 \times 1.5 \times 1.05$
13. $3.618 \div 0.6$	14. $3.60 \div 0.03$	15. $4.40 \div 0.22$
16. $6.5 \div 0.05$	17. $0.0405 \div 0.9$	18. $0.378 \div 0.003$

19. $3 \div 0.05$	20. $10 \div 0.001$	21. $4 \div 0.01$
22. $2.59 \div 70$	23. $44.22 \div 6.7$	24. $104.2 \div 0.0320$
25. $484 \div 0.8$	26. 6.05×2.3	27. 0.0027×1.4
28. $0.0783 \div 0.27$	29. $0.00456 \div 0.095$	30. $800 \div 0.25$
31. $324 \div 0.0072$	32. $0.08322 \div 228$	33. $0.0092 \div 115$
34. 0.047×0.024	35. 0.0056×0.065	36. $0.02 \times 0.06 \times 0.04$
37. $0.008 \times 0.4 \times 0.03$	38. 123.4×0.45	39. 0.062×27.5

B. Round the following numbers as indicated.

1. 42.875 (Two decimal places)   2. 0.5728 (Nearest tenth)

3. 6.54 (One decimal place)   4. 117.6252 (Three decimal places)

5. 79.135 (Nearest hundredth)   6. 1462.87 (Nearest whole number)

7. 3.64937 (Four decimal places)   8. 19.4839 (Nearest hundredth)

9. 0.2164 (Nearest thousandth)   10. 56.826 (Two decimal places)

C. Divide and round as indicated.
Round to two decimal digits.

1. $10 \div 3$	2. $5 \div 6$	3. $2.0 \div 0.19$
4. $3 \div 0.081$	5. $0.023 \div 0.19$	6. $12.3 \div 4.7$
7. $2.37 \div 0.07$	8. $6.5 \div 1.31$	

Round to the nearest tenth.

9. $100 \div 3$	10. $21.23 \div 98.7$	11. $1 \div 4$
12. $100 \div 9$	13. $0.006 \div 0.04$	14. $1.008 \div 3$

Round to three decimal places.

15. $10 \div 70$	16. $0.09 \div 0.402$	17. $0.091 \div 0.0014$
18. $3.41 \div 0.257$	19. $6.001 \div 2.001$	20. $123.21 \div 0.1111$

D. Word Problems

1. **Painting**   A power-spraying outfit is advertised for $899. It can also be bought "on time" for 24 payments of $39.75 each. How much extra do you pay by purchasing it on the installment plan?

2. **Allied Health**   How much total medication will a patient receive if he is given four 0.075-mg tablets each day for 12 days?

3. **Metalworking**   Find the average weight of five castings that weigh 17.0, 21.0, 12.0, 20.6, and 23.4 lb.

4. **Office Services**   If you work $8\frac{1}{4}$ hours on Monday, 10.1 hours on Tuesday, 8.5 hours on Wednesday, 9.4 hours on Thursday, $6\frac{1}{2}$ hours on Friday, and 4.2 hours on Saturday, what is the average number of hours worked per day?

5. **Machine Trades**   For the following four machine parts, find W, the number of pounds per part; C, the cost of the metal per part; and T, the total cost.

Metal Parts	Number of Inches Needed	Number of Pounds per Inch	Cost per Pound	Pounds (W)	Cost per Part (C)
A	44.5	0.38	$0.98		
B	122.0	0.19	$0.89		
C	108.0	0.08	$1.05		
D	9.5	0.32	$2.15		
					$T =$

6. **Machine Trades** How much does 15.7 sq ft of No. 16 gauge steel weigh if 1 sq ft weighs 2.55 lb?

7. **Construction** A 4-ft by 8-ft sheet of $\frac{1}{4}$-in. plywood has an area of 32 sq ft. If the weight of $\frac{1}{4}$-in. plywood is 1.5 lb/sq ft (pounds per square foot), what is the weight of the sheet?

8. **Manufacturing** Each inch of 1-in.-diameter cold-rolled steel weighs 0.22 lb. How much would a piece weigh that was 38 in. long?

9. **Transportation** A truck can carry a load of 5000 lb. Assuming that it could be cut to fit, how many feet of steel beam weighing 32.6 lb/ft (pounds per foot) can the truck carry?

10. **Fire Protection** One gallon of water weighs 8.34 lb. How much weight is added to a fire truck when its tank is filled with 750 gal of water?

11. **Electrical Trades** Voltage values are often stated in millivolts [1 millivolt (mV) equals 0.001 volt] and must be converted to volts to be used in Ohm's law. If a voltage is given to be 75 mV, how many volts is this?

12. **Industrial Technology** An industrial engineer must estimate the cost of building a storage tank. The tank requires 208 sq ft of material at $9.29 per square foot. It also requires 4.5 hours of labor at $26.40 per hour plus 1.6 hours of labor at $19.60 per hour. Find the total cost of the tank.

13. **Sheet Metal Trades** How many sheets of metal are in a stack 5.00 in. high if each sheet is 0.0149 in. thick?

14. **Metalworking** A casting weighs 3.68 lb. How many castings are contained in a load weighing 5888.00 lb?

15. **Agriculture** A barrel partially filled with liquid fertilizer weighs 267.75 lb. The empty barrel weighs 18.00 lb, and the fertilizer weighs 9.25 lb/gal (pounds per gallon). How many gallons of fertilizer are in the barrel?

16. **Construction** A 4-ft width of $\frac{1}{4}$-in. gypsum costs $6.15. Using this material, how much would it cost to drywall a room with 52 ft of total wall width?

17. **Flooring and Carpeting** A room addition requires 320 sq ft of wood floor. Hardwood costs $4.59 per sq ft, while laminate costs $2.79 per sq ft. What is the total difference in cost between the two types of flooring?

18. **Office Services** The Happy Hacker electronics warehouse purchases 460 USB flash drives, which are then packed in sets of four selling for $24.95 per set. Calculate the total income from the sale of these flash drives.

19. **Electronics** An oscillator is a device that generates an ac signal at some specified frequency. If the oscillator's output frequency is 7500 cycles per second (hertz), how many cycles does it generate in a 0.35-sec interval?

20. **Trades Management** A shop owner needs to move to a larger shop. He has narrowed his choice down to two possibilities: a 2400-sq-ft shop renting for $3360 per month and a 2800-sq-ft shop renting for $4060 per month. Which shop has the lower cost per square foot?

21. **Office Services** A shop tech earns a base pay of $19.88 per hour, plus "time-and-a-half" for overtime (time exceeding 40 hours). If he works 43.5 hours during a particular week, what is his gross pay?

22. **Automotive Trades** To calculate the torque at the rear wheels of an automobile, we multiply the torque of the engine by the drive ratio. If the engine torque is 305 foot-pounds and the drive ratio is 8.68 (to 1), find the torque at the rear wheels. Round to the nearest ten foot-pounds.

23. **Agriculture** Fertilizer is used at the Tofu Soy Farms at the rate of 3.5 lb per acre. How much fertilizer should be spread on a 760-acre field?

24. **Flooring and Carpeting** A carpet installer is paid $0.75 per square foot of carpet laid. How many square feet would he need to install in a week in order to earn $6000?

25. **HVAC** Regulations require a ventilation rate for houses of 0.35 air changes per hour—that is, 0.35 of the total volume of air must be changed. For a house with a volume of 24,000 cu ft, how many cubic feet of air must be changed per hour?

26. **Allied Health** A litter of 7 puppies has been brought into a veterinary clinic to receive antibiotic for treatment of an infection. Each puppy should receive 0.3 milliliters (mL) of the antibiotic solution per kilogram (kg) of body weight. If the puppies weigh an average of 6.7 kg each, how much antibiotic solution should the veterinary technician prepare for the 7 puppies? (Round to the nearest milliliter.)

27. **Flooring and Carpeting** Carpet installers often are paid by the square foot rather than by the hour. If a certain installer is paid $0.75 per square foot, how much would he earn on a day when he installed 864 sq ft of carpet?

28. **Sports and Leisure** To calculate the distance traveled by a bicycle with each pedal stroke, multiply the gear ratio by the circumference of the tires. If the gear ratio is 2.96 and the circumference of the tires is 81.5 in., how far does the bike roll during one pedal stroke? (Round to the nearest inch.)

29. **Graphic Design** Artists and architects often use the "golden rectangle" (also called the "golden ratio") to determine the most visually pleasing proportions for their designs. The proportions of this rectangle specify that the longer side should be 1.618 times the shorter side. An artist wants to use this proportion for a rectangular design. If the shorter side of the design must be 3.5 in. long, what should the longer side measure? (Round to the nearest tenth.)

30. **Automotive Trades** A 2008 Saturn Astra is EPA-rated at 24 mi/gal in the city and 32 mi/gal on the highway.

    (a) How many gallons of gas would be used on a trip consisting of 66 miles of city driving and 280 miles of highway driving?

    (b) At $1.90 per gallon, how much would the gas cost for this trip?

31. **Office Services** Because of a drought emergency, a water district began charging its customers according to the following block rate system:

Block	Quantity Used	Rate per HCF*
1	The first 20 HCF	$3.90
2	The next 40 HCF (From 21—60 HCF)	$4.15
3	The next 60 HCF (From 61—120 HCF)	$4.90
4	Usage in excess of 120 HCF	$5.90
*HCF = Hundred cubic feet		

What would a customer be charged for using:

(a) 38 HCF (b) 96 HCF (c) 132 HCF (*Hint:* See Example 3 on page 50.)

E. **Calculator Problems**

1. Divide. 9.87654321 ÷ 1.23456789.

   Notice anything interesting? (Divide it out to eight decimal digits.) You should be able to get the correct answer even if your calculator will not accept a nine-digit number.

2. Divide. (a) $\dfrac{1}{81}$ (b) $\dfrac{1}{891}$ (c) $\dfrac{1}{8991}$

   Notice a pattern? (Divide them out to about eight decimal places.)

3. **Metalworking** The outside diameter of a steel casting is measured six times at different positions with a digital caliper. The measurements are 4.2435, 4.2426, 4.2441, 4.2436, 4.2438, and 4.2432 in. Find the average diameter of the casting.

4. **Printing** To determine the thickness of a sheet of paper, five batches of 12 sheets each are measured with a digital micrometer. The thickness of each of the five batches of 12 is 0.7907, 0.7914, 0.7919, 0.7912, and 0.7917 mm. Find the average thickness of a single sheet of paper.

5. **Construction** The John Hancock Towers office building in Boston had a serious problem when it was built: when the wind blew hard, the pressure caused its windows to pop out! The only reasonable solution was to replace all 10,344 windows in the building at a cost of $6,000,000. What was the replacement cost per window? Round to the nearest dollar.

6. **Culinary Arts** In purchasing food for his chain of burger shops, José pays $1.89 per pound for cheese to be used in cheeseburgers. He calculates that an order of 5180 lb will be enough for 165,000 cheeseburgers.

   (a) How much cheese will he use on each cheeseburger? Round to three decimal places.

   (b) What is the cost, to the nearest cent, of the cheese used on a cheeseburger?

7. **Plumbing** The pressure in psi (pounds per square inch) in a water system at a point 140.8 ft below the water level in a storage tank is given by the formula

$$P = \frac{62.4 \text{ lb per cu ft} \times 140.8 \text{ ft} \times 0.43}{144 \text{ sq in. per sq ft}}$$

Find P to the nearest tenth.

8. **Flooring and Carpeting** A carpet cleaner charges $50 for two rooms and $120 for five rooms. Suppose a two-room job contained 320 sq ft and a five-room job contained 875 sq ft. Calculate the difference in cost per square foot between these two jobs. Round to the nearest tenth of a cent.

9. **Electronics** Radar operates by broadcasting a high-frequency radio wave pulse that is reflected from a target object. The reflected signal is detected at the original source. The distance of the object from the source can be determined by multiplying the speed of the pulse by the elapsed time for the round trip of the pulse and dividing by 2. A radar signal traveling at 186,000 miles per second is reflected from an object. The signal has an elapsed round trip time of 0.0001255 sec. Calculate the distance of the object from the radar antenna.

10. **Plumbing** Normal air pressure at sea level is about 14.7 psi. Additional pressure exerted by a depth of water is given by the formula

Pressure in psi $\approx$ 0.434 $\times$ depth of water in ft

Calculate the total pressure on the bottom of a diving pool filled with water to a depth of 12 ft. (The sign $\approx$ means "approximately equal to.")

11. **Life Skills** A total of 12.4 gal was needed to fill a car's gas tank. The car had been driven 286.8 miles since the tank was last filled. Find the fuel economy in miles per gallon since this car was last filled. Round to one decimal place.

12. **Aviation** In November 2005, a Boeing 777-200LR set the world record for the longest nonstop flight by a commercial airplane. The jet flew east from Hong Kong to London, covering 13,422 miles in 22.7 hours. What was the jet's average speed in miles per hour? Round to the nearest whole number.

13. **Aviation** The jet referred to in problem 12 burned 52,670 gal of fuel during the 13,422-mile trip.

    (a) What was the fuel economy of the jet in miles per gallon? Round to three decimal places.

    (b) A more meaningful measure of fuel economy is "people-miles per gallon," which is found by multiplying actual miles per gallon by the number of passengers on the plane. If there were 35 passengers on this historic flight, what was the fuel economy in people-miles per gallon? Round to two decimal places.

14. **Flooring and Carpeting** A certain type of prefinished oak flooring comes in 25-sq-ft cartons and is priced at $101.82 per carton. Assuming that you cannot purchase part of a carton, (a) how many cartons will you need to lay 860 sq ft of floor, and (b) what will the cost of the flooring be?

15. **Office Services** A gas company bills a customer according to a unit called "therms." They calculate the number of therms by multiplying the number of CCF (hundred cubic feet) used by a billing factor that varies according to geographical location. If a customer used 122 CCF during a particular month, and the customer's billing factor is 1.066, how many therms did the customer consume?

16. **Life Skills** In the summer of 2008, a gallon of regular gas cost an average of $4.11 in the United States and $8.96 in England. How much more did it cost to fill up a 16-gal tank in England than in the United States? (Source: AIRINC)

17. **Office Services** Once the number of therms is established on a customer's gas bill (see problem 15), the monthly bill is calculated as follows:

Item	Charge
Customer charge	$0.16438 per day
Regulatory fee	$0.05012 per therm
Gas usage	$0.91071 per therm for the first 56 therms
	$1.09365 per therm for each therm over 56 therms

Suppose that during a 31-day month a customer used 130 therms. What would the customer's total bill be?

18. **Automotive Trades** Most people measure the fuel efficiency of a vehicle using the number of miles per gallon. But to compare the effectiveness of different grades of gasoline—which, of course, have different prices—it may be more meaningful to calculate and compare the number of "miles per dollar" that different grades give you under similar driving conditions. Perform this calculation for parts (a), (b), and (c) and then answer part (d). (Round to the nearest hundredth.) (Source: Mike Allen, Senior Editor Automotive, Popular Mechanics)

(a) 278 miles on 12.4 gal of regular at $1.79 per gallon

(b) 304 miles on 13.1 gal of mid-grade at $1.89 per gallon

(c) 286 miles on 12.0 gal of premium at $1.99 per gallon

(d) Which grade gave the highest number of miles per dollar?

When you have finished these exercises, check your answers to the odd-numbered problems in the Appendix, then continue in Section 3-3.

3-3 Decimal Fractions

Because decimal numbers are fractions, they may be used, as fractions are used, to represent a part of some quantity. For example, recall that

"$\frac{1}{2}$ of 8 equals 4"   means $\frac{1}{2} \times 8 = 4$

and therefore,

"0.5 of 8 equals 4"   means $0.5 \times 8 = 4$

The word *of* used in this way indicates multiplication, and decimal numbers are often called "decimal fractions."

It is very useful to be able to convert any number from fraction form to decimal form. Simply divide the top by the bottom of the fraction. If the division has no remainder, the decimal number is called a *terminating* decimal.

Example 1

$$\frac{5}{8} = \underline{\quad ? \quad}$$

$$\begin{array}{r} .625 \\ 8{\overline{\smash{\big)}\,5.000}} \\ \underline{4\ 8} \\ 20 \\ \underline{16} \\ 40 \\ \underline{40} \\ 0 \end{array}$$ ← Attach as many zeros as needed.

← Zero remainder; therefore, the decimal terminates or ends.

$$\frac{5}{8} = 0.625$$

If the decimal does not terminate, you may round it to any desired number of decimal digits.

Example 2

$$\frac{2}{13} = \underline{\quad ? \quad}$$

$$\begin{array}{r} .1538 \\ 13{\overline{\smash{\big)}\,2.0000}} \\ \underline{1\ 3} \\ 70 \\ \underline{65} \\ 50 \\ \underline{39} \\ 110 \\ \underline{104} \\ 6 \end{array}$$ ← Attach zeros.

← Remainder is not equal to zero.

$$\frac{2}{13} = 0.154 \quad \text{rounded to three decimal places}$$

→ **Your Turn**

Convert the following fractions to decimal form and round if necessary to two decimal places.

(a) $\dfrac{4}{5}$ (b) $\dfrac{2}{3}$ (c) $\dfrac{17}{7}$ (d) $\dfrac{5}{6}$ (e) $\dfrac{7}{16}$ (f) $\dfrac{5}{9}$

→ **Solutions**

(a) $\dfrac{4}{5} = \underline{\quad ? \quad}$ $\begin{array}{r} .8 \\ 5{\overline{\smash{\big)}\,4.0}} \end{array}$

$\dfrac{4}{5} = 0.8$

(b) $\dfrac{2}{3} = $ _____?_____

$$
\begin{array}{r}
.666\ldots \\
3\overline{)2.000} \\
\underline{1\,8} \\
20 \\
\underline{18} \\
20 \\
\underline{18} \\
2
\end{array}
$$

$\dfrac{2}{3} = 0.67$ rounded to two decimal places

Notice that to round to *two* decimal places, we must carry the division out to at least *three* decimal digits.

(c) $\dfrac{17}{7} = $ _____?_____

$$
\begin{array}{r}
2.428\ldots \\
7\overline{)17.000} \\
\underline{14} \\
3\,0 \\
\underline{2\,8} \\
20 \\
\underline{14} \\
60 \\
\underline{56} \\
4
\end{array}
$$

$\dfrac{17}{7} = 2.43$ rounded to two decimal digits

(d) $\dfrac{5}{6} = $ _____?_____

$$
\begin{array}{r}
.833\ldots \\
6\overline{)5.000} \\
\underline{4\,8} \\
20 \\
\underline{18} \\
20 \\
\underline{18} \\
2
\end{array}
$$

$\dfrac{5}{6} = 0.83$ rounded to two decimal places

(e) $\dfrac{7}{16} = $ _____?_____

$$
\begin{array}{r}
.4375 \\
16\overline{)7.0000} \\
\underline{6\,4} \\
60 \\
\underline{48} \\
120 \\
\underline{112} \\
80 \\
\underline{80} \\
0
\end{array}
$$

$\dfrac{7}{16} = 0.4375$ or 0.44 rounded to two decimal digits

(f) $\dfrac{5}{9} =$ _____?_____

$$
\begin{array}{r}
.555 \\
9\overline{)5.000} \\
\underline{4\,5} \\
50 \\
\underline{45} \\
50 \\
\underline{45} \\
5
\end{array}
$$

$\dfrac{5}{9} = 0.555\ldots$ or 0.56 rounded

Repeating Decimals

Decimal numbers that do not terminate will repeat a sequence of digits. This type of decimal number is called a *repeating* decimal. For example,

$$\frac{1}{3} = 0.3333\ldots$$

where the three dots are read "and so on," and they tell us that the digit 3 continues without end.

Similarly, $\dfrac{2}{3} = 0.6666\ldots$ and $\dfrac{3}{11}$ is

$$
\begin{array}{r}
.2727 \\
11\overline{)3.0000} \\
\underline{2\,2} \\
80 \\
\underline{77} \\
30 \\
\underline{22} \\
80 \\
\underline{77} \\
3
\end{array}
$$

The remainder 3 is equal to the original dividend. This tells us that the decimal quotient repeats itself.

$$\frac{3}{11} = 0.272727\ldots$$

If you did the division in problem (c) with a calculator, you may have noticed that

$$\frac{17}{7} = 2.\overparen{428571}\overparen{428571}\ldots \qquad \text{The digits 428571 repeat endlessly.}$$

Mathematicians often use a shorthand notation to show that a decimal repeats.

Write $\dfrac{1}{3} = 0.\overline{3}$ or $\dfrac{2}{3} = 0.\overline{6}$

where the bar means that the digits under the bar repeat endlessly.

$\dfrac{3}{11} = 0.\overline{27}$ means $0.272727\ldots$ and $\dfrac{17}{7} = 2.\overline{428571}$

→ **Your Turn**

Write $\dfrac{41}{33}$ as a repeating decimal using the "bar" notation.

$$\begin{array}{r} 1.24 \\ 33\overline{)41.00} \\ \underline{33} \\ 8\,0 \\ \underline{6\,6} \\ 1\,40 \\ \underline{1\,32} \\ 8 \end{array}$$

These remainders are the same, so we know that further division will produce a repeat of the digits 24 in the answer.

$$\frac{41}{33} = 1.242424\ldots = 1.\overline{24}$$

Terminating Decimals

Some fractions are used so often in practical work that it is important for you to know their decimal equivalents. Study the table on page 158, then continue below.

→ **More Practice**

Quick now, without looking at the table, fill in the blanks in the problems shown.

$\dfrac{1}{2} = \underline{\hspace{2cm}}$　　$\dfrac{3}{4} = \underline{\hspace{2cm}}$　　$\dfrac{3}{8} = \underline{\hspace{2cm}}$　　$\dfrac{7}{8} = \underline{\hspace{2cm}}$

$\dfrac{1}{4} = \underline{\hspace{2cm}}$　　$\dfrac{5}{8} = \underline{\hspace{2cm}}$　　$\dfrac{3}{16} = \underline{\hspace{2cm}}$　　$\dfrac{7}{16} = \underline{\hspace{2cm}}$

$\dfrac{1}{8} = \underline{\hspace{2cm}}$　　$\dfrac{5}{16} = \underline{\hspace{2cm}}$　　$\dfrac{9}{16} = \underline{\hspace{2cm}}$　　$\dfrac{13}{16} = \underline{\hspace{2cm}}$

$\dfrac{1}{16} = \underline{\hspace{2cm}}$　　$\dfrac{11}{16} = \underline{\hspace{2cm}}$　　$\dfrac{15}{16} = \underline{\hspace{2cm}}$

Check your work against the table when you are finished, then continue.

→ **More Practice**

Convert each of the following fractions into decimal form.

(a) $\dfrac{95}{100}$　　(b) $\dfrac{1}{20}$　　(c) $\dfrac{7}{10}$　　(d) $\dfrac{4}{1000}$　　(e) $\dfrac{11}{1000}$

(f) $\dfrac{327}{10000}$　　(g) $\dfrac{473}{1000}$　　(h) $\dfrac{3}{50}$　　(i) $\dfrac{1}{25}$

(j) $\dfrac{27}{64}$ in. (to the nearest thousandth of an inch)

→ **Answers**

(a)　0.95　　(b)　0.05　　(c)　0.7　　(d)　0.004　　(e)　0.011

(f)　0.0327　　(g)　0.473　　(h)　0.06　　(i)　0.04　　(j)　0.422 in.

DECIMAL–FRACTION EQUIVALENTS

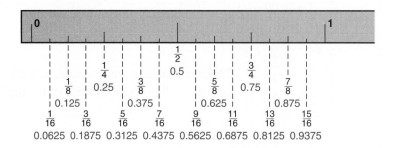

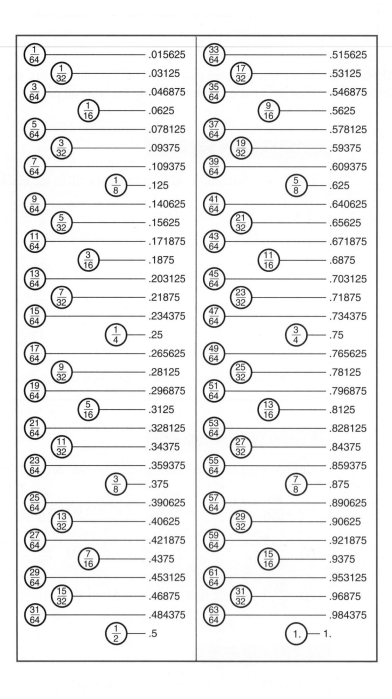

$\frac{1}{64}$		.015625	$\frac{33}{64}$		.515625
	$\frac{1}{32}$	.03125		$\frac{17}{32}$	.53125
$\frac{3}{64}$		.046875	$\frac{35}{64}$		.546875
	$\frac{1}{16}$	.0625		$\frac{9}{16}$	.5625
$\frac{5}{64}$		.078125	$\frac{37}{64}$		.578125
	$\frac{3}{32}$	.09375		$\frac{19}{32}$	.59375
$\frac{7}{64}$		.109375	$\frac{39}{64}$		.609375
	$\frac{1}{8}$	.125		$\frac{5}{8}$	.625
$\frac{9}{64}$		.140625	$\frac{41}{64}$		.640625
	$\frac{5}{32}$	.15625		$\frac{21}{32}$	.65625
$\frac{11}{64}$		.171875	$\frac{43}{64}$		.671875
	$\frac{3}{16}$	.1875		$\frac{11}{16}$	.6875
$\frac{13}{64}$		.203125	$\frac{45}{64}$		.703125
	$\frac{7}{32}$	.21875		$\frac{23}{32}$	.71875
$\frac{15}{64}$		.234375	$\frac{47}{64}$		.734375
	$\frac{1}{4}$	.25		$\frac{3}{4}$	.75
$\frac{17}{64}$		.265625	$\frac{49}{64}$		.765625
	$\frac{9}{32}$	.28125		$\frac{25}{32}$	.78125
$\frac{19}{64}$		.296875	$\frac{51}{64}$		.796875
	$\frac{5}{16}$	.3125		$\frac{13}{16}$	.8125
$\frac{21}{64}$		.328125	$\frac{53}{64}$		.828125
	$\frac{11}{32}$	.34375		$\frac{27}{32}$	.84375
$\frac{23}{64}$		.359375	$\frac{55}{64}$		.859375
	$\frac{3}{8}$	.375		$\frac{7}{8}$	.875
$\frac{25}{64}$		.390625	$\frac{57}{64}$		.890625
	$\frac{13}{32}$	.40625		$\frac{29}{32}$	.90625
$\frac{27}{64}$		.421875	$\frac{59}{64}$		.921875
	$\frac{7}{16}$	.4375		$\frac{15}{16}$	.9375
$\frac{29}{64}$		.453125	$\frac{61}{64}$		.953125
	$\frac{15}{32}$	.46875		$\frac{31}{32}$	.96875
$\frac{31}{64}$		.484375	$\frac{63}{64}$		.984375
	$\frac{1}{2}$	.5		$1.$	1.

You will do many of these calculations more quickly if you remember that dividing by a power of ten is equivalent to shifting the decimal point to the left. To divide by a power of ten, move the decimal point to the left as many digits as there are zeros in the power of ten. For example,

$$\frac{95}{100} = 0.95$$

| Two zeros | Move the decimal point two digits to the left |

and

$$\frac{11}{1000} = 0.011$$

| 3 zeros | Move the decimal point three digits to the left |

HOW TO WRITE A REPEATING DECIMAL AS A FRACTION

A repeating decimal is one in which some sequence of digits is endlessly repeated. For example, $0.333\ldots = 0.\overline{3}$ and $0.272727\ldots = 0.\overline{27}$ are repeating decimals. The bar over the number is a shorthand way of showing that those digits are repeated.

What fraction is equal to $0.\overline{3}$? To answer this, form a fraction with numerator equal to the repeating digits and denominator equal to a number formed with the same number of 9s.

$0.\overline{3} = \frac{3}{9} = \frac{1}{3}$

$0.\overline{27} = \frac{27}{99} = \frac{3}{11}$ Two digits in $0.\overline{27}$; therefore, use 99 as the denominator.

$0.\overline{123} = \frac{123}{999} = \frac{41}{333}$ Three digits in $0.\overline{123}$; therefore, use 999 as the denominator.

This procedure works only when *all* of the decimal part repeats.

Now turn to Exercises 3-3 for a set of problems on decimal fractions.

Exercises 3-3   Decimal Fractions

A. Write as decimal numbers. Round to two decimal digits if necessary.

1. $\frac{1}{4}$ 2. $\frac{2}{3}$ 3. $\frac{3}{4}$ 4. $\frac{2}{5}$

5. $\frac{4}{5}$ 6. $\frac{5}{6}$ 7. $\frac{2}{7}$ 8. $\frac{4}{7}$

9. $\frac{6}{7}$ 10. $\frac{3}{8}$ 11. $\frac{6}{8}$ 12. $\frac{1}{10}$

13. $\frac{3}{10}$ 14. $\frac{2}{12}$ 15. $\frac{5}{12}$ 16. $\frac{3}{16}$

17. $\frac{6}{16}$ 18. $\frac{9}{16}$ 19. $\frac{13}{16}$ 20. $\frac{3}{20}$

21. $\frac{7}{32}$ 22. $\frac{13}{20}$ 23. $\frac{11}{24}$ 24. $\frac{39}{64}$

25. $\frac{3}{100}$ 26. $\frac{213}{1000}$ 27. $\frac{19}{1000}$ 28. $\frac{17}{50}$

B. Calculate in decimal form. Round to agree with the number of decimal digits in the decimal number.

1. $2\frac{3}{5} + 1.785$ 2. $\frac{1}{5} + 1.57$ 3. $3\frac{7}{8} - 2.4$

4. $1\frac{3}{16} - 0.4194$ 5. $2\frac{1}{2} \times 3.15$ 6. $1\frac{3}{25} \times 2.08$

7. $3\frac{4}{5} \div 2.65$ 8. $3.72 \div 1\frac{1}{4}$ 9. $2.76 + \frac{7}{8}$

10. $16\frac{3}{4} - 5.842$ 11. $3.14 \times 2\frac{7}{16}$ 12. $1.17 \div 1\frac{3}{32}$

C. Practical Problems

1. **Allied Health** If one tablet of calcium pantothenate contains 0.5 gram, how much is contained in $2\frac{3}{4}$ tablets? How many tablets are needed to make up 2.6 grams?

2. **Flooring and Carpeting** Estimates of matched or tongue-and-groove (T&G) flooring—stock that has a tongue on one edge—must allow for the waste from milling. To allow for this waste, $\frac{1}{4}$ of the area to be covered must be added to the estimate when 1-in. by 4-in. flooring is used. If 1-in. by 6-in. flooring is used, $\frac{1}{6}$ of the area must be added to the estimate. (*Hint:* The floor is a rectangle. Area of a rectangle = length × width.)

   (a) A house has a floor size 28 ft long by 12 ft wide. How many square feet of 1-in. by 4-in. T&G flooring will be required to lay the floor?

   (b) A motel contains 12 units or rooms, and each room is 22 ft by 27 ft or 594 sq ft. In five rooms the builder is going to use 4-in. stock and in the other seven rooms 6-in. stock—both being T&G. How many square feet of each size will be used?

   (c) A contractor is building a motel with 24 rooms. Eight rooms will be 16 ft by 23 ft, nine rooms 18 ft by 26 ft, and the rest of the rooms 14 ft by 20 ft. How much did she pay for flooring, using 1-in. by 4-in. T&G at $4920 per 1000 sq ft?

3. **Roofing** If a roofer lays $10\frac{1}{2}$ squares of shingles in $4\frac{1}{2}$ days, how many squares does he do in one day?

4. **Machine Trades** A bearing journal measures 1.996 in. If the standard readings are in $\frac{1}{32}$-in. units, what is its probable standard size as a fraction?

5. **Interior Design** Complete the following invoice for upholstery fabric.

   (a) $5\frac{1}{4}$ yd @ $12.37 per yd = _____

   (b) $23\frac{3}{4}$ yd @ $16.25 per yd = _____

   (c) $31\frac{5}{6}$ yd @ $9.95 per yd = _____

   (d) $16\frac{2}{3}$ yd @ $17.75 per yd = _____

                    Total   _____

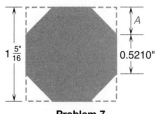

Problem 7

6. **Construction** A land developer purchased a piece of land containing 437.49 acres. He plans to divide it into a 45-acre recreational area and lots of $\frac{3}{4}$ acre each. How many lots will he be able to form from this piece of land?

7. **Machine Trades** Find the corner measurement A needed to make an octagonal end on a square bar as shown in the diagram.

8. **Automotive Trades** A cylinder is normally $3\frac{7}{16}$ in. in diameter. It is rebored 0.040 in. larger. What is the size of the rebored cylinder? (Give your answer in decimal form.)

9. Plumbers deal with measurements in inches and common fractions of an inch, while surveyors use feet and decimal fractions of a foot. Often one trade needs to interpret the measurements of the other.

   (a) **Construction** A drain has a run of 52 ft at a grade of $\frac{1}{8}$ ft./ft. The high end of the drain has an elevation of 126.70 ft. What is the elevation at the low end?

   (b) **Plumbing** The elevation at one end of a lot is 84.25 ft, and the elevation at the other end is 71.70 ft. Express the difference in elevation in feet and inches and round to the nearest $\frac{1}{8}$ in.

10. **Plumbing** To fit a 45 degree connection, a plumber can approximate the diagonal length by multiplying the offset length by 1.414. Find the diagonal length of a 45 degree fitting if the offset length is $13\frac{11}{16}$ in. Express the answer as a fraction to the nearest $\frac{1}{16}$ in.

11. **Sports and Leisure** In the 2006 Nathan's Famous Hot Dog Eating Contest, Takeru Kobayashi ate $53\frac{3}{4}$ hot dogs in 12 minutes. To the nearest tenth, how many hot dogs did he consume per minute?

D. Calculator Problems

1. **Sheet Metal Trades** In the table on page 162 the first column lists the U.S. Standard sheet metal gauge number. These tables were at one time used to give the equivalent sheet metal thickness in fractions of an inch, as shown in the second column. Today, these thicknesses are listed as decimal numbers. Complete column three to reflect this change. Round to the nearest thousandth of an inch.

2. **Construction** Calculate the total cost of the following building materials.

8024 sq ft of flooring	@ $521.51 per 100 sq ft	_____
1825 ft of 6-in. redwood siding	@ $173.48 per 100 ft	_____
24,390 bricks	@ $461.17 per 1000	_____
19,200 sq ft of shingles	@ $93.74 per 100 sq ft	_____
	Total	_____

3. **Office Services** Machine helpers at the CALMAC Tool Co. earn $10.72 per hour. The company pays time-and-a-half for hours over 40 per week. Sunday hours are paid at double time. How much would you earn if you worked the following hours: Monday, 9; Tuesday, 8; Wednesday, $9\frac{3}{4}$; Thursday, 8; Friday, $10\frac{1}{2}$; Saturday, $5\frac{1}{4}$; Sunday, $4\frac{3}{4}$?

When you have finished these exercises, check your answers to the odd-numbered problems in the Appendix, and then turn to Problem Set 3 on page 165 for practice working with decimal numbers. If you need a quick review of the topics in this chapter, visit the chapter Summary on page 163 first.

Gauge No.	Fraction Thickness (in.)	Decimal Thickness (in.)	Gauge No.	Fraction Thickness (in.)	Decimal Thickness (in.)
7-0	$\frac{1}{2}$		14	$\frac{5}{64}$	
6-0	$\frac{15}{32}$		15	$\frac{9}{128}$	
5-0	$\frac{7}{16}$		16	$\frac{1}{16}$	
4-0	$\frac{13}{32}$		17	$\frac{9}{160}$	
3-0	$\frac{3}{8}$		18	$\frac{1}{20}$	
2-0	$\frac{11}{32}$		19	$\frac{7}{160}$	
0	$\frac{5}{16}$		20	$\frac{3}{80}$	
1	$\frac{9}{32}$		21	$\frac{11}{320}$	
2	$\frac{17}{64}$		22	$\frac{1}{32}$	
3	$\frac{1}{4}$		23	$\frac{9}{320}$	
4	$\frac{15}{64}$		24	$\frac{1}{40}$	
5	$\frac{7}{32}$		25	$\frac{7}{320}$	
6	$\frac{13}{64}$		26	$\frac{3}{160}$	
7	$\frac{3}{16}$		27	$\frac{11}{640}$	
8	$\frac{11}{64}$		28	$\frac{1}{64}$	
9	$\frac{5}{32}$		29	$\frac{9}{640}$	
10	$\frac{9}{64}$		30	$\frac{1}{80}$	
11	$\frac{1}{8}$		31	$\frac{7}{640}$	
12	$\frac{7}{64}$		32	$\frac{13}{1280}$	
13	$\frac{3}{32}$		—	—	—

Objective

Review

Write decimal numbers in words and translate words to numbers. (p. 123)

The word "and" represents the decimal point. In written form, the decimal portion of the number is followed by the place value of the far right digit.

Example: (a) The number 26.035 is written in words as "twenty-six and thirty-five thousandths."

(b) The number "one hundred six and twenty-seven ten-thousandths" is written in decimal form as 106.0027.

Add and subtract decimal numbers. (pp. 127–130)

Write the numbers in column format with the decimal points aligned vertically. Then add or subtract as with whole numbers.

Example:

$$
\text{(a)} \quad 5.82 + 0.096 \Rightarrow \begin{array}{r} 5.820 \\ + 0.096 \\ \hline 5.916 \end{array}
$$

Attach a zero

$$
\text{(b)} \quad 27 - 4.03 \Rightarrow \begin{array}{r} 27.00 \\ - 4.03 \end{array} \Rightarrow \begin{array}{r} \overset{6\ 9\ 10}{2\cancel{7}.\cancel{0}\cancel{0}} \\ - 4.03 \\ \hline 22.97 \end{array}
$$

Attach a decimal point and two zeros.

Multiply decimal numbers. (p. 135)

First, multiply, ignoring decimal points—they need not be aligned. Then, to find the total number of decimal digits in the product, count the total number of decimal digits in the factors. Attach additional zeros if needed.

Example: $7.25 \times 0.301 =$

$$
\begin{array}{r}
7.25 \quad \longleftarrow \textit{two decimal digits} \\
\times\, 0.301 \quad \longleftarrow \textit{three decimal digits} \\
\hline
2.18225 \quad \longleftarrow \textit{five decimal digits}
\end{array}
$$

Divide decimal numbers. (p. 139)

Shift the decimal point in the divisor to the right so as to make it a whole number. Shift the decimal point in the dividend the same amount. Attach zeros if necessary. Divide as with whole numbers, placing the decimal point directly above the new position in the dividend.

Example: $104.96 \div 0.032 = 0.032\overline{)104.96}$

$$
0.032\overline{)104.960.} = 32\overline{)104960.}
$$

$$
\begin{array}{r}
3280. \\
32\overline{)104960.} \\
\underline{96} \\
89 \\
\underline{64} \\
256 \\
\underline{256} \\
0 \\
0 \\
\underline{0} \\
0
\end{array}
$$

Objective	Review

Calculate an average.
(p. 146)

To find the average of a set of numbers, divide the sum of the numbers by the number of numbers in the set. Round the answer to agree in precision with the least precise number in the set.

Example: The average of the numbers 4.2, 4.8, 5.7, 2.5, 3.6, and 5.0 is

$$\text{Average} = \frac{\text{sum}}{\text{number}} = \frac{25.8}{6} = 4.3$$

Round decimal numbers.
(p. 144)

Follow the procedure for rounding whole numbers, except after rounding to a decimal place, drop all digits to the right of that place.

Example: Round 16.2785 to the nearest hundredth
16.27$\underset{\wedge}{}$85 = 16.28

Convert fractions to decimal numbers.
(p. 153)

Divide the numerator by the denominator and round if necessary. Use the over-bar notation for repeating decimals.

Example:

$$\frac{3}{16} \Rightarrow 16\overline{)3.0000} = 0.1875$$

```
      .1875
16)3.0000
   16
   140
   128
    120
    112
     80
     80
```

$$\frac{5}{6} \Rightarrow 6\overline{)5.00} = 0.8\bar{3}$$

```
    .833...
6)5.00
  4 8
   20
   18
   20
   18
```

Solve practical problems involving decimal numbers.

Read the problem statement carefully. Look for key words or phrases that indicate which operation(s) to use and then perform the calculation.

Example: A box weighs 61.3 lb and contains 175 bolts. If the container itself weighs 1.8 lb, how much does each bolt weigh?

First, find the weight of the bolts by subtracting the weight of the container:

61.3 lb − 1.8 lb = 59.5 lb

Then, divide this amount by the total number of bolts to find the weight of each bolt:

59.5 lb ÷ 175 bolts = 0.34 lb per bolt

Decimal Numbers

Answers to odd-numbered problems are given in the Appendix.

A. **Write in words.**

1. 0.91
2. 0.84
3. 23.164
4. 63.219
5. 9.3
6. 3.45
7. 10.06
8. 15.037

Write as decimal numbers.

9. Seven hundredths
10. Eighteen thousandths
11. Two hundred and eight tenths
12. Sixteen and seventeen hundredths
13. Sixty-three and sixty-three thousandths
14. One hundred ten and twenty-one thousandths
15. Five and sixty-three ten-thousandths
16. Eleven and two hundred eighteen ten-thousandths

B. **Add or subtract as shown.**

1. $4.39 + 18.8$
2. $18.8 + 156.16$
3. $\$7.52 + \$11.77$
4. $26 + 0.06$
5. $3.68 - 1.74$
6. $\$12.46 - \$8.51$
7. $104.06 - 15.80$
8. $16 - 3.45$
9. $264.3 + 12.804$
10. $0.232 + 5.079$
11. $165.4 + 73.61$
12. $245.94 + 7.07$
13. $116.7 - 32.82$
14. $4.07 - 0.085$
15. $0.42 + 1.452 + 31.8$
16. $\frac{1}{2} + 4.21$
17. $3\frac{1}{5} + 1.08$
18. $1\frac{1}{4} - 0.91$
19. $3.045 - 1\frac{1}{8}$
20. $8.1 + 0.47 - 1\frac{4}{5}$

C. **Calculate as shown.**

1. 0.004×0.02
2. 0.06×0.05
3. 1.4×0.6
4. 3.14×12
5. $0.2 \times 0.6 \times 0.9$
6. 5.3×0.4
7. 6.02×3.3
8. $3.224 \div 2.6$

Name

Date

Course/Section

9. $187.568 \div 3.04$

10. $0.078 \div 0.3$

11. $0.6 \times 3.15 \times 2.04$

12. $3.78 + 4.1 \times 6.05$

13. $0.008 - 0.001 \div 0.5$

14. $3.1 \times 4.6 - 2.7 \div 0.3$

15. $\dfrac{4.2 + 4.6 \times 1.2}{2.73 \div 21 + 0.41}$

16. $\dfrac{7.2 - 3.25 \div 1.3}{3.5 + 5.7 \times 4.0 - 2.8}$

17. $1.2 + 0.7 \times 2.2 + 1.6$

18. $1.2 \times 0.7 + 2.2 \times 1.6$

Round to two decimal digits.

19. $0.007 \div 0.03$

20. $3.005 \div 2.01$

21. $17.8 \div 6.4$

22. $0.0041 \div 0.019$

23. 0.0371

24. 16.8449

Round to three decimal places.

25. $0.04 \div 0.076$

26. $234.1 \div 465.8$

27. $17.6 \div 0.082$

28. $0.051 \div 1.83$

29. 27.0072

30. 1.1818

Round to the nearest tenth.

31. $0.08 \div 0.053$

32. $3.05 \div 0.13$

33. $18.76 \div 4.05$

34. $0.91 \div 0.97$

35. 47.233

36. 123.7666

D. Write as a decimal number.

1. $\frac{1}{16}$

2. $\frac{7}{8}$

3. $\frac{5}{32}$

4. $\frac{7}{16}$

5. $\frac{11}{8}$

6. $1\frac{3}{4}$

7. $2\frac{11}{20}$

8. $1\frac{1}{6}$

9. $2\frac{2}{3}$

10. $1\frac{1}{32}$

11. $2\frac{5}{16}$

Write in decimal form, calculate, round to two decimal digits.

12. $4.82 \div \frac{1}{4}$

13. $11.5 \div \frac{3}{8}$

14. $1\frac{3}{16} \div 0.62$

15. $2\frac{3}{4} \div 0.035$

16. $0.45 \times 2\frac{1}{8}$

17. $0.068 \times 1\frac{7}{8}$

E. Practical Problems

1. **Machine Trades**   A $\frac{5}{16}$-in. bolt weighs 0.43 lb. How many bolts are there in a keg containing 125 lb of bolts?

2. **Metalworking**   Twelve equally spaced holes are to be drilled in a metal strip $34\frac{1}{4}$-in. long, with 2 in. remaining on each end. What is the distance, to the nearest hundredth of an inch, from center to center of two consecutive holes?

3. **Plumbing**   A plumber finds the length of pipe needed to complete a bend by performing the following multiplication:

   Length needed $= 0.01745 \times$ (radius of the bend) $\times$ (angle of bend)

   What length of pipe is needed to complete a 35° bend with a radius of 16 in.?

4. **Office Services** During the first six months of the year the Busy Bee Printing Co. spent the following amounts on paper:

January	$1070.16	February	$790.12	March	$576.77
April	$600.09	May	$1106.45	June	$589.21

What is the average monthly expenditure for paper?

5. **Machine Trades** The *feed* of a drill is the distance it advances with each revolution. If a drill makes 310 rpm and drills a hole 2.125 in. deep in 3.25 minutes, what is the feed? Round to four decimal places.

$$\left(Hint\text{: Feed} = \frac{\text{total depth}}{(\text{rpm}) \times (\text{time of drilling in minutes})}. \right)$$

6. **Electrical Trades** The resistance of an armature while it is cold is 0.208 ohm. After running for several minutes, the resistance increases to 1.340 ohms. Find the increase in resistance of the armature.

7. **Painting** The Ace Place Paint Co. sells paint in steel drums each weighing 36.4 lb empty. If 1 gallon of paint weighs 9.06 lb, what is the total weight of a 50-gal drum of paint?

8. **Electrical Trades** The diameter of No. 12 bare copper wire is 0.08081 in., and the diameter of No. 15 bare copper wire is 0.05707 in. How much larger is No. 12 wire than No. 15 wire?

9. **Machine Trades** What is the decimal equivalent of each of the following drill bit diameters?
    (a) $\frac{3}{16}$ in.　　　(b) $\frac{5}{32}$ in.　　　(c) $\frac{3}{8}$ in.　　　(d) $\frac{13}{64}$ in.

10. What is the closest fractional equivalent in 16ths of an inch of the following dimensions? (*Hint:* Use the table on page 158.)
    (a) 0.185 in.　　(b) 0.127 in.　　(c) 0.313 in.　　(d) 0.805 in.

11. **Sheet Metal Trades** The following table lists the thickness in inches of several sizes of sheet steel:

U.S. Gauge	Thickness (in.)
35	0.0075
30	0.0120
25	0.0209
20	0.0359
15	0.0673
10	0.1345
5	0.2092

(a) What is the difference in thickness between 30 gauge and 25 gauge sheet?

(b) What is the difference in thickness between seven sheets of 25 gauge and four sheets of 20 gauge sheet?

(c) What length of $\frac{3}{16}$-in.-diameter rivet is needed to join one thickness of 25 gauge sheet to a strip of $\frac{1}{4}$-in. stock? Add $1\frac{1}{2}$ times the diameter of the rivet to the length of the rivet to ensure that the rivet is long enough that a proper rivet head can be formed.

(d) What length of $\frac{5}{32}$-in. rivet is needed to join two sheets of 20 gauge sheet steel? (Don't forget to add the $1\frac{1}{2}$ times rivet diameter.)

12. **Trades Management** When the owners of the Better Builder Co. completed a small construction job, they found that the following expenses had been incurred: labor, $972.25; gravel, $86.77; sand, $39.41; cement, $280.96; and

bricks, $2204.35. What total bill should they give the customer if they want to make $225 profit on the job?

13. **Construction** One linear foot of 12-in. I-beam weighs 25.4 lb. What is the length of a beam that weighs 444.5 lb?

14. **Sheet Metal Trades** To determine the average thickness of a metal sheet, a sheet metal worker measures it at five different locations. Her measurements are 0.0401, 0.0417, 0.0462, 0.0407, and 0.0428 in. What is the average thickness of the sheet?

15. **Drafting** A dimension in a technical drawing is given as $2\frac{1}{2}$ in. with a tolerance of ±0.025 in. What are the maximum and minimum permissible dimensions written in decimal form?

16. **Trades Management** The four employees of the Busted Body Shop earned the following amounts last week: $811.76, $796.21, $808.18, and $876.35. What is the average weekly pay for the employees of the shop?

17. **Electrical Trades** If 14-gauge Romex cable sells for 19 cents per foot, what is the cost of 1210 ft of this cable?

18. **Life Skills** In Lucy's first week as an electrician, she earned $1593.75 for 37.5 hours of work. What is her hourly rate of pay?

19. **Welding** A welder finds that 2.1 cu ft of acetylene gas is needed to make one bracket. How much gas will be needed to make 27 brackets?

20. **Automotive Trades** What is the outside diameter d of a 6.50 × 14 tire? (See the figure.)

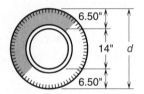
Problem 20

21. **Metalworking** How many 2.34-in. spacer blocks can be cut from a 2-in. by 2-in. square bar 48 in. long? Allow $\frac{1}{4}$ in. waste for each saw cut.

22. **Automotive Trades** At the Fixum Auto Shop, a mechanic is paid $22.50 per hour for a motor overhaul and is allotted 23 hours of labor time for the job. The mechanic completes that job in 18.5 hours, but he is still paid as if he had worked for 23 hours. Calculate his actual hourly compensation under these circumstances.

23. **Machine Trades** A machinist estimates the following times for fabricating a certain part: 0.6 hour for setup, 2.4 hours of turning, 5.2 hours of milling, 1.4 hours of grinding, and 1.8 hours of drilling. What is the total time needed to make the part?

24. **Printing** A 614-page book was printed on paper specified as 0.00175 in. thick and finished with a cover 0.165 in. thick. What was the total thickness of the bound book? (Be sure to count the cover twice and remember that there are two pages of the book for every one sheet of paper.)

25. **Allied Health** The primary care physician has directed that his patient should receive a total weekly dose of 0.35 mg codeine sulfate. How many 0.035-mg tablets should she be given during a week?

26. **Fire Protection** One gallon of water weighs approximately 8.34 lb. When sent to a particularly rough terrain, a Forestry Service truck is allowed to carry only 1.5 tons of weight. How many gallons of water can it carry? (1 ton = 2000 lb.)

27. **Plumbing** To con struct a certain pipe system, the following material is needed:

6 1-in. elbows at $0.45 each
5 tees at $0.99 each
3 couplings at $0.49 each
4 pieces of pipe each 6 ft long
6 pieces of pipe each 2 ft long
4 pieces of pipe each 1.5 ft long

The pipe is schedule 40 1-in. PVC pipe, costing $0.48 per linear foot. Find the total cost of the material for the system.

28. **Masonry**   A tile setter purchases the following supplies for the day:

1 bag of thin-set mortar	@ $13.49 per bag
44 sq ft of tile	@ $5.50 per sq ft
2 boxes of sanded grout mix	@ $8.99 per box
2 tubes of caulk	@ $3.79 per tube
3 containers of grout sealer	@ $6.99 per container
3 containers of grout and tile cleaner	@ $5.99 per container
4 scrub pads	@ $2.79 each
1 finishing trowel	@ $33.87 each
2 packages of tile spacers	@ $3.99 each
1 grout bag	@ $2.49 each
1 grout float	@ $8.95 each

What is the total cost of these items before tax is added?

29. **Trades Management**   One way in which the owner of an auto shop evaluates the success of his business is to keep track of "production per square foot." To calculate this indicator, he divides total revenue by the area (in square feet) of his shop. Last year, a 2260-sq-ft shop generated total revenue of $586,435. This year, the shop expanded to 2840 sq ft and generated total revenue of $734,592. During which year was the production per square foot the greater and by how much?

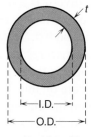

Problem 30

30. **Plumbing**   The inside diameter (I.D.) and outside diameter (O.D.) of a pipe are shown in the figure. The wall thickness of the pipe is the dimension labeled t. Calculate the wall thickness of schedule 120 pipe if its I.D. is 0.599 in. and its O.D. is 1.315 in.

31. **Electronics**   In the circuit shown, a solar cell is connected to a 750-ohm load. If the circuit current is 0.0045 ampere, calculate the voltage output of the solar cell by multiplying the load by the current.

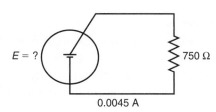

32. **Office Services**   A shop technician earns $18.26 per hour plus time-and-a-half for overtime (time exceeding 40 hours). If he worked 42.5 hours during a particular week, what would be his gross pay?

33. **Life Skills**   A driver refueling her car fills the tank with 15.6 gal. If the odometer currently reads 63589.2, and it read 63186.7 when the tank was last filled, what was her fuel economy between refuelings? (Round to one decimal place.)

34. **Automotive Trades**   An auto repair job consisted of a new timing belt, an oil change, and a brake job. The timing belt cost $44.95 and required 4.6 hours of labor. The oil and filter cost $40 and required 1.5 hours of labor. The brake pads cost $67.95 and required 1.8 hours of labor. If the labor rate is $95 per hour, find the total cost of all repairs before tax.

35. **Automotive Trades**   The *taper* of an automobile cylinder is caused by uneven wearing between the top and bottom of the cylinder. The size of this taper is calculated by subtracting the bore at the top of the cylinder from the bore at the bottom. If the bore at the top of a cylinder is 4.258 in. and the bore at the bottom is 4.276 in., what is the taper of this cylinder?

36. **Office Services** A water district charges residential customers a monthly meter service charge of $51.60, plus an additional charge of $3.75 per HCF (hundred cubic feet) of water usage. What would be the total bill if a customer used 4600 cubic feet of water during a particular month?

37. **Automotive Trades** An automotive technician, who is paid on a flat rate system, has logged the following daily hours during one particular week:

Monday	7.2	Thursday	8.8
Tuesday	8.4	Friday	7.4
Wednesday	6.6		

If he is paid $22.40 per hour, how much did he earn during the week?

38. **HVAC** In June 2006, home heating oil sold for an average of $2.50 per gal in the northeast, In June 2008, it sold for $4.61 per gal. A typical heating oil tank holds 250 gal. How much more did it cost to fill this tank in 2008 than in 2006?

39. **Sports and Leisure** In the 2008 Olympics, China's Chen Ruolin performed a nearly perfect dive in the final round of the women's 10-meter platform competition to overtake Canada's Emilie Heymans for the gold medal. The dive was a back $2\frac{1}{2}$ somersault with $1\frac{1}{2}$ twists, which was rated at a 3.4 degree of difficulty. Her scores from the seven judges were: 10.0, 10.0, 10.0, 9.5, 10.0, 9.5, and 9.5. To calculate the total score of a dive, the two best scores and the two worst scores are all thrown out, the remaining three are added, and that sum is multiplied by the degree of difficulty. Use this procedure to calculate Chen Ruolin's total score on this dive.

40. **Trades Management** A business owner is looking for a new commercial space for his shop. An 8000-sq-ft space is advertised for a base price of $1.35 per square foot per month plus additional costs for property taxes and insurance. Suppose that property taxes and insurance are estimated to be $8856 per year.
    (a) What is the cost per month for property taxes and insurance?
    (b) Use your answer to part (a) to calculate the monthly cost per square foot for property taxes and insurance.
    (c) Add your answer to part (b) to the base price to determine the total cost per square foot per month.

41. **Office Services** As shown in Example 3 on page 50, many public utilities charge customers according to a "tier" or "block" system in order to encourage conservation. For example, a certain electric utility divides a customer's usage of kilowatt-hours (kWh) into five tiers according to the following table:

Tier	Usage	Cost per kWh
1	first 200 kWh	$0.02930
2	next 60 kWh	$0.05610
3	next 140 kWh	$0.16409
4	next 200 kWh	$0.21260
5	remaining kWh	$0.26111

Use this table to calculate the cost of using (a) 350 kWh and (b) 525 kWh. (Round to the nearest cent.)

USING A CALCULATOR, III: DECIMALS

To perform addition, subtraction, multiplication, and division of decimal numbers on a calculator, use the same procedures outlined at the end of Chapter 1, and press the ⊡ key to enter the decimal point. This key is normally located along the bottom row beneath the "2" key.

Example 1

To calculate $243.78 + 196.1 \times 2.75$ use the keystroke sequence

243 · **78** ⊕ **196** · **1** ⊗ **2** · **75** ⊜ → ▓ *783.055* ▓

The decimal point key was shown as a special symbol here for emphasis, but in future calculator sequences in this book we will save space by not showing the decimal point as a separate key.

Notice that in the preceding calculation, the numbers and operations were entered, and the calculator followed the standard order of operations.

When you work with decimal numbers, remember that for a calculator with a ten-digit display if an answer is a nonterminating decimal or a terminating decimal with more than ten digits, the calculator will display a number rounded to ten digits. Knowledge of rounding is especially important in such cases because the result must often be rounded further.

→ **Your Turn**

Work the following problems for practice and round as indicated.

(a) $\$28.75 + \$161.49 - \$37.60$

(b) 2.8×0.85 (Round to the nearest tenth.)

(c) $2347.68 \div 12.9$ (Round to the nearest whole number.)

(d) $46.8 - 27.3 \times 0.49$ (Round to the nearest hundredth.)

(e) $\dfrac{16500 + 3700}{12 \times 68}$ (Round to the nearest tenth.)

(f) $(3247.9 + 868.7) \div 0.816$ (Round to the nearest ten.)

(g) $6 \div 0.07 \times 0.8 + 900$ (Round to the nearest hundred.)

→ **Answers**

(a) **28.75** ⊕ **161.49** ⊖ **37.6** ⊜ → ▓ *152.64* ▓

⟋ Final zeros to the right of the decimal point may be omitted.

(b) **2.8** ⊗ **.85** ⊜ ▓ *2.38* ▓ or 2.4 rounded

∟ We need not enter the leading zero in 0.85

(c) **2347.68** ⊘ **12.9** ⊜ → ▓ *181.9906977* ▓ or 182 rounded

(d) **46.8** ⊖ **27.3** ⊗ **.49** ⊜ → ▓ *33.423* ▓ or 33.42 rounded

(e) **16500** ⊕ **3700** ⊜ ⊘ ⦅ **12** ⊗ **68** ⦆ → ▓ *24.75490196* ▓ or 24.8 rounded

(f) **3247.9** ⊕ **868.7** ⊜ ⊘ **.816** ⊜ → ▓ *5044.852941* ▓ or 5040 rounded

(g) **6** ⊘ **.07** ⊗ **.8** ⊕ **900** ⊜ → ▓ *968.5714286* ▓ or 1000 rounded

Scientific calculators display decimal numbers between 0 and 1 as they are written in textbooks, with a leading 0 written before the decimal point. As mentioned earlier in the chapter, this helps prevent you from overlooking the decimal point. However, graphing calculators do not normally display this zero, so you must be extra careful when examining results on graphing calculators. ◄

If you need to convert a decimal answer to fraction form or vice versa, look for a second function key labeled $\boxed{F \leftrightarrow D}$. If your calculator does not have this key, simply press the $\boxed{a^b_c}$ key to convert back and forth between fraction and decimal form.

Example 2

(a) To calculate $\dfrac{2}{3} + \dfrac{4}{5}$ and convert the result to decimal form, enter whichever one of the following sequences works for your calculator:

$2\;\boxed{A^b_c}\;3\;\boxed{+}\;4\;\boxed{A^b_c}\;5\;\boxed{=}\;\boxed{F\leftrightarrow D}\;\boxed{=}\;\rightarrow\;\boxed{1.466666667}$

(For single-line calculators, the $\boxed{=}$ is not required after $\boxed{F\leftrightarrow D}$.)

or $2\;\boxed{a^b_c}\;3\;\boxed{+}\;4\;\boxed{a^b_c}\;5\;\boxed{=}\;\boxed{a^b_c}\;\rightarrow\;\boxed{1.466666667}$

(b) To calculate $5.67 - 3.22$ and convert the result to fraction form, enter whichever one of the following sequences works for your calculator:

$5.67\;\boxed{-}\;3.22\;\boxed{=}\;\boxed{F\leftrightarrow D}\;\boxed{=}\;\rightarrow\;\boxed{2\llcorner9/20}$

or $5.67\;\boxed{-}\;3.22\;\boxed{=}\;\boxed{a^b_c}\;\rightarrow\;\boxed{2\llcorner9/20}$

The sequences shown in the text will use the $\boxed{F\leftrightarrow D}$ key.

→ Your Turn

Calculate the following and then convert the result to the form indicated:

(a) $3\dfrac{7}{8} \div 1\dfrac{3}{16}$ (Decimal, nearest hundredth)

(b) 7.2×4.3 (Fraction, mixed number)

→ Solutions

(a) $3\;\boxed{A^b_c}\;7\;\boxed{A^b_c}\;8\;\boxed{\div}\;1\;\boxed{A^b_c}\;3\;\boxed{A^b_c}\;16\;\boxed{=}\;\boxed{F\leftrightarrow D}\;\boxed{=}\;\rightarrow\;\boxed{3.263157895} \approx 3.26$

(b) $7.2\;\boxed{\times}\;4.3\;\boxed{=}\;\boxed{F\leftrightarrow D}\;\boxed{=}\;\rightarrow\;\boxed{30\llcorner24/25}$ or $30\dfrac{24}{25}$

4

Ratio, Proportion, and Percent

Objective	**Sample Problems**	**For help, go to**

When you finish this chapter you will be able to:

1. Calculate ratios.

(a) Find the ratio of the pulley diameters.

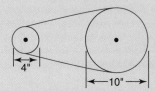

_____ Page 178

(b) **Automotive Trades** A small gasoline engine has a maximum cylinder volume of 520 cu cm and a compressed volume of 60 cu cm. Find the compression ratio.

_____ Page 179

2. Solve proportions.

(a) Solve for x.

$$\frac{8}{x} = \frac{12}{15}$$

_____ Page 182

(b) Solve for y.

$$\frac{4.4}{2.8} = \frac{y}{9.1}$$

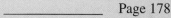

3. Solve problems involving proportions.

(a) **Landscaping** A mixture of plant food must be prepared by combining three parts of a concentrate with every 16 parts water. How much water should be added to 12 ounces of concentrate?

_____ Page 185

Name

_____

Date

_____

Course/Section

Objective	Sample Problems	For help, go to
	(b) **Drafting** An architectural drawing of the living room of a house is $5\frac{1}{2}$ in. long and $4\frac{1}{2}$ in. wide. If the actual length of the living room is 22 ft, what is the actual width?	_____ Page 192
	(c) **Manufacturing** A large production job is completed by four machines in 6 hours. The same size job must be finished in 2 hours the next time it is ordered. How many machines should be used to accomplish this task?	_____ Page 196
4. Write fractions and decimal numbers as percents.	(a) Write $\frac{1}{4}$ as a percent.	_____ Page 208
	(b) Write 0.46 as a percent.	_____ Page 208
	(c) Write 5 as a percent.	_____
	(d) Write 0.075 as a percent.	_____
5. Convert percents to decimal numbers and fractions.	(a) Write 35% as a decimal number.	_____ Page 211
	(b) Write 0.25% as a decimal number.	_____
	(c) Write 112% as a fraction.	_____ Page 212
6. Solve problems involving percent.	(a) Find $37\frac{1}{2}$% of 600.	_____ Page 215
	(b) Find 120% of 45.	_____
	(c) What percent of 80 is 5?	_____ Page 218
	(d) 12 is 16% of what number?	_____ Page 219
	(e) If a measurement is given as 2.778 ± 0.025 in., state the tolerance as a percent.	_____ Page 237
	(f) **Machine Trades** If a certain kind of solder is 52% tin, how many pounds of tin are needed to make 20 lb of solder?	_____
	(g) **Painting** The paint needed for a redecorating job originally cost $144.75, but is discounted by 35%. What is its discounted price?	_____ Page 225

Preview

(h) **Machine Trades**   A mower motor rated at 2.0 hp is found to deliver only 1.6 hp when connected to a transmission system. What is the efficiency of the transmission?

_____   Page 236

(i) **Electrical Trades**   If the voltage in a circuit is increased from 70 volts to 78 volts, what is the percent increase in voltage?

_____   Page 239

(Answers to these preview problems are given in the Appendix. Also, worked solutions to many of these problems appear in the chapter Summary.)

If you are certain that you can work *all* these problems correctly, turn to page 253 for a set of practice problems. If you cannot work one or more of the preview problems, turn to the page indicated to the right of the problem. Those who wish to master this material with the greatest success should turn to Section 4-1 and begin work there.

Chapter 4

Ratio, Proportion, and Percent

DON'T FEEL BAD IF YOU ONLY GOT A 3% RAISE; I ONLY GOT 2% MYSELF.

CAN WE FEEL BAD THAT 2% OF YOUR PAY IS BIGGER THAN 3% OF OUR PAY?

DON'T GET ALL MATHY ON ME.

DILBERT: © Scott Adams/Dist. by United Feature Syndicate, Inc

4-1   Ratio and Proportion

Machinists, mechanics, carpenters, and other trades workers use the ideas of ratio and proportion to solve very many technical problems. The compression ratio of an automobile engine, the gear ratio of a machine, scale drawings, the pitch of a roof, the mechanical advantage of a pulley system, and the voltage ratio in a transformer are all practical examples of the ratio concept.

Ratio   A *ratio* is a comparison, using division, of two quantities of the same kind, both expressed in the same units. For example, the steepness of a hill can be written as the ratio of its height to its horizontal extent.

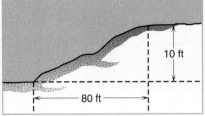

Steepness $= \dfrac{10 \text{ ft}}{80 \text{ ft}} = \dfrac{1}{8}$

The ratio is usually written as a fraction in lowest terms, and you would read this ratio as either "one-eighth" or "one to eight."

The *gear ratio* of a gear system is defined as

$$\text{Gear ratio} = \frac{\text{number of teeth on the driving gear}}{\text{number of teeth on the driven gear}}$$

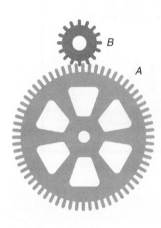

Example 1

General Trades   Find the gear ratio of the system shown if *A* is the driving gear and *B* is the driven gear, and where *A* has 64 teeth and *B* has 16 teeth.

Gear ratio = ?

The ratio of the number of teeth on the driving gear to the number of teeth on the driven gear is

$$\text{Gear ratio} = \frac{64 \text{ teeth}}{16 \text{ teeth}} = \frac{4}{1}$$

Always write the fraction in lowest terms.

The gear ratio is 4 to 1. In technical work this is sometimes written as 4:1 and is read "four to one."

Typical gear ratios on a passenger car are

First gear: 3.54:1

Second gear: 1.90:1

Third gear: 1.31:1

Reverse: 3.25:1

A gear ratio of 3.54:1 means that the engine turns 3.54 revolutions for each revolution of the drive shaft. If the drive or rear-axle ratio is 3.72, the engine needs to make 3.54 × 3.72 or approximately 13.2 revolutions in first gear to turn the wheels one full turn.

Here are a few important examples of the use of ratios in practical work.

Pulley Ratios   A pulley is a device that can be used to transfer power from one system to another. A pulley system can be used to lift heavy objects in a shop or to connect a power source to a piece of machinery. The ratio of the pulley diameters will determine the relative pulley speeds.

Example 2

General Trades   Find the ratio of the diameter of pulley *A* to the diameter of pulley *B* in the following drawing.

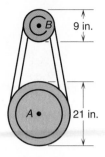

$$\frac{\text{diameter of pulley } A}{\text{diameter of pulley } B} = \frac{21 \text{ in.}}{9 \text{ in.}}$$ ⬅ Same units

$$= \frac{7}{3}$$

The ratio is 7 to 3 or 7 : 3.

Notice that the units, inches, cancel from the ratio. A ratio is a fraction or decimal number, and it has no units.

Compression Ratio   In an automobile engine there is a large difference between the volume of the cylinder space when a piston is at the bottom of its stroke and when it is at the top of its stroke. This difference in volumes is called the *engine displacement.* Automotive mechanics find it very useful to talk about the compression ratio of an automobile engine. The *compression ratio* of an engine compares the volume of the cylinder at maximum expansion to the volume of the cylinder at maximum compression.

$$\text{Compression ratio} = \frac{\text{expanded volume}}{\text{compressed volume}}$$

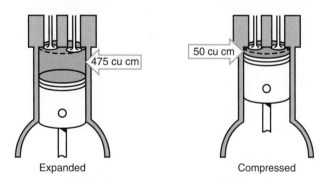

Expanded   Compressed

Example 3

Automotive Trades   Find the compression ratio of a gasoline engine if each cylinder has a maximum volume of 475 cu cm and a minimum or compression volume of 50 cu cm.

$$\text{Compression ratio} = \frac{475 \text{ cu cm}}{50 \text{ cu cm}} = \frac{19}{2}$$

$$= 9.5$$

Compression ratios are always written so that the second number in the ratio is 1. This compression ratio would be written as $9\frac{1}{2}$ to 1.

Now, for some practice in calculating ratios, work the following problems.

(a)

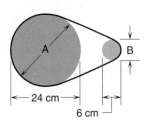

(b)

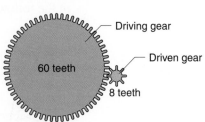

Ratio of pulley diameters $= \dfrac{A}{B} =$

Gear ratio $=$

(c) **Roofing**

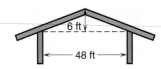

$\text{Pitch} = \dfrac{\text{rise}}{\text{run}} =$

(d) **Automotive Trades**   Find the rear-axle ratio of a car if the ring gear has 54 teeth and the pinion gear has 18 teeth.

$$\text{Rear-axle ratio} = \dfrac{\text{number of teeth on ring gear}}{\text{number of teeth on pinion gear}}$$

(e) **Automotive Trades**   A gasoline engine has a maximum cylinder volume of 47 cu in. and a compressed volume of 5.0 cu in. Find its compression ratio.

(f) **Business and Finance**   Stock in Amazing.com sells for $18 per share. The company has earnings of $0.50 per share. Find its P-E or price to earnings ratio.

→ **Solutions**

(a)   Ratio of pulley diameters $= \dfrac{24 \text{ cm}}{6 \text{ cm}}$ or   4 to 1.

(b)   Gear ratio $= \dfrac{60 \text{ teeth}}{8 \text{ teeth}} = \dfrac{15}{2}$ or   15 to 2.

(c)   Pitch $= \dfrac{6\text{-ft rise}}{24\text{-ft run}} = \dfrac{6}{24}$ or $\dfrac{3}{12}$.

The pitch is 3:12. Recall that roofers usually express pitch with a denominator of 12.

(d)   Rear-axle ratio $= \dfrac{54 \text{ teeth}}{18 \text{ teeth}} = \dfrac{3}{1}$.

The rear-axle ratio is 3 to 1.

(e)   Compression ratio $= \dfrac{47 \text{ cu in.}}{5.0 \text{ cu in.}} = \dfrac{47}{5.0}$ or   9.4 to 1.

(f)   P-E $= \dfrac{\$18}{\$0.50} = \dfrac{\$18 \times \boxed{2}}{\$0.50 \times \boxed{2}} = \dfrac{\$36}{\$1}$ or   36 to 1.

Simple Equations   To use ratios to solve a variety of problems, you must be able to solve a simple kind of algebraic equation. Consider this puzzle: "I'm thinking of a number. When I multiply my number by 3, I get 15. What is my number?" Solve the puzzle and check your solution with ours.

If you answered "five," you're correct. Think about how you worked it out. Most people take the answer 15 and do the "reverse" of multiplying by 3. That is, they divide by 3. In symbols the problem would look like this:

If $3 \times \square = 15$

then $\square = 15 \div 3 = 5$

Example 4

Try another one: A number multiplied by 40 gives 2200. Find the number.

Using symbols again,

$40 \times \square = 2200$

so $\square = 2200 \div 40 = \dfrac{2200}{40} = 55$

We have just solved a simple algebraic equation. But in algebra, instead of drawing boxes to represent unknown quantities, we use letters of the alphabet. For example, $40 \times \square$ could be written as $40 \times N$. Furthermore, we may signify multiplication in any one of these additional three ways:

- A raised dot

  For example, $40 \cdot N$ or $8 \cdot 5$

- Writing a multiplier next to a letter.

  For example, $40N$

- Enclosing in parentheses one or both of the numbers being multiplied. (This is especially useful when one or both of the numbers are fractions or decimals.)

  For example, $8(5.6)$

  or $\left(2\dfrac{1}{3}\right)\left(4\dfrac{5}{8}\right)$

Example 5

$40 \times \square = 2200$

can be written $40 \cdot N = 2200$

or $40N = 2200$

To solve this type of equation, we divide 2200 by the multiplier of N, which is 40.

→ Your Turn

For practice, solve these equations.

(a) $25n = 275$ (b) $6M = 96$

(c) $100x = 550$ (d) $12y = 99$

(e) $88 = 8P$ (f) $15 = 30a$

(a) $n = \dfrac{275}{25} = 11$ (b) $M = \dfrac{96}{6} = 16$

(c) $x = \dfrac{550}{100} = 5.5$ (d) $y = \dfrac{99}{12} = 8.25$

(e) Notice that P, the unknown quantity, is on the *right* side of the equation. You must divide the number by the multiplier of the unknown, so

$$P = \dfrac{88}{8} = 11$$

(f) Again, a, the unknown quantity, is on the right side, so

$$a = \dfrac{15}{30} = 0.5$$

Careful   In problems like (f) some students mistakenly think that they must always divide the larger number by the smaller. That is not always correct. Remember, for an equation of this simple form, always divide by the number that multiplies the unknown. ◄

Other types of equations will be solved in Chapter 7, but only this type is needed for our work with proportions and percent.

Proportions   A *proportion* is an equation stating that two ratios are equal. For example,

$$\dfrac{1}{3} = \dfrac{4}{12} \quad \text{is a proportion}$$

Notice that the equation is true because the fraction $\frac{4}{12}$ expressed in lowest terms is equal to $\frac{1}{3}$.

Note   When you are working with proportions, it is helpful to have a way of stating them in words. For example, the proportion

$$\dfrac{2}{5} = \dfrac{6}{15} \quad \text{can be stated in words as}$$

"Two is to five as six is to fifteen." ◄

When one of the four numbers in a proportion is unknown, it is possible to find the value of that number. In the proportion

$$\dfrac{1}{3} = \dfrac{4}{12}$$

notice what happens when we multiply diagonally:

$$\dfrac{1}{3} = \dfrac{4}{12} \rightarrow 1 \cdot 12 = 12 \qquad \dfrac{1}{3} = \dfrac{4}{12} \rightarrow 3 \cdot 4 = 12$$

These diagonal products are called the *cross-products* of the proportion. If the proportion is a true statement, the cross-products will always be equal. Here are more examples:

$$\dfrac{5}{8} = \dfrac{10}{16} \rightarrow 5 \cdot 16 = 80 \text{ and } 8 \cdot 10 = 80$$

$$\dfrac{9}{12} = \dfrac{3}{4} \rightarrow 9 \cdot 4 = 36 \text{ and } 12 \cdot 3 = 36$$

$$\dfrac{10}{6} = \dfrac{5}{3} \rightarrow 10 \cdot 3 = 30 \text{ and } 6 \cdot 5 = 30$$

This very important fact is called the cross-product rule.

THE CROSS-PRODUCT RULE

$$\text{If } \frac{a}{b} = \frac{c}{d} \quad \text{then } a \cdot d = b \cdot c.$$

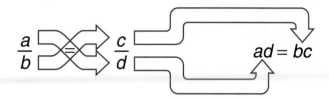

Example 6

Find the cross-products for the proportion $\frac{3}{5} = \frac{12}{20}$.

The cross-products are $3 \cdot 20 = 60$ and $5 \cdot 12 = 60$.

The cross-product rule can be used to solve proportions, that is, to find the value of an unknown number in the proportion. For example, in the proportion

$$\frac{x}{4} = \frac{12}{16}$$

to solve the proportion means to find the value of the unknown quantity x that makes the equation true. To do this:

First, use the cross-product rule.

If $\quad \frac{x}{4} = \frac{12}{16} \quad$ then $\quad 16 \cdot x = 4 \cdot 12 \quad$ or $\quad 16x = 48$

Then, solve this equation using the division technique.

$$x = \frac{48}{16} = 3$$

 Finally, check your answer by replacing x with 3 in the original proportion equation.

$$\frac{3}{4} = \frac{12}{16}$$

Check 1: Find the cross-products.

$3 \cdot 16 = 48 \qquad$ and $\qquad 4 \cdot 12 = 48 \qquad$ The answer is correct.

Check 2: Write $\frac{12}{16}$ in lowest terms.

$$\frac{12}{16} = \frac{4 \cdot 3}{4 \cdot 4} = \frac{3}{4}$$

→ Your Turn

Now try this one:

$\frac{6}{N} = \frac{15}{10} \qquad$ Solve this proportion for N.

Step 1 Apply the cross-product rule.

If $\dfrac{6}{N} = \dfrac{15}{10}$ then $6 \cdot 10 = 15N$

or $15N = 60$

Step 2 Solve the equation.

$$N = \dfrac{60}{15} = 4$$

✓ **Step 3** Substitute 4 for N in the original proportion.

$$\dfrac{6}{4} = \dfrac{15}{10}$$

The cross-products are equal: $6 \cdot 10 = 60$ and $4 \cdot 15 = 60$.

Notice that, in lowest terms,

$$\dfrac{6}{4} = \dfrac{3}{2} \quad \text{and} \quad \dfrac{15}{10} = \dfrac{3}{2}$$

Note The unknown quantity can appear in any one of the four positions in a proportion. No matter where the unknown appears, solve the proportion the same way: write the cross-products and solve the resulting equation. ◄

→ **More Practice**

Here are more practice problems. Solve each proportion.

(a) $\dfrac{28}{40} = \dfrac{x}{100}$ (b) $\dfrac{12}{y} = \dfrac{8}{50}$ (c) $\dfrac{n}{6} = \dfrac{7}{21}$

(d) $\dfrac{12}{9} = \dfrac{32}{M}$ (e) $\dfrac{Y}{7} = \dfrac{3}{4}$ (f) $\dfrac{6}{5} = \dfrac{2}{T}$

(g) $\dfrac{2\frac{1}{2}}{3\frac{1}{2}} = \dfrac{w}{2}$ (h) $\dfrac{12}{E} = \dfrac{0.4}{1.5}$

→ **Answers**

(a) $\dfrac{28}{40} = \dfrac{x}{100}$ $\quad 28 \cdot 100 = 40 \cdot x$ or $\quad\quad 40x = 2800$

$x = 2800 \div 40$

$x = 70$

(b) $\dfrac{12}{y} = \dfrac{8}{50}$ $\quad 600 = 8y$ or $\quad\quad y = \dfrac{600}{8} = 75$

(c) $\dfrac{n}{6} = \dfrac{7}{21}$ $\quad 21n = 42$ $\quad\quad n = \dfrac{42}{21} = 2$

(d) $\dfrac{12}{9} = \dfrac{32}{M}$ $\quad 12M = 288$ $\quad\quad M = \dfrac{288}{12} = 24$

$$(e) \quad \frac{Y}{7} = \frac{3}{4} \qquad\qquad 4Y = 21 \qquad\qquad Y = \frac{21}{4} = 5\tfrac{1}{4} \text{ or } 5.25$$

$$(f) \quad \frac{6}{5} = \frac{2}{T} \qquad\qquad 6T = 10 \qquad\qquad T = \frac{10}{6} = 1\tfrac{2}{3}$$

$$(g) \quad \frac{2\tfrac{1}{2}}{3\tfrac{1}{2}} = \frac{w}{2} \qquad\qquad 3.5w = 5 \qquad\qquad w = \frac{5}{3.5} = 1\tfrac{3}{7}$$

$$(h) \quad \frac{12}{E} = \frac{0.4}{1.5} \qquad\qquad 18 = 0.4E \qquad\qquad E = \frac{18}{0.4} = 45$$

A Closer Look   You may have noticed that the two steps for solving a proportion can be simplified into one step. For example, in problem (a), to solve

$$\frac{28}{40} = \frac{x}{100} \qquad \text{we can write the answer directly as}$$

$$x = \frac{28 \cdot 100}{40}$$

Always divide by the number that is diagonally opposite the unknown, and multiply the two remaining numbers that are diagonally opposite each other.

This shortcut comes in handy when using a calculator. For example, with a calculator, problem (h) becomes

12 $\boxed{\times}$ 1.5 $\boxed{\div}$ .4 $\boxed{=}$ → 45.

The numbers 12 and 1.5 are diagonally opposite each other in the proportion.

0.4 is diagonally opposite the unknown E. ◄

→ Your Turn

Practice this one-step process by solving the proportion

$$\frac{120}{25} = \frac{6}{Q}$$

→ Solution

$$Q = \frac{6 \cdot 25}{120} = 1.25 \qquad \text{Check it.}$$

Problem Solving Using Proportions   We can use proportions to solve a variety of problems involving ratios. If you are given the value of a ratio and one of its terms, it is possible to find the other term.

Example 7

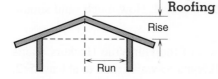

↓ **Roofing**   If the pitch of a roof is supposed to be 2 : 12, and the run is 20 ft, what must be the rise?

$$\text{Pitch} = \frac{\text{rise}}{\text{run}}$$

A ratio of $2:12$ is equivalent to the fraction $\frac{2}{12}$; therefore, the equation becomes

$$\frac{2}{12} = \frac{\text{rise}}{20} \qquad \text{or} \qquad \frac{2}{12} = \frac{R}{20}$$

Find the cross-products. $\qquad\qquad 12R = 40$

Then solve for R. $\qquad\qquad\qquad R = \dfrac{40}{12}$

The rise is 3 ft 4 inches. $\left(\dfrac{1}{3} \text{ ft} = 4 \text{ in.}\right) R = 3\dfrac{1}{3} \text{ ft}$

The algebra you learned earlier in this chapter will enable you to solve any ratio problem of this kind. Here is another example.

Example 8

Allied Health   For medications provided in a liquid formulation, the amount of liquid given to a patient depends on the concentration of drug in the fluid. Medication Y is available only in a liquid formulation with a concentration of 100 mg/mL (milligrams per milliliter). Suppose a physician orders 225 mg of Medication Y. To determine the amount of liquid required, first note that 100 mg/mL means that one milliliter of liquid contains 100 mg of the medication. Therefore, we can set up the following proportion:

$$\frac{1 \text{ mL}}{x \text{ mL}} = \frac{100 \text{ mg}}{225 \text{ mg}}$$

Solving for x, we have $\qquad 100x = 225$

$$x = 2.25 \text{ mL}$$

Therefore, the patient should take 2.25 mL of the liquid in order to receive 225 mg of Medication Y.

→ Your Turn

Try these problems.

(a)   **Manufacturing**   If the gear ratio on a mixing machine is $6:1$ and the driven gear has 12 teeth, how many teeth are on the driving gear?

(b)   **Manufacturing**   The pulley system of an assembly belt has a pulley diameter ratio of 4. If the larger pulley has a diameter of 15 in., what is the diameter of the smaller pulley?

(c)   **Automotive Trades**   The compression ratio of a classic Datsun 280Z is 8.3 to 1. If the compressed volume of the cylinder is 36 cu cm, what is the expanded volume of the cylinder?

(d)   **Construction**   On a certain construction job, concrete is made using a volume ratio of 1 part cement to $2\frac{1}{2}$ parts sand and 4 parts gravel. How much sand should be mixed with 3 cu ft of cement?

(e)   **Sports and Leisure**   Manager Sparky Spittoon of the Huntville Hackers wants his ace pitcher Lefty Groove to improve his strikeouts-to-walks ratio to at least $5:2$.

If he has 65 strikeouts so far this season, what is the maximum number of walks he should have to keep Sparky happy?

(f) **Roofing**   If the pitch of a roof must be 5 : 12, how much rise should there be over a run of 54 ft?

(g) **Allied Health**   Medication TTQ is available only in a liquid formulation with a concentration of 25 μg/mL (micrograms per milliliter). How much of this liquid formulation should be given to a patient if 300 μg of TTQ is ordered by the physician?

→ **Solutions**

(a)   Gear ratio $= \dfrac{\text{number of teeth on driving gear}}{\text{number of teeth on driven gear}}$

A ratio of 6 : 1 is equivalent to the fraction $\dfrac{6}{1}$.

Therefore,

$$\frac{6}{1} = \frac{x}{12}$$

or $72 = x$ The driving gear has 72 teeth.

(b)   Pulley ratio $= \dfrac{\text{diameter of larger pulley}}{\text{diameter of smaller pulley}}$

$$4 = \frac{15 \text{ in.}}{D} \qquad \text{or} \qquad \frac{4}{1} = \frac{15}{D}$$

$$4D = 15$$

$$D = \frac{15}{4} \qquad \text{or} \qquad D = 3\frac{3}{4} \text{ in.}$$

(c)   Compression ratio $= \dfrac{\text{expanded volume}}{\text{compressed volume}}$

$$8.3 = \frac{V}{36 \text{ cu cm}} \qquad \text{or} \qquad \frac{83}{10} = \frac{V}{36}$$

$$V = \frac{83 \cdot 36}{10} = 298.8 \text{ cu cm}$$

 83 ⊗ 36 ⊕ 10 ⊜ → [298.8]

In any practical situation this answer would be rounded to 300 cu cm.

(d)   Ratio of cement to sand $= \dfrac{\text{volume of cement}}{\text{volume of sand}}$

$$\frac{1}{2\frac{1}{2}} = \frac{3 \text{ cu ft}}{S}$$

Therefore,

$$S = 3\left(2\tfrac{1}{2}\right)$$

$$S = 7\tfrac{1}{2} \text{ cu ft}$$

(e) $\dfrac{\text{Strikeouts}}{\text{Walks}} = \dfrac{5}{2} = \dfrac{65}{x}$

$5x = 130$

$x = 26$ Lefty should have no more than 26 walks.

(f) $\text{Pitch} = \dfrac{\text{Rise}}{\text{Run}}$

$\dfrac{5}{12} = \dfrac{R}{54}$

$12R = 270$

$R = 22.5 \text{ ft}$

(g) **First,** set up a proportion. $\dfrac{1 \text{ mL}}{x \text{ mL}} = \dfrac{25 \ \mu g}{300 \ \mu g}$

Then, solve. $25x = 300$

$x = 12 \text{ mL}$

Therefore, 12 milliliters of liquid should be given to the patient.

Now turn to Exercises 4-1 for more practice on ratio and proportion.

Exercises 4-1   Ratio and Proportion

A. Complete the following tables.

1.

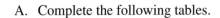

	Teeth on Driving Gear A	Teeth on Driven Gear B	Gear Ratio, $\dfrac{A}{B}$
(a)	35	5	
(b)	12	7	
(c)		3	2 to 1
(d)	21		$3\frac{1}{2}$ to 1
(e)	15		1 to 3
(f)		18	1 to 2
(g)		24	2:3
(h)	30		3:5
(i)	27	18	
(j)	12	30	

2.

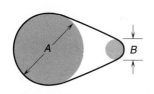

	Diameter of Pulley A	Diameter of Pulley B	Pulley Ratio, $\dfrac{A}{B}$
(a)	16 in.	6 in.	
(b)	15 in.	12 in.	
(c)		8 in.	2 to 1
(d)	27 cm		4.5 to 1
(e)		10 cm	4 to 1
(f)	$8\frac{1}{8}$ in.	$3\frac{1}{4}$ in.	
(g)	8.46 cm	11.28 cm	
(h)	20.41 cm		3.14 to 1
(i)		12.15 cm	1 to 2.25
(j)	4.45 cm		0.25 to 1

3.

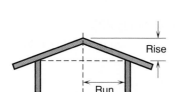

	Rise	Run	Pitch
(a)	8 ft	6 ft	
(b)		24 ft	4:12
(c)	7 ft		3:12
(d)	14 ft 4 in.	25 ft	
(e)	9 ft	15 ft	
(f)		20 ft	2.4:12
(g)	3 ft		1.8:12
(h)		30 ft 6 in.	2:12

B. Solve these proportion equations.

1. $\dfrac{3}{2} = \dfrac{x}{8}$ 2. $\dfrac{6}{R} = \dfrac{5}{72}$

3. $\dfrac{y}{60} = \dfrac{5}{3}$ 4. $\dfrac{2}{15} = \dfrac{8}{H}$

5. $\dfrac{5}{P} = \dfrac{30}{7}$ 6. $\dfrac{1}{6} = \dfrac{17}{x}$

7. $\dfrac{A}{2.5} = \dfrac{13}{10}$ 8. $\dfrac{27}{M} = \dfrac{3}{0.8}$

9. $\dfrac{2}{5} = \dfrac{T}{4.5}$ 10. $\dfrac{0.12}{N} = \dfrac{2}{7}$

11. $\dfrac{138}{23} = \dfrac{18}{x}$ 12. $\dfrac{3.25}{1.5} = \dfrac{A}{0.6}$

13. $\dfrac{x}{34.86} = \dfrac{1.2}{8.3}$

14. $\dfrac{2\frac{1}{2}}{R} = \dfrac{1\frac{1}{4}}{3\frac{1}{4}}$

15. $\dfrac{2 \text{ ft } 6 \text{ in.}}{4 \text{ ft } 3 \text{ in.}} = \dfrac{L}{8 \text{ ft } 6 \text{ in.}}$

16. $\dfrac{6.2 \text{ cm}}{x} = \dfrac{1.2 \text{ in.}}{11.4 \text{ in.}}$

17. $\dfrac{3 \text{ ft } 4 \text{ in.}}{4 \text{ ft } 2 \text{ in.}} = \dfrac{3.2 \text{ cm}}{x}$

18. $\dfrac{3\frac{1}{2} \text{ in.}}{W} = \dfrac{1.4}{0.05}$

C. Solve.

1. **Automotive Trades** The compression ratio in a certain engine is 9.6 to 1. If the expanded volume of a cylinder is 48 cu in., what is the compressed volume?

2. **Painting** If 1 gal of paint covers 360 sq ft, how many gallons will be needed to cover 2650 sq ft with two coats? (Assume that you cannot buy a fraction of a gallon.)

3. **Machine Trades** If 28 tapered pins can be machined from a steel rod 12 ft long, how many tapered pins can be made from a steel rod 9 ft long?

4. **Carpentry** If 6 lb of nails are needed for each thousand lath, how many pounds are required for 4250 lath?

5. **Masonry** For a certain kind of plaster work, 1.5 cu yd of sand are needed for every 100 sq yd of surface. How much sand will be needed for 350 sq yd of surface?

6. **Printing** The paper needed for a printing job weighs 12 lb per 500 sheets. How many pounds of paper are needed to run a job requiring 12,500 sheets?

7. **Machine Trades** A cylindrical oil tank 8 ft deep holds 420 gallons when filled to capacity. How many gallons remain in the tank when the depth of oil is $5\frac{1}{2}$ ft?

8. **Agriculture** A liquid fertilizer must be prepared by using one part of concentrate for every 32 parts of water. How much water should be added to 7 oz of concentrate?

9. **Welding** A 10-ft bar of I-beam weighs 208 lb. What is the weight of a 6-ft length?

10. **Photography** A photographer must mix a chemical in the ratio of 1 part chemical for every 7 parts of water. How many ounces of chemical should be used to make a 3-qt *total* mixture? (1 qt = 32 oz)

11. **General Trades** If you earn $684.80 for a 32-hour work week, how much would you earn for a 40-hour work week at the same hourly rate?

12. **Machine Trades** A machinist can produce 12 parts in 40 min. How many parts can the machinist produce in 4 hours?

13. **Automotive Trades** The headlights on a car are set so the light beam drops 2 in. for each 25 ft measured horizontally. If the headlights are mounted 30 in. above the ground, how far ahead of the car will they hit the ground?

14. **Machine Trades** A machinist creates $2\frac{3}{4}$ lb of steel chips in fabricating 16 rods. How many pounds of steel chips will be created in producing 120 rods?

15. **Agriculture** To prepare a pesticide spray, 3.5 lb of BIOsid is added to 30 gal of water. How much BIOsid should be added to a spray tank holding 325 gal? Round to the nearest 0.1 lb.

16. **Painting**   A painter must thin some paint for use in a sprayer. If the recommended rate is $\frac{1}{2}$ pint of water per gallon of paint, how many pints of water should be added to $5\frac{1}{2}$ gallons of paint?

17. **Allied Health**   The label on a concentrated drug solution indicates that it contains 85 mg of medication in 5 mL. If the patient is to receive 220 mg of solution, how much of the solution should be given? Round to the nearest tenth.

18. **Automotive Trades**   In winter weather, fuel-line antifreeze must be added at a rate of one can per 8 gal of fuel. How many cans should be added for an 18-gal fuel tank?

19. **Culinary Arts**   One small 0.5-kg package of Fromage de Cernex cheese contains 2000 calories. How many calories are contained in a large 10-kg package?

20. **Automotive Trades**   The air to fuel ratio of an engine helps determine how efficiently the engine is running. In most cases, a ratio of 14.7:1 is ideal. A larger ratio indicates that the engine is running *lean*, while a smaller ratio indicates that it is running *rich*. Suppose a certain engine draws 160 lb of air in burning 12 lb of fuel. Find the air to fuel ratio, and state whether the engine is running lean or rich.

21. **Automotive Trades**   In problem 20, it was stated that the ideal air to fuel ratio for an engine is 14.7:1. If a vehicle burns 9 lb of fuel, how many pounds of air should it draw to achieve the ideal ratio? Round to the nearest pound.

22. **Sports and Leisure**   Power hitter Sammy Sockitome strikes out too much to suit manager Sparky Spittoon. Sammy is promised an incentive bonus of a million dollars if he can reduce his strikeout to home run ratio to 3:1. If Sammy hits 56 home runs, what is the maximum number of strikeouts he can have and still earn the bonus?

23. **Landscaping**   A landscape architect is seeding a 6000-sq-ft lawn. If the seed manufacturer recommends using 6 lb of seed per 1500 sq ft, how many pounds of seed will be needed?

24. **Masonry**   A mason needs to purchase mortar mix for a retaining wall consisting of 640 blocks. The guidelines for this mix suggest that 12 bags are required for every 100 blocks. Assuming that he cannot purchase part of a bag, how many bags will he need?

25. **Allied Health**   Medication Q is available only in a liquid form with a concentration of 30 μg/mL. Determine how much of the liquid should be given to a patient when the following amounts of Medication Q are ordered by the physician:

(a) 15 μg   (b) 10 μg   (c) 200 μg

(Round to two decimal digits if necessary.)

26. **Construction**   Concrete requires cement, sand, and gravel in a ratio of 1 to 2 to 3 by volume for optimum strength. PTO Construction is pouring a small concrete foundation for which 38 bags of sand will be used. (a) How many bags of cement should be used? (b) How many bags of gravel should be used?

27. **Allied Health**   Each tablet of Medication Z contains 50 micrograms (μg) of drug. Determine how many tablets (or fractions of tablets) should be given when the following amounts of Medication Z are ordered by the physician:

(a) 100 μg   (b) 75 μg   (c) 230 μg

(Round to the nearest half-tablet.)

28. **Trades Management**   A business owner is currently renting her shop space for $8460 per month. Her lease agreement states that her rent will be adjusted

each July according to the CPI (Consumer Price Index). Specifically, the ratio of this June's CPI to the previous June's CPI will be multiplied by the current monthly rent to determine the new monthly rent. If this June's CPI is 226.765, and last June's CPI was 217.273, what will be her new monthly rent beginning in July? (Round your answer to the nearest cent.)

29. **Trades Management**   A small welding shop employs four welders and one secretary. A workers' compensation insurance policy charges a premium of $10.59 per $100 of gross wages for the welders and $1.38 per $100 of gross wages for the secretary. If each welder earns $36,000 per year, and the secretary earns $28,000 per year, what is the total annual premium for this insurance?

30. **Construction**   On a crisp fall day, a builder wants to drive from his house to a work site in the mountains to pour the concrete foundations for a cabin. The minimum temperature at which the concrete will set with adequate strength is 40°F. The temperature at his house is 60°F, and the cabin is at an altitude that is 5900 ft higher than the town where the builder lives. If temperature decreases by about 4°F for every 1000 ft of altitude increase, should the builder bother to drive to the work site?

31. **General Trades**   A trades worker is considering a job offer in another city. Her current job pays $2950 per month and is in a city with a cost of living index of 98.3. The cost of living index in the new location is 128.5. If she were to maintain her present lifestyle, how much would she need to earn per month in the new job? (*Hint:* Cost of living must be directly proportional to salary).

Check your answers to the odd-numbered problems in the Appendix, then turn to Section 4-2.

4-2   Special Applications of Ratio and Proportion

In the previous section, we used the concepts of ratio and proportion to solve some simple technical applications. In this section we will learn about additional applications of ratio and proportion in the trades and technical areas.

Scale Drawings   Proportion equations are found in a wide variety of practical situations. For example, when a drafter makes a drawing of a machine part, building layout, or other large structure, he or she must *scale it down*. The drawing must represent the object accurately, but it must be small enough to fit on the paper. The draftsperson reduces every dimension by some fixed ratio.

Example 1

Drafting   Drawings that are larger than life involve an expanded scale.

For the automobile shown, the ratio of the actual length to the scale-drawing length is equal to the ratio of the actual width to the scale-drawing width.

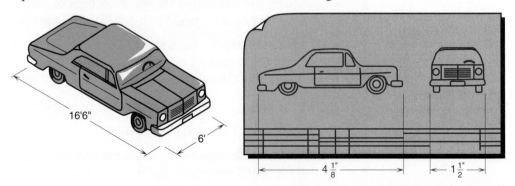

$$\frac{\text{actual length}}{\text{drawing length}} = \frac{\text{actual width}}{\text{drawing width}}$$

$$\frac{16\,\text{ft}\,6\,\text{in.}}{4\frac{1}{8}\,\text{in.}} = \frac{6\,\text{ft}}{1\frac{1}{2}\,\text{in.}}$$

This equation is a proportion

Ratio of lengths

Ratio of widths

Rewrite all quantities in the same units: (1 ft = 12 in.)

$$\frac{198\,\text{in.}}{4\frac{1}{8}\,\text{in.}} = \frac{72\,\text{in.}}{1\frac{1}{2}\,\text{in.}}$$

You should notice first of all that each side of this equation is a ratio. Each side is a ratio of *like* quantities: lengths on the left and widths on the right.

Second, notice that the ratio $\dfrac{198\,\text{in.}}{4\frac{1}{8}\,\text{in.}}$ is equal to $\dfrac{48}{1}$.

Divide it out: $198 \div 4\frac{1}{8} = 198 \div \dfrac{33}{8}$

$$= \frac{198}{1} \times \frac{8}{33}$$

$$= 48$$

198 ✕ **8** ÷ **33** = → ⬛ *48.*

Notice also that the ratio $\dfrac{72\,\text{in.}}{1\frac{1}{2}\,\text{in.}}$ is equal to $\dfrac{48}{1}$.

The common ratio $\dfrac{48}{1}$ is called the *scale* factor of the drawing.

Example 2

Architecture   In the following problem, one of the dimensions is unknown:

Suppose that a rectangular room has a length of 18 ft and a width of 12 ft. An architectural scale drawing of this room is made so that on the drawing the length of the room is 4.5 in. What will be the width of the room on the drawing? What is the scale factor?

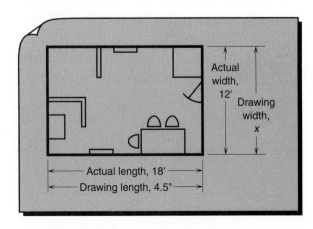

Actual width, 12'

Drawing width, x

Actual length, 18'

Drawing length, 4.5"

First, set up a ratio of lengths and a ratio of widths.

Length ratio $= \dfrac{\text{actual length}}{\text{drawing length}}$

$= \dfrac{18 \text{ ft}}{4.5 \text{ in.}} = \dfrac{216 \text{ in.}}{4.5 \text{ in.}}$ ← Convert 18 ft to inches so that the top and the bottom of the fraction have the same units.

Width ratio $= \dfrac{\text{actual width}}{\text{drawing width}}$

$= \dfrac{12 \text{ ft}}{x \text{ in.}} = \dfrac{144 \text{ in.}}{x \text{ in.}}$ ← Change 12 ft to 144 in.
← Let x = the drawing width

Second, write a proportion equation.

$\dfrac{216}{4.5} = \dfrac{144}{x}$

Third, solve this proportion. Cross-multiply to get

$216x = (4.5)144$

$216x = 648$

$x = 3 \text{ in.}$ The width of the room on the drawing is 3 in.

For the room drawing shown, the scale factor is

Scale factor $= \dfrac{18 \text{ ft}}{4.5 \text{ in.}} = \dfrac{216 \text{ in.}}{4.5 \text{ in.}}$

$= 48$ or   48 to 1

One inch on the drawing corresponds to 48 in. or 4 ft on the actual object. We can express this as 1 in. = 4 ft. A draftsperson would divide by 4 and write it as $\frac{1}{4}$ in. = 1 ft.

→ **Your Turn**

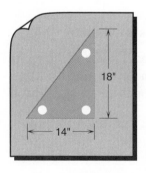

Drafting   The drawing of the triangular plate shown has a height of 18 in. and a base length of 14 in.

(a)   What will be the corresponding height in a reduced copy of this drawing if the base length in the copy is $2\frac{3}{16}$ in.?

(b)   Find the scale factor of the reduction.

→ **Solutions**

(a) $\dfrac{18 \text{ in.}}{x} = \dfrac{14 \text{ in.}}{2\frac{3}{16} \text{ in.}}$

Cross-multiply $18\left(2\frac{3}{16}\right) = 14x$

$14x = 39\frac{3}{8}$

$x = 2\frac{13}{16} \text{ in.}$

*C4-1   2 $\boxed{A^b_c}$ 3 $\boxed{A^b_c}$ 16 $\boxed{\times}$ 18 $\boxed{\div}$ 14 $\boxed{=}$ → 2⎵13/16 $\boxed{F \leftrightarrow D}$ $\boxed{=}$ → 2.8125

(b) Scale factor $= \dfrac{18 \text{ in.}}{2\frac{13}{16} \text{ in.}}$ or $\dfrac{14 \text{ in.}}{2\frac{3}{16} \text{ in.}}$

$\qquad\qquad\qquad = 6\frac{2}{5}$ or 6.4 to 1

*C4-2   18 $\boxed{\div}$ 2 $\boxed{A^b_c}$ 13 $\boxed{A^b_c}$ 16 $\boxed{=}$ → 6⎵2/5 $\boxed{F \leftrightarrow D}$ $\boxed{=}$ → 6.4

*__Reminder:__ The reference codes C4-1 and C4-2 indicate that, for those who use calculators with single-line displays, alternative key sequences are provided in the Calculator Appendix.

Similar Figures

In general, two geometric figures that have the same shape but are not the same size are said to be *similar* figures. The blueprint drawing and the actual object are a pair of similar figures. An enlarged photograph and the smaller original are similar.

  and are similar triangles.

A [⬚] B and C [⬚] D are similar rectangles.

and are similar figures.

In any similar figures, all parts of corresponding dimensions have the same scale ratio. For example, in the preceding rectangles

$$\dfrac{A}{C} = \dfrac{B}{D}$$

In the irregular figures above,

$$\dfrac{x}{p} = \dfrac{y}{q} = \dfrac{z}{s} = \dfrac{w}{t} \text{ and so on}$$

The triangles shown here are *not* similar:

Example 3

__Landscaping__   A landscaper is designing a garden for a house that is not yet built. He needs to determine how long a shadow the house will cast into the garden area. To determine this, the landscaper, who is 6 ft tall, measures his shadow to be 8 ft long at a particular time on a summer afternoon.

If the roofline of the house will be 15 ft high where the landscaper was standing, how far from the edge of the house will its shadow extend at the same time of day?

To solve this problem we draw a sketch of the situation.

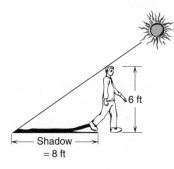

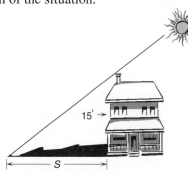

At the same time of day, the triangles formed by the objects, the sun's rays, and the shadows are similar. Therefore,

$$\frac{6 \text{ ft}}{15 \text{ ft}} = \frac{8 \text{ ft}}{S \text{ ft}}$$

$$S = \frac{15(8)}{6} = 20 \text{ ft}$$

The shadow cast by the house will extend 20 ft into the garden from the edge of the house.

→ Your Turn

Find the missing dimension in each of the following pairs of similar figures.

(a)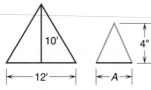

(b)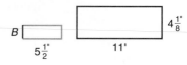

(c)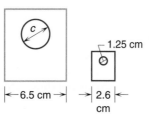

(d)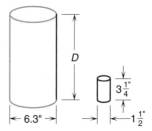

→ Solutions

(a) $\dfrac{A}{144} = \dfrac{4}{120}$ Convert ft to in.

$$A = \frac{4(144)}{120} = 4.8 \text{ in.}$$

(b) $\dfrac{B}{4\frac{1}{8}} = \dfrac{5\frac{1}{2}}{11}$

$$B = \frac{\left(5\frac{1}{2}\right)\left(4\frac{1}{8}\right)}{11} = 2\frac{1}{16} \text{ in.}$$

(c) $\dfrac{c}{1.25} = \dfrac{6.5}{2.6}$

$$c = \frac{6.5(1.25)}{2.6} = 3.125 \text{ cm}$$

(d) $\dfrac{D}{3\frac{1}{4}} = \dfrac{6.3}{1\frac{1}{2}}$

$$D = \frac{6.3\left(3\frac{1}{4}\right)}{1\frac{1}{2}} = 13.65 \text{ in.}$$

Work problem (d) on a calculator this way:

 3 $\boxed{A\frac{b}{c}}$ 1 $\boxed{A\frac{b}{c}}$ 4 $\boxed{\times}$ 6.3 $\boxed{\div}$ 1.5 $\boxed{=}$ →   *13.65*

Direct and Inverse Proportion   Many trade problems can be solved by setting up a proportion involving four related quantities. But it is important that you recognize that there are *two* kinds of proportions—direct and inverse. Two quantities are said to be *directly proportional* if an increase in one quantity leads to a proportional increase in the other quantity, or if a decrease in one leads to a decrease in the other.

> **Direct proportion:** increase → increase
> or decrease → decrease

Example 4

Electrical Trades The electrical resistance of a wire is directly proportional to its length—the longer the wire, the greater the resistance. If 1 ft of Nichrome heater element wire has a resistance of 1.65 ohms, what length of wire is needed to provide a resistance of 19.8 ohms?

Resistance = 1.65 ohms |← 1 ft →| Resistance = 19.8 ohms |←——— L ———→|

First, recognize that this problem involves a *direct* proportion. As the length of wire increases, the resistance increases proportionally.

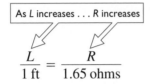

$$\frac{L}{1 \text{ ft}} = \frac{R}{1.65 \text{ ohms}}$$

Both ratios increase in size when L increases.

Second, set up a direct proportion and solve.

$$\frac{L}{1 \text{ ft}} = \frac{19.8 \text{ ohms}}{1.65 \text{ ohms}}$$

$$L = 12 \text{ ft}$$

→ **Your Turn**

Solve each of the following problems by setting up a direct proportion.

(a) **Manufacturing** If a widget machine produces 88 widgets in 2 hours, how many will it produce in $3\frac{1}{2}$ hours?

(b) **Painting** If 1 gal of paint covers 825 sq ft, how much paint is needed to cover 2640 sq ft?

(c) **Automotive Trades** What is the cost of six air filters if eight filters cost $55.92?

(d) **Transportation** A diesel truck was driven 273 miles on 42 gal of fuel. How much fuel is needed for a trip of 600 miles? Round to the nearest gallon.

(e) **Industrial Technology** A cylindrical oil tank holds 450 gal when it is filled to its full height of 8 ft. When it contains oil to a height of 2 ft 4 in., how many gallons of oil are in the tank?

→ **Solutions**

(a) A direct proportion—the more time spent, the more widgets produced:

$$\frac{88}{x} = \frac{2 \text{ hours}}{3\frac{1}{2} \text{ hours}}$$

Cross-multiply:

$2x = 308$
$x = 154$ widgets

 88 $\boxed{\times}$ **3.5** $\boxed{\div}$ **2** $\boxed{=}$ → 　　　　　 *154.*

(b) A direct proportion—the more paint, the greater the area that can be covered:

$$\frac{1 \text{ gal}}{x \text{ gal}} = \frac{825 \text{ sq ft}}{2640 \text{ sq ft}}$$

$x = 3.2$ gal

(c) A direct proportion—the less you get, the less you pay:

$$\frac{6 \text{ filters}}{8 \text{ filters}} = \frac{x}{\$55.92}$$

$x = \$41.94$

(d) A direct proportion—the more miles you drive, the more fuel it takes:

$$\frac{273 \text{ mi}}{600 \text{ mi}} = \frac{42 \text{ gal}}{x \text{ gal}}$$

$x \approx 92$ gallons (*Reminder:* $\approx$ means "approximately equal to")

(e) A direct proportion—the volume is directly proportional to the height:

$$\frac{450 \text{ gal}}{x \text{ gal}} = \frac{8 \text{ ft}}{2\frac{1}{3} \text{ ft}} \quad \Longleftarrow \boxed{2'\,4'' = 2\frac{4}{12}' = 2\frac{1}{3}'}$$

$x = 131\frac{1}{4}$ gallons

Two quantities are said to be *inversely proportional* if an increase in one quantity leads to a proportional decrease in the other quantity, or if a decrease in one leads to an increase in the other.

> **Inverse Proportion:**　increase → decrease
> 　　　　　　　or　decrease → increase

For example, the time required for a trip of a certain length is *inversely* proportional to the speed of travel. The faster you go (*increase* in speed), the quicker you get there (*decrease* in time).

Example 5

If a certain trip takes 2 hours at 50 mph, how long will it take at 60 mph?

The correct proportion equation is

$$\frac{50 \text{ mph}}{60 \text{ mph}} \qquad = \qquad \frac{x \text{ hours}}{2 \text{ hours}}$$

| Ratio of speeds | Increase in speed . . . leads to a . . . Decrease in time | Inverse ratio of times |

By inverting the time ratio, we have set it up so that both sides of the equation are in balance—both ratios decrease as speed increases.

Before attempting to solve the problem, make an estimate of the answer. We expect that the time to make the trip at 60 mph will be *less* than the time at 50 mph. The correct answer should be less than 2 hours.

Now solve it by cross-multiplying:

$$60x = 2 \cdot 50$$

$$x = 1\tfrac{2}{3} \text{ hours}$$

Learning Help Remember, in a *direct* proportion

These terms go together. If B and D are fixed, then an increase in A goes with an increase in C.

In an *inverse* proportion

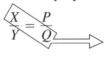

These terms go together. If P and Y are fixed, then an increase in X goes with a decrease in Q. ◄

→ Your Turn

Automotive Trades   In an automobile cylinder, the pressure is inversely proportional to the volume if the temperature does not change. If the volume of gas in the cylinder is 300 cu cm when the pressure is 20 psi, what is the volume when the pressure is increased to 80 psi?

Set this up as an inverse proportion and solve.

→ Solution

Pressure is inversely proportional to volume. If the pressure increases, we expect the volume to decrease. The answer should be less than 300 cu cm.

Set up an inverse proportion. $\qquad\qquad\qquad \dfrac{P_1}{P_2} = \dfrac{V_2}{V_1}$

Substitute the given information. $\qquad\qquad \dfrac{20 \text{ psi}}{80 \text{ psi}} = \dfrac{V}{300 \text{ cu cm}}$

Cross-multiply. $\qquad\qquad\qquad\qquad\quad 80V = 20 \cdot 300$

$$V = 75 \text{ cu cm}$$

Gears and Pulleys

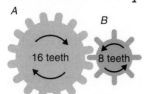

A particularly useful kind of inverse proportion involves the relationship between the size of a gear or pulley and the speed with which it rotates.

In the figure in the margin, A is the driving gear.

Because gear A has twice as many teeth as gear B, when A turns one turn, B will make two turns. If gear A turns at 10 turns per second, gear B will turn at 20 turns per second. The speed of the gear is inversely proportional to the number of teeth.

$$\frac{\text{speed of gear } A}{\text{speed of gear } B} = \frac{\text{teeth in gear } B}{\text{teeth in gear } A}$$

This ratio has the *A* term on top	This ratio has the *B* term on top

In this proportion, gear speed is measured in revolutions per minute, abbreviated rpm.

Example 6

For the gear assembly shown on the previous page, if gear *A* is turned by a shaft at 40 rpm, what will be the speed of gear *B*?

Because the relation is an inverse proportion, the smaller gear moves with the greater speed. We expect the speed of gear *B* to be faster than 40 rpm.

$$\frac{40 \text{ rpm}}{B} = \frac{8 \text{ teeth}}{16 \text{ teeth}}$$

$$8B = 640$$

$$B = 80 \text{ rpm}$$

On an automobile the speed of the drive shaft is converted to rear axle motion by the ring and pinion gear system.

$$\frac{\text{drive shaft speed}}{\text{rear axle speed}} = \frac{\text{teeth in ring gear on axle}}{\text{teeth in pinion gear on drive shaft}}$$

This ratio has the drive shaft term on top	This ratio has the drive shaft term on the bottom

Again, gear speed is inversely proportional to the number of teeth on the gear.

→ **Your Turn**

Automotive Trades   If the pinion gear has 9 teeth and the ring gear has 40 teeth, what is the rear axle speed when the drive shaft turns at 1200 rpm?

→ **Solution**

$$\frac{1200 \text{ rpm}}{R} = \frac{40 \text{ teeth}}{9 \text{ teeth}}$$

Cross-multiply: $40R = 9 \cdot 1200$

$$R = 270 \text{ rpm}$$

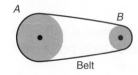

A   *B*

Belt

Pulleys transfer power in much the same way as gears. For the pulley system shown in the margin, the speed of a pulley is inversely proportional to its diameter.

If pulley A has a diameter twice that of pulley B, when pulley A makes one turn, pulley B will make two turns, assuming of course that there is no slippage of the belt.

$$\frac{\text{speed of pulley } A}{\text{speed of pulley } B} = \frac{\text{diameter of pulley } B}{\text{diameter of pulley } A}$$

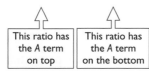

| This ratio has the A term on top | This ratio has the A term on the bottom |

→ **Your Turn**

If pulley B is 16 in. in diameter and is rotating at 240 rpm, what is the speed of pulley A if its diameter is 20 in.?

→ **Solution**

$$\frac{A}{240 \text{ rpm}} = \frac{16 \text{ in.}}{20 \text{ in.}}$$

Cross-multiply: $20A = 16 \cdot 240$

$$A = 192 \text{ rpm}$$

→ **More Practice**

Solve each of the following problems by setting up an inverse proportion.

(a) **Machine Trades**   A 9-in. pulley on a drill press rotates at 960 rpm. It is belted to a 5-in. pulley on an electric motor. Find the speed of the motor shaft.

(b) **Automotive Trades**   A 12-tooth gear mounted on a motor shaft drives a larger gear. The motor shaft rotates at 1450 rpm. If the speed of the large gear is to be 425 rpm, how many teeth must be on the large gear?

(c) **Physics**   For gases, pressure is inversely proportional to volume if the temperature does not change. If 30 cu ft of air at 15 psi is compressed to 6 cu ft, what is the new pressure?

(d) **Manufacturing**   If five assembly machines can complete a given job in 3 hours, how many hours will it take for two assembly machines to do the same job?

(e) **Physics**   The forces and lever arm distances for a lever obey an inverse proportion.

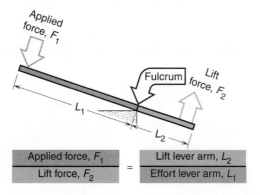

$$\frac{\text{Applied force, } F_1}{\text{Lift force, } F_2} = \frac{\text{Lift lever arm, } L_2}{\text{Effort lever arm, } L_1}$$

If a 100-lb force is applied to a 22-in. crowbar pivoted 2 in. from the end, what lift force is exerted?

(a) $\dfrac{9 \text{ in.}}{5 \text{ in.}} = \dfrac{x}{960 \text{ rpm}}$ An inverse proportion: the larger pulley turns more slowly.

$5x = 9 \cdot 960$

$x = 1728$

(b) $\dfrac{12 \text{ teeth}}{x \text{ teeth}} = \dfrac{425 \text{ rpm}}{1450 \text{ rpm}}$ An inverse proportion:
the larger gear turns more slowly.

$425x = 12 \cdot 1450$

$x = 40.941 \ldots$ or 41 teeth, rounding to the nearest whole number. We can't have a part of a gear tooth!

$x \approx 41$

(c) $\dfrac{30 \text{ cu ft}}{6 \text{ cu ft}} = \dfrac{P}{15 \text{ psi}}$ An inverse proportion: the higher the pressure, the smaller the volume.

$6P = 30 \cdot 15$

$P = 75 \text{ psi}$

(d) Careful on this one! An inverse proportion should be used. The *more* machines used, the *fewer* the hours needed to do the job.

$\dfrac{5 \text{ machines}}{2 \text{ machines}} = \dfrac{x \text{ hours}}{3 \text{ hours}}$

$2x = 15$

$x = 7\frac{1}{2} \text{ hours}$ Two machines will take much longer to do the job than will five machines.

(e) $\dfrac{100 \text{ lb}}{F} = \dfrac{2 \text{ in.}}{20 \text{ in.}}$ If the entire bar is 22 in. long, and $L_2 = 2$ in., then $L_1 = 20$ in.

$2F = 20 \cdot 100$

$F = 1000 \text{ lb}$

Now turn to Exercises 4-2 for a set of practice problems on these special applications of ratio and proportion.

Exercises 4-2   Special Applications of Ratio and Proportion

A. Complete the following tables.

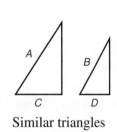

Similar triangles

1.

	A	B	C	D
(a)	$5\frac{1}{2}$ in.	$1\frac{1}{4}$ in.	$2\frac{3}{4}$ in.	
(b)		23.4 cm	20.8 cm	15.6 cm
(c)	12 ft		9 ft	6 ft
(d)	4.5 m	3.6 m		2.4 m

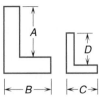

Similar figures

2.

	A	B	C	D
(a)	4 in.	5 in.	$3\frac{1}{2}$ in.	
(b)	5 ft		5 ft	2 ft
(c)	6.4 m	5.6 m		1.6 m
(d)		36 m	12 cm	14 cm

3.

	Number of Teeth on Gear 1	Number of Teeth on Gear 2	RPM of Gear 1	RPM of Gear 2
(a)	20	48	240	
(b)	25		150	420
(c)		40	160	100
(d)	32	40		1200

4.

	Diameter of Pulley 1	Diameter of Pulley 2	RPM of Pulley 1	RPM of Pulley 2
(a)	18 in.	24 in.	200	
(b)	12 in.		300	240
(c)		5 in.	400	640
(d)	14 in.	6 in.	300	

B. Practical Problems

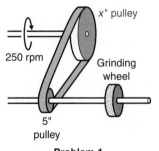

x" pulley

250 rpm

Grinding wheel

5" pulley

Problem 1

1. **Metalworking** A line shaft rotating at 250 rpm is connected to a grinding wheel by the pulley assembly shown. If the grinder shaft must turn at 1200 rpm, what size pulley should be attached to the line shaft?

2. **Architecture** The architectural drawing of an outside deck is $3\frac{1}{2}$ in. wide by $10\frac{7}{8}$ in. long. If the deck will actually be 14 ft wide, calculate the following:
   (a) The actual length of the deck
   (b) The scale factor

3. **Automotive Trades** Horsepower developed by an engine varies directly as its displacement. How many horsepower will be developed by an engine with a displacement of 240 cu in. if a 380-cu in. engine of the same kind develops 220 hp?

4. **Transportation** The distance necessary to stop a subway train at a given speed is inversely proportional to its deceleration. If a train traveling at 30 mph requires 180 ft to stop when decelerating at 0.18 g, what is the stopping distance at the same speed when it is decelerating at 0.15 g?

5. **Physics** A crowbar 28 in. long is pivoted 6 in. from the end. What force must be applied at the long end in order to lift a 400-lb object at the short end?

6. **Industrial Technology**   If 60 gal of oil flow through a certain pipe in 16 minutes, how long will it take to fill a 450-gal tank using this pipe?

7. **Automotive Trades**   If the alternator-to-engine drive ratio is 2.45 to 1, what rpm will the alternator have when the engine is idling at 400 rpm? (*Hint:* Use a direct proportion.)

8. **General Trades**   The length of a wrench is inversely proportional to the amount of force needed to loosen a bolt. A wrench 6 in. long requires a force of 240 lb to loosen a rusty bolt. How much force would be required to loosen the same bolt using a 10-in. wrench?

9. **Construction**   The Santa Barbara Planning Commission recently voted to restrict the size of home remodels by limiting the floor area to lot area ratio to a maximum of 0.45 to 1. Under these guidelines,
   (a)   What would be the maximum allowable size of a remodel on an 11,800-sq-ft lot?
   (b)   What size lot would be required in order to create a 3960-sq-ft remodel?

10. **Machine Trades**   A pair of belted pulleys have diameters of 20 in. and 16 in., respectively. If the larger pulley turns at 2000 rpm, how fast will the smaller pulley turn?

11. **Manufacturing**   A 15-tooth gear on a motor shaft drives a larger gear having 36 teeth. If the motor shaft rotates at 1200 rpm, what is the speed of the larger gear?

12. **Electronics**   The power gain of an amplifier circuit is defined as

$$\text{Power gain} = \frac{\text{output power}}{\text{input power}}$$

If the audio power amplifier circuit has an input power of 0.72 watt and a power gain of 30, what output power will be available at the speaker?

13. **Industrial Technology**   If 12 assemblers can complete a certain job in 4 hours, how long will the same job take if the number of assemblers is cut back to 8?

14. **Electronics**   A 115-volt power transformer has 320 turns on the primary. If it delivers a secondary voltage of 12 volts, how many turns are on the secondary? (*Hint:* Use a direct proportion.)

15. **Automotive Trades**   The headlights of a car are mounted at a height of 3.5 ft. If the light beam drops 1 in. per 35 ft, how far ahead in the road will the headlights illuminate?

16. **Transportation**   A truck driver covers a certain stretch of the interstate in $4\frac{1}{2}$ hours traveling at the posted speed limit of 55 mph. If the speed limit is raised to 65 mph, how much time will the same trip require?

17. **Machine Trades**   It is known that a cable with a cross-sectional area of 0.60 sq in. has a capacity to hold 2500 lb. If the capacity of the cable is proportional to its cross-sectional area, what size cable is needed to hold 4000 lb?

18. **Masonry**   Cement, sand, and gravel are mixed to a proportion of $1:3:6$ for a particular batch of concrete. How many cubic yards of each should be used to mix 125 cubic yards of concrete?

19. **Roofing**   A cylindrical vent 6 in. in diameter must be cut at an angle to fit on a gable roof with a $\frac{2}{3}$ pitch. This means that for the vent itself the ratio of rise to run will be $2:3$. Find the height x of the cut that must be made on the cylinder to make it fit the slope of the roof. (See the figure.)

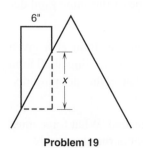

Problem 19

Chapter 4 Ratio, Proportion, and Percent

20. **Architecture** A wrought-iron gate for a new house will be 18 ft 9 in. long and 9 ft 6 in. high. An architect makes a drawing of the gate using a scale factor of $\frac{1}{4}$ in. = 1 ft. What will be the dimensions of the gate on the drawing?

21. **Automotive Trades** When a tire is inflated, the air pressure is inversely proportional to the volume of the air. If the pressure of a certain tire is 28 psi when the volume is 120 cu in., what is the pressure when the volume is 150 cu in.?

22. **Electrical Trades** The electrical resistance of a given length of wire is inversely proportional to the square of the diameter of the wire:

$$\frac{\text{resistance of wire } A}{\text{resistance of wire } B} = \frac{(\text{diameter of } B)^2}{(\text{diameter of } A)^2}$$

If a certain length of wire with a diameter of 34.852 mils has a resistance of 8.125 ohms, what is the resistance of the same length of the same composition wire with a diameter of 45.507 mils?

23. **Life Skills** If you are paid $238.74 for $21\frac{1}{2}$ hours of work, what amount should you be paid for 34 hours of work at this same rate of pay?

24. **Sheet Metal Trades** If the triangular plate shown is cut into eight pieces along equally spaced dashed lines, find the height of each cut. Round to two decimal digits.

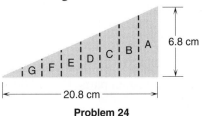

$A =$ _____ $B =$ _____

$C =$ _____ $D =$ _____

$E =$ _____ $F =$ _____

$G =$ _____

Problem 24

Problem 27

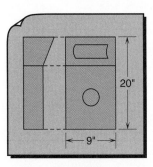

Problem 28

25. **Wastewater Technology** The time required for an outlet pipe to empty a tank is inversely proportional to the cross-sectional area of the pipe. A pipe with a cross-sectional area of 113.0 sq in. requires 6.4 hours to empty a certain tank. If the pipe was replaced with one with a cross-sectional area of 50.25 sq in., how long would it take this pipe to empty the same tank?

26. **Physics** A gas has a volume of 2480 cu cm at a pressure of 63.5 psi. What is the pressure when the gas is compressed to 1830 cu cm? (Round to the nearest tenth.)

27. **Machine Trades** The base of the flange shown in the drawing is actually 8 in. Its drawing width is 3 in. (a) What will be the actual dimension A if A is 2.25 in. on the drawing? (b) Find the scale factor of the drawing.

28. **Drafting** The drawing in the margin shows the actual length and width of a display panel. (a) What will be the corresponding drawing length of the panel if the drawing width is $2\frac{1}{4}$ in.? (b) Find the scale factor for this drawing.

29. **Plumbing** If a 45-gal hot water tank holds 375 lb of water, what weight of water will a 55-gal tank hold? (Round to the nearest pound.)

30. **Automotive Trades** When different-size tires are put on a vehicle, the speedometer gear must be changed so that the mechanical speedometer will continue to give the correct reading. The number of teeth on the gear is inversely proportional to the size of the tires. Suppose that a driver decides to replace 24-in. tires with $26\frac{1}{2}$-in. ones. If the old speedometer gear had 18 teeth,

how many teeth should the replacement gear have? (*Remember:* You cannot have a fraction of a tooth on a gear.)

When you have completed these exercises, check your answers to the odd-numbered problems in the Appendix, then turn to Section 4-3.

4-3 Introduction to Percent

In many practical calculations, it is helpful to be able to compare two numbers, and it is especially useful to express the comparison in terms of a percent. Percent calculations are a very important part of any work in business, technical skills, or the trades.

The word *percent* comes from the Latin phrase *per centum* meaning "by the hundred" or "for every hundred." A number written as a percent is being compared with a second number called the standard or *base*.

Example 1

What part of the base length is length *A*?

Base length	Length *A*

We could answer the question with a fraction or ratio, a decimal, or a percent. First, divide the base into 100 equal parts. Use a 100-mm metric rule to help visualize this.

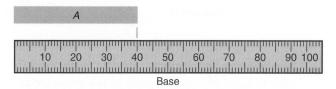

Then compare length *A* with it.

The length of *A* is 40 parts out of the 100 parts that make up the base.

A is $\dfrac{40 \text{ mm}}{100 \text{ mm}}$ or 0.40 or 40% of *A*.

$$\frac{40}{100} = \boxed{40}\% \qquad \boxed{40}\% \text{ means } \boxed{40} \text{ parts in 100 or } \frac{40}{100}$$

→ Your Turn

For the following diagram, what part of the base length is length *B*?

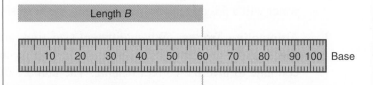

Answer with a percent.

B is $\dfrac{60}{100}$ or 60% of the base.

The compared number may be larger than the base. In such cases, the percent will be larger than 100%.

Example 2

What percent of the base length is length C?

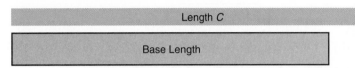

In this case, divide the base into 100 parts and extend it in length.

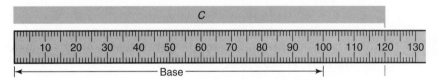

The length of C is 120 of the 100 parts that make up the base.

C is $\dfrac{120}{100}$ or 120 % of the base.

Ratios, decimals, and percents are all alternative ways to compare two numbers. For example, in the following drawing what fraction of the rectangle is shaded? What part of 12 is 3?

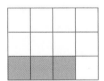

First, we can write the answer as a ratio:

$$\dfrac{3 \text{ shaded squares}}{12 \text{ squares}} = \dfrac{3}{12} = \dfrac{1}{4}$$

Second, by dividing 4 into 1 we can write this ratio or fraction as a decimal:

$$\dfrac{1}{4} = 0.25 \qquad 4\overline{)1.00}^{\,.25}$$

Finally, we can rewrite the fraction with a denominator of 100:

$$\dfrac{1}{4} = \dfrac{1 \times 25}{4 \times 25}$$

$$\dfrac{1}{4} = \dfrac{25}{100} \qquad \text{or} \qquad 25\%$$

We will now learn in more detail how to convert back and forth among these different comparisons.

Changing Decimal Numbers to Percents

Converting a number from fraction or decimal form to a percent is a very useful skill.

> To write a decimal number as a percent, multiply the decimal number by 100%.

Example 3

(a) $0.60 = 0.60 \times 100\% = 60\%$

Multiply by 100%

(b) $0.375 = 0.375 \times 100\% = 37.5\%$

(c) $3.4 = 3.4 \times 100\% = 340\%$

(d) $0.02 = 0.02 \times 100\% = 2\%$

> **Learning Help**
>
> Notice that multiplying by 100 is equivalent to shifting the decimal point two places to the right. For example,
>
> $0.60 = 0.60 = 60.\%$ or 60%
>
> Shift decimal point two places right
>
> $0.056 = 0.056 = 5.6\%$ ◄

> **→ Your Turn**
>
> Rewrite the following decimal numbers as percents.
>
> (a) 0.75   (b) 1.25   (c) 0.064
>
> (d) 3   (e) 0.05   (f) 0.004

> **→ Solutions**
>
> (a) $0.75 = 0.75 \times 100\% = 75\%$:   0.75 becomes 75%
>
> (b) $1.25 = 1.25 \times 100\% = 125\%$ 1.25 becomes 125%
>
> (c) $0.064 = 0.064 \times 100\% = 6.4\%$ 0.064 becomes 6.4%
>
> (d) $3 = 3 \times 100\% = 300\%$ $3 = 3.00$ or   300%
>
> (e) $0.05 = 0.05 \times 100\% = 5\%$ 0.05 becomes 5%
>
> (f) $0.004 = 0.004 \times 100\% = 0.4\%$ 0.004 becomes 0.4%

Changing Fractions to Percents

> To rewrite a fraction as a percent, first change to decimal form by dividing, then multiply by 100%.

Example 4

(a) $\frac{1}{2}$ is 1 divided by 2, or $2\overline{)1.0}^{\,0.5}$ so that

$$\frac{1}{2} = 0.5 = 0.5 \times 100\% = 50\%$$

⌐ Multiply by 100%

└ Change to a decimal

(b) $\frac{3}{4} = 0.75 = 0.75 \times 100\% = 75\%$

(c) $\frac{3}{20} = 0.15 = 0.15 \times 100\% = 15\%$ since $\frac{3}{20} = 20\overline{)3.00}^{\,0.15}$

(d) $1\frac{7}{20} = \frac{27}{20} = 1.35 = 1.35 \times 100\% = 135\%$ since $\frac{27}{20} = 20\overline{)27.00}^{\,1.35}$

→ **Your Turn**

Rewrite $\frac{5}{16}$ as a percent.

→ **Solution**

$\frac{5}{16}$ means $16\overline{)5.0000}^{\,0.3125}$ so that $\frac{5}{16} = 0.3125 = 31.25\%$

This is often written as $31\frac{1}{4}\%$.

Some fractions cannot be converted to an exact decimal. For example, $\frac{1}{3} = 0.3333\ldots$, where the 3s continue endlessly. We can round to get an approximate percent.

$\frac{1}{3} \approx 0.3333 \approx 33.33\%$ or $33\frac{1}{3}\%$

The fraction $\frac{1}{3}$ is roughly equal to 33.33% and exactly equal to $33\frac{1}{3}\%$.

Example 5

Landscaping   In landscaping or road construction, the slope of a hill or *grade* is often expressed as a percent, converted from the following fraction:

$$\text{Slope} = \frac{\text{vertical distance}}{\text{horizontal distance}}$$

The hillside in the figure on the next page has a slope of:

$$\frac{20 \text{ ft}}{70 \text{ ft}} = \frac{2}{7}$$

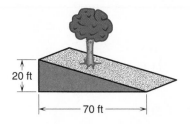

20 ft

70 ft

To express this as a percent,

$\frac{2}{7}$ is $7\overline{)2.0000}$ $\overset{0.2857\ldots}{} = 0.2857\ldots$ or $28.57\ldots\%$

Thus, the grade is approximately 29%.

Learning Help Here is an easy way to change certain fractions into percents. If the denominator of the fraction divides exactly into 100, change the given fraction into an equivalent fraction with a denominator of 100. The new numerator is the percent. For example,

$$\frac{3}{20} = \frac{3 \times 5}{20 \times 5} = \frac{15}{100} \qquad so \qquad \frac{3}{20} = 15\% \;\blacktriangleleft$$

→ More Practice

Rewrite the following fractions as percents.

(a) $\frac{4}{5}$ (b) $\frac{2}{3}$ (c) $3\frac{1}{8}$

(d) $\frac{5}{12}$ (e) $\frac{5}{6}$ (f) $1\frac{1}{3}$

(g) **Landscaping** A driveway 44 ft long has a vertical rise of 2.5 ft. Express the slope of the driveway as a percent to the nearest tenth.

→ Solutions

(a) $\frac{4}{5} = 0.80 = 80\%$

Shift the decimal point two places to the right

Change to a decimal by dividing

(b) $\frac{2}{3} \approx 0.6667 \approx 66.67\%$ or exactly $66\frac{2}{3}\%$

(c) $\frac{1}{8}$ is $8\overline{)1.000}$ $\overset{0.125}{}$ so $3\frac{1}{8} = 3.125$

$3\frac{1}{8} = 3.125 = 312.5\%$

(d) $\dfrac{5}{12}$ is $12\overline{)5.000}^{\,0.4166\ldots}$ so

$$\dfrac{5}{12} = 0.4166\ldots \approx 41.67\% \text{ or exactly } 41\tfrac{2}{3}\%$$

(e) $\dfrac{5}{6}$ is $6\overline{)5.000}^{\,0.833\ldots}$ so

$$\dfrac{5}{6} \approx 0.833 \approx 83.3\% \text{ or exactly } 83\tfrac{1}{3}\%$$

(f) $1\dfrac{1}{3} = 1.333\ldots \approx 133.3\% \text{ or exactly } 133\tfrac{1}{3}\%$

(g) Expressed as a fraction, the slope is

$$\dfrac{2.5 \text{ ft}}{44 \text{ ft}}$$

Converting to a percent, we have

$$44\overline{)2.50000}^{\,0.0568\ldots} = 0.0568\ldots \quad 5.7\%$$

Changing Percents to Decimal Numbers

To use percent numbers when you solve practical problems, it is often necessary to change a percent to a decimal number.

> To change a percent to a decimal number, divide by 100%.

Example 6

(a) $50\% = \dfrac{50\%}{100\%} = \dfrac{50}{100} = 0.5$

Divide by 100%

(b) $6\% = \dfrac{6\%}{100\%} = \dfrac{6}{100} = 0.06$

(c) $0.2\% = \dfrac{0.2\%}{100\%} = \dfrac{0.2}{100} = 0.002$

▶ **Learning Help** Notice that in each of these examples division by 100% is the same as moving the decimal point two digits to the left. For example,

$50\% = 50.\% = 0.50 \text{ or } 0.5$

Shift the decimal point two places left

$6\% = 06.\% = 0.06$

$0.2\% = 00.2\% = 0.002$ ◀

If a fraction is part of the percent number, write it as a decimal number before dividing by 100%.

Example 7

$$8\frac{1}{2}\% = 8.5\% = \frac{8.5\%}{100\%} = 0.085 \quad \text{or} \quad 8.5\% = 08.5\% = 0.085$$

→ Your Turn

Now try these. Write each percent as a decimal number.

(a) 4%   (b) 112%   (c) 0.5%

(d) $9\frac{1}{4}\%$ (e) 45%   (f) $12\frac{1}{3}\%$

→ Solutions

(a) $4\% = \dfrac{4\%}{100\%} = \dfrac{4}{100} = 0.04$ since $100\overline{)4.00}^{\,0.04}$

This may also be done by shifting the decimal point: $4\% = 04.\% = 0.04$.

(b) $112\% = 112.\% = 1.12$

(c) $0.5\% = 00.5\% = 0.005$

(d) $9\frac{1}{4}\% = 9.25\% = 09.25\% = 0.0925$

(e) $45\% = 45.\% = 0.45$

(f) $12\frac{1}{3}\% \approx 12.33\% \approx 0.1233$

A Closer Look   Did you notice that in problem (b) a percent greater than 100% gives a decimal number greater than 1?

$100\% = 1$

$200\% = 2$

$300\% = 3$ and so on ◄

Changing Percents to Fractions

It is often necessary to rewrite a percent number as a fraction in lowest terms.

> To write a percent number as a fraction, form a fraction with the percent number as numerator and 100 as the denominator. Then rewrite this fraction in lowest terms.

Example 8

(a) $40\% = \dfrac{40}{100} = \dfrac{40 \div 20}{100 \div 20} = \dfrac{2}{5}$

(b) $68\% = \dfrac{68}{100} = \dfrac{68 \div 4}{100 \div 4} = \dfrac{17}{25}$

(c) $125\% = \dfrac{125}{100} = \dfrac{125 \div 25}{100 \div 25} = \dfrac{5}{4}$ or $1\dfrac{1}{4}$

Write each percent as a fraction in lowest terms.

(a)   8%   (b)   65%   (c)   140%

(d)   73%   (e)   0.2%   (f) $16\frac{2}{3}\%$

→ Solutions

(a) $8\% = \dfrac{8}{100} = \dfrac{8 \div 4}{100 \div 4} = \dfrac{2}{25}$

(b) $65\% = \dfrac{65}{100} = \dfrac{65 \div 5}{100 \div 5} = \dfrac{13}{20}$

(c) $140\% = \dfrac{140}{100} = \dfrac{140 \div 20}{100 \div 20} = \dfrac{7}{5}$ or $1\frac{2}{5}$

(d) $73\% = \dfrac{73}{100}$ in lowest terms

(e) $0.2\% = \dfrac{0.2}{100} = \dfrac{0.2 \times 5}{100 \times 5} = \dfrac{1}{500}$

(f) $16\frac{2}{3}\% = \dfrac{16\frac{2}{3}}{100} = \dfrac{\frac{50}{3}}{100} = \dfrac{50}{3} \times \dfrac{1}{100} = \dfrac{1}{6}$

A Closer Look   Notice in (e) that we had to multiply numerator and denominator by 5 to avoid having a decimal number within a fraction. ◀

Here is a table of the most often used fractions with their percent equivalents.

Study the table, then go to Exercises 4-3 for a set of problems involving percent conversion.

Percent Equivalents

Percent	Decimal	Fraction	Percent	Decimal	Fraction
5%	0.05	$\frac{1}{20}$	50%	0.50	$\frac{1}{2}$
$6\frac{1}{4}\%$	0.0625	$\frac{1}{16}$	60%	0.60	$\frac{3}{5}$
$8\frac{1}{3}\%$	$0.08\overline{3}$	$\frac{1}{12}$	$62\frac{1}{2}\%$	0.625	$\frac{5}{8}$
10%	0.10	$\frac{1}{10}$	$66\frac{2}{3}\%$	$0.\overline{6}$	$\frac{2}{3}$
$12\frac{1}{2}\%$	0.125	$\frac{1}{8}$	70%	0.70	$\frac{7}{10}$
$16\frac{2}{3}\%$	$0.1\overline{6}$	$\frac{1}{6}$	75%	0.75	$\frac{3}{4}$
20%	0.20	$\frac{1}{5}$	80%	0.80	$\frac{4}{5}$
25%	0.25	$\frac{1}{4}$	$83\frac{1}{3}\%$	$0.8\overline{3}$	$\frac{5}{6}$
30%	0.30	$\frac{3}{10}$	$87\frac{1}{2}\%$	0.875	$\frac{7}{8}$
$33\frac{1}{3}\%$	$0.\overline{3}$	$\frac{1}{3}$	90%	0.90	$\frac{9}{10}$
$37\frac{1}{2}\%$	0.375	$\frac{3}{8}$	100%	1.00	$\frac{10}{10}$
40%	0.40	$\frac{2}{5}$			

A. Convert to a percent.

1. 0.32   2. 1   3. 0.5   4. 2.1

5. $\frac{1}{4}$ 6. 3.75   7. 40   8. 0.675

9. 2   10. 0.075   11. $\frac{1}{2}$ 12. $\frac{1}{6}$

13. 0.335   14. 0.001   15. 0.005   16. $\frac{3}{10}$

17. $\frac{3}{2}$ 18. $\frac{3}{40}$ 19. $3\frac{3}{10}$ 20. $\frac{1}{5}$

21. 0.40   22. 0.10   23. 0.95   24. 0.03

25. 0.3   26. 0.015   27. 0.60   28. 7.75

29. 1.2   30. 4   31. 6.04   32. 9

33. $\frac{5}{4}$ 34. $\frac{1}{5}$ 35. $\frac{7}{20}$ 36. $\frac{3}{8}$

37. $\frac{5}{6}$ 38. $2\frac{3}{8}$ 39. $3\frac{7}{10}$ 40. $1\frac{4}{5}$

B. Convert to a decimal number.

1. 6%   2. 45%   3. 1%   4. 33%

5. 71%   6. 456%   7. $\frac{1}{4}$%   8. 0.05%

9. $6\frac{1}{4}$%   10. $8\frac{3}{4}$%   11. 30%   12. 2.1%

13. 800%   14. 8%   15. 0.25%   16. $16\frac{1}{3}$%

17. 7%   18. 3%   19. 56%   20. 15%

21. 1000%   22. $7\frac{1}{2}$%   23. 90%   24. 0.3%

25. 150%   26. $1\frac{1}{2}$%   27. $6\frac{3}{4}$%   28. $\frac{1}{2}$%

29. $12\frac{1}{4}$%   30. $125\frac{1}{5}$%   31. 1.2%   32. 240%

C. Rewrite each percent as a fraction in lowest terms.

1. 5%   2. 20%   3. 250%   4. 16%

5. 53%   6. 0.1%   7. 92%   8. 2%

9. 45%   10. 175%   11. $8\frac{1}{3}$%   12. 37.5%

13. 24%   14. 119%   15. 0.05%   16. 15%

17. 480%   18. $83\frac{1}{3}$%   19. 62.5%   20. 1.2%

When you have had the practice you need, check your answers to the odd-numbered problems in the Appendix, then turn to Section 4-4.

In all your work with percent you will find that there are three basic types of problems. These three are related to all percent problems that arise in business, technology, or the trades. In this section we show you how to solve any percent problem, and we examine each of the three types of problems.

All percent problems involve three quantities:

B, the *base* or whole or total amount, a standard used for comparison

P, the *percentage* or part being compared with the base

R, the *rate* or *percent,* a percent number

> **Note** To avoid possible confusion between the words "percentage" and "percent," we will refer to the percentage P as the "part." ◄

For any percent problem, these three quantities are related by the proportion

$$\frac{P}{B} = \frac{R}{100}$$

For example, the proportion

$$\frac{3}{4} = \frac{75}{100}$$

can be translated to the percent statement

"*Three is 75% of four.*"

Here, the part P is 3, the base B is 4, and the percent R is 75.

To solve any percent problem, we need to identify which of the quantities given in the problem is P, which is B, and which is R. Then we can write the percent proportion and solve for the missing or unknown quantity.

When P Is Unknown

Consider these three problems:

What is 30% of 50?

Find 30% of 50.

30% of 50 is what number?

These three questions are all forms of the same problem. They are all asking you to find P, the part.

Example 1

Now let's solve these problems.

We know that 30 is the percent R because it has the % symbol attached to it. The number 50 is the base B. To solve, write the percent proportion, substituting 50 for B and 30 for R.

$$\frac{P}{50} = \frac{30}{100}$$

Now solve using cross-products and division.

$$100P = 30 \cdot 50$$

$$100P = 1500$$

$$P = \frac{1500}{100} = 15$$

15 is 30% of 50.

Substitute the answer back into the proportion.

$$\frac{15}{50} = \frac{30}{100} \qquad \text{or} \qquad 15 \cdot 100 = 30 \cdot 50 \qquad \text{which is correct}$$

The answer is reasonable: 30% is roughly one-third, and 15 is roughly one-third of 50.

Learning Help   The percent number R is easy to identify because it always has the % symbol attached to it. If you have trouble distinguishing the part P from the base B, notice that B is usually associated with the word "of" and P is usually associated with the word "is."

What is   30%   of 50?

"is" indicates the part **P**. It is unknown in this case.

"%" indicates the percent or rate **R**.

"of" indicates the base **B**.

Therefore, the proportion $\dfrac{P}{B} = \dfrac{R}{100}$ can be thought of as $\dfrac{\text{is}}{\text{of}} = \dfrac{\%}{100}$. ◄

→ Your Turn

Now try this problem to test yourself.

Find $8\frac{1}{2}\%$ of 160.

→ Solution

Step 1 If your mental math skills are good, estimate the answer first. In this case, $8\frac{1}{2}\%$ of 160 should be a little less than 10% of 160, or 16.

Step 2 Now identify the three quantities, P, B, and R. R is obviously $8\frac{1}{2}$. The word "is" does not appear, but "of" appears with 160, so 160 is B. Therefore, P must be the unknown quantity.

Step 3 Set up the percent proportion.

$$\frac{P}{B} = \frac{R}{100} \longrightarrow \frac{P}{160} = \frac{8\frac{1}{2}}{100}$$

Step 4 Solve.

$$100P = \left(8\frac{1}{2}\right)160$$

$$100P = 1360$$

$$P = 13.6$$

Step 5 Check your answer by substituting into the original proportion equation and comparing cross-products.

Notice that the calculated answer, 13.6, is a little less than 16, our rough estimate.

→ **More Practice**

Solve the following percent problems.

(a) Find 2% of 140 lb.

(b) 35% of $20 is equal to what amount?

(c) What is $7\frac{1}{4}$% of $1000?

(d) Calculate $16\frac{2}{3}$% of 66.

(e) **Carpentry** To account for waste, a carpenter needs to order 120% of the total wood used in making a cabinet. If 15 ft of white oak is actually used, how much should he order?

→ **Solutions**

(a) $\dfrac{P}{140} = \dfrac{2}{100}$

$100P = 2 \cdot 140 = 280$

$P = 2.8 \text{ lb}$

✓ $\dfrac{2.8}{140} = \dfrac{2}{100}$

$280 = 280$

(b) $\dfrac{P}{20} = \dfrac{35}{100}$

$100P = 700$

$P = \$7$

✓ $\dfrac{7}{20} = \dfrac{35}{100}$

$700 = 700$

(c) $\dfrac{P}{1000} = \dfrac{7\frac{1}{4}}{100}$

$100P = 1000\left(7\frac{1}{4}\right) = 7250$

$P = \$72.50$

✓ $\dfrac{72.50}{1000} = \dfrac{7\frac{1}{4}}{100}$

$7250 = 7250$

(d) $\dfrac{P}{66} = \dfrac{16\frac{2}{3}}{100}$ $\qquad 16\frac{2}{3} = \dfrac{50}{3}$

$100P = 66 \cdot \dfrac{50}{3} = \dfrac{\overset{22}{\cancel{66}}}{1} \cdot \dfrac{50}{\underset{1}{\cancel{3}}} = 22 \cdot 50$

$100P = 1100$

$P = 11 \qquad$ Check it.

16 Ⓐᵇ𝒸 2 Ⓐᵇ𝒸 3 ✕ 66 ÷ 100 = → ▮▮▮▮▮▮▮▮▮ *11.*

(e) The key phrases in this problem are

"120% of the total . . ." and "15 ft . . . is used."

We are finding 120% of 15 ft. $B = 15$ ft

$$\frac{P}{15} = \frac{120}{100}$$

$$100P = 1800$$

$$P = 18 \text{ ft}$$

> **Learning Help**   Notice that in all problems where P is to be found, we end by dividing by 100 in the last step. Recall that the quick way to divide by 100 is to move the decimal point two places to the left. In problem (c), for example, $100P = 7250$. The decimal point is after the zero in 7250. Move it two places to the left.
>
> $7250 \div 100 = 7250. = 72.50$ ◀

When R is Unknown   Consider the following problems:

5 is what percent of 8?

Find what percent 5 is of 8.

What percent of 8 is 5?

Once again, these statements represent three different ways of asking the same question. In each statement the percent R is unknown. We know this because neither of the other two numbers has a % symbol attached.

Example 2

To solve this kind of problem, first identify P and B. The word "of" is associated with the base B, and the word "is" is associated with the part P. In this case, $B = 8$ and $P = 5$.

Next, set up the percent proportion.

$$\frac{P}{B} = \frac{R}{100} \rightarrow \frac{5}{8} = \frac{R}{100}$$

Finally, solve for R.

$$8R = 5 \cdot 100 = 500$$

$$R = 62.5\%$$

> **Note**   When you solve for R, remember to include the % symbol with your answer. ◀

> → **Your Turn**
>
> Now try these problems for practice.
>
> (a)   What percent of 40 lb is 16 lb?
>
> (b)   65 is what percent of 25?
>
> (c)   Find what percent $9.90 is of $18.00.
>
> (d)   **Machine Trades**   During reshaping, 6 lb of metal is removed from a casting originally weighing 80 lb. What percent of the metal is removed?
>
> (e)   **Allied Health**   Twelve milliliters (mL) of pure acetic acid are mixed with 38 mL of water. Calculate the percent concentration of the resulting solution.

(a) $\dfrac{16}{40} = \dfrac{R}{100}$

$40R = 1600$

$R = 40\%$

(b) $\dfrac{65}{25} = \dfrac{R}{100}$

$25R = 6500$

$R = 260\%$

(c) $\dfrac{\$9.90}{\$18.00} = \dfrac{R}{100}$

$18R = 990$

$R = 55\%$

(d) $\dfrac{6}{80} = \dfrac{R}{100}$

$80R = 600$

$R = 7.5\%$

(e) In this problem, the *part* is the 12 mL of acid. The *base* is the total amount of solution, the sum of the amounts of acid and water.

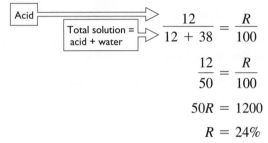

$$\dfrac{12}{12 + 38} = \dfrac{R}{100}$$

$$\dfrac{12}{50} = \dfrac{R}{100}$$

$$50R = 1200$$

$$R = 24\%$$

A Closer Look Notice that in problem (b) the percentage is larger than the base, resulting in a percent larger than 100%. Some students mistakenly believe that the percentage is always smaller than the base. This is not always true. When in doubt use the "is–of" method to determine P and B. ◄

Learning Help In all problems where R is unknown, we end up multiplying by 100. Recall that the quick way to do this is to move the decimal point two places to the right. If there is no decimal point, attach two zeros. ◄

When B is Unknown This third type of percent problem requires that you find the total or base when the percent and the percentage are given. Problems of this kind can be stated as

8.7 is 30% of what number?

30% of what number is 8.7?

Find a number such that 30% of it is 8.7.

Example 3

We can solve this problem in the same way as the previous ones.

First, identify P, B, and R. Because it carries the % symbol, we know that $R = 30$. Because the word "is" is associated with the number 8.7, we know that $P = 8.7$. The base B is unknown.

Next, set up the percent proportion.

$$\dfrac{P}{B} = \dfrac{R}{100} \rightarrow \dfrac{8.7}{B} = \dfrac{30}{100}$$

Finally, solve the proportion.

$$30B = 870$$

$$B = 29 \qquad ✓ \qquad \frac{8.7}{\boxed{29}} = \frac{30}{100}$$

$$8.7(100) = 30 \cdot 29 \quad \text{which is correct}$$

→ **Your Turn**

Ready for a few practice problems? Try these.

(a) 16% of what amount is equal to $5.76?

(b) $2 is 8% of the cost. Find the cost.

(c) Find a distance such that $12\frac{1}{2}\%$ of it is $26\frac{1}{4}$ ft.

(d) **Trades Management** The $2800 actually spent on a construction job was 125% of the original estimate. What was the original estimate?

(e) **Construction** A builder told his crew that the site preparation and foundation work on a house would represent about 5% of the total building time. If it took them 21 days to do this preliminary work, how long will the entire construction take?

→ **Solutions**

(a) $\dfrac{\$5.76}{B} = \dfrac{16}{100}$

$$16B = 576$$

$$B = \$36$$

(b) $\dfrac{2}{B} = \dfrac{8}{100}$

$$8B = 200$$

$$B = \$25$$

(c) $\dfrac{26\frac{1}{4}}{B} = \dfrac{12\frac{1}{2}}{100}$

$$B = \frac{26.25(100)}{12.5} = 210 \text{ ft}$$

(d) $\dfrac{2800}{B} = \dfrac{125}{100}$

$$B = \frac{2800 \cdot 100}{125} = \$2240$$

For problem (c): $\boxed{26.25}\ \boxed{\times}\ \boxed{100}\ \boxed{\div}\ \boxed{12.5}\ \boxed{=}\ \rightarrow\ \boxed{210.}$

(e) We are asked to find the total construction time, and we are told that preliminary work is 5% of the total. Because preliminary work takes 21 days, we can restate the problem as

21 is 5% of the total

or $\dfrac{21}{B} = \dfrac{5}{100}$

$$B = \frac{21 \cdot 100}{5} = 420 \text{ days}$$

So far we have solved sets of problems that were all of the same type. The key to solving percent problems is to be able to identify correctly the quantities *P*, *B*, and *R*.

→ **More Practice**

Use the hints you have learned to identify these three quantities in the following problems, then solve each problem.

(a) What percent of 25 is 30?

(b) 12 is $66\frac{2}{3}\%$ of what amount?

(c) Find 6% of 2400.

(d) **Life Skills** Television City asked you to make a 15% down payment on your purchase of a new LCD TV set. If the down payment is $210, what is the cost of the set?

(e) **Printing** A printer gives a 5% discount to preferred customers. If the normal cost of a certain job is $3400, what discount would a preferred customer receive?

(f) **Industrial Technology** Twelve computer chips were rejected out of a run of 400. What percent were rejected?

→ **Solutions**

(a) <u>What percent</u>   of <u>25</u>   is <u>30</u>? $\dfrac{30}{25} = \dfrac{R}{100}$

           R B P $25R = 3000$

                                         $R = 120\%$

(b) <u>12</u>   is   <u>$66\frac{2}{3}\%$</u>   of <u>what?</u> $\dfrac{12}{B} = \dfrac{66\frac{2}{3}}{100}$

    P R B $66\frac{2}{3} \cdot B = 1200$

                                      $B = 18$

 12 $\times$ **100** $\div$ **66** $\boxed{A\frac{b}{c}}$ **2** $\boxed{A\frac{b}{c}}$ **3** $=$ →   *18.*

(c) <u>Find</u>   <u>6%</u>   of <u>$2400.</u> $\dfrac{P}{2400} = \dfrac{6}{100}$

       P R B $100P = 6 \cdot 2400 = 14400$

                                        $P = \$144$

(d) In this problem, 15% is the rate, so $210 must be either the part or the base. Becasuse $210 is the down payment, a part of the whole price, $210 must be P.

$$\frac{210}{B} = \frac{15}{100}$$

$$15B = 21{,}000$$

$$B = \$1400 \qquad \text{The total cost of the TV set is } \$1400.$$

(e) 5% is the rate R, so $3400 must be either P or B. The discount is a part of the normal cost, so P is unknown and B is $3400.

$$\frac{P}{3400} = \frac{5}{100}$$

$$100P = 5 \cdot 3400 = 17000$$

$$P = \$170 \qquad \text{The preferred customer receives a discount of } \$170.$$

(f) The percent R is not given. The 12 chips are a part (P) of the whole (B) run of 400.

$$\frac{12}{400} = \frac{R}{100}$$

$$400R = 1200$$

$$R = 3\% \qquad 3\% \text{ of the chips were rejected.}$$

Now go to Exercises 4-4 for more practice on the three basic kinds of percent problems.

Exercises 4-4 Percent Problems

A. Solve.

1. 4 is _____ % of 5.
2. What percent of 25 is 16?
3. 20% of what number is 3?
4. 8 is what percent of 8?
5. 120% of 45 is _____.
6. 8 is _____ % of 64.
7. 3% of 5000 = _____.
8. 2.5% of what number is 2?
9. What percent of 54 is 36?
10. 60 is _____ % of 12.
11. 17 is 17% of _____.
12. 13 is what percent of 25?
13. $8\frac{1}{2}$% of $250 is _____.
14. 12 is _____ % of 2.
15. 6% of 25 is _____.
16. 60% of what number is 14?
17. What percent of 16 is 7?
18. 140 is _____ % of 105?
19. Find 24% of 10.
20. 45 is 12% of what number?
21. 30% of what number is equal to 12?
22. What is 65% of 5?
23. Find a number such that 15% of it is 750.
24. 16% of 110 is what number?

B. Solve.

1. 75 is $33\frac{1}{3}$% of _____.
2. What percent of 10 is 2.5?
3. 6% of $3.29 is _____.
4. 63 is _____ % of 35.
5. 12.5% of what number is 20?
6. $33\frac{1}{3}$% of $8.16 = _____.
7. 9.6 is what percent of 6.4?
8. $0.75 is _____ % of $37.50.
9. $6\frac{1}{4}$% of 280 is _____.
10. 1.28 is _____ % of 0.32.
11. 42.7 is 10% of _____.
12. 260% of 8.5 is _____.
13. $\frac{1}{2}$ is _____ % of 25.
14. $7\frac{1}{4}$% of 50 is _____.
15. 287.5% of 160 is _____.
16. 0.5% of _____ is 7.
17. 112% of _____ is 56.
18. $2\frac{1}{4}$% of 110 is _____.

C. Word Problems

1. **General Interest**   If you answered 37 problems correctly on a 42-question test, what percent of the problems did you answer correctly?

2. **Trades Management**   The profits from Ed's Plumbing Co. increased by $14,460, or 30%, this year. What profit did it earn last year?

3. **Machine Trades**   A casting weighed 146 lb out of the mold. It weighed 138 lb after finishing. What percent of the weight was lost in finishing?

4. **Masonry**   On the basis of past experience, a contractor expects to find 4.5% broken bricks in every truck load. If he orders 2000 bricks, will he have enough to complete a job requiring 1900 bricks?

5. **Automotive Trades**   The weight distribution of a vehicle specifies that the front wheels carry 54% of the weight. If a certain vehicle weighs 3860 lb, what weight is carried by the front wheels? Round to the nearest pound.

6. **Life Skills**   If state sales tax is 6%, what is the price before tax if the tax is $7.80?

7. **General Trades**   A worker earns $28.50 per hour. If she receives a $7\frac{1}{2}$% pay raise, what is the amount of her hourly raise?

8. **General Trades**   If 7.65% of your salary is withheld for Social Security and Medicare, what amount is withheld from monthly earnings of $1845.00?

9. **Printing**   If an 8-in. by 10-in. photocopy is enlarged to 120% of its size, what are its new dimensions?

10. **Police Science**   Year-end statistics for a community revealed that in 648 of the 965 residential burglaries, the burglar entered through an unlocked door or window. What percent of the entries were made in this way?

11. **Carpentry**   Preparing an estimate for a decking job, a carpenter assumes about a 15% waste factor. If the finished deck contains 1230 linear feet of boards, how many linear feet should the carpenter order? (*Hint:* If the waste is 15%, then the finished deck is 100% − 15% or 85% of the total order.)

12. **Printing**   Jiffy Print needs to produce 4000 error-free copies of a brochure. For this type of run there is usually a 2% spoilage rate. What is the minimum number of brochures that should be printed to ensure 4000 good copies?

13. **Office Services**   The Bide-A-While Hotel manager is evaluated on the basis of the occupancy rate of the hotel, that is, the average daily percent of the available rooms that are actually occupied. During one particular week daily occupancy totals were 117, 122, 105, 96, 84, 107, and 114. If the hotel has 140 rooms available each night, what was the occupancy rate for that week?

14. **Culinary Arts**   The manager of the Broccoli Palace restaurant needs 25 lb of fresh broccoli for an evening's business. Because of damage and spoilage, he expects to use only about 80% of the broccoli delivered by his supplier. What is the minimum amount he should order?

15. **Construction**   The BiltWell Construction Company has submitted a bid on construction of a spa deck. If their $8320 bid includes $2670 for lumber, what percent of the bid was the lumber? Round to the nearest percent.

16. **Flooring and Carpeting**   White oak flooring boards 3 in. wide are used to construct a hallway. If the flooring is laid at an angle, approximately 7200 linear feet of boards are needed for the job. How many linear feet should be ordered if an allowance of 18% is made for waste? Round to the nearest foot. (*Hint:* If there is 18% waste, then the amount needed is 100% − 18% = 82% of the amount that should be ordered.)

17. **Construction**   Quality Homes Construction Company orders only lumber that is at least 80% *clear*—free of defects, including knots and edge defects. In an order containing 65,500 linear feet, what minimum amount of lumber will be free of defects?

18. **Electrical Trades**   A electrician needs fuses for a motor being installed. If the motor current is 50 amperes and the fuse must be rated at 175% of the motor current, what size fuse will the electrician need?

19. **Allied Health**   A typical 35-year-old man has an age-adjusted maximum heart rate of 185 bpm (beats per minute). To maintain top conditioning, an athlete should exercise so that his or her heart rate is between 80% and 90% of his or her maximum heart rate for at least 30 minutes a day, 3 days a week. Calculate the range of pulse rate (bpm) for this level of exercise.

20. **Trades Management**   A masonry contractor estimates that a certain job will require eight full-time (8-hour) working days. He plans to use a crew consisting of three laborers earning $12 per hour each and three masons earning $20 per hour each. He estimates that materials will cost $3360. If he wishes to make a profit equal to 25% of his total cost for labor and materials, how much should his bid price be?

21. **Automotive Trades**   The volumetric efficiency of a certain engine is 88.2%. This means that the actual airflow is 88.2% of the theoretic airflow. If the actual airflow is 268.4 cfm (cubic feet per minute), what is the theoretic airflow of this engine? Round to the nearest 0.1 cfm.

22. **Allied Health**   If 5 oz of alcohol are mixed with 12 oz of water, what is the percent concentration of alcohol in the resulting solution? Round to the nearest percent. [*Hint:* See part (e) of "Your Turn" on page 218.]

23. **Sports and Leisure**   During the 2008 Wimbledon Final, Roger Federer got 128 out of 195 first serves in, and his opponent Rafael Nadal got 159 out of 218 first serves in. Which player had a higher percent of first serves in? (Round to the nearest percent.)

24. **Electronics**   A technician setting up a wireless network for an office is performing a survey to evaluate signal coverage. The technician finds that, after passing through a steel door in a brick wall, the signal power from a 62-mW wireless router is only 4 mW (milliwatts). What percent of the signal passes through the door? (Round to the nearest tenth of a percent.)

25. **Life Skills**   When sales of a particular hybrid automobile reach 60,000 units federal income tax credits begin to phase out. One year after reaching this threshold, buyers are still eligible for 25% of the original $3000 credit. What tax credit would buyers receive if they purchased the vehicle at this time?

26. **Culinary Arts**   Wolfgang runs a small lunchtime cafeteria that serves about 120 customers each day, or a total of 600 meals in the normal working week. The cafeteria's weekly income is about $3680, 25% of which is set aside for taxes. An additional $1200 per week is paid for supplies, rent, and for the salary of Wolfgang's assistant. On average, what is the cafeteria's remaining profit for each meal served?

27. **Landscaping**   An arborist is preparing the fuel mix for her chainsaw before climbing a tree to remove a dead limb. The two-stroke chainsaw requires a 2% mix of engine oil to gas—that is, the amount of oil should be 2% of the amount of gas. How much engine oil should be added to a jerry can that contains 2.5 gal of gas?

28. **Automotive Trades**   An independent mechanic purchases parts from her distributor at a 20% discount from the retail price. She then makes part of her profit on the repair job by charging her customer the retail price. If the 20% discounted price for a set of brake pads was $55.20, what is the retail price that the mechanic will charge her customer?

29. **Allied Health** A pharmaceutical sales representative makes a base salary of $66,200 per year. In addition, she can earn an annual bonus of up to $22,500 if she meets her prescription targets for the two medications she sells. She receives 70% of the bonus if she meets her target for Medication A, and she receives 30% of the bonus if she meets her target for Medication B. If she only meets her target for Medication A, what would her total compensation (salary plus bonus) be for the year?

30. **Allied Health** A pharmaceutical sales representative can earn a maximum bonus of $11,250 for meeting 100% of his annual prescription target of 4000 scrips for a certain medication. However, he can earn a portion of that bonus by meeting more than 80% of his target according to the following scale:

Percent of Target Reached	Percent of Bonus Earned
81%	5%
82%	10%
83%	15%
as this column increases by 1% . . .	this column increases by 5% . . .
98%	90%
99%	95%
100%	100%

What would the dollar amount of his bonus be for generating

(a)   3480 scrips?   (b)   3840 scrips?

When you have had the practice you need, check your answers to the odd-numbered problems in the Appendix, then continue with the study of some special applications of percent to practical problems.

4-5   Special Applications of Percent Calculations

Now that you have seen and solved the three basic percent problems, you need some help in applying what you have learned to practical situations.

Discount   An important kind of percent problem, especially in business, involves the idea of *discount*. To stimulate sales, a merchant may offer to sell some item at less than its normal price. A discount is the amount of money by which the normal price is reduced. It is a percentage or part of the normal price. The *list price* is the normal or original price before the discount is subtracted. The *net price* is the new, reduced price. The net price is sometimes called the *discounted* price or the *sale* price. It is always less than the list price.

Net price = list price − discount

The discount rate (R) is a percent number that enables you to calculate the discount (the percentage or part P) as a part of the list price (the base B). Therefore, the percent proportion

$$\frac{\text{part}}{\text{base}} = \frac{\text{rate}}{100}$$

becomes

$$\frac{\text{discount } D}{\text{list price } L} = \frac{\text{discount rate } R}{100}$$

Example 1

General Trades   The list price of a tool is $18.50. On a special sale it is offered at 20% off. What is the net price?

Step 1   Write the proportion formula.

$$\frac{\text{discount}}{\text{list price}} = \frac{\text{discount rate}}{100}$$

Step 2   Substitute the given numbers.

$$\frac{D}{18.50} = \frac{20}{100}$$

Step 3   Solve for D.

$$100D = 18.50 \times 20 = 370$$
$$D = \$3.70$$

Step 4   Calculate the net price.

$$\text{Net price} = \text{list price} - \text{discount}$$
$$= \$18.50 - \$3.70$$
$$= \$14.80$$

Think of it this way:

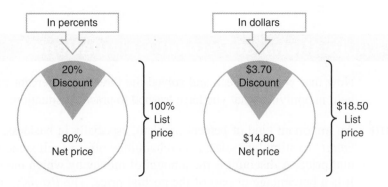

Ready for another problem?

→ Your Turn

Carpentry   After a 25% discount, the net price of a 5-in. random orbit sander is $54. What was its list price?

We can use a pie chart diagram to help set up the correct percent proportion.

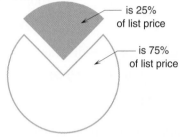

The net price, $54, is 75% of the list price. Therefore, the correct proportion is

$$\frac{\text{net price}}{\text{list price}} = \frac{\text{net price rate}}{100}$$

or $\quad \dfrac{54}{L} = \dfrac{75}{100}$

Solving for L, we have

$75L = 5400$

$L = \$72$ The list price is $72.

 $\dfrac{\text{discount}}{\text{list price}} = \dfrac{\text{discount rate}}{100}$

$\dfrac{D}{72} = \dfrac{25}{100}$

$100D = 1800$

$D = \$18$ The discount is $18.

Net price = list price − discount

        = \$72 − \$18

        = \$54   which is the correct list price as given in the original problem.

→ **More Practice**

Here are a few problems to test your understanding of the idea of discount.

(a) **Carpentry** A preinventory sale advertises all tools 20% off. What would be the net price of a rotary hammer drill that cost $199.95 before the sale?

(b) **Painting** An airless sprayer is on sale for $1011.12 and is advertised as "12% off regular price." What was its regular price?

(c) **Office Services** A set of four 205/65/15 automobile tires is on sale for 15% off list price. What is the sale price if their list price is $79.50 each?

(d) **Plumbing** A kitchen sink that retails for $625 is offered to a plumbing contractor for $406.25. What discount rate is the contractor receiving?

→ **Solutions**

(a) $\dfrac{D}{L} = \dfrac{R}{100}$ becomes $\dfrac{D}{199.95} = \dfrac{20}{100}$

$$100D = (199.95)20 = 3999$$

$$D = \$39.99$$

Net price = list price − discount

$$= \$199.95 - \$39.99$$

$$= \$159.96$$

A pie chart provides a quicker way to do this problem.

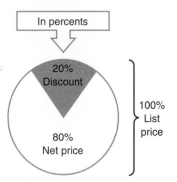

If the discount is 20% of the list price, then the net price is 80% of the list price.

$$\frac{\text{net price}}{\text{list price}} = \frac{80\%}{100\%}$$

or $\dfrac{N}{199.95} = \dfrac{80}{100}$

$$100N = 15996$$

$$N = \$159.96$$

(b) Pie charts will help you set up the correct proportions.

Use the net price proportion, substitute, and solve.

$$\frac{\text{net price}}{\text{list price}} = \frac{\text{net price rate}}{100}$$

$$\frac{1011.12}{L} = \frac{88}{100}$$

$$88L = 101112$$

$$L = \$1149.00$$

(c) Use the discount proportion.

$$\frac{D}{L} = \frac{R}{100} \qquad \text{becomes} \qquad \frac{D}{4(79.50)} = \frac{15}{100}$$

Four tires at $79.50 each $\longrightarrow$

$$\frac{D}{318} = \frac{15}{100}$$

$$100D = 4770$$

$$D = \$47.70$$

Net (sale) price = list price − discount

$$= \$318 - \$47.70$$

$$= \$270.30$$

Using a calculator, first find the discount.

4 $\times$ 79.50 $\times$ 15 $\div$ 100 $=$ $\rightarrow$ [47.7]

Then subtract it from the list price, $318, using the "last answer" key.

318 $-$ [ANS] $=$ $\rightarrow$ [270.3]

(d) **First,** subtract to find the discount.

Discount = list price − net price

$$= \$625 - \$406.25$$

$$= \$218.75$$

Then, set up the discount proportion.

$$\frac{D}{L} = \frac{R}{100} \qquad \text{becomes} \qquad \frac{218.75}{625} = \frac{R}{100}$$

$$625R = 21875$$

$$R = 35\% \qquad \text{The contractor receives a 35\% discount.}$$

A Quick Method for Calculating Percentage Using the proportion method for solving percent problems makes it easier to distinguish among the different types of problems. However, many practical applications involve finding the percentage, and there is a quicker way to do this, especially using a calculator. Simply change the percent to a decimal number and multiply by the base.

Example 2

Find 20% of $640.

Step 1 Change the percent to a decimal number.

$$20\% = 20.\% = 0.20 = 0.2$$

Step 2 Multiply this decimal equivalent by the base.

$$0.2 \times \$640 = \$128$$

This is the same answer you would get using a proportion.

$$\frac{P}{640} = \frac{20}{100} \qquad \text{or} \qquad P = \frac{640 \cdot 20}{100} = 640(0.2)$$

Use this quick method to solve the following problems.

(a) Find 35% of 80.

(b) What is 4.5% of 650?

(c) What is 125% of 18?

(d) Find 0.3% of 5000.

→ **Solutions**

(a) $35\% = 35.\% = 0.35$
     $0.35 \times 80 = 28$

(b) $4.5\% = 04.5\% = 0.045$
     $0.045 \times 650 = 29.25$

(c) $125\% = 125.\% = 1.25$
     $1.25 \times 18 = 22.5$

(d) $0.3\% = 00.3\% = 0.003$
     $0.003 \times 5000 = 15$

Sales Tax *Taxes* are almost always calculated as a percent of some total amount. *Property taxes* are written as some fraction of the value of the property involved. *Income taxes* are most often calculated from complex formulas that depend on many factors. We cannot consider either income or property taxes here.

A *sales tax* is an amount calculated from the actual list price of a purchase. It is then added to the list price to determine the total cost. Retail sales tax rates are set by the individual states in the United States and vary from 0 to 8.25% of the sales price. A sales tax of 6% is often stated as "6 cents on the dollar," because 6% of $1.00 equals 6 cents.

Using a proportion equation, we can write the formulas for sales tax and total cost as follows:

$$\frac{\text{sales tax}}{\text{list price}} = \frac{\text{tax rate}}{100}$$

Total cost = list price + sales tax

Example 3

Life Skills If the retail sales tax rate is 7.75% in a certain city, how much would you pay for a pair of shoes with a list price of $76.50?

Using a proportion equation, we have

$$\frac{\text{sales tax}}{\text{list price}} = \frac{\text{tax rate}}{100}$$

or $\dfrac{\text{sales tax } T}{\$76.50} = \dfrac{7.75}{100}$

$100T = (7.75)(76.50)$

$100T = 592.875$

$T = \$5.93$ rounded

Using the equivalent quick method, we have

Sales tax = 7.75% of $76.50 $7.75\% = 07.75 = 0.0775$
= (0.0775)($76.50)
= $5.92875 = $5.93 rounded

Total cost = list price + sales tax

$$= \$76.50 + \$5.93$$

$$= \$82.43$$

.0775 $\times$ 76.5 $=$ → ▨▨▨▨ 5.92875 $+$ 76.5 $=$ → ▨▨▨▨ 82.42875

Many stores provide their salesclerks with computerized cash registers that calculate sales tax automatically. However, all consumers and most small business or shop owners still need to be able to do the sales tax calculations.

→ **Your Turn**

Here are a few problems in calculating sales tax.

General Trades   Find the tax and total cost for each of the following if the tax rate is as shown.

(a)   A roll of plastic electrical tape at $2.79   (5%).

(b)   A random orbit sander priced at $67.50   (6%).

(c)   A tape measure priced at $21.99 $\left(4\frac{1}{2}\%\right)$.

(d)   A new car priced at $23,550   (4%).

(e)   A band saw priced at $346.50 $\left(6\frac{1}{2}\%\right)$.

(f)   A toggle switch priced at $10.49 $\left(7\frac{3}{4}\%\right)$.

→ **Answers**

	Tax	Total Cost	Quick Calculation
(a)	14 cents	$2.93	$0.05 \times \$2.79 = 13.95$ cents ≈ 14 cents
(b)	$4.05	$71.55	$0.06 \times \$67.50 = \$4.05$
(c)	$0.99	$22.98	$0.045 \times \$21.99 = \$0.98955 \approx \$0.99$
(d)	$942.00	$24,492	$0.04 \times \$23,550 = \$942.00$
(e)	$22.52	$369.02	$0.065 \times \$346.50 = \$22.5225 \approx \$22.52$
(f)	$0.81	$11.30	$0.0775 \times \$10.49 = \$0.812975 \approx \$0.81$

Interest   In modern society we have set up complex ways to enable you to use someone else's money. A *lender,* with money beyond his or her needs, supplies cash to a *borrower,* whose needs exceed his money. The money is called a *loan.*

Interest is the amount the lender is paid for the use of his money. Interest is the money you pay to use someone else's money. The more you use and the longer you use it, the more interest you must pay. *Principal* is the amount of money lent or borrowed.

When you purchase a house with a bank loan, a car or a refrigerator on an installment loan, or gasoline on a credit card, you are using someone else's money, and you pay interest for that use. If you are on the other end of the money game, you may earn interest for money you invest in a savings account or in shares of a business.

Simple interest* can be calculated from the following formulas.

$$\frac{\text{annual interest}}{\text{principal}} = \frac{\text{annual interest rate}}{100}$$

$$\text{Total interest} = \text{annual interest} \times \text{time in years}$$

Example 4

Life Skills Suppose that you have $5000 in a 12-month CD (certificate of deposit) earning 4.25% interest annually in your local bank. How much interest do you receive in a year?

$$\frac{\text{annual interest } I}{\$5000} = \frac{4.25}{100}$$

$$100I = 21250$$

$$I = \$212.50$$

The total interest for one year is $212.50.

Using the quick calculation method to calculate 4.25% of $5000,

$$\text{Annual interest } I = \$5000(4.25\%)$$

$$= 5000(0.0425)$$

$$= \$212.50$$

By placing your $5000 in a bank CD, you allow the bank to use it in various profitable ways, and you are paid $212.50 per year for its use.

→ Your Turn

Trades Management Most of us play the money game from the other side of the counter. Suppose that your small business is in need of cash and you arrange to obtain a loan from a bank. You borrow $12,000 at 9% per year for 3 months. How much interest must you pay?

Try to set up and solve this problem exactly as we did in the previous problem.

→ Solution

$$\frac{\text{annual interest } I}{\$12,000} = \frac{9}{100}$$

$$I = \frac{12,000 \times 9}{100} = \$1080$$

The time of the loan is 3 months, so the time in years is

$$\frac{3 \text{ months}}{12 \text{ months}} = \frac{3}{12} \text{ yr}$$

$$= 0.25 \text{ yr}$$

---

*In many applications, compound interest is used instead of simple interest. The formula for compound interest is beyond the scope of this chapter.

Total interest = annual interest × time in years

$$= \$1080(0.25)$$

$$= \$270$$

By the quick calculation method,

$$\text{Total interest} = \$12{,}000(9\%)\left(\frac{3}{12}\,\text{yr}\right)$$

$$= 12{,}000(0.09)(0.25)$$

$$= \$270$$

 12000 ⊗ **.09** ⊗ **3** ⊘ **12** ⊜ →  *270.*

Depending on how you and the bank decide to arrange it, you may be required to pay the total principal ($12,000) plus interest ($270) all at once, at the end of three months, or according to some sort of regular payment plan—for example, pay $4090 each month for three months.

HOW DOES A CREDIT CARD WORK?

Many forms of small loans, such as credit card loans, charge interest by the month. These are known as "revolving credit" or "charge account" plans. (If you borrow very much money this way, *you* do the revolving and may run in circles for years trying to pay it back!) Essentially, you buy now and pay later. Generally, if you re-pay the full amount borrowed within 25 or 30 days, there is no charge for the loan.

After the first pay period of 25 or 30 days, you pay a percent of the unpaid balance each month, usually between $\frac{1}{2}$ and 2%. In addition, you must pay a yearly fee and some minimum amount each month, usually $10 or 10% of the unpaid balance, whichever is larger. Often, you are also charged a small monthly amount for insurance premiums. (The credit card company insures itself against your defaulting on the loan or disappearing, and you pay for their insurance.)

Let's see how it works. Suppose that you go on a short vacation and pay for gasoline, lodging, and meals with your Handy Dandy credit card. A few weeks later you receive a bill for $1000. You can pay it within 30 days and owe no interest or you can pay over a longer period of time. Suppose you agree to pay $100 per month. The calculations look like this:

Month 1 $1\frac{1}{2}\%$ of $1000 = $1000 × 0.015 = $15.00   owe $1015.00
pay $100 and carry $915.00 over to next month

Month 2 $1\frac{1}{2}\%$ of $915 = $13.73   owe $928.73
pay $100 and carry $828.73 over to next month

. . . and so on.

Eleven months later you will have repaid the $1000 loan and all interest.

The $1\frac{1}{2}\%$ per month interest rate seems small, but it is equivalent to between 15 and 18%. You pay at a high rate for the convenience of using the credit card and the no-questions-asked ease of getting the loan.

Commission The simplest practical use of percent is in the calculation of a part or percentage of some total. For example, salespersons are often paid on the basis of their success at selling and receive a *commission* or share of the sales receipts. Commission is usually described as a percent of sales income.

$$\frac{\text{commission}}{\text{sales}} = \frac{\text{rate}}{100}$$

Example 5

Life Skills Suppose that your job as a lumber broker pays 12% commission on all sales. How much do you earn from a sale of $4000?

$$\frac{\text{commission } C}{4000} = \frac{12}{100}$$

$$C = \frac{12 \cdot 4000}{100}$$

$$C = \$480 \qquad \text{You earn \$480 commission.}$$

By the quick calculation method,

Commission = ($4000)(12%) = ($4000)(0.12) = $480

→ Your Turn

Try this one yourself.

Office Services At the Happy Bandit Used Car Company each salesperson receives a 6% commission on sales. What would a salesperson earn if she sold a 1960 Airedale for $1299.95?

→ Solution

$$\frac{C}{\$1299.95} = \frac{6}{100}$$

$$C = \frac{6(1299.95)}{100}$$

$$C = \$77.9970 \qquad \text{or} \qquad \$78 \text{ rounded}$$

or

Commission = ($1299.95)(6%) = ($1299.95)(0.06)
$$\approx \$78$$

Some situations require that you determine the total sales or the rate of commission.

→ More Practice

In the following two problems, concentrate on identifying each quantity in the commission proportion.

(a) **Life Skills** A sales representative generated $360,000 in sales last year and was paid a salary of $27,000. If she were to switch to straight commission compensation, what rate of commission would be needed for her commission to match her salary?

(b) **Life Skills** A computer salesperson is paid a salary of $800 per month plus a 4% commission on sales. What amount of sales must he generate in order to earn $40,000 per year in total compensation?

→ **Solutions**

(a) Sales = $360,000   Commission = $27,000   Rate is unknown.

$$\frac{\$27,000}{\$360,000} = \frac{R}{100}$$

$$R = \frac{27,000 \cdot 100}{360,000}$$

$$R = 7.5\%$$ She needs to be paid a 7.5% commission in order to match her salary.

(b) Salary = $800 per month × 12 months
= $9600 per year

Commission = total compensation − salary
= $40,000 − $9600 = $30,400

The commission needed is $30,400. Write the commission proportion. The sales, S, is not known.

$$\frac{\$30,400}{S} = \frac{4}{100}$$

$$S = \frac{30,400 \cdot 100}{4}$$

$$S = \$760,000$$

He must sell $760,000 worth of computers to earn a total of $40,000 per year.

 40000 ⊖ 800 ⊗ 12 ⊜ → **30400.** ⊗ 100 ⊘ 4 ⊜ → **760000.**

SMALL LOANS

Sooner or later everyone finds it necessary to borrow money. When you do, you will want to know beforehand how the loan process works. Suppose that you borrow $200 and the loan company specifies that you repay it at $25 per month plus interest at 3% per month on the unpaid balance. What interest do you actually pay?

Month 1 3% of $200 = $ 6.00	you pay	$25 + $6.00 = $ 31.00
Month 2 3% of $175 = $ 5.25	you pay	$25 + $5.25 = $ 30.25
Month 3 3% of $150 = $ 4.50	you pay	$25 + $4.50 = $ 29.50
Month 4 3% of $125 = $ 3.75	you pay	$25 + $3.75 = $ 28.75
Month 5 3% of $100 = $ 3.00	you pay	$25 + $3.00 = $ 28.00
Month 6 3% of $ 75 = $ 2.25	you pay	$25 + $2.25 = $ 27.25

(continued)

Month 7	3% of $ 50 = $ 1.50	you pay	$25 + $1.50 = $ 26.50
Month 8	3% of $ 25 = $ 0.75	you pay	$25 + $0.75 = $ 25.75
	$ 27.00		$227.00

Total interest is $27.00

Total of eight loan payments

They might also set it up as eight equal payments of $227.00 ÷ 8 = $28.375 or $28.38 per month.

The 3% monthly interest rate seems small, but it amounts to about 20% per year.

A bank loan for $200 at 12% for 8 months would cost you

$$(12\% \text{ of } \$200)\left(\frac{8}{12}\right)$$

or

$$(\$24)\left(\frac{8}{12}\right)$$

or $16.00   Quite a difference for a $200 loan.

The loan company demands that you pay more, and in return they are less worried about your ability to meet the payments. For a bigger risk, they want a higher rate of interest.

Efficiency

When energy is converted from one form to another in any machine or conversion process, it is useful to talk about the *efficiency* of the machine or the process. The efficiency of a process is a fraction comparing the energy or power output to the energy or power input. It is usually expressed as a percent.

$$\frac{\text{output}}{\text{input}} = \frac{\text{efficiency}}{100}$$

Example 6

Automotive Trades   An auto engine is rated at 175 hp and is found to deliver only 140 hp to the transmission. What is the efficiency of the process?

$$\frac{\text{output}}{\text{input}} = \frac{\text{efficiency } E}{100}$$

$$\frac{140}{175} = \frac{E}{100}$$

$$E = \frac{140 \cdot 100}{175}$$

$$E = 80\%$$

→ Your Turn

General Trades   A gasoline shop engine rated at 65 hp is found to deliver 56 hp through a belt drive to a pump. What is the efficiency of the drive system?

→ **Solution**

output = 56 hp,   input = 65 hp

$$\frac{56}{65} = \frac{E}{100}$$

$$E = \frac{56 \cdot 100}{65} = 86\% \text{ rounded}$$

56 ⊗ 100 ⊙ 65 ⊜ →   `86.15384615`

The efficiency of any practical system will always be less than 100%—you can't produce an output greater than the input!

Tolerances   On technical drawings or other specifications, measurements are usually given with a *tolerance* showing the allowed error. For example, the fitting shown in the margin has a length of

1.370 ± 0.015 in.

This means that the dimension shown must be between

1.370 + 0.015 in.   or   1.385 in.

and

1.370 − 0.015 in.   or   1.355 in.

The *tolerance limits* are 1.385 in. and 1.355 in.

Very often the tolerance is written as a percent.

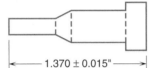

1.370 ± 0.015"

$$\frac{\text{tolerance}}{\text{measurement}} = \frac{\text{percent tolerance}}{100}$$

Example 7

To rewrite 1.370 ± 0.015 in. with a percent tolerance, substitute into the proportion equation and solve as follows:

$$\frac{0.015}{1.370} = \frac{\text{percent tolerance } R}{100}$$

$$R = \frac{(0.015)(100)}{1.370}$$

$$R = 1.0948\ldots \quad \text{or} \quad 1.1\% \text{ rounded}$$

The dimension would be written 1.370 in. ± 1.1%.

→ **Your Turn**

Rewrite the following dimension with a percent tolerance:

1.426 ± 0.010 in. = _____

→ **Solution**

The tolerance is 0.010 in. and the measurement is 1.426 in. Substituting into the tolerance proportion, we have

$$\frac{0.010}{1.426} = \frac{\text{percent tolerance } R}{100}$$

$$R = \frac{(0.010)(100)}{1.426}$$

$$R = 0.70126\ldots \qquad \text{or} \qquad 0.70\% \text{ rounded}$$

Rewriting the dimension with a percent tolerance, we have

$$1.426 \pm 0.010 \text{ in.} = 1.426 \text{ in.} \pm 0.70\%$$

If the dimension is given in **metric units,** the procedure is exactly the same.

Example 8

The dimension

$$315 \pm 0.25 \text{ mm}$$

can be converted to percent tolerance.

$$\boxed{\text{tolerance}} \longrightarrow \frac{0.25}{315} = \frac{R}{100}$$
$$\boxed{\text{measurement}} \longrightarrow$$

$$R = \frac{(0.25)(100)}{315}$$

$$R = 0.07936\ldots \qquad \text{or} \qquad 0.08\% \text{ rounded}$$

The dimension is $315 \text{ mm} \pm 0.08\%$.

 .25 $\times$ 100 $\div$ 315 $=$ → `0.079365079`

If the tolerance is given as a **percent,** it is easy to use the proportion on page 237 to calculate the tolerance as a dimension value.

Example 9

To convert 2.450 in. $\pm$ 0.2% to a dimension value, note that the measurement is 2.450 in., the percent tolerance is 0.2%, and the actual tolerance T is unknown. Therefore

$$\frac{\text{tolerance } T}{2.450} = \frac{0.2}{100}$$

$$T = \frac{(0.2)(2.45)}{100}$$

$$T = 0.0049 \text{ in.} \qquad \text{or} \qquad 0.005 \text{ in. rounded}$$

We can rewrite the dimension without a percent as follows:

$$2.450 \text{ in.} \pm 0.2\% = 2.450 \text{ in.} \pm 0.005 \text{ in.}$$

 .2 $\times$ 2.45 $\div$ 100 $=$ → `0.0049`

Now, for some practice in working with tolerances, convert the measurement tolerances below to percents, and vice versa. Round the tolerances to three decimal places and round the percents to two decimal places.

Measurement	Tolerance	Percent Tolerance
2.345 in.	± 0.001 in.	
	± 0.002 in.	
	± 0.005 in.	
	± 0.010 in.	
	± 0.125 in.	
274 mm	± 0.10 mm	
	± 0.20 mm	
	± 0.50 mm	
3.475 in.		± 0.10%
		± 0.20%
123 mm		± 0.05%
		± 0.15%

→ **Answers**

Measurement	Tolerance	Percent Tolerance
2.345 in.	± 0.001 in.	± 0.04%
	± 0.002 in.	± 0.09%
	± 0.005 in.	± 0.21%
	± 0.010 in.	± 0.43%
	± 0.125 in.	± 5.33%
274 mm	± 0.10 mm	± 0.04%
	± 0.20 mm	± 0.07%
	± 0.50 mm	± 0.18%
3.475 in.	± 0.003 in.	± 0.10%
	± 0.007 in.	± 0.20%
123 mm	± 0.062 mm	± 0.05%
	± 0.185 mm	± 0.15%

Percent Change   In many situations in technical work and the trades, you may need to find a *percent increase* or *percent decrease* in a given quantity. Use the following formula to calculate percent increase or decrease.

$$\frac{\text{amount of increase or decrease}}{\text{original amount}} = \frac{\text{percent of increase or decrease}}{100}$$

Example 10

Electrical Trades   Suppose that the output of a certain electrical circuit is 20 amperes (A), and the output increases by 10%. What is the new value of the output?

Step 1 Write the percent change proportion.

$$\frac{\text{amount of increase}}{\text{original amount}} = \frac{\text{percent increase}}{100}$$

Step 2 Substitute. The original amount is 20 A and the percent increase is 10%.

$$\frac{\text{amount of increase } i}{20} = \frac{10}{100}$$

Step 3 Solve.

$$100i = 200$$
$$i = 2 \text{ A}$$

Step 4 Calculate the new value.

New value = original amount + increase

$$= 20 \text{ A} + 2 \text{ A} = 22 \text{ A}$$

▶ **Learning Help** A quicker way to solve the problem is to add 100% to the percent increase and use this total for the rate R. The revised proportion becomes.

$$\frac{\text{new value } V}{\text{original amount}} = \frac{100\% + \text{percent increase}}{100} \quad ◄$$

For this example,

$$\frac{V}{20} = \frac{100\% + 10\%}{100}$$

$$\frac{V}{20} = \frac{110}{100}$$

$$V = \frac{20 \cdot 110}{100}$$

$$V = 22A$$

Percent decrease can be calculated in a very similar way.

Example 11

Manufacturing Suppose that the output of a certain machine at the Acme Gidget Co. is 600 gidgets per day. If the output decreases by 8%, what is the new output?

Percent *decrease* involves subtracting, and, as with percent increase, we can subtract either at the beginning or at the end of the calculation. Subtracting at the end, we have

Step 1 Write the percent change proportion.

$$\frac{\text{amount of decrease}}{\text{original amount}} = \frac{\text{percent decrease}}{100}$$

Step 2 Substitute: Original amount = 600; percent decrease = 8%.

$$\frac{\text{amount of decrease } d}{600} = \frac{8}{100}$$

Step 3 Solve.

$$d = \frac{8 \cdot 600}{100}$$

$$d = 48 \text{ gidgets per day}$$

Step 4 Now subtract to find the new value.

New value = original amount − amount of decrease

= 600 − 48 = 552 gidgets per day

Learning Help We can also solve the problem by subtracting at the beginning of the problem.

$$\frac{\text{new value } V}{\text{original amount}} = \frac{100\% - \text{percent decrease}}{100} \quad \blacktriangleleft$$

For this example,

$$\frac{\text{new value } V}{600} = \frac{100\% - 8\%}{100}$$

$$\frac{V}{600} = \frac{92}{100}$$

$$100V = 55200$$

$$V = 552 \text{ gidgets per day}$$

→ Your Turn

Try these problems involving percent change. In each, add or subtract the percents at the beginning of the problem.

(a) **Automotive Trades** The markup, or profit, on a part is the amount a dealer adds to his cost to determine the retail price, or the amount that the customer pays. If an air cleaner cost a dealer $5.12 and the markup is 40%, what retail price does he charge the customer?

(b) **Electrical Trades** Normal line voltage, 115 volts, drops 3.5% during a system malfunction. What is the reduced voltage?

(c) **Trades Management** Because of changes in job specifications, the cost of a small construction job is increased 25% from the original cost of $2475. What is the cost of the job now?

(d) **Welding** By using automatic welding equipment, the time for a given job can be reduced by 40% from its original 30 hours. How long will the job take using the automatic equipment?

→ Solutions

(a) The markup is the percent increase.

$$\frac{\text{retail price}}{\$5.12} = \frac{140}{100} \quad \Longleftarrow \boxed{100\% + 40\%}$$

$$\text{Retail price} = \frac{140(5.12)}{100} = \$7.168 \quad \text{or} \quad \$7.17 \text{ rounded}$$

(b) $\dfrac{\text{reduced voltage}}{115 \text{ volts}} = \dfrac{96.5}{100}$ ⇐ 100% − 3.5%

$\text{Reduced voltage} = \dfrac{115(96.5)}{100} = 110.975 \quad \text{or} \quad 111 \text{ volts rounded}$

115 [×] **96.5** [÷] **100** [=] → 110.975

(c) $\dfrac{\text{new cost}}{\$2475} = \dfrac{125}{100}$

$\text{New cost} = \dfrac{2475 \cdot 125}{100} = \$3093.75$

(d) $\dfrac{\text{new time}}{30} = \dfrac{60}{100}$

$\text{New time} = \dfrac{30 \cdot 60}{100} = 18 \text{ hours}$

In trade and technical work, changes in a measured quantity are often specified as a percent increase or decrease.

Example 12

General Trades If the reading on a pressure valve increases from 30 psi to 36 psi, the percent increase can be found by using the following proportion:

$$\begin{array}{l} \text{The change} \rightarrow \\ \text{Original value} \longrightarrow \end{array} \dfrac{36 \text{ psi} - 30 \text{ psi}}{30 \text{ psi}} = \dfrac{\text{percent increase}}{100}$$

$$\frac{6}{30} = \frac{R}{100}$$

$$R = \frac{100 \cdot 6}{30} = 20\%$$

The reading on the pressure valve increased by 20%.

100 [×] [(] **36** [−] **30** [)] [÷] **30** [=] → 20.

Example 13

HVAC The length of a heating duct is reduced from 110 in. to 96 in. because of design changes. The percent decrease is calculated as follows:

$$\begin{array}{l} \text{The change} \rightarrow \\ \text{Original value} \longrightarrow \end{array} \dfrac{110 \text{ in.} - 96 \text{ in.}}{110 \text{ in.}} = \dfrac{\text{percent decrease}}{100}$$

$$\frac{14}{110} = \frac{R}{100}$$

$$R = \frac{100 \cdot 14}{110} = 12.7272\ldots \quad \text{or} \approx 13\%$$

The length of the duct was reduced by about 13%.

Students often are confused about which number to use for the base in percent change problems. The difficulty usually arises when they think in terms of which is the bigger or smaller value. Always use the *original* value as the base. In percent increase problems, the original or base number will be the smaller number, and in percent decrease problems, the original or base number will be the larger number. ◄

→ Your Turn

Now try these problems to sharpen your ability to work with percent changes.

(a) **Carpentry**   If 8 in. is cut from a 12-ft board, what is the percent decrease in length?

(b) **Electrical Trades**   What is the percent increase in voltage when the voltage increases from 65 volts to 70 volts?

(c) **Industrial Technology**   The measured value of the power output of a motor is 2.7 hp. If the motor is rated at 3 hp, what is the percent difference between the measured value and the expected value?

(d) **Automotive Trades**   A water pump cost a parts department \$98.50. The pump was sold to a customer for \$137.90. What was the percent markup of the pump?

(e) **Life Skills**   A house purchased for \$400,000 during the housing boom decreased in value by 30% during the recent housing slump. By what percent must it now increase in value in order to be worth its original purchase price? (Round your answer to one decimal digit.)

→ Solutions

(a)   Use 12 ft = 144 in.

$$\text{The change} \longrightarrow \frac{8 \text{ in.}}{144 \text{ in.}} = \frac{\text{percent decrease}}{100} \longleftarrow \text{Original value}$$

$$R = \frac{100 \cdot 8}{144} = 5.555 \ldots \quad \text{or} \quad 5.6\% \text{ rounded}$$

The length of the board decreased by about 5.6%.

(b)   To find the amount of increase, subtract 65 volts from 70 volts.

$$\frac{70 \text{ volts} - 65 \text{ volts}}{\text{Original value} \longrightarrow 65 \text{ volts}} = \frac{\text{percent increase}}{100}$$

$$\frac{5}{65} = \frac{R}{100}$$

$$R = \frac{100 \cdot 5}{65} = 7.6923 \ldots \quad \text{or } 7.7\% \text{ rounded}$$

The voltage increased by about 7.7%.

(c)   To find the amount of decrease, subtract 2.7 hp from 3 hp.

$$\frac{3 \text{ hp} - 2.7 \text{ hp}}{3 \text{ hp}} = \frac{\text{percent difference}}{100}$$

The expected value is used as the base.

$$\frac{0.3}{3} = \frac{R}{100}$$

$$R = \frac{100(0.3)}{3} = 10\%$$

There is a 10% difference between the measured value and the expected value.

(d) To find the amount of increase or markup, subtract the cost from the retail price.

Amount of markup = \$137.90 − \$98.50 = \$39.40

The percent markup is the percent increase over the *cost*.

$$\text{Amount of markup}\rightarrow \frac{\$39.40}{\$98.50}\leftarrow \text{Cost} = \frac{\text{percent markup } R}{100}$$

$$R = \frac{100(39.40)}{\$98.50} = 40\%$$

(e) Many people mistakenly believe that 30% is the answer to this problem: a 30% increase should make up for the 30% decrease, right? However, the percent decrease is based on the original price of \$400,000, whereas the percent increase is based on the price after devaluation (\$280,000). As you will see if you perform the calculation, 30% of 400,000 is quite a bit more than 30% of 280,000. Therefore, the required increase must be larger than 30%.

First, determine the dollar amount D of the decrease in value.

$$\frac{D}{400,000} = \frac{30}{100}$$

$$100D = 12,000,000$$

$$D = \$120,000$$

Then, find the new value V of the house. $V = \$400,000 − \$120,000 = \$280,000$

Finally, note that the house must increase in value by \$120,000 in order to be worth its original purchase price. But the percent P by which it must increase must be calculated using the current value of \$280,000 as the base:

$$\frac{P}{100} = \frac{120,000}{280,000}$$

$$280,000P = 12,000,000$$

$$P = 43\%, \text{rounded}$$

▶ **A Closer Look**

Here is a quick calculator method for doing problem (e):

First, multiply the original value by 0.70 to calculate the decreased value. (100% − 30% = 70% = 0.70)

400000 $\times$ .7 $=$ → `280000.`

Then, divide the original value by this decreased value.

400000 $\div$ (ANS) $=$ → `1.428571429`

Finally, this tells us that the original value is about 1.43 times the decreased value, or about 143% of it. This means that the decreased value must increase by approximately 43% in order to return to the original price. ◀

▶ **Note**

We do not actually need to know the original price in order to work this problem. The final answer will be the same no matter what the original amount is. For example, if we use \$100 as the original price, then the first step becomes 100 $\times$ .7 $=$ → `70.` The second step becomes 100 $\div$ (ANS) $=$ → `1.428571429`, which gives us the same answer, approximately 43%. ◀

Now, turn to Exercises 4-5 for a set of problems involving these special applications of percent calculations.

Exercises 4-5 Special Applications of Percent Calculations

1. **Electronics** A CB radio is rated at 7.5 watts, and actual measurements show that it delivers 4.8 watts to its antenna. What is its efficiency?

2. **Electrical Trades** An electric motor uses 6 kW (kilowatts) at an efficiency of 63%. How much power does it deliver?

3. **Electrical Trades** An engine supplies 110 hp to an electric generator, and the generator delivers 70 hp of electrical power. What is the efficiency of the generator?

4. **Electronics** Electrical resistors are rated in ohms and color coded to show both their resistance and percent tolerance. For each of the following resistors, find its tolerance limits and actual tolerance.

Resistance (ohms)	Limits (ohms)	Tolerance (ohms)
5300 ± 5%	_____ to _____	_____
2750 ± 2%	_____ to _____	_____
6800 ± 10%	_____ to _____	_____
5670 ± 20%	_____ to _____	_____

5. **Automotive Trades** A 120-hp automobile engine delivers only 81 hp to the driving wheels of the car. What is the efficiency of the transmission and drive mechanism?

6. **Electronics** An electrical resistor is rated at 500 ohms ±10%. What is the highest value its resistance could have within this tolerance range?

7. **Roofing** On a roofing job, 21 of 416 shingles had to be rejected for minor defects. What percent is this?

8. **Life Skills** If you earn 12% commission on sales of $4200, what actual amount do you earn?

9. **Metalworking** On a cutting operation 2 sq ft of sheet steel is wasted for every 16 sq ft used. What is the percent waste?

10. **Metalworking** Four pounds of a certain bronze alloy is one-sixth tin, 0.02 zinc, and the rest copper. Express the portion of each metal in (a) percents and (b) pounds.

11. **Metalworking** An iron casting is made in a mold with a hot length of 16.40 in. After cooling, the casting is 16.25 in. long. What is the shrinkage in percent?

12. **Manufacturing** Because of friction, a pulley block system is found to be only 83% efficient. What actual load can be raised if the theoretical load is 2000 lb?

13. **Manufacturing** A small gasoline shop engine develops 65 hp at 2000 rpm. At 2400 rpm its power output is increased by 25%. What actual horsepower does it produce at 2400 rpm?

14. **Carpentry** An online tool supplier offers a 15% discount on orders of $100 or more. If a carpenter orders a bevel setter priced at $33.50, an electronic caliper priced at $39.50, and set-up blocks priced at $42.50, what would be the total cost after the discount is applied?

15. **Machine Trades** Specifications call for a hole in a machined part to be 2.315 in. in diameter. If the hole is measured to be 2.318 in., what is the machinist's percent error?

16. **Trades Management** A welding shop charges the customer 185% of its labor and materials costs to cover overhead and profit. If a bill totals $822.51, what is the cost of labor and materials?

17. Complete the following table.

Measurement	Tolerance	Percent Tolerance
3.425 in.	±0.001 in.	(a)
3.425 in.	±0.015 in.	(b)
3.425 in.	(c)	±0.20%

18. **General Trades** If you receive a pay increase of 12%, what is your new pay rate, assuming that your old rate was $16.80 per hour?

19. **Manufacturing** The pressure in a hydraulic line increases from 40 psi to 55 psi. What is the percent increase in pressure?

20. **Painting** The cost of the paint used in a redecorating job is $123.20. This is a reduction of 20% from its initial cost. What was the original cost?

21. **Life Skills** What sales tax, at a rate of $6\frac{1}{2}$%, must you pay on the purchase of a computer hard disk drive costing $256.75?

22. **Interior Design** An interior designer working for a department store is paid a weekly salary of $350 plus a 10% commission on total sales. What would be his weekly pay if his total sales were $6875.00 for the week?

23. **Electronics** A one-terabyte external hard drive is on sale at a 25% discount. If the original price was $349, what was the sale price?

24. **Trades Management** Chris paid for the initial expenses in setting up her upholstery shop with a bank loan for $39,000 for 5 years at $8\frac{1}{2}$% annual interest.
    (a) What total interest will she pay?
    (b) If she pays off the interest only in equal monthly installments, what will be her monthly payments?

25. **Carpentry** A 7-in. medium-angle grinder is on sale for $101.50 and is marked as being 30% off. What was the original list price?

26. **Automotive Trades** Driving into a 20-mph breeze cuts gas mileage by about 12%. What will be your gas mileage in a 20-mph wind if it is 24 mi/gal with no wind?

27. **Automotive Trades** Supplier A offers a mechanic a part at 35% off the retail price of $68.40. Supplier B offers him the same part at 20% over the wholesale cost of $35.60. Which is the better deal, and by how much?

28. **Automotive Trades** An auto mechanic purchases the parts necessary for a repair for $126.40 and sells them to the customer at a 35% markup. How much does the mechanic charge the customer?

29. **Automotive Trades** A new tire had a tread depth of $\frac{3}{8}$ in. After one year of driving the tire had $\frac{9}{32}$ in. of tread remaining. What percent of the tread was worn?

30. **Printing** A printer agrees to give a nonprofit organization a 10% discount on a printing job. The normal price to the customer is $1020, and the printer's cost is $850. What is the printer's percent profit over cost after the discount is subtracted?

31. **Wastewater Technology** A treatment plant with a capacity of 20 MGD (million gallons per day) has a normal daily flow equal to 60% of capacity.

When it rains the flow increases by 30% of the normal flow. What is the flow on a rainy day?

32. **Police Science**   Last year a community had 23 homicides. This year they had 32 homicides. What was the percent of increase?

33. **Fire Protection**   The pressure reading coming out of a pump is 180 lb. Every 50-ft section reduces the pressure by about 3%. What will the nozzle pressure be at the end of six 50-ft sections? (*Hint:* One reduction of 18% is *not* equivalent to six reductions of 3% each. You must do this the long way.)

34. **Machine Trades**   Complete the following table of tolerances for some machine parts.

Measurement	Tolerance	Maximum	Minimum	Percent Tolerance
1.58 in.	±0.002 in.			
	±0.005 in.			
				±0.15%
0.647 in.	±0.004 in.			
	±0.001 in.			
				±0.20%
165.00 mm	±0.15 mm			
	±0.50 mm			
				±0.05%
35.40 mm	±0.01 mm			
	±0.07 mm			
				±0.05%
				±0.08%
				±0.10%

35. **Electrical Trades**   An electrician purchases some outdoor lighting for a customer at a wholesale price of $928.80. The customer would have paid the retail price of $1290. What was the wholesale discount?

36. **Life Skills**   In problem 35, what will the total retail price be after 7.75% tax is added?

37. **Trades Management**   A contractor takes a draw of $46,500 from a homeowner to start a remodeling job. He uses $22,450 to purchase materials and deposits the remaining money in a three-month CD paying 2.80% *annual* interest. How much interest does he earn on the remaining funds during the three-month term?

38. **Culinary Arts**   A wholesale food sales rep is paid a base salary of $1200 per month plus a $3\frac{1}{2}$% commission on sales. How much sales must she generate in order to to earn $35,000 per year in total compensation?

39. **Life Skills**   A sales rep for roofing materials figures he can generate about $1,800,000 in business annually. What rate of commission does he need in order to earn $50,000?

40. **Carpentry**   A power planer originally selling for $139.50 is on sale for $111.60. By what percent was it discounted?

41. **Trades Management**   Last year, the owner of the Fixum Auto Repair Shop paid a premium of $9.92 per $100 of gross wages for workmen's compensation insurance. This year his premium will increase by 4%. If he has 8 employees averaging $22 per hour, 40 hours per week for 50 weeks, what will be his total premium for this year?

42. **Trades Management**   Because they have a history of paying their bills on time, the Radius Electric Company receives an 8.25% discount on all purchases of electrical supplies from their supplier. What would they pay for cable and fixtures originally costing $4760?

43. **Automotive Trades**   When a turbocharger is added to a certain automobile engine, horsepower increases by 15%. If the non-turbocharged engine produces 156 peak horsepower, what peak horsepower does the turbocharged engine produce? Round to the nearest whole number.

44. **Life Skills**   The National Center for Real Estate Research released a study showing how property characteristics affect the selling price of a home. For example, they found that having central air conditioning increased the selling price of a house by 5%. If an identical house without air conditioning sold for $380,000, what would be the selling price of the air-conditioned house?

45. **HVAC**   The following table shows the recommended cooling capacity (in Btu per hour) of a room air conditioner based on the area (in square feet) of the room. If more than two people regularly occupy the room, add 600 Btu/hr per additional person; if the room is particularly sunny, add 10%; and if the room gets very little sun, subtract 10% from the listed capacity.

Area (sq ft)	Capacity (Btu/hr)
250–300	7,000
300–350	8,000
350–400	9,000
400–450	10,000

Determine the recommended cooling capacity of a room air conditioner for:
(a)   A sunny 320-sq-ft room normally occupied by one or two people.
(b)   A shady 440-sq-ft room normally occupied by four people.

46. **Sports and Leisure**   In 1976 new world records in the high jump were set for both men and women. Dwight Stones set the men's record at 2.32 meters, and Rosemarie Ackerman set the women's record at 1.96 meters. The record in the year 2008 was 2.45 meters for men (Javier Sotomayor) and 2.09 meters for women (Stefka Kostadinova). Which record, men's or women's, has increased by the larger percent since 1976? Find the percent increase for each to the nearest tenth.

47. **Allied Health**   For maximum benefit during exercise, one's heart rate should increase by 50% to 85% above resting heart rate.
(a)   For a woman with a resting heart rate of 64 bpm, what would her heart rate be if she increased it by 75% during exercise?
(b)   A man with a resting heart rate of 70 bpm increased it to 105 bpm during exercise. By what percent did he increase his resting heart rate?

48. **Transportation** An aerodynamically designed truck can improve fuel economy by as much as 45% over older, square-nosed rigs. If a trucker currently gets 4.5 mi/gal in his old truck, what mileage could he potentially achieve in an aerodynamically designed model? (Round to the nearest tenth.)

49. **Office Services** A water district has been charging its customers a flat rate of $3.75 per HCF. Because of a drought emergency, it is instituting a block rate of $3.90 for the first 20 HCF and $4.15 for the next 40 HCF. If a family uses an average of 52 HCF per month, by what percent would their average bill increase under the block system? (Round to the nearest tenth.)

50. **General Trades** Due to an economic downturn, a trades worker had to take a 10% pay cut from her $20 per hour wage. By what percent must her wage now increase in order to regain the original level? (Round your answer to one decimal digit.)

51. **Printing** A poster was reduced in size by 25%. By what percent must it then be increased in size in order to restore it to its original size? (*Hint:* See the Note on page 244. Pick any arbitrary number to represent the original size and then work the problem.)

52. **Hydrology** During a drought year, a customer's normal monthly water allocation of 4600 cubic feet was reduced by 15%. When the drought is over, by what percent must the allocation be increased to restore it to its original level?

53. **Allied Health** A patient originally weighing 180 lb took a course of medication with a side effect that caused his weight to increase by 15%.
    (a) How much weight did he gain?
    (b) What percent of this resulting weight must he now lose in order to return to his original weight?

When you have completed these exercises, check your answers to the odd-numbered problems in the Appendix, then turn to Problem Set 4 on page 253 for a set of practice problems on ratio, proportion, and percent. If you need a quick review of the topics in this chapter, visit the chapter Summary first.

Summary → Ratio, Proportion, and Percent

Objective	Review
Calculate ratios. (p. 178)	A ratio is a comparison, using division, of two quantities of the same kind, both expressed in the same units. Final answers should be expressed either as a fraction in lowest terms or as a comparison to the number 1 using the word "to" or a colon (:).

> **Example:** Find the compression ratio C for an engine with maximum cylinder volume of 520 cu cm and minimum cylinder volume of 60 cu cm.
>
> $$C = \frac{520 \text{ cu cm}}{60 \text{ cu cm}} = \frac{26}{3} \quad \text{or} \quad 8\tfrac{2}{3} \text{ to } 1 \quad \text{or} \quad 8\tfrac{2}{3}{:}1$$

Objective	Review
Solve proportions. (p. 182)	A proportion is an equation stating that two ratios are equal: $$\frac{a}{b} = \frac{c}{d}$$ To solve a proportion use the cross-product rule: $$a \cdot d = b \cdot c$$

Example: Solve: $\dfrac{8}{x} = \dfrac{12}{15}$

$$12x = 120$$
$$x = 10$$

Solve problems involving proportions. (pp. 185, 196)	Determine whether the problem involves a direct or an inverse proportion. Then set up the ratios accordingly and solve.

Example:

(a) An architectural drawing of a living room is $5\frac{1}{2}$ in. long and $4\frac{1}{2}$ in. wide. If the actual length of the room is 22 ft, find the actual width. Use a direct proportion.

$$\frac{5\frac{1}{2}''}{4\frac{1}{2}''} = \frac{22'}{x}$$

$$5\frac{1}{2} \cdot x = 99$$

$$x = 18'$$

(b) A large production job is completed by four machines working for 6 hours. The same size job must be finished in 2 hours the next time it is ordered. How many machines should be used to accomplish this task? Use an inverse proportion.

$$\frac{4 \text{ machines}}{x \text{ machines}} = \frac{2 \text{ hr}}{6 \text{ hr}}$$

$$2x = 24$$

$$x = 12 \text{ machines}$$

Change decimal numbers to percents. (p. 208)	Multiply the decimal number by 100%, or shift the decimal point two places to the right.

Example:
(a) $5 = 5 \times 100\% = 500\%$

(b) $0.075 = 0.075 = 7.5\%$

Change fractions to percents. (p. 208)	First change the fraction to decimal form. Then proceed as in the previous objective.

Example: $\frac{1}{4} = 0.25 = 0.25 = 25\%$

Change percents to decimal numbers. (p. 211)	Divide by 100%, or shift the decimal point two places to the left.

Example:
(a) $35\% = \dfrac{35\%}{100\%} = 0.35$

(b) $0.25\% = 00.25\% = 0.0025$

Objective	Review

Change percents to fractions. (p. 212)

Write a fraction with the percent number as the numerator and 100 as the denominator. Then rewrite this fraction in lowest terms.

Example: $112\% = \dfrac{112}{100} = \dfrac{28}{25}$

Solve problems involving percent. (p. 215)

Set up the proportion.

$$\frac{\text{part }(P)}{\text{base }(B)} = \frac{\text{rate or percent }(R)}{100} \quad \text{or} \quad \frac{\text{is}}{\text{of}} = \frac{\%}{100}$$

Then solve for the missing quantity.

Example: (a) Find 120% of 45.

$$\frac{P}{45} = \frac{120}{100}$$
$$100P = 5400$$
$$P = 54$$

(b) 12 is 16% of what number?

$$\frac{12}{B} = \frac{16}{100}$$
$$16B = 1200$$
$$B = 75$$

Solve applications involving percent. Specific applications may involve discount (p. 225), sales tax (p. 230), interest (p. 231), commission (p. 234), efficiency (p. 236), tolerances (p. 237), or percent change (p. 239).

In each case, read the problem carefully to identify *P, B,* and *R.* Use "is," "of," and the "%" symbol to help you. Then proceed as in the previous objective.

Example: (a) The paint needed for a redecorating job originally cost $144.75, but it is now discounted by 35%. What is its discounted price?

$$\frac{\text{net price}}{144.75} = \frac{100 - 35}{100} \longrightarrow \frac{N}{144.75} = \frac{65}{100}$$
$$100N = 9408.75$$
$$N \approx \$94.09$$

(b) If the voltage in a circuit is increased from 70 volts to 78 volts, what is the percent increase in voltage?

$$\frac{78 - 70}{70} = \frac{R}{100} \longrightarrow \frac{8}{70} = \frac{R}{100}$$
$$70R = 800$$
$$R \approx 11.4\%$$

4

Ratio, Proportion, and Percent

Answers to odd-numbered problems are given in the Appendix.

A. Complete the following tables.

1.

	Diameter of Pulley A	Diameter of Pulley B	Pulley Ratio, $\dfrac{A}{B}$
(a)	12 in.		2 to 5
(b)	95 cm	38 cm	
(c)		15 in.	5 to 3

2.

	Teeth on Driving Gear A	Teeth on Driven Gear B	Gear Ratio, $\dfrac{A}{B}$
(a)	20	60	
(b)		10	6.5
(c)	56		4 to 3

B. Solve the following proportions.

1. $\dfrac{5}{6} = \dfrac{x}{42}$ 2. $\dfrac{8}{15} = \dfrac{12}{x}$ 3. $\dfrac{x}{12} = \dfrac{15}{9}$

4. $\dfrac{3\frac{1}{2}}{x} = \dfrac{5\frac{1}{4}}{18}$ 5. $\dfrac{1.6}{5.2} = \dfrac{4.4}{x}$ 6. $\dfrac{x}{12.4} = \dfrac{4\text{ ft }6\text{ in.}}{6\text{ ft }3\text{ in.}}$

C. Write each number as a percent.

1. 0.72   2. 0.06   3. 0.6   4. 0.358   5. 1.3

6. 3.03   7. 4   8. $\dfrac{7}{10}$ 9. $\dfrac{1}{6}$ 10. $2\frac{3}{5}$

D. Write each percent as a decimal number.

1. 4%   2. 37%   3. 11%   4. 94%

5. $1\frac{1}{4}\%$ 6. 0.09%   7. $\frac{1}{5}\%$ 8. 1.7%

9. $3\frac{7}{8}\%$ 10. 8.02%   11. 115%   12. 210%

E. Write each of the following percents as a fraction in lowest terms.

1. 28%   2. 375%   3. 81%   4. 4%

5. 0.5%   6. 70%   7. 14%   8. $41\frac{2}{3}\%$

Name _____

Date _____

Course/Section _____

F. Solve.

1. 3 is _____ % of 5.
2. 5% of $120 is _____.
3. 25% of what number is 1.4?
4. 16 is what percent of 8?
5. 105% of 40 is _____.
6. 1.38 is _____ % of 1.15?
7. $7\frac{1}{4}$% of _____ is $2.10.
8. 250% of 50 is _____.
9. 0.05% of _____ is 4.
10. $8\frac{1}{4}$% of 1.2 is _____.

G. Solve.

1. **Metalworking**   Extruded steel rods shrink 12% in cooling from furnace temperature to room temperature. If a standard tie rod is 34 in. exactly when it is formed, how long will it be after cooling?
2. **Metalworking**   Cast iron contains up to 4.5% carbon, and wrought iron contains up to 0.08% carbon. How much carbon is in a 20-lb bar of each metal?
3. **Life Skills**
   (a) A real estate saleswoman sells a house for $664,500. Her company pays her 70% of the 1.5% commission on the sale. How much does she earn on the sale?
   (b) All salespeople in the Ace Department Store receive $360 per week plus a 6% commission. If you sold $3975 worth of goods in a week, what would be your income?
   (c) A salesman at the Wasteland TV Company sold five identical LCD TV sets last week and earned $509.70 in commissions. If his commission is 6%, what does one of these models cost?
4. **Carpentry**   What is the net price of a 10-in. bench table saw with a list price of $165.95 if it is on sale at a 35% discount?
5. **Life Skills**   If the retail sales tax in your state is 6%, what would be the total cost of each of the following items?
   (a) An $18.99 pair of pliers.
   (b) A $17.49 adjustable wrench.
   (c) 69 cents worth of washers.
   (d) A $118.60 textbook.
   (e) A $139.99 five-drawer shop cabinet rollaway.
6. **Life Skills**   A computer printer sells for $376 after a 12% discount. What was its original or list price?
7. **Construction**   How many running feet of matched 1-in. by 6-in. boards will be required to lay a subfloor in a house that is 28 ft by 26 ft? Add 20% to the area to allow for waste and matching.
8. **Sheet Metal Trades**   In the Easy Does It Metal Shop, one sheet of metal is wasted for every 25 purchased. What percent of the sheets are wasted?
9. **Machine Trades**   Complete the following table.

Measurement	Tolerance	Percent Tolerance
1.775 in.	±0.001 in.	(a)
1.775 in.	±0.05 in.	(b)
1.775 in.	(c)	±0.50%
310 mm	±0.1 mm	(d)
310 mm	(e)	±0.20%

10. **Masonry** A mason must purchase enough mortar mix to lay 1820 bricks. The guidelines for this mix suggest that about 15 bags are required for every 400 bricks.
    (a) Assuming that he cannot purchase a fraction of a bag, how many bags will he need?
    (b) At $6.40 per bag, what will be the total cost of the mix?

11. **Manufacturing** A production job is bid at $6275, but cost overruns amount to 15%. What is the actual job cost?

12. **Metalworking** When heated, a metal rod expands 3.5%, to 15.23 cm. What is its cold length?

13. **Electrical Trades** An electrical resistor is rated at 4500 ohms $\pm$ 3%. Express this tolerance in ohms and state the actual range of resistance.

14. **Automotive Trades** A 140-hp automobile engine delivers only 126 hp to the driving wheels of the car. What is the efficiency of the transmission and drive mechanism?

15. **Metalworking** A steel casting has a hot length of 26.500 in. After cooling, the length is 26.255 in. What is the shrinkage expressed as a percent? Round to one decimal place.

16. **Automotive Trades** The parts manager for an automobile dealership can buy a part at a 25% discount off the retail price. If the retail price is $46.75, how much does he pay?

17. **Manufacturing** An electric shop motor rated at 2.5 hp is found to deliver 1.8 hp to a vacuum pump when it is connected through a belt drive system. What is the efficiency of the drive system?

18. **Trades Management** To purchase a truck for his mobile welding service, Jerry arranges a 36-month loan of $45,000 at 4.75% annual interest.
    (a) What total interest will he pay?
    (b) What monthly payment will he make? (Assume equal monthly payments on interest and principal.)

19. **HVAC** The motor to run a refrigeration system is rated at 12 hp, and the system has an efficiency of 76%. What is the effective cooling power output of the system?

20. **Interior Design** An interior designer is able to purchase a sofa from a design center for $1922. His client would have paid $2480 for the same sofa at a furniture store. What rate of discount was the designer receiving off the retail price?

21. **Life Skills** The commission rate paid to a manufacturer's rep varies according to the type of equipment sold. If his monthly statement showed $124,600 in sales, and his total commission was $4438, what was his average rate of commission to the nearest tenth of a percent?

22. **Masonry** Six square feet of a certain kind of brick wall contains 78 bricks. How many bricks are needed for 150 sq ft?

23. **Transportation** If 60 mph (miles per hour) is equivalent to 88 fps (feet per second), express
    (a) 45 mph in feet per second.
    (b) 22 fps in miles per hour.

24. **Automotive Trades** The headlights on a car are mounted at the height of 28 in. The light beam must illuminate the road for 400 ft. That is, the beam must hit the ground 400 ft ahead of the car.
    (a) What should be the drop ratio of the light beam in inches per foot?
    (b) What is this ratio in inches per 25 ft?

25. **Printing** A photograph measuring 8 in. wide by 12 in. long must be reduced to a width of $3\frac{1}{2}$ in. to fit on a printed page. What will be the corresponding length of the reduced photograph?

26. **Machine Trades** If 250 ft of wire weighs 22 lb, what will be the weight of 100 ft of the same wire?

27. **Agriculture**   A sprayer discharges 600 cc of herbicide in 45 sec. For what period of time should the sprayer be discharged in order to apply 2000 cc of herbicide?

28. **Trades Management**   An auto mechanic currently rents a 3460-sq-ft shop for $5363 per month. How much will his monthly rent be if he moves to a comparably priced 2740-sq-ft shop? Round to the nearest dollar.

29. **Machine Trades**   Six steel parts weigh 1.8 lb. How many of these parts are in a box weighing 142 lb if the box itself weighs 7 lb?

30. **Manufacturing**   Two belted pulleys have diameters of 24 in. and 10 in.
   (a)   Find the pulley ratio.
   (b)   If the larger pulley turns at 1500 rpm, how fast will the smaller pulley turn?

31. **Printing**   If six printing presses can do a certain job in $2\frac{1}{2}$ hours, how long will it take four presses to do the same job? (*Hint:* Use an inverse proportion.)

32. **Drafting**   Drafters usually use a scale of $\frac{1}{4}$ in. = 1 ft on their architectural drawings. Find the actual length of each of the following items if their blueprint length is given.
   (a)   A printing press $2\frac{1}{4}$ in. long.   (b)   A building $7\frac{1}{2}$ in. long.
   (c)   A rafter $\frac{11}{16}$ in. long.   (d)   A car $1\frac{7}{8}$ in. long.

33. **Construction**   A certain concrete mix requires one sack of cement for every 650 lb of concrete. How many sacks of cement are needed for 3785 lb of concrete? (*Remember:* You cannot buy a fraction of a sack.)

34. **Automotive Trades**   The rear axle of a car turns at 320 rpm when the drive shaft turns at 1600 rpm. If the pinion gear has 12 teeth, how many teeth are on the ring gear?

35. **Physics**   If a 120-lb force is applied to a 36-in. crowbar pivoted 4 in. from the end, what lift force is exerted?

36. **Machine Trades**   How many turns will a pinion gear having 16 teeth make if a ring gear having 48 teeth makes 120 turns?

37. **Electrical Trades**   The current in a certain electrical circuit is inversely proportional to the line voltage. If a current of 0.40 A is delivered at 440 volts, what is the current of a similar system operating at 120 volts? Round to the nearest 0.01 A.

38. **Printing**   A printer needs 12,000 good copies of a flyer for a particular job. Previous experience has shown that she can expect no more than a 5% spoilage rate on this type of job. How many should she print to ensure 12,000 clean copies?

39. **General Trades**   If a mechanic's helper is paid $456.28 for $38\frac{1}{4}$ hours of work, how much should she receive for $56\frac{1}{2}$ hours at the same rate?

40. **Manufacturing**   A pair of belted pulleys have diameters of $8\frac{3}{4}$ in. and $5\frac{5}{8}$ in. If the smaller pulley turns at 850 rpm, how fast will the larger one turn?

41. **Trades Management**   Each production employee in a plant requires an average of 120 sq ft of work area. How many employees will be able to work in an area that measures $8\frac{1}{16}$ in. by $32\frac{5}{8}$ in. on a blueprint if the scale of the drawing is $\frac{1}{4}$ in. = 1 ft?

42. **General Trades**   Find the missing dimension in the following pair of similar figures.

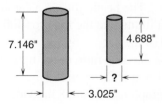

43. **Physics**   In a closed container, the pressure is inversely proportional to the volume when the temperature is held constant. Find the pressure of a gas compressed to 0.386 cu ft if the pressure is 12.86 psi at 2.52 cu ft.

44. **Automotive Trades**   The mechanical efficiency of a certain engine is specified to be 82.5%. This means that 82.5% of its indicated or theoretical horsepower becomes brake horsepower. If the indicated horsepower of the engine is 172, what is the brake horsepower? Round to the nearest whole number.

45. **Automotive Trades**   An automobile engine loses approximately 3% of its power for every 1000 ft of elevation gain. A certain engine generates 180 horsepower (hp) at sea level. How much power will it generate at an elevation of 4000 ft? Round to the nearest whole number.

46. **Allied Health**   Calculate the percent concentration of a solution created by adding 15 mL of a medication to 160 mL of water. Round your answer to the nearest tenth of a percent.

47. **Construction**   The National Center for Real Estate Research released a study showing that houses advertised as "fixer-uppers" sell for an average of 24% less than comparable houses. How much would a fixer-upper sell for if a comparable house that was in good condition sold for $485,000?

48. **Automotive Trades**   A headlight costs an auto service parts department $9.25. If the markup is 60%, what retail price will the customer be charged?

49. **Automotive Trades**   During a compression check, the cylinder with the highest compression was measured at 156 psi (pounds per square inch), while the cylinder with the lowest compression was measured at 136 psi. If the maximum allowable reduction from the highest compression is 15%, does this engine fall within the allowable limit?

50. **Allied Health**   A pre-op antiseptic solution is to be used to deliver 0.3 mg of atropine to the patient. There is, on hand, a solution of atropine containing 0.4 mg per 0.5 mL. (a) How much of the solution should be used to obtain the amount of atropine needed? (b) Suppose the only measuring device available is calibrated in minims (1 mL = 16 minims). How much of the solution is needed in minims?

51. **Sports and Leisure**   Use the following five steps to determine the NFL passing rating for quarterback Bubba Grassback.
    (a) Bubba completed 176 passes in 308 attempts. Calculate his pass completion rate as a percent. Subtract 30 from this number and multiply the result by 0.05.
    (b) He has 11 touchdown passes in 308 attempts. Calculate the percent of his attempts that resulted in touchdowns. Multiply that number by 0.2.
    (c) He has thrown 13 interceptions. Calculate the percent of his 308 attempts that have resulted in interceptions. Multiply this percent by 0.25, and then subtract the result from 2.375.
    (d) Bubba passed for 2186 yards. Calculate the average number of yards gained per passing attempt. Subtract 3 from this number, and then multiply the result by 0.25.
    (e) Add the four numbers obtained in parts (a), (b), (c), and (d).
    (f) Divide the total in part (e) by 6 and then multiply by 100. This is Bubba's NFL quarterback passing rating.

    (*Note:* Ratings for starting quarterbacks usually range from a low of about 50 to a high of about 110.)

52. **Allied Health**   Medication Q is available only in tablets containing 30 mg of the drug. How many tablets (or fractions of tablets) should be administered if the physician orders 75 mg of Medication Q?

53. **Allied Health**   Medication TTQ is available only in a liquid formulation with a concentration of 25 μg/mL (micrograms per milliliter). How much of this liquid formulation should be administered if the following amounts are ordered by the physician?
    (a) 300 μg   (b) 150 μg   (c) 20 μg

54. **Transportation** For every 1-mph reduction of average speed below 65 mph, a trucker will save 1.5% in fuel consumption. If a trucker reduces her average speed from 65 mph to 60 mph, how much will she save if she currently spends $4000 per week on fuel?

55. **Life Skills** A working person who needs an emergency loan, but who has no home or business to offer as security, may require the services of a company offering short-term paycheck advances. One such company offers a 14-day maximum advance of $255 for a fee of $45.

(a) What percent of the loan amount does the fee represent? (Round to the nearest hundredth of a percent.)

(b) Use a proportion to calculate the annual (365-day) percentage rate of this fee. (Round to the nearest percent.)

Preview

Measurement

Preview

Objective	Sample Problems		For help, go to

When you finish this chapter, you will be able to:

1. Determine the precision and accuracy of measurement numbers.

Determine the precision of each of the following measurement numbers. Which one is most precise?
(a) 4.27 psi   (b) 6758 psi   (c) 350 psi

(a) _____   Page 263

(b) _____

(c) _____

Determine the number of significant digits in each of the following measurement numbers.
(d) 9.6 kg   (e) 458 kg   (f) 6000 kg

(d) _____   Page 262

(e) _____

(f) _____

2. Add and subtract measurement numbers. (Round to the correct precision.)

(a)   38.26 sec + 5.9 sec =

(b)   8.37 lb − 2.435 lb =

(a) _____   Page 265

(b) _____

3. Multiply and divide measurement numbers. (Round to the proper accuracy.)

(a)   16.2 ft × 5.8 ft =

(b)   480 mi ÷ 32.5 mph =

(c)   240 lb/sq in. × 52.5 sq in. =

(a) _____   Page 267

(b) _____

(c) _____

Name _____

Date _____

Course/Section _____

		For help,
Objective	**Sample Problems**	**go to**

4. Convert units within the English system. (Round to the correct number of significant digits.)

    (a) 65.0 lb = _____ oz   Page 276

    (b) 8200 yd = _____ mi

    (c) 3.50 sq ft = _____ sq in.

5. Think metric.

    (a) Estimate the weight in pounds of a casting weighing 45 kg.   (a) _____   Page 294

    (b) Estimate the volume in quarts   (b) _____
           of 80 liters.

6. Convert units within the metric system. (Round to the correct number of significant digits.)

    (a) 6500 mg = _____ g   Page 296

    (b) 0.45 km = _____ m

    (c) 0.285 ha = _____ sq m

7. Convert between English and metric system units.

    (a) 48.0 liters = _____ gal   Page 297

    (b) 45 mi/hr = _____ km/h

    (c) 42°C = _____ °F

8. Use common technical measuring instruments.

Correctly read a length rule, micrometer, vernier caliper, protractor, and meters, and combine gauge blocks properly.   Page 314

(Answers to these preview problems are given in the Appendix. Also, worked solutions to many of these problems appear in the chapter Summary.)

If you are certain that you can work *all* these problems correctly, turn to page 339 for a set of practice problems. If you cannot work one or more of the preview problems, turn to the page indicated to the right of the problem. For those who wish to master this material with the greatest success, turn to Section 5-1 and begin work there.

Chapter 5

Measurement

© Zits Partnership. Reprinted with special Permission of King Features Syndicate.

5-1 Working with Measurement Numbers

Most of the numbers used in practical or technical work either come from measurements or lead to measurements. A carpenter must measure a wall to be framed; a machinist may weigh materials; an electronic technician may measure circuit resistance or voltage output. At one time or another each of these technicians must make measurements, do calculations based on measurements, perhaps round measurement numbers, and worry about the units of measurement. Of course, they must also work with measurement numbers from drawings, blueprints, or other sources.

Units A measurement number usually has units associated with it, and the unit name for that measurement number *must* be included when you use that number, write it, or talk about it. For example, we know immediately that 20 is a number, 20 ft is a length, 20 lb is a weight or force, and 20 sec is a time interval.

Suppose that the length of a piece of pipe is measured to be 8 ft. Two bits of information are given in this measurement: the size of the measurement (8) and the units of measurement (ft). The measurement unit compares the size of the quantity being measured to some standard.

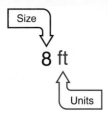

Size

8 ft

Units

Example 1

If the piece of pipe is measured to be 8 ft in length, it is 8 times the standard 1-ft length.

Length = 8 ft = 8 × 1 ft

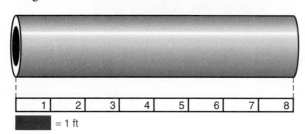

= 1 ft

→ **Your Turn**

Complete the following.

(a) 6 in. = 6 × _____

(b) 10 lb = _____ × _____

(c) $5 = 5 × _____

(d) 14.7 psi = _____ × _____

→ **Answers**

(a) 6 in. = 6 × 1 in.

(b) 10 lb = 10 × 1 lb

(c) $5 = 5 × $1

(d) 14.7 psi = 14.7 × 1 psi

Significant Digits

Every measuring instrument and every person making measurements has limitations. Measurement numbers—numbers used in technical work and in the trades—have a built-in uncertainty, and this uncertainty is not stated explicitly but is expressed by the number of digits used to communicate the measurement information.

In technical work, when we perform calculations with measurement numbers, we must round the result according to agreed-upon rules involving **significant digits.**

The significant digits in a number are those that represent an actual measurement. To determine the number of significant digits in a measurement number, follow these rules:

Rule 1 Digits other than zero are always significant.

Rule 2 A zero **is** significant when

    (a) it appears between two significant digits,

    (b) it is at the right end of a decimal number,

    (c) it is marked as significant with an overbar.

Rule 3 A zero **is not** significant when

    (a) it is at the right end of a whole number,

    (b) it is at the left end of a number.

Example 2

General Trades

- A tachometer reading of 84.2 rpm has three significant digits. All three digits in the number are nonzero. (Rule 1)

- A pressure reading of 3206 torr has four significant digits. The zero is significant because it appears between two nonzero digits. (Rule 2a)
- A voltage reading of 240 volts has two significant digits. The zero is not a significant digit because it is at the right end of a whole number. (Rule 3a)
- A voltage reading of $24\overline{0}$ volts has three significant digits. The overbar tells us that the zero marked is significant. (Rule 2c) The person making the measurement had an instrument that could measure to the nearest unit.
- A voltage reading of 24.0 volts has three significant digits. The zero is significant because it appears at the right end of a decimal number. (Rule 2b)
- A current reading of 0.025 amp has two significant digits. The zeros are not significant because they are at the left end of the number. (Rule 3b)

→ **Your Turn**

Determine the number of significant digits in each of the following measurement numbers.

(a) 3.04 lb   (b) 38,000 mi   (c) 15.730 sec

(d) 0.005 volt   (e) $18,4\overline{0}0$ gal   (f) 200.0 torr

→ **Answers**

(a) Three significant digits. The zero is significant because it is between two nonzero digits.

(b) Two significant digits. The zeros are not significant because they are at the right end of a whole number.

(c) Five significant digits. The zero is significant because it is at the right end of a decimal number.

(d) One significant digit. The zeros are not significant because they are at the left end of the number.

(e) Four significant digits. The zero with the overbar is significant, but the right-end zero is not significant.

(f) Four significant digits. The right-end digit is significant because it is at the right end of a decimal number. The middle two zeros are significant because they are in between two significant digits.

Precision and Accuracy

Because measurement numbers are never exact we need to be careful that the significant digits we specify in the number actually reflect the situation.

Look at this line:

If we measured the length of this line using a rough ruler, we would write that its length is

Length = 2 in.

To the nearest inch

But with a better measuring device, we might be able to say with confidence that its length is

Length = 2.1 in.

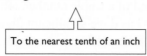

To the nearest tenth of an inch

Using a caliper, we might find that the length is

Length = 2.134 in.

To the nearest thousandth of an inch

We could go on using better and better measuring devices until we found ourselves looking through a microscope trying to decide where the ink line begins. If someone told us that the length of that line was

Length = 2.1304756 in.

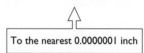

To the nearest 0.0000001 inch

we would know immediately that the length had been measured with incredible precision—they couldn't get a number like that by using a carpenter's rule!

To work with measurement numbers, we define precision and accuracy* as follows:

> The **precision** of a measurement number is indicated by the place value of its right-most significant digit.
>
> The **accuracy** of a measurement number refers to the number of significant digits that it contains.

Example 3

State the precision and accuracy of each of the following measurement numbers using the rules given.

(a)   4.01 sec   (b)   40 sec   (c)   420.1 sec

Measurement (a) is precise to the nearest hundredth of a second. It is accurate to three significant digits.

In (b) the zero is not significant. Therefore the number is precise to the nearest 10 seconds. It is accurate to one significant digit.

Measurement (c) is precise to the nearest tenth of a second. It is accurate to four significant digits.

---

*In engineering, *accuracy* is the degree of closeness between the measured value of some quantity and its true, or nominal, value. It has to do with the ability of an instrument to make measurements with small uncertainty. *Precision* is the degree of consistency and agreement among independent measurements of a quantity under the same conditions. For practical purposes we define them as shown in the box.

Notice that measurement (a) is considered the most precise because it has the most decimal digits. However, measurement (c) is considered the most accurate because it contains the most significant digits.

→ **Your Turn**

Determine the precision and accuracy of each of the following measurement numbers. Then indicate which is the most precise and which is the most accurate.

(a) 34.1 lb   (b) 0.568 lb   (c) 2735 lb   (d) 5.09 lb   (e) 400 lb

→ **Answers**

Measurement	Precise to the nearest . . .	Accurate to this number of significant digits
(a)	tenth of a pound	3
(b)	thousandth of a pound	3
(c)	pound	4
(d)	hundredth of a pound	3
(e)	hundred pounds	1

Measurement (b) is the most precise and measurement (c) is the most accurate.

TOLERANCE

Very often the technical measurements and numbers shown in drawings and specifications have a *tolerance* attached so that you will know the precision needed. For example, if a dimension on a machine tool is given as 2.835 ± 0.004 in., this means that the dimension should be 2.835 in. to within 4 thousandths of an inch. In other words, the dimension may be no more than 2.839 in. and no less than 2.831 in. These are the limits of size that are allowed or tolerated.

Addition and Subtraction of Measurement Numbers

When measurement numbers are added or subtracted, the resulting answer cannot be more precise than the least precise measurement number in the calculation. Therefore, use the following rule when adding or subtracting measurement numbers:

Always round a sum or difference to agree in precision with the least precise measurement number being added or subtracted.

Example 4

To add

4.1 sec + 3.5 sec + 1.27 sec = ____?____

First, check to be certain that all numbers have the same units. In the given problem all numbers have units of seconds. If one or more of the numbers are given in other units, you will need to convert so that all have the same units.

Second, add the numerical parts.

$$
\begin{array}{r}
4.1 \\
3.5 \\
+ \ 1.27 \\
\hline
8.87
\end{array}
$$
Notice that we line up the decimal points and add as we would with any decimal numbers.

4.1 $\boxed{+}$ **3.5** $\boxed{+}$ **1.27** $\boxed{=}$ → *8.87*

Third, attach the common units to the sum.

8.87 sec

Fourth, round the sum to agree in precision with the least precise number in the calculation. We must round 8.87 sec to the nearest tenth, or 8.9 seconds. Our answer agrees in precision with 4.1 sec and 3.5 sec, the least precise measurements in the calculation.

→ **Your Turn**

Perform the following subtraction and round your answer to the correct precision.

7.425 in. − 3.5 in. = _____?_____

→ **Solution**

First, check to be certain that both numbers have the same units. In this problem both numbers have inch units. If one of the numbers was given in other units, we would convert so that they have the same units. (You'll learn more about converting measurement units later in this chapter.)

Second, subtract the numerical parts.

$$
\begin{array}{r}
7.425 \\
- \ 3.500 \\
\hline
3.925
\end{array}
$$
⟵ Attach zeros if necessary

7.425 $\boxed{-}$ **3.5** $\boxed{=}$ → *3.925*

Third, attach the common units to the difference.   3.925 in.

Fourth, round the answer to agree in precision with the least precise number in the original problem. The least precise number is 3.5 in., which is precise to the nearest tenth. We should round our answer, 3.925 in., to the nearest tenth, or 3.9 in.

Example 5

Here are a few more examples to help fix the concept in your mind.

(a)   4.6 gal + 2.145 gal = 6.745 gal ≈ 6.7 gal

Round to the nearest tenth to agree with the least precise measurement number, 4.6 gal, which is stated to the nearest tenth.

(b) $0.24 \text{ mi} - 0.028 \text{ mi} = 0.212 \text{ mi} \approx 0.21 \text{ mi}$

Round the answer to the nearest hundredth to agree with the least precise number, 0.24 mi.

(c) $2473 \text{ lb} + 321.2 \text{ lb} = 2794.2 \text{ lb} \approx 2794 \text{ lb}$

Round to the nearest whole number to agree with the least precise measurement, 2473 lb.

→ **More Practice**

Try these problems for practice in adding and subtracting measurement numbers. Be careful to round your answers correctly.

(a) 10.5 in. + 8.72 in.

(b) 2.46 cu ft + 3.517 cu ft + 4.8 cu ft

(c) 47.5 sec − 21.844 sec

(d) 56 lb − 4.8 lb

(e) 325 sq ft + 11.8 sq ft + 86.06 sq ft

(f) 0.027 volt + 0.18 volt + 0.009 volt

→ **Answers**

(a) $10.5 \text{ in.} + 8.72 \text{ in.} = 19.22 \text{ in.} \approx 19.2 \text{ in.}$

(b) $2.46 \text{ cu ft} + 3.517 \text{ cu ft} + 4.8 \text{ cu ft} = 10.777 \text{ cu ft} \approx 10.8 \text{ cu ft}$

(c) $47.5 \text{ sec} - 21.844 \text{ sec} = 25.656 \text{ sec} \approx 25.7 \text{ sec}$

(d) $56 \text{ lb} - 4.8 \text{ lb} = 51.2 \text{ lb} \approx 51 \text{ lb}$

(e) $325 \text{ sq ft} + 11.8 \text{ sq ft} + 86.06 \text{ sq ft} = 422.86 \text{ sq ft} \approx 423 \text{ sq ft}$

(f) $0.027 \text{ volt} + 0.18 \text{ volt} + 0.009 \text{ volt} = 0.216 \text{ volt} \approx 0.22 \text{ volt}$

Note   If some of the numbers to be added are in fraction form, you should convert them to decimal form before adding. For example, the sum $3.2 \text{ lb} + 1\frac{3}{4} \text{ lb} + 2.4 \text{ lb}$ becomes

$$
\begin{array}{r}
3.2 \ \text{lb} \\
1.75 \ \text{lb} \\
+2.4 \ \text{lb} \\
\hline
7.35 \ \text{lb}
\end{array}
$$

If we do not know the true precision of the fractional measurement, the usual procedure is to round to the precision of the original *decimal* measurements. The least precise of these, 3.2 lb and 2.4 lb, are both given to the nearest tenth of a pound. Our answer, 7.35 lb, should be rounded to 7.4 lb. ◄

Multiplication and Division of Measurement Numbers

When multiplying or dividing measurement numbers, we need to pay special attention to both the proper units and the proper rounding of our final answer. First, let's consider the units.

With both addition and subtraction of measurement numbers, the answer to the arithmetic calculation has the same units as the numbers being added or subtracted. This is not true when we multiply or divide measurement numbers.

Example 6

To multiply

3 ft × 2 ft = _____?_____

First, multiply the numerical parts: $3 \times 2 = 6$

Second, multiply the units: $1 \text{ ft} \times 1 \text{ ft} = 1 \text{ ft}^2$ or 1 sq ft

The answer is 6 ft^2 or 6 sq ft—both indicate 6 square feet.

▶ Learning Help | It may help if you remember that this multiplication is actually

$$3 \text{ ft} \times 2 \text{ ft} = (3 \times 1 \text{ ft}) \times (2 \times 1 \text{ ft})$$
$$= (3 \times 2) \times (1 \text{ ft} \times 1 \text{ ft})$$
$$= 6 \times 1 \text{ sq ft} = 6 \text{ sq ft}$$

The rules of arithmetic say that we can do the multiplications in any order we wish, so we multiply the units part separately from the number part.

Notice that the product 1 ft × 1 ft can be written either as 1 ft^2 or as 1 sq ft. ◀

Example 7

To multiply

2 in. × 3 in. × 5 in. = _____?_____

First, multiply the numbers: $2 \times 3 \times 5 = 30$

Second, multiply the units: $1 \text{ in.} \times 1 \text{ in.} \times 1 \text{ in.} = 1 \text{ in.}^3$ or 1 cu in. (1 cubic inch).

The answer is 30 cu in.

VISUALIZING UNITS

It is helpful to have a visual understanding of measurement units and the "dimension" of a measurement.

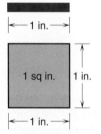

The length unit **inch,** in., specifies a **one-**dimensional or linear measurement. It gives the length of a straight line.

The area unit **square inch,** sq in., specifies a **two-**dimensional measurement. Think of a square inch as giving the area of a square whose sides are one inch in length.

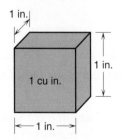

The volume unit **cubic inch,** cu in., specifies a **three-**dimensional measurement. Think of a cubic inch as the volume of a cube whose edges are each one inch in length.

Division with measurement numbers requires that you take the same care with the units that is needed in multiplication.

Example 8

To divide

$$24 \text{ mi} \div 1.5 \text{ hr} = \underline{\quad ? \quad}$$

First, divide the numerical parts as usual.

$$24 \div 1.5 = 16$$

Second, divide the units.

$$1 \text{ mi} \div 1 \text{ hr} = \frac{1 \text{ mi}}{1 \text{ hr}} \quad \text{or} \quad 1 \text{ mph} \quad \text{or} \quad 1 \text{ mi/hr}$$

In this case the unit $\frac{1 \text{ mi}}{1 \text{ hr}}$ is usually written mi/hr or mph—miles per hour—and the answer to the problem is 16 mph. Remember that

$$24 \text{ mi} \div 1.5 \text{ hr} = \frac{24 \times 1 \text{ mi}}{1.5 \times 1 \text{ hr}} = \left(\frac{24}{1.5}\right) \times \left(\frac{1 \text{ mi}}{1 \text{ hr}}\right)$$

$$= 16 \times 1 \text{ mph} = 16 \text{ mph}$$

Units such as sq ft or mph, which are made up of a combination of simpler units, are called *compound* units.

> **Note** Both h and hr are acceptable abbreviations for the unit *hour.* The standard metric abbreviation is h, and hr is in common usage in the United States. ◄

> **→ Your Turn**
>
> $$36 \text{ lb} \div 1.5 \text{ sq in.} = \underline{\quad\quad}$$
>
> **→ Solution**
>
> **First,** divide the numbers: $36 \div 1.5 = 24$
>
> **Second,** divide the units: $1 \text{ lb} \div 1 \text{ sq in.} = \dfrac{1 \text{ lb}}{1 \text{ sq in.}} = 1 \text{ lb/sq in.} = 1 \text{ psi}$
>
> $$\left(\text{The unit } psi \text{ means "pounds per square inch" or } \frac{\text{lb}}{\text{sq in.}} . \right)$$
>
> The answer is 24 psi.

When multiplying or dividing quantities having compound units, it is sometimes possible to "cancel" or divide out common units. This procedure is identical to the cancellation of common numerical factors when multiplying fractions.

Example 9

(a) To multiply

$$45 \text{ mph} \times 4 \text{ hr} = \underline{\quad ? \quad}$$

set it up this way:

$$45\frac{\text{mi}}{\text{hr}} \times 4\,\text{hr} = (45 \times 4) \times \left(\frac{\text{mi}}{\cancel{\text{hr}}} \times \cancel{\text{hr}}\right)$$

$$= 180\,\text{mi}$$

Notice that the "hr" units cancelled just as common numerical factors do.

(b) To divide

$$42\,\text{sq yd} \div 6\,\text{yd} = \underline{\quad ? \quad}$$

work with the numbers and the units separately as follows.

$$\left(\frac{42}{6}\right) \times \left(\frac{\text{sq yd}}{\text{yd}}\right) = \left(\frac{42}{6}\right) \times \left(\frac{\text{yd} \times \cancel{\text{yd}}}{\cancel{\text{yd}}}\right) = 7\,\text{yd}$$

→ More Practice

Before we discuss rounding in multiplication and division, try the following practice problems. Be careful with units.

(a) 5 ft × 4 ft   (b) 4 ft ÷ 2 sec

(c) 2 yd × 6 yd × 0.5 yd   (d) 24 sq in. ÷ 3 in.

(e) 30 mph × 2 hr   (f) 70 ft ÷ 35 ft/sec

→ Answers

(a) 20 sq ft   (b) 2 ft/sec

(c) 6 cu yd   (d) 8 in.

(e) 60 mi   (f) 2 sec

A Closer Look   Did you have difficulty with (d), (e), or (f)? Do them this way:

(d) $24\,\text{sq in.} \div 3\,\text{in.} = \left(\dfrac{24}{3}\right) \times \left(\dfrac{\text{sq in.}}{\text{in.}}\right)$

$$= 8 \times \left(\frac{\text{in.}^2}{\text{in.}}\right) = 8 \times \left(\frac{\text{in.} \times \cancel{\text{in.}}}{\cancel{\text{in.}}}\right) = 8\,\text{in.}$$

(e) $30\,\text{mph} \times 2\,\text{hr} = 30\dfrac{\text{mi}}{\text{hr}} \times 2\,\text{hr}$

$$= (30 \times 2) \times \left(\frac{\text{mi}}{\cancel{\text{hr}}} \cdot \cancel{\text{hr}}\right)$$

$$= 60\,\text{mi}$$

(f) $70\,\text{ft} \div 35\,\text{ft/sec} = \left(\dfrac{70}{35}\right) \times (\text{ft} \div \text{ft/sec})$

$$= 2 \times \left(\cancel{\text{ft}} \times \frac{\text{sec}}{\cancel{\text{ft}}}\right)$$

$$= 2\,\text{sec} \blacktriangleleft$$

Rounding Products and Quotients

When measurement numbers are multiplied or divided, the resulting answer cannot be expressed with greater accuracy than the least accurate measurement number in the calculation. The result of a calculation should not be written to reflect greater accuracy than any number used in that calculation.

> Always round a product or quotient to the same number of significant digits as the least accurate number used in the calculation.

Example 10

Multiply: 4.35 ft $\times$ 3.6 ft.

First, multiply the numerical factors: $4.35 \times 3.6 = 15.66$

Second, multiply the units: $1 \text{ ft} \times 1 \text{ ft} = 1 \text{ ft}^2$ or 1 sq ft

Third, round the answer to agree in accuracy with the least accurate number in the calculation.

4.35 ft is accurate to three significant digits.

3.6 ft is accurate to two significant digits.

The least accurate factor has two significant digits, so the answer must be rounded to two significant digits.

15.66 sq ft $\approx$ 16 sq ft

→ Your Turn

Divide: 375 mi $\div$ 65 mph.

→ Solution

First, divide the numerical parts: $375 \div 65 = 5.7692\ldots$

Second, divide the units:

$$\text{mi} \div \frac{\text{mi}}{\text{hr}} = \cancel{\text{mi}} \times \frac{\text{hr}}{\cancel{\text{mi}}} = \text{hr}$$

Third, round the answer to two significant digits to agree in accuracy with 65, the least accurate measurement number in the calculation.

375 mi $\div$ 65 mph $\approx$ 5.8 hr

Note Numbers that result from simple counting should be considered to be exact. They do not affect the accuracy of the calculation. For example, if a flywheel turns 48 rotations in 1.25 minutes, the speed of the flywheel is calculated as $48 \div 1.25 = 38.4$ rpm. We should assume that the 48 is an exact number—rotations are counted directly—so we express the answer as 38.4 with three significant digits to agree with 1.25, the least accurate measurement number in the calculation. ◄

Perform the following calculations. Round each answer to the proper accuracy.

(a) 7.2 in. × 0.48 in.   (b) 0.18 ft × 3.15 ft × 20 ft

(c) 82.45 ft ÷ 3.65 sec   (d) 28 sq yd ÷ 1.25 yd

(e) 2450 ton/sq ft × 3178 sq ft   (f) 148 mi ÷ 15 mph

→ **Answers**

(a) 7.2 in. × 0.48 in. = 3.456 sq in. ≈ 3.5 sq in.

(b) 0.18 ft × 3.15 ft × 20 ft = 11.34 cu ft ≈ 10 cu ft

(c) 82.45 ft ÷ 3.65 sec = 22.58904... ft/sec ≈ 22.6 ft/sec

(d) 28 sq yd ÷ 1.25 yd = 22.4 yd ≈ 22 yd

(e) 2450 ton/sq ft × 3178 sq ft = 7786100 ton ≈ 7,790,000 ton

(f) 148 mi ÷ 15 mph = 9.8666... hr ≈ 9.9 hr

Careful   When dividing by hand it is important that you carry the arithmetic to one digit more than will be retained after rounding. ◀

Follow these rounding rules for all calculations shown in this book unless specific instructions for rounding are given.

Decimal Equivalents   Some of the measurements made by technical workers in a shop are made in fractions. But on many shop drawings and specifications, the dimensions may be given in decimal form as shown in the figure.

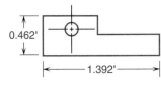

The steel rule used in the shop or in other technical work may be marked in 8ths, 16ths, 32nds, or even 64ths of an inch. A common problem is to rewrite the decimal number to the nearest 32nd or 64th of an inch and to determine how much error is involved in using the fraction number rather than the decimal number. The following example illustrates how to do this.

Example 11

Machine Trades   In the preceding drawing, what is the fraction, to the nearest 32nd of an inch, equivalent to 0.462 in.?

Step 1   Multiply the decimal number by the fraction $\frac{32}{32}$.

$$0.462 \text{ in.} = 0.462 \text{ in.} \times \boxed{\frac{32}{32}}$$

$$= \frac{0.462 \text{ in.} \times 32}{32}$$

$$= \frac{14.784}{32} \text{ in.}$$

Step 2   Round to the nearest 32nd of an inch by rounding the numerator to the nearest whole number.

$$\approx \frac{15}{32} \text{ in.}$$

Step 3 The error involved is the difference between

$$\frac{15}{32} \text{ in. and } \frac{14.784}{32} \text{ in., or } \frac{15 - 14.784}{32} \text{ in. } = \frac{0.216}{32} \text{ in.}$$

$$= 0.00675 \text{ in.}$$

$$\approx 0.0068 \text{ in. rounded}$$

The error involved in using $\frac{15}{32}$ in. instead of 0.462 in. is about 68 ten-thousandths of an inch. In problems of this kind we would usually round the error to the nearest ten-thousandth.

Fraction: **.462** ⊠ **32** ⊜ → 　　14.784　　 Round to $\frac{15}{32}$.

C5-1

Error: **15** ⊖ (ANS) ⊜ ⊘ **32** ⊜ → 　　0.00675　　 about 0.0068 in.

▶ **A Closer Look**　Pressing the (ANS) key retrieves the last answer—in this case, 14.784. ◀

→ **Your Turn**

Change 1.392 in. to a fraction expressed in 64ths of an inch, and find the error.

→ **Solution**

$$1.392 \text{ in.} = 1.392 \text{ in.} \times \frac{64}{64}$$

$$= \frac{1.392 \times 64}{64} \text{ in.}$$

$$= \frac{89.088}{64} \text{ in.}$$

$$\approx \frac{89}{64} \text{ in.} \quad \boxed{\text{Rounded to the nearest unit}}$$

$$\approx 1\frac{25}{64} \text{ in.}$$

The error in using the fraction rather than the decimal is

$$\frac{89.088}{64} \text{ in.} - \frac{89}{64} \text{ in.} = \frac{0.088}{64} \text{ in.}$$

$$\approx 0.0014 \text{ in.} \qquad \text{or about fourteen ten-thousandths of an inch.}$$

Fraction: **1.392** ⊠ **64** ⊜ → 　　89.088　　 Round to $\frac{89}{64}$.

Error: ⊖ **89** ⊜ ⊘ **64** ⊜ → 　　0.001375　　 about 0.0014 in.

→ **More Practice**

(a) Express to the nearest 16th of an inch.

　(1)　0.438 in.　　　　(2)　0.30 in.

　(3)　0.18 in.　　　　(4)　2.70 in.

(b) Express to the nearest 32nd of an inch.

    (1)   1.650 in.   (2)   0.400 in.

    (3)   0.720 in.   (4)   2.285 in.

(c) Express to the nearest 64th of an inch and find the error, to the nearest ten-thousandth, involved in using the fraction in place of the decimal number.

    (1)   0.047 in.   (2)   2.106 in.

    (3)   1.640 in.   (4)   0.517 in.

→ **Answers**

(a) (1) $\frac{7}{16}$ in.   (2) $\frac{5}{16}$ in.   (3) $\frac{3}{16}$ in.   (4) $2\frac{11}{16}$ in.

(b) (1) $1\frac{21}{32}$ in.   (2) $\frac{13}{32}$ in.   (3) $\frac{23}{32}$ in.   (4) $2\frac{9}{32}$ in.

(c) (1) $\frac{3}{64}$ in. The error is about 0.0001 in.

    (2) $2\frac{7}{64}$ in. The error is about 0.0034 in.

    (3) $1\frac{41}{64}$ in. The error is about 0.0006 in.

    (4) $\frac{33}{64}$ in. The error is about 0.0014 in.

A DIFFERENT WAY TO ROUND NUMBERS

Some engineers and technicians use the following rounding rule rather than the ones we presented in Chapter 3.

Examples

1. Determine the place to which the number is to be rounded. Mark it with a $\wedge$.

    Round   4.786 to three digits.

            4.78$\wedge$6

2. When the digit to the right of the mark is greater than 5, increase the digit to the left by 1.

    4.78$\wedge$6   becomes   4.79

    1.701$\wedge$72   becomes   1.702

3. When the digit to the right of the mark is less than 5, drop the digits to the right or replace them with zeros.

    3.81$\wedge$2   becomes   3.81

    14$\wedge$39   becomes   1400

4. When the digit to the right of the mark is equal to 5, and there are no other nonzero digits to the right of the 5, round the left digit to the nearest *even* value.

    1.41$\wedge$5   becomes   1.42

    37.00$\wedge$5   becomes   37.00

    53$\wedge$50   becomes   5400

5. In step 4, if there are any nonzero digits after the 5, always round up.

    2.62$\wedge$52   becomes   2.63

    487$\wedge$56   becomes   48800

For a set of practice problems over this section of Chapter 5, turn to Exercises 5-1.

Working with Measurement Numbers

A. State the precision and the accuracy of each of the following measurement numbers.

1. 4.7 gal
2. 0.008 ft
3. 562 psi
4. 3400 yd
5. 7.024 lb
6. 360 volts
7. $19,\overline{0}00$ tons
8. 19.65 oz
9. 10.0 in.
10. 0.20 sec

B. Add or subtract as shown. Round your answer to the correct precision.

1. 7 in. + 14 in.
2. 628 lb + 400 lb
3. 3.25 in. + 1.7 in. + 4.6 in.
4. 16.2 psi + 12.64 psi
5. 0.28 gal + 1.625 gal
6. 64.5 mi − 12.72 mi
7. 1.565 oz − 0.38 oz
8. 12.372 sec − 2.1563 sec
9. 28.3 psi + 16 psi + 5.9 psi
10. 6.4 tons + 8.26 tons − 5.0 tons

C. Multiply or divide as shown. Round to the appropriate number of significant digits.

1. 6 ft × 7 ft
2. 3.1 in. × 1.7 in.
3. 3.0 ft × 2.407 ft
4. 215 mph × 1.2 hr
5. $1\frac{1}{4}$ ft × 1.5 ft
6. 48 in. × 8.0 in.
7. 2.1 ft × 1.7 ft × 1.3 ft
8. 18 sq ft ÷ 2.1 ft
9. 458 mi ÷ 7.3 hr
10. $1.37 ÷ 3.2 lb
11. $40\frac{1}{2}$ mi ÷ 25 mph
12. 2.0 sq ft ÷ 1.073 ft

D. Convert as shown.

1. Write each of the following fractions as a decimal number rounded to two decimal places.

    (a) $1\frac{7}{8}$ in.
    (b) $4\frac{3}{64}$ in.
    (c) $3\frac{1}{8}$ sec
    (d) $\frac{1}{16}$ in.
    (e) $\frac{3}{32}$ in.
    (f) $1\frac{3}{16}$ lb
    (g) $2\frac{17}{64}$ in.
    (h) $\frac{19}{32}$ in.
    (i) $\frac{3}{8}$ lb

2. Write each of the following decimal numbers as a fraction rounded to the nearest 16th of an inch.

    (a) 0.921 in.
    (b) 2.55 in.
    (c) 1.80 in.
    (d) 3.69 in.
    (e) 0.802 in.
    (f) 0.306 in.
    (g) 1.95 in.
    (h) 1.571 in.
    (i) 0.825 in.

3. Write each of the following decimal numbers as a fraction rounded to the nearest 32nd of an inch.

    (a) 1.90 in.
    (b) 0.85 in.
    (c) 2.350 in.
    (d) 0.666 in.
    (e) 2.091 in.
    (f) 0.285 in.
    (g) 0.600 in.
    (h) 0.685 in.
    (i) 1.525 in.

4. Write each of the following decimal numbers as a fraction rounded to the nearest 64th of an inch, and find the error, to the nearest ten-thousandth, involved in using the fraction in place of the decimal number.

   (a) 0.235 in.   (b) 0.515 in.   (c) 1.80 in.   (d) 2.420 in.

   (e) 3.175 in.   (f) 2.860 in.   (g) 1.935 in.   (h) 0.645 in.

   (i) 0.480 in.

E. Practical Problems

   1. **Machine Trades**   Specifications call for drilling a hole 0.637 ± 0.005 in. in diameter. Will a $\frac{41}{64}$-in. hole be within the required tolerance?

   2. **Metalworking**   According to standard American wire size tables, 3/0 wire is 0.4096 in. in diameter. Write this size to the nearest 32nd of an inch. What error is involved in using the fraction rather than the decimal?

   3. **Automotive Trades**   An auto mechanic converts a metric part to 0.473 in. The part comes only in fractional sizes given to the nearest 64th of an inch. What would be the closest size?

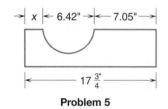

Problem 5

   4. **General Trades**   What size wrench, to the nearest 32nd of an inch, will fit a bolt with a 0.748-in. head?

   5. **Drafting**   Find the missing distance x in the drawing.

   6. **Carpentry**   A truckload of knotty pine paneling weighs 180 lb. If this kind of wood typically has a density of 38 lb/cu ft, find the volume of wood in the load. (*Hint:* Divide the weight by the density.) Round to the nearest 0.1 cu ft.

   7. **Automotive Trades**   The wear on a cylinder has increased its bore to 3.334 in. The most likely original bore was the nearest 32nd-in. size less than this. What was the most likely original bore?

   8. **Interior Design**   If the rectangular floor of a room has an area of 455 sq ft and a length of 26 ft, what is its width?

   9. **Transportation**   A truck, moving continuously, averages 52.5 mph for 6.75 hr. How far does it travel during this time?

   10. **Industrial Technology**   The volume of a cylindrical tank can be calculated as its base area multiplied by its height. If the tank has a capacity of 2760 cu ft of liquid and has a base area of 120 sq ft, what is its height?

   11. **Electronics**   If an electromagnetic signal moves at the speed of light, about 186,282 miles per second, how long will it take this signal to travel from New Jersey to California, approximately 2500 miles?

When you have completed these problems, check your answers to the odd-numbered problems in the Appendix, then turn to Section 5-2 for some information on units and unit conversion.

5-2 English Units and Unit Conversion

When you talk about a 4-ft by 8-ft wall panel, a $1\frac{1}{4}$-in. bolt, 3 lb of solder, or a gallon of paint, you are comparing an object with a *standard*. The units or standards used in science, technology, or the trades have two important characteristics: They are convenient in size, and they are standardized. Common units such as the foot, pound, gallon, or hour were originally chosen because they were related either to natural quantities or to body measurements.

A pinch of salt, a drop of water, a handful of sand, or a tank of gas are all given in *natural* units. The width of a finger, the length of a foot, or a stride length are all distances given in *body-related* units. In either case, there is a need to define and standardize these units before they are useful for business, science, or technology.

In ancient times, the *digit* was a length unit defined as the width of a person's finger. Four digits was called a *hand,* a unit still used to measure horses. Four hands was a *cubit,* the distance from fingertip to elbow, about 18 in. One *foot* was defined as the length of a man's foot. A *yard* was the distance from nose to tip of outstretched arm. Of course all these lengths depend on whose finger, hand, arm, or foot is used.

Standardization in the English system of units began in the fifteenth century, and one yard was defined as the length of a standard iron bar kept in a London vault. One foot was defined as exactly one-third the length of the bar, and one inch was one-twelfth of a foot. (The word *inch* comes from Latin and Anglo-Saxon words meaning "one-twelfth.") Today the inch is legally defined worldwide in terms of metric or SI units.

We will consider the familiar English system of units in this section and metric units in Section 5-3.

Because there are a great many units available for writing the same quantity, it is necessary that you be able to convert a given measurement from one unit to another. For example, a length of exactly one mile can be written as

1 mile = 1760 yd, for the traffic engineer
        = 5280 ft, for a landscape architect
        = 63,360 in., in a science problem
        = 320 rods, for a surveyor
        = 8 furlongs, for a horse racing fan
        = 1609.344 meters, in metric units

Here are some basic conversion factors for length, weight, and liquid capacity in the English system of units.

LENGTH UNITS

1 foot (ft) = 12 inches (in.)
1 yard (yd) = 3 ft = 36 in.
1 rod (rd) = $16\frac{1}{2}$ ft
1 mile (mi) = 5280 ft = 1760 yd

WEIGHT UNITS

1 pound (lb) = 16 ounces (oz)
1 ton = 2000 lb

LIQUID CAPACITY UNITS

1 quart (qt) = 2 pints (pt)
1 gallon (gal) = 4 qt

In this section we will show you a quick and mistake-proof way to convert measurements from one unit to another. Whatever the units given, if a new unit is defined, you should be able to convert the measurement to the new units.

Example 1

Consider the following simple unit conversion. Convert 4 yards to feet.

4 yd = _____ ft

From the table we know that 1 yard is defined as exactly 3 feet; therefore,

4 yd = 4 × 3 ft = 12 ft

Easy.

→ **Your Turn**

Try another, more difficult, conversion. Surveyors use a unit of length called a *rod,* defined as exactly $16\frac{1}{2}$ ft.

Convert: 11 yd = _____ rods.

→ **Solution**

The correct answer is 2 rods.

You might have thought to first convert 11 yd to 33 ft and then noticed that 33 ft divided by $16\frac{1}{2}$ equals 2. There are 2 rods in a 33-ft length.

Unity Fractions

Most people find it difficult to reason through a problem in this way. To help you, we have devised a method of solving *any* unit conversion problem quickly with no chance of error. This is the *unity fraction* method.

The first step in converting a number from one unit to another is to set up a unity fraction from the definition linking the two units.

Example 2

To convert a distance from yard units to foot units, take the equation relating yards to feet:

1 yd = 3 ft

and form the fractions $\dfrac{1 \text{ yd}}{3 \text{ ft}}$ and $\dfrac{3 \text{ ft}}{1 \text{ yd}}$.

These fractions are called unity fractions because they are both equal to 1. Any fraction whose top and bottom are equal has the value 1.

→ **Your Turn**

Practice this step by forming pairs of unity fractions from each of the following definitions.

(a) 12 in. = 1 ft   (b) 1 lb = 16 oz   (c) 4 qt = 1 gal

(a) 12 in. = 1 ft. The unity fractions are $\dfrac{12 \text{ in.}}{1 \text{ ft}}$ and $\dfrac{1 \text{ ft}}{12 \text{ in.}}$.

(b) 1 lb = 16 oz. The unity fractions are $\dfrac{1 \text{ lb}}{16 \text{ oz}}$ and $\dfrac{16 \text{ oz}}{1 \text{ lb}}$.

(c) 4 qt = 1 gal. The unity fractions are $\dfrac{4 \text{ qt}}{1 \text{ gal}}$ and $\dfrac{1 \text{ gal}}{4 \text{ qt}}$.

The second step in converting units is to multiply the original number by one of the unity fractions. Always choose the unity fraction that allows you to cancel out the units you do not want in your answer. The units you do not want will be in the denominator of the appropriate unity fraction.

Example 3

To convert the distance 4 yd to foot units, multiply this way:

$$4 \text{ yd} = (4 \times 1 \, \cancel{\text{yd}}) \times \left(\frac{3 \text{ ft}}{1 \, \cancel{\text{yd}}} \right) \quad \boxed{\text{Choose the unity fraction that cancels yd and replaces it with ft}}$$

$$= 4 \times 3 \text{ ft}$$

$$= 12 \text{ ft}$$

You must choose one of the two fractions. Multiply by the fraction that allows you to cancel out the *yard* units you do not want and to keep the *feet* units you do want.

Did multiplying by a fraction give you any trouble? Remember that any number A can be written as $\dfrac{A}{1}$, so that we have

$$A \times \frac{B}{C} = \frac{A}{1} \times \frac{B}{C} \text{ or } \frac{A \times B}{C}$$

This   means   this

Example 4

$$6 \times \frac{3}{4} = \frac{\overset{3}{\cancel{6}}}{1} \times \frac{3}{\underset{2}{\cancel{4}}}$$

$$= \frac{9}{2} \text{ or } 4\frac{1}{2}$$

(If you need to review the multiplication of fractions, turn back to page 81.)

Now try this problem. Use unity fractions to convert 44 oz to pounds. Remember that 1 lb = 16 oz.

→ Solution

Use the equation 1 lb $=$ 16 oz to set up the unity fractions $\dfrac{1\text{ lb}}{16\text{ oz}}$ and $\dfrac{16\text{ oz}}{1\text{ lb}}$.

Multiply by the first fraction to convert ounces to pounds.

$$44\text{ oz} = (44\ \cancel{\text{oz}}) \times \left(\frac{1\text{ lb}}{16\ \cancel{\text{oz}}}\right)$$

| Choose the unity fraction that cancels oz and replaces it with lb |

$$= \frac{44 \times 1\text{ lb}}{16}$$

$$= \frac{44}{16}\text{ lb or } 2\frac{3}{4}\text{ lb}$$

Note When measurement units are converted by division, the remainder often is expressed as the number of original units. For example, instead of giving this last answer as $2\frac{3}{4}$ lb, we can express it as 2 lb 12 oz. ◄

It may happen that several unity fractions are needed in the same problem.

Example 5

If you know that

$$1\text{ rod} = 16\tfrac{1}{2}\text{ ft} \quad\text{and}\quad 1\text{ yd} = 3\text{ ft}$$

then to convert 11 yd to rods multiply as follows:

| First cancel yd and keep ft | | Then cancel ft and keep rod |

$$11\text{ yd} = (11\ \cancel{\text{yd}}) \times \left(\frac{3\ \cancel{\text{ft}}}{1\ \cancel{\text{yd}}}\right) \times \left(\frac{1\text{ rod}}{16\frac{1}{2}\ \cancel{\text{ft}}}\right)$$

$$= \frac{11 \times 3 \times 1\text{ rod}}{16\frac{1}{2}}$$

$$= \frac{33}{16\frac{1}{2}}\text{ rod}$$

$$11\text{ yd} = 2\text{ rods}$$

To work with unity fractions on a calculator, simply multiply by any quantity in a numerator and divide by any quantity in a denominator. For this example we would enter

11 ⊗ 3 ⊘ 16.5 ⊜ → ░░░░░░ *2.*

Be careful to set up the unity fraction so that the unwanted units cancel. At first it helps to write both fractions and then choose the one that fits. As you become an expert at unit conversion, you will be able to write down the correct fraction at first try.

Conversion factors are usually definitions and therefore are considered to be **exact** numbers rather than approximate measurement numbers. When converting units, round all answers to the same number of significant digits as the original measurement number. ◀

→ **Your Turn**

Try these practice problems. Use unity fractions to convert. Be sure to round your answers properly.

(a) 6.25 ft = _____ in.

(b) $5\frac{1}{4}$ yd = _____ ft

(c) 2.1 mi = _____ yd

(d) 12 gallons = _____ pints

(e) 32 psi = _____ atm (1 atm = 14.7 psi)

(f) 87 in. = _____ ft _____ in.

(g) 9 lb 7 oz = _____ oz (Round to the nearest ounce.)

(h) **Automotive Trades**   The circumference of each tire on a truck is approximately 98 in., which means that the truck travels 98 in. with each revolution of the tire. How many miles will the truck travel if the wheels make 8500 revolutions (rev)?

→ **Solutions**

(a) $6.25 \text{ ft} = (6.25 \text{ ft}) \times \left(\dfrac{12 \text{ in.}}{1 \text{ ft}}\right)$

$= 6.25 \times 12 \text{ in.}$

$= 75 \text{ in.}$

(b) $5\frac{1}{4} \text{ yd} = (5\frac{1}{4} \text{ yd}) \times \left(\dfrac{3 \text{ ft}}{1 \text{ yd}}\right)$

$= 5\frac{1}{4} \times 3 \text{ ft}$

$= 15\frac{3}{4} \text{ ft}$

(c) $2.1 \text{ mi} = (2.1 \text{ mi}) \times \left(\dfrac{1760 \text{ yd}}{1 \text{ mi}}\right)$

$= 2.1 \times 1760 \text{ yd}$

$= 3696 \text{ yd}$

$\approx 3700 \text{ yd}$ rounded to 2 significant digits

(d) $12 \text{ gal} = (12 \text{ gal}) \times \left(\dfrac{4 \text{ qt}}{1 \text{ gal}} \times \dfrac{2 \text{ pt}}{1 \text{ qt}}\right)$

$= 12 \times 4 \times 2 \text{ pt}$

$= 96 \text{ pt}$

(e) $32 \text{ psi} = (32 \text{ psi}) \times \left(\dfrac{1 \text{ atm}}{14.7 \text{ psi}}\right)$

$= \dfrac{32}{14.7} \text{ atm}$

$\approx 2.176870748 \text{ atm}$

$\approx 2.2 \text{ atm}$ rounded to 2 significant digits

(f) $87 \text{ in.} = (87 \text{ in.}) \times \left(\dfrac{1 \text{ ft}}{12 \text{ in.}}\right)$

$= \dfrac{87}{12} \text{ ft}$

$= 7\text{R}3 \text{ ft} = 7 \text{ ft } 3 \text{ in.}$

(g) **First,** convert the pounds to ounces:

$9 \text{ lb} = (9 \text{ lb}) \times \left(\dfrac{16 \text{ oz}}{1 \text{ lb}}\right)$

$= 9 \times 16 \text{ oz}$

$= 144 \text{ oz}$

Then, add the ounces portion of the original measurement:

$$9 \text{ lb } 7 \text{ oz} = 144 \text{ oz} + 7 \text{ oz}$$
$$= 151 \text{ oz}$$

(h) **First,** convert revolutions to inches:

$$8500 \text{ rev} = (8500 \text{ rev}) \times \left(\frac{98 \text{ in.}}{1 \text{ rev}} \right)$$

$$= 8500 \times 98 \text{ in.}$$

$$= 833{,}000 \text{ in.}$$

Then, convert inches to miles:

$$833{,}000 \text{ in.} = (833{,}000 \text{ in.}) \times \left(\frac{1 \text{ ft}}{12 \text{ in.}} \right) \times \left(\frac{1 \text{ mi}}{5280 \text{ ft}} \right)$$

$$= \frac{833{,}000}{12 \times 5280} \text{ mi}$$

$$\approx 13.147 \text{ mi}$$

$$\approx 13 \text{ mi rounded to two significant digits}$$

Compound Units

In the English system of units commonly used in the United States in technical work, the basic units for length, weight, and time are named with a single word or abbreviation: ft, lb, and sec. We often define units for other similar quantities in terms of these. For example, 1 mile = 5280 ft, 1 ton = 2000 lb, 1 hr = 3600 sec, and so on. We may also define units for more complex quantities using these basic units. For example, the common unit of speed is miles per hour or mph, and it involves both distance (miles) and time (hour) units. Units named using the product or quotient of two or more simpler units are called *compound units.*

Example 6

If a car travels 75 miles at a constant rate in 1.5 hours, it is moving at a speed of

$$\frac{75 \text{ miles}}{1.5 \text{ hours}} = \frac{75 \times 1 \text{ mi}}{1.5 \times 1 \text{ hr}}$$

$$= \frac{75}{1.5} \times \boxed{\frac{1 \text{ mi}}{1 \text{ hr}}}$$

$$= 50 \text{ mi/hr}$$

We write this speed as 50 mi/hr or 50 mph, and read it as "50 miles per hour." Ratios and rates are often written with compound units.

→ Your Turn

Which of the following are expressed in compound units?

(a) Diameter, 4.65 in.

(b) Density, 12 lb/cu ft

(c) Gas usage, mi/gal or mpg

(d) Current, 4.6 amperes

(e) Rotation rate, revolution/min or rpm

(f) Pressure, lb/sq in. or psi

(g)   Cost ratio, cents/lb

(h)   Length, 5 yd 2 ft

(i)   Area, sq ft

(j)   Volume, cu in.

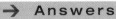

Answers

Compound units are used in quantities (b), (c), (e), (f), (g), (i), and (j). Two different units are used in (h), but they have not been combined by multiplying or dividing.

In Chapters 8 and 9, we study the geometry needed to calculate the area and volume of the various plane and solid figures used in technical work. Here we look at the many different units used to measure area and volume.

Area

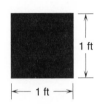

When the area of a surface is found, the measurement or calculation is given in "square units." If the lengths are measured in inches, the area is given in square inches or sq in.; if the lengths are measured in feet, the area is given in square feet or sq ft.

$$\text{Area} = 1\text{ ft} \times 1\text{ ft}$$
$$= 1\text{ sq ft}$$

To see the relation between square feet area units and square inch area units, rewrite the above area in inch units.

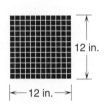

$$\text{Area} = 12\text{ in.} \times 12\text{ in.}$$
$$= 144 \times 1\text{ in.} \times 1\text{ in.}$$
$$= 144\text{ sq in.}$$

Therefore,

$$1\text{ sq ft} = 144\text{ sq in.}$$

> **Note**   The area units "sq ft," "sq in.," and so on are often expressed as ft², in.², and so on. The reason for this will become clear when we study exponents in Chapter 6. ◄

Example 7

Let's use this information to do the following conversion:

15 sq ft = _____ sq in.

$$= (15 \times 1 \text{ sq ft}) \times \left(\frac{144\text{ sq in.}}{1 \text{ sq ft}}\right) \quad \text{This unity fraction cancels sq ft and keeps sq in.}$$

$$= 15 \times 144\text{ sq in.}$$

$$= 2160\text{ sq in.} \quad \text{or} \quad 2200\text{ sq in.} \quad \text{rounded}$$

 15 ☒ **144** ═ → 　　　　　*2160.*

If you do not remember the conversion number, 144, try it this way:

$$15\text{ sq ft} = (15 \times 1\text{ ft} \times 1\text{ ft}) \times \left(\frac{12\text{ in.}}{1\text{ ft}}\right) \times \left(\frac{12\text{ in.}}{1\text{ ft}}\right)$$

$$= 15 \times 1 \times 1 \times 12\text{ in.} \times 12\text{ in.}$$

$$= 2160\text{ sq in.}$$

$$= 2200\text{ sq in.} \quad \text{rounded}$$

A number of other convenient area units have been defined.

AREA UNITS

1 square foot (sq ft or ft^2) $\quad= 144$ square inches (sq in. or in.^{2})

1 square yard (sq yd or yd^2) $= 9$ sq ft $= 1296$ sq in.

1 square rod (sq rod) $\qquad\quad = 30.25$ sq yd

1 acre $= 160$ sq rod $\qquad\quad = 4840$ sq yd $= 43,560$ sq ft

1 sq mile (sq mi or mi^2) $\qquad = 640$ acres

→ **Your Turn**

Solve the following conversion problems.

(a) 4800 acres = _____ sq mi

(b) 2.50 sq yd = _____ sq in.

(c) 1210 sq yd = _____ acre

(d) **Interior Design** Carpeting used to be sold strictly by the square yard, but now many companies are pricing it by the square foot. Suppose Carpet City offered a certain carpet at $42.95 per square yard, and a competitor, Carpeterium, offered a comparable carpet at $4.95 per square foot. Which is the better buy?

→ **Solutions**

(a) $4800 \text{ acres} = (4800 \times 1 \text{ acre}) \times \left(\dfrac{1 \text{ sq mi}}{640 \text{ acres}} \right)$

$= \dfrac{4800}{640} \text{ sq mi}$

$= 7.5 \text{ sq mi}$

(b) $2.50 \text{ sq yd} = (2.50 \times 1 \text{ sq yd}) \times \left(\dfrac{1296 \text{ sq in.}}{1 \text{ sq yd}} \right)$

$= 3240 \text{ sq in.}$

(c) $1210 \text{ sq yd} = (1210 \times 1 \text{ sq yd}) \times \left(\dfrac{1 \text{ acre}}{4840 \text{ sq yd}} \right)$

$= \dfrac{1210}{4840} \text{ acre}$

$= 0.25 \text{ acre}$

(d) To compare prices, we will convert Carpeterium's price to cost per square yard.

$\$4.95 \text{ per sq ft} = \left(\dfrac{\$4.95}{1 \text{ sq ft}} \right) \times \left(\dfrac{9 \text{ sq ft}}{1 \text{ sq yd}} \right) = \dfrac{\$44.55}{1 \text{ sq yd}} = \$44.55 \text{ per sq yd}$

Carpeterium's cost is $44.55 per square yard compared to Carpet City's cost of $42.95 per square yard. Carpet City has the better deal. Notice that we could also have converted Carpet City's price to cost per square foot and reached the same conclusion.

Volume

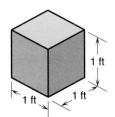

Volume units are given in "cubic units." If the lengths are measured in inches, the volume is given in cubic inches or cu in.; if the lengths are measured in feet, the volume is given in cubic feet or cu ft.

$$\text{Volume} = 1 \text{ ft} \times 1 \text{ ft} \times 1 \text{ ft}$$
$$= 1 \text{ cu ft}$$

In inch units, this same volume is 12 in. $\times$ 12 in. $\times$ 12 in. = 1728 cu in., so that 1 cu ft = 1728 cu in.

> **Note**   Volume units such as cu ft and cu in. can also be written as ft^3 and in.^{3} ◄

Example 8

To convert 0.25 cu ft to cu in., proceed as follows:

$$0.25 \text{ cu ft} = (0.25 \times 1 \, \cancel{\text{cu ft}}) \times \left(\frac{1728 \text{ cu in.}}{1 \, \cancel{\text{cu ft}}} \right)$$

$$= (0.25 \times 1728) \text{ cu in.}$$

$$= 432 \text{ cu in.}$$

A very large number of special volume units have been developed for various uses.

VOLUME UNITS

1 cubic foot (cu ft or ft^3)   = 1728 cubic inches (cu in. or in.^{3})

1 cubic yard (cu yd or yd^3) = 27 cu ft = 46,656 cu in.

1 gallon (gal)   = 231 cu in.   or   1 cu ft $\approx$ 7.48 gal

1 bushel (bu)   = 2150.42 cu in.

1 fluid ounce (fl oz)   = 1.805 cu in.

1 pint (pt)   = 28.875 cu in. (liquid measure)

Dozens of other volume units are in common use: cup, quart, peck, teaspoonful, and others. The cubic inch, cubic foot, and gallon are most used in trades work.

→ **Your Turn**

Convert the following volume measurements for practice. Round to the proper number of significant digits.

(a) 94 cu ft = _____ cu yd

(b) 12.0 gal = _____ cu in.

(c) 10,500 cu in. = _____ bu

(a) $94 \text{ cu ft} = (94 \times 1 \text{ cu ft}) \times \left(\dfrac{1 \text{ cu yd}}{27 \text{ cu ft}} \right)$

$= \left(\dfrac{94}{27} \right) \text{cu yd}$

$= 3.481 \ldots \text{cu yd} \approx 3.5 \text{ cu yd}$

(b) $12.0 \text{ gal} = (12.0 \times 1 \text{ gal}) \times \left(\dfrac{231 \text{ cu in.}}{1 \text{ gal}} \right)$

$= 12.0 \times 231 \text{ cu in.}$

$= 2772 \text{ cu in.} \approx 2770 \text{ cu in.}$

(c) $10,500 \text{ cu in.} = (10,500 \times 1 \text{ cu in.}) \times \left(\dfrac{1 \text{ bu}}{2150.42 \text{ cu in.}} \right)$

$= \left(\dfrac{10,500}{2150.42} \right) \text{bu}$

$= 4.882 \ldots \text{bu} \approx 4.88 \text{ bu}$

Lumber Measure

Carpenters and workers in the construction trades use a special unit to measure the amount of lumber. The volume of lumber is measured in *board feet*. One board foot (bf) of lumber is a piece having an area of 1 sq ft and a thickness of 1 in. or less.

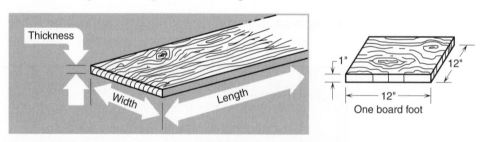

Thickness · Width · Length · One board foot

> Number of board feet = thickness in inches × width in feet × length in feet

To find the number of board feet in a piece of lumber, multiply the length in feet by the width in feet by the thickness in inches. A thickness of less than 1 in. should be counted as 1 in. If the lumber is dressed or finished, use the full size or rough stock dimension to calculate board feet.

Example 9

Construction   A 2-by-4 used in framing a house would actually measure $1\frac{1}{2}$ in. by $3\frac{1}{2}$ in., but we would use the dimensions 2 in. by 4 in. in calculating board feet. A 12-ft length of 2 in. by 4 in. would have a volume of

$2 \text{ in.} \times \frac{4}{12} \text{ ft} \times 12 \text{ ft} = 8 \text{ board ft or 8 bf}$

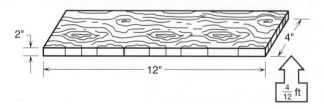

| **Note** | In lumber measure the phrase "per foot" means "per board foot." The phrase "per running foot" means "per foot of length." ◄ |

→ **Your Turn**

Flooring and Carpeting To floor a small building requires 245 boards, each $1\frac{1}{2}$ in. by 12 in. by 12 ft long. How many board feet should be ordered? (*Hint:* Use 2 in. for the thickness to calculate board feet.)

→ **Solution**

The volume of each board is

$$2 \text{ in.} \times \frac{12}{12} \text{ ft} \times 12 \text{ ft} = 24 \text{ bf}$$

For 245 boards, we must order

$$245 \times 24 \text{ bf} = 5880 \text{ bf}$$

→ **More Practice**

1. **Construction** Find the number of board feet in each of the following pieces of lumber. Be sure to convert the middle dimension, width, to feet.

   (a) $\frac{3}{4}$ in. $\times$ 6 in. $\times$ 4 ft   (b) 2 in. $\times$ 12 in. $\times$ 16 ft

   (c) 1 in. $\times$ 8 in. $\times$ 12 ft   (d) 4 in. $\times$ 4 in. $\times$ 8 ft

2. **Construction** How many board feet are there in a shipment containing 80 boards, 2 in. by 6 in., each 16 ft long?

→ **Answers**

1. (a) 2 bf   (b) 32 bf   (c) 8 bf   (d) $10\frac{2}{3}$ bf

2. **2** ⊗ **6** ⊘ **12** ⊗ **16** ⊗ **80** ⊜ → ▨ *1280.*

There are 1280 board feet.

| **A Closer Look** | In part 1(a), remember to round the thickness up to 1 in. ◄ |

Speed Speed is usually specified by a compound unit that consists of a distance unit divided by a time unit. For example, a speed given as 60 mph can be written as

$$60 \text{ mi/hr or } 60 \times \frac{1 \text{ mi}}{1 \text{ hr}}$$

To convert speed units within the English system of units, use the length units from page 277 and the following time conversion factors.

> **TIME UNITS**
>
> 1 minute (min) = 60 seconds (sec)
>
> 1 hour (hr or h) = 60 min = 3600 sec
>
> 1 day = 24 hr or 24 h

Converting speeds often involves converting both the distance and the time units. This requires two separate unity fractions.

Example 10

To convert 55 mph to ft/sec, use two conversion factors: 1 mi = 5280 ft (to convert miles to feet) and 1 hr = 3600 sec (to convert hours to seconds). Set up the unity fractions like this:

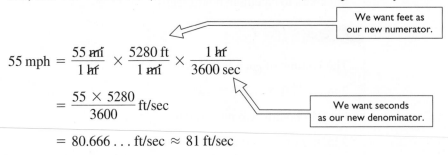

$$55 \text{ mph} = \frac{55 \text{ mi}}{1 \text{ hr}} \times \frac{5280 \text{ ft}}{1 \text{ mi}} \times \frac{1 \text{ hr}}{3600 \text{ sec}}$$

We want feet as our new numerator.

We want seconds as our new denominator.

$$= \frac{55 \times 5280}{3600} \text{ ft/sec}$$

$$= 80.666 \ldots \text{ ft/sec} \approx 81 \text{ ft/sec}$$

→ **Your Turn**

Convert the following speed measurements.

(a) 3.8 in./sec = _____ ft/min

(b) 98.5 ft/sec = _____ mi/hr

(c) 160 rpm = _____ rev/sec

→ **Solutions**

This replaces in. with ft in the numerator

(a) $3.8 \text{ in./sec} = \frac{3.8 \text{ in.}}{1 \text{ sec}} \times \frac{1 \text{ ft}}{12 \text{ in.}} \times \frac{60 \text{ sec}}{1 \text{ min}}$

This replaces sec with min in the denominator

$$= 19 \text{ ft/min}$$

This replaces ft with mi in the numerator

(b) $98.5 \text{ ft/sec} = \frac{98.5 \text{ ft}}{1 \text{ sec}} \times \frac{1 \text{ mi}}{5280 \text{ ft}} \times \frac{3600 \text{ sec}}{1 \text{ hr}}$

This replaces sec with hr in the denominator

$$= 67.159 \ldots \text{ mi/hr} \approx 67.2 \text{ mi/hr}$$

98.5 ÷ 5280 × 3600 = → **67.15909091**

(c) $160 \text{ rpm} = 160 \text{ rev/min} = \frac{160 \text{ rev}}{1 \text{ min}} \times \frac{1 \text{ min}}{60 \text{ sec}}$

$$= 2.666 \ldots \text{ rev/sec} \approx 2.7 \text{ rev/sec}$$

We can use the cancellation of units method to solve problems involving unique conversion factors stated in the problem.

Example 11

Automotive Trades   The 2009 Toyota Camry Hybrid has an EPA fuel economy rating of 33.5 mi/gal (miles per gallon; combined city and highway driving). The comparable 6-cylinder nonhybrid has an average fuel economy of 23.5 mi/gal, but it costs $2500 less. If gas costs an average of $4 per gallon, we can calculate the amount of money the hybrid would save over a 160,000-mi lifespan.

First, use the cancellation of units method to calculate the cost of gas over the life of each car.

For the hybrid: $\dfrac{160,000 \, \cancel{mi}}{1 \text{ lifespan}} \times \dfrac{1 \, \cancel{gal}}{33.5 \, \cancel{mi}} \times \dfrac{\$4}{1 \, \cancel{gal}}$ $\approx \$19,104$ rounded to the nearest dollar

For the nonhybrid: $\dfrac{160,000 \, \cancel{mi}}{1 \text{ lifespan}} \times \dfrac{1 \, \cancel{gal}}{23.5 \, \cancel{mi}} \times \dfrac{\$4}{1 \, \cancel{gal}} \approx \$27,234$ to the nearest dollar

Then, subtract to find the savings on gas.

The hybrid would save $27,234 − $19,104 = $8130 on gas.

Finally, because the hybrid costs $2500 more than the nonhybrid, we subtract this from the gas savings to find the total savings.

The hybrid saves a total of $8130 − $2500 = $5630 over a 160,000-mi lifespan.

160000 ÷ 23.5 ✕ 4 ⊖ 160000 ÷ 33.5 ✕ 4 ⊖ 2500 = → **5629.564941**

<u>   </u>   <u>   </u>   ↑
   Gas cost for nonhybrid   Gas cost for hybrid   Price difference

---

The following table summarizes all English conversion factors given in this section.

Length	Weight
1 ft = 12 in. 1 yd = 3 ft = 36 in. 1 rod (rd) = $16\frac{1}{2}$ ft 1 mi = 5280 ft = 1760 yd	1 lb = 16 oz 1 ton = 2000 lb
Liquid Capacity	**Area**
1 qt = 2 pt 1 gal = 4 qt	1 sq ft = 144 sq in. 1 sq yd = 9 sq ft = 1296 sq in. 1 sq rod = 30.25 sq yd 1 acre = 160 sq rod = 4840 sq yd 1 acre = 43,560 sq ft 1 sq mi = 640 acres
Volume	**Time**
1 cu ft = 1728 cu in. 1 gal = 231 cu in. 1 bu = 2150.42 cu in. 1 pt = 28.875 cu in. 1 cu yd = 27 cu ft = 46,656 cu in. 1 cu ft ≈ 7.48 gal 1 fl oz = 1.805 cu in.	1 min = 60 sec 1 hr or 1 h = 60 min = 3600 sec 1 day = 24 hr

Now, turn to Exercises 5-2 for a set of problems involving unit conversion in the English system.

A. Convert the units as shown. (Round your answer to the same number of significant digits as the given measurement.)

1. 4.25 ft = _____ in.

2. 33 yd = _____ ft

3. 3.4 mi = _____ ft

4. 17.3 lb = _____ oz

5. 126 gal = _____ qt

6. 8.5 mi = _____ yd

7. 46 psi = _____ atm (1 atm = 14.7 psi)

8. 7.5 lb = _____ oz

9. 18.4 pt = _____ gal

10. 32 in. = _____ ft

11. 4640 ft = _____ mi

12. 19 yd = _____ rd

13. 26 ft = _____ yd

14. $23\overline{0}$ oz = _____ lb

15. 5.10 gal = _____ cu in.

16. 114 gal = _____ cu in.

17. 1400 yd = _____ mi

18. 0.85 cu ft = _____ gal

19. 6 ft 11 in. = _____ in.

20. 13.5 lb = _____ oz

21. 326 oz = _____ lb _____ oz

22. 29 ft = _____ yd _____ ft

B. Convert the units as shown. (Round as needed.)

1. 6.0 sq ft = _____ sq in.

2. 2.50 acre = _____ sq ft

3. 17.25 cu yd = _____ cu ft

4. 385 sq ft = _____ sq yd

5. 972 cu in. = _____ cu ft

6. 325 rev/min = _____ rev/sec

7. $60 \dfrac{\text{lb}}{\text{cu ft}} = \underline{\qquad} \dfrac{\text{lb}}{\text{cu in.}}$

8. $3\overline{0}$ cu ft = _____ cu in.

9. 14,520 sq ft = _____ acre

10. $2500 \dfrac{\text{ft}}{\text{min}} = \underline{\qquad} \dfrac{\text{in.}}{\text{sec}}$

11. 20.0 sq mi = _____ acre

12. 65 mph = _____ ft/sec

13. 2.1 sq yd = _____ sq in.

14. 42 sq yd = _____ sq ft

15. 14,000 cu in. = _____ cu yd

16. 8.3 cu yd = _____ cu ft

C. Solve:

1. **Sports and Leisure** The Manitou Incline, which leads to the top of Pike's Peak in Colorado, climbs from an elevation of 6574 feet to 8585 feet above sea level. The fastest reported time for running up the incline is 16 minutes and 42 seconds, by Mark Fretta. How many vertical feet per second did Mark climb during his record-breaking performance? (Round to 1 decimal digit.)

2. **Carpentry** Find the number of board feet in each of the following quantities of lumber.

(a) 24 pieces of 2 in. × 10 in. × 16 ft

(b) 36 pieces of 1 in. × 8 in. × 10 ft

(c) 50 pieces of 2 in. × 4 in. × 12 ft

(d) 72 pieces of $\frac{7}{8}$ in. × 6 in. × 8 ft

3. **Construction** The hole for a footing needs to be 5 ft deep. If it is currently 2 ft 8 in. deep, how much deeper does it need to be dug? Give the answer in inches.

4. **Painting** The area covered by a given volume of paint is determined by the consistency of the paint and the nature of the surface. The measure of covering power is called the *spreading rate* of the paint and is expressed in units of square feet per gallon. The following table gives some typical spreading rates.

Paint	Surface	Spreading Rate	
		One Coat	Two Coats
Oil-based paint	Wood, smooth	580	320
	Wood, rough	360	200
	Plaster, smooth	300	200
	Plaster, rough	250	160
Latex paint	Wood	500	350
Varnish	Wood	400	260

Use the table to answer the following questions. (Assume that you cannot buy a fraction of a gallon.)

(a) How many gallons of oil-based paint are needed to cover 2500 sq ft of rough wood with two coats?

(b) At $34.39 per gallon, how much would it cost to paint 850 sq ft of wood with one coat of latex paint?

(c) How many square feet of rough plaster will 8 gal of oil-based paint cover with one coat?

(d) How many gallons of varnish are needed to cover 600 sq ft of wood cabinets with two coats?

(e) At $41.99 per gallon, how much will it cost to paint 750 sq ft of smooth plaster with two coats of oil-based paint?

5. **General Interest** The earliest known unit of length to be used in a major construction project is the "megalithic yard" used by the builders of Stonehenge in southwestern Britain about 2600 B.C. If 1 megalithic yard = 2.72 ± 0.05 ft, convert this length to inches. Round to the nearest tenth of an inch.

6. **General Interest** In the Bible (Genesis, Chapter 7), Noah built an ark 300 cubits long, 50 cubits wide, and 30 cubits high. In I Samuel, Chapter 17, it is reported that the giant Goliath was "six cubits and a span" in height. If 1 cubit = 18 in. and 1 span = 9 in.:

(a) What were the dimensions of the ark in English units?

(b) How tall was Goliath?

7. **Flooring and Carpeting** A homeowner needing carpeting has calculated that $140\frac{5}{9}$ sq yd of carpet must be ordered. However, carpet is sold by the square foot. Make the necessary conversion for her.

8. **Metalworking** Cast aluminum has a density of 160.0 lb/cu ft. Convert this density to units of lb/cu in. and round to one decimal place.

9. **Machine Trades** A $\frac{5}{16}$-in. twist drill with a periphery speed of 50.0 ft/min has a cutting speed of 611 rpm (revolutions per minute). Convert this speed to rps (revolutions per second) and round to one decimal place.

10. **Welding** How many pieces $8\frac{3}{8}$ in. long can be cut from a piece of 20-ft stock? Allow for $\frac{3}{16}$ in. kerf (waste due to the width of a cut).

11. **Wastewater Technology** A sewer line has a flow rate of 1.25 MGD (million gallons per day). Convert this to cfs (cubic feet per second).

12. **Hydrology** A reservoir has a capacity of 9000 cu ft. How long will it take to fill the reservoir at the rate of 250 gallons per minute?

13. **Sheet Metal Trades** A machinist must cut 12 strips of metal that are each 3 ft $2\frac{7}{8}$ in. long. What is the total length needed in inches?

14. **Marine Technology** In the petroleum industry, a barrel is defined to be 42 gal. A scientist wishes to treat a 430,000-barrel oil spill with a bacterial culture. If the culture must be applied at the rate of 1 oz of culture per 100 cu ft of oil, how many ounces of culture will the scientist need? Round to two significant digits.

15. **Painting** A painter must thin some paint for use in a sprayer. If the recommended rate is $\frac{1}{2}$ pint of water per gallon of paint, how many total gallons will there be after thinning 8 gal of paint?

16. **Interior Design** A designer is pricing a certain wool carpet for a new house. She finds it listed for $139.50 per square yard at Carpet City and $16.65 per square foot at Carpetorium. (a) Which is the better buy? (b) How much will she save on 166 sq yd?

17. **Landscaping** A 2800-sq-ft house is built on a flat and bare $\frac{1}{2}$-acre parcel of land. If an additional 600 sq ft are needed for a driveway, how many square feet remain for landscaping?

18. **Transportation** A car at rest begins traveling at a constant acceleration of 18 feet per second per second (18 ft/sec^2). After 6.0 sec, the car is traveling at 108 ft/sec. Convert this final speed to miles per hour. (Round to two significant digits.)

19. **Automotive Trades** If the circumference of each tire on a vehicle is approximately 88 in., how many miles will the vehicle travel when the tires make 12,000 revolutions? (See problem (h) on page 281.)

20. **Automotive Trades** A drum of oil contains 55 gal and costs $164.80. What is the cost per quart?

21. **Automotive Trades** Suppose a hybrid automobile has an average fuel economy of 40 mi/gal, and a comparable nonhybrid averages 25 mi/gal. If the hybrid model costs $6000 more than the nonhybrid, and gasoline averages $3.75 per gallon, which model would be less expensive over a lifespan of 125,000 miles? (See Example 11 on page 289.)

22. **Architecture** Calculate the height of the church tower shown.

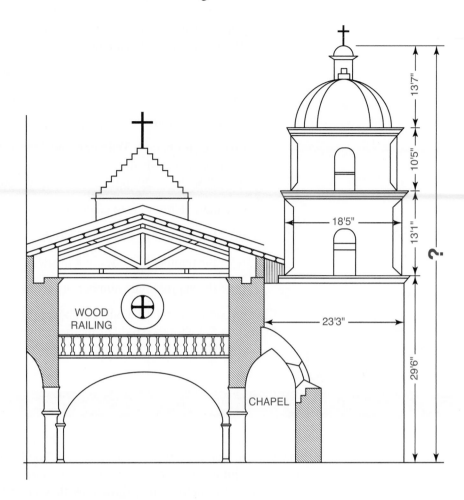

23. **Hydrology** Hydrological engineers often measure very large amounts of water in acre-feet. 1 acre-ft of water fills a rectangular volume of base 1 acre and height 1 ft. Convert 1 acre-ft to

(a) _____ cu ft   (b) _____ gal   (c) _____ cu yd

24. **Hydrology** The average daily usage of water from Bradbury Dam at Lake Cachuma during a particular week in the summer was 134.2 acre-feet. To the nearest thousand, how many gallons is this? (See problem 23.)

25. **Automotive Trades** The 2008 Ford Escape hybrid is EPA-rated at 34 mi/gal (miles per gallon) for city driving and 30 mi/gal for highway driving. The Toyota Yaris is rated at 29 mi/gal city and 35 mi/gal highway. How many gallons of gas would each vehicle consume for:

(a) 300 mi of city driving plus 100 mi of highway driving.

(b) 100 mi of city driving plus 300 mi of highway driving.

26. **Life Skills** A trades worker has a job in a large city. She can either rent an apartment in the city for $1400 per month or a comparable one in the suburbs for $1200 per month. If she lives in the city, she can walk or bike to work. If she lives in the suburbs, she must drive 60 miles per day round trip an average of 22 days per month. Her car averages 24 mi/gal and gas costs $2.80 per gallon. If she decides to live in the suburbs, would the total monthly cost of gas be more or less than the savings in rent?

27. **General Interest** Researchers have recently "cracked the code" of the Aztec system of measurement. Their basic unit of length was the rod, which is approximately 8 ft. (Notice that this is not the same as the modern definition of a rod.) Other units are defined in terms of the rod as follows:

5 hands = 3 rods   3 arms = 1 rod

2 arrows = 1 rod   5 bones = 1 rod

5 hearts = 2 rods

(Source: Barbara J. Williams and Maria del Carmen Jorge y Jorge, "Aztec Arithmetic Revisited: Land-Area Algorithms and Acolhua Congruence Arithmetic," *Science* 320, pp. 72–77, 4 April 2008.)

Use these definitions to calculate:

(a) the number of hands in 15 rods   (b) the number of rods in 35 hearts.

(c) the number of bones in 8 arrows. (d) the number of hands in 12 arms.

(e) the approximate number of feet in 20 hearts.

Check your answers to the odd-numbered problems in the Appendix, then continue in Section 5-3 with the study of metric units.

5-3 Metric Units

For all technical and many trades workers, the ability to use metric units is important. In many ways, working with meters, kilograms, and liters is easier and more logical than using our traditional units of feet, pounds, or quarts. In this section we define and explain the most useful metric units, show how they are related to the corresponding English units, explain how to convert from one unit to another, and provide practice problems designed to help you use the metric system and to "think metric."

The most important common units in the metric system are those for length or distance, speed, weight, volume, area, and temperature. Time units—year, day, hour, minute, second—are the same in the metric as in the English system.

Length The basic unit of length in the metric system is the *meter,* pronounced *meet-ur* and abbreviated *m.* (The word is sometimes spelled *metre* but pronounced exactly the same.) One meter is roughly equal to one yard.

The meter is the appropriate unit to use in measuring your height, the width of a room, length of lumber, or the height of a building.

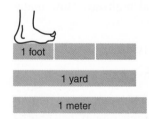

→ **Your Turn**

Estimate the following lengths in meters:

(a) The length of the room in which you are now sitting = _____ m.

(b) Your height = _____ m.

(c) The length of a Ping-Pong table = _____ m.

(d) The length of a football field = _____ m.

Guess as closely as you can before looking at the solution.

(a)   A small room might be 3 or 4 m long, and a large one might be 6 or 8 m long.

(b)   Your height is probably between $1\frac{1}{2}$ and 2 m. Note that 2 m is roughly 6 ft 7 in.

(c)   A Ping-Pong table is a little less than 3 m in length.

(d)   A football field is about 100 m long.

All other length units used in the metric system are defined in terms of the meter, and these units differ from one another by multiples of ten. For example, the *centimeter,* pronounced *cent-a-meter* and abbreviated *cm,* is defined as exactly one-hundredth of a meter. The *kilometer,* pronounced *kill-AH-meter* and abbreviated *km,* is defined as exactly 1000 meters. The *millimeter,* abbreviated *mm,* is defined as exactly $\frac{1}{1000}$ meter.

Because metric units increase or decrease in multiples of ten, they may be named using prefixes attached to a basic unit. The following table summarizes the most commonly used metric prefixes.

Metric Prefix	Multiplier	Common Example
tera	1,000,000,000,000 (10^{12})	terabyte: one trillion bytes
giga	1,000,000,000 (10^{9})	gigahertz: one billion hertz
mega	1,000,000 (10^{6})	megawatt: one million watts
kilo	1,000 (10^{3})	kilopascal: one thousand pascal
centi	0.01 (10^{-2})	centimeter: one hundredth of a meter
milli	0.001 (10^{-3})	milliliter: one thousandth of a liter
micro	0.000001 (10^{-6})	microgram: one millionth of a gram
nano	0.000000001 (10^{-9})	nanosecond: one billionth of a second
pico	0.000000000001 (10^{-12})	picofarad: one trillionth of a farad

Note   The prefixes "hecto" (multiplier of 100), "deca" (multiplier of 10), and "deci" (multiplier of one tenth) were not included in the table because they are rarely used in the trades or even in technical work. ◀

The following length conversion factors are especially useful for workers in the trades.

METRIC LENGTH UNITS

1 centimeter (cm) = 10 millimeters (mm)

1 meter (m) = 100 cm = 1000 mm

1 kilometer (km) = 1000 m

The millimeter (mm) unit of length is very often used on technical drawings, and the centimeter (cm) is handy for shop measurements.

1 cm is roughly the width of a large paper clip.

1 mm is roughly the thickness of the wire in a paper clip.

1 meter is roughly 10% more than a yard.

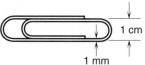

Metric to Metric Conversion

One very great advantage of the metric system is that we can easily convert from one metric unit to another. There are no hard-to-remember conversion factors: 36 in. in a yard, 5280 ft in a mile, 220 yd in a furlong, or whatever.

For example, with English units, if we convert a length of 137 in. to feet or yards, we find

$$137 \text{ in.} = 11\tfrac{5}{12} \text{ ft} = 3\tfrac{29}{36} \text{ yd}$$

Here we must divide by 12 and then by 3 to convert the units. But in the metric system we simply shift the decimal point.

Example 1

Convert 348 cm to meters like this:

$$\boxed{100 \text{ cm} = 1 \text{ m}}$$

$$348 \text{ cm} = 3.48 \text{ m}$$

> To divide by 100, shift the decimal point two digits to the left

Of course we may also use unity fractions:

$$348 \text{ cm} = 348 \text{ cm} \times \left(\frac{1 \text{ m}}{100 \text{ cm}}\right)$$

> Choose the unity fraction that cancels cm and replaces it with m

$$= \frac{348}{100} \text{ m}$$

$$= 3.48 \text{ m}$$

→ Your Turn

Try it. Complete the following unit conversions.

(a) 147 cm = _____ m

(b) 3.1 m = _____ cm

(c) 2.1 km = _____ m

(d) 307 m = _____ km

(e) 7 cm = _____ mm

(f) 20.5 mm = _____ cm

(a) $147 \text{ cm} = 147 \text{ cm} \times \left(\dfrac{1 \text{ m}}{100 \text{ cm}} \right)$

$\qquad\qquad = \dfrac{147}{100} \text{ m}$

$\qquad\qquad = 1.47 \text{ m} \qquad$ The decimal point shifts two places to the left.

(b) $3.1 \text{ m} = 3.1 \text{ m} \times \left(\dfrac{100 \text{ cm}}{1 \text{ m}} \right)$

$\qquad\qquad = 3.1 \times 100 \text{ cm}$

$\qquad\qquad = 310 \text{ cm} \qquad$ The decimal point shifts two places right.

(c) $2.1 \text{ km} = 2.1 \text{ km} \times \left(\dfrac{1000 \text{ m}}{1 \text{ km}} \right)$

$\qquad\qquad = 2100 \text{ m}$

(d) $307 \text{ m} = 307 \text{ m} \times \left(\dfrac{1 \text{ km}}{1000 \text{ m}} \right)$

$\qquad\qquad = 0.307 \text{ km}$

(e) $7 \text{ cm} = 7 \text{ cm} \times \left(\dfrac{10 \text{ mm}}{1 \text{ cm}} \right)$

$\qquad\qquad = 70 \text{ mm}$

(f) $20.5 \text{ mm} = 20.5 \text{ mm} \times \left(\dfrac{1 \text{ cm}}{10 \text{ mm}} \right)$

$\qquad\qquad = 2.05 \text{ cm}$

Metric–English Conversion Because metric units are the only international units, all English units are defined in terms of the metric system. Use the following table for metric–English length conversions.

1 inch

1 centimeter

METRIC–ENGLISH LENGTH CONVERSIONS

1 in. = 2.54 cm ⟵ This is the exact legal definition of the inch.

1 ft = 30.48 cm

1 yd = 0.9144 m

1 mi ≈ 1.6093 km

To shift from English to metric units or from metric to English units, use either unity fractions or the conversion factors given in the table.

Example 2

Convert 4.3 yd to meters.

From the conversion table, 1 yd = 0.9144 m.

Using unity fractions,

$$4.3 \text{ yd} \times \frac{0.9144 \text{ m}}{1 \text{ yd}} = 3.93192 \text{ m} \approx 3.9 \text{ m rounded to two significant digits.}$$

Sometimes more than one unity fraction is needed. This is demonstrated in the next example.

Example 3

Convert 2.50 in. to millimeters.

The conversion table does not list an inch–millimeter conversion factor.

Therefore we must use two factors: 1 in. = 2.54 cm and 1 cm = 10 mm

$$2.50 \text{ in.} = 2.5 \text{ in.} \times \frac{2.54 \text{ cm}}{1 \text{ in.}} \times \frac{10 \text{ mm}}{1 \text{ cm}} = 63.5 \text{ mm}$$

| This unity fraction cancels in. and leaves cm | This unity fraction cancels cm and leaves mm |

The answer agrees in accuracy with the three significant digits of the original measurement.

2.5 ⊗ **2.54** ⊗ **10** ⊜ →   *63.5*

→ More Practice

Convert the following measurements as shown. Round answers to agree in accuracy with the original measurement.

(a) 16.0 in. = _____ cm   (b) 6.0 km = _____ mi

(c) 1.2 m = _____ ft   (d) 37.0 cm = _____ in.

(e) 4.2 m = _____ yd   (f) 140 cm = _____ ft

(g) 18 ft 6 in. = _____ m

(h) **Automotive Trades** An automobile tire is designated as P205/55R16. This means that the overall width of the tire is 205 mm, the aspect ratio is 55%, and the diameter of the rim is 16 in. To determine the overall diameter of the tire in inches, we must first convert the overall width to inches. Perform this conversion.

→ Answers

(a) 40.6 cm   (b) 3.7 mi

(c) 3.9 ft   (d) 14.6 in.

(e) 4.6 yd   (f) 4.6 ft

(g) 5.6 m   (h) 8.07 in.

In problem (c), $1.2 \not{m} \times \dfrac{1 \not{yd}}{0.9144 \not{m}} \times \dfrac{3 \text{ ft}}{1 \not{yd}} = 3.937 \text{ ft} \ldots \approx 3.9 \text{ ft}.$

1.2 ÷ .9144 × 3 = → **3.937007874** ◄

DUAL DIMENSIONING

Some companies involved in international trade use "dual dimensioning" on their technical drawings and specifications. With dual dimensioning, both inch and metric dimensions are given. For example, a part might be labeled like this:

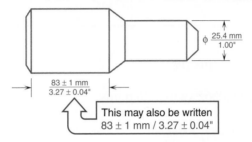

Notice that the metric measurement is written first or on top of the fraction bar. Diameter dimensions are marked with the symbol ϕ.

Weight

The weight of an object is the gravitational pull of the earth exerted on that object. The mass of an object is related to how it behaves when pushed or pulled. For scientists the difference between mass and weight can be important. For all practical purposes they are the same.

The basic unit of mass or metric weight in the International Metric System (SI) is the *kilogram,* pronounced *kill-o-gram* and abbreviated *kg.* The kilogram is defined as the mass of a standard platinum–iridium cylinder kept at the International Bureau of Weights and Measures near Paris. By law in the United States, the pound is defined as exactly equal to the weight of 0.45359237 kg, so that 1 kg weighs about 2.2046 lb.

One kilogram weighs about 10% more than 2 lb.

The kilogram was originally designed so that it was almost exactly equal to the weight of one liter of water.

Two smaller metric units of mass are also defined. The *gram,* abbreviated *g,* is equal to 0.001 kg and the *milligram,* abbreviated *mg,* is equal to 0.001 g.

The following conversion factors will make it easy for you to convert weight units within the metric system.

METRIC WEIGHT/MASS UNITS

1 milligram (mg) = 1000 micrograms (µg)

1 gram (g) = 1000 milligrams (mg)

1 kilogram (kg) = 1000 g

1 metric ton = 1000 kg

The gram is a very small unit of weight, equal to roughly $\frac{1}{500}$ of a pound, or the weight of a paper clip. The milligram is often used by druggists and nurses to measure very small amounts of medication or other chemicals, and it is also used in scientific measurements.

Metric to Metric Conversion

To convert weights within the metric system, use the conversion factors in the table to form unity fractions.

Example 4

To convert 4500 mg to grams, use the conversion factor 1 g = 1000 mg.

$$4500 \text{ mg} = 4500 \text{ mg} \times \frac{1 \text{ g}}{1000 \text{ mg}} = 4.5 \text{ g}$$

→ Your Turn

Convert 0.0062 kg to milligrams.

→ Solution

Here we need to use two conversion factors: 1 kg = 1000 g and 1 g = 1000 mg

$$0.0062 \text{ kg} = 0.0062 \text{ kg} \times \frac{1000 \text{ g}}{1 \text{ kg}} \times \frac{1000 \text{ mg}}{1 \text{ g}} = 6200 \text{ mg}$$

Cancels kg and leaves g

Cancels g and leaves mg

→ More Practice

Convert the following measurements as shown.

(a) 21.4 g = _____ mg   (b) 156,000 mg = _____ kg

(c) 0.56 kg = _____ g   (d) 467 µg = _____ mg

→ Answers

(a) 21,400 mg   (b) 0.156 kg   (c) 560 g   (d) 0.467 mg

Metric–English Conversion

To convert weights from metric to English or English to metric units, use the conversion factors in the following table.

METRIC–ENGLISH WEIGHT CONVERSIONS

1 ounce (oz) ≈ 28.35 grams (g)

1 pound (lb) ≈ 0.4536 kilogram (kg)

1 ton (T) ≈ 907.2 kilograms (kg)

Example 5

To convert 75 g to ounces, use the conversion factor 1 ounce = 28.35 grams to form a unity fraction.

$$75 \text{ g} = 75 \text{ g} \times \frac{1 \text{ oz}}{28.35 \text{ g}} = 2.6455 \ldots \approx 2.6 \text{ oz, rounded}$$

Example 6

Culinary Arts Cactus sweetener sells for $7.20 per pound. What is its cost in cents per gram?

First, express "$7.20 per pound" as a fraction. $\quad$ $\text{Cost} = \dfrac{\$7.20}{1 \text{ lb}}$

Next, convert dollars to cents. $\quad = \dfrac{\$7.20}{1 \text{ lb}} \times \dfrac{100 \cancel{\text{¢}}}{\$1} = \dfrac{720 \text{ ¢}}{1 \text{ lb}}$

Now, we must replace pounds with grams in the denominator.

Use two conversion factors: 1 lb = 0.4536 kg and 1 kg = 1000 g

$$\text{Cost} = \frac{720 \text{ ¢}}{1 \cancel{\text{lb}}} \times \frac{1 \cancel{\text{lb}}}{0.4536 \cancel{\text{kg}}} \times \frac{1 \cancel{\text{kg}}}{1000 \text{ g}}$$

Cancels lb, leaves kg $\qquad$ Cancels kg, leaves g

Finally, perform the calculations. $\quad = \dfrac{720 \text{ ¢}}{0.4536 \times 1000 \text{ g}}$

$$= 1.587 \text{ ¢/g} \approx 1.6 \text{ ¢/g}$$

 720 ÷ **.4536** ÷ **1000** = → `1.587301587`

→ **Your Turn**

Convert (a) 0.075 lb to grams.

$\qquad$ (b) 1.40 T to kg

→ **Solutions**

(a) $\quad$ Using two unity fractions we have

$$0.075 \text{ lb} = 0.075 \cancel{\text{lb}} \times \frac{0.4536 \text{ kg}}{1 \cancel{\text{lb}}} \times \frac{1000 \text{ g}}{1 \cancel{\text{kg}}}$$

$$= 34.02 \text{ g} \approx 34 \text{ g, rounded}$$

 .075 × **.4536** × **1000** = → `34.02`

(b) $\quad 1.40 \text{ T} = 1.40 \cancel{\text{T}} \times \dfrac{907.2 \text{ kg}}{1 \cancel{\text{T}}}$

$$= 1.40 \times 907.2 \text{ kg}$$

$$\approx 1270 \text{ kg}$$

Convert the following measurements as shown. Round to the same number of significant digits.

(a) 150 lb = _____ kg   (b) 6.0 oz = _____ g

(c) 2.50 T = _____ kg   (d) $10\overline{0}$ kg = _____ lb

(e) 1.50 kg = _____ oz   (f) 35,600 kg = _____ T

(g) 150 g = _____ lb   (h) 4 lb 5 oz = _____ kg

→ **Answers**

(a) 68 kg   (b) 170 g   (c) 2270 kg

(d) 220 lb   (e) 52.9 oz   (f) 39.2 tons

(g) 0.33 lb   (h) 2.0 kg

A Closer Look

In problem (e) $1.50 \; \cancel{kg} \times \dfrac{1 \; \cancel{lb}}{0.4536 \; \cancel{kg}} \times \dfrac{16 \; oz}{1 \; \cancel{lb}} = 52.91 \ldots \approx 52.9 \; oz$

In problem (h) the answer 1.96 is 2.0 when rounded to two significant digits. ◄

Area Because an area is essentially the product of two lengths, it is specified by a compound unit. When the area of a surface is calculated, the units will be given in "square length units." If the lengths are measured in feet, the area will be given in square feet; if the lengths are measured in meters, the area will be given in square meters.

For example, the area of a carpet 2 m wide and 3 m long would be

$2 \; m \times 3 \; m = 6$ square meters
$\qquad\qquad\quad = 6 \; sq \; m$

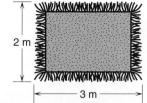

Areas roughly the size of carpets and room flooring are usually measured in square meters. Larger areas are usually measured in *hectares,* a metric surveyor's unit. One hectare (ha) is the area of a square 100 m on each side. The hectare is roughly $2\frac{1}{2}$ acres, or about double the area of a football field.

METRIC AREA UNITS

1 hectare (ha) = 100 m × 100 m

$\qquad\qquad\qquad = 10,000 \; sq \; m \; (or \; m^2)$

Example 7

Convert 456,000 sq m to hectares.

Using a unity fraction.

$456,000 \; sq \; m = 456,000 \; \cancel{sq \; m} \times \dfrac{1 \; ha}{10,000 \; \cancel{sq \; m}} = 45.6 \; ha$

→ **Your Turn**

Convert 0.85 hectare to square meters.

→ **Solution**

$$0.85 \text{ ha} = 0.85 \,\cancel{\text{ha}} \times \frac{10{,}000 \text{ sq m}}{1 \,\cancel{\text{ha}}} = 8500 \text{ sq m}$$

Area conversions within the metric system, such as square meters to square kilometers, are seldom used in the trades and we will not cover them here. To convert area units from metric to English or English to metric units, use the following conversion factors.

METRIC–ENGLISH AREA CONVERSIONS

1 square inch (sq in. or in.^{2}) $\approx$ 6.452 sq cm (or cm^2)

1 square foot (sq ft or ft^2) $\approx$ 0.0929 sq m (or m^2)

1 square yard (sq yd or yd^2) $\approx$ 0.836 sq m

→ **More Practice**

Use this information to solve the following problems. Round carefully.

(a)   15 sq ft = _____ sq m   (b)   0.21 sq m = _____ sq ft

(c) $1\overline{0}$ sq cm = _____ sq in.   (d)   78 sq yd = _____ sq m

→ **Answers**

(a)   1.4 sq m   (b)   2.3 sq ft

(c)   1.6 sq in.   (d)   65 sq m

Volume   When the volume of an object is calculated or measured, the units will normally be given in "cubic length units." If the lengths are measured in feet, the volume will be given in cubic feet; if the lengths are measured in meters, the volume will be given in cubic meters.

For example, the volume of a box 2 m high by 4 m wide by 5 m long would be

$$2 \text{ m} \times 4 \text{ m} \times 5 \text{ m} = 40 \text{ cu m or } 40 \text{ m}^3$$

The cubic meter is an appropriate unit for measuring the volume of large amounts of water, sand, or gravel, or the volume of a room. However, a more useful metric volume unit for most practical purposes is the *liter,* pronounced *leet-ur.* You may find this volume unit spelled either *liter* in the United States or *litre* elsewhere in the world.

One liter (L) is defined as 1000 cu cm, the volume of a cube 10 cm on each edge.

Because 1000 cubic centimeters are needed to make up 1 liter, a cubic centimeter is $\frac{1}{1000}$ of a liter or a *milliliter.*

Volumes smaller than one liter are usually measured in units of cubic centimeters or milliliters. The following table summarizes the most useful metric volume relationships.

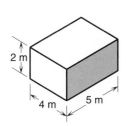

2 m

4 m   5 m

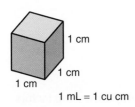

1 cm

1 cm

1 cm

1 mL = 1 cu cm

METRIC VOLUME UNITS

1 cubic centimeter (cu cm or cc or cm^3) = 1 milliliter (mL)

1 milliliter (mL) = 1000 microliters (μL)

1 liter (L) = 1000 cu cm

Example 8

To convert 0.24 liter to cubic centimeters, use the conversion factor, 1 liter = 1000 cu cm. Using a unity fraction,

$$0.24 \text{ liter} = 0.24 \text{ liter} \times \frac{1000 \text{ cu cm}}{1 \text{ liter}} = 240 \text{ cu cm}$$

→ **Your Turn**

(a) Convert 5780 mL to liters.   (b) Convert 0.45 mL to microliters.

→ **Solution**

(a) Because 1 mL = 1 cu cm, we need only one unity fraction.

$$5780 \text{ mL} = 5780 \text{ cu cm} \times \frac{1 \text{ liter}}{1000 \text{ cu cm}} = 5.78 \text{ liters}$$

(b) We use the conversion factor 1 mL = 1000 μL.

$$0.45 \text{ mL} = 0.45 \text{ mL} \times \frac{1000 \text{ μL}}{1 \text{ mL}} = 450 \text{ μL}$$

Metric–English Conversion

In the United States, one quart is legally defined as 0.94635295 liter. This means that a volume of one liter is slightly larger than 1 qt.

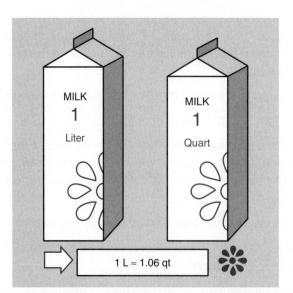

One liter is roughly 6% more than 1 qt.

The following table summarizes the most useful metric–English volume conversion factors.

METRIC–ENGLISH VOLUME CONVERSIONS

1 cubic inch (cu in. or in.^{3}) ≈ 16.387 cubic centimeters (cu cm or cm^3)

1 cubic yard (cu yd or yd^3) ≈ 0.765 cubic meters (cu m or m^3)

1 fluid ounce ≈ 29.574 cu cm

1 quart (qt) ≈ 0.946 liter

1 gallon (gal) ≈ 3.785 liters

Example 9

To convert 5.5 gal to liters, use the conversion factor 1 gal = 3.785 L. Using unity fractions,

$$5.5 \text{ gal} \times \frac{3.785 \text{ L}}{1 \text{ gal}} = 20.8175 \text{ liters} \approx 21 \text{ liters}$$

The final answer is rounded to the same number of significant digits as the original measurement.

→ Your Turn

Convert 4.2 cu in. to cubic centimeters.

→ Solution

The only conversion factor needed is 1 cu in. ≈ 16.387 cu cm.

$$4.2 \text{ cu in.} = 4.2 \text{ cu in.} \times \frac{16.387 \text{ cu cm}}{1 \text{ cu in.}} = 68.8254 \text{ cu cm} \approx 69 \text{ cu cm}$$

→ More Practice

Convert the following measurements as shown. Round the answer to the same number of significant digits as in the original number.

(a) 13.4 gal = _____ liter

(b) 9.1 fl oz = _____ cu cm

(c) 16.0 cu yd = _____ cu m

(d) 5.1 liters = _____ qt

(e) 140 cu cm = _____ cu in.

(f) 24.0 cu m = _____ cu yd

(g) 64.0 liters = _____ gal

(h) 12.4 qt = _____ liter

→ Answers

(a) 50.7 liters   (b) 270 cu cm   (c) 12.2 cu m   (d) 5.4 qt

(e) 8.5 cu in.   (f) 31.4 cu yd   (g) 16.9 gal   (h) 11.7 liters

Speed In the metric system, ordinary highway speeds are measured in kilometers per hour, abbreviated km/h. To "think metric" while you drive, remember that

100 km/h is approximately equal to 62 mph.

To convert speed units, either within the metric system or between metric and English units, use the appropriate combination of time and distance unity fractions.

Example 10

To convert 2.4 ft/sec to m/min, use the following conversion factors:

For converting time in the denominator, use 1 min = 60 sec.

For converting distance in the numerator, use 1 ft = 30.48 cm and 1 m = 100 cm.

Using unity fractions,

$$2.4 \text{ ft/sec} = \frac{2.4 \text{ ft}}{1 \text{ sec}} \times \frac{30.48 \text{ cm}}{1 \text{ ft}} \times \frac{1 \text{ m}}{100 \text{ cm}} \times \frac{60 \text{ sec}}{1 \text{ min}}$$

| These unity fractions combine to convert ft to m | This one replaces sec with min in the denominator |

$$= 2.4 \times 30.48 \div 100 \times 60 \text{ m/min}$$

$$= 43.8912 \text{ m/min} \approx 44 \text{ m/min}$$

→ Your Turn

Convert 27.5 km/h to cm/sec.

→ Solution

First, determine the necessary conversion factors:

For time, 1 h = 3600 sec. For distance, 1 km = 1000 m, and 1 m = 100 cm

Then, set up the appropriate unity fractions:

$$27.5 \text{ km/h} = \frac{27.5 \text{ km}}{1 \text{ h}} \times \frac{1000 \text{ m}}{1 \text{ km}} \times \frac{100 \text{ cm}}{1 \text{ m}} \times \frac{1 \text{ h}}{3600 \text{ sec}}$$ ← To convert h to sec

To convert km to cm

$$= \frac{27.5 \times 1000 \times 100}{3600} \text{ cm/sec}$$

$$= 763.88 \ldots \text{ cm/sec} \approx 764 \text{ cm/sec}$$

 27.5 ⊗ 1000 ⊗ 100 ÷ 3600 = → **763.8888889**

→ More Practice

(a) 35 mph = _____ km/h

(b) 30.0 m/sec = _____ km/h

(c) 120 cm/sec = _____ in./sec

(d) 18 in./sec = _____ cm/sec

(e) $28\overline{0}$ km/h = _____ mph

(f) 775 m/min = _____ cm/sec

(g) **Machine Trades**   The cutting speed for a soft steel part in a lathe is 165 ft/min. Express this in cm/sec.

Temperature The Fahrenheit temperature scale is commonly used in the United States for weather reports, cooking, and other practical work. On this scale, water boils at 212°F and freezes at 32°F. The metric or *Celsius* temperature scale is a simpler scale originally designed for scientific work but now used worldwide for most temperature measurements. On the Celsius scale, water boils at 100°C and freezes at 0°C.

Notice that the range from freezing to boiling is covered by 180 degrees on the Fahrenheit scale and 100 degrees on the Celsius scale. The size and meaning of a temperature degree is different for the two scales. A temperature of zero does not mean "no temperature"—it is simply another point on the scale. Temperatures less than zero are possible, and they are labeled with negative numbers.

Because zero does not correspond to the same temperature on both scales, converting Celsius to Fahrenheit or Fahrenheit to Celsius is not as easy as converting length or weight units. The simplest way to convert from one scale to the other is to use the chart on this page.

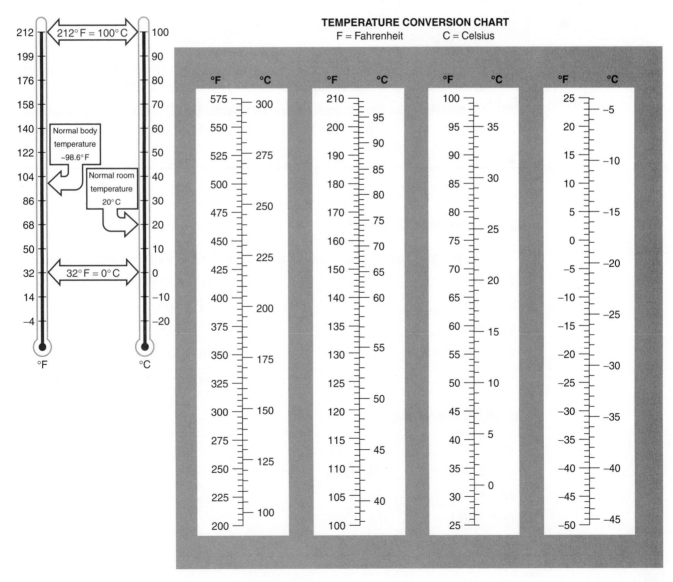

TEMPERATURE CONVERSION CHART
F = Fahrenheit   C = Celsius

Example 11

On this chart equal temperatures are placed side by side. To convert 50°F to Celsius:

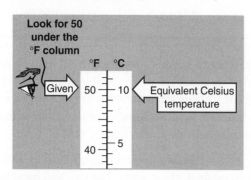

$$50° \, F = 10° \, C$$

→ **Your Turn**

Use this temperature conversion chart to convert

(a)   140°F to Celsius

(b)   20°C to Fahrenheit

→ **Solutions**

(a)   On the second chart from the left, locate 140° under the °F column. Notice that 60 in the °C column is exactly aligned with 140°F. Therefore, 140°F = 60°C.

(b)   On the third chart from the left, locate 20°C under the °C column. Notice that 68 in the °F column is approximately aligned with 20°C, to the nearest degree. Therefore, 20°C ≈ 68°F.

→ **More Practice**

Now use this temperature conversion chart to determine the following temperatures.

(a)   375°F = _____ °C   (b)   −20°F = _____ °C

(c)   14°C = _____ °F   (d)   80°C = _____ °F

(e)   37°C = _____ °F   (f)   68°F = _____ °C

(g)   −40°C = _____ °F   (h)   525°F = _____ °C

(i)   80°F = _____ °C   (j)   −14°C = _____ °F

→ **Answers**

(a)   191°C   (b)   −29°C

(c)   57.5°F   (d)   176°F

(e)   98.6°F   (f)   20°C

(g)   −40°F   (h)   274°C

(i)   27°C   (j)   7°F

If you must convert a temperature with more accuracy than this chart allows, or if you are converting a temperature not on the chart, algebraic conversion formulas should be used. These temperature conversion formulas will be explained in Chapter 7 when we study basic algebra.

This table summarizes conversions involving metric units.

Metric–Metric	Metric–English
Length	
1 cm = 10 mm 1 m = 100 cm = 1000 mm 1 km = 1000 m	1 in. = 2.54 cm 1 ft = 30.48 cm 1 yd = 0.9144 m 1 mi $\approx$ 1.6093 km
Weight/Mass	
1 mg = 1000 μg 1 g = 1000 mg 1 kg = 1000 g 1 metric ton = 1000 kg	1 oz = 28.35 g 1 lb = 0.4536 kg 1 T = 907.2 kg
Area	
1 ha = 10,000 sq m (m^2)	1 sq in. (in.^{2}) $\approx$ 6.452 sq cm (cm^2) 1 sq ft (ft^2) $\approx$ 0.0929 sq m (m^2) 1 sq yd (yd^2) $\approx$ 0.836 sq m
Volume	
1 mL = 1000 μL 1 cu cm (cm^3) = 1 mL 1 liter (L) = 1000 cu cm	1 cu in. (in.^{3}) $\approx$ 16.387 cu cm (cm^3) 1 cu yd (yd^3) $\approx$ 0.765 cu m (m^3) 1 fluid ounce $\approx$ 29.574 cu cm 1 qt $\approx$ 0.946 liter 1 gal $\approx$ 3.785 liters

Now turn to Exercises 5-3 for a set of practice problems designed to help you use the metric system and to "think metric."

Exercises 5-3 Metric Units

A. **Think Metric.** For each problem, circle the measurement closest to the first one given. No calculations are needed.

Remember: (1) A meter is a little more (about 10%) than a yard.
(2) A kilogram is a little more (about 10%) than 2 pounds.
(3) A liter is a little more (about 6%) than a quart.

1. 30 cm   (a) 30 in.   (b) 75 in.   (c) 1 ft

2. 5 ft   (a) 1500 cm   (b) 1.5 m   (c) 2 m

3. 1 yd   (a) 90 cm   (b) 110 cm   (c) 100 cm

4. 2 m   (a) 6 ft 6 in.   (b) 6 ft   (c) 2 yd

5. 3 km (a) 3 mi (b) 2 mi (c) 1 mi

6. 200 km (a) 20 mi (b) 100 mi (c) 120 mi

7. 50 km/h (a) 30 mph (b) 50 mph (c) 60 mph

8. 55 mph (a) 30 km/h (b) 60 km/h (c) 90 km/h

9. 100 m (a) 100 yd (b) 100 ft (c) 1000 in.

10. 400 lb (a) 800 kg (b) 180 kg (c) 250 kg

11. 6 oz (a) 1.7 g (b) 17 g (c) 170 g

12. 5 kg (a) 2 lb (b) 5 lb (c) 10 lb

13. 50 liters (a) 12 gal (b) 120 gal (c) 50 gal

14. 6 qt (a) 2 liters (b) 3 liters (c) 6 liters

15. 14 sq ft (a) 1.3 sq m (b) 13 sq m (c) 130 sq m

16. 8 sq in. (a) 50 sq cm (b) 5 sq cm (c) 500 sq cm

17. 100°F (a) 38°C (b) 212°C (c) 32°C

18. 60°C (a) 20°F (b) 140°F (c) 100°F

19. 212°F (a) 100°C (b) 400°C (c) 50°C

20. 0°C (a) 100°F (b) 32°F (c) −30°F

B. **Think Metric.** Choose the closest estimate.

1. Diameter of a penny (a) 3 cm (b) 1.5 cm (c) 5 cm

2. Length of a man's foot (a) 3 m (b) 3 cm (c) 30 cm

3. Tank of gasoline (a) 50 liters (b) 5 liters (c) 500 liters

4. Volume of a wastebasket (a) 10 liters (b) 100 liters (c) 10 cu cm

5. One-half gallon of milk (a) 1 liter (b) 2 liters (c) $\frac{1}{2}$ liter

6. Hot day in Phoenix, Arizona (a) 100°C (b) 30°C (c) 45°C

7. Cold day in Minnesota (a) −80°C (b) −10°C (c) 20°C

8. 200-lb barbell (a) 400 kg (b) 100 kg (c) 40 kg

9. Length of paper clip (a) 10 cm (b) 35 mm (c) 3 mm

10. Your height (a) 17 m (b) 17 cm (c) 1.7 m

11. Your weight (a) 80 kg (b) 800 kg (c) 8 kg

12. Boiling water (a) 100°C (b) 212°C (c) 32°C

C. Perform the following metric–metric conversions. Round to the appropriate number of significant digits.

1. 5.6 cm = _____ mm

2. 0.08 g = _____ mg

3. 0.25 ha = _____ sq m

4. 64,600 cu cm = _____ liters

5. 0.045 kg = _____ g

6. 12.5 km/h = _____ cm/sec

7. 1.25 liters = _____ cu cm

8. 720,000 g = _____ kg

9. 4.2 m = _____ cm

10. 1026 mm = _____ cm

11. 0.16 m/sec = _____ cm/sec

12. 598 cm = _____ m

13. 9.62 km = _____ m

14. 785 mm = _____ m

15. 56,500 m = _____ km

16. 78,000 sq m = _____ ha

17. 9,500 mg = _____ g

18. 870 m/min = _____ km/h

19. 580 ml = _____ liters

20. 42 cm/sec = _____ m/min

21. 1400 μL = _____ mL

22. 42.5 mL = _____ μL

23. 650 cu cm = _____ liters

24. 27 km = _____ m

25. 246 km/h = _____ m/min

26. 0.2 L/kg = _____ mL/g

D. Convert to the units shown. Round to the appropriate number of significant digits.

1. 3.0 in. = _____ cm

2. 4.2 ft = _____ cm

3. 2.0 mi = _____ km

4. 20 ft 6 in. = _____ m
   (three sig. digits)

5. $9\frac{1}{4}$ in. = _____ cm
   (two sig. digits)

6. 31 kg = _____ lb

7. 152 lb = _____ kg

8. 3 lb 4 oz = _____ kg
   (two sig. digits)

9. 8.5 kg = _____ lb

10. 10.4 oz = _____ g

11. 3.15 gal = _____ liters

12. 2.5 liters = _____ qt

13. 6.5 qt = _____ liters

14. 61 mph = _____ km/h

15. 8.0 km/h = _____ mph

16. 230°F = _____ °C

17. 10°C = _____ °F

18. 35 in./sec = _____ cm/sec

19. 63 ft/sec = _____ m/sec

20. 2.0 m = _____ ft

21. 285 sq ft = _____ sq m

22. 512 cu in. = _____ cu cm

23. 2.75 m^3 = _____ cu yd

24. 46.2 sq cm = _____ sq in.

25. 687 km = _____ mi

26. 260 mL = _____ oz

27. 2400 sq ft = _____ sq m

28. 62.0 m/sec = _____ ft/sec

29. 31.0 kg/sq m = _____ lb/sq ft

30. 4.50 metric tons = _____ lb

E. Practical Problems. Round as indicated.

1. **Welding**   Welding electrode sizes are presently given in inches. Convert the following set of electrode sizes to millimeters. (1 in. = 25.4 mm) Round to the nearest one-thousandth of a millimeter.

in.	mm
0.030	
0.035	
0.040	
0.045	

in.	mm
$\frac{1}{16}$	
$\frac{5}{64}$	
$\frac{3}{32}$	
$\frac{1}{8}$	

in.	mm
$\frac{5}{32}$	
$\frac{3}{16}$	
$\frac{3}{8}$	
$\frac{11}{64}$	

2. **Hydrology**   One cubic foot of water weighs 62.4 lb. (a) What is the weight in pounds of 1 liter of water? (b) What is the weight in kilograms of 1 cu ft of water? (c) What is the weight in kilograms of 1 liter of water? Round to one decimal place.

3. **Sports and Leisure** At the 2008 Beijing Olympic Games, Sammy Wanjiru of Kenya won the 26.2-mile marathon race in 2 hr 6.32 min. What was his average speed in mph and km/h? Round to the nearest tenth.

4. **Physics** The metric unit of pressure is the *pascal,* where

   1 lb/sq in. (psi) ≈ 6894 pascal (Pa), and 1 lb/sq ft (psf) ≈ 47.88 Pa

   For very large pressures, the kilopascal (kPa) is the appropriate unit.

   1 kPa = 1000 Pa and therefore

   $$1 \frac{lb}{sq\ in.} \approx 6.894\ kPa \quad and \quad 1 \frac{lb}{sq\ ft} \approx 0.04788\ kPa$$

   Convert the following pressures to metric units.

   (a) 15.0 psi = _____ Pa   (b) 100 psi = _____ kPa

   (c) 40 psf = _____ Pa   (d) 2500 psf = _____ kPa

5. Find the difference between

   (a) 1 mile and 1 kilometer   _____ km

   (b) 3 miles and 5000 meters   _____ ft

   (c) 120 yards and 110 meters   _____ ft

   (d) 1 quart and 1 liter   _____ qt

   (e) 10 pounds and 5 kilograms   _____ lb

6. **Flooring and Carpeting** Standard American carpeting comes in 12-ft widths, but imported European carpeting comes in 4-m widths. What would this European width be in feet and inches? (Round to the nearest half-inch.)

7. **Automotive Trades** Which is performing more efficiently, a car getting 12 km per liter of gas or one getting 25 miles per gallon of gas?

8. **Painting** Wood lacquer sells for $38.29 per gallon. At this rate, what would 1 liter cost?

9. **Roofing** In roofing, the unit "one square" is sometimes used to mean 100 sq ft. Convert this unit to the metric system. Round to one decimal place.

   1 square = _____ sq m

10. **General Interest** A *cord* of wood is a volume of wood 8 ft long, 4 ft wide, and 4 ft high.
    (a) What are the equivalent metric dimensions for a "metric cord" of wood?
    (b) What is the volume of a cord in cubic meters? Round to the nearest tenth. (*Hint:* volume = length × width × height.)

11. **General Interest** Translate these well-known phrases into metric units.

    (a) A miss is as good as _____ km. (1 mile = _____ km)

    (b) An _____ cm by _____ cm sheet of paper. ($8\frac{1}{2}$ in. = _____ cm; 11 in. = _____ cm)

    (c) He was beaten within _____ cm of his life. (1 in. = _____ cm)

    (d) Race in the Indy _____ km. (500 miles = _____ km)

    (e) Take it with _____ grams of salt. (*Hint:* 437.5 grains = 1 oz)

    (f) _____ grams of prevention is worth _____ grams of cure.

         (1 oz = _____ g; 1 lb = _____ g)

(g) Peter Piper picked _____ liters of pickled peppers. (*Hint:* 1 peck = 8 quarts)

(h) A cowboy in a _____ liter hat. (10 gal = _____ liter)

12. **Automotive Trades** In an automobile, cylinder displacement is measured in either cubic inches or liters. What is the cylinder displacement in liters of a 400-cu in. engine? Round to the nearest liter.

13. **Electronics** Convert the following quantities as shown.

(a) 18,000 volts = _____ kilovolts

(b) 435 millivolts = _____ volts

14. **Culinary Arts** Fried crickets, served as an appetizer, sell for 5400 Cambodian riels per kg. If 4200 riels are equal to one dollar, how much will 5 lb of crickets cost in U.S. dollars?

15. **Sports and Leisure** Nutrition scientists have determined that for long-distance running in the desert, a drink containing 5.0 g of glycerol per 100 mL of water is ideal. Convert this to ounces per quart. Round to two significant digits.

16. **Automotive Trades** A BMW X5 has a fuel tank capacity of 85.0 liters. How many gallons does the fuel tank hold? (Round to the nearest tenth.)

17. **Culinary Arts** A British recipe for Beef Wellington requires a cooking temperature of 170°C. Determine the equivalent of this temperature in °F.

18. **Sports and Leisure** By changing the rear derailleur of a racing bike, a mechanic is able to reduce the overall weight of the bike by 60 g. If the bike weighs 7.50 kg before changing the derailleur, what is its weight in kilograms after the derailleur is changed?

19. **Aviation** The normal cruising speed of the Airbus A380 is about 250 m/sec. What is this speed in miles per hour?

20. **Allied Health** For a baby's milk formula, 1 oz of formula should be mixed with 0.5 qt of water. Express this baby formula concentration in grams per liter (g/L).

21. **Culinary Arts** A caterer for a wedding reception wants to use a Hungarian recipe for eggplant salad that recommends using 1.5 kg of eggplant for 8 guests. However, in her area, eggplant is only sold by the pound. If 150 guests are expected, how many pounds of eggplant should the caterer purchase? (Round to the nearest pound.)

22. **Culinary Arts** The standard recipe for chocolate chip cookies calls for 12 oz of chocolate chips to make approximately 60 cookies. An American visiting Europe wants to make 200 cookies for a Christmas party. Because chocolate chips are difficult to find in Europe, he decides to chop up some chocolate bars to replace the chips. How many 100-g chocolate bars will he need to purchase? (Assume that you cannot buy a fraction of a bar.)

23. **Construction** The Taipei 101 in Taipei, Taiwan, is 509 m tall. Its elevators are the fastest in the world, rising at 60.48 km/hr. How many seconds would it take for an elevator to rise the entire height of the building, nonstop?

24. **Sports and Leisure** In the 2008 Olympics, Usain Bolt of Jamaica set world records of 9.69 sec in the 100-m run and 19.30 sec in the 200-m run. Determine his average speed in miles per hour for (a) the 100-m run, and (b) the 200-m run. (Round your answer to two decimal digits.)

When you have completed these exercises, check your answers to the odd-numbered problems in the Appendix, then turn to Section 5-4 to learn about direct measurement.

5-4 Direct Measurements

The simplest kind of measuring device to use is one in which the output is a digital display. For example, in a digital clock or digital voltmeter, the measured quantity is translated into electrical signals, and each digit of the measurement number is displayed.

A digital display is easy to read, but most technical work involves direct measurement where the numerical reading is displayed on some sort of scale. Reading the scale correctly requires skill and the ability to estimate or *interpolate* accurately between scale divisions.

Length Measurements

Length measurement involves many different instruments that produce either fraction or decimal number readouts. Carpenters, electricians, plumbers, roofers, and machinists all use rulers, yardsticks, or other devices to make length measurements. Each instrument is made with a different scale that determines the precision of the measurement that can be made. The more precise the measurement has to be, the smaller the scale divisions needed.

For example, on the carpenter's rule shown, scale A is graduated in eighths of an inch. Scale B is graduated in sixteenths of an inch.

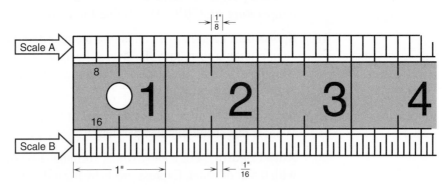

A carpenter or roofer may need to measure no closer than $\frac{1}{8}$ or $\frac{1}{16}$ in., but a machinist will often need a scale graduated in 32nds or 64ths of an inch. The finely engraved steel rule shown is a machinist's rule.

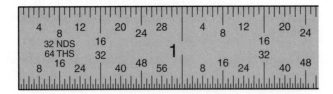

Some machinists' rules are marked in tenths or hundredths of an inch on one side and may be marked in metric units on the other side.

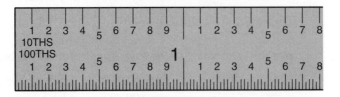

Notice that on all these rules only a few scale markings are actually labeled with numbers. It would be easy to read a scale marked like this:

Any useful scale has divisions so small that this kind of marking is impossible. An expert in any technical field must be able to identify the graduations on the scale correctly even though the scale is *not* marked. To use any scale, first determine the size of the smallest division by counting the number of spaces in a 1-in. interval.

Example 1

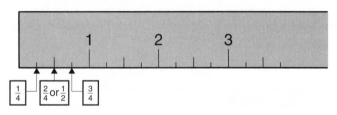

There are four spaces in the first 1-in. interval. The smallest division is therefore $\frac{1}{4}$ in., and we can label the scale like this:

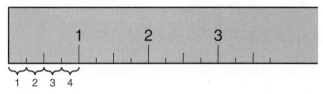

→ Your Turn

Try it. Label all marks on the following scales. The units are inches.

(a)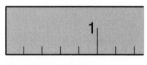

Smallest division = _____

(b)

Smallest division = _____

(c)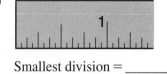

Smallest division = _____

(d)

Smallest division = _____

→ Answers

(a)   Smallest division = $\frac{1}{8}$ in.

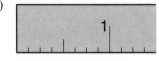

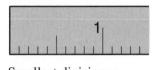

(b)   Smallest division = $\frac{1}{5}$ in.

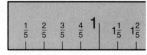

(c) Smallest division = $\frac{1}{10}$ in.

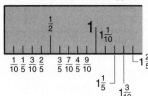

(d) Smallest division = $\frac{1}{16}$ in.

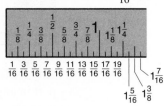

To use the rule once you have mentally labeled all scale markings, first count inches, then count the number of smallest divisions from the last whole-inch mark.

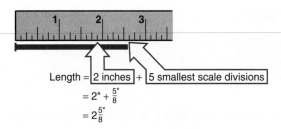

Length = 2 inches + 5 smallest scale divisions

$= 2" + \frac{5"}{8}$

$= 2\frac{5"}{8}$

If the length of the object being measured falls between scale divisions, record the nearest scale division.

Example 2

In this measurement:

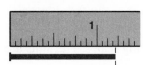

the endpoint of the object falls between $1\frac{3}{16}$ and $1\frac{4}{16}$ in.

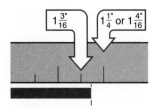

To the nearest scale division, the length is $1\frac{3}{16}$ in. When a scale with $\frac{1}{16}$-in. divisions is used, we usually write the measurement to the nearest $\frac{1}{16}$ in., but we may estimate it to the nearest half of a scale division or $\frac{1}{32}$ in.

→ Your Turn

For practice in using length rulers, find the dimensions given on the following inch rules. Express all fractions in lowest terms.

(a)

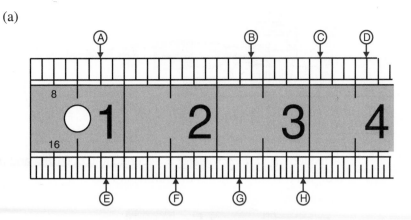

(b)

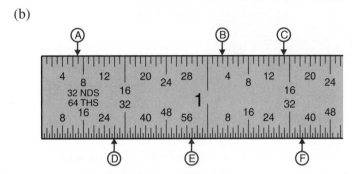

(c) Express answers in decimal form.

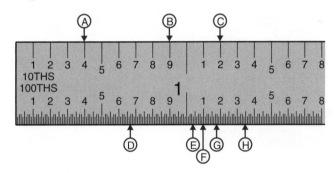

→ **Answers**

(a) A. $\frac{3}{4}$ in.   B. $2\frac{3}{8}$ in.   C. $3\frac{1}{8}$ in.   D. $3\frac{5}{8}$ in.

     E. $\frac{13}{16}$ in.   F. $1\frac{9}{16}$ in.   G. $2\frac{1}{4}$ in.   H. $2\frac{15}{16}$ in.

(b) A. $\frac{7}{32}$ in.   B. $1\frac{3}{32}$ in.   C. $1\frac{15}{32}$ in.

     D. $\frac{28}{64}$ in. $= \frac{7}{16}$ in.   E. $\frac{29}{32}$ in.   F. $1\frac{37}{64}$ in.

(c) A. $\frac{4}{10}$ in. $= 0.4$ in.   B. $\frac{9}{10}$ in. $= 0.9$ in.   C. $1\frac{2}{10}$ in. $= 1.2$ in.

     D. $\frac{67}{100}$ in. $= 0.67$ in.   E. $1\frac{4}{100}$ in. $= 1.04$ in.   F. $1\frac{10}{100}$ in. $= 1.10$ in.

     G. $1\frac{18}{100}$ in. $= 1.18$ in.   H. $1\frac{35}{100}$ in. $= 1.35$ in.

Micrometers A steel rule can be used to measure lengths of $\pm\frac{1}{32}$ in. or $\pm\frac{1}{64}$ in., or as fine as $\pm\frac{1}{100}$ in. To measure lengths with greater accuracy than this, more specialized instruments are needed. A *micrometer* can be used to measure lengths to ± 0.001 in., one one-thousandth of an inch.

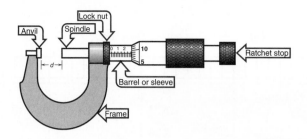

A modern *digital micrometer* is accurate to 0.0001 in., and the measurement is given as a digital readout in inches or millimeters.

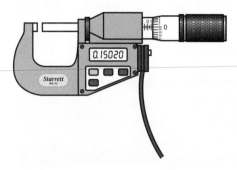

Two scales are used on a micrometer. The first scale, marked on the sleeve, records the movement of a screw machined accurately to 40 threads per inch. The smallest divisions on this scale record movements of the spindle of one-fortieth of an inch or 0.025 in.

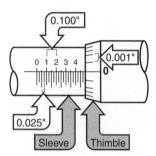

The second scale is marked on the rotating spindle. One complete turn of the thimble advances the screw one turn or $\frac{1}{40}$ in. The thimble scale has 25 divisions, so that each mark on the thimble scale represents a spindle movement of $\frac{1}{25} \times \frac{1}{40}$ in. $= \frac{1}{1000}$ in. or 0.001 in.

To read a micrometer, follow these steps.

Step 1 Read the largest numeral visible on the sleeve and multiply this number by 0.100 in.

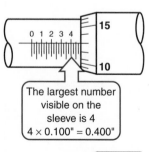

The largest number visible on the sleeve is 4
$4 \times 0.100" = 0.400"$

Step 2 Read the number of additional scale spaces visible on the sleeve. (This will be 0, 1, 2, or 3.) Multiply this number by 0.025 in.

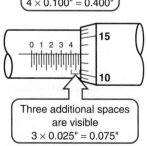

Three additional spaces are visible
$3 \times 0.025" = 0.075"$

Step 3   Read the number on the thimble scale opposite the horizontal line on the sleeve and multiply this number by 0.001 in.

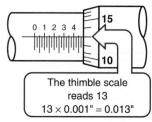

The thimble scale reads 13
$13 \times 0.001" = 0.013"$

Step 4   Add these three products to find the measurement.

$$\begin{array}{r} 0.400 \text{ in.} \\ 0.075 \text{ in.} \\ + 0.013 \text{ in.} \\ \hline 0.488 \text{ in.} \end{array}$$

> **Careful**   To count as a "space" in step 2, the right end marker must be visible. ◄

Here are more examples.

Example 3

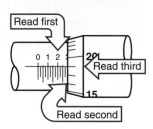

Read first
Read third
Read second

Step 1   The largest number completely visible on the sleeve is 2.

2×0.100 in. $= 0.200$ in.

Step 2   Three additional spaces are visible on the sleeve.

3×0.025 in. $= 0.075$ in.

Step 3   The thimble scale reads 19.

19×0.001 in. $= 0.019$ in.

Step 4   Add:
$$\begin{array}{r} 0.200 \text{ in.} \\ 0.075 \text{ in.} \\ 0.019 \text{ in.} \\ \hline \text{Length } = 0.294 \text{ in.} \end{array}$$

Example 4

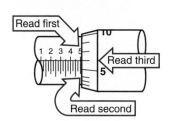

Read first
Read third
Read second

Step 1 5×0.100 in. $= 0.500$ in.

Step 2   No additional spaces are visible on the sleeve.

0×0.025 in. $= 0.000$ in.

Step 3 6×0.001 in. $= 0.006$ in.

Step 4   Add:
$$\begin{array}{r} 0.500 \text{ in.} \\ 0.000 \text{ in.} \\ 0.006 \text{ in.} \\ \hline \text{Length } = 0.506 \text{ in.} \end{array}$$

→ **Your Turn**

For practice, read the following micrometers.

(a)
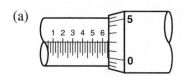

(b)
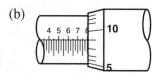

(c)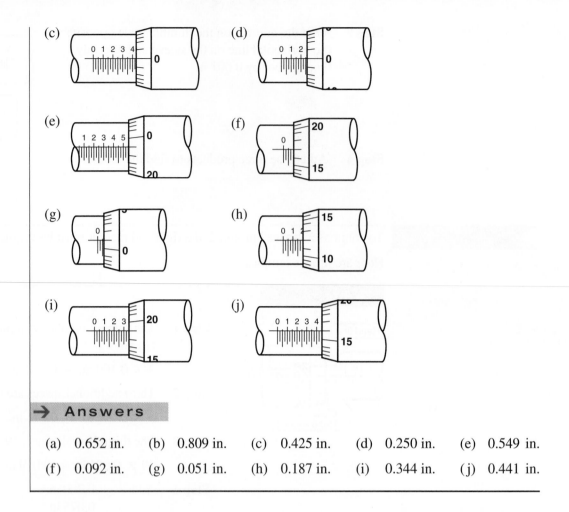

(d)

(e)

(f)

(g)

(h)

(i)

(j)

→ **Answers**

(a) 0.652 in. (b) 0.809 in. (c) 0.425 in. (d) 0.250 in. (e) 0.549 in.

(f) 0.092 in. (g) 0.051 in. (h) 0.187 in. (i) 0.344 in. (j) 0.441 in.

Metric micrometers are easier to read. Each small division on the sleeve represents 0.5 mm. The thimble scale is divided into 50 spaces; therefore, each division is $\frac{1}{50} \times 0.5$ mm $= \frac{1}{100}$ mm or 0.01 mm.

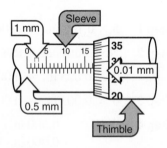

Turning the thimble scale of a metric micrometer by one scale division advances the spindle one one-hundredth of a millimeter, or 0.001 cm.

To read a metric micrometer, follow these steps.

Step 1 Read the largest mark visible on the sleeve and multiply this number by 1 mm.

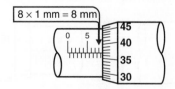

Step 2 Read the number of additional half spaces visible on the sleeve. (This number will be either 0 or 1.) Multiply this number by 0.5 mm.

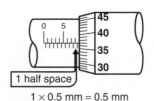

1 half space

1×0.5 mm $= 0.5$ mm

Step 3 Read the number on the thimble scale opposite the horizontal line on the sleeve and multiply this number by 0.01 mm.

The thimble scale reads 37.
37×0.01 mm $= 0.37$ mm

Step 4 Add these three products to find the measurement value.

8.00 mm
0.50 mm
$\underline{0.37 \text{ mm}}$
8.87 mm

Example 5

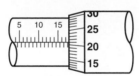

17×1 mm $= 17$ mm
1×0.5 mm $= 0.5$ mm
21×0.01 mm $= 0.21$ mm
Length $= 17$ mm $+ 0.5$ mm $+ 0.21$ mm
$ = 17.71$ mm

→ Your Turn

Read these metric micrometers.

(a)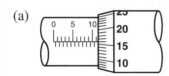

(b)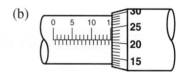

(c)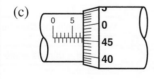

(d)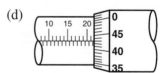

→ Answers

(a) 11.16 mm (b) 15.21 mm

(c) 7.96 mm (d) 21.93 mm

Vernier Micrometers When more accurate length measurements are needed, the *vernier* micrometer can be used to provide an accuracy of $\pm$ 0.0001 in.

On the vernier micrometer, a third or vernier scale is added to the sleeve. This new scale appears as a series of ten lines parallel to the axis of the sleeve above the usual scale.

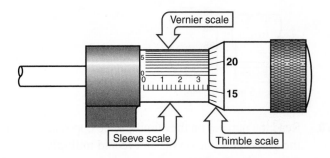

The ten vernier marks cover a distance equal to nine thimble scale divisions—as shown on this spread-out diagram of the three scales.

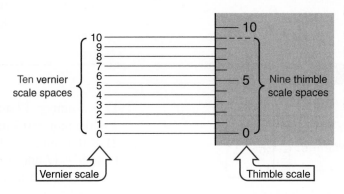

The difference between a vernier scale division and a thimble scale division is one-tenth of a thimble scale division, or $\frac{1}{10}$ of 0.001 in. = 0.0001 in.

To read the vernier micrometer, follow the first three steps given for the ordinary micrometer, then:

Step 4 Find the number of the line on the vernier scale that exactly lines up with any line on the thimble scale. Multiply this number by 0.0001 in. and add this amount to the other three distances.

Example 6

For the vernier micrometer shown:

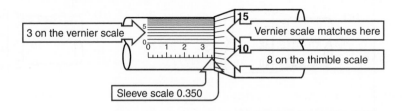

Sleeve scale: 3×0.100 in. = 0.300 in.　When adding these four lengths,
　　　　　　　2×0.025 in. = 0.050 in.　be careful to align the decimal
Thimble scale: 8×0.001 in. = 0.008 in.　digits correctly.
Vernier scale: 3×0.0001 in. = 0.0003 in.
　　　　Length　　 = 0.3583 in.

The vernier scale provides a way of measuring accurately the distance to an additional tenth of the finest scale division.

→ **Your Turn**

Now for some practice in using the vernier scale, read each of the following.

(a)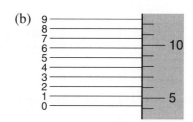

(b)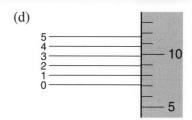

(c)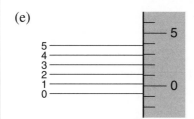

(d)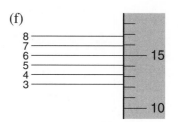

(e)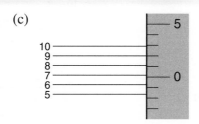

(f)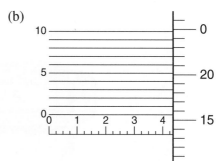

→ **Answers**

(a)	7	(b)	3	(c)	9
(d)	1	(e)	3	(f)	6

→ **More Practice**

Now read the following vernier micrometers.

(a)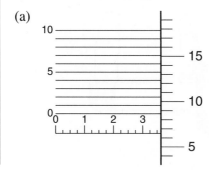

(b)

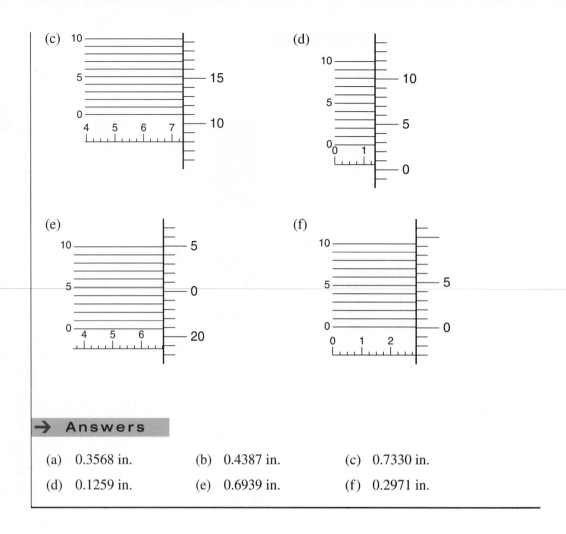

(c) 10 5 0 | 4 5 6 7 | 15 10

(d) 10 5 0 | 0 1 | 10 5 0

(e) 10 5 0 | 4 5 6 | 5 0 20

(f) 10 5 0 | 0 1 2 | 5 0

→ **Answers**

(a) 0.3568 in.   (b) 0.4387 in.   (c) 0.7330 in.

(d) 0.1259 in.   (e) 0.6939 in.   (f) 0.2971 in.

Vernier Calipers

The vernier caliper is another length measuring instrument that uses the vernier principle and can measure lengths to ± 0.001 in.

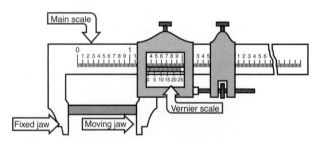

Modern direct *digital calipers,* battery or solar-powered, give measurements accurate to 0.0001 in. They convert automatically between English and metric units. If the jaws of the measuring device cannot access both sides of the part being measured, an ultrasonic digital readout thickness gauge will do the job.

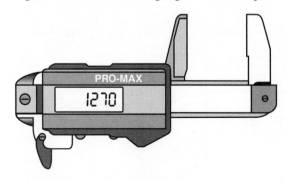

Notice that on the manual instrument each inch of the main scale is divided into 40 parts, so that each smallest division on the scale is 0.025 in. Every fourth division or tenth of an inch is numbered. The numbered marks represent 0.100 in., 0.200 in., 0.300 in., and so on.

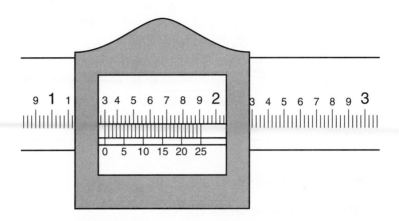

The sliding vernier scale is divided into 25 equal parts numbered by 5s. These 25 divisions on the vernier scale cover the same length as 24 divisions on the main scale.

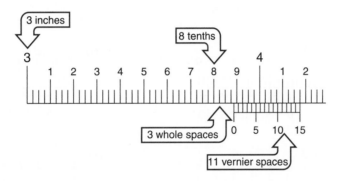

To read the vernier caliper above, follow these steps.

Step 1 Read the number of whole-inch divisions on the main scale to the left of the vernier zero. This gives the number of whole inches.

3 in. = 3.000 in.

Step 2 Read the number of tenths on the main scale to the left of the vernier zero. Multiply this number by 0.100 in.

8 tenths
8 × 0.100 in. = 0.800 in.

Step 3 Read the number of additional whole spaces on the main scale to the left of the vernier zero. Multiply this number by 0.025 in.

3 spaces
3 × 0.025 in. = 0.075 in.

Step 4 On the sliding vernier scale, find the number of the line that exactly lines up with any line on the main scale. Multiply this number by 0.001 in.

11 lines up exactly.
11 × 0.001 in. = 0.011 in.

Step 5 Add the four numbers.

$$\begin{array}{r} 3.000 \text{ in.} \\ 0.800 \text{ in.} \\ 0.075 \text{ in.} \\ + \underline{0.011 \text{ in.}} \\ 3.886 \text{ in.} \end{array}$$

Example 7

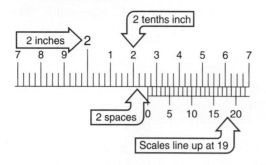

2 in.   = 2.000 in.
2   × 0.100 in. = 0.200 in.
2   × 0.025 in. = 0.050 in.
19 × 0.001 in. = 0.019 in.
                  2.269 in.

→ **Your Turn**

Read the following vernier calipers.

(a)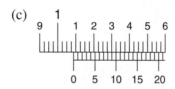

(b)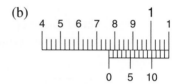

(c)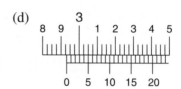

(d)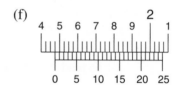

(e)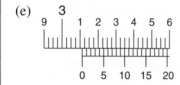

(f)

→ **Answers**

(a)   2.478 in.   (b)   0.763 in.   (c)   1.084 in.

(d)   2.931 in.   (e)   3.110 in.   (f)   1.470 in.

Protractors   A *protractor* is used to measure and draw angles. The simplest kind of protractor, shown here, is graduated in degrees with every tenth degree labeled with a number.

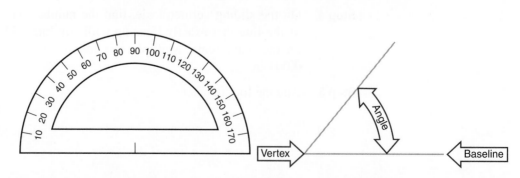

In measuring or drawing angles with a protractor, be certain that the vertex of the angle is exactly at the center of the protractor base. One side of the angle should be placed along the 0°–180° baseline.

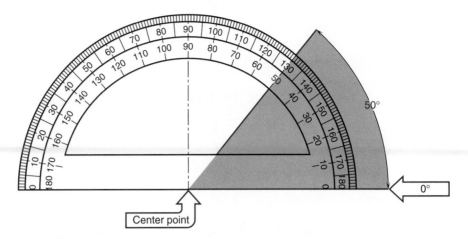

Notice that the protractor can be read from the 0° point on either the right or left side. Measuring from the right side, the angle is read as 50° on the inner scale. Measuring from the left side, the angle is 130° on the outer scale.

The *bevel protractor* is very useful in shopwork because it can be adjusted with an accuracy of ±0.5°. The angle being measured is read directly from the movable scale.

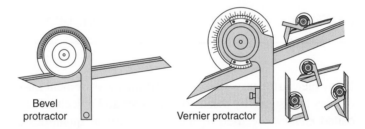

For very accurate angle measurements, the *vernier protractor* can be used with an accuracy of ±$\frac{1}{60}$ of a degree or 1 minute, because 60 minutes = 1 degree.

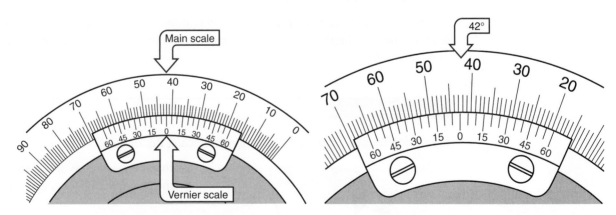

Example 8

To read the vernier protractor shown, follow these steps.

Step 1   Read the number of whole degrees on the main scale between the zero on the main scale and the zero on the vernier scale.   The mark on the main scale just to the right of 0° is 42°.

Step 2 Find the vernier scale line on the left of the vernier zero that lines up exactly with a main scale marking. The vernier protractor in the example reads 42°45′.

The vernier scale lines up with a mark on the main scale at 45. This gives the fraction of a degree in minutes.

→ Your Turn

Read the following vernier protractors.

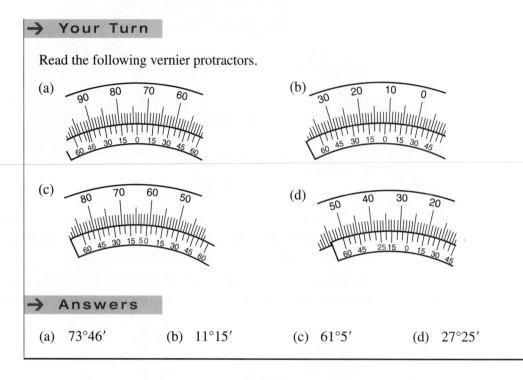

(a)

(b)

(c)

(d)

→ Answers

(a) 73°46′   (b) 11°15′   (c) 61°5′   (d) 27°25′

Meters Many measuring instruments convert the measurement into an electrical signal and then display that signal by means of a pointer and a decimal scale. Instruments of this kind, called *meters,* are used to measure electrical quantities such as voltage, current, or power, and other quantities such as air pressure, flow rates, or speed. As with length scales, not every division on the meter scale is labeled, and practice is required in order to read meters quickly and correctly.

The *range* of a meter is its full-scale or maximum reading. The *main* divisions of the scale are the numbered divisions. The *small* divisions are the smallest marked portions of a main division.

Example 9

Electrical Trades   On the following meter scale, the *range* is 10 volts;

the *main* divisions are 1 volt;
the *smallest* divisions are 0.5 volt.

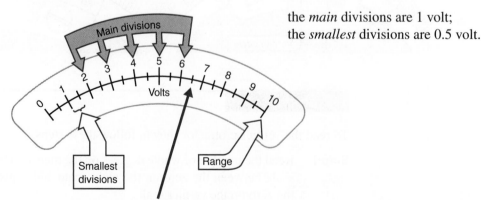

Chapter 5 Measurement

A *dual-scale meter* has an upper and lower scale and allows the user to switch ranges as needed.

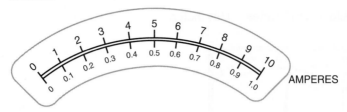

AMPERES

→ **Your Turn**

General Trades   Find the range, main divisions, and smallest divisions for each of the following meter scales.

(a) Range = _____
     Main divisions = _____
     Smallest divisions = _____

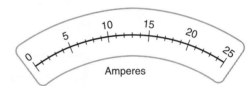

Volts

(b) Range = _____
     Main divisions = _____
     Smallest divisions = _____

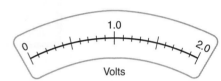

Amperes

(c) Range = _____
     Main divisions = _____
     Smallest divisions = _____

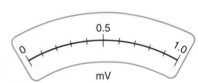

mV

(d) Range = _____
     Main divisions = _____
     Smallest divisions = _____

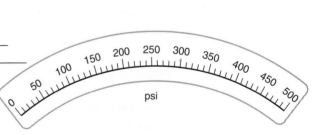

psi

→ **Answers**

(a) Range = 2.0 V   (b) Range = 25 A
     Main divisions = 1.0 V   Main divisions = 5 A
     Smallest divisions = 0.1 V   Smallest divisions = 1 A

(c) Range = 1.0 mV   (d) Range = 500 psi
     Main divisions = 0.5 mV   Main divisions = 50 psi
     Smallest divisions = 0.1 mV   Smallest divisions = 10 psi

To read a meter scale, either we can choose the small scale division nearest to the pointer, or we can estimate between scale markers.

Example 10

Electrical Trades   This meter

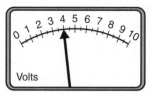

reads exactly 4.0 volts.

But this meter

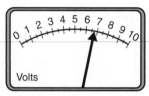

reads 6.5 volts to the nearest scale marker, or about 6.7 volts if we estimate between scale markers. The smallest division is 0.5 volt.

On the following meter scale

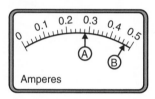

the smallest scale division is $\frac{1}{5}$ of 0.1 ampere or 0.02 ampere. The reading at A is therefore 0.2 plus 4 small divisions or 0.28 ampere. The reading at B is 0.4 plus $3\frac{1}{2}$ small divisions or 0.47 ampere.

→ **Your Turn**

General Trades   Read the following m ters. Estimate between scale divisions where necessary.

(a)

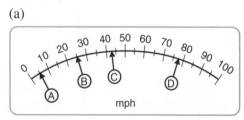

(b)

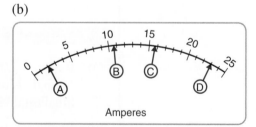

(c)

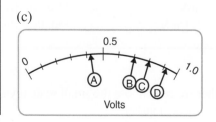

(d)

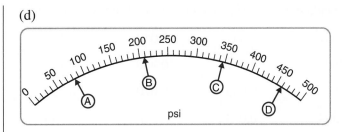

(e)

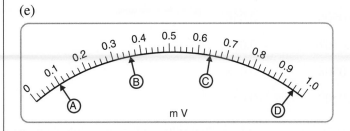

(f)

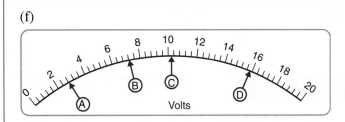

→ **Answers**

Note: Your estimated answers may differ slightly from ours.

(a)	A. 5	B. 25	C. 43	D. 78 mph
(b)	A. 2	B. 10.6	C. 15.8	D. 23 amperes
(c)	A. 0.42	B. 0.7	C. 0.8	D. 0.93 volt
(d)	A. 82	B. 207	C. 346	D. 458 psi
(e)	A. 0.1	B. 0.36	C. 0.64	D. 0.96 mV
(f)	A. 2.8	B. 7.2	C. 10.2	D. 15.7 volts

A Closer Look Notice in meter (f) that the smallest division is $\frac{1}{5}$ of 2 volts or 0.4 volt. If your answers differ slightly from ours, the difference is probably due to the difficulty of estimating the position of the arrows on the diagrams. ◄

Gauge Blocks *Gauge blocks* are extremely precise rectangular steel blocks used for making, checking, or inspecting tools and other devices. Opposite faces of each block are ground and lapped to a high degree of flatness, so that when two blocks are fitted together they will cling tightly. Gauge blocks are sold in standard sets of from 36 to 81 pieces in precisely measured widths. These pieces may be combined to give any decimal dimension within the capacity of the set.

A standard 36-piece set would consist of blocks of the following sizes (in inches):

0.0501	0.051	0.050	0.110	0.100	1.000
0.0502	0.052	0.060	0.120	0.200	2.000
0.0503	0.053	0.070	0.130	0.300	
0.0504	0.054	0.080	0.140	0.400	
0.0505	0.055	0.090	0.150	0.500	
0.0506	0.056				
0.0507	0.057				
0.0508	0.058				
0.0509	0.059				

To combine gauge blocks to produce any given dimension, work from digit to digit, right to left. This procedure will produce the desired length with a minimum number of blocks.

Example 11

Combine gauge blocks from the set given to equal 1.8324 in.

Step 1 To get the 4 in 1.8324 in., select block 0.0504 in. It is the only block in the set with a 4 in the fourth decimal place.

$$\begin{array}{r} 1.8324 \text{ in.} \\ -0.0504 \text{ in.} \\ \hline 1.7820 \text{ in.} \end{array}$$

The target length
The first block
The remainder

Step 2 To get the 2 in 1.782 in., select block 0.052 in. It is the only block in the set with a 2 in the third decimal place.

$$\begin{array}{r} 1.782 \text{ in.} \\ -0.052 \text{ in.} \\ \hline 1.730 \text{ in.} \end{array}$$

The new target length
The second block
The remainder

Step 3 To get the 3 in 1.73 in., select block 0.130 in.

$$\begin{array}{r} 1.73 \text{ in.} \\ -0.13 \text{ in.} \\ \hline 1.60 \text{ in.} \end{array}$$

The new target length
The third block
The remainder

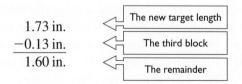

Step 4   To get 1.6, select blocks 1.000, 0.500, and 0.100 in. The six blocks can be combined or "stacked" to total the required 1.8324-in. length.

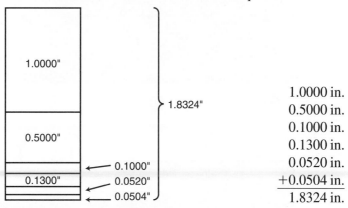

1.0000 in.	
0.5000 in.	
0.1000 in.	
0.1300 in.	
0.0520 in.	
+0.0504 in.	
1.8324 in.	the required length

→ **Your Turn**

Stack sets of gauge blocks that total each of the following.

(a)   2.0976 in.   (b)   1.8092 in.   (c)   3.6462 in.

(d)   0.7553 in.   (e)   1.8625 in.   (f)   0.7474 in.

→ **Answers**

(a)   1.000 in. + 0.500 in. + 0.400 in. + 0.090 in. + 0.057 in. + 0.0506 in.

(b)   1.000 in. + 0.500 in. + 0.200 in. + 0.059 in. + 0.0502 in.

(c)   1.000 in. + 2.000 in. + 0.400 in. + 0.140 in. + 0.056 in. + 0.0502 in.

(d)   0.500 in. + 0.100 in. + 0.050 in. + 0.055 in. + 0.0503 in.

(e)   1.000 in. + 0.500 in. + 0.200 in. + 0.060 in. + 0.052 in. + 0.0505 in.

(f)   0.500 in. + 0.140 in. + 0.057 in. + 0.0504 in.

Now turn to Exercises 5-4 for a set of practice problems designed to help you become more expert in making direct measurements.

Exercises 5-4   Direct Measurements

A. Find the lengths marked on the following rules.

1.

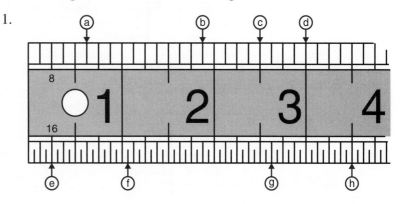

2.

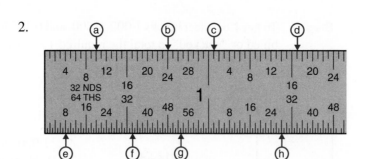

3.

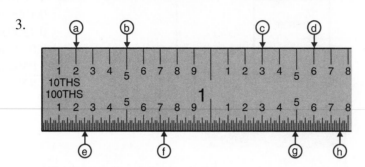

B. Read the following micrometers.

1.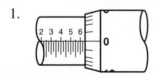

2.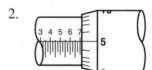

3.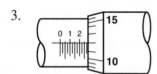

4.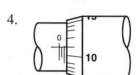

5.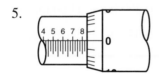

6.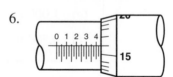

7.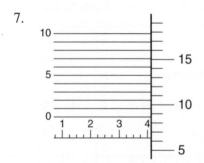

8.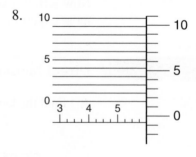

9.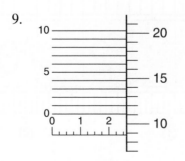

10.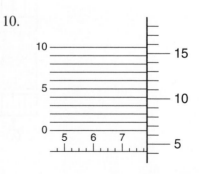

11.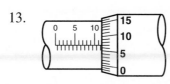

12.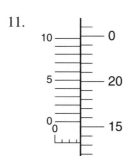

13.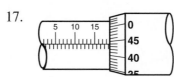

14.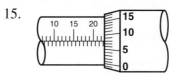

15.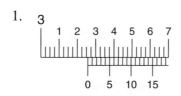

16.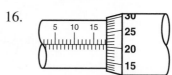

17.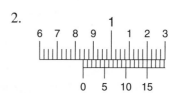

18.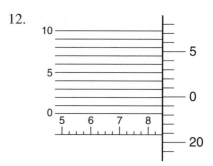

C. Read the following vernier calipers.

1.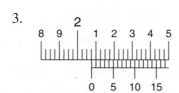

2.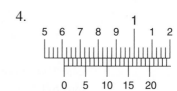

3.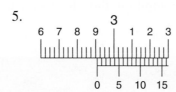

4.

5.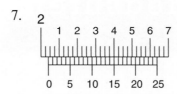

6.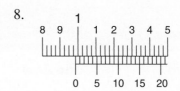

7.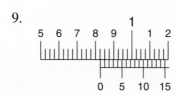

8.

9.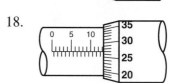

10.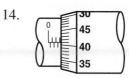

D. Read the following vernier protractors.

1.

2.

3.

4.

5.

6.

7.

8.

E. Find combinations of gauge blocks that produce the following lengths. (Use the 36-block set given on page 332.)

1.	2.3573 in.	2.	1.9467 in.
3.	0.4232 in.	4.	3.8587 in.
5.	1.6789 in.	6.	2.7588 in.
7.	0.1539 in.	8.	3.0276 in.
9.	2.1905 in.	10.	1.5066 in.

Check your answers to the odd-numbered problems in the Appendix, then turn to Problem Set 5 on page 339 for practice on working with measurement numbers. If you need a quick review of the topics in this chapter, visit the chapter Summary first.

Summary Measurement

Objective

Determine the precision and accuracy of measurement numbers.
(p. 263)

Review

The precision of a measurement number is indicated by the place value of its right-most significant digit. The accuracy of a measurement number refers to the number of significant digits it contains.

> **Example:** A pressure reading of 4.27 psi is precise to the nearest hundredth. The measurement of a mass given as 6000 kg is accurate to one significant digit.

Objective	Review
Add and subtract measurement numbers. (p. 265)	**First,** make certain that all numbers have the same units. **Then,** add or subtract the numerical parts and attach the common units. **Finally,** round the answer to agree in precision with the least precise number in the problem.

> **Example:** $38.26 \text{ sec} + 5.9 \text{ sec} = 44.16 \text{ sec}$
> $$= 44.2 \text{ sec, rounded}$$
>
> The final answer has been rounded to one decimal digit to agree with 5.9 sec, the least precise number in the sum.

Multiply and divide measurement numbers. (p. 267)	Multiply or divide the given numbers, then round the answer to the same number of significant digits as the least accurate number used in the calculation. The units of the answer will be the product or quotient of the units in the original problem.

> **Example:** $16.2 \text{ ft} \times 5.8 \text{ ft} = 93.96 \text{ sq ft}$
> $$= 94 \text{ sq ft, rounded}$$
>
> The final answer is rounded to two significant digits to agree with 5.8 ft, the least accurate of the original factors.
>
> Note that ft $\times$ ft = sq ft.

Convert units: (a) within the English system (p. 276), **(b) within the metric system** (p. 294), **and (c) between the English and metric systems** (p. 297)	Use the basic conversion factors given in the text, along with the unity fraction method, for all conversions. Round your answer to the same number of significant digits as the original number.

> **Example:**
>
> (a) $65.0 \text{ lb} = (65.0 \text{ lb}) \times \left(\dfrac{16 \text{ oz}}{1 \text{ lb}} \right) = 1040 \text{ oz}$
>
> (b) $0.45 \text{ km} = (0.45 \text{ km}) \times \left(\dfrac{1000 \text{ m}}{1 \text{ km}} \right) = 450 \text{ m}$
>
> (c) $45 \text{ mph} = \dfrac{45 \text{ mi}}{1 \text{ hr}} \times \dfrac{1.6093 \text{ km}}{1 \text{ mi}} = 72.4185 \text{ km/h} \approx 72 \text{ km/h}$
>
> The final answer has been rounded to two significant digits to agree in accuracy with the original speed, 45.

Think metric. (p. 294)	Use appropriate conversion factors to perform English to metric or metric to English conversions mentally.

> **Example:** Estimate the weight, in pounds, of a casting weighing 45 kg. Remember that a kilogram is about 10% more than 2 lb.
>
> Our first estimate is that 45 kg is about 2×45 or 90 lb. For a better estimate add 10% to 90: $90 + 9 = 99$, or about 100 lb.

Use common technical measuring instruments. (p. 314)	Correctly read a length rule, micrometer, vernier caliper, protractor, and meters, and combine gauge blocks properly.

Measurement

Answers to odd-numbered problems are given in the Appendix.

A. State the accuracy and precision of each of the following measurement numbers.

1. 8.3 lb
2. 960 ft
3. 3.50 in.
4. 8775 gal

5. 0.04 kg
6. 17 psi
7. 4000 sq yd
8. 1.375 liters

B. Perform the following calculations with measurement numbers. Round to the correct precision or accuracy.

1. 5.75 sec − 2.3 sec
2. 3.1 ft × 2.2 ft
3. 4.56 sq in. ÷ 6.1 in.
4. 3.2 cm × 1.26 cm
5. $7\frac{1}{2}$ in. + $6\frac{1}{4}$ in. + $3\frac{1}{8}$ in.
6. 9 lb 6 oz − 5 lb 8 oz
7. 765 mi ÷ 36.2 gal
8. 1.5 cm + 2.38 cm + 5.8 cm
9. 34.38 psi − 16.732 psi
10. $7\frac{1}{4}$ ft − 4.1 ft
11. 7.064 in. − 3.19 in.
12. 22.0 psi × 18.35 sq in.
13. 8 ft 7 in. + 11 ft 11 in. + 9 in.
14. 2.6 in. × 3.85 in. × 14 in.

C. Convert the following measurement numbers to new units as shown. Round to the same number of significant digits as the original measurement.

1. 6.0 in. = _____ cm
2. 54 mph = _____ ft/sec
3. 22°C = _____ °F
4. 2.6 liters = _____ qt
5. 0.82 kg = _____ g
6. 1.62 cm = _____ mm
7. $26\frac{1}{2}$ cu ft = _____ cu in.
8. 41°F = _____ °C
9. 14.3 kg = _____ lb
10. 123.7 psi = _____ atm
11. 24.0 km/h = _____ cm/sec
12. 127,000 sq m = _____ ha
13. $14\frac{1}{2}$ mi = _____ km
14. 350 cu in. = _____ cu cm
15. 14.3 cu yd = _____ cu ft
16. 160 sq ft = _____ sq m
17. 1400 mL = _____ liters
18. 5650 m/min = _____ km/h
19. 24,680 cu in. = _____ bu
20. 6.9 liters = _____ gal
21. 6.8 sq ft = _____ sq in.
22. $12\frac{1}{4}$ qt = _____ liters
23. 0.095 g = _____ mg
24. 596,000 m = _____ km
25. 3.40 ft = _____ cm
26. 45.0 in./sec = _____ cm/sec

Name _____

Date _____

Course/Section _____

27. 1.85 liters = _____ cu cm   28. 4,000,000 mg = _____ g

29. 85 km/h = _____ mph   30. 28 lb = _____ kg

D. Think metric. Choose the closest estimate.

1. Weight of a pencil   (a) 4 kg   (b) 4 g   (c) 40 g

2. Size of a house   (a) 6 sq m   (b) 200 sq m   (c) 1200 sq m

3. Winter in New York   (a) −8°C   (b) −40°C   (c) 20°C

4. Volume of a 30-gal trash can   (a) 0.1 cu m   (b) 2.0 cu m   (c) 5.0 cu m

5. Height of an oak tree   (a) 500 mm   (b) 500 cm   (c) 50 km

6. Capacity of a 50-gal water heater   (a) 25 liters   (b) 100 liters   (c) 180 liters

E. Read the following measuring devices.

1. Rulers

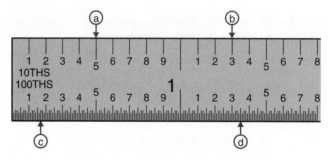

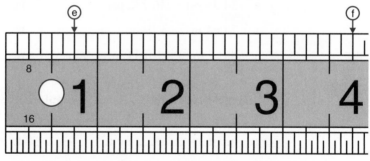

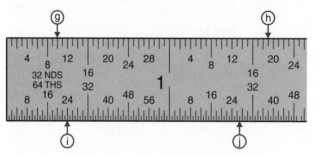

2. Micrometers. Problems (a) and (b) are English, (c) and (d) are metric, and (e) and (f) are English with a vernier scale.

(a) 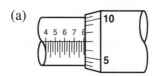   (b)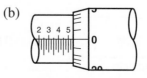

(c)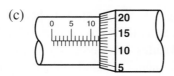

(d)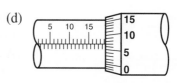

(e)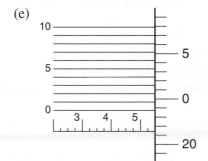

(f)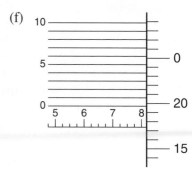

3. Vernier calipers.

(a)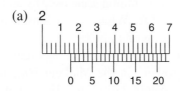

(b)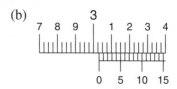

(c)

(d)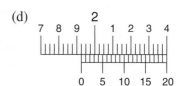

(e)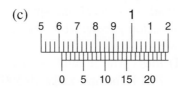

(f)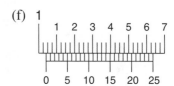

F. Practical Problems. Round appropriately.

Problem 4

1. **Machine Trades**   A piece $6\frac{1}{16}$ in. long is cut from a steel bar $28\frac{5}{8}$ in. long. How much is left?

2. **Life Skills**   An American traveling in Italy notices a road sign saying 85 km/h. What is this speed in miles per hour?

3. **Automotive Trades**   An automotive technician testing cars finds that a certain bumper will prevent damage up to 9.0 ft/sec. Convert this to miles per hour.

4. **Machine Trades**   The gap in the piece of steel shown in the figure should be 2.4375 in. wide. Which of the gauge blocks in the set on page 332 should be used for testing this gap?

5. **Carpentry**   What size wrench, to the nearest 32nd of an inch, will fit a 0.455-in. bolt head?

6. **Metalworking**   The melting point of a casting alloy is 260°C. What temperature is this on the Fahrenheit scale?

7. **Sheet Metal Trades**   What area of sheet metal in square inches is needed to construct a vent with a surface area of 6.50 sq ft?

8. **Life Skills**   How many gallons of gas are needed to drive a car exactly 350 miles if the car normally averages 18.5 mi/gal?

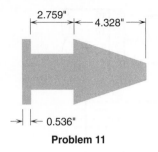

2.759" ← **4.328"** →

← **0.536"**

Problem 11

9. **Metalworking** A certain alloy is heated to a temperature of 560°F. What Celsius temperature does this correspond to? Round to the nearest degree.

10. **Automotive Trades** A Nissan sports car has a 2680-cc engine. What would this be in cubic inches? Round to the nearest cubic inch.

11. **Metalworking** Find the total length of the piece of steel shown in the figure.

12. **Plumbing** The Uniform Plumbing Code is being converted to the metric system. Convert the following measurements.
    (a) The amount of gas used by a furnace (136 cu ft/hr) to cubic meters per hour.
    (b) The length of a gas line (25.0 ft) to meters.
    (c) The diameter of a water pipe (2.50 in.) to centimeters.

13. **Metalworking** Wrought iron has a density of 480 lb/cu ft. Convert this to pounds per cubic inch.

14. **Hydrology** A cubic foot of water contains 7.48 gal. What volume is this in liters?

15. **Carpentry** What are the dimensions of a metric "2-by-4" (a piece of wood $1\frac{1}{2}$ in. by $3\frac{1}{2}$ in.) in centimeters?

16. **Architecture** A common scale used by architects and drafters in making blueprints is $\frac{1}{4}$ in. = 1 ft. What actual length in meters would correspond to a drawing dimension of $1\frac{1}{4}$ in.?

17. **Forestry** The world's largest living organism is a fungus known as the Honey Mushroom, growing in the Malheur National Forest in Oregon. The fungus covers approximately 2200 acres. How many square miles is this?

18. **Automotive Trades** A mechanic spends 45 min on an oil change. Convert this to hours and determine the labor charge given a rate of $62 per hour. Do not round the answer.

19. **Automotive Trades** The circumference of each tire on a motor vehicle is approximately 81.7 in. If the tires make 10,500 revolutions, how many miles will the vehicle travel?

20. **Automotive Trades** An automobile tire is designated as P205/65R15. This means that the overall width of the tire is 205 mm, the sidewall height of the tire is 65% of the width, and the diameter of the rim is 15.0 in. Find the overall diameter of the tire in inches by adding the rim diameter to twice the sidewall height.

21. **Automotive Trades** An auto repair shop allots 1.2 hr to replace a fan belt. The mechanic actually spends 1 hr 20 min doing the job. By how many minutes did the mechanic exceed the allotted time?

22. **Automotive Trades** A customer buying a new car has narrowed her decision down to two finalists: a hybrid selling for $32,000 and an economical nonhybrid selling for $22,000. The average fuel economy of the hybrid is 48 mi/gal, compared to 30 mi/gal for the nonhybrid. By purchasing the hybrid, the customer will receive a one-time tax credit of $1500 and will save about $500 over the life of the vehicle on discounted auto insurance. Assume that the customer would keep each car for 120,000 miles and that the cost of fuel will average $3.80 per gallon over that time period.
    (a) What will be the total cost of fuel for the nonhybrid for 120,000 miles?
    (b) To find the comparative cost of the nonhybrid, add the answer to part (a) to the purchase price.
    (c) What is the total cost of fuel for the hybrid for 120,000 miles?
    (d) To find the comparative cost of the hybrid, first add the answer to part (c) to the purchase price. Then subtract the tax credit and the insurance savings.
    (e) If total cost were the only consideration, which vehicle would the customer choose?

23. **Construction**   A remodeling job requires the following quantities of lumber: 36 boards, 2 in. by 10 in., each 8 ft long; and 48 boards, 2 in. by 4 in., each 12 ft long. What is the total number of board feet required?

24. **Hydrology**   The average daily evaporation from Bradbury Dam at Lake Cachuma during a particular week in the summer was 63.6 acre-feet. To the nearest thousand, how many gallons is this? (See Exercises 5-2, problem C23.)

25. **Automotive Trades**   Europeans commonly express gas mileage in "liters per 100 km" (L/100 km), whereas North Americans express mileage in miles per gallon (mi/gal). If a European car has a mileage rating of 6.5 L per 100 km, how would this be expressed in North America?

26. **Automotive Trades**   Tire pressure is expressed in pounds per square inch (psi) in the United States, and in kiloPascals (kPa) in countries using the metric system. What is the pressure in pounds per square inch of a tire with a pressure of 220 kPa? (Use 1 psi = 6.895 kPa and round your final answer to the nearest tenth.)

27. **Sports and Leisure**   Kenenisa Bekele (Ethiopia) holds the world record for both the men's 5000-m run and the 10,000-m run. His record time for 5000 m is 12 min 37.35 sec. His record time for 10,000 m is 26 min 17.53 sec. To solve the following problems, convert his times from minutes and seconds to only seconds, and round each answer to two decimal places.

   (a)   What was Bekele's average speed (in meters per second) during his record-breaking 5000-m run?

   (b)   What was Bekele's average speed (in meters per second) during his record-breaking 10,000-m run?

   (c)   If Bekele could maintain his average 10,000-m speed for an entire marathon (42,195 m), what would his total time be for the marathon? (Express the final answer in hours, minutes, seconds.)

Pre-Algebra

Objective	Sample Problems	For help, go to
When you finish this chapter, you will be able to:		
1. Understand the meaning of signed numbers.	(a) Which is smaller, -12 or -14? _____	Page 347
	(b) Mark the following points on a number line: $-5\frac{1}{4}$, $+1.0$, -0.75, $+3\frac{1}{2}$. _____	
	(c) Represent a loss of \$250 as a signed number. _____	
2. Add signed numbers.	(a) $(-3) + (-5)$ _____	Page 349
	(b) $18\frac{1}{2} + \left(-6\frac{3}{4}\right)$ _____	
	(c) $-21.7 + 14.2$ _____	
	(d) $(-3) + 15 + 8 + (-20) + 6 + (-9)$ _____	
3. Subtract signed numbers.	(a) $4 - 7$ _____	Page 358
	(b) $-11 - (-3)$ _____	
	(c) $6.2 - (-4.7)$ _____	
	(d) $-\frac{5}{8} - 1\frac{1}{4}$ _____	

_____
Name

_____
Date

_____
Course/Section

Objective	Sample Problems		For help, go to

4. Multiply and divide signed numbers.

(a) $(-12) \times 6$ _____ Page 363

(b) $(-54) \div (-9)$ _____

(c) $(-4.73)(-6.5)$ (two significant digits) _____

(d) $\frac{3}{8} \div \left(-2\frac{1}{2}\right)$ _____

(e) $60 \div (-0.4) \times (-10)$ _____

5. Work with exponents.

(a) 4^3 _____ Page 368

(b) $(2.05)^2$ _____

6. Use the order of operations.

(a) $2 + 3^2$ _____ Page 371

(b) $(4 \times 5)^2 + 4 \times 5^2$ _____

(c) $6 + 8 \times 2^3 \div 4$ _____

7. Find square roots.

(a) $\sqrt{169}$ _____ Page 373

(b) $\sqrt{14.5}$ (to two decimal places) _____

(Answers to these preview problems are given in the Appendix. Worked solutions to many of these problems appear in the chapter Summary.)

If you are certain that you can work *all* these problems correctly, turn to page 381 for a set of practice problems. If you cannot work one or more of the preview problems, turn to the page indicated to the right of the problem. Those who wish to master this material should turn to Section 6-1 and begin work there.

Chapter 6

Pre-Algebra

Reprinted with permission of Universal Press Syndicate.

6-1 Addition of Signed Numbers

The Meaning of Signed Numbers

What kind of number would you use to name each of the following: a golf score two strokes below par, a $2 loss in a poker game, a debt of $2, a loss of 2 yards on a football play, a temperature 2 degrees below zero?

The answer is that they are all *negative numbers,* all equal to -2.

We can better understand the meaning of negative numbers if we can picture them on a number line. First we will display the familiar whole numbers as points along a number line:

0   1   2   3   4...

Notice that the numbers become greater as you move to the right along the number line, and they become lesser as you move to the left. The numbers to the *right* of zero, *greater* than zero, are considered to be *positive numbers.* The negative numbers are all the numbers *less than* zero, and we must extend the number line to the left in order to picture them:

Think of the negative part of the number line as a reflection of the positive part through an imaginary mirror placed through zero. This means that ordering the negative numbers is the opposite of ordering the positive numbers.

To indicate the ordering or relative size of numbers, we often use the following symbols:

The symbol $>$ means "greater than." For example, $a > b$ signifies that "a is greater than b."

The symbol $<$ means "less than." For example, $c < d$ signifies that "c is less than d."

Example 1

(a) On the number line, 4 is to *the right* of 1 so that 4 is *greater than* 1, or $4 > 1$.

(b) On the negative side, -4 is to *the left* of -1 so that -4 is *less than* -1, or $-4 < -1$.

→ **Your Turn**

Make each statement true by replacing the $\square$ with either the $>$ or the $<$ symbol.

(a) $-5 \ \square \ -3$ (b) $9 \ \square \ 6$ (c) $-15 \ \square \ -22$

(d) $12 \ \square \ 31$ (e) $-8 \ \square \ 2$ (f) $43 \ \square \ -6$

→ **Answers**

(a) $-5 < -3$ (b) $9 > 6$ (c) $-15 > -22$

(d) $12 < 31$ (e) $-8 < 2$ (f) $43 > -6$

Note   In algebra, we also use the symbol $\geq$ to signify "greater than or equal to," and the symbol $\leq$ to signify "less than or equal to." ◄

To help in solving the preceding problems, you may extend the number line in both directions. This is perfectly legitimate, because the number line is infinitely long. Notice we have marked the positions of only the positive and negative whole numbers. These numbers are called the *integers.* The positions of fractions and decimals, both positive and negative, can be located between the integers.

→ **Your Turn**

Try marking the following points on a number line.

(a) $-3\frac{1}{2}$ (b) $\frac{1}{4}$ (c) -2.4 (d) 1.75 (e) -0.6

→ **Answers**

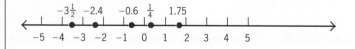

When we work with both positive and negative numbers and need to use names to distinguish them from each other, we often refer to them as *signed numbers*. Signed numbers are a bookkeeping concept. They were used in bookkeeping by the ancient Greeks, Chinese, and Hindus more than 2000 years ago. Chinese merchants wrote positive numbers in black and negative numbers in red in their account books. (Being "in the red" still means losing money or having a negative income.) The Hindus used a dot or circle to show that a number was negative. We use a minus sign (−) to show that a number is negative.

For emphasis, the "+" symbol is sometimes used to denote a positive number, although the absence of any symbol automatically implies that a number is positive. Because the + and − symbols are used also for addition and subtraction, we will use parentheses when writing problems involving addition and subtraction of signed numbers.

→ **More Practice**

Before we attempt these operations, test your understanding of signed numbers by representing each of the following situations with the appropriate signed number.

(a) Two seconds before liftoff

(b) A profit of eight dollars

(c) An elevator going up six floors

(d) Diving 26 feet below the surface of the ocean

(e) A debt of $50

(f) Twelve degrees above zero

→ **Answers**

(a) −2 sec	(b) +$8	(c) +6
(d) −26 ft	(e) −$50	(f) +12°

Addition of Signed Numbers

Addition of signed numbers may be pictured using the number line. A positive number may be represented on the line by an arrow to the right or positive direction. Think of a positive arrow as a "gain" of some quantity such as money. The sum of two numbers is the net "gain."

Example 2

The sum of 4 + 3 or (+4) + (+3) is

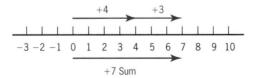

The numbers +4 and +3 are represented by arrows directed to the right. The +3 arrow begins where the +4 arrow ends. The sum 4 + 3 is the arrow that begins at the start of the +4 arrow and ends at the tip of the +3 arrow. In terms of money, a "gain" of $4 followed by a "gain" of $3 produces a net gain of $7.

A negative number may be represented on the number line by an arrow to the left or negative direction. Think of a negative arrow as a loss of some quantity.

Example 3

The sum $(-4) + (-3)$ is -7.

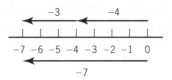

The numbers -4 and -3 are represented by arrows directed to the left. The -3 arrow begins at the tip of the -4 arrow. Their sum, -7, is the arrow that begins at the start of the -4 arrow and ends at the tip of the -3 arrow. A loss of $4 followed by another loss of $3 makes a total loss of $7.

→ **Your Turn**

Try setting up number line diagrams for the following two sums.

(a) $4 + (-3)$ (b) $(-4) + 3$

→ **Solutions**

(a) $4 + (-3) = 1$

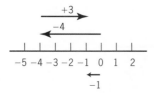

The number 4 is represented by an arrow directed to the right. The number -3 is represented by an arrow starting at the tip of $+4$ and directed to the left. A gain of $4 followed by a loss of $3 produces a net gain of $1.

(b) $(-4) + 3 = -1$

The number -4 is represented by an arrow directed to the left. The number 3 is represented by an arrow starting at the tip of -4 and directed to the right. A loss of $4 followed by a gain of $3 produces a net loss of $1.

ANOTHER WAY TO PICTURE ADDITION OF SIGNED NUMBERS

If you are having trouble with the number line, you might find it easier to picture addition of signed numbers using ordinary poker chips. Suppose that we designate white chips to represent positive integers and blue chips to represent negative integers. The sum of $(+4) + (+3)$ can be found by taking four white chips and combining them with three more white chips. The result is seven white chips or $+7$:

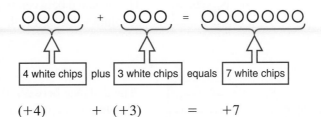

$$(+4) \quad + \quad (+3) \quad = \quad +7$$

Similarly, $(-4) + (-3)$ can be illustrated by taking four *blue* chips and combining them with three additional *blue* chips. We end up with seven blue chips or -7:

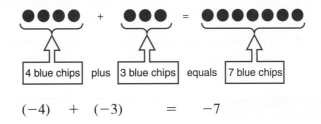

$$(-4) \quad + \quad (-3) \quad = \quad -7$$

To illustrate $4 + (-3)$, take four white chips and three blue chips. Realizing that a white and a blue will "cancel" each other out $[(+1) + (-1) = 0]$, we rearrange them as shown below, and the result is one white chip or $+1$:

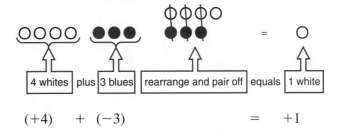

$$(+4) \quad + \quad (-3) \quad = \quad +1$$

Finally, to represent $(-4) + 3$, picture the following:

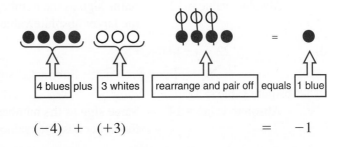

$$(-4) \quad + \quad (+3) \quad = \quad -1$$

Number lines and poker chips provide a good way to see what is happening when we add signed numbers, but we need a nice easy rule that will allow us to do the addition more quickly.

Absolute Value To add or subtract signed numbers you should work with their absolute value. The **absolute value** of a number is the distance along the number line between that number and the zero point. We indicate the absolute value by enclosing the number with vertical lines. The absolute value of any number is always a nonnegative number. For example,

$|4| = 4$ 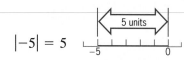   The distance between 0 and 4 is 4 units, so the absolute value of 4 is 4.

$|-5| = 5$ The distance between 0 and -5 is 5 units, so the absolute value of -5 is 5.

To add two signed numbers, follow this three-step rule:

Step 1 Find the absolute value of each number.

Step 2 If the two numbers have the **same sign,** add their absolute values and give the sum the sign of the original numbers.

Step 3 If the two numbers have **opposite signs,** subtract the absolute values and give the difference the sign of the number with the larger absolute value.

Example 4

(a) $9 + 3 = +12$

Absolute value $= 9$
Absolute value $= 3$ Add absolute values: $9 + 3 = 12$
            Sign is positive because both numbers are positive

(b) $(-6) + (-9) = -15$

Absolute value $= 6$
Absolute value $= 9$ Add absolute values: $6 + 9 = 15$
            Sign is negative because both numbers are negative

(c) $7 + (-12) = -5$

Absolute value $= 7$
Absolute value $= 12$ Subtract absolute values: $12 - 7 = 5$
            Same sign as the number with the larger absolute value (-12)

(d) $22 + (-14) = +8$

Absolute value $= 22$
Absolute value $= 14$ Subtract absolute values: $22 - 14 = 8$
            Same sign as the number with the larger absolute value $(+22)$

Try it. Use this method to do the following calculations.

(a) $7 + (-2)$ (b) $-6 + (-14)$ (c) $(-15) + (-8)$

(d) $23 + 13$ (e) $-5 + 8 + 11 + (-21) + (-9) + 12$

→ Solutions

(a) $7 + (-2) = +5$ Signs are opposite; subtract absolute values. Answer has the sign of the number with the larger absolute value (7).

(b) $-6 + (-14) = -20$ Signs are the same; add absolute values. Answer has the sign of the original numbers.

(c) $(-15) + (-8) = -23$ Signs are the same; add absolute values. Answer has the sign of the original numbers.

(d) $23 + 13 = +36$ Signs are the same; add absolute values. Answer has the sign of the original numbers.

(e) When more than two numbers are being added, rearrange them so that numbers with like signs are grouped together:

$$-5 + 8 + 11 + (-21) + (-9) + 12 = \underbrace{(8 + 11 + 12)}_{\text{Positives}} + \underbrace{[(-5) + (-21) + (-9)]}_{\text{Negatives}}$$

Rearrange:

Find the positive total: $8 + 11 + 12 = +31$

Find the negative total: $(-5) + (-21) + (-9) = -35$

Add the positive total to the negative total: $31 + (-35) = -4$

The procedure is exactly the same if the numbers being added are fractions or decimal numbers.

Example 5

(a) To add $6\frac{3}{8} + \left(-14\frac{3}{4}\right)$, **first** note that the signs are opposite.

We must subtract absolute values:

$$14\frac{3}{4}$$
$$-6\frac{3}{8}$$

The common denominator is 8.

$$14\frac{6}{8} \quad \Longleftarrow \quad \boxed{\frac{3}{4} = \frac{3 \times 2}{4 \times 2} = \frac{6}{8}}$$
$$-6\frac{3}{8}$$
$$\overline{8\frac{3}{8}}$$

The answer has the sign of the number with the larger absolute value $\left(-14\frac{3}{4}\right)$;

Answer: $-8\frac{3}{8}$

(b) To add $-21.6 + (-9.8)$, notice that the signs are the same.

We must add absolute values:

$$\begin{array}{r} 21.6 \\ + 9.8 \\ \hline 31.4 \end{array}$$

The original numbers were both negative.

Answer: -31.4

→ **Your Turn**

Find each sum.

(a) $-1\frac{2}{3} + \left(-2\frac{3}{4}\right)$

(b) $-7.2 + 11.6$

→ **Solutions**

(a) The signs are the same so we must add absolute values.

The common denominator is 12:

$$1\frac{2}{3} \longrightarrow 1\frac{8}{12} \Longleftarrow \boxed{\frac{2\times\boxed{4}}{3\times\boxed{4}}}$$

$$+2\frac{3}{4} \longrightarrow +2\frac{9}{12} \Longleftarrow \boxed{\frac{3\times\boxed{3}}{4\times\boxed{3}}}$$

$$3\frac{17}{12} = 3 + 1\frac{5}{12} = 4\frac{5}{12}$$

The original numbers were both negative.   Answer: $-4\frac{5}{12}$

(b) The signs are opposite so we must subtract absolute values.

$$\begin{array}{r} 11.6 \\ - 7.2 \\ \hline 4.4 \end{array}$$

The answer is $+4.4$ because the number with the larger absolute value ($+11.6$) is positive.

Using a Calculator To work with signed numbers on a calculator, we need only one new key, the $\boxed{(-)}$ key. This key allows us to enter a negative number on our display.

For example, to enter -4, simply press $\boxed{(-)}$ followed by 4.

▶ **Note** Scientific calculators with a one-line display normally have a key marked $\boxed{+/-}$ for entering negative numbers. For these models, enter the number key first and then press the $\boxed{+/-}$ key. If you are using a one-line display model, be sure to consult the calculator appendix for alternate key sequences. ◀

Example 6

C6-1

To add $(-587) + 368$, enter the following sequence:

$\boxed{(-)}$ **587** $\boxed{+}$ **368** $\boxed{=}$ → ░░░░ *-219.* ░░░░

→ More Practice

Use a calculator to do the following addition problems.

(a) $(-2675) + 1437$ (b) $(-6.975) + (-5.2452)$

(c) $0.026 + (-0.0045)$ (d) $(-683) + (-594) + 438 + (-862)$

(e) $\left(-\frac{5}{8}\right) + 2\frac{2}{3} + \left(-3\frac{1}{2}\right)$

→ Solutions

C6-2

(a) $\boxed{(-)}$ **2675** $\boxed{+}$ **1437** $\boxed{=}$ → ░░░ *-1238.* ░░░

(b) $\boxed{(-)}$ **6.975** $\boxed{+}$ $\boxed{(-)}$ **5.2452** $\boxed{=}$ → ░░░ *-12.2202* ░░░

(c) **.026** $\boxed{+}$ $\boxed{(-)}$ **.0045** $\boxed{=}$ → ░░░ *0.0215* ░░░

(d) $\boxed{(-)}$ **683** $\boxed{+}$ $\boxed{(-)}$ **594** $\boxed{+}$ **438** $\boxed{+}$ $\boxed{(-)}$ **862** $\boxed{=}$ → ░░░ *-1701.* ░░░

(e) $\boxed{(-)}$ **5** $\boxed{A\frac{b}{c}}$ **8** $\boxed{+}$ **2** $\boxed{A\frac{b}{c}}$ **2** $\boxed{A\frac{b}{c}}$ **3** $\boxed{+}$ $\boxed{(-)}$ **3** $\boxed{A\frac{b}{c}}$ **1** $\boxed{A\frac{b}{c}}$ **2** $\boxed{=}$ → ░░░ *-1⌐11/24* ░░░

For more practice in addition of signed numbers, turn to Exercises 6-1 for a set of problems.

Exercises 6-1 **Addition of Signed Numbers**

A. Make each statement true by replacing the $\square$ symbol with either the $>$ or $<$ symbol.

1. $-4 \ \square \ 7$	2. $-17 \ \square \ -14$	3. $23 \ \square \ 18$
4. $13 \ \square \ -15$	5. $-9 \ \square \ -10$	6. $-31 \ \square \ 41$
7. $-65 \ \square \ -56$	8. $47 \ \square \ 23$	9. $-47 \ \square \ -23$
10. $-3 \ \square \ 0$	11. $6 \ \square \ 0$	12. $-1.4 \ \square \ -1.39$
13. $2.7 \ \square \ -2.8$	14. $-3\frac{1}{4} \ \square \ -3\frac{1}{8}$	15. $-11.4 \ \square \ -11\frac{1}{4}$
16. $-0.1 \ \square \ -0.001$		

B. For each problem, make a separate number line and mark the points indicated.

1. $-8.7, -9.4, -8\frac{1}{2}, -9\frac{1}{2}$

2. $-1\frac{3}{4}, +1\frac{3}{4}, -\frac{1}{4}, +\frac{1}{4}$

3. $-0.2, -1\frac{1}{2}, -2.8, -3\frac{1}{8}$

4. $+4.1, -1.4, -3.9, +1\frac{1}{4}$

C. Represent each of the following situations with a signed number.

1. A temperature of six degrees below zero

2. A debt of three hundred dollars

3. An airplane rising to an altitude of 12,000 feet

4. The year A.D. 240

5. A golf score of nine strokes below par

6. A quarterback getting sacked for a seven-yard loss

7. Diving eighty feet below the surface of the ocean

8. Winning fifteen dollars at poker

9. Taking five steps backward

10. Ten seconds before takeoff

11. A profit of six thousand dollars

12. An elevation of thirty feet below sea level

13. A checking account overdrawn by $17.50

14. A 26-point drop in the NAZWAC stock index

D. Add.

1. $-1 + 6$
2. $-11 + 7$

3. $-13 + (-2)$
4. $-6 + (-19)$

5. $31 + (-14)$
6. $23 + (-28)$

7. $21 + 25$
8. $12 + 19$

9. $-8 + (-8)$
10. $-7 + (-9)$

11. $-34 + 43$
12. $-56 + 29$

13. $45 + (-16)$
14. $22 + (-87)$

15. $-18 + (-12)$
16. $-240 + 160$

17. $-3.6 + 2.7$
18. $5.7 + (-7.9)$

19. $-12.43 + (-15.66)$
20. $-90.15 + (-43.69)$

21. $-\frac{5}{12} + \left(-\frac{11}{12}\right)$
22. $\frac{3}{5} + \left(-\frac{4}{5}\right)$

23. $2\frac{5}{8} + \left(-3\frac{7}{8}\right)$
24. $-4\frac{1}{4} + 6\frac{3}{4}$

25. $-9\frac{1}{2} + \left(-4\frac{1}{2}\right)$
26. $-6\frac{2}{3} + \left(-8\frac{2}{3}\right)$

27. $-2\frac{1}{2} + 1\frac{3}{4}$
28. $-5\frac{5}{8} + 8\frac{3}{4}$

29. $9\frac{7}{16} + \left(-7\frac{7}{8}\right)$
30. $-13\frac{3}{10} + \left(-8\frac{4}{5}\right)$

31. $6\frac{5}{6} + \left(-11\frac{1}{4}\right)$
32. $\left(-4\frac{3}{8}\right) + \left(-3\frac{7}{12}\right)$

33. $-6 + 4 + (-8)$
34. $22 + (-38) + 19$

35. $-54 + 16 + 22$
36. $11 + (-6) + (-5)$

37. $-7 + 12 + (-16) + 13$
38. $23 + (-19) + 14 + (-21)$

39. $-36 + 45 + 27 + (-18)$
40. $18 + (-51) + (-8) + 26$

41. $-250 + 340 + (-450) + (-170) + 320 + (-760) + 880 + 190 + (-330)$

42. $-1 + 3 + (-5) + 7 + (-9) + 11 + (-13) + (15) + (-17) + 19$

43. $23 + (-32) + (-14) + 55 + (-66) + 44 + 28 + (-31)$

44. $-12.64 + 22.56 + 16.44 + (-31.59) + (-27.13) + 44.06 + 11.99$

45. $(-2374) + (5973)$

46. $(6.875) + (-11.765)$

47. $(-46,720) + (-67,850) + 87,950$

48. $(0.0475) + (-0.0875) + (-0.0255)$

E. Word Problems

1. **Printing** The Jiffy Print Shop has a daily work quota that it tries to meet. During the first two weeks of last month its daily comparisons to this quota showed the following: $+\$30$, $-\$27$, $-\$15$, $+\$8$, $-\$6$, $+\$12$, $-\$22$, $+\$56$, $-\$61$, $+\$47$. (In each case "+" means they were above the quota and "−" means they were below.) How far below or above their quota was the two-week total?

2. **Sports and Leisure** During a series of plays in a football game, a running back carried the ball for the following yardages: 7, -3, 2, -1, 8, -5, and -2. Find his total yards and his average yards per carry.

3. **Meteorology** On a cold day in Fairbanks the temperature stood at 12 degrees below zero $(-12°)$ at 2 P.M. Over the next 5 hours the temperature rose 4 degrees, dropped 2 degrees, rose 5 degrees, dropped 1 degree, and finally, dropped 3 degrees. What was the temperature at 7 P.M.?

4. **Life Skills** Pedro's checking account contained $489 on June 1. He then made the following transactions: deposited $150, withdrew $225, withdrew $34, deposited $119, withdrew $365, and deposited $750. What was his new balance after these transactions?

5. **Business and Finance** A struggling young company reported the following results for the four quarters of the year: A loss of $257,000, a loss of $132,000, a profit of $87,000, a profit of $166,000. What was the net result for the year?

6. **Plumbing** In determining the fitting allowance for a pipe configuration, a plumber has negative allowances of $-2\frac{1}{2}$ in. and $-1\frac{7}{16}$ in. and two positive allowances of $\frac{1}{2}$ in. each. What is the net fitting allowance for the job?

7. **Electronics** What is the sum of currents flowing into Node N in the circuit below?

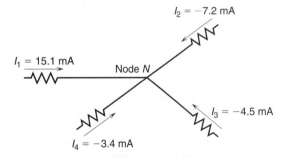

8. **Electrical Trades** Excess energy generated by solar panels can often be sold back to the power company. Over a 6-month period, a household consumed 212, 150, -160, -178, -148, and 135 kWh (kilowatt-hours) of electricity from the power company, where a negative number indicates that the household generated more solar energy than it consumed. What was the net amount of electricity consumed by the household? (Express your answer as a signed number.)

When you have completed these exercises, check your answers to the odd-numbered problems in the Appendix, then turn to Section 6-2 to learn how to subtract signed numbers.

Opposites

When the number line is extended to the left to include the negative numbers, every positive number can be paired with a negative number whose absolute value, or distance from zero, is the same. These pairs of numbers are called *opposites*. For example, -3 is the opposite of $+3$, $+5$ is the opposite of -5, and so on.

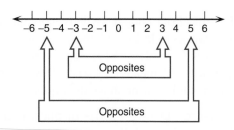

> **Note** Zero has no opposite because it is neither positive nor negative. ◄

The concept of opposites will help us express a rule for subtraction of signed numbers.

Subtraction of Signed Numbers

We know from basic arithmetic that $5 - 2 = 3$. From the rules for addition of signed numbers, we know that $5 + (-2) = 3$ also. Therefore,

$$(+5) - (+2) = (+5) + (-2) = 3$$

This means that subtracting $(+2)$ is the same as adding its opposite, (-2). Our knowledge of arithmetic and signed numbers verifies that subtracting and adding the opposite are equivalent for all the following cases:

$$5 - 5 = 5 + (-5) = 0$$
$$5 - 4 = 5 + (-4) = 1$$
$$5 - 3 = 5 + (-3) = 2$$
$$5 - 2 = 5 + (-2) = 3$$
$$5 - 1 = 5 + (-1) = 4$$
$$5 - 0 \qquad\qquad = 5 \quad \text{(zero has no opposite)}$$

But what happens with a problem like $5 - 6 = ?$ If we rewrite this subtraction as addition of the opposite like the others, we have

$$5 - 6 = 5 + (-6) = -1$$

Notice that this result fits right into the pattern established by the previous results. The answers *decrease* as the numbers being subtracted *increase*: $5 - 3 = 2, 5 - 4 = 1$, $5 - 5 = 0$, so it seems logical that $5 - 6 = -1$.

What happens if we subtract a negative? For example, how can we solve the problem $5 - (-1) = ?$ If we rewrite this subtraction as addition of the opposite, we have

$$5 - (-1) = 5 + (+1) = 6$$

| subtracting -1 | becomes | adding the opposite of -1 or $+1$ |

Notice that this too fits in with the pattern established by the previous results. The answers *increase* as the numbers being subtracted *decrease*: $5 - 2 = 3, 5 - 1 = 4$, $5 - 0 = 5$, so again it seems logical that $5 - (-1) = 6$.

We can now state a rule for subtraction of signed numbers. To subtract two signed numbers, follow this two-step rule:

Step 1   Rewrite the subtraction as an addition of the opposite.

Step 2   Use the rules for addition of signed numbers to add.

Example 1

(a) $3 - 7 = (+3) \underbrace{- (+7)}_{\text{subtracting } +7} = 3 \underbrace{+ (-7)}_{\text{adding } -7} = -4$

$\underbrace{\qquad\qquad\qquad}_{\text{is the same as}}$

(b) $-2 \underbrace{- (-6)}_{\text{subtracting } -6} = -2 \underbrace{+ (+6)}_{\text{adding } +6} = +4$

$\underbrace{\qquad\qquad\qquad}_{\text{is the same as}}$

(c) $-8 \underbrace{- (-3)}_{\text{subtracting } -3} = -8 \underbrace{+ (+3)}_{\text{adding } +3} = -5$

$\underbrace{\qquad\qquad\qquad}_{\text{is the same as}}$

(d) $-4 - 9 = -4 \underbrace{- (+9)}_{\text{subtracting } +9} = (-4) \underbrace{+ (-9)}_{\text{adding } -9} = -13$

$\underbrace{\qquad\qquad\qquad}_{\text{is the same as}}$

> **Careful**   Never change the sign of the first number. You change the sign of the *second* number only *after* changing subtraction to addition. ◄

→ **Your Turn**

Try these:

(a) $-7 - 5$

(b) $8 - 19$

(c) $4 - (-4)$

(d) $-12 - (-8)$

(e) $\left(-\dfrac{3}{4}\right) - \left(\dfrac{5}{16}\right)$

(f) $-2 - (-3) + (-1)$

(g) $-3 - 9 - (-7) + (-8) + (+2)$

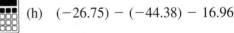 (h) $(-26.75) - (-44.38) - 16.96$

→ **Solutions**

(a) $-7 - 5 = -7 + (-5) = -12$

(b) $8 - 19 = (+8) + (-19) = -11$

(c) $4 - (-4) = (+4) + (+4) = +8$

(d) $-12 - (-8) = -12 + (+8) = -4$

(e) $\left(-\dfrac{3}{4}\right) - \left(\dfrac{5}{16}\right) = \left(-\dfrac{3}{4}\right) + \left(-\dfrac{5}{16}\right)$

$$= \left(-\dfrac{12}{16}\right) + \left(-\dfrac{5}{16}\right)$$

$$= -\dfrac{17}{16} = -1\dfrac{1}{16}$$

(f) $-2 - (-3) + (-1) = \underbrace{-2 + (+3)} + (-1)$

$$= (+1) + (-1)$$

$$= 0$$

(g) $-3 - 9 - (-7) + \underbrace{(-8) + (+2)} = -3 + (-9) + (+7) + \underbrace{(-8) + (+2)} = -11$

> These last two operations are additions, so they remain unchanged when the problem is rewritten.

C6-3   (h) $\boxed{(-)}$ **26.75** $\boxed{-}$ $\boxed{(-)}$ **44.38** $\boxed{-}$ **16.96** $\boxed{=}$ → $\quad$ *0.67*

▶ **Careful**

When using a calculator, make sure to use the $\boxed{-}$ key only for subtraction and the $\boxed{(-)}$ key only to enter a negative number. Using the incorrect key will result in either an "Error" message or a wrong answer. ◀

SUBTRACTION WITH CHIPS

If you are having trouble visualizing subtraction of signed numbers, try the poker chip method that we used to show addition. Subtraction means "taking away" chips. First, a simple example: $7 - 3$ or $(+7) - (+3)$.

Begin with 7 white (positive) chips.

Take away 3 white chips.

Four white chips remain.

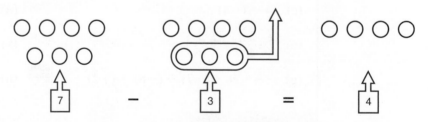

What about the problem: $(+3) - (+7) = ?$ How can you remove 7 white chips if you have only 3 white chips? Here we must be very clever. Remember that a

(continued)

Chapter 6 Pre-Algebra

positive chip and a negative chip "cancel" each other [$(+1) + (-1) = 0$]. So if we add both a white (positive) and a blue (negative) chip to a collection, we do not really change its value. Do $(+3) - (+7)$ this way:

| Start with 3 white chips. | Add in 4 more whites and 4 more blues | Now we can remove 7 white chips. | Four blue chips remain. |

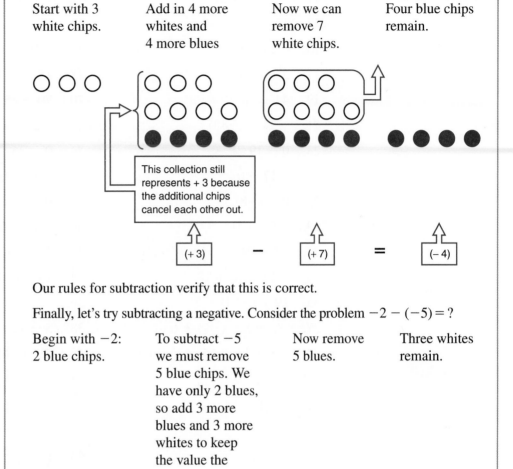

This collection still represents + 3 because the additional chips cancel each other out.

$(+ 3)$ $-$ $(+ 7)$ $=$ $(- 4)$

Our rules for subtraction verify that this is correct.

Finally, let's try subtracting a negative. Consider the problem $-2 - (-5) = ?$

| Begin with -2: 2 blue chips. | To subtract -5 we must remove 5 blue chips. We have only 2 blues, so add 3 more blues and 3 more whites to keep the value the same. | Now remove 5 blues. | Three whites remain. |

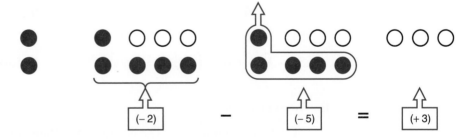

$(- 2)$ $-$ $(- 5)$ $=$ $(+ 3)$

Our rules for subtraction verify that this is correct.

Now turn to Exercises 6-2 for a set of problems on subtraction of signed numbers.

Exercises 6-2 Subtraction of Signed Numbers

A. Find the value of the following.

1. $3 - 5$ 　　　　　　　　　　　　　2. $7 - 20$

3. $-4 - 12$ 　　　　　　　　　　　　4. $-9 - 15$

5. $6 - (-3)$ 　　　　　　　　　　　　6. $14 - (-7)$

7. $-11 - (-5)$

8. $-22 - (-10)$

9. $12 - 4$

10. $23 - 9$

11. $-5 - (-16)$

12. $-14 - (-17)$

13. $12 - 30$

14. $30 - 65$

15. $-23 - 8$

16. $-18 - 11$

17. $2 - (-6)$

18. $15 - (-12)$

19. $44 - 18$

20. $70 - 25$

21. $-34 - (-43)$

22. $-27 - (-16)$

23. $0 - 19$

24. $0 - 26$

25. $-17 - 0$

26. $-55 - 0$

27. $0 - (-8)$

28. $0 - (-14)$

29. $-56 - 31$

30. $-45 - 67$

31. $22 - 89$

32. $17 - 44$

33. $-12 - 16 + 18$

34. $23 - (-11) - 30$

35. $19 + (-54) - 7$

36. $-21 + 6 - (-40)$

37. $-2 - 5 - (-7) + 6$

38. $5 - 12 + (-13) - (-7)$

39. $-3 - 4 - (-5) + (-6) + 7$

40. $2 - (-6) - 10 + 4 + (-11)$

41. $\frac{2}{9} - \frac{7}{9}$

42. $\frac{5}{12} - \left(-\frac{7}{12}\right)$

43. $\left(-\frac{3}{8}\right) - \frac{1}{4}$

44. $2\frac{1}{2} - \left(-3\frac{1}{4}\right)$

45. $6\frac{9}{16} - 13\frac{1}{16}$

46. $\left(-\frac{5}{6}\right) - \left(-\frac{2}{3}\right)$

47. $\left(-2\frac{3}{4}\right) - 5\frac{5}{6}$

48. $\left(-1\frac{7}{8}\right) - \left(-3\frac{1}{6}\right)$

49. $6.5 - (-2.7)$

50. $(-0.275) - 1.375$

51. $(-86.4) - (-22.9)$

52. $0.0025 - 0.075$

53. $22.75 - 36.45 - (-54.72)$

54. $(-7.35) - (-4.87) - 3.66$

55. $3\frac{1}{2} - 6.6 - \left(-2\frac{1}{4}\right)$

56. $(-5.8) - \left(-8\frac{1}{8}\right) - 3.9$

B. Word Problems

1. **General Interest** The height of Mt. Rainier is $+14{,}410$ ft. The lowest point in Death Valley is -288 ft. What is the difference in altitude between these two points?

2. **Meteorology** The highest temperature ever recorded on Earth was $+136°F$ at Libya in North Africa. The coldest was $-127°F$ at Vostok in Antarctica. What is the difference in temperature between these two extremes?

3. **Meteorology** At 8 P.M. the temperature was $-2°F$ in Anchorage, Alaska. At 4 A.M. it was $-37°F$. By how much did the temperature drop in that time?

4. **General Interest** A vacation trip involved a drive from a spot in Death Valley 173 ft below sea level (-173 ft) to the San Bernardino Mountains 4250 ft above sea level ($+4250$ ft). What was the vertical ascent for this trip?

5. **Trades Management** Dave's Auto Shop was $2465 in the "red" ($-2465) at the end of April—they owed this much to their creditors. At the end of

December they were $648 in the red. How much money did they make in profit between April and December to achieve this turnaround?

6. **Sports and Leisure** A golfer had a total score of six strokes over par going into the last round of a tournament. He ended with a total score of three strokes under par. What was his score relative to par during the final round?

7. **Trades Management** Pacific Roofing was $3765 in the "red" (see problem 5) at the end of June. At the end of September they were $8976 in the "red." Did they make or lose money between June and September? How much?

8. **Sheet Metal Trades** To establish the location of a hole relative to a fixed zero point, a machinist must make the following calculation:

$$y = 5 - (3.750 - 0.500) - 2.375$$

Find y.

9. **Electronics** The peak-to-peak voltage of an ac circuit is found by subtracting the negative peak voltage from the positive peak voltage. Calculate the peak-to-peak voltage V in the diagram shown.

10. **Automotive Trades** During an alignment, the caster angle on a vehicle was adjusted from $-0.25°$ to $+1.75°$. What was the amount of the adjustment.

Check your answers to the odd-numbered problems in the Appendix, then turn to Section 6-3 to learn about multiplication and division of signed numbers.

6-3 Multiplication and Division of Signed Numbers

Multiplication of Signed Numbers

As we learned earlier, multiplication is the shortcut for repeated addition. For example,

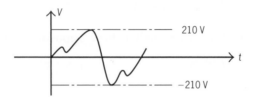

Notice that the *product of two positive numbers* is always a *positive* number.

We can use repeated addition to understand multiplication involving negative numbers. Consider the problem $(+5) \times (-3)$.

$$(+5) \times (-3) = \underbrace{(-3) + (-3) + (-3) + (-3) + (-3)}_{} = -15$$

Five times negative three   negative three added five times

Notice that the product of a *positive number* and a *negative number* is a *negative* number. The order in which you multiply two numbers does not matter, so a negative times a positive is also a negative. For example,

$$(-5) \times (+3) = (+3) \times (-5) = (-5) + (-5) + (-5) = -15$$

To understand the product of a negative times a negative, look at the pattern in the following sequence of problems:

$(-5) \times (+3) = -15$
$(-5) \times (+2) = -10$
$(-5) \times (+1) = -5$
$(-5) \times \quad 0 = 0$
$(-5) \times (-1) = ?$

Each time we *decrease* the factor on the right by 1, we *increase* the product by 5. So to preserve this pattern, the product of $(-5) \times (-1)$ must be $+5$. In other words, *a negative times a negative* is a *positive*.

Rules for Multiplication

In summary, to multiply two signed numbers, follow this two-step rule:

Step 1 Find the product of their absolute values.

Step 2 Attach a sign to the answer:

- If both numbers have the **same** sign, the answer is **positive.**

$$(+) \times (+) = + \qquad (-) \times (-) = +$$

- If the two numbers have **opposite** signs, the answer is **negative.**

$$(+) \times (-) = - \qquad (-) \times (+) = -$$

Example 1

Multiply:

(a) $(+3) \times (-2)$

    Step 1 The product of their absolute values is $3 \times 2 = 6$

    Step 2 The original numbers have opposite signs, so the product is negative. $(+3) \times (-2) = -6$

(b) $(-4) \times (-8)$

    Step 1 The product of their absolute values is $4 \times 8 = 32$

    Step 2 The original numbers have the same sign, so the product is positive. $(-4) \times (-8) = +32$

(c) $\left(-\dfrac{1}{4}\right) \times \left(+\dfrac{2}{3}\right)$

    Step 1 The product of their absolute values is $\dfrac{1}{\overset{2}{4}} \times \dfrac{\overset{1}{2}}{3} = \dfrac{1}{6}$

    Step 2 The original fractions have opposite signs, so the product is negative. $\left(-\dfrac{1}{4}\right) \times \left(+\dfrac{2}{3}\right) = -\dfrac{1}{6}$

→ Your Turn

Use these rules on the following problems.

(a) $(-4) \times (+7)$ (b) $(-8) \times (-6)$ (c) $(+5) \times (+9)$

(d) $(+10) \times (-3)$ (e) $(-3) \times (-3)$ (f) $(-2) \times (-2) \times (-2)$

(g) $(-2) \times (-3) \times (+4) \times (-5) \times (-6)$ (h) $\left(-2\frac{1}{3}\right) \times \left(\frac{9}{14}\right)$

(i) $(-3.25) \times (-1.40)$

→ **Solutions**

(a) $(-4) \times (+7) = -28$

opposite signs ⟶ answer is negative

(b) $(-8) \times (-6) = +48$

same sign ⟶ answer is positive

(c) $+45$ (same sign, answer is positive)

(d) -30 (opposite signs, answer is negative)

(e) $+9$

(f) -8

(g) $+720$

(h) $-1\frac{1}{2}$

C6-4 (i) ⟮(−)⟯ **3.25** ⟮×⟯ ⟮(−)⟯ **1.4** ⟮=⟯ → | **4.55** |

A Closer Look   Did you discover an easy way to do problems (f) and (g) where more than two signed numbers are being multiplied? You could go left to right, multiply two numbers, then multiply the answer by the next number. You would then need to apply the sign rules for each separate product. However, because every *two* negative factors results in a positive, an *even* number of negative factors gives a positive answer, and an *odd* number of negative factors gives a negative answer. In a problem with more than two factors, count the total number of negative signs to determine the sign of the answer.

In problem (f) there are three negative factors—an odd number—so the answer is negative. Multiply the absolute values as usual to get the numerical part of the answer, 8. In problem (g) there are four negative factors—an even number—so the answer is positive. The product of the absolute values of all five numbers gives us the numerical part of the answer, 720. ◀

Division of Signed Numbers   Recall that division is the reverse of the process of multiplication. The division

$24 \div 4 = \square$ can be written as the multiplication $24 = 4 \times \square$

and in this case the number $\square$ is 6. Notice that *a positive divided by a positive equals a positive.* We can use this "reversing" process to verify that the rest of the sign rules for division are exactly the same as those for multiplication. To find $(-24) \div (+4)$, rewrite the problem as follows:

$(-24) \div (+4) = \square$ $-24 = (+4) \times \square$

From our knowledge of signed number multiplication, $\square = -6$. Therefore, *a negative divided by a positive equals a negative.* Similarly,

$(+24) \div (-4) = \square$ means $(+24) = (-4) \times \square$ so that $\square = -6$

A positive divided by a negative is also a negative. Finally, let's try a negative divided by a negative:

$$(-24) \div (-4) = \square \quad \text{means} \quad (-24) = (-4) \times \square \quad \text{so that} \quad \square = +6$$

A negative divided by a negative equals a positive.

When dividing two signed numbers, determine the sign of the answer the same way you do for multiplication:

- If both numbers have the **same** sign, the answer is **positive.**

$$(+) \div (+) = + \qquad (-) \div (-) = +$$

- If the two numbers have **opposite** signs, the answer is **negative.**

$$(+) \div (-) = - \qquad (-) \div (+) = -$$

Divide the absolute values of the two numbers to get the numerical value of the quotient.

Example 2

Divide:

(a) $(+28) \div (-4)$

    Step 1 The quotient of their absolute values is $28 \div 4 = 7$

    Step 2 The original numbers have opposite signs, so the quotient is negative. $(+28) \div (-4) = -7$

(b) $(-45) \div (-5)$

    Step 1 The quotient of their absolute values is $45 \div 5 = 9$

    Step 2 The original numbers have the same sign, so the quotient is positive. $(-45) \div (-5) = +9$

(c) $\left(-\dfrac{2}{5}\right) \div \left(+\dfrac{7}{10}\right)$

    Step 1 The quotient of their absolute values is $\dfrac{2}{5} \div \dfrac{7}{10} = \dfrac{2}{\cancel{5}_{1}} \times \dfrac{\cancel{10}^{2}}{7} = \dfrac{4}{7}$

    Step 2 The original fractions have opposite sign, so the quotient is negative. $\left(-\dfrac{2}{5}\right) \div \left(+\dfrac{7}{10}\right) = -\dfrac{4}{7}$

→ Your Turn

Try these division problems:

(a) $(-36) \div (+3)$ (b) $(-72) \div (-9)$ (c) $(+42) \div (-6)$

(d) $(-5.4) \div (-0.9)$ (e) $\left(\frac{3}{8}\right) \div \left(-\frac{1}{4}\right)$ (f) $(-3.75) \div (6.4)$

    (Round to two decimal places.)

→ Answers

(a) -12 (b) $+8$ (c) -7 (d) $+6$ (e) $-1\frac{1}{2}$ (f) -0.59

Enter (f) this way on a calculator:

C6-5 $\boxed{(-)}\ 3.75\ \boxed{\div}\ 6.4\ \boxed{=} \rightarrow$ -0.5859375

For more practice in multiplication and division of signed numbers, turn to Exercises 6-3 for a set of problems.

Exercises 6-3 Multiplication and Division of Signed Numbers

A. Multiply or divide as indicated.

1. $(-7) \times (9)$
2. $8 \times (-5)$
3. $(-7) \times (-11)$
4. $(-4) \times (-9)$
5. $28 \div (-7)$
6. $(-64) \div 16$
7. $(-72) \div (-6)$
8. $(-44) \div (-11)$
9. $(-8.4) \div 2.1$
10. $(-10.8) \div (-1.2)$
11. $\left(-4\frac{1}{2}\right) \times \left(-2\frac{3}{4}\right)$
12. $\frac{2}{3} \times \left(-\frac{3}{4}\right)$
13. $(-3.4) \times (-1.5)$
14. $(-4) \div (16)$
15. $(-5) \div (-25)$
16. $(-13) \times (-4)$
17. $15 \times (-6)$
18. $(-18) \times 3$
19. $(-12) \times (-7)$
20. $(-120) \div 6$
21. $140 \div (-5)$
22. $(-200) \div (-10)$
23. $(-350) \div (-70)$
24. $2.3 \times (-1.5)$
25. $(-1.6) \times (5.0)$
26. $\left(-6\frac{2}{5}\right) \times \left(-1\frac{1}{4}\right)$
27. $\left(-\frac{5}{8}\right) \times \left(-\frac{4}{15}\right)$
28. $(-2.25) \div 0.15$
29. $8\frac{1}{2} \div \left(-3\frac{1}{3}\right)$
30. $\left(-\frac{5}{6}\right) \div \left(-\frac{3}{4}\right)$
31. $(-4.8) \div (-12)$
32. $(-2) \times \left(-5\frac{7}{8}\right)$
33. $(-3) \times \left(6\frac{1}{2}\right)$
34. $(-0.07) \times (-1.1)$
35. $3.2 \times (-1.4)$
36. $(-0.096) \div 1.6$
37. $(-5) \times (-6) \times (-3)$
38. $(-8) \times 4 \times (-7)$
39. $(-24) \div (-6) \times (-5)$
40. $35 \div (-0.5) \times (-10)$
41. $(-12) \times (-0.3) \div (-9)$
42. $0.2 \times (-5) \div (-0.02)$

B. Multiply or divide using a calculator and round as indicated.

1. $(-3.87) \times (4.98)$ — (three significant digits)
2. $(-5.8) \times (-9.75)$ — (two significant digits)
3. $(-0.075) \times (-0.025)$ — (two significant digits)
4. $(648) \times (-250)$ — (two significant digits)
5. $(-4650) \div (1470)$ — (three significant digits)
6. $(-14.5) \div (-3.75)$ — (three significant digits)
7. $(-0.58) \div (-2.5)$ — (two significant digits)
8. $0.0025 \div (-0.084)$ — (two significant digits)
9. $(-4.75) \times 65.2 \div (-6.8)$ — (two significant digits)
10. $78.4 \div (-8.25) \times (-22.6)$ — (three significant digits)
11. $(-2400) \times (-450) \times (-65)$ — (two significant digits)
12. $(-65,000) \div (2.75) \times 360$ — (two significant digits)

C. Practical Problems

1. **Meteorology** To convert 17°F to degrees Celsius, it is necessary to perform the calculation

$$°C = \frac{5}{9} \times (°F - 32) = \frac{5}{9} \times (17 - 32)$$

What is the Celsius equivalent of 17°F?

2. **Business and Finance** In the first four months of the year, a certain country had an average monthly trade deficit (a *negative* quantity) of $16.5 billion. During the final eight months of the year, the country had an average monthly trade surplus (a *positive* quantity) of $5.3 billion. What was the overall trade balance for the year? (Use the appropriate sign to indicate an overall deficit or surplus.)

3. **Industrial Technology** Due to the relative sizes of two connected pulleys A and B, pulley A will turn $1\frac{1}{2}$ times more than pulley B and in the opposite direction. If positive numbers are used to indicate counterclockwise revolutions and negative numbers are used for clockwise revolutions, express as a signed number the result for:
   (a) Pulley A when pulley B turns 80 revolutions counterclockwise.
   (b) Pulley B when pulley A turns 60 revolutions clockwise.

4. **Machine Trades** A long metal rod changes in length by 0.4 mm for every 1°C change in temperature. If it expands when heated and contracts when cooled, express its change in length as a signed number as the temperature changes from 12°C to 4°C.

5. **Aviation** An airplane descends from 42,000 ft to 18,000 ft in 12 minutes. Express its rate of change in altitude in feet per minute as a signed number.

6. **Meteorology** To convert −8°C to °F, you must do the following calculation:

$$°F = \frac{9 \times °C}{5} + 32 = \frac{9 \times (-8)}{5} + 32$$

What is the Fahrenheit equivalent of −8°C?

Check your answers to the odd-numbered problems in the Appendix, then turn to Section 6-4 to learn about exponents and square roots.

6-4 Exponents and Square Roots

Exponents   When the same number appears many times in a multiplication, writing the product may become monotonous, tiring, and even inaccurate. It is easy, for example, to miscount the twos in

$$32{,}768 = 2 \times 2 \times 2 \times 2 \times 2 \times 2 \times 2 \times 2 \times 2 \times 2 \times 2 \times 2 \times 2 \times 2 \times 2$$

or the tens in

$$100{,}000{,}000{,}000 = 10 \times 10 \times 10 \times 10 \times 10 \times 10 \times 10 \times 10 \times 10 \times 10 \times 10$$

Products of this sort are usually written in a shorthand form as 2^{15} and 10^{11}. In this *exponential* form the raised integer 15 shows the number of times that 2 is to be used as a factor in the multiplication.

Example 1

$2 \times 2 = 2^2$ Product of <u>two</u> factors of 2

$2 \times 2 \times 2 = 2^3$ Product of <u>three</u> factors of 2

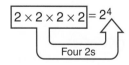   Product of <u>four</u> factors of 2

Four 2s

Write $3 \times 3 \times 3 \times 3 \times 3$ in exponential form.

$3 \times 3 \times 3 \times 3 \times 3 = $ _____

→ Solution

Counting 3s,

$$\underbrace{3 \times 3 \times 3 \times 3 \times 3}_{\text{Five factors of 3}} = 3^5$$

In this expression, 3 is called the *base,* and the integer 5 is called the *exponent.* The exponent 5 tells you how many times the base 3 must be used as a factor in the multiplication.

Example 2

Find the value of $4^3 = $ _____

The exponent 3 tells us to use 4 as a factor 3 times.

$4^3 = 4 \times 4 \times 4 = (4 \times 4) \times 4 = 16 \times 4 = 64$

It is important that you be able to read exponential forms correctly.

2^2 is read "two to the second power" or "two squared"

2^3 is read "two to the third power" or "two cubed"

2^4 is read "two to the fourth power"

2^5 is read "two to the fifth power" and so on

Students studying basic electronics, or those going on to another mathematics or science class, will find that it is important to understand exponential notation.

→ More Practice

Do the following problems for practice in using exponents.

(a) Write in exponential form.

$5 \times 5 \times 5 \times 5$ = ____	base = ____	exponent = ____
7×7 = ____	base = ____	exponent = ____
$10 \times 10 \times 10 \times 10 \times 10$ = ____	base = ____	exponent = ____
$3 \times 3 \times 3 \times 3 \times 3 \times 3 \times 3$ = ____	base = ____	exponent = ____
$9 \times 9 \times 9$ = ____	base = ____	exponent = ____
$1 \times 1 \times 1 \times 1$ = ____	base = ____	exponent = ____

(b) Write as a product of factors and multiply out.

2^6	= _____ = ____	base = ____	exponent = ____
10^7	= _____ = ____	base = ____	exponent = ____
$(-3)^4$	= _____ = ____	base = ____	exponent = ____

5^2	= _____	= ____	base = ____	exponent = ____
$(-6)^3$	= _____	= ____	base = ____	exponent = ____
4^5	= _____	= ____	base = ____	exponent = ____
12^3	= _____	= ____	base = ____	exponent = ____
1^5	= _____	= ____	base = ____	exponent = ____

→ **Solutions**

(a) $5 \times 5 \times 5 \times 5$ = 5^4 base = 5 exponent = 4

7×7 = 7^2 base = 7 exponent = 2

$10 \times 10 \times 10 \times 10 \times 10$ = 10^5 base = 10 exponent = 5

$3 \times 3 \times 3 \times 3 \times 3 \times 3 \times 3$ = 3^7 base = 3 exponent = 7

$9 \times 9 \times 9$ = 9^3 base = 9 exponent = 3

$1 \times 1 \times 1 \times 1$ = 1^4 base = 1 exponent = 4

(b) $2^6 = 2 \times 2 \times 2 \times 2 \times 2 \times 2$ = 64   base = 2 exponent = 6

$10^7 = 10 \times 10 \times 10 \times 10 \times 10 \times 10 \times 10 = 10{,}000{,}000$

                                                base = 10 exponent = 7

$(-3)^4 = (-3) \times (-3) \times (-3) \times (-3) =$ +81   base = −3 exponent = 4

$5^2 = 5 \times 5$ = 25   base = 5 exponent = 2

$(-6)^3 = (-6) \times (-6) \times (-6)$ = −216   base = −6 exponent = 3

> **Note**   Notice that $(-3)^4 = +81$, but $(-6)^3 = -216$. The rules for multiplying signed numbers tell us that a negative number raised to an **even** power is positive, but a negative number raised to an **odd** power is negative. ◄

$4^5 = 4 \times 4 \times 4 \times 4 \times 4$ = 1024   base = 4 exponent = 5

$12^3 = 12 \times 12 \times 12$ = 1728   base = 12 exponent = 3

$1^5 = 1 \times 1 \times 1 \times 1 \times 1$ = 1   base = 1 exponent = 5

> **Careful**   Be careful when there is a negative sign in front of the exponential expression. For example,

$$3^2 = 3 \times 3 = 9 \quad \text{and} \quad (-3)^2 = (-3) \times (-3) = 9$$

but

$$-3^2 = -(3^2) = -(3 \times 3) = -9$$

To calculate $-(-2)^3$ first calculate

$$(-2)^3 = (-2) \times (-2) \times (-2) = -8$$

Then

$-(-2)^3 = -(-8) = 8$. A useful way of remembering this: exponents act only on what they touch. ◄

Any power of 1 is equal to 1, of course.

$1^2 = 1 \times 1 = 1$
$1^3 = 1 \times 1 \times 1 = 1$
$1^4 = 1 \times 1 \times 1 \times 1 = 1$ and so on

Notice that when the base is ten, the product is easy to find.

$10^2 = 10 \times 10 = 100$
$10^3 = 10 \times 10 \times 10 = 1000$
$10^4 = 10,000$
$10^5 = \underbrace{100,000}$

5 zeros

The exponent number is always exactly equal to the number of zeros in the final product when the base is 10.

Continue the pattern to find the value of 10^0 and 10^1.

$10^5 = 100,000$ $10^2 = 100$
$10^4 = 10,000$ $10^1 = 10$
$10^3 = 1000$ $10^0 = 1$

For any base, powers of 1 or 0 are easy to find.

$2^1 = 2$ $2^0 = 1$
$3^1 = 3$ $3^0 = 1$
$4^1 = 4$ $4^0 = 1$ and so on

Order of Operations with Exponents

In Chapter 1 we discussed the order of operations when you add, subtract, multiply, or divide. To evaluate an arithmetic expression containing exponents, first calculate the value of the exponential factor, then perform the other operations using the order of operations already given. If the calculations are to be performed in any other order, parentheses will be used to show the order. The revised order of operations is:

> **First,** simplify all operations within **parentheses.**
>
> **Second,** simplify all **exponents,** left to right.
>
> **Third,** perform all **multiplications** and **divisions,** left to right.
>
> **Finally,** perform all **additions** and **subtractions,** left to right.

Learning Help

If you think you will have trouble remembering this order of operations, try memorizing this phrase: **P**lease **E**xcuse **M**y **D**ear **A**unt **S**ally. The first letter of each word should remind you of **P**arentheses, **E**xponents, **M**ultiply/**D**ivide, **A**dd/**S**ubtract. ◄

Here are some examples showing how to apply this expanded order of operations.

Example 3

(a) To calculate $12 + 54 \div 3^2$

     First, simplify the exponent factor. $= 12 + 54 \div 9$
     Second, divide. $= 12 + 6$
     Finally, add. $= 18$

(b) To calculate $237 - (2 \times 3)^3$

    First, work within parentheses. $= 237 - (6)^3$
    Second, simplify the exponent term. $= 237 - 216$
    Finally, subtract. $= 21$

→ Your Turn

Calculate.

(a) $2^4 \times 5^3$ (b) $6 + 3^2$ (c) $(6 + 3)^2$

(d) $(24 \div 6)^3$ (e) $2^2 + 4^2 \div 2^3$ (f) $4 + 3 \times 2^4 \div 6$

→ Solutions

(a) $2^4 \times 5^3 = 16 \times 125 = 2000$ (b) $6 + 3^2 = 6 + 9 = 15$

(c) $(6 + 3)^2 = (9)^2 = 81$ (d) $(24 \div 6)^3 = (4)^3 = 64$

(e) $2^2 + 4^2 \div 2^3 = 4 + 16 \div 8 = 4 + 2 = 6$ (Do the division before adding.)

(f) $4 + 3 \times 2^4 \div 6 = 4 + 3 \times 16 \div 6 = 4 + 48 \div 6 = 4 + 8 = 12$

Finding Powers on a Calculator   To square a number using a multi-line calculator, enter the number, then press the $\boxed{x^2}$ key followed by the $\boxed{=}$ key. If you have a single-line model, it is not necessary to press the $\boxed{=}$ key.

Example 4

C6-6

The value of 38^2 can be found by entering

38 $\boxed{x^2}$ $\boxed{=}$ →   *1444.*

To raise a number to any other power, enter the base number, press the power key (labeled $\boxed{\wedge}$ or $\boxed{y^x}$ or $\boxed{x^y}$), then enter the exponent number, and finally press the $\boxed{=}$ key.

Example 5

To calculate 3.8^5, enter

3.8 $\boxed{\wedge}$ **5** $\boxed{=}$ →   *792.35168*

Note   Many calculators have an $\boxed{x^2}$ key for raising a number to the third power. ◄

To raise a *negative* number to a power on a multi-line calculator, you must enclose the number in parentheses. (Parentheses are not required with a single-line model.)

Example 6

C6-7

(a) To calculate $(-25.4)^2$, enter

$\boxed{(}$ $\boxed{(-)}$ **25.4** $\boxed{)}$ $\boxed{x^2}$ $\boxed{=}$ →   *645.16*

C6-8

(b) To calculate $(-0.85)^4$, enter

$\boxed{(}$ $\boxed{(-)}$ **.85** $\boxed{)}$ $\boxed{\wedge}$ **4** $\boxed{=}$ →   *0.52200625*

→ Your Turn

A bit of practice will help. Calculate the following:

(a) 68^2 (b) 65^3 (c) $(-38)^2$

(d) 2.75^4 (Round to three significant digits.)

(e) $(-0.475)^5$ (Round to three significant digits.)

→ Solutions

C6-9

(a) **68** $\boxed{x^2}$ $\boxed{=}$ → | 4624. |

(b) **65** $\boxed{\wedge}$ **3** $\boxed{=}$ → | 274625. |

(c) $\boxed{(}$ $\boxed{(-)}$ **38** $\boxed{)}$ $\boxed{x^2}$ $\boxed{=}$ → | 1444. |

(d) **2.75** $\boxed{\wedge}$ **4** $\boxed{=}$ → | 57.19140625 | or 57.2 rounded

(e) $\boxed{(}$ $\boxed{(-)}$ **.475** $\boxed{)}$ $\boxed{\wedge}$ **5** $\boxed{=}$ → | -0.024180654 | or −0.0242 rounded

Square Roots   What is interesting about the numbers

1, 4, 9, 16, 25, 36, 49, 64, 81, 100, . . . ?

Do you recognize them?

These numbers are the squares or second powers of the counting numbers,

$1^2 = 1$
$2^2 = 4$
$3^2 = 9$
$4^2 = 16$, and so on. 1, 4, 9, 16, 25, . . . are called *perfect squares*.

PERFECT SQUARES

$1^2 = 1$	$6^2 = 36$	$11^2 = 121$	$16^2 = 256$
$2^2 = 4$	$7^2 = 49$	$12^2 = 144$	$17^2 = 289$
$3^2 = 9$	$8^2 = 64$	$13^2 = 169$	$18^2 = 324$
$4^2 = 16$	$9^2 = 81$	$14^2 = 196$	$19^2 = 361$
$5^2 = 25$	$10^2 = 100$	$15^2 = 225$	$20^2 = 400$

If you have memorized the multiplication table for one-digit numbers, you will recognize them immediately. The number 3^2 is read "three squared." What is "square" about $3^2 = 9$? The name comes from an old Greek idea about the nature of numbers. Ancient Greek mathematicians called certain numbers "square numbers" or "perfect squares" because they could be represented by a square array of dots.

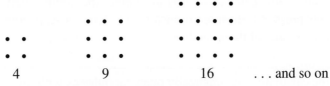

4   9   16   . . . and so on

The number of dots along the side of the square was called the "root" or origin of the square number. We call it the *square root*. When you square this number, or multiply it by itself, you obtain the original number.

Example 7

(a) The square root of 16 is 4, because 4^2 or $4 \times 4 = 16$.

(b) The square root of 64 is equal to 8 because 8^2 or $8 \times 8 = 64$.

   The sign $\sqrt{}$ is used to indicate the square root.

(c) $\sqrt{16} = 4$ read it "square root of 16"

(d) $\sqrt{169} = 13$ read it "square root of 169"

→ Your Turn

Find $\sqrt{81} = $ _____ $\sqrt{361} = $ _____ $\sqrt{289} = $ _____

Try using the table of perfect squares if you do not recognize these.

→ Solutions

$\sqrt{81} = 9$ ✓ $9 \times 9 = 81$

$\sqrt{361} = 19$ ✓ $19 \times 19 = 361$

$\sqrt{289} = 17$ ✓ $17 \times 17 = 289$

Always check your answer as shown.

How do you find the square root of any number? Using a calculator is the easiest way to find the square or other power of a number, or to find a square root.

To find the square root of a number on a multi-line calculator, simply press the $\boxed{\sqrt{}}$ key and then enter the number followed by the $\boxed{=}$ key. On some calculators, $\boxed{\sqrt{}}$ is a second function of the $\boxed{x^2}$ key.

Example 8

C6-10

To calculate $\sqrt{237}$, enter

$\boxed{\sqrt{}}$ **237** $\boxed{=}$ → $\boxed{15.39480432}$

You will usually need to round your answer. Rounded to three significant digits, $\sqrt{237} \approx 15.4$.

For single-line calculators, you must enter the number first and then press the $\boxed{\sqrt{}}$ key. Do not press $\boxed{=}$. Be sure to consult the Calculator Appendix for alternate key sequences if you use one of these models.

Note Some calculators automatically open parentheses with the $\boxed{\sqrt{}}$ function, and our key sequences will be based on ones that do. However, unless there are additional calculations after the square root is taken, it is not necessary to close parentheses before pressing $\boxed{=}$. Therefore the sequence shown in Example 8 will work for all multi-line calculators. ◄

→ **Your Turn**

Perform the following calculations. Round your answer to three significant digits.

(a) $\sqrt{760}$ (b) $\sqrt{5.74}$

→ **Answers**

C6-11

(a) ☑ **760** ⊟ → **27.5680975** $\sqrt{760} \approx 27.6$

(b) ☑ **5.74** ⊟ → **2.39582971** $\sqrt{5.74} \approx 2.40$

Careful Notice that you cannot take the square root of a negative number. There is no number □ such that □ × □ = −81.

If □ is positive, then □ × □ is positive, and if □ is negative, then □ × □ is also positive. You cannot multiply a number by itself and get a negative result. (To perform higher-level algebra, mathematicians have invented imaginary numbers, which are used to represent the square roots of negative numbers. Students who go on to higher-level courses will learn about these numbers later.) ◄

Order of Operations with Square Roots Square roots have the same place in the order of operations as exponents. Simplify square roots before performing multiplication, division, addition, or subtraction.

The expanded order of operations now reads:

> **First,** simplify all operations within **parentheses.**
>
> **Second,** simplify all **exponents and roots,** left to right.
>
> **Next,** perform all **multiplications and divisions,** left to right.
>
> **Finally,** perform all **additions and subtractions,** left to right.

Example 9

(a) $3\sqrt{49} = 3 \times \sqrt{49} = 3 \times 7 = 21$

   Note that $3\sqrt{49}$ is a shorthand way of writing $3 \times \sqrt{49}$.

(b) $2 + \sqrt{64} = 2 + 8 = 10$

(c) $\dfrac{\sqrt{81}}{3} = \dfrac{9}{3} = 3$

(d) If an arithmetic operation appears under the square root symbol, perform this operation before taking the square root. For example,

   $\sqrt{9 + 16} = \sqrt{25} = 5$

Careful Notice $\sqrt{9 + 16}$ is not the same as $\sqrt{9} + \sqrt{16} = 3 + 4 = 7$. ◄

Your calculator is programmed to follow the order of operations. When square roots are combined with other operations, you may enter the calculation in the order that it is written, but you must be aware of the correct use of parentheses. If your calculator automatically opens parentheses with square root, then you must close parentheses if another operation follows the square root. Remember that our key sequences are based on models that do automatically open parentheses.

Example 10

(a) To calculate $\dfrac{38^2\sqrt{3}}{4}$, enter the following sequence:

C6-12

38 $\boxed{x^2}$ $\boxed{\times}$ $\boxed{\sqrt{}}$ **3** $\boxed{)}$ $\boxed{\div}$ **4** $\boxed{=}$ $\rightarrow$ $\boxed{\textbf{625.2703415}}$

If your calculator does not open parentheses with square root, then do not press $\boxed{)}$ after entering 3.

(b) To calculate $\sqrt{17^2 - 15^2}$, enter

C6-13

$\boxed{\sqrt{}}$ **17** $\boxed{x^2}$ $\boxed{-}$ **15** $\boxed{x^2}$ $\boxed{=}$ $\rightarrow$ $\boxed{\textbf{8.}}$

On this problem, if your calculator does not automatically open parentheses with square root, then you must press $\boxed{(}$ after $\boxed{\sqrt{}}$ in order to take the square root of the entire expression. If you wish to avoid parentheses, use this alternate key sequence instead:

17 $\boxed{x^2}$ $\boxed{-}$ **15** $\boxed{x^2}$ $\boxed{=}$ $\boxed{\sqrt{}}$ $\boxed{\text{ANS}}$ $\boxed{=}$ $\rightarrow$ $\boxed{\textbf{8.}}$

Recall that "ANS" stands for "last answer," so that $\boxed{\sqrt{}}$ $\boxed{\text{ANS}}$ takes the square root of the answer to $17^2 - 15^2$.

→ Your Turn

Calculate:

(a) $100 - \sqrt{100}$ (b) $\dfrac{4^2 \times \sqrt{36}}{3}$ (c) $\sqrt{169 - 144}$ (d) $\sqrt{64} + \sqrt{121}$

(e) $6 + 4\sqrt{25}$ (f) $49 - 3^2\sqrt{4}$ (g) $\dfrac{\sqrt{225} - \sqrt{81}}{\sqrt{36}}$ (h) $\sqrt{\dfrac{225 - 81}{36}}$

(i) $\sqrt{12.4^2 + 21.6^2}$ (j) $\dfrac{3(8.4)^2\sqrt{3}}{2}$

→ Solutions

(a) $100 - \sqrt{100} = 100 - 10 = 90$

(b) $\dfrac{4^2 \times \sqrt{36}}{3} = \dfrac{16 \times (6)}{3} = \dfrac{96}{3} = 32$

(c) $\sqrt{169 - 144} = \sqrt{25} = 5$

(d) $\sqrt{64} + \sqrt{121} = 8 + 11 = 19$

(e) $6 + 4\sqrt{25} = 6 + 4 \times 5 = 6 + 20 = 26$

(f) $49 - 3^2\sqrt{4} = 49 - 9\sqrt{4} = 49 - 9 \times 2 = 49 - 18 = 31$

(g) $\dfrac{\sqrt{225} - \sqrt{81}}{\sqrt{36}} = \dfrac{15 - 9}{6} = \dfrac{6}{6} = 1$

(h) $\sqrt{\dfrac{225 - 81}{36}} = \sqrt{\dfrac{144}{36}} = \sqrt{4} = 2$

(i) $\boxed{\sqrt{}}$ **12.4** $\boxed{x^2}$ $\boxed{+}$ **21.6** $\boxed{x^2}$ $\boxed{=}$ → $\boxed{24.90622412}$

or **12.4** $\boxed{x^2}$ + **21.6** $\boxed{x^2}$ $\boxed{=}$ $\boxed{\sqrt{}}$ $\boxed{\text{ANS}}$ $\boxed{=}$ → $\boxed{24.90622412}$

(j) **3** $\boxed{\times}$ **8.4** $\boxed{x^2}$ $\boxed{\times}$ $\boxed{\sqrt{}}$ $\boxed{3}$ $\boxed{)}$ $\boxed{\div}$ **2** $\boxed{=}$ → $\boxed{183.3202575}$

Now go to Exercises 6-4 for a set of problems on exponents and square roots.

Exercises 6-4   Exponents and Square Roots

A. Find the value of these.

1.	2^4	2.	3^2	3.	4^3
4.	5^3	5.	10^3	6.	7^2
7.	2^8	8.	6^2	9.	8^3
10.	3^4	11.	5^4	12.	10^5
13.	$(-2)^3$	14.	$(-3)^5$	15.	9^3
16.	6^0	17.	5^1	18.	1^4
19.	2^5	20.	8^2	21.	$(-14)^2$
22.	$(-21)^2$	23.	15^3	24.	16^2
25.	$2^2 \times 3^3$	26.	$2^6 \times 3^2$	27.	$3^2 \times 5^3$
28.	$2^3 \times 7^3$	29.	$6^2 \times 5^2 \times 3^1$	30.	$2^{10} \times 3^2$
31.	$2 + 4^2$	32.	$14 - 2^3$	33.	$(8 - 2)^3$
34.	$(14 \div 2)^2$	35.	$3^3 - 8^2 \div 2^3$	36.	$12 - 5 \times 2^5 \div 2^4$
37.	$5 \times 6^2 - 4^3$	38.	$64 \div 4^2 + 18$	39.	$(28 \div 7)^2 - 14$
40.	$3(8^2 - 3^3)$	41.	$26 + 40 \div 2^3 \times 3^2$	42.	$18 + 4 \times 5^2 - 3^4$

B. Calculate. Round to two decimal places if necessary.

1.	$\sqrt{81}$	2.	$\sqrt{144}$	3.	$\sqrt{36}$
4.	$\sqrt{16}$	5.	$\sqrt{25}$	6.	$\sqrt{9}$
7.	$\sqrt{256}$	8.	$\sqrt{400}$	9.	$\sqrt{225}$
10.	$\sqrt{49}$	11.	$\sqrt{324}$	12.	$\sqrt{121}$
13.	$\sqrt{4.5}$	14.	$\sqrt{500}$	15.	$\sqrt{12.4}$
16.	$\sqrt{700}$	17.	$\sqrt{210}$	18.	$\sqrt{321}$
19.	$\sqrt{810}$	20.	$\sqrt{92.5}$	21.	$\sqrt{1000}$
22.	$\sqrt{2000}$	23.	$\sqrt{25000}$	24.	$\sqrt{2500}$
25.	$\sqrt{150}$	26.	$\sqrt{300}$	27.	$\sqrt{3000}$
28.	$\sqrt{30000}$	29.	$\sqrt{1.25}$	30.	$\sqrt{1.008}$
31.	$3\sqrt{20}$	32.	$4 + 2\sqrt{5}$	33.	$55 - \sqrt{14}$
34.	$\dfrac{3^2\sqrt{24}}{3 \times 2}$	35.	$\sqrt{2^5 - 3^2}$	36.	$5.2^2 - 3\sqrt{52 - 5^2}$
37.	$\sqrt{105^2 + 360^2}$	38.	$\left(68.5 + 27.4 + \sqrt{68.5 \times 27.4}\right)\left(\dfrac{8}{3}\right)$		

C. Practical Problems

1. **General Trades**   Find the length of the side of a square whose area is 184.96 sq ft. (*Hint:* The area of a square is equal to the square of one of its sides. The side of a square equals the square root of its area.)

2. **Construction**   A square building covers an area of 1269 sq ft. What is the length of each side of the building? Round to the nearest tenth.

3. **Mechanical Engineering**   The following calculation must be performed to find the stress (in pounds per square inch) on a plunger. Find this stress to the nearest thousand psi.

$$\frac{3000\sqrt{34}}{0.196}$$

4. **Electronics**   The resistance (in ohms) of a certain silicon diode is found using the expression

$$\sqrt{15 \times 10^6 \times 24}$$

Calculate this resistance to the nearest thousand ohms.

5. **Fire Protection**   The velocity (in feet per second) of water discharged from a hose with a nozzle pressure of 62 psi is given by

$$12.14\sqrt{62}$$

Calculate and round to the nearest whole number.

6. **Police Science**   Skid marks can be used to determine the maximum speed of a car involved in an accident. If a car leaves a skid mark of 175 ft on a road with a 35% coefficient of friction, its speed (in miles per hour) can be estimated using

$$\sqrt{30 \times 0.35 \times 175}$$

Find this speed to the nearest whole number.

7. **Allied Health**   The Fit-4-U weight loss center calculates the body mass index for each prospective client.

$$\text{BMI} = \frac{703 \times (\text{weight in pounds})}{(\text{height in inches})^2}$$

A person with a BMI greater than 30 is considered to be obese, and a person with a BMI over 25 is considered overweight. Calculate the BMI of a person 5 ft 6 in. tall weighing 165 lb.

8. **Roofing**   The length L of a rafter on a shed roof is calculated using the formula

$$L = \sqrt{(H_2 - H_1)^2 + W^2}$$

If H_2, the maximum roof height, is 13 ft, H_1, the minimum wall height, is 7.5 ft, and W, the width of the shed, is 12 ft, calculate the rafter length, not including allowance for overhang. Round to the nearest tenth.

9. **Construction**   The Golden Gate bridge is about 1 mi long. On a warm day it expands about 2 ft in length. If there were no expansion joints to compensate for the expansion, how high a bulge would this produce? Use the following formula for your calculation, and round to the nearest foot.

$$\text{Height} \atop (\text{in ft}) = \sqrt{\left(\frac{5280}{2} + 1\right)^2 - \left(\frac{5280}{2}\right)^2}$$

Check your answers to the odd-numbered problems in the Appendix, then turn to Problem Set 6 on page 381 for practice problems on Chapter 6. If you need a quick review of the topics in this chapter, visit the chapter Summary first.

Objective	Review
Understand the meaning of signed numbers. (p. 347)	Negative numbers are used to represent real-world situations such as debts, losses, temperatures below zero, and locations below sea level. The negative part of the number line is a reflection of the positive part of the line. Therefore, the order of the negative numbers is the opposite of the order of the positive numbers.

> **Example:**   14 is greater than 12 ($14 > 12$), but -14 is less than -12 ($-14 < -12$). A loss of \$250 can be represented as the signed number $-\$250$.

Add signed numbers. (p. 349)

To add two signed numbers, follow these steps.

Step 1   Find the absolute value of each number.

Step 2   If the two numbers have the same sign, add their absolute values and give the sum the sign of the original numbers.

Step 3   If the two numbers have opposite signs, subtract the absolute values and give the difference the sign of the number with the larger absolute value.

> **Example:**   (a) $(-3) + (-5) = -8$ ← Add absolute values: $3 + 5 = 8$
> Absolute value $= 3$
> Absolute value $= 5$ — Sign of the sum is negative because both numbers are negative.
>
> (b) $-21.7 + 14.2 = -7.5$ ← Subtract absolute values: $21.7 - 14.2 = 7.5$
> — Same sign as number with the larger absolute value (-21.7)

Subtract signed numbers. (p. 358)

Rewrite the subtraction as an addition of the opposite signed number.

> **Example:**   (a) $4 - 7 = 4 + (-7) = -3$
>
> (b) $-11 - (-3) = -11 + (+3) = -8$

Multiply and divide signed numbers. (p. 363)

To multiply or divide two signed numbers, first find the product or quotient of their absolute values. If the numbers have the same sign, the answer is positive. If they have opposite signs, the answer is negative.

> **Example:**   (a) $(-12) \times 6 = -72$ Opposite signs, answer is negative
>
> (b) $(-54) \div (-9) = +6$ Same sign, answer is positive

Work with exponents. (p. 368)

In exponential form, the exponent indicates the number of times the base is multiplied by itself to form the number.

> **Example:**   Exponent
> Base ⟶ $4^3 = \underbrace{4 \times 4 \times 4}_{} = 64$
> — 4 multiplied by itself 3 times

Objective

Use the order of operations. (p. 371)

Find square roots. (p. 373)

Review

1. Simplify all operations within **parentheses.**
2. Simplify all **exponents,** left to right.
3. Perform all **multiplications** and **divisions,** left to right.
4. Perform all **additions** and **subtractions,** left to right.

Example: Simplify: $(4 \times 5)^2 + 4 \times 5^2$

First, simplify within parentheses $= (20)^2 + 4 \times 5^2$

Second, simplify exponents $\qquad = 400 + 4 \times 25$

Third, multiply $\qquad\qquad = 400 + 100$

Finally, add $\qquad\qquad\qquad = 500$

The square root of a positive number is that number such that when it is squared (multiplied by itself), it produces the original number. Square roots occupy the same place in the order of operations as exponents.

Example: $\sqrt{169} = 13$ because 13^2 or $13 \times 13 = 169$

Problem Set

Pre-Algebra

Answers to odd-numbered problems are given in the Appendix.

A. Rewrite each group of numbers in order from smallest to largest.

1. $-13, 7, 4, -8, -2$

2. $-0.6, 0, 0.4, -0.55, -0.138, -0.96$

3. $-150, -140, -160, -180, -120, 0$

4. $-\dfrac{5}{8}, \dfrac{3}{4}, \dfrac{1}{2}, -\dfrac{1}{2}, -\dfrac{3}{4}, \dfrac{3}{8}$

B. Add or subtract as indicated.

1. $-16 + 6$

2. $-5 + (-21)$

3. $7 - 13$

4. $-4 - (-4)$

5. $23 + (-38)$

6. $-13 - 37$

7. $20 - (-5)$

8. $11 + (-3)$

9. $-2.6 + 4.9$

10. $2\frac{1}{4} - 4\frac{3}{4}$

11. $-\$5.26 - \$3.89$

12. $-2\frac{3}{8} + \left(-7\frac{1}{2}\right)$

13. $-9 - 6 + 4 + (-13) - (-8)$

14. $-12.4 + 6\frac{1}{2} - \left(-1\frac{3}{4}\right) + (-18) - 5.2$

C. Multiply or divide as indicated.

1. $6 \times (-10)$

2. $(-7) \times (-5)$

3. $-48 \div (-4)$

4. $(-320) \div 8$

5. $\left(-1\frac{1}{8}\right) \times \left(-6\frac{1}{4}\right)$

6. $\left(-5\frac{1}{4}\right) \div 7$

7. $6.4 \div (-16)$

8. $(-2.4) \times 15$

9. $(-4) \times (-5) \times (-6)$

10. $72 \div (-4) \times (-3)$

11. $(2.4735) \times (-6.4)$ (two significant digits)

12. $(-0.675) \div (-2.125)$ (three significant digits)

D. Find the numerical value of each expression.

1. $12.5 \times (-6.3 + 2.8)$

2. $(-3) \times 7 - 4 \times (-9)$

3. $\dfrac{14 - 46}{(-4) \times (-2)}$

4. $\left(-4\frac{1}{2} - 3\frac{3}{4}\right) \div 8$

5. 2^6

6. 3^5

7. 17^2

8. 23^3

Name

_____

Date

_____

Course/Section

9. 0.5^2

10. 1.2^3

11. 0.02^2

12. 0.03^3

13. 0.001^3

14. 2.01^2

15. 4.02^2

16. $(-3)^3$

17. $(-2)^4$

18. $\sqrt{1369}$

19. $\sqrt{784}$

20. $\sqrt{4.41}$

21. $\sqrt{0.16}$

22. $\sqrt{5.29}$

23. $5.4 + 3.3^2$

24. $2.1^2 + 3.1^3$

25. 4.5×5.2^2

26. $4.312 \div 1.4^2$

27. $(3 \times 6)^2 + 3 \times 6^2$

28. $28 - 4 \times 3^2 \div 6$

29. $2 + 2\sqrt{16}$

30. $5\sqrt{25} - 4.48$

31. $\sqrt{0.49} - \sqrt{0.04}$

32. $10 - 3\sqrt{1.21}$

Round to two decimal digits.

33. $\sqrt{80}$

34. $\sqrt{106}$

35. $\sqrt{310}$

36. $\sqrt{1.8}$

37. $\sqrt{4.20}$

38. $\sqrt{3.02}$

39. $\sqrt{1.09}$

40. $\sqrt{0.08}$

41. $7 - 3\sqrt{2}$

42. $\sqrt{6} - \sqrt{5}$

43. $3\sqrt{2} - 2\sqrt{3}$

44. $3\sqrt{2} + 2\sqrt{3}$

45. $\sqrt{9.65^2 - 5.73^2}$

46. $\left(13^2 + 18^2 + \sqrt{13^2 \cdot 18^2}\right)\left(\frac{4}{3}\right)$

E. Practical Problems

1. **Electrical Trades** The current in a circuit changes from -2 A (amperes) to 5 A. What is the change in current?

2. **General Trades** Find the side of a square whose area is 225 sq ft. (The side of a square is the square root of its area.)

3. **Meteorology** In Hibbing, Minnesota, the temperature dropped 8°F overnight. If the temperature read -14°F before the plunge, what did it read afterward?

4. **Trades Management** Morgan Plumbing had earnings of $3478 in the last week of January and expenses of $4326. What was the net profit or loss? (Indicate the answer with the appropriate sign.)

5. **Electrical Trades** The current capacity (in amperes) of a service using a three-phase system for 124 kW of power and a line voltage of 440 volts is given by the calculation

$$\frac{124{,}000}{440\sqrt{3}}$$

Determine this current to three significant digits.

6. **Fire Protection** The flow rate (in gallons per minute) of water from a $1\frac{1}{2}$-in. hose at 65 psi is calculated from the expression

$$29.7 \times (1.5)^2\sqrt{65}$$

Find the flow rate to the nearest gallon per minute.

7. **Construction** The crushing load (in tons) for a square pillar 8 in. thick and 12 ft high is given by the formula

$$L = \frac{25 \times (8)^4}{12^2}$$

Find this load to the nearest hundred tons.

8. **Trades Management**   An auto shop was $14,560 in the "red" at the beginning of the year and $47,220 in the "black" at the end of the year. How much profit did the shop make during the year?

9. **Electronics**   In determining the bandwidth of a high-fidelity amplifier, a technician would first perform the following calculation to find the rise time of the signal:

$$T = \sqrt{3200^2 - 22.5^2}$$

Calculate this rise time to the nearest hundred. The time will be in nanoseconds.

10. **Electronics**   The following calculation is used to determine one of the voltages for a sweep generator:

$$V = 10 - 9.5 \div 0.03$$

Calculate this voltage. Round to one significant digit.

11. **Meteorology**   Use the formula

$$°C = \frac{5}{9} \times (°F - 32)$$

to convert the following Fahrenheit temperatures to Celsius, rounded to the nearest degree.

(a)   −4°F   (b)   24°F

12. **Meteorology**   Use the formula

$$°F = \frac{9 \times °C}{5} + 32$$

to convert the following Celsius temperatures to Fahrenheit.

(a)   −21°C   (b)   −12°C

13. **Electronics**   The value of a 1000-ohm resistor decreases by 0.2 ohm for every degree Celsius of temperature decrease. By how many ohms does the resistance decrease if the temperature drops from 125°C to −40°C?

14. **Meteorology**   The fastest temperature drop in recorded history in the United States occurred on January 10, 1911, in Rapid City, South Dakota. Between 7 A.M. and 7:15 A.M., the temperature plummeted from 55°F to 8°F. To the nearest hundredth, what was the average change in temperature per minute? Express the answer as a signed number.

Basic Algebra

Objective	Sample Problems	For help, go to
When you finish this chapter, you will be able to:		
1. Evaluate formulas and literal expressions.	If $x = 2$, $y = 3$, $a = 5$, $b = 6$, find the value of	Page 392

    (a) $2x + y$ _____

    (b) $\dfrac{1 + x^2 + 2a}{y}$ _____

    (c) $A = x^2y$ $A = $ _____

    (d) $T = \dfrac{2(a + b + 1)}{3x}$ $T = $ _____

    (e) $P = abx^2$ $P = $ _____

2. Perform the basic algebraic operations.   (a) $3ax^2 + 4ax^2 - ax^2$ $= $ _____   Page 403

    (b) $5x - 3y - 8x + 2y$ $= $ _____

    (c) $5x - (x + 2)$ $= $ _____

    (d) $3(x^2 + 5x) - 4(2x - 3)$ $= $ _____

Name _____

Date _____

Course/Section _____

Objective	Sample Problems		For help, go to

3. Solve linear equations in one unknown and solve formulas.

(a) $3x - 4 = 11$ $x =$ _____   Page 420

(b) $2x = 18$ $x =$ _____   Page 411

(c) $2x + 7 = 43 - x$ $x =$ _____   Page 433

(d) Solve the following formula

for *N:* $S = \dfrac{N}{2} + 26$ $N =$ _____   Page 437

(e) Solve the following formula

for *A:* $M = \dfrac{(A - B)L}{8}$ $A =$ _____

4. Translate simple English phrases and sentences into algebraic expressions and equations.

Write each phrase as an algebraic expression or equation.

(a) Four times the area.   _____   Page 444

(b) Current squared times resistance.   _____

(c) Efficiency is equal to 100 times the output divided by the input.   _____

(d) Resistance is equal to 12 times the length divided by the diameter squared.   _____

5. Solve word problems

(a) **Trades Management**   An electrician collected $1265.23 for a job that included $840 labor, which was not taxed, and the rest for parts, which were taxed, at 6%. Determine how much of the total was tax.   _____   Page 449

(b) **General Trades**   A plumber's helper earns $18 per hour plus $27 per hour overtime. If he works 40 regular hours during a week, how much overtime does he need in order to earn $1000?   _____

6. Multiply and divide simple algebraic expressions.

(a) $2y \cdot 3y$ _____   Page 458

(b) $(6x^4y^2)(-2xy^2)$ _____

(c) $3x(y - 2x)$ _____

Objective	Sample Problems	For help, go to

(d) $\dfrac{10x^7}{-2x^2}$ _____ Page 460

(e) $\dfrac{6a^2b^4}{18a^5b^2}$ _____

(f) $\dfrac{12m^4 - 9m^3 + 15m^2}{3m^2}$ _____

7. Use scientific notation.

Write in scientific notation.

(a) 0.000184 _____ Page 465

(b) 213,000 _____

Calculate.

(c) $(3.2 \times 10^{-6}) \times (4.5 \times 10^2)$ _____ Page 468

(d) $(1.56 \times 10^{-4}) \div (2.4 \times 10^3)$ _____

(Answers to these preview problems are given in the Appendix. Also, worked solutions to many of these problems appear in the chapter Summary.)

If you are certain that you can work *all* these problems correctly, turn to page 477 for the set of practice problems. If you cannot work one or more of the preview problems, turn to the page indicated to the right of the problem. Those who wish to master this material should turn to Section 7-1 and begin work there.

DRABBLE: © Kevin Fagan/Dist. by United Feature Syndicate, Inc.

In this chapter you will study algebra, but not the very formal algebra that deals with theorems, proofs, sets, and abstract problems. Instead, we shall study practical or applied algebra as actually used by technical and trades workers. Let's begin with a look at the language of algebra and algebraic formulas.

7-1 Algebraic Language and Formulas

The most obvious difference between algebra and arithmetic is that in algebra letters are used to replace or to represent numbers. A mathematical statement in which letters are used to represent numbers is called a *literal* expression. Algebra is the arithmetic of literal expressions—a kind of symbolic arithmetic.

Any letters will do, but in practical algebra we use the normal lower- and uppercase letters of the English alphabet. It is helpful in practical problems to choose the letters to be used on the basis of their memory value: t for time, D for diameter, C for cost, A for area, and so on. The letter used reminds you of its meaning.

Multiplication

Most of the usual arithmetic symbols have the same meaning in algebra that they have in arithmetic. For example, the addition ($+$) and subtraction ($-$) signs are used in exactly the same way. However, the multiplication sign ($\times$) of arithmetic looks like the letter x and to avoid confusion we have other ways to show multiplication in algebra. The product of two algebraic quantities a and b, "a times b," may be written using

A raised dot	$a \cdot b$
Parentheses	$a(b)$ or $(a)b$ or $(a)(b)$
Nothing at all	ab

Obviously, this last way of showing multiplication won't do in arithmetic; we cannot write "two times four" as "24"—it looks like twenty-four. But it is a quick and easy way to write a multiplication in algebra.

Placing two quantities side by side to show multiplication is not new and it is not only an algebra gimmick; we use it every time we write 20 cents or 4 feet.

$$20 \text{ cents} = 20 \times 1 \text{ cent}$$
$$4 \text{ feet} \quad = 4 \times 1 \text{ foot}$$

→ **Your Turn**

Write the following multiplications using no multiplication symbols.

(a) 8 times a = _____

(b) m times p = _____

(c) 2 times s times t = _____

(d) 3 times x times x = _____

→ **Answers**

(a) $8a$ (b) mp (c) $2st$ (d) $3x^2$

A Closer Look

Did you notice in problem (d) that powers are written just as in arithmetic:

$$x \cdot x = x^2$$
$$x \cdot x \cdot x = x^3$$
$$x \cdot x \cdot x \cdot x = x^4 \text{ and so on.} \blacktriangleleft$$

Parentheses

Parentheses ( ) are used in arithmetic to show that some complicated quantity is to be treated as a unit. For example,

$$2 \cdot (13 + 14 - 6)$$

means that the number 2 multiplies *all* of the quantity in the parentheses.

In exactly the same way in algebra, parentheses ( ), brackets [ ], or braces { } are used to show that whatever is enclosed in them should be treated as a single quantity. An expression such as

$$(3x^2 - 4ax + 2by^2)^2$$

should be thought of as (something)^2. The expression

$$(2x + 3a - 4) - (x^2 - 2a)$$

should be thought of as (first quantity) − (second quantity). Parentheses are the punctuation marks of algebra. Like the period, comma, or semicolon in regular sentences, they tell you how to read an expression and get its correct meaning.

Division In arithmetic we would write "48 divided by 2" as

$$2\overline{)48} \quad \text{or} \quad 48 \div 2 \quad \text{or} \quad \frac{48}{2}$$

But the first two ways of writing division are used very seldom in algebra. Division is usually written as a fraction.

"x is divided by y" is written $\dfrac{x}{y}$

"$(2n + 1)$ divided by $(n - 1)$" is written $\dfrac{(2n + 1)}{(n - 1)}$ or $\dfrac{2n + 1}{n - 1}$

→ **Your Turn**

Write the following using algebraic notation.

(a) 8 times $(2a + b)$ = _____

(b) $(a + b)$ times $(a - b)$ = _____

(c) x divided by y^2 = _____

(d) $(x + 2)$ divided by $(2x - 1)$ = _____ _

→ **Answers**

(a) $8(2a + b)$

(b) $(a + b)(a - b)$

(c) $\dfrac{x}{y^2}$

(d) $\dfrac{x + 2}{2x - 1}$

Algebraic Expressions The word "expression" is used very often in algebra. An *expression* is a general name for any collection of numbers and letters connected by arithmetic signs. For example,

$$x + y \qquad 2x^2 + 4 \qquad 3(x^2 - 2ab)$$
$$\frac{D}{T} \qquad \sqrt{x^2 + y^2} \qquad \text{and} \qquad (b - 1)^2$$

are all algebraic expressions.

If the algebraic expression has been formed by multiplying quantities, each multiplier is called a *factor* of the expression.

Expression	**Factors**
ab	a and b
$2x(x + 1)$	$2x$ and $(x + 1)$ or 2, x, and $(x + 1)$
$(R - 1)(2R + 1)$	$(R - 1)$ and $(2R + 1)$

The algebraic expression can also be a sum or difference of simpler quantities or *terms*.

Expression	Terms
$x + 4y$	x and $4y$
$2x^2 + xy$	$2x^2$ and xy
$A - R$	A and R

The first term is x
The second term is $4y$

→ **Your Turn**

Now let's check to see if you understand the difference between terms and factors. In each algebraic expression below, tell whether the portion being named is a term or a factor.

(a) $2x^2 - 3xy$ y is a _____

(b) $7x - 4$ $7x$ is a _____

(c) $4x(a + 2b)$ $4x$ is a _____

(d) $-2y(3y - 5)$ $(3y - 5)$ is a _____

(e) $3x^2y + 8y^2 - 9$ $3x^2y$ is a _____

→ **Answers**

(a) factor   (b) term   (c) factor   (d) factor   (e) term

Evaluating Formulas

One of the most useful algebraic skills for any technical or practical work involves finding the value of an algebraic expression when the letters are given numerical values. A *formula* is a rule for calculating the numerical value of one quantity from the values of the other quantities. The formula or rule is usually written in mathematical form because algebra gives a brief, convenient to use, and easy to remember form for the rule. Here are a few examples of rules and formulas used in the trades.

1. **Rule:** Ohm's law: The voltage V (in volts) across a simple resistor is equal to the product of the current i (in amperes) through the resistor and the value of its resistance R (in ohms).

    Formula: $V = iR$

2. **Rule:** The cost C of setting type is equal to the total characters set T in the job multiplied by the hourly wage H divided by the rate E at which the type is set, in characters per hour.

    Formula: $C = \dfrac{TH}{E}$

3. **Rule:** The number of standard bricks N needed to build a wall is about 21 times the volume of the wall.

    Formula: $N = 21\,LWH$

Evaluating a formula or algebra expression means to find its value by substituting numbers for the letters in the expression.

To be certain you do it correctly, follow this two-step process:

Step 1 Place the numbers being substituted in parentheses, and then substitute them in the formula.

Step 2 Do the arithmetic carefully *after* the numbers are substituted.

The order of operations for arithmetic calculations must be used when evaluating formulas. Remember:

First, do any operations **inside parentheses.**

Second, find all **powers** and **roots.**

Next, do all **multiplications** or **divisions** left to right.

Finally, do all **additions** or **subtractions** left to right.

Example 1

In construction, the following formula is used to calculate the number of joists J needed to cover a distance L when spaced 16 in. on center:

$$J = \tfrac{3}{4}L + 1$$

Suppose you know that $L = 44$ ft. You can determine J as follows:

First, substitute 44 for L. $\qquad\qquad\qquad\qquad\qquad J = \tfrac{3}{4}(44) + 1$

Then, do the arithmetic according to the order of operations.

$$\text{Multiply:} \qquad J = 33 + 1$$
$$\text{Then add:} \qquad J = 34$$

34 joists will be needed for the given length.

→ Your Turn

Automotive Trades   Automotive engineers use the following formula to calculate the SAE rating (R) of an engine.

$$R = \frac{D^2 N}{2.5} \qquad \text{where } D \text{ is the diameter of a cylinder in inches and } N \text{ is the number of cylinders.}$$

Find R when $D = 3\tfrac{1}{2}$ in. and $N = 6$.

→ Solution

Step 1   Substitute the given values, in parentheses, into the formula.

$$R = \frac{(3.5)^2(6)}{2.5}$$

Step 2   Perform the calculation using the order of operations.

First, find the power: $R = \dfrac{(12.25)(6)}{2.5}$

Next, multiply: $= \dfrac{73.5}{2.5}$

Finally, divide: $= 29.4$

 3.5 $\boxed{x^2}$ $\boxed{\times}$ **6** $\boxed{\div}$ **2.5** $\boxed{=}$ → $\boxed{29.4}$

$R = 29.4$ or roughly 29

→ More Practice

Evaluate the following formulas using the two-step process.

(a) $D = 2R$ for $R = 3.45$ in.

(b) $W = T - C$ for $T = 1420$ lb, $C = 385$ lb

(c) $P = 0.433H$ for $H = 11.4$ in. Round to three significant digits.

(d) $A = bh - 2$ for $b = 1.75$ in., $h = 4.20$ in.

(e) **Allied Health**   The basal metabolic rate (BMR) for a woman can be estimated in an algebraic formula.

$$BMR = 655 + 4.35W + 4.7H - 4.7A$$

where W is her weight in pounds, H is her height in inches, and A is her age in years.

Calculate the BMR for a woman who weighs 145 lb, is 5 ft 6 in. tall, and is 55 years old. Round to three significant digits.

→ Solutions

(a) $D = 2(3.45) = 6.90$ in.   (b) $W = (1420) - (385)$
                                                  $= 1420 - 385$
                                                  $= 1035$ lb

(c) $P = 0.433(11.4)$ (d) $A = (1.75)(4.2) - 2$
      $= 4.9362$ or 4.94 in., rounded $= 7.35 - 2$
                                                      $= 5.35$ sq in.

For problem (d),

 1.75 $\boxed{\times}$ **4.2** $\boxed{-}$ **2** $\boxed{=}$ → $\boxed{5.35}$

(e) First, convert 5 ft 6 in. to 66 in. and then substitute for H, W, and A.

$$BMR = 655 + 4.35(145) + 4.7(66) - 4.7(55)$$
$$= 655 + 630.75 + 310.2 - 258.5$$
$$= 1337.45$$
$$\approx 1340$$

 655 $\boxed{+}$ **4.35** $\boxed{\times}$ **145** $\boxed{+}$ **4.7** $\boxed{\times}$ **66** $\boxed{-}$ **4.7** $\boxed{\times}$ **55** $\boxed{=}$ → $\boxed{1337.45}$

When negative numbers must be substituted into a formula, it is especially important to place them in parentheses.

Example 2

To evaluate the formula $B = x^2 - y$ for $x = 4$ and $y = -3$, follow these steps:

Step 1 Substitute the numbers in parentheses. $\qquad B = (4)^2 - (-3)$

Step 2 Do the arithmetic. $\qquad\qquad\qquad\qquad\quad = 16 - (-3)$

$$= 16 + 3 = 19$$

Using parentheses in this way may seem like extra work for you, but it is the key to avoiding mistakes when evaluating formulas.

→ **More Practice**

1. Evaluate the following formulas using the standard order of operations.

$a = 3, b = 4, c = -6$

(a) $3ab$ $\quad = \underline{\hspace{1cm}}$ (b) $2a^2c$ $\quad = \underline{\hspace{1cm}}$

(c) $2a^2 - b^2$ $\quad = \underline{\hspace{1cm}}$ (d) $a + 3(2b - c)$ $\quad = \underline{\hspace{1cm}}$

(e) $(c^2 - 1) - (2 + b) = \underline{\hspace{1cm}}$

2. To convert Fahrenheit F to Celsius C temperature, use the following formula:

$$°C = \frac{5(°F - 32)}{9}$$

To convert Celsius C to Fahrenheit F temperature, use this formula:

$$°F = \frac{9°C}{5} + 32$$

Use these formulas to find the following temperatures. Round to the nearest degree.

(a) $650°F = \underline{\hspace{1cm}} °C$ (b) $-160°C = \underline{\hspace{1cm}} °F$

(c) $-10°F = \underline{\hspace{1cm}} °C$ (d) $52.5°C = \underline{\hspace{1cm}} °F$

→ **Solutions**

1. (a) $3(3)(4) = 36$

(b) $2(3)^2(-6) = 2(9)(-6) = -108$

(c) $2(3)^2 - (4)^2 = 2(9) - 16$
$\qquad\qquad\qquad = 18 - 16 = 2$

(d) $(3) + 3[2(4) - (-6)] = 3 + 3[8 - (-6)]$
$\qquad\qquad\qquad\qquad\quad = 3 + 3[14]$
$\qquad\qquad\qquad\qquad\quad = 3 + 42 = 45$

(e) $[(-6)^2 - 1] - [2 + (4)] = [36 - 1] - [6]$
$\qquad\qquad\qquad\qquad\qquad = 35 - 6 = 29$

2. (a) $C = \dfrac{5(650 - 32)}{9}$

$= \dfrac{5(618)}{9}$

$= \dfrac{3090}{9}$

$\approx 343°C$

(b) $F = \dfrac{9(-160)}{5} + 32$

$= \dfrac{-1440}{5} + 32$

$= -288 + 32$

$= -256°F$

(c) $C = \dfrac{5[(-10) - 32]}{9}$

$= \dfrac{5[-42]}{9}$

$= \dfrac{-210}{9}$

$\approx -23°C$

(d) $F = \dfrac{9(52.5)}{5} + 32$

$= \dfrac{472.5}{5} + 32$

$= 94.5 + 32$

$\approx 127°F$

On a calculator, do 2(a) like this:

5 $\boxed{\times}$ $\boxed{(}$ 650 $\boxed{-}$ 32 $\boxed{)}$ $\boxed{\div}$ 9 $\boxed{=}$ → $\boxed{343.3333333}$

and do 2(b) like this:

C7-1 9 $\boxed{\times}$ $\boxed{(-)}$ 160 $\boxed{\div}$ 5 $\boxed{+}$ 32 $\boxed{=}$ → $\boxed{-256.}$

Careful Avoid the temptation to combine steps when you evaluate formulas. Take it slowly and carefully, follow the standard order of operations, and you will arrive at the correct answer. Rush through problems like these and you usually make mistakes. ◀

Now turn to Exercises 7-1 for a set of practice problems in evaluating formulas.

Exercises 7-1 ## Algebraic Language and Formulas

A. Write the following using algebraic notation.

1. 6 times y times x

2. $2x$ divided by 5

3. $(a + c)$ times $(a - c)$

4. $4x$ times $(x + 6)$

5. $(n + 5)$ divided by $4n$

6. $7y$ divided by $(y - 4)$

B. For each algebraic expression, tell whether the portion named is a term or a factor.

1. $5x - 2y$ x is a _____

2. $6x^2 + 4$ 4 is a _____

3. $5n^3 - 6mn$ n^3 is a _____

4. $2y^2 - 3y + 6$ $2y^2$ is a _____

5. $4abc + d$ b is a _____

6. $5m(m + n)$ $(m + n)$ is a _____

7. $2y(3y^2 - 6)$ $2y$ is a _____

8. $4x^2y + 6xy^2 + 2y^3$ $6xy^2$ is a _____

C. Find the value of each of these formulas for $x = 2$, $y = 3$, $z = -4$, $R = 5$.

1. $A = 3x$
2. $D = 2R - y$
3. $T = x^2 + y^2$
4. $H = 2x + 3y - z$
5. $K = 3z - x^2$
6. $Q = 2xyz - 10$
7. $F = 2(x + y^2) - 3$
8. $W = 3(y - 1)$
9. $L = 3R - 2(y^2 - x)$
10. $A = R^2 - y^2 - xz$
11. $B = 3R - y + 1$
12. $F = 3(R - y + 1)$

D. Find the value of each of the following formulas. Round to the nearest whole number if necessary.

1. $A = 2(x + y) - 1$ for $x = 2$, $y = 4$
2. $V = (L + W)(2L + W)$ for $L = 7.5$ ft, $W = 5.0$ ft
3. $I = PRT$ for $P = 150$, $R = 0.05$, $T = 2$
4. $H = 2(a^2 + b^2)$ for $a = 2$ cm, $b = 1$ cm
5. $T = \dfrac{(A + B)H}{2}$ for $A = 3.26$ m, $B = 7.15$ m, $H = 4.4$ m
6. $V = \dfrac{\pi D^2 H}{4}$ for $\pi = 3.14$, $D = 6.25$ in., $H = 7.2$ in.
7. $P = \dfrac{NR(T + 273)}{V}$ for $N = 5$, $R = 0.08$, $T = 27$, $V = 3$
8. $W = D(AB - \pi R^2)H$ for $D = 9$ lb/in.^{3}, $A = 6.3$ in., $B = 2.7$ in., $\pi = 3.14$, $R = 2$ in., $H = 1.0$ in.
9. $V = LWH$ for $L = 16.25$ m, $W = 3.1$ m, $H = 2.4$ m
10. $V = \pi R^2 A$ for $\pi = 3.14$, $R = 3.2$ ft, $A = 0.425$ ft

E. Practical Problems

1. **General Trades** The perimeter of a rectangle is given by the formula $P = 2L + 2W$, where L is the length of the rectangle and W is its width. Find P when L is $8\frac{1}{2}$ in. and W is 11 in.

2. **Electronics** The current in amperes in a simple electrical circuit is given by Ohm's law $i = \frac{V}{R}$, where V is the voltage and R is the resistance of the circuit. Find the current in a circuit whose resistance is 10 ohms and which is connected across a 120-volt power source.

3. **Electrical Trades** Find the power used in an electric light bulb, $P = i^2 R$, if the current $i = 0.80$ ampere and the resistance $R = 150$ ohms. P will be in watts.

4. **Electrical Trades** When charging a battery with solar cells connected in series, the number N of solar cells required is determined by the following formula:

    $$N = \frac{\text{Voltage of the battery}}{\text{Voltage of each cell}}$$

    How many solar cells, providing 0.5 volt each, must be connected in series to provide sufficient voltage to charge a 12-volt battery?

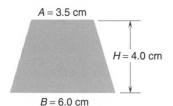
A = 3.5 cm

H = 4.0 cm

B = 6.0 cm

Problem 9

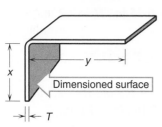
T→
←d→
←—D—→

Problem 10

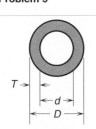
x
Dimensioned surface
→‖← T

Problem 11

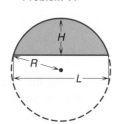
H
R
L

Problem 12

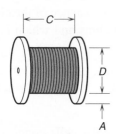
←C→
D
A

Problem 13

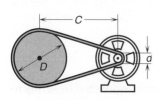
←C→
D
d

Problem 14

5. **Meteorology** The Fahrenheit temperature °F is related to the Celsius temperature °C by the formula $°F = \frac{9}{5}°C + 32$. Find the Fahrenheit temperature when °C = 40°.

6. **Machine Trades** The volume of a round steel bar depends on its length L and diameter D according to the formula $V = \frac{\pi D^2 L}{4}$. Find the volume of a bar 20.0 in. long and 3.0 in. in diameter. Use $\pi \approx 3.14$ and round to the nearest 10 cu in.

7. **Business and Finance** If D dollars is invested at p percent simple interest for t years, the amount A of the investment is

$$A = D\left(1 + \frac{pt}{100}\right)$$

Find A if D = $10,000, p = 5%, and t = 5 years.

8. **Electronics** The total resistance R of two resistances a and b connected in parallel is $R = \frac{ab}{a + b}$. What is the total resistance if a = 200 ohms and b = 300 ohms?

9. **General Trades** The area of a trapezoid is $T = \frac{(A + B)H}{2}$, where A and B are the lengths of its parallel sides and H is the height. Find the area of the trapezoid shown.

10. **Plumbing** Use the formula $T = \frac{1}{2}(D - d)$ to find the wall thickness T of tubing having the following dimensions. (See the figure.)

D, outside diameter   d, inside diameter

(a)   2.125 in.   1.500 in.
(b)   0.785 in.   0.548 in.
(c)   1.400 cm   0.875 cm
(d) $\frac{15}{16}$ in. $\frac{5}{8}$ in.

11. **Sheet Metal Trades** To make a right-angle inside bend in sheet metal, the length of sheet used is given by the formula $L = x + y + \frac{1}{2}T$. Find L when $x = 6\frac{1}{4}$ in., $y = 11\frac{7}{8}$ in., $T = \frac{1}{4}$ in. (See the figure.)

12. **Sheet Metal Trades** The length of a chord of a circle is given by the formula $L = 2RH - H^2$, where R is the radius of the circle and H is the height of the arc above the chord. If you have a portion of a circular disk for which $R = 4\frac{1}{4}$ in. and $H = 5\frac{1}{2}$ in., what is the length of the chord? (See the figure.)

13. **Construction** The rope capacity of a drum is given by the formula $L = ABC(A + D)$. How many feet of $\frac{1}{2}$-in. rope can be wound on a drum where A = 6 in., C = 30 in., D = 24 in., and B = 1.05 for $\frac{1}{2}$-in. rope? Round to the nearest 100 in. (See the figure.)

14. **Manufacturing** A millwright uses the following formula as an approximation to find the required length of a pulley belt. (See the figure.)

$$L = 2C + 1.57(D + d) + \frac{(D + d)}{4C}$$

Find the length of belt needed if

C = 36 in. between pulley centers
D = 24-in. follower
d = 4-in. driver

Round to the nearest 10 in.

15. **Electrical Trades**   An electrician uses a bridge circuit to locate a ground in an underground cable several miles long. The formula

$$\frac{R_1}{L - x} = \frac{R_2}{x} \qquad \text{or} \qquad x = \frac{R_2 L}{R_1 + R_2}$$

is used to find x, the distance to the ground. Find x if $R_1 = 750$ ohms, $R_2 = 250$ ohms, and $L = 4000$ ft.

16. **Machine Trades**   The cutting speed of a lathe is the rate, in feet per minute, that the revolving workpiece travels past the cutting edge of the tool. Machinists use the following formula to calculate cutting speed:

$$\text{Cutting speed, } C = \frac{\pi DN}{12}$$

where $\pi \approx 3.1416$, D is the diameter in inches of the work, and N is the turning rate in rpm. Find the cutting speed if a steel shaft 3.25 in. in diameter is turned at 210 rpm. Round to the nearest whole number.

17. **Electronics**   Find the power load P in kilowatts of an electrical circuit that takes a current I of 12 amperes at a voltage V of 220 volts if

$$P = \frac{VI}{1000}$$

18. **Aviation**   A jet engine developing T pounds of thrust and driving an airplane at V mph has a thrust horsepower, H, given approximately by the formula

$$H = \frac{TV}{375} \qquad \text{or} \qquad V = \frac{375H}{T}$$

Find the airspeed V if $H = 16{,}000$ hp and $T = 10{,}000$ lb.

19. **Electronics**   The resistance R of a conductor is given by the formula

$$R = \frac{PL}{A} \qquad \text{or} \qquad L = \frac{AR}{P}$$

where P = coefficient of resistivity
L = length of conductor
A = cross-sectional area of the conductor

Find the length in cm of No. 16 Nichrome wire needed to obtain a resistance of 8 ohms. For this wire $P = 0.000113$ ohm-cm and $A = 0.013$ cm^2. Round to the nearest centimeter.

20. **Automotive Trades**   The pressure P and total force F exerted on a piston of diameter D are approximately related by the equation

$$P = \frac{4F}{\pi D^2} \qquad \text{or} \qquad F = \frac{\pi PD^2}{4}$$

Find the total force on a piston of diameter 3.25 in. if the pressure exerted on it is 150 lb/sq in. Use $\pi \approx 3.14$ and round to the nearest 50 lb.

21. **Printing**   The formula for the number of type lines L set solid in a form is

$$L = \frac{12D}{T}$$

where D is the depth in picas and T is the point size of the type. How many lines of 10-point type can be set in a space 30 picas deep?

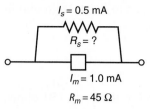

$I_s = 0.5$ mA

$R_s = ?$

$I_m = 1.0$ mA

$R_m = 45\ \Omega$

Problem 22

22. **Electronics**   In this circuit, the shunt resistance (R_s) has been selected to pass a current of 0.5 mA. Determine the value of R_s using the formula given. (See the figure.)

$$R_s = \frac{R_m \cdot I_m}{I_s}$$

23. **Automotive Trades**   The overall valve lift L_v for a vehicle is given by $L_v = L_c R$ where L_c is the lift at the cam and R is the rocker-arm ratio. Find the overall valve lift if the lift at the cam is 0.350 in. and the rocker arm ratio is 1.50 to 1.

24. **Plumbing**   To determine the number N of smaller pipes that provide the same total flow as a larger pipe, use the formula

$$N = \frac{D^2}{d^2}$$

where D is the diameter of the larger pipe and d is the diameter of each smaller pipe. How many $1\frac{1}{2}$-in. pipes will it take to produce the same flow as a $2\frac{1}{2}$-in. pipe?

25. **Automotive Trades**   Find the volume of a spherical ball bearing of radius $R = 1.0076$ in.:

$$\text{Volume } V = \frac{4\pi R^3}{3}$$

Use 3.1416 for π and round to 0.0001 cu in.

26. **Machine Trades**   How many minutes will it take a lathe to make 17 cuts each 24.5 in. in length on a steel shaft if the tool feed F is 0.065 in. per revolution and the shaft turns at 163 rpm? Use the formula

$$T = \frac{LN}{FR}$$

where T is the cutting time (min), N is the number of cuts, L the length of cut (in.), F the tool feed rate (in./rev), and R the rpm rate of the workpiece. Round to 0.1 min.

27. **Machine Trades**   Find the area of each of the following circular holes using the formula

$$A = \frac{\pi D^2}{4}$$

(a) $D = 1.04$ in.   (b) $D = \frac{7}{8}$ in.   (c) $D = 4.13$ in.   (d) $D = 2.06$ cm

Use 3.14 for π and round to three significant digits.

28. **Automotive Trades**   For modern automotive engines, horsepower (hp) is calculated as

$$\text{hp} = \frac{\text{torque (in lb} \cdot \text{ft)} \ \times \text{ engine speed (in rpm)}}{5252}$$

In the 2004 Mercedes-Benz E55 AMG, at 6100 rpm the engine produces 404 lb · ft of torque. Calculate the horsepower at this engine speed. Round to two significant digits.

29. **Automotive Trades**   The theoretic airflow of an engine is given by

$$T = \frac{DS}{3456}$$

where D is the displacement in cubic inches and S is the engine speed in rpm (revolutions per minute). The theoretic airflow will be in cubic feet per minute, or cfm. Find the theoretic airflow if the displacement is 190 in.^{3} and the engine is running at 3800 rpm. Round to the nearest tenth.

30. **Business and Finance**   A company's net profit margin M, expressed as a percent, is determined using the formula

$$M = \frac{100I}{R}$$

where I represents net income and R represents net revenue. Calculate the net profit margin for a company with a net income of \$254,000 and a net revenue of \$7,548,000. (Round to the nearest hundredth of a percent.)

31. **Electronics**   For a sinusoidal voltage, the root mean square voltage, V_{rms}, is defined as $V_{rms} = \dfrac{V_{peak}}{\sqrt{2}}$. If V_{peak} is 2.50 volts what is V_{rms}? (Round to the nearest hundredth.)

32. **Allied Health**   The body mass index (BMI), which is used to determine whether a person is overweight, underweight, or in the normal range, is calculated with metric units as:

$$BMI = \frac{\text{weight in kg}}{(\text{height in m})^2}$$

An adult with a BMI of less than 18.5 is considered to be underweight, and an adult with a BMI above 25.0 is considered to be overweight. A BMI in between 18.5 and 25.0 indicates weight in the normal range. Find the BMI for each person described and determine whether they are overweight, underweight, or in the normal range.

(a)   A person 1.87 m tall and weighing 95.0 kg.

(b)   A person 1.58 m tall and weighing 55.0 kg.

(c)   A person 1.70 m tall and weighing 52.0 kg.

33. **General Interest**   Suppose that a steel band was placed tightly around the earth at the equator. If the temperature of the steel is raised 1°F, the metal will expand 0.000006 in. each inch. How much space would there be between the earth and the steel band if the temperature was raised 1°F? Use the formula

$$D \text{ (in ft)} = \frac{(0.000006)(\text{diameter of the earth})(5280)}{\pi}$$

where diameter of the earth = 7917 miles. Use $\pi \approx 3.1415927$ and round to two decimal places.

34. **Plumbing**   When a cylindrical container is lying on its side, the following formula can be used for calculating the volume of liquid in the container:

$$V = \frac{4}{3}h^2L\sqrt{\frac{d}{h} - 0.608}$$

Use this formula to calculate the volume of water in such a tank 6.0 ft long (L), 2.0 ft in diameter (d), and filled to a height (h) of 0.75 ft. Round to two significant digits.

35. **Meteorology** The following formula is used to calculate the wind chill factor W in degrees Celsius:

$$W = 33 - \frac{(10.45 + 10\sqrt{V} - V)(33 - T)}{22.04}$$

where T is the air temperature in degrees Celsius and V is the wind speed in meters per second. Determine the wind chill for the following situations:

(a) Air temperature 7°C and wind speed 20 meters per second.

(b) Air temperature 28°F and wind speed 22 mph. (*Hint:* Use the conversion factors from Section 5-3 and the temperature formulas given on page 395 in this chapter to convert to the required units. Convert your final answer to degrees Fahrenheit.)

$C_1 = 4\ \mu F$

$C_T = \dfrac{C_1\, C_2}{C_1 + C_2}$

$C_2 = 14\ \mu F$

Problem 36

36. **Electronics** Calculate the total capacitance C_T in the circuit shown. Round to the nearest 0.1 μF (microfarad).

37. **Allied Health** The *vital capacity* is the amount of air that can be exhaled after one deep inhalation. Vital capacity is usually measured directly, but it can be estimated using the following formula:

$$V = (21.78 - 0.101a)h$$

where a is the patient's age in years, h is his height in cm, and V is his vital capacity in cc (cubic centimeters).

If Bob is 185 cm tall and 60.5 years old, calculate his vital capacity. Round to two significant digits.

38. **Sheet Metal Trades** A sheet metal worker uses the following formula for calculating bend allowance, BA, in inches:

$$BA = N(0.01743R + 0.0078T)$$

where $\quad N =$ number of degrees in the bend
$\qquad\quad R =$ inside radius of the bend, in inches
$\qquad\quad T =$ thickness of the metal, in inches

Find BA for each of the following situations. (You'll want to use a calculator on this one.)

	N	R	T
(a)	50°	$1\frac{1}{4}$ in.	0.050 in.
(b)	65°	0.857 in.	0.035 in.
(c)	40°	1.025 in.	0.0856 in.

39. **Automotive Trades** The fuel economy F of a certain hybrid electric car at speeds greater than or equal to 50 mph is given by the formula

$$F = 0.005V^2 - 1.25V + 96$$

where F is in miles per gallon and V is the speed in miles per hour. What is the fuel economy at a speed of (a) 50 mph? (b) 80 mph? (Round to the nearest whole number.)

When you have completed these exercises, check your answers to the odd-numbered problems in the Appendix, then turn to Section 7-2 to learn how to add and subtract algebraic expressions.

In the preceding section, we learned how to find the value of an algebraic expression after substituting numbers for the letters. There are other useful ways of using algebra where we must manipulate algebraic expressions *without* first substituting numbers for letters. In this section we learn how to simplify algebraic expressions by adding and subtracting terms.

Combining Like Terms

Two algebraic terms are said to be *like terms* if they contain exactly the same literal part. For example, the terms

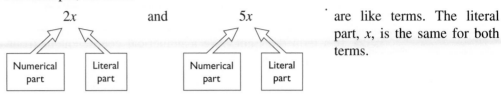

$2x$ and $5x$ are like terms. The literal part, x, is the same for both terms.

The number multiplying the letters is called the *numerical coefficient* of the term.

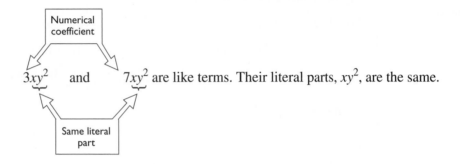

$3xy^2$ and $7xy^2$ are like terms. Their literal parts, xy^2, are the same.

But

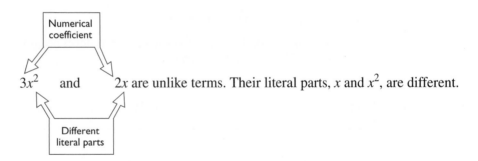

$3x^2$ and $2x$ are unlike terms. Their literal parts, x and x^2, are different.

You can add and subtract like terms but not unlike terms. To add or subtract like terms, add or subtract their numerical coefficients and keep the same literal part.

Example 1

(a) Add $2x + 3x$ like this:

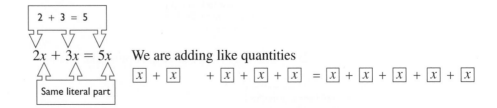

$2x + 3x = 5x$ We are adding like quantities

$\boxed{x} + \boxed{x} \quad + \boxed{x} + \boxed{x} + \boxed{x} = \boxed{x} + \boxed{x} + \boxed{x} + \boxed{x} + \boxed{x}$

(b) Subtract $8a^2x - 3a^2x$
like this:

$$8 - 3 = 5$$

$$8a^2x - 3a^2x = 5a^2x$$

Same literal part

(c) But $6n + 4n^2 - 3mn = 6n + 4n^2 - 3mn$

Different literal parts

These are unlike terms. They cannot be added.

> **Note**   If a term does not have a numerical coefficient in front of it, the coefficient is 1: x means $1x$. ◄

→ **Your Turn**

Try these problems for practice.

(a) $12d^2 + 7d^2$ = _____

(b) $2ax - ax - 5ax$ = _____

(c) $3(y + 1) + 9(y + 1)$ = _____

(d) $8x^2 + 2xy - 2x^2$ = _____

(e) $x - 6x + 2x$ = _____

(f) $4xy - xy + 3xy$ = _____

→ **Answers**

(a) $12d^2 + 7d^2 = 19d^2$

(b) $2ax - ax - 5ax = -4ax$ The term ax is equal to $1 \cdot ax$.

(c) $3(y + 1) + 9(y + 1) = 12(y + 1)$

(d) $8x^2 + 2xy - 2x^2 = 6x^2 + 2xy$

We cannot combine the x^2-term and the xy-term because the literal parts are not the same. They are unlike terms.

(e) $x - 6x + 2x = 1x - 6x + 2x = -3x$

(f) $4xy - xy + 3xy = 4xy - 1xy + 3xy = 6xy$

In general, to simplify a series of terms being added or subtracted, first group together like terms, then add or subtract.

Example 2

$3x + 4y - x + 2y + 2x - 8y$

Be careful not to lose the negative sign on 8y.

becomes

$(3x - x + 2x) + (4y + 2y - 8y)$ after grouping like terms,

Be careful not to lose the negative sign on x.

$$(3x - 1x + 2x) + (4y + 2y - 8y) = 4x + (-2y) \quad \text{or} \quad 4x - 2y$$

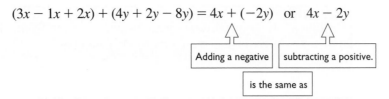

It is simpler to write the final answer as a subtraction rather than as an addition.

→ Your Turn

Simplify the following expressions by adding and subtracting like terms.

(a) $5x + 4xy - 2x - 3xy$

(b) $3ab^2 + a^2b - ab^2 + 3a^2b - a^2b$

(c) $x + 2y - 3z - 2x - y + 5z - x + 2y - z$

(d) $17pq - 9ps - 6pq + ps - 6ps - pq$

(e) $4x^2 - x^2 + 2x + 2x^2 + x$

→ Solutions

(a) $(5x - 2x) + (4xy - 3xy) = 3x + xy$

(b) $(3ab^2 - ab^2) + (a^2b + 3a^2b - a^2b) = 2ab^2 + 3a^2b$

(c) $(x - 2x - x) + (2y - y + 2y) + (-3z + 5z - z) = -2x + 3y + z$

(d) $(17pq - 6pq - pq) + (-9ps + ps - 6ps) = 10pq - 14ps$

(e) $(4x^2 - x^2 + 2x^2) + (2x + x) = 5x^2 + 3x$

Expressions with Parentheses Parentheses are used in algebra to group together terms that are to be treated as a unit. Adding and subtracting expressions usually involves working with parentheses. There are three main rules for removing parentheses.

Rule 1 If the parenthesis has a plus sign in front, simply remove the parentheses.

Example 3

(a) $1 + (3x + y) = 1 + 3x + y$

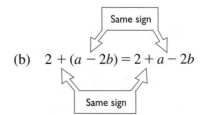

(b) $2 + (a - 2b) = 2 + a - 2b$

> **Rule 2** If the parenthesis has a negative sign in front, change the sign of each term inside, then remove the parentheses.

Example 4

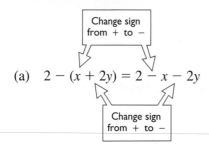

(a) $2 - (x + 2y) = 2 - x - 2y$

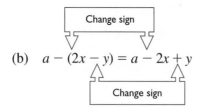

(b) $a - (2x - y) = a - 2x + y$

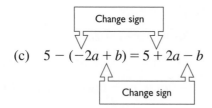

(c) $5 - (-2a + b) = 5 + 2a - b$

Note In using Rule 2 you are simply rewriting subtraction as addition of the opposite. However, you must add the opposite of *all* terms inside parentheses. ◄

→ **More Practice**

Simplify the following expressions by using these two rules to remove parentheses.

(a) $x + (2y - a^2)$ (b) $4 - (x^2 - y^2)$

(c) $-(x + 1) + (y + a)$ (d) $ab - (a - b)$

(e) $(x + y) - (p - q)$ (f) $-(-x - 2y) - (a + 2b)$

(g) $3 + (-2p - q^2)$ (h) $-(x - y) + (-3x^2 + y^2)$

→ **Answers**

(a) $x + 2y - a^2$ (b) $4 - x^2 + y^2$

(c) $-x - 1 + y + a$ (d) $ab - a + b$

(e) $x + y - p + q$ (f) $x + 2y - a - 2b$

(g) $3 - 2p - q^2$ (h) $-x + y - 3x^2 + y^2$

A third rule is needed when a multiplier is in front of the parentheses.

> **Rule 3** If the parenthesis has a multiplier in front, multiply each term inside the parentheses by the multiplier.

Example 5

(a) $+2(a + b) = +2a + 2b$

Think of this as $(+2)(a + b) = (+2)a + (+2)b$
$$= +2a + 2b$$

Each term inside the parentheses is multiplied by $+2$.

(b) $-2(x + y) = -2x - 2y$

Think of this as $(-2)(x + y) = (-2)x + (-2)y$

Each term inside the parentheses is multiplied by -2.

(c) $-(x - y) = (-1)(x - y)$
$$= (-1)(x) + (-1)(-y) = -x + y$$

(d) $2 - 5(3a - 2b) = 2 + (-5)(3a - 2b) = 2 + (-5)(3a) + (-5)(-2b)$
$$= 2 + (-15a) + (+10b) \quad \text{or} \quad 2 - 15a + 10b$$

> ► **Careful** In the last example it would be incorrect first to subtract the 5 from the 2. The order of operations requires that we multiply before we subtract. ◄

When you multiply negative numbers, you may need to review the arithmetic of negative numbers starting on page 347.

Notice that we must multiply *every* term inside the parentheses by the number outside the parentheses. Once the parentheses have been removed, you can add and subtract like terms, as explained previously.

→ Your Turn

Simplify the following expressions by removing parentheses. Use the three rules.

(a) $2(2x - 3y)$ (b) $1 - 4(x + 2y)$ (c) $a - 2(b - 2x)$

(d) $x^2 - 3(x - y)$ (e) $p - 2(-y - 2x)$ (f) $3(x - y) - 2(2x^2 + 3y^2)$

→ Solutions

(a) $2(2x - 3y) = 2(2x) + 2(-3y) = 4x + (-6y)$ or $4x - 6y$

(b) $1 - 4(x + 2y) = 1 + (-4)(x) + (-4)(2y) = 1 + (-4x) + (-8y)$ or $1 - 4x - 8y$

(c) $a - 2(b - 2x) = a + (-2)(b) + (-2)(-2x) = a + (-2b) + (+4x)$ or $a - 2b + 4x$

(d) $x^2 - 3(x - y) = x^2 + (-3)(x) + (-3)(-y) = x^2 + (-3x) + (+3y)$ or $x^2 - 3x + 3y$

(e) $p - 2(-y - 2x) = p + (-2)(-y) + (-2)(-2x) = p + (+2y) + (+4x)$ or $p + 2y + 4x$

(f) $3(x - y) - 2(2x^2 + 3y^2) = 3(x) + 3(-y) + (-2)(2x^2) + (-2)(3y^2)$
$$= 3x + (-3y) + (-4x^2) + (-6y^2) \text{ or } 3x - 3y - 4x^2 - 6y^2$$

Once you can simplify expressions by removing parentheses, it is easy to add and subtract like terms.

Example 6

$(3x - y) - 2(x - 2y) = 3x - y - 2x + 4y$ Simplify by removing parentheses.

$= \underbrace{3x - 2x} \; \underbrace{- y + 4y}$ Group like terms.

$= x + 3y$ Combine like terms.

→ **Your Turn**

Try these problems for practice.

(a) $(3y + 2) + 2(y + 1)$ (b) $(2x + 1) + 3(4 - x)$

(c) $(a + b) - (a - b)$ (d) $2(a + b) - 2(a - b)$

(e) $2(x - y) - 3(y - x)$ (f) $2(x^3 + 1) - 3(x^3 - 2)$

(g) $(x^2 - 2x) - 2(x - 2x^2)$ (h) $-2(3x - 5) - 4(x - 1)$

→ **Answers**

(a) $5y + 4$ (b) $-x + 13$ (c) $2b$ (d) $4b$

(e) $5x - 5y$ (f) $-x^3 + 8$ (g) $5x^2 - 4x$ (h) $-10x + 14$

Some step-by-step solutions:

(a) $(3y + 2) + 2(y + 1) = 3y + 2 + 2y + 2$
                           $= 3y + 2y + 2 + 2$
                           $= 5y + 4$

(b) $(2x + 1) + 3(4 - x) = 2x + 1 + 12 - 3x$
                           $= 2x - 3x + 1 + 12$
                           $= -x + 13$

(c) $(a + b) - (a - b) = a + b - a - (-b)$
                        $= a + b - a + b$
                        $= a - a + b + b = 0 + 2b = 2b$

(h) $-2(3x - 5) - 4(x - 1) = -6x - 2(-5) - 4x - 4(-1)$
                           $= -6x + 10 - 4x + 4$
                           $= -6x - 4x + 10 + 4 = -10x + 14$

Now turn to Exercises 7-2 for additional practice on addition and subtraction of algebraic expressions.

Exercises 7-2   Adding and Subtracting Algebraic Expressions

A.   Simplify by adding or subtracting like terms.

    1. $3y + y + 5y$ 2. $4x^2y + 5x^2y$

    3. $E + 2E + 3E$ 4. $ax - 5ax$

5. $9B - 2B$ 6. $3m - 3m$

7. $3x^2 - 5x^2$ 8. $4x + 7y + 6x + 9y$

9. $6R + 2R^2 - R$ 10. $1.4A + 0.05A - 0.8A^2$

11. $x - \frac{1}{2}x - \frac{1}{4}x - \frac{1}{8}x$ 12. $x + 2\frac{1}{2}x - 5\frac{1}{2}x$

13. $2 + W - 4.1W - \frac{1}{2}$ 14. $q - p - 1\frac{1}{2}p$

15. $2xy + 3x + 4xy$ 16. $ab + 5ab - 2ab$

17. $x^2 + x^2y + 4x^2 + 3x$ 18. $1.5p + 0.3pq + 3.1p$

B. Simplify by removing parentheses and, if possible, combining like terms.

1. $3x^2 + (2x - 5)$ 2. $6 + (-3a + 8b)$

3. $8m + (4m^2 + 2m)$ 4. $9x + (2x - 5x^2)$

5. $2 - (x + 5y)$ 6. $7x - (4 + 2y)$

7. $3a - (8 - 6b)$ 8. $5 - (w - 6z)$

9. $4x - (10x^2 + 7x)$ 10. $12m - (6n + 4m)$

11. $15 - (3x - 8)$ 12. $-12 - (5y - 9)$

13. $-(x - 2y) + (2x + 6y)$ 14. $-(3 + 5m) + (11 - 4m)$

15. $-(14 + 5w) - (2w - 3z)$ 16. $-(16x - 8) - (-2x + 4y)$

17. $3(3x - 4y)$ 18. $4(5a + 6b)$

19. $-8(7m + 6)$ 20. $-2(6x - 3)$

21. $x - 5(3 + 2x)$ 22. $4y - 2(8 + 3y)$

23. $9m - 7(-2m + 6)$ 24. $w - 5(4w - 3)$

25. $3 - 4(2x + 3y)$ 26. $6 - 2(3a - 5b)$

27. $12 - 2(3w - 8)$ 28. $2 - 11(7 + 5a)$

29. $2(3x + 4y) - 4(6x^2 - 5y^2)$ 30. $-4(5a - 6b) + 7(2ab - 4b^2)$

31. $(22x - 14y) + 3(8y - 6x)$ 32. $8(3w - 5z) + (9z - 6w)$

33. $6(x + y) - 3(x - y)$ 34. $-5(2x - 3x^2) + 2(9x^2 + 8x)$

35. $4(x^2 - 6x + 8) - 6(x^2 + 3x - 5)$

36. $2(3a - 5b + 6ab) - 8(2a + b - 4ab)$

When you have completed these exercises, check your answers to the odd-numbered problems in the Appendix, then turn to Section 7-3 to learn how to solve algebraic equations.

7-3   Solving Simple Equations

An arithmetic equation such as $3 + 2 = 5$ means that the number named on the left $(3 + 2)$ is the same as the number named on the right (5).

An algebraic equation such as $x + 3 = 7$ is a statement that the sum of some number x and 3 is equal to 7. If we choose the correct value for x, the number $x + 3$ will be equal to 7.

x is a *variable,* a symbol that stands for a number in an equation, a blank space to be filled. Many numbers might be put in the space, but only one makes the equation a true statement.

Find the missing numbers in the following arithmetic equations.

(a) $37 + \underline{\hspace{1cm}} = 58$ (b) $\underline{\hspace{1cm}} - 15 = 29$

(c) $4 \times \underline{\hspace{1cm}} = 52$ (d) $28 \div \underline{\hspace{1cm}} = 4$

→ **Answers**

(a) $37 + \boxed{21} = 58$ (b) $\boxed{44} - 15 = 29$

(c) $4 \times \boxed{13} = 52$ (d) $28 \div \boxed{7} = 4$

We could have written these equations as follows, with variables instead of blanks.

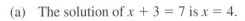

$$37 + A = 58 \qquad B - 15 = 29 \qquad 4C = 52 \qquad \frac{28}{D} = 4$$

Of course, any letters would do in place of A, B, C, and D in these algebraic equations.

How did you solve these equations? You probably "eyeballed" them—mentally juggled the other information in the equation until you found a number that made the equation true. Solving algebraic equations is very similar except that we can't "eyeball" it entirely. We need certain and systematic ways of solving the equation that will produce the correct answer quickly every time.

In this section you will learn first what a solution to an algebraic equation is—how to recognize it if you stumble over it in the dark—then how to solve linear equations.

Solution   Each value of the variable that makes an equation true is called a *solution* of the equation.

Example 1

(a)   The solution of $x + 3 = 7$ is $x = 4$.

   $(4) + 3 = 7$

(b)   The solution of the equation $2x - 9 = 18 - 7x$ is $x = 3$.

$$
\begin{aligned}
2(3) - 9 &= 18 - 7(3) \\
6 - 9 &= 18 - 21 \\
-3 &= -3
\end{aligned}
$$

For certain equations more than one value of the variable may make the equation true.

Example 2

The equation $x^2 + 6 = 5x$ is true for $x = 2$,

$$
\begin{aligned}
(2)^2 + 6 &= 5(2) \\
4 + 6 &= 5 \cdot 2 \\
10 &= 10
\end{aligned}
$$

and it is also true for $x = 3$.

$$
\begin{aligned}
(3)^2 + 6 &= 5(3) \\
9 + 6 &= 5 \cdot 3 \\
15 &= 15
\end{aligned}
$$

Determine whether each given value of the variable is a solution to the equation.

(a)   For $4 - x = -3$ is $x = -7$ a solution?

(b)   For $3x - 8 = 22$ is $x = 10$ a solution?

(c)   For $-2x + 15 = 5 + 3x$ is $x = -2$ a solution?

(d)   For $3(2x - 8) = 5 - (6 - 4x)$ is $x = 11.5$ a solution?

(e)   For $3x^2 - 5x = -2$ is $x = \dfrac{2}{3}$ a solution?

→ Answers

(a)   No—the left side equals 11.

(b)   Yes.   (c)   No—the left side equals 19, the right side equals -1.

(d)   Yes.   (e)   Yes.

Here is the check for (e): $3\left(\dfrac{2}{3}\right)^2 - 5\left(\dfrac{2}{3}\right) = -2$

$$3\left(\dfrac{4}{9}\right) - 5\left(\dfrac{2}{3}\right) = -2$$

$$\dfrac{4}{3} - \dfrac{10}{3} = -2$$

$$-\dfrac{6}{3} = -2$$

$$-2 = -2$$

Equivalent Equations

Equations with the exact same solution are called *equivalent* equations. The equations $2x + 7 = 13$ and $3x = 9$ are equivalent because substituting the value 3 for x makes them both true.

We say that an equation with the variable x is *solved* if it can be put in the form

$x = \square$ where $\square$ is some number

For example, the solution to the equation

$2x - 1 = 7$ is $x = 4$

because $2(4) - 1 = 7$

   or $8 - 1 = 7$

is a true statement.

Solving Equations

Equations as simple as the previous one are easy to solve by guessing, but guessing is not a very dependable way to do mathematics. We need some sort of rule that will enable us to rewrite the equation to be solved ($2x - 1 = 7$, for example) as an equivalent solution equation ($x = 4$).

The general rule is to treat every equation as a balance of the two sides.

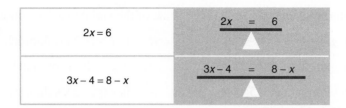

Any changes made in the equation must not disturb this balance.

Any operation performed on one side of the equation must also be performed on the other side.

Two kinds of balancing operations may be used.

1. Adding or subtracting a number on both sides of the equation does not change the balance

   Original equation: $a = b$

   $a + 2 = b + 2$

   $a - 2 = b - 2$

2. Multiplying or dividing both sides of the equation by a number (but not zero) does not change the balance.

   $2 \cdot a = 2 \cdot b$

   $\dfrac{a}{3} = \dfrac{b}{3}$

Example 3

Let's work through an example.

Solve: $x - 4 = 2$.

Step 1 We want to change this equation to an equivalent equation with only x on the left, so we add 4 to each side of the equation.

$$x - 4 \boxed{+ 4} = 2 \boxed{+ 4}$$

Step 2 Combine terms.

$$x \underbrace{- 4 \boxed{+ 4}}_{0} = 2 \boxed{+ 4}$$

$$x = 6 \quad \text{Solution}$$

 $(6) - 4 = 2$
$2 = 2$

→ Your Turn

Use these balancing operations to solve the equation

$8 + x = 14$

Solution

Solve: $8 + x = 14$

Step 1 We want to change this equation to an equivalent equation with only x on the left, so we subtract 8 from each side of the equation.

$$8 + x \boxed{-8} = 14 \boxed{-8}$$

Step 2 Combine terms.

$$x + \underbrace{8 \boxed{-8}}_{0} = 14 \boxed{-8} \quad \text{where } 8 + x = x + 8$$

$$x = 6 \quad \text{Solution}$$

$$8 + (6) = 14$$
$$14 = 14$$

→ **More Practice**

Solve these in the same way.

(a) $x - 7 = 10$ (b) $12 + x = 27$ (c) $x + 6 = 2$

(d) $8.4 = 3.1 + x$ (e) $6.7 + x = 0$ (f) $\frac{1}{4} = x - \frac{1}{2}$

(g) $-11 = x + 5$ (h) $x - 5.2 = -3.7$

→ **Solutions**

(a) **Solve:** $x - 7 = 10$

Add 7 to each side. $x - 7 \boxed{+7} = 10 \boxed{+7}$

Combine terms. $x \underbrace{- 7 \boxed{+7}}_{0} = 10 \boxed{+7}$

Solution: $x = 17$

$$(17) - 7 = 10$$
$$10 = 10$$

(b) **Solve:** $12 + x = 27$

Subtract 12 from each side. $12 + x \boxed{-12} = 27 \boxed{-12}$

Combine terms.
(Note that $12 + x = x + 12$.) $x + \underbrace{12 \boxed{-12}}_{0} = 27 \boxed{-12}$

Solution: $x = 15$

$$12 + (15) = 27$$
$$27 = 27$$

(c) **Solve:** $x + 6 = 2$

Subtract 6 from each side. $x + 6 \boxed{-6} = 2 \boxed{-6}$

Combine terms. $x + \underbrace{6 \boxed{-6}}_{0} = 2 \boxed{-6}$

The solution is a negative number. $x = -4$
Remember, any number, positive or negative, may be the solution of an equation.

$(-4) + 6 = 2$
$\qquad 2 = 2$

(d) **Solve:** $8.4 = 3.1 + x$

The variable x is on the right side of this equation.

Subtract 3.1 from each side. $\qquad 8.4 \;\boxed{-\; 3.1} = 3.1 + x \;\boxed{-\; 3.1}$

Combine terms. $\qquad\qquad\qquad 8.4 - 3.1 = x$

Decimal numbers often appear $\qquad\qquad 5.3 = x$
in practical problems.

$5.3 = x$ is the same as $x = 5.3$. $\qquad x = 5.3$

$8.4 = 3.1 + (5.3)$
$8.4 = 8.4$

(e) **Solve:** $6.7 + x = 0$

Subtract 6.7 from each side. $\qquad 6.7 + x \;\boxed{-\; 6.7} = 0 \;\boxed{-\; 6.7}$

Solution: $\qquad\qquad\qquad\qquad\qquad x = -6.7$

$6.7 + (-6.7) = 0$
$\quad 6.7 - 6.7 = 0$
$\qquad\qquad\; 0 = 0$

(f) **Solve:** $\dfrac{1}{4} = x - \dfrac{1}{2}$

Add $\dfrac{1}{2}$ to each side.

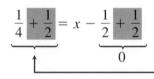

Solution: $\qquad\qquad \dfrac{3}{4} = x \qquad$ or $\qquad x = \dfrac{3}{4}$

$\dfrac{1}{4} = \left(\dfrac{3}{4}\right) - \dfrac{1}{2}$

$\dfrac{1}{4} = \dfrac{1}{4}$

(g) **Solve:** $-11 = x + 5$

Subtract 5 from each side. $\qquad -11 \;\boxed{-\; 5} = x + 5 \;\boxed{-\; 5}$

Solution: $\qquad\qquad\qquad\qquad -16 = x \quad$ or $\quad x = -16$

$-11 = (-16) + 5$
$-11 = -11$

(h) **Solve:** $x - 5.2 = -3.7$

Add 5.2 to each side. $\qquad x - 5.2 \;\boxed{+\; 5.2} = -3.7 \;\boxed{+\; 5.2}$

Solution: $\qquad\qquad\qquad\qquad x = 1.5$

$(1.5) - 5.2 = -3.7$
$\qquad\quad -3.7 = -3.7$

Notice that to solve these equations, we always do the opposite of whatever operation is being performed on the variable. If a number is being added to x in the equation, we must subtract that number in order to solve. If a number is being subtracted from x in the equation, we must add that number in order to solve. Finally, whatever we do to one side of the equation, we must also do to the other side of the equation. ◄

In the equations solved so far, the variable has appeared as simply x. In some equations the variable may be multiplied or divided by a number, so that it appears as $2x$ or $\frac{x}{3}$, for example. To solve such an equation, do the operation opposite of the one being performed on the variable. If the variable appears as $2x$, divide by 2; if the variable appears as $\frac{x}{3}$, multiply by 3, and so on.

Example 4

To solve the equation $27 = -3x$ follow these steps:

$27 = -3x$ Notice that the x-term is on the right side of the equation. It is x multiplied by -3. To change $-3x$ into $1x$ or x, we must perform the opposite operation, that is, divide by -3.

Step 1 Divide each side by -3. $\dfrac{27}{-3} = \dfrac{-3x}{-3}$

Step 2 Simplify. $-9 = \left(\dfrac{-3}{-3}\right)x$

Solution: $-9 = x$ or $x = -9$

$27 = -3(-9)$
$27 = 27$

Example 5

Here is a slightly different one to solve: $\dfrac{x}{3} = 13$

In this equation the variable x is divided by 3. We want to change this equation to an equivalent equation with x alone on the left. Therefore, we must perform the opposite operation, multiplying by 3.

Step 1 Multiply both sides by 3 $\dfrac{x}{3} \cdot 3 = 13 \cdot 3$

Step 2 Simplify. $x = 39$

$$\dfrac{x}{3} \cdot \dfrac{3}{1} = \dfrac{3x}{3} = \left(\dfrac{3}{3}\right)x = 1x = x$$

$\dfrac{(39)}{3} = 13$

$13 = 13$

Example 6

To solve $-\dfrac{x}{4} = 5$

Step 1 Multiply both sides by -4. $\left(-\dfrac{x}{4}\right)(-4) = 5\,(-4)$

Step 2 $\dfrac{(-x)(-4)}{4} = -20$ because $-\dfrac{x}{4} = \dfrac{-x}{4}$

$\dfrac{4x}{4} = -20$ because $(-x)(-4) = +4x$

$x = -20$ because $\dfrac{4x}{4} = \left(\dfrac{4}{4}\right)x = 1x = x$

$-\dfrac{(-20)}{4} = 5$

$-(-5) = 5$

$5 = 5$

▶ **A Closer Look** When a negative sign precedes a fraction, we may move it either to the numerator or to the denominator, but not to both. In the previous problem, we placed the negative sign in the numerator in Step 2. If we had moved it to the denominator instead, we would have the following:

$$\left(\dfrac{x}{-4}\right)(-4) = -20 \qquad \text{or} \qquad \dfrac{-4x}{-4} = -20 \qquad \text{or} \qquad x = -20$$

which is exactly the same solution. ◀

→ **Your Turn**

Try the following problems for more practice with one-step multiplication and division equations.

(a) $-5x = -25$ (b) $\dfrac{x}{6} = 6$ (c) $-16 = 2x$

(d) $7x = 35$ (e) $-\dfrac{x}{2} = 14$ ▦ (f) $2.4x = 0.972$

→ **Solutions**

(a) **Solve:** $-5x = -25$ The variable x appears multiplied by -5.

   Step 1 Divide both sides by -5. $\dfrac{-5x}{-5} = \dfrac{-25}{-5}$

   Step 2 Simplify: $\dfrac{-5x}{-5} = \left(\dfrac{-5}{-5}\right)x = 1x = x$ $x = 5$

$-5(5) = -25$

$-25 = -25$

(b) **Solve:** $\dfrac{x}{6} = 6$ The variable x appears divided by 6.

   Step 1 Multiply both sides by 6. $\left(\dfrac{x}{6}\right)(6) = 6(6)$

   Step 2 Simplify: $\left(\dfrac{x}{6}\right)(6) = x\left(\dfrac{6}{6}\right) = x(1) = x$ $x = 36$

$\dfrac{(36)}{6} = 6$

$6 = 6$

(c) **Solve:** $-16 = 2x$

    Step 1 Divide each side by 2.

$$\frac{-16}{\boxed{2}} = \frac{2x}{\boxed{2}}$$

    Step 2 Simplify: $\frac{2x}{2} = \left(\frac{2}{2}\right)x = 1x = x$ $-8 = x$ or $x = -8$

$-16 = 2(-8)$
$-16 = -16$

(d) **Solve:** $7x = 35$

    Step 1 Divide both sides by 7.

$$\frac{7x}{\boxed{7}} = \frac{35}{\boxed{7}}$$

    Step 2 Simplify: $\frac{7x}{7} = \left(\frac{7}{7}\right)x = x$ $x = 5$

$7 \cdot (5) = 35$
$\ \ \ \ 35 = 35$

(e) **Solve:** $-\dfrac{x}{2} = 14$

    Step 1 Multiply both sides by -2.

$$\left(-\frac{x}{2}\right)\boxed{(-2)} = 14\ \boxed{(-2)}$$

    Step 2   Simplify: $\boxed{-\frac{x}{2} = -\frac{1}{2}x = x\left(-\frac{1}{2}\right)}\ \Longrightarrow\ x\left(-\frac{1}{2}\right)(-2) = 14(-2)$

$$\boxed{\left(-\frac{1}{2}\right)(-2) = \frac{2}{2} = 1}\ \Longrightarrow\qquad x \cdot 1 = -28$$

$$\boxed{x \cdot 1 = x}\ \Longrightarrow\qquad\qquad x = -28$$

$-\dfrac{(-28)}{2} = 14$

$\dfrac{28}{2} = 14$

$14 = 14$

(f) **Solve:** $2.4x = 0.972$

          Divide by 2.4. $x = 0.405$ **Solution**

    .972 ⊙ 2.4 ⊜ →   **0.405**

We have now covered the four basic single-operation equations. Before moving on, try the following set of problems covering these four operations. For each problem, think carefully whether you must add, subtract, multiply, or divide in order to make the equation read $x = $ _____ or _____ $= x$.

→ **More Practice**

Solve:

(a) $14 = x - 8$ (b) $\dfrac{y}{6} = 3$ (c) $n + 11 = 4$

(d) $3x = 33$ (e) $7 = -\dfrac{a}{5}$ (f) $z - 2 = -10$

(g) $-8 = -16x$ (h) $26 = y + 5$

Each solution states the operation you should have performed on both sides and then gives the final answer.

(a) Add 8 to get $x = 22$.

(b) Multiply by 6 to get $y = 18$.

(c) Subtract 11. $n = -7$.

(d) Divide by 3. $x = 11$.

(e) Multiply by -5. $a = -35$.

(f) Add 2. $z = -8$.

(g) Divide by -16. $x = \dfrac{1}{2}$.

(h) Subtract 5. $y = 21$.

Now turn to Exercises 7-3 for practice on solving simple equations.

Exercises 7-3 Solving Simple Equations

A. Solve the following equations.

1. $x + 4 = 13$

2. $23 = A + 6$

3. $17 - x = 41$

4. $z - 18 = 29$

5. $6 = a - 2\dfrac{1}{2}$

6. $73 + x = 11$

7. $y - 16.01 = 8.65$

8. $11.6 - R = 3.7$

9. $-39 = 3x$

10. $-9y = 117$

11. $13a = 0.078$

12. $\dfrac{x}{3} = 7$

13. $\dfrac{z}{1.3} = 0.5$

14. $\dfrac{N}{2} = \dfrac{3}{8}$

15. $m + 18 = 6$

16. $-34 = x - 7$

17. $-6y = -39$

18. $\dfrac{a}{-4} = 9$

19. $22 - T = 40$

20. $79.2 = 2.2y$

21. $-5.9 = -6.6 + Q$

22. $\dfrac{5}{4} = \dfrac{Z}{8}$

23. $12 + x = 37$

24. $66 - y = 42$

25. $\dfrac{K}{0.5} = 8.48$

26. $-12z = 3.6$

B. Practical Problems

1. **Electrical Trades**   Ohm's law is often written in the form

$$I = \frac{E}{R}$$

where I is the current in amperes (A), E is the voltage, and R is the resistance in ohms. What is the voltage necessary to push a 0.80-A current through a resistance of 450 ohms?

2. **Aviation**   The formula

$$D = RT$$

is used to calculate the distance D traveled by an object moving at a constant average speed R during an elapsed time T. How long would it take a pilot to fly 1240 miles at an average speed of 220 mph?

3. **Wastewater Technology**   A wastewater treatment operator uses the formula

$$A = 8.34FC$$

to determine the amount A of chlorine in pounds to add to a basin. The flow F through the basin is in millions of gallons per day, and the desired concentration C of chlorine is in parts per million.

If 1800 pounds of chlorine was added to a basin with a flow rate of 7.5 million gallons per day, what would be the resulting concentration?

4. **Sheet Metal Trades**   The water pressure P in pounds per square foot is related to the depth of water D in feet by the formula

$$P = 62.4D$$

where 62.4 lb is the weight of 1 cu ft of water. If the material forming the bottom of a tank is made to withstand 800 pounds per square foot of pressure, what is the maximum safe height for the tank?

5. **Physics**   The frequency f (in waves per second) and the wavelength w (in meters) of a sound traveling in air are related by the equation

$$fw = 343$$

where 343 m/s is the speed of sound in air. Calculate the wavelength of a musical note with a frequency of 200 waves per second.

6. **Physics**   At a constant temperature, the pressure P (in psi) and the volume V (in cu ft) of a particular gas are related by the equation

$$PV = 1080$$

If the volume is 60 cu ft, calculate the pressure.

7. **Electrical Trades**   For a particular transformer, the voltage E in the circuits is related to the number of windings W of wire around the core by the equation

$$E = 40W$$

How many windings will produce a voltage of 840 V?

8. **Metalworking**   The surface speed S in fpm (feet per minute) of a rotating cylindrical object is

$$S = \frac{\pi dn}{12}$$

where d is the diameter of the object in inches and n is the speed of rotation in rpm (revolutions per minute). If an 8-in. grinder must have a surface speed of 6000 fpm, what should the speed of rotation be? Use $\pi \approx 3.14$, and round to the nearest hundred rpm.

9. **Construction**   The amount of lumber in board feet (bf) can be expressed by the formula

$$\text{bf} = \frac{TWL}{12}$$

where T is the thickness of a board in inches, W is its width in inches, and L is its length in feet. What total length of 1 in. by 6 in. boards is needed for a total of 16 bf?

10. **Automotive Trades** For modern automotive engines, horsepower is defined as

$$\text{hp} = \frac{\text{torque (in lb} \cdot \text{ft)} \times \text{engine speed (in rpm)}}{5252}$$

If an engine has a peak horsepower of 285 hp at 4800 rpm, what is the torque of the engine at this speed? Round to two significant digits.

11. **Business and Finance** The formula for a company's net profit margin M expressed as a percent is given by

$$M = \frac{100I}{R}$$

where I represents net income and R represents net revenue.

If a company had a net profit margin of 2.85% on net revenue of $4,625,000, what was its net income? (Round to the nearest hundred dollars.)

12. **Allied Health** In problem 32 on page 401, the metric formula for BMI (body mass index) was given as

$$\text{BMI} = \frac{\text{weight in kg}}{(\text{height in m})^2}$$

If a patient has a BMI of 23.8 and is 1.77 m tall, how much does the patient weigh?

13. **Construction** On a construction site, a winch is used to wind cable onto the drum of the winch. For the first layer of cable, the line pull P of the winch is related to the diameter D of the drum by the formula

$$P = \frac{k}{D}$$

where k is a constant. If $P = 5,400$ Ib and $D = 4$ in., find the value of the constant k in foot-pounds. (*Hint:* Convert D to feet first.)

When you have completed these exercises, check your answers to the odd-numbered problems in the Appendix, then continue in Section 7-4.

7-4 Solving Two-Step Equations

Solving many simple algebraic equations involves both kinds of operations: addition/subtraction and multiplication/division.

Example 1

Solve: $2x + 6 = 14$

Step 1 We want to change this equation to an equivalent equation with only x or terms that include x on the left.

Subtract 6 from both sides.	$2x + 6 \boxed{- 6} = 14 \boxed{- 6}$
Combine terms. (Be careful. You cannot add $2x$ and 6—they are unlike terms.)	$2x + \underbrace{6 - 6}_{0} = 8$
Now this is an equivalent equation with only an x-term on the left.	$2x = 8$

Step 2 Divide both sides of the equation by 2.

$$\dfrac{2x}{2} = \dfrac{8}{2}$$

$$\dfrac{2x}{2} = x \qquad x = 4$$

$$2(4) + 6 = 14$$
$$8 + 6 = 14$$
$$14 = 14$$

→ **Your Turn**

Try this one to test your understanding of the process.

Solve: $3x - 7 = 11$

→ **Solution**

Solve: $3x - 7 = 11$

Step 1 Add 7 to each side.

$$3x - 7 \boxed{+ 7} = 11 \boxed{+ 7}$$

$$3x \underbrace{- 7 \boxed{+ 7}}_{0} = 11 \boxed{+ 7}$$

$$3x = 18$$

Step 2 Divide both sides of the equation by 3.

$$\dfrac{3x}{3} = \dfrac{18}{3}$$

Solution: $\qquad\qquad x = 6$

$$3(6) - 7 = 11$$
$$18 - 7 = 11$$
$$11 = 11$$

→ **More Practice**

Here is another two-step equation.

Solve: $23 = 9 - \dfrac{y}{3}$

→ **Solution**

Solve: $23 = 9 - \dfrac{y}{3}$

Step 1 Notice that the variable y is on the right side of the equation. We must first eliminate the 9 by subtracting 9 from both sides. Be sure to keep the negative sign in front of the $\dfrac{y}{3}$ term.

$$23 - 9 = 9 - \frac{y}{3} - 9$$

$$\boxed{9 - 9 = 0}$$

$$14 = -\frac{y}{3}$$

Step 2 Multiply both sides by -3.

$$(14)(-3) = \left(-\frac{y}{3}\right)(-3)$$

$$-42 = (-3)\left(-\frac{1}{3}\right)y \quad \Longleftarrow \quad \boxed{-\frac{y}{3} = -\frac{1}{3}y}$$

$$-42 = 1y \quad \Longleftarrow \quad \boxed{(-3)\left(-\frac{1}{3}\right) = 1}$$

Solution: $y = -42$

$$23 = 9 - \frac{(-42)}{3}$$
$$23 = 9 - (-14)$$
$$23 = 23$$

Note In these two-step problems, we did the addition or subtraction in Step 1 and the multiplication or division in Step 2. We could have reversed this order and arrived at the correct solution, but the problem might have become more complicated to solve. Always add or subtract first, and when you multiply or divide, do so to *all* terms. ◀

If more than one variable term appears on the same side of an equation, combine these like terms before performing any operation to both sides.

Example 2

Solve: $2x + 5 + 4x = 17$

Step 1 Combine the *x*-terms on the left side.

$$(2x + 4x) + 5 = 17$$
$$6x + 5 = 17$$

Step 2 Subtract 5 from each side.

$$6x + 5 - 5 = 17 - 5$$
$$6x = 12$$

Step 3 Divide each side by 6.

$$\frac{6x}{6} = \frac{12}{6}$$

Solution:

$$x = 2$$

Substitute 2 for *x* in each *x*-term.

$$2(2) + 5 + 4(2) = 17$$
$$4 + 5 + 8 = 17$$
$$17 = 17$$

→ Your Turn

Now you try one.

Solve: $32 = x - 12 - 5x$

Solution

Step 1   Combine the x-terms on the right.

$$32 = (x - 5x) - 12$$
$$32 = -4x - 12$$

Step 2   Add 12 to both sides.

$$32 \boxed{+ 12} = -4x - 12 \boxed{+ 12}$$
$$44 = -4x$$

Step 3   Divide both sides by -4.

$$\frac{44}{-4} = \frac{-4x}{-4}$$

Solution: $-11 = x$ or $x = -11$

$$32 = (-11) - 12 - 5(-11)$$
$$32 = -11 - 12 + 55$$
$$32 = -23 + 55$$
$$32 = 32$$

The next example involves the two steps of multiplication and division.

Example 3

Solve: $\dfrac{5x}{3} = 25$

Step 1   Multiply both sides by 3.

$$\frac{5x}{3} \cdot \boxed{3} = 25 \cdot \boxed{3}$$

$$\boxed{\frac{5x}{3} \cdot 3 = \frac{5x}{\cancel{3}} \cdot \frac{\cancel{3}}{1} = 5x} \Longrightarrow 5x = 75$$

Step 2   Divide both sides by 5.

$$\frac{5x}{\boxed{5}} = \frac{75}{\boxed{5}}$$

Solution: $x = 15$

A Closer Look   In the previous example, you could solve for x in one step if you multiply both sides by the fraction $\frac{3}{5}$.

$$\frac{5x}{3} \cdot \boxed{\frac{3}{5}} = 25 \cdot \boxed{\frac{3}{5}}$$

$$\frac{{}^{1}\cancel{5}x}{{}_{1}\cancel{3}} \cdot \frac{{}^{1}\cancel{3}}{{}_{1}\cancel{5}} = \frac{25}{1} \cdot \frac{3}{5}$$

$$x = 15 \blacktriangleleft$$

More Practice

Here are a few more two-operation equations for practice.

(a) $7x + 2 = 51$ (b) $18 - 5x = 3$ (c) $15.3 = 4x - 1.5$

(d) $\dfrac{x}{5} - 4 = 6$ (e) $11 - x = 2$ (f) $5 = 7 - \dfrac{x}{4}$

(g) $\dfrac{x + 2}{3} = 4$ (h) $2x - 9.4 = 0$ (i) $2.75 = 14.25 - 0.20x$

(j) $3x + x = 18$ (k) $12 = 9x + 4 - 5x$ (l) $\dfrac{3x}{4} = -15$

(m) **Electronics**   The resistance R in ohms (Ω) of a particular circuit element is given by the formula

$R = 0.02T + 5.00$

where T is the temperature of the element in degrees Celsius (°C). At what temperature is the resistance 5.48 Ω?

→ **Solutions**

(a) **Solve:** $7x + 2 = 51$

Change this equation to an equivalent equation with only an x-term on the left.

Step 1   Subtract 2 from each side. $7x + 2 \boxed{-\,2} = 51 \boxed{-\,2}$

Combine like terms. $7x + \underbrace{2 \boxed{-\,2}}_{0} = 51 \boxed{-\,2}$

$7x = 49$

Step 2   Divide both sides by 7. $\dfrac{\boxed{7}x}{\boxed{7}} = \dfrac{49}{\boxed{7}}$

$x = \dfrac{49}{7}$

Solution: $x = 7$

✓
$7(7) + 2 = 51$
$49 + 2 = 51$
$51 = 51$

(b) **Solve:** $18 - 5x = 3$

Step 1   Subtract 18 from each side. $18 - 5x \boxed{-\,18} = 3 \boxed{-\,18}$

Rearrange terms and combine like terms. $-5x + \underbrace{18 - 18}_{0} = 3 - 18$

$-5x = -15$

Step 2   Divide both sides by -5. $\dfrac{-5x}{-5} = \dfrac{-15}{-5}$

Solution: $x = 3$

✓
$18 - 5(3) = 3$
$18 - 15 = 3$
$3 = 3$

(c) **Solve:** $15.3 = 4x - 1.5$

Step 1   Add 1.5 to each side and combine like terms. $15.3 \boxed{+\,1.5} = 4x \underbrace{-\,1.5 \boxed{+\,1.5}}_{0}$

$16.8 = 4x$

$4x = 16.8$

Step 2   Divide both sides by 4.

$$\frac{4x}{4} = \frac{16.8}{4}$$

Solution:

$$x = 4.2$$

Decimal number solutions are common in practical problems.

 15.3 ⊞ 1.5 ⊟ ⊡ 4 ⊟ → [████████ 4.2]

$15.3 = 4(4.2) - 1.5$

$15.3 = 16.8 - 1.5$

$15.3 = 15.3$

(d) **Solve:** $\dfrac{x}{5} - 4 = 6$

Step 1   Add 4 to both sides.

$$\frac{x}{5} - 4 \;\boxed{+\,4} = 6 \;\boxed{+\,4}$$

$$\frac{x}{5} = 10$$

Step 2   Multiply both sides by 5.

$$\left(\frac{x}{5}\right) \cdot \boxed{5} = 10 \cdot \boxed{(5)}$$

Solution:

$$x = 50$$

$$\frac{(50)}{5} - 4 = 6$$

$$10 - 4 = 6$$

$$6 = 6$$

(e) **Solve:** $11 - x = 2$

Step 1   Subtract 11 from each side.

$$11 - x \;\boxed{-\,11} = 2 \;\boxed{-\,11}$$

Combine terms.

$$-x + \underbrace{11 \;\boxed{-\,11}}_{0} = 2 \;\boxed{-\,11}$$

Step 2   Multiply each side by -1.

$$-x = -9$$

Solution:

$$x = 9$$

$11 - (9) = 2$

$2 = 2$

(f) **Solve:** $5 = 7 - \dfrac{x}{4}$

Step 1   Subtract 7 from both sides.

$$5 \;\boxed{-\,7} = 7 - \frac{x}{4} \;\boxed{-\,7}$$

$$\boxed{\,7 - \frac{x}{4} - 7 = (7 - 7) - \frac{x}{4} = -\frac{x}{4}\,}$$

$$-2 = -\frac{x}{4}$$

Step 2   Multiply both sides by -4.

$$-2 \;\boxed{(-4)} = \left(-\frac{x}{4}\right)\boxed{(-4)}$$

$$\boxed{\,\left(-\frac{x}{4}\right)(-4) = (-x)\left(\frac{-4}{4}\right) = (-x)(-1) = x\,}$$

$$8 = x$$

Solution:

$$x = 8$$

$$5 = 7 - \frac{(8)}{4}$$

$$5 = 7 - 2$$

$$5 = 5$$

(g) **Solve:** $\dfrac{x + 2}{3} = 4$

Multiply both sides by 3. $\qquad\qquad \dfrac{x + 2}{3} \cdot \boxed{3} = 4 \cdot \boxed{3}$

This eliminates the fraction on the left side. $\qquad \dfrac{x + 2}{{}_1\cancel{3}} \cdot \dfrac{{}^1\cancel{3}}{1} = 4 \cdot 3$

$$x + 2 = 12$$

Subtract 2 from both sides. $\qquad\qquad x + 2 \boxed{- 2} = 12 \boxed{- 2}$

$$x = 10$$

$$\frac{10 + 2}{3} = 4$$

$$\frac{12}{3} = 4$$

$$4 = 4$$

(h) **Solve:** $2x - 9.4 = 0$

Step 1   Add 9.4 to each side. $\qquad\qquad 2x - 9.4 \boxed{+ 9.4} = 0 \boxed{+ 9.4}$

$$2x = 9.4$$

Step 2   Divide each side by 2. $\qquad\qquad\qquad x = 4.7 \quad$ Solution

$$2(4.7) - 9.4 = 0$$

$$9.4 - 9.4 = 0$$

(i) **Solve:** $2.75 = 14.25 - 0.20x$

Step 1   Subtract 14.25 from both sides. $2.75 \boxed{- 14.25} = 14.25 - 0.20x \boxed{- 14.25}$

$$-11.5 = -0.20x$$

Step 2   Divide both sides by -0.20 $\qquad \dfrac{-11.5}{\boxed{-0.20}} = \dfrac{-0.20x}{\boxed{-0.20}}$

$$57.5 = x$$

or $\qquad x = 57.5$

C7-2   2.75 ⊟ 14.25 ⊜ ÷ ⊡ .2 ⊜ → $\boxed{57.5}$

$$2.75 = 14.25 - 0.20(57.5)$$

$$2.75 = 14.25 - 11.5$$

$$2.75 = 2.75$$

(j) **Solve:** $3x + x = 18$

Step 1   Combine like terms on the left side. $\qquad\qquad 4x = 18$

Step 2   Divide each side by 4. $\qquad\qquad\qquad \dfrac{4x}{\boxed{4}} = \dfrac{18}{\boxed{4}}$

$$x = 4.5$$

✓ $3(4.5) + (4.5) = 18$

$13.5 + 4.5 = 18$

$18 = 18$

(k) **Solve:** $12 = 9x + 4 - 5x$

Step 1 Combine like terms. $\qquad\qquad\qquad 12 = 4x + 4$

Step 2 Subtract 4 from each side. $\qquad\quad 12 \boxed{-4} = 4x + 4 \boxed{-4}$

$8 = 4x$

Step 3 Divide each side by 4. $\qquad\qquad \dfrac{8}{\boxed{4}} = \dfrac{4x}{\boxed{4}}$

$2 = x \qquad \text{or} \qquad x = 2$

The check is left to you.

(l) **Solve:** $\dfrac{3x}{4} = -15$

Step 1 Multiply both sides by 4. $\qquad\qquad \dfrac{3x}{4} \cdot \boxed{4} = -15 \cdot \boxed{4}$

$3x = -60$

Step 2 Divide both sides by 3. $\qquad\qquad \dfrac{3x}{\boxed{3}} = \dfrac{-60}{\boxed{3}}$

Solution: $\qquad\qquad\qquad\qquad\qquad x = -20$

Don't forget to check your solution.

(m) **First,** substitute 5.48 for R:

$5.48 = 0.02T + 5.00$

Then, solve:

Subtract 5.00 from both sides. $\qquad 5.48 \boxed{-5.00} = 0.02T + 5.00 \boxed{-5.00}$

$0.48 = 0.02T$

Divide both sides by 0.02. $\qquad\qquad \dfrac{0.48}{\boxed{0.02}} = \dfrac{0.02T}{\boxed{0.02}}$

$24 = T$

The resistance is 5.48 Ω at a temperature of 24°C.

Be sure to check your solution.

Now turn to Exercises 7-4 for more practice on solving equations.

Exercises 7-4 Solving Two-Step Equations

A. Solve.

1. $2x - 3 = 17$ $\qquad\qquad\qquad$ 2. $4x + 6 = 2$

3. $\dfrac{x}{5} = 7$ $\qquad\qquad\qquad\quad$ 4. $-8y + 12 = 32$

5. $4.4m - 1.2 = 9.8$

6. $\dfrac{3}{4} = 3x + \dfrac{1}{4}$

7. $14 - 7n = -56$

8. $38 = 58 - 4a$

9. $3z - 5z = 12$

10. $17 = 7q - 5q - 3$

11. $23 - \dfrac{x}{4} = 11$

12. $9m + 6 + 3m = -60$

13. $-15 = 12 - 2n + 5n$

14. $2.6y - 19 - 1.8y = 1$

15. $\dfrac{1}{2} + 2x = 1$

16. $3x + 16 = 46$

17. $-4a + 45 = 17$

18. $\dfrac{x}{2} + 1 = 8$

19. $-3Z + \dfrac{1}{2} = 17$

20. $2x + 6 = 0$

21. $1 = 3 - 5x$

22. $23 = 17 - \dfrac{x}{4}$

23. $-5P + 18 = 3$

24. $5x - 2x = 24$

25. $6x + 2x = 80$

26. $x + 12 - 6x = -18$

27. $27 = 2x - 5 + 4x$

28. $-13 = 22 - 3x + 8x$

29. $\dfrac{x - 5}{4} = -2$

30. $7 = \dfrac{x + 6}{2}$

31. $\dfrac{2x}{3} = -4$

32. $6 = -\dfrac{5x}{4}$

B. Practical Problems

1. **Office Services**   A repair service charges $45 for a house call and an additional $60 per hour of work. The formula

   $T = 45 + 60H$

   represents the total charge T for H hours of work. If the total bill for a customer was $375, how many hours of actual labor were there?

2. **Meteorology**   The air temperature T (in degrees Fahrenheit) at an altitude h (in feet) above a particular area can be approximated by the formula

   $T = -0.002h + G$

   where G is the temperature on the ground directly below. If the ground temperature is 76°F, at what altitude will the air temperature drop to freezing, 32°F?

3. **Sports and Leisure**   Physical fitness experts sometimes use the following formula to approximate the maximum target heart rate R during exercise based on a person's age A:

   $R = -0.8A + 176$

   At what age should the heart rate during exercise not exceed 150 beats per minute?

4. **Life Skills**   The formula $A = p + prt$ is used to determine the total amount of money A in a bank account after an amount p is invested for t years at a rate of interest r. What rate of interest is needed for $8000 to grow to $12,000 after

5 years? Be sure to convert your decimal answer to a percent and round to the nearest tenth of a percent.

5. **Automotive Trades** The formula

$$P + 2T = C$$

gives the overall diameter C of the crankshaft gear for a known pitch diameter P of the small gear and the height T of teeth above the pitch diameter circle. If $P = 2.875$ in. and $C = 3.125$ in., find T.

6. **Sheet Metal Trades** The allowance A for a Pittsburgh lock is given by

$$A = 2w + \frac{3}{16} \text{ in.}$$

where w is the width of the pocket. If the allowance for a Pittsburgh lock is $\frac{11}{16}$ in., what is the width of the pocket?

7. **Trades Management** A plumber's total bill A can be calculated using the formula

$$A = RT + M$$

where R is his hourly rate, T is the total labor time in hours, and M is the cost of materials. A plumber bids a particular job at $2820. If materials amount to $980, and his hourly rate is $64 per hour, how many hours should the job take for the estimate to be accurate?

8. **Machine Trades** The formula

$$L = 2d + 3.26(r + R)$$

can be used under certain conditions to approximate the length L of belt needed to connect two pulleys of radii r and R if their centers are a distance d apart. How far apart can two pulleys be if their radii are 8 in. and 6 in., and the total length of the belt connecting them is 82 in.? Round to the nearest inch.

9. **Forestry** The fire damage potential D of a class 5 forest is given by the formula

$$D = 2A + 5$$

where A is the average age (in years) of the brush in the forest. If $D = 20$, find the average age of the brush.

When you have completed these exercises, check your answers to the odd-numbered problems in the Appendix, then continue in Section 7-5.

7-5 Solving More Equations and Formulas

Parentheses in Equations In Section 7-2 you learned how to deal with algebraic expressions involving parentheses. Now you will learn to solve equations containing parentheses by using these same skills.

Example 1

Solve: $2(x + 4) = 27$

Step 1 Use Rule 3 on page 407. Multiply each term inside the parentheses by 2.

$$2x + 8 = 27$$

Now solve this equation using the techniques of the previous section.

Step 2   Subtract 8 from both sides. $2x + 8 \boxed{- 8} = 27 \boxed{- 8}$

$$2x = 19$$

Step 3   Divide both sides by 2. $\dfrac{2x}{2} = \dfrac{19}{2}$

Solution: $x = 9.5$

$$2[(9.5) + 4] = 27$$
$$2(13.5) = 27$$
$$27 = 27$$

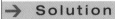 **Your Turn**

Try this similar example. Solve for y.

$$-3(y - 4) = 36$$

→ Solution

Solve: $-3(y - 4) = 36$

Step 1   Multiply each term inside parentheses by -3.

$\boxed{-3(y - 4) = -3(y) + (-3)(-4)}$ $\Longrightarrow$ $-3y + 12 = 36$

Step 2   Subtract 12 from each side. $-3y + 12 \boxed{- 12} = 36 \boxed{- 12}$

$$-3y = 24$$

Step 3   Divide each side by -3. $\dfrac{-3y}{\boxed{-3}} = \dfrac{24}{\boxed{-3}}$

$$y = -8$$

$$-3[(-8) - 4] = 36$$
$$-3(-12) = 36$$
$$36 = 36$$

A Closer Look   In each of the last two examples, you could have first divided both sides by the number in front of parentheses.

Here is how each solution would have looked.

First Example	**Second Example**
$\dfrac{2(x + 4)}{2} = \dfrac{27}{2}$	$\dfrac{-3(y - 4)}{-3} = \dfrac{36}{-3}$
$x + 4 = 13.5$	$y - 4 = -12$
$x + 4 - 4 = 13.5 - 4$	$y - 4 + 4 = -12 + 4$
$x = 9.5$	$y = -8$

Some students may find this technique preferable, especially when the right side of the equation is exactly divisible by the number in front of the parentheses. ◀

Here is a more difficult equation involving parentheses.

Example 2

Solve: $5x - (2x - 3) = 27$

Step 1 Use Rule 2 on page 406. Change the sign of each term inside parentheses and then remove them.

$$\boxed{-(2x - 3) = -2x + 3} \Longrightarrow \quad 5x - 2x + 3 = 27$$

Step 2 Combine the like terms on the left.

$$\boxed{5x - 2x = 3x} \Longrightarrow \quad 3x + 3 = 27$$

Step 3 Subtract 3 from both sides. $\qquad 3x + 3 \;\boxed{- 3} = 27 \;\boxed{- 3}$

$$3x = 24$$

Step 4 Divide both sides by 3. $\qquad \dfrac{3x}{3} = \dfrac{24}{3}$

$$x = 8$$

$$5(8) - [2(8) - 3] = 27$$
$$40 - (16 - 3) = 27$$
$$40 - 13 = 27$$
$$27 = 27$$

→ Your Turn

Try this problem.

Solve: $8 - 3(2 - 3x) = 34$

→ Solution

Step 1 Be very careful here. Some students are tempted to subtract the 3 from the 8. However, the order of operations rules on page 371 specify that multiplication must be performed before addition or subtraction. Therefore, your first step is to multiply the expression in parentheses by -3.

$$8 \;\boxed{-\; 3}(2 - 3x) = 8 + \boxed{(-3)}(2) + \boxed{(-3)}(-3x) \qquad 8 - 6 + 9x = 34$$

Step 2 Combine the like terms on the left in Step 1. $\qquad 2 + 9x = 34$

Step 3 Subtract 2 from each side. $\qquad 2 + 9x \;\boxed{- 2} = 34 \;\boxed{- 2}$

$$9x = 32$$

Step 4 Divide each side by 9. $\qquad \dfrac{9x}{9} = \dfrac{32}{9}$

$$x = 3\tfrac{5}{9}$$

Be sure to check your answer.

Here are more equations with parentheses for you to solve.

(a) $4(x - 2) = 26$

(b) $11 = -2(y + 5)$

(c) $23 = 6 - (3n + 4)$

(d) $4x - (6x - 9) = 41$

(e) $7 + 3(5x + 2) = 38$

(f) $20 = 2 - 5(9 - a)$

(g) $(3m - 2) - (5m - 3) = 19$

(h) $2(4x + 1) + 5(3x + 2) = 58$

→ **Answers**

(a) $x = 8.5$ (b) $y = -10.5$ (c) $n = -7$ (d) $x = -16$

(e) $x = 1\frac{2}{3}$ (f) $a = 12.6$ (g) $m = -9$ (h) $x = 2$

Here are worked solutions to (b), (d), (f), and (h). The checks are left to you.

(b) **Solve:** $11 = -2(y + 5)$

Step 1 Multiply each term inside parentheses by -2.

$$11 = -2y - 10$$

Step 2 Add 10 to each side.

$$11 \boxed{+ 10} = -2y - 10 \boxed{+ 10}$$

$$21 = -2y$$

Step 3 Divide each side by -2.

$$\frac{21}{-2} = \frac{-2y}{-2}$$

$$-10.5 = y \qquad \text{or} \qquad y = -10.5$$

(d) **Solve:** $4x - (6x - 9) = 41$

Step 1 Change the sign of each term inside parentheses and then remove parentheses.

$$4x - 6x + 9 = 41$$

Step 2 Combine like terms on the left side in Step 1.

$$-2x + 9 = 41$$

Step 3 Subtract 9 from each side.

$$-2x + 9 \boxed{- 9} = 41 \boxed{- 9}$$

$$-2x = 32$$

Step 4 Divide each side by -2.

$$\frac{-2x}{-2} = \frac{32}{-2}$$

$$x = -16$$

(f) **Solve:** $20 = 2 - 5(9 - a)$

Step 1 Multiply both terms inside parentheses by -5 and then remove parentheses.

$$\boxed{\begin{array}{l} 2 - 5(9 - a) \\ = 2 + (-5)(9) + (-5)(-a) \end{array}}$$

$$20 = 2 - 45 + 5a$$

Step 2 Combine like terms on the right side in Step 1.

$$20 = -43 + 5a$$

Step 3 Add 43 to both sides.

$$20 \boxed{+ 43} = \boxed{-43} + 5a + 43$$

$$63 = 5a$$

			$\dfrac{63}{5} = \dfrac{5a}{5}$

Step 4 Divide both sides by 5.

$$12.6 = a \qquad \text{or} \qquad a = 12.6$$

(h) **Solve:** $2(4x + 1) + 5(3x + 2) = 58$

Step 1 Multiply each term inside the first parentheses by 2 and each term inside the second parentheses by 5.

$$8x + 2 + 15x + 10 = 58$$

Step 2 Combine both pairs of like terms on the left side in Step 1. $23x + 12 = 58$

Step 3 Subtract 12 from both sides. $23x + 12 - 12 = 58 - 12$

$$23x = 46$$

Step 4 Divide both sides by 23. $\dfrac{23x}{23} = \dfrac{46}{23}$

$$x = 2$$

Variable on Both Sides

In all the equations we have solved so far, the variable has been on only one side of the equation. Sometimes it is necessary to solve equations with variable terms on both sides.

Example 3

To solve

$$3x - 4 = 8 - x$$

First, move all variable terms to the left side by adding x to both sides.

$$3x - 4 + x = 8 - x + x$$
$$4x - 4 = 8$$

$3x + x = 4x$
$-x + x = 0$

Next, proceed as before.

Add 4 to both sides. $4x - 4 + 4 = 8 + 4$

$$4x = 12$$

Divide both sides by 4. $\dfrac{4x}{4} = \dfrac{12}{4}$

$$x = 3$$

Finally, check your answer.

$$3(3) - 4 = 8 - (3)$$
$$9 - 4 = 5$$
$$5 = 5$$

→ Your Turn

Ready to attempt one yourself? Solve this equation for y, then check it with our solution.

$$5y - 21 = 8y$$

Did you move the variable terms to the left? Here you can save a step by moving them to the right instead of to the left.

Solve: $5y - 21 = 8y$

Step 1 Subtract $5y$ from both sides. $\quad 5y - 21 \;\boxed{-\,5y} = 8y \;\boxed{-\,5y}$

$$-21 = 3y$$

Step 2 Divide both sides by 3.

$$\frac{-21}{3} = \frac{3y}{3}$$

$$-7 = y \quad \text{or} \quad y = -7$$

$$5(-7) - 21 = 8(-7)$$
$$-35 - 21 = -56$$
$$-56 = -56$$

Learning Help If a variable term is already by itself on one side of the equation, move all variable terms to this side. As in the last example, this will save a step. ◀

→ **More Practice**

Now try these problems for practice in solving equations in which the variable appears on both sides.

(a) $\;x - 6 = 3x$ (b) $\;5(x - 2) = x + 4$

(c) $\;2(x - 1) = 3(x + 1)$ (d) $\;4x + 9 = 7x - 15$

(e) $\;4n + 3 = 18 - 2n$ (f) $\;2A = 12 - A$

(g) $\;6y = 4(2y + 7)$ (h) $\;8m - (2m - 3) = 3(m - 4)$

(i) $\;3 - (5x - 8) = 6x + 22$ (j) $\;-2(3x - 5) = 7 - 5(2x + 3)$

→ **Solutions**

(a) **Solve:** $\;x - 6 = 3x$

     Subtract x from each side, so that $x - 6 \;\boxed{-\,x} = 3x \;\boxed{-\,x}$
     x-terms will appear only on the right.

     Combine like terms. $-6 = 2x$

     or $2x = -6$

     Divide by 2. $x = -3$

$$(-3) - 6 = 3(-3)$$
$$-9 = -9$$

(b) **Solve:** $\;5(x - 2) = x + 4$

     Multiply each term inside the $5x - 10 = x + 4$
     parentheses by 5.

     Subtract x from each side. $5x - 10 \;\boxed{-\,x} = x + 4 \;\boxed{-\,x}$

     Combine terms $4x - 10 = 4$

Add 10 to each side. $4x - 10 \; +\,10 = 4 \; +\,10$

$$4x = 14$$

Divide each side by 4. $x = 3\tfrac{1}{2}$

$$5\left(3\tfrac{1}{2} - 2\right) = \left(3\tfrac{1}{2}\right) + 4$$
$$5\left(1\tfrac{1}{2}\right) = 7\tfrac{1}{2}$$
$$7\tfrac{1}{2} = 7\tfrac{1}{2}$$

(c) **Solve:** $2(x - 1) = 3(x + 1)$

Remove parentheses by multiplying. $2x - 2 = 3x + 3$

Subtract $3x$ from each side. $2x - 2 \; -\,3x = 3x + 3 \; -\,3x$

Combine like terms. $-2 - x = 3$

Add 2 to each side. $-2 - x \; +\,2 = 3 \; +\,2$

$$-x = 5$$

Multiply both sides by -1. $x = -5$

Solve for x, not $-x$.

$$2(-5 - 1) = 3(-5 + 1)$$
$$2(-6) = 3(-4)$$
$$-12 = -12$$

(d) **Solve:** $4x + 9 = 7x - 15$

Subtract $7x$ from each side. $4x + 9 \; -\,7x = 7x - 15 \; -\,7x$

Combine like terms. $-3x + 9 = -15$

Subtract 9 from each side. $-3x + 9 \; -\,9 = -15 \; -\,9$

$$-3x = -24$$

Divide each side by -3. $x = 8$

$$4(8) + 9 = 7(8) - 15$$
$$32 + 9 = 56 - 15$$
$$41 = 41$$

(e) **Solve:** $4n + 3 = 18 - 2n$

Add $2n$ to both sides. $4n + 3 \; +\,2n = 18 - 2n \; +\,2n$

Now the n-terms appear only on the left side. $6n + 3 = 18$

Subtract 3 from both sides. $6n + 3 \; -\,3 = 18 \; -\,3$

$$6n = 15$$

Divide by 6. $\dfrac{6n}{6} = \dfrac{15}{6}$

Solution: $n = 2.5$

$$4(2.5) + 3 = 18 - 2(2.5)$$
$$10 + 3 = 18 - 5$$
$$13 = 13$$

(f) **Solve:** $2A = 12 - A$

Add A to each side. $2A + A = 12 - A + A$

The A-terms now appear only on the left. $3A = 12$

Divide by 3. $\dfrac{3A}{3} = \dfrac{12}{3}$

Solution: $A = 4$

☑ $2(4) = 12 - (4)$

$8 = 8$

(g) **Solve:** $6y = 4(2y + 7)$

Multiply to remove parentheses. $6y = 8y + 28$

Subtract $8y$ from both sides. $6y - 8y = 8y + 28 - 8y$

$-2y = 28$

Divide by -2. $\dfrac{-2y}{-2} = \dfrac{28}{-2}$

Solution: $y = -14$

Check the solution.

(h) **Solve:** $8m - (2m - 3) = 3(m - 4)$

Change signs to remove parentheses on the left. $8m - 2m + 3 = 3(m - 4)$

Multiply by 3 to remove parentheses on the right. $8m - 2m + 3 = 3m - 12$

Combine like terms. $6m + 3 = 3m - 12$

Subtract $3m$ from both sides. $6m + 3 - 3m = 3m - 12 - 3m$

$3m + 3 = -12$

Subtract 3 from both sides. $3m + 3 - 3 = -12 - 3$

$3m = -15$

Divide by 3. $\dfrac{3m}{3} = \dfrac{-15}{3}$

Solution: $m = -5$

Check the solution.

(i) **Solve:** $3 - (5x - 8) = 6x + 22$

Remove parentheses by changing the signs of both terms. $3 - 5x + 8 = 6x + 22$

Combine terms. $11 - 5x = 6x + 22$

Subtract $6x$ from both sides. $11 - 5x - 6x = 6x + 22 - 6x$

$11 - 11x = 22$

Subtract 11 from both sides. $11 - 11x - 11 = 22 - 11$

$-11x = 11$

Divide by -11. $x = -1$

Check it.

(j) **Solve:** $-2(3x - 5) = 7 - 5(2x + 3)$

Multiply by -2 to remove parentheses on the left.

$$-6x + 10 = 7 - 5(2x + 3)$$

Multiply by -5 to remove parentheses on the right (leave the 7 alone!).

$$-6x + 10 = 7 - 10x - 15$$

Combine like terms on the right.

$$-6x + 10 = -8 - 10x$$

Add $10x$ to both sides.

$$-6x + 10 + 10x = -8 - 10x + 10x$$

$$4x + 10 = -8$$

Subtract 10 from both sides.

$$4x + 10 - 10 = -8 - 10$$

$$4x = -18$$

Divide by 4.

$$x = -4.5$$

Check it.

Remember:

1. Do only legal operations: Add or subtract the same quantity from both sides of the equation; multiply or divide both sides of the equation by the same nonzero quantity.

2. Remove all parentheses carefully.

3. Combine like terms when they are on the same side of the equation.

4. Use legal operations to change the equation so that you have only x by itself on one side of the equation and a number on the other side of the equation.

5. Always check your answer.

Solving Formulas

To *solve a formula* for some letter means to rewrite the formula as an equivalent formula with that letter isolated on the left of the equals sign.

For example, the area of a triangle is given by the formula

$$A = \frac{BH}{2}$$ where A is the area, B is the length of the base, and H is the height.

Solving for the base B gives the equivalent formula

$$B = \frac{2A}{H}$$

Solving for the height H gives the equivalent formula

$$H = \frac{2A}{B}$$

Solving formulas is a very important practical application of algebra. Very often a formula is not written in the form that is most useful. To use it you may need to rewrite the formula, solving it for the letter whose value you need to calculate.

To solve a formula, use the same balancing operations that you used to solve equations. You may add or subtract the same quantity on both sides of the formula and you may multiply or divide both sides of the formula by the same nonzero quantity.

Example 4

To solve the formula

$$S = \frac{R + P}{2} \qquad \text{for } R$$

First, multiply both sides of the equation by 2 to eliminate the fraction.

$$2 \cdot S = 2 \cdot \left(\frac{R + P}{2} \right)$$

$$2S = R + P$$

Second, subtract P from both sides of the equation to isolate R on one side.

$$2S - P = R + \underbrace{P - P}_{0}$$

$$2S - P = R$$

This formula can be reversed to read

$$R = 2S - P \qquad \text{We have solved the original formula for } R.$$

→ **Your Turn**

Solve the following formulas for the variable indicated.

(a) $V = \dfrac{3K}{T}$ for K

(b) $Q = 1 - R + T$ for R

(c) $V = \pi R^2 H - AB$ for H

(d) $P = \dfrac{T}{A - B}$ for A

→ **Solutions**

(a) $V = \dfrac{3K}{T}$

First, multiply both sides by T to get $\qquad VT = 3K$

Second, divide both sides by 3 to get $\qquad \dfrac{VT}{3} = K$

Solved for K, the formula is $\qquad K = \dfrac{VT}{3}$

(b) $Q = 1 - R + T$

First, subtract T from both sides to get $\qquad Q - T = 1 - R$

Second, subtract 1 from both sides to get $\qquad Q - T - 1 = -R$

This is equivalent to $\qquad -R = Q - T - 1$

$\qquad$ or $\quad R = -Q + T + 1$

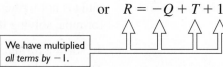

We have multiplied all terms by -1.

$\qquad$ or $\quad R = 1 - Q + T$

(c) $V = \pi R^2 H - AB$

First, add AB to both sides to get $\qquad\qquad V + AB = \pi R^2 H$

Second, divide both sides by πR^2 to get $\qquad \dfrac{V + AB}{\pi R^2} = H$

Notice that we divide *all* of the left side by πR^2.

Solved for H, the formula is $\qquad\qquad\qquad H = \dfrac{V + AB}{\pi R^2}$

(d) $P = \dfrac{T}{A - B}$

First, multiply each side by $(A - B)$ to get $\qquad P(A - B) = T$

Second, multiply to remove the parentheses. $\qquad PA - PB = T$

Next, add PB to each side. $\qquad\qquad\qquad PA = T + PB$

Finally, divide each side by P. $\qquad\qquad\qquad A = \dfrac{T + PB}{P}$

▶ **Careful** Remember, when using the multiplication/division rule, you must multiply or divide *all* of both sides of the formula by the same quantity. ◀

→ **More Practice**

Practice solving formulas with the following problems.

Solve:

(a) $P = 2A + 3B$ $\qquad$ for A $\qquad\qquad$ (b) $E = MC^2$ $\qquad\qquad$ for M

(c) $S = \dfrac{A - RT}{1 - R}$ $\qquad$ for A $\qquad\qquad$ (d) $S = \dfrac{1}{2} gt^2$ $\qquad\qquad$ for g

(e) $P = i^2 R$ $\qquad$ for R $\qquad\qquad$ (f) $I = \dfrac{V}{R + a}$ $\qquad\qquad$ for R

(g) $A = \dfrac{2V - W}{R}$ $\qquad$ for V $\qquad\qquad$ (h) $F = \dfrac{9C}{5} + 32$ $\qquad$ for C

(i) $A = \dfrac{\pi R^2 S}{360}$ $\qquad$ for S $\qquad\qquad$ (j) $P = \dfrac{t^2 dN}{3.78}$ $\qquad\qquad$ for d

(k) $C = \dfrac{AD}{A + 12}$ $\qquad$ for D $\qquad\qquad$ (l) $V = \dfrac{\pi L T^2}{6} + 2$ $\qquad$ for L

→ **Answers**

(a) $A = \dfrac{P - 3B}{2}$ $\qquad\qquad$ (b) $M = \dfrac{E}{C^2}$

(c) $A = S - SR + RT$ $\qquad$ (d) $g = \dfrac{2S}{t^2}$

(e) $R = \dfrac{P}{i^2}$

(f) $R = \dfrac{V - aI}{I}$

(g) $V = \dfrac{AR + W}{2}$

(h) $C = \dfrac{5F - 160}{9}$

(i) $S = \dfrac{360A}{\pi R^2}$

(j) $d = \dfrac{3.78P}{t^2 N}$

(k) $D = \dfrac{CA + 12C}{A}$

(l) $L = \dfrac{6V - 12}{\pi T^2}$

Note   The equations you learned to solve in this chapter are all *linear* equations. The variable appears only to the first power—no x^2 or x^3 terms appear in the equations. You will learn how to solve more difficult algebraic equations in Chapter 11. ◄

Now turn to Exercises 7-5 for a set of practice problems on solving equations and formulas.

Exercises 7-5   Solving More Equations and Formulas

A. Solve the following equations.

1. $5(x - 3) = 30$

2. $22 = -2(y + 6)$

3. $3(2n + 4) = 41$

4. $-6(3a - 7) = 21$

5. $2 - (x - 5) = 14$

6. $24 = 5 - (3 - 2m)$

7. $6 + 2(y - 4) = 13$

8. $7 - 11(2z + 3) = 18$

9. $8 = 5 - 3(3x - 4)$

10. $7 + 9(2w + 3) = 25$

11. $6c - (c - 4) = 29$

12. $9 = 4y - (y - 2)$

13. $5x - 3(2x - 8) = 31$

14. $6a + 2(a + 7) = 8$

15. $9t - 3 = 4t - 2$

16. $7y + 5 = 3y + 11$

17. $12x = 4x - 16$

18. $22n = 16n - 18$

19. $8y - 25 = 13 - 11y$

20. $6 - 2p = 14 - 4p$

21. $9x = 30 - 6x$

22. $12 - y = y$

23. $2(3t - 4) = 10t + 7$

24. $5A = 4(2 - A)$

25. $2 - (3x - 20) = 4(x - 2)$

26. $2(2x - 5) = 6x - (5 - x)$

27. $8 + 3(6 - 5x) = 11 - 10x$

28. $2(x - 5) - 3(2x - 8) = 16 - 6(4x - 3)$

B. Solve the following formulas for the variable shown.

1. $S = LW$ for L

2. $A = \dfrac{1}{2}BH$ for B

3. $V = IR$ for I

4. $H = \dfrac{D - R}{2}$ for D

5. $S = \dfrac{W}{2}(A + T)$ for T

6. $V = \pi R^2 H$ for H

7. $P = 2A + 2B$ for B

8. $H = \dfrac{R}{2} + 0.05$ for R

9. $T = \dfrac{RP}{R + 2}$ for P

10. $I = \dfrac{E + V}{R}$ for V

C. Practical Problems

Problem 1

1. **Sheet Metal Trades**   The length of the arc of a sector of a circle is given by the formula

$$L = \frac{2\pi Ra}{360}$$

R is the radius of the circle and a is the central angle in degrees (see art at left).

(a) Solve for a.   (b) Solve for R.

(c) Find L when $R = 10$ in. and $a = 30°$. Use $\pi \approx 3.14$.

2. **Sheet Metal Trades**   The area of the sector shown in problem 1 is $A = \pi R^2 a/360$.

(a) Solve this formula for a.

(b) Find A if $R = 12$ in., $a = 45°$, $\pi \approx 3.14$. Round to two significant digits.

3. **Electronics**   When two resistors R_1 and R_2 are put in series with a battery giving V volts, the current through the resistors is

$$i = \frac{V}{R_1 + R_2}$$

(a) Solve for R_1.

(b) Find R_2 if $V = 100$ volts, $i = 0.4$ ampere, $R_1 = 200$ ohms.

4. **Machine Trades**   Machinists use a formula known as Pomeroy's formula to determine roughly the power required by a metal punch machine.

$$P \approx \frac{t^2 dN}{3.78}$$

where P = power needed, in horsepower
   t = thickness of the metal being punched
   d = diameter of the hole being punched
   N = number of holes to be punched at one time

(a) Solve this formula for N.

(b) Find the power needed to punch six 2-in.-diameter holes in a sheet $\frac{1}{8}$ in. thick. Round to one significant digit.

5. **Physics**   When a gas is kept at constant temperature and the pressure on it is changed, its volume changes in accord with the pressure–volume relationship known as Boyle's law:

$$\frac{V_1}{V_2} = \frac{P_2}{P_1}$$

where P_1 and V_1 are the beginning volume and pressure and P_2 and V_2 are the final volume and pressure.

(a) Solve for V_1.   (b) Solve for V_2.

(c) Solve for P_1.   (d) Solve for P_2.

(e) Find P_1 when $V_1 = 10$ cu ft, $V_2 = 25$ cu ft, and $P_2 = 120$ psi.

6. **Sports and Leisure**   The volume of a football is roughly $V = \frac{\pi L T^2}{6}$, where L is its length and T is its thickness.

   (a) Solve for L.

   (b) Solve for T^2.

7. **Allied Health**   Nurses use a formula known as Young's rule to determine the amount of medicine to give a child under 12 years of age when the adult dosage is known.

   $$C = \frac{AD}{A + 12}$$

   C is the child's dose; A is the age of the child in years; D is the adult dose.

   (a) Work backward and find the adult dose in terms of the child's dose. Solve for D.

   (b) Find D if $C = 0.05$ gram and $A = 7$.

8. **Carpentry**   The projection or width P of a protective overhang of a roof is determined by the height T of the window, the height H of the header above the window, and a factor F that depends on the latitude of the construction site.

   $$P = \frac{T + H}{F}$$ Solve this equation for the header height H.

9. **Electronics**   For a current transformer, $\dfrac{i_L}{i_S} = \dfrac{T_P}{T_S}$

   where i_L = line current
   i_S = secondary current
   T_P = number of turns of wire in the primary coil
   T_S = number of turns of wire in the secondary coil

   (a) Solve for i_L.

   (b) Solve for i_S.

   (c) Find i_L when $i_S = 1.5$ amperes, $T_P = 1500$, $T_S = 100$.

10. **Electronics**   The electrical power P dissipated in a circuit is equal to the product of the current I and the voltage V, where P is in watts, I is in amperes, and V is in volts.

    (a) Write an equation giving P in terms of I and V.

    (b) Solve for I.

    (c) Find V when $P = 15,750$ watts, $I = 42$ amperes.

11. **Roofing**   The formula

    $$L = U(R + H)$$

    is used to determine rafter length (L) of a roof where R is the run and H is the overhang. The quantity U represents unit line length. When the run and the overhang are in feet, the rafter length will be in inches. If a rafter 247 in. long is used on a roof with a run of 16.5 ft and $U = 13.0$, how long will the overhang be?

12. **Automotive Trades**   The compression ratio R of a cylinder in a diesel engine is given by

    $$R = \frac{S + C}{C}$$

where S is the swept volume of the cylinder and C is the clearance volume of the cylinder.

(a) Solve for S.

(b) Find the swept volume if the clearance volume is 6.46 in.³ and the compression ratio is 17.4 to 1. Round to three significant digits.

13. **Machine Trades**   Suppose that, on the average, 3% of the parts produced by a particular machine have proven to be defective. Then the formula

$$N - 0.03N = P$$

will give the number of parts N that must be produced in order to manufacture a total of P nondefective ones. How many parts should be produced by this machine in order to end up with 7500 nondefective ones?

14. **Life Skills**   The formula

$$A = P(1 + rt)$$

is used to find the total amount A of money in an account when an original amount or principal P is invested at a rate of simple interest r for t years. How long would it take $8000 to grow to $10,000 at 8% simple interest?

15. **Construction**   The formula

$$I = 0.000014L(T - t)$$

gives the expansion I of a particular highway of length L at a temperature of T degrees Fahrenheit. The variable t stands for the temperature at which the highway was built. If a 2-mile stretch of highway was built at an average temperature of 60°F, what is the maximum temperature it can withstand if expansion joints allow for 7.5 ft of expansion? (*Hint:* The units of L must be the same as the units of I.)

16. **Automotive Trades**   The 2007 Mercedes Benz S550 generates 382 hp at 6000 rpm. Use the following formula to solve for the torque created at this engine speed:

$$hp = \frac{\text{torque (in lb} \cdot \text{ft)} \times \text{engine speed (in rpm)}}{5252}$$

Round to the nearest whole number.

17. **Automotive Trades**   The total engine displacement D (in inches) is given by

$$D = \frac{\pi B^2 SN}{4}$$

where B is the bore, S is the stroke, and N is the number of cylinders. What is the stroke of a 6-cylinder engine with a bore of 3.50 in. and a total displacement of 216 in.³? Use $\pi \approx 3.14$.

18. **Wastewater Technology**   The pressure P (in pounds per square inch or psi) on a submerged body in wastewater is given by

$$P = 0.520D + 14.7$$

where D is the depth in feet. At what depth is the pressure 25.0 psi?

When you have completed these exercises, check your answers to the odd-numbered problems in the Appendix, then turn to Section 7-6 to learn about word problems.

Translating English to Algebra

Algebra is a very useful tool for solving real problems. But in order to use it you may find it necessary to translate simple English sentences and phrases into mathematical expressions or equations. In technical work especially, the formulas to be used are often given in the form of English sentences, and they must be rewritten as algebraic formulas before they can be used.

Example 1

Transportation The statement

Horsepower required to overcome vehicle air resistance is equal to the cube of the vehicle speed in miles per hour multiplied by the frontal area in square feet divided by 150,000

translates to the formula

$$hp = \frac{mph^3 \cdot area}{150,000}$$

or

$$P = \frac{v^3 A}{150,000} \qquad \text{in algebraic form, where } v \text{ is the vehicle speed.}$$

In the next few pages of this chapter we show you how to translate English statements into algebraic formulas. To begin, try the following problem.

→ Your Turn

Automotive Trades An automotive technician found the following statement in a manual:

The pitch diameter of a cam gear is twice the diameter of the crank gear.

Translate this sentence into an algebraic equation.

→ Answer

The equation is $P = 2C$ where P is the pitch diameter of the cam gear and C is the diameter of the crank gear.

You may use any letters you wish of course, but we have chosen letters that remind you of the quantities they represent: P for *pitch* and C for *crank*.

Notice that the phrase "twice C" means "two times C" and is written as $2C$ in algebra.

Certain words and phrases appear again and again in statements to be translated. They are signals alerting you to the mathematical operations to be used. Here is a handy list of the *signal words* and their mathematical translations.

English Term	**Math Translation**	**Example**
Equals	$=$	$A = B$
Is, is equal to, was, are, were		
The same as . . .		
What is left is . . .		
The result is . . .		
Gives, makes, leaves, having		
Plus, sum of	$+$	$A + B$
Increased by, more than		
Minus B, subtract B	$-$	$A - B$
Less B		
Decreased by B, take away B		
Reduced by B, diminished by B		
B less than A		
B subtracted from A		
Difference between A and B		
Times, multiply, of	$\times$	AB
Multiplied		
Product of		
Divide, divided by B	$\div$	$A \div B$ or $\dfrac{A}{B}$
Quotient of		
Twice, twice as much	$\times 2$	$2A$
Double		
Squared, square of A		A^2
Cubed, cube of A		A^3

→ **Your Turn**

Translate the phrase "length plus 3 inches" into an algebraic expression. Try it, then check your answer with ours.

→ **Solution**

First, make a word equation by using parentheses.

(length)   (plus)   (3 inches)

Second, substitute mathematical symbols.

(length)   (plus)   (3 inches)

   ↓   ↓   ↓

   L $+$ 3   or $L + 3$

Notice that signal words, such as "plus," are translated directly into math symbols. Unknown quantities are represented by letters of the alphabet, chosen to remind you of their meaning.

→ **More Practice**

General Trades   Translate the following phrases into math expressions.

(a) Weight divided by 13.6 _____

(b) $6\frac{1}{4}$ in. more than the width _____

(c) One-half of the original torque _____

(d) The sum of the two lengths _____

(e) The voltage decreased by 5 _____

(f) Five times the gear reduction _____

(g) 8 in. less than twice the height _____

→ **Solutions**

(a)   (weight)   (divided by)   (13.6)

$\qquad$ W $\qquad$ $\div$ $\qquad$ 13.6 or $\dfrac{W}{13.6}$

(b) $(6\frac{1}{4}$ in.)   (more than)   (the width)

$\qquad$ $6\frac{1}{4}$ $\qquad$ $+$ $\qquad$ W $\qquad$ or $6\frac{1}{4} + W$

(c)   (one-half)   (of)   (the original torque)

$\qquad$ $\dfrac{1}{2}$ $\qquad$ $\times$ $\qquad$ T $\qquad$ or $\dfrac{1}{2} T$ or $\dfrac{T}{2}$

(d)   (the sum of)   (the two lengths)

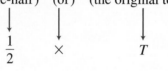

$\qquad\qquad L_1 + L_2$

(e)   (the voltage)   (decreased by)   (5)

$\qquad$ V $\qquad\qquad$ $-$ $\qquad$ 5 or $V - 5$

(f)   (five)   (times)   (the gear reduction)

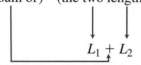

$\qquad$ 5 $\qquad$ $\times$ $\qquad$ G $\qquad\qquad$ or $5G$

(g)   8 less than twice the height means

$\qquad$ (twice the height)   (less)   (8)

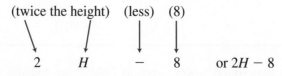

$\qquad\qquad$ 2 $\quad$ H $\qquad$ $-$ $\quad$ 8 $\qquad$ or $2H - 8$

Of course, any letters could be used in place of the ones used above.

The phrase "less than" has a meaning very different from the word "less."

"8 *less* 5" means $8 - 5$

"8 *less than* 5" means $5 - 8$ ◄

Translating Sentences to Equations

So far we have translated only phrases, pieces of sentences, but complete sentences can also be translated. An English phrase translates into an algebraic expression, and an English sentence translates into an algebraic formula or equation.

Example 2

HVAC The sentence

The height of the duct ⎵ is equal to ⎵ its width

translates to H $=$ W or $H = W$

Each word or phrase in the sentence becomes a mathematical term, letter, number, expression, or arithmetic operation sign.

→ **Your Turn**

Machine Trades Translate the following sentence into algebraic form as we did above.

The size of a drill for a tap is equal to the tap diameter minus the depth.

→ **Solution**

$S = T - D$

Follow these steps when you must translate an English sentence into an algebraic equation or formula.

Step 1 Cross out all unnecessary words.

~~The~~ size ~~of a drill for a tap~~ is equal to ~~the~~ tap diameter minus ~~the~~ depth.

Step 2 Make a word equation using parentheses.

(Size) (is equal to) (tap diameter) (minus) (depth)
↓   ↓   ↓   ↓   ↓

Step 3 Substitute a letter or an arithmetic symbol for each parentheses.

S $=$ T $-$ D

Step 4 Combine and simplify.

$S = T - D$

In most formulas the units for the quantities involved must be given. In the formula above, T and D are in inches.

Translating English sentences or verbal rules into algebraic formulas requires that you read the sentences very differently from the way you read stories or newspaper articles. Very few people are able to write out the math formula after reading the problem only once. You should expect to read it several times, and you'll want to read it slowly. No speed reading here! ◄

The ideas in technical work and formulas are usually concentrated in a few key words, and you must find them. If you find a word you do not recognize, stop reading and look it up in a dictionary, textbook, or manual. It may be important. Translating and working with formulas is one of the skills you must have if you are to succeed at any technical occupation.

Example 3

Electrical Trades   Here is another example of translating a verbal rule into an algebraic formula:

> The electrical resistance of a length of wire is equal to the resistivity of the metal times the length of the wire divided by the square of the wire diameter.

Step 1   Eliminate all but the key words.

"~~The electrical~~ resistance ~~of a length of wire~~ is equal to ~~the~~ resistivity ~~of the metal~~ times ~~the~~ length ~~of the wire~~ divided by ~~the~~ square of ~~the wire~~ diameter."

Step 2   Make a word equation.

(Resistance) (is equal to) (resistivity) (times) (length) (divided by) (square of diameter)

Step 3   Substitute letters and symbols.

$$R = \frac{rL}{D^2}$$

If the resistivity r has units of ohms times inches, L and D will be in inches.

→ More Practice

The more translations you do, the easier it gets. Translate each of the following technical statements into algebraic formulas.

(a) **Sheet Metal Trades**   A sheet metal worker measuring a duct cover finds that the width is $8\frac{1}{2}$ in. less than the height.

(b) **General Trades**   One quarter of a job takes $3\frac{1}{2}$ days.

(c) **Electrical Trades**   One half of a coil of wire weighs $16\frac{2}{3}$ lb.

(d) **Industrial Technology**   The volume of an elliptical tank is approximately 0.7854 times the product of its height, length, and width.

(e) **Automotive Trades**   The engine speed is equal to 168 times the overall gear reduction multiplied by the speed in miles per hour and divided by the rolling radius of the tire.

(f) **Transportation**   The air resistance force in pounds acting against a moving vehicle is equal to 0.0025 times the square of the speed in miles per hour times the frontal area of the vehicle.

(g) **Electrical Trades**   Two pieces of wire have a combined length of 24 in. The longer piece is five times the length of the shorter piece. (*Hint:* Write two separate equations.)

(h) **Automotive Trades** Two shims are to have a combined thickness of 0.090 in. The larger shim must be 3.5 times as thick as the smaller shim. (*Hint:* Write two separate equations.)

→ **Answers**

(a) $W = H - 8\frac{1}{2}$

(b) $\frac{1}{4}J = 3\frac{1}{2}$ or $\frac{J}{4} = 3\frac{1}{2}$

(c) $\frac{1}{2}C = 16\frac{2}{3}$ or $\frac{C}{2} = 16\frac{2}{3}$

(d) $V = 0.7854HLW$

(e) $S = \dfrac{168Gv}{R}$ where v is in mph.

(f) $R = 0.0025v^2A$ where v is in mph.

(g) $24 = L + S$ and $L = 5S$

(h) $S + L = 0.090$ and $L = 3.5S$

A Closer Look   Notice in problems (g) and (h) that two equations can be written. These can be combined to form a single equation.

(g) $24 = L + S$ and $L = 5S$ give $24 = 5S + S$

(h) $S + L = 0.090$ and $L = 3.5S$ give $S + 3.5S = 0.090$ ◀

General Word Problems   Now that you can translate English phrases and sentences into algebraic expressions and equations, you should be able to solve many practical word problems.

The following six-step method will help you to solve any word problems.

1. **Read** the problem carefully, noting what information is given and what is unknown.

2. **Assign a variable** to represent the unknown. If there are additional unknowns, try to represent these with an expression involving the same variable.

3. **Write an equation** using the unknown(s) and some given fact about the unknown(s) in the problem statement.

4. **Solve** the equation.

5. **Answer** the original question, being careful to state the value of all unknown quantities asked for in the problem.

6. **Check** your solution.

Example 4

Machine Trades   Consider this problem:

A machinist needs to use two shims with a combined thickness of 0.084 in. One shim is to be three times as thick as the other. What are the thicknesses of the two shims?

Read the problem carefully, then note that there are two unknowns—a thinner shim and a thicker shim. We know their total thickness and we know that one is three times thicker than the other. We use this last fact to assign variable expressions to the unknowns.

Let x = thickness of the thinner shim

then $3x$ = thickness of the thicker shim

Use their combined thickness to write an equation. $3x + x = 0.084$ in.

To solve the equation, first combine terms. $4x = 0.084$ in.

Then divide each side by 4. $x = 0.021$ in.

$$3x = 3(0.021 \text{ in.}) = 0.063 \text{ in.}$$

Notice that two answers are required here.

The thin shim is 0.021 in. and the thicker one is 0.063 in.

0.021 in. $+ 3(0.021$ in.$) = 0.084$ in.
  0.021 in. $+ 0.063$ in. $= 0.084$ in.   which is correct

Example 5

Life Skills   A family prepares a budget that allows for $220 per month to be spent on gas. Suppose that gas costs $2.50 per gallon and their vehicles have an average gas mileage of 25 mi/gal. The family wishes to determine the maximum number of miles, m, they can drive and stay within this budget. We can solve this problem in two stages.

First, set up and solve an equation to determine the number of gallons, g, they can purchase.

$$\$2.50g = \$220$$
$$g = 88 \text{ gal}$$

Then, multiply this result by their gas mileage.

Notice that the gallons unit cancels out.

$$m = (88 \text{ gal})(25 \text{ mi/gal})$$
$$m = \frac{88 \text{ gal}}{1 \text{ month}} \times \frac{25 \text{ mi}}{1 \text{ gal}}$$

Finally, this leaves us with the maximum number of miles they can drive in a month.

$$m = \frac{2200 \text{ mi}}{1 \text{ month}}$$
or 2200 miles per month

→ Your Turn

Use your knowledge of algebra to translate the following problem to an algebraic equation and then solve it.

Life Skills   Bob's small appliance repair service charges a fixed cost of $65 plus $50 per hour for the total repair time. Frank's small appliance repair service charges a simple rate of $70 per hour.

(a)   After how many hours would the total repair bill be the same for Bob and Frank?

(b)   Who would be less expensive for a repair job lasting 2 hours?

→ Solution

(a)   There is one unknown:

Let x = the repair time for which the two shops would charge the same amount

Use the fact that the two amounts would be the same to write an equation:

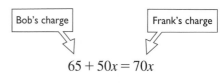

$$65 + 50x = 70x$$

To solve, subtract $50x$ from both sides.
$$65 = 20x$$

Then divide each side by 20.
$$3.25 = x$$

Bob and Frank would charge the same amount for a repair job involving 3.25 hours of work.

(b) For a repair job of 2 hours:

Bob's charge is $\$65 + \$50(2) = \$165$
Frank's charge is $\$70(2) = \$140$

Frank would be less expensive. (This would be the case for any job involving up to 3.25 hours, after which Bob would be less expensive.)

Some of the more difficult percent problems can be simplified using algebraic equations. One such problem, backing the tax out of a total, is commonly encountered by all those who own their own businesses.

Example 6

Trades Management   Suppose that an auto mechanic has collected \$1468.63 for parts, including 6% tax. For accounting purposes he must determine exactly how much of the total is sales tax. If we let x stand for the dollar amount of the parts before tax was added, then the tax is 6% of that or $0.06x$, and we have

$$x \qquad + \qquad 0.06x \qquad = \qquad \$1468.63$$

| Cost of the parts | 6% Tax on the parts | Total amount |

To solve the equation $\qquad\qquad\qquad x + 0.06x = \$1468.63$

First, note that $x = 1x = 1.00x \qquad 1.00x + 0.06x = \$1468.63$

Next, combine like terms $\qquad\qquad\qquad 1.06x = \$1468.63$

Now, divide by 1.06 to get $\qquad\qquad\qquad x = \$1385.50$

x is the cost of the parts. To calculate the tax,

multiply x by 0.06 $\qquad\Longrightarrow\qquad 0.06x = 0.06(\$1385.50) = \$83.13$

or subtract x from \$1468.63 $\qquad\Longrightarrow\qquad \$1468.63 - \$1385.50 = \$83.13$

Example 7

Automotive Trades   When fuel costs increase, more people consider the purchase of a hybrid car, and they need to make informed decisions as to whether such a purchase will save them money in the long run. Consider a vehicle that is available in a hybrid and a nonhybrid model. The following table shows us several pertinent facts about the two models:

Vehicle	Purchase price	Combined mileage rating	Tax credit
Hybrid	$29,500	28 mi/gal	$2200
Nonhybrid	$25,000	20 mi/gal	none

Let's assume that the average price of gas will be $1.95 per gallon. We wish to determine how many miles of driving it will take before the fuel savings of the hybrid will make up for its greater initial cost.

First, determine the difference in their initial costs. The tax credit on the hybrid reduces its initial cost to $29,500 − $2200, or $27,300, so the actual difference in their initial costs is $27,300 − $25,000, or $2300.

Next, determine the cost per mile for each of the two vehicles. To do so, divide cost per gallon by miles per gallon. Note that:

$$\frac{\$}{gal} \div \frac{mi}{gal} = \frac{\$}{\cancel{gal}} \times \frac{\cancel{gal}}{mi} = \frac{\$}{mi}$$

In our example,

Cost per mile of the hybrid: $1.95 per gallon ÷ 28 mi/gal ≈ $0.0696 per mile

Cost per mile of the nonhybrid: $1.95 per gallon ÷ 20 mi/gal = $0.0975 per mile

Next, set up an equation. Let x = the number of miles one must drive to make up for the difference in the initial cost: At what value of x will the difference between the cost of driving the nonhybrid ($0.0975x$) and the cost of driving the hybrid ($0.0696x$) be equal to the difference in the initial cost ($2300)?

$$0.0975x - 0.0696x = 2300$$

Finally, solve the equation:

Combine like terms. $0.0279x = 2300$

Divide by 0.0279 $x = 82,437.2 \ldots$

Therefore, it will take approximately 82,400 miles of driving before the fuel savings of the hybrid will make up for its greater initial cost.

Note While the method of the previous example gives us a fairly accurate picture of the "payback" time for a hybrid, there may be other cost differences involved, such as insurance or maintenance, that may also affect the cost comparison. ◀

→ **More Practice**

Practice makes perfect. Work the following set of word problems.

(a) **Trades Management**   Geeks R Us, a computer repair service, charges $90 for the first hour of labor and $75 for each additional hour. The total labor charge for a repair job was $540. Set up an equation to find the total repair time, then solve the equation.

(b) **Metalworking**   A 14-ft-long steel rod is cut into two pieces. The longer piece is $2\frac{1}{2}$ times the length of the shorter piece. Find the length of each piece. (Ignore waste.)

(c) **General Trades** A tech is paid $22 per hour or $33 per hour for overtime (weekly time in excess of 40 hours). How much total time must the tech work to earn $1078?

(d) **Machine Trades** Find the dimensions of a rectangular cover plate if its length is 6 in. longer than its width and if its perimeter is 68 in.
(*Hint:* Perimeter = 2 · length + 2 · width.)

(e) **Trades Management** Mike and Jeff are partners in a small manufacturing firm. Because Mike provided more of the initial capital for the business, they have agreed that Mike's share of the profits should be one-fourth greater than Jeff's. The total profit for the first quarter of this year was $17,550. How should they divide it?

(f) **Trades Management** A plumber collected $2784.84 during the day, $1650 for labor, which is not taxed, and the rest for parts, which includes 5% sales tax. Determine the total amount of tax that was collected.

(g) **Printing** A printer knows from past experience that about 2% of a particular run of posters will be spoiled. How many should she print in order to end up with 1500 usable posters?

(h) **Construction** A contractor needs to order some Buell Flat gravel for a driveway. Acme Building Materials sells it for $51 per cubic yard plus a $42 delivery charge. The Stone Yard sells it for $45 per cubic yard plus a $63 delivery charge. (1) For an order of 2 cu yd, which supplier is less expensive? (2) For an order of 5 cu yd, which is less expensive? (3) For what size order would both suppliers charge the same amount?

→ **Solutions**

(a) Let H = the total number of hours

then $H - 1$ = the number of additional hours charged at $75 per hour

$$90 + 75(H - 1) = 540$$

The cost of the first hour

The cost of $H - 1$ additional hours

$$90 + 75H - 75 = 540$$
$$75H + 15 = 540$$
$$75H = 525$$
$$H = 7$$

The total repair time was 7 hours.

(b) Let x = length of the shorter piece

then $\left(2\frac{1}{2}\right)x$ = length of the longer piece

The equation is $\qquad x + \left(2\frac{1}{2}\right)x = 14$ ft

Add like terms. ($x = 1x$) $\qquad \left(3\frac{1}{2}\right)x = 14$ ft

or $\qquad\qquad\qquad\qquad\qquad \dfrac{7x}{2} = 14$

Multiply by 2. $\qquad\qquad\qquad\qquad 7x = 28$

Divide by 7. $\qquad\qquad\qquad\qquad\quad x = 4$ ft

Longer piece $= 2\frac{1}{2}x = 2\frac{1}{2}(4) = 10$ ft.

The shorter piece is 4 ft long and the longer piece is 10 ft long.

(c) Let t = total time worked
then $t - 40$ = hours of overtime worked

For the first 40 hours, the tech's pay is

$22(40) = \$880$

For $t - 40$ overtime hours, the tech's pay is

$33(t - 40)$

Therefore the equation for total pay is

$\$880 + \$33(t - 40) = \$1078$

Solving for t:

$$880 + 33t - 1320 = 1078$$
$$33t - 440 = 1078$$
$$33t = 1518$$
$$t = 46$$

The tech must work 46 total hours to earn \$1078.

(d) Let width = W
then length = $W + 6$
and perimeter = 68 in. so that $\qquad$ $68 = 2W + 2(W + 6)$

Remove parentheses, $\qquad\qquad\qquad$ $68 = 2W + 2W + 12$
$2(W + 6) = 2W + 12.$
Combine terms. $\qquad\qquad$ $4W + 12 = 68$
Subtract 12 from each side. $\qquad$ $4W = 56$
Divide by 4. $\qquad\qquad\qquad\qquad$ $W = 14$ in.
$\qquad\qquad\qquad\qquad$ length = $W + 6 = 20$ in.

(e) Let Jeff's share = J
then Mike's share = $J + \frac{1}{4}J$

The equation is $\qquad\qquad\qquad$ $\$17{,}550 = J + \left(J + \frac{1}{4}J\right)$

Combine terms; remember $J = 1 \cdot J$. $\qquad$ $\$17{,}550 = \left(2\frac{1}{4}\right)J$

Write the mixed number $\qquad\qquad$ $\$17{,}550 = \frac{9}{4}J = \frac{9J}{4}$
as an improper fraction.

or $\qquad\qquad\qquad\qquad\qquad\qquad$ $\frac{9J}{4} = 17{,}550$

Multiply by 4. $\qquad\qquad\qquad\qquad$ $9J = 70{,}200$

Divide by 9. $\qquad\qquad\qquad\qquad$ $J = \$7800$ $\quad$ Jeff's share

Mike's share = $\$7800 + \dfrac{\$7800}{4}$

$\qquad\qquad\qquad = \$7800 + \$1950$

$\qquad\qquad\qquad = \$9750$

(f) Let x = cost of the parts
then $0.05x$ = amount of the tax

Total cost with tax

Subtract out the labor to
get the total cost of the parts.

The equation is $\qquad\quad$ $x + 0.05x = \$2784.84 - \$1650$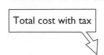

Combine terms. $1.05x = \$1134.84$ (Remember: $x = 1x = 1.00x$)

Divide by 1.05. $x = \$1080.80$ This is the cost of the parts.

The amount of the tax is $0.05(\$1080.80) = \$54.04$.

(g) Let x = number of posters she needs to print
then $0.02x$ = number that will be spoiled

$$x \qquad - \qquad 0.02x \qquad = \qquad 1500$$

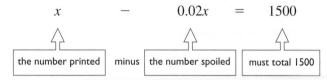

the number printed | minus | the number spoiled | must total 1500

We can rewrite this equation as $1.00x - 0.02x = 1500$

Now subtract like terms. $0.98x = 1500$

Divide by 0.98. $x = 1531$ rounded

She must print approximately 1531 posters to end up with 1500 unspoiled ones.

(h) (1) For 2 cu yd
Acme price: $\$51(2) + \$42 = \$144$
Stone Yard price: $\$45(2) + \$63 = \$153$
Acme is less expensive.

(2) For 5 cu yd
Acme price: $\$51(5) + \$42 = \$297$
Stone Yard price: $\$45(5) + \$63 = \$288$
Stone Yard price is less expensive.

(3) Let x = the amount of gravel for which each supplier's price
will be the same.

Then if Acme's price = Stone Yard's price
$$\$51x + \$42 = \$45x + \$63$$

Subtract $45x$. $\qquad 6x + 42 = 63$
Subtract 42. $\qquad\qquad 6x = 21$
Divide by 6. $\qquad\qquad x = 3.5$ cu yd

The total cost for each supplier is the same for an order of 3.5 cu yd.

Now turn to Exercises 7-6 for a set of word problems.

Exercises 7-6 Solving Word Problems

A. Translate the following into algebraic equations:

1. **Wastewater Technology** The height of the tank is 1.4 times its width.

2. **General Trades** The sum of the two weights is 167 lb.

3. **General Trades** The volume of a cylinder is equal to $\frac{1}{4}$ of its height times π times the square of its diameter.

4. **General Trades** The weight in kilograms is equal to 0.454 times the weight in pounds.

5. **Machine Trades** The volume of a solid bar is equal to the product of the cross-sectional area and the length of the bar.

6. **General Trades** The volume of a cone is $\frac{1}{3}$ times π times the height times the square of the radius of the base.

7. **Machine Trades** The pitch diameter D of a spur gear is equal to the number of teeth on the gear divided by the pitch.

8. **Machine Trades** The cutting time for a lathe operation is equal to the length of the cut divided by the product of the tool feed rate and the revolution rate of the workpiece.

9. **Machine Trades** The weight of a metal cylinder is approximately equal to 0.785 times the height of the cylinder times the density of the metal times the square of the diameter of the cylinder.

10. **Electrical Trades** The voltage across a simple circuit is equal to the product of the resistance of the circuit and the current flowing in the circuit.

B. Set up equations and solve.

1. **Wastewater Technology** Two tanks must have a total capacity of 400 gallons. If one tank needs to be twice the size of the other, how many gallons should each tank hold?

2. **Machine Trades** Two metal castings weigh a total of 84 lb. One weighs 12 lb more than the other. How much does each one weigh?

3. **Plumbing** The plumber wants to cut a 24-ft length of pipe into three sections. The largest piece should be twice the length of the middle piece, and the middle piece should be twice the length of the smallest piece. How long should each piece be?

4. **Electrical Trades** The sum of three voltages in a circuit is 38 volts. The middle-sized one is 2 volts more than the smallest. The largest is 6 volts more than the smallest. What is the value of each voltage?

5. **Masonry** There are 156 concrete blocks available to make a retaining wall. The bottom three rows will all have the same number of blocks. The next six rows will each have two blocks fewer than the row below it. How many blocks are in each row?

6. **Construction** A concrete mix is in volume proportions of 1 part cement, 2 parts water, 2 parts aggregate, and 3 parts sand. How many cubic feet of each ingredient are needed to make 54 cu ft of concrete?

7. **Trades Management** Jo and Ellen are partners in a painting business. Because Jo is the office manager, she is to receive one-third more than Ellen when the profits are distributed. If their profit is $85,400, how much should each of them receive?

8. **Construction** A total of $1,120,000 is budgeted for constructing a roadway. The rule of thumb for this type of project is that the pavement costs twice the amount of the base material, and the sidewalk costs one-fourth the amount of the pavement. Using these figures, how much should each item cost?

9. **Photography** A photographer has enough liquid toner for fifty 5-in. by 7-in. prints. How many 8-in. by 10-in. prints will this amount cover? (*Hint:* The toner covers the *area* of the prints: Area = length × width.)

10. **Construction** A building foundation has a length of 82 ft and a perimeter of 292 ft. What is the width? (*Hint:* Perimeter = 2 × length + 2 × width.)

11. **General Trades** A roofer earns $24 per hour plus $36 per hour overtime. If he puts in 40 hours of regular time during a certain week and he wishes to earn $1200, how many hours of overtime should he work?

12. **Agriculture**   An empty avocado crate weighs 4.2 kg. How many avocados weighing 0.3 kg each can be added before the total weight of the crate reaches 15.0 kg?

13. **Machine Trades**   One-tenth of the parts tooled by a machine are rejects. How many parts must be tooled to ensure 4500 acceptable ones?

14. **Office Services**   An auto mechanic has total receipts of $1861.16 for a certain day. He determines that $1240 of this is labor. The remaining amount is for parts plus 6% sales tax on the parts only. Find the amount of the sales tax he collected.

15. **Office Services**   A travel agent receives a 7% commission on the base fare of a customer—that is, the fare *before* sales tax is added. If the total fare charged to a customer comes to $753.84 *including* 5.5% sales tax, how much commission should the agent receive?

16. **Office Services**   A newly established carpenter wishes to mail out letters advertising her services to the families of a small town. She has a choice of mailing the letters first class or obtaining a bulk-rate permit and mailing them at the cheaper bulk rate. The bulk-rate permit and mailing fee cost $160 total and each piece of bulk mail then costs 23.4 cents. If the first-class rate is 39 cents, how many pieces would she have to mail in order to make bulk rate the cheaper way to go?

17. **Trades Management**   An appliance repair shop charges $75 for the first hour of labor and $60 for each additional hour. The total labor charge for an air conditioning system installation was $495. Write an equation that allows you to solve for the total repair time; then solve the equation.

18. **Trades Management**   Four partners in a plumbing business decide to sell the business. Based on their initial investments, Al will receive five shares of the selling price, Bob will receive three shares, and Cody and Dave will receive two shares each. If the business sells for $840,000, how much is each share worth?

19. **Automotive Trades**   An automotive technician has two job offers. Shop A would pay him a straight salary of $720 per week. Shop B would pay him $20 per hour to work flat rate, where his hours would vary according to how many jobs he completes. How many hours would he have to log during a week at Shop B in order to match the salary at Shop A?

20. **Culinary Arts**   Laurie runs a small cafeteria. She must set aside 25% of her revenue for taxes and spends another $1300 per week on other expenses. How much total revenue must she generate in order to clear a profit of $2000 per week?

21. **Life Skills**   Suppose you budgeted $1800 for fuel expenses for the year. How many miles could you drive if gas were $2.25 per gallon and your vehicle averaged 28 mi/gal? (See Example 5 on page 450.)

22. **Automotive Trades**   A certain vehicle has both a hybrid and a nonhybrid model. Their purchase prices, combined mileage ratings, and eligible tax credits are shown in the following table.

Vehicle	Purchase price	Combined mileage rating	Tax credit
Hybrid	$32,000	34 mi/gal	$500
Nonhybrid	$28,000	23 mi/gal	none

How many miles of driving will it take until the savings on gas makes up for the difference in cost between the two models if the average price of gas is $1.90 per gallon? (Round to the nearest hundred miles.) (See Example 7 on page 451.)

23. **Automotive Trades** In problem 22, suppose the average price of gas has increased to $2.90 per gallon by the time you purchase your vehicle.

   (a) How many miles would you have to drive in order to offset the difference in the initial cost between the two vehicles? (Round to the nearest hundred.)

   (b) Choose the correct word to fill in the underlined portion of the quote: "The higher the price of gas, the (more/fewer) miles it takes for the savings on gas to make up for the difference in the initial cost."

24. **Landscaping** A landscaper must order some topsoil for a job. The Dirt Yard sells it for $60 per cubic yard plus a $75 delivery charge. Mud City sells it for $68 per cubic yard plus a $55 delivery charge. Set up and solve an equation to determine the number of cubic yards at which both suppliers would charge the same total amount. (See problem (h) on page 453.)

Check your answers to the odd-numbered problems in the Appendix, then turn to Section 7-7 to learn more algebra.

7-7 Multiplying and Dividing Algebraic Expressions

Multiplying Simple Factors

Earlier we learned that only like algebraic terms, those with the same literal part, can be added and subtracted. However, any two terms, like or unlike, can be multiplied.

To multiply two terms such as $2x$ and $3xy$, first remember that $2x$ means 2 times x. Second, recall from arithmetic that the order in which you do multiplications does not make a difference. For example, in arithmetic

$$2 \cdot 3 \cdot 4 = (2 \cdot 4) \cdot 3 = (3 \cdot 4) \cdot 2$$

and in algebra

$$a \cdot b \cdot c = (a \cdot c) \cdot b = (c \cdot b) \cdot a \qquad \text{or} \qquad 2 \cdot x \cdot 3 \cdot x \cdot y = 2 \cdot 3 \cdot x \cdot x \cdot y$$

Remember that $x \cdot x = x^2, x \cdot x \cdot x = x^3$, and so on. Therefore,

$$2x \cdot 3xy = 2 \cdot 3 \cdot x \cdot x \cdot y = 6x^2y$$

The following examples show how to multiply two terms.

Example 1

(a) $a \cdot 2a = a \cdot 2 \cdot a$

$= 2 \cdot \underbrace{a \cdot a}$ Group like factors together.

$= 2 \cdot a^2$

$= 2a^2$

(b) $2x^2 \cdot 3xy = 2 \cdot x \cdot x \cdot 3 \cdot x \cdot y$

$= \underbrace{2 \cdot 3} \cdot \underbrace{x \cdot x \cdot x} \cdot y$ Group like factors together.

$= 6 \cdot x^3 \cdot y$

$= 6x^3y$

(c) $\quad 3x^2yz \cdot 2xy = 3 \cdot x^2 \cdot y \cdot z \cdot 2 \cdot x \cdot y$

$$= \underbrace{3 \cdot 2} \cdot \underbrace{x^2 \cdot x} \cdot \underbrace{y \cdot y} \cdot z$$

$$= 6 \cdot x^3 \cdot y^2 \cdot z$$

$$= 6x^3y^2z$$

Remember to group like factors together before multiplying.

If you need to review exponents, return to page 368.

→ **Your Turn**

Now try the following problems.

(a) $\quad x \cdot y \qquad =$ _____   (b) $\quad 2x \cdot 3x \qquad =$ _____

(c) $\quad 2x \cdot 5xy \qquad =$ _____   (d) $\quad 4a^2b \cdot 2a \qquad =$ _____

(e) $\quad 3x^2y \cdot 4xy^2 \qquad =$ _____   (f) $\quad 5xyz \cdot 2ax^2 \qquad =$ _____

(g) $\quad 3x \cdot 2y^2 \cdot 2x^2y =$ _____   (h) $\quad x^2y^2 \cdot 2x \cdot y^2 \qquad =$ _____

(i) $\quad 2x^2(x + 3x^2) \qquad =$ _____   (j) $\quad -2a^2b(a^2 - 3b^2) =$ _____

→ **Answers**

(a) $\quad xy$ (b) $\quad 6x^2$

(c) $\quad 10x^2y$ (d) $\quad 8a^3b$

(e) $\quad 12x^3y^3$ (f) $\quad 10ax^3yz$

(g) $\quad 12x^3y^3$ (h) $\quad 2x^3y^4$

(i) $\quad 2x^3 + 6x^4$ (j) $\quad -2a^4b + 6a^2b^3$

▶ **A Closer Look**   Were the last two problems tricky for you? Recall from Rule 3 on page 407 that when there is a multiplier in front of parentheses, every term inside must be multiplied by this factor. Try it this way:

(i) $\quad 2x^2(x + 3x^2) = (2x^2)(x) + (2x^2)(3x^2)$

$$= 2 \cdot \underbrace{x^2 \cdot x} + \underbrace{2 \cdot 3} \cdot \underbrace{x^2 \cdot x^2}$$

$$= 2 \cdot x^3 + 6 \cdot x^4$$

$$= 2x^3 + 6x^4$$

(j) $\quad -2a^2b(a^2 - 3b^2) = (-2a^2b)(a^2) + (-2a^2b)(-3b^2)$

$$= (-2) \cdot \underbrace{a^2 \cdot a^2} \cdot b + \underbrace{(-2) \cdot (-3)} \cdot a^2 \cdot \underbrace{b \cdot b^2}$$

$$= -2 \cdot a^4 \cdot b + 6 \cdot a^2 \cdot b^3$$

$$= -2a^4b + 6a^2b^3 \quad ◀$$

Rule for Multiplication

As you did the preceding problems, you may have noticed that when you multiply like factors together, you actually add their exponents. For example,

$$x^2 \cdot x^3 = x \cdot x \cdot x \cdot x \cdot x = x^5$$

In general, we can state the following rule:

Rule 1   To multiply numbers written in exponential form having the same base, add the exponents.

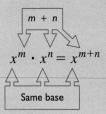

$$x^m \cdot x^n = x^{m+n}$$

Same base

Example 2

(a) $x^2 \cdot x^3 = x^{2+3} = x^5$

(b) $3^5 \times 3^2 = 3^{5+2} = 3^7$

(c) $2a^6 \cdot 8a \cdot a^5 = 2 \cdot 8 \cdot a^6 \cdot a^1 \cdot a^5 = 16a^{6+1+5} = 16a^{12}$

Careful

1. Notice in part (b) that the like bases remain unchanged when you multiply—that is, $3^5 \times 3^2 = 3^7$, not 9^7.

2. Exponents are added during multiplication, but numerical factors are multiplied as usual. For example,

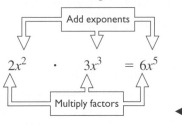

Add exponents

$$2x^2 \quad \cdot \quad 3x^3 \quad = 6x^5$$

Multiply factors ◀

→ Your Turn

Use Rule 1 to simplify.

(a) $m^4 \cdot m^3$

(b) $5^3 \cdot 5^6$

(c) $4y^4 \cdot 6y \cdot y^2$

→ Solutions

(a) $m^4 \cdot m^3 = m^{4+3} = m^7$

(b) $5^3 \cdot 5^6 = 5^{3+6} = 5^9$

(c) $4y^4 \cdot 6y \cdot y^2 = 4 \cdot 6 \cdot y^4 \cdot y^1 \cdot y^2 = 24y^{4+1+2} = 24y^7$

Dividing Simple Factors

To divide a^5 by a^2, think of it this way:

$$\frac{a^5}{a^2} = \frac{\cancel{a} \cdot \cancel{a} \cdot a \cdot a \cdot a}{\cancel{a} \cdot \cancel{a}} = a^3$$

Cancel common factors

Notice that the final exponent, 3, is the difference between the original two exponents, 5 and 2. Remember that multiplication and division are reverse operations. It makes sense that if you add exponents when multiplying, you would subtract exponents when dividing. We can now state the following rule:

> **Rule 2** To divide numbers written in exponential form having the same base, subtract the exponents.
>
> $$\frac{x^m}{x^n} = x^{m-n}$$

Example 3

(a) $\dfrac{x^7}{x^3} = x^{7-3} = x^4$

(b) $\dfrac{4^5}{4^2} = 4^{5-2} = 4^3$

(c) $\dfrac{6a^6}{2a^3} = \dfrac{6}{2} \cdot \dfrac{a^6}{a^3} = 3a^3$

Careful

1. As with multiplication, the like bases remain unchanged when dividing with exponents. This is especially important to keep in mind when the base is a number, as in part (b) of Example 3.

2. When dividing expressions that contain both exponential expressions and numerical factors, *subtract* the exponents but *divide* the numerical factors. For example,

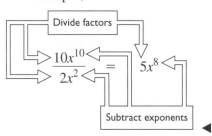

→ Your Turn

Try out the division rule on these problems:

(a) $\dfrac{x^6}{x^3}$ (b) $\dfrac{6a^4}{2a}$ (c) $\dfrac{-12m^3n^5}{4m^2n^2}$ (d) $\dfrac{5^8}{5^2}$ (e) $\dfrac{4x^5 - 8x^4 + 6x^3}{2x^2}$

→ Solutions

(a) $\dfrac{x^6}{x^3} = x^{6-3} = x^3$

(b) $\dfrac{6a^4}{2a} = \dfrac{6}{2} \cdot \dfrac{a^4}{a^1} = 3a^3$

(c) $\dfrac{-12m^3n^5}{4m^2n^2} = \dfrac{-12}{4} \cdot \dfrac{m^3}{m^2} \cdot \dfrac{n^5}{n^2} = -3mn^3$

(d) $\dfrac{5^8}{5^2} = 5^{8-2} = 5^6$

(e) $\dfrac{4x^5 - 8x^4 + 6x^3}{2x^2} = \dfrac{4x^5}{2x^2} - \dfrac{8x^4}{2x^2} + \dfrac{6x^3}{2x^2} = 2x^3 - 4x^2 + 3x$

▶ A Closer Look   Did you have trouble with (e)? Unlike the other problems, which have only one term in the numerator (dividend), problem (e) has *three* terms in the numerator. Just imagine that the entire numerator is enclosed in parentheses and divide each of the three terms separately by the denominator (divisor). ◀

▶ Note   The rule for division confirms the fact stated in Section 6-4 that $a^0 = 1$. We know from arithmetic that any quantity divided by itself is equal to 1, so

$$\frac{a^n}{a^n} = 1 \qquad \text{(for } a \neq 0\text{)}$$

But according to the division rule,

$$\frac{a^n}{a^n} = a^{n-n} = a^0 \qquad \text{Therefore, } a^0 \text{ must be equal to 1. ◀}$$

Negative Exponents

If we apply the division rule to a problem in which the divisor has a larger exponent than the dividend, the answer will contain a negative exponent. For example,

$$\frac{x^4}{x^6} = x^{4-6} = x^{-2}$$

If we use the "cancellation method" to do this problem, we have

$$\frac{x^4}{x^6} = \frac{\cancel{x} \cdot \cancel{x} \cdot \cancel{x} \cdot \cancel{x}}{\cancel{x} \cdot \cancel{x} \cdot \cancel{x} \cdot \cancel{x} \cdot x \cdot x} = \frac{1}{x^2} \qquad \text{Therefore, } \quad x^{-2} = \frac{1}{x^2}.$$

In general, we define negative exponents as follows:

$$x^{-n} = \frac{1}{x^n}$$

Example 4

(a) To divide $-\dfrac{12a}{4a^5}$

Think of this as the product of two separate quotients:

$$-\frac{12a}{4a^5} = -\frac{12}{4} \cdot \frac{a^1}{a^5} \quad \Longleftarrow \boxed{a = a^1}$$

First, divide the numerical coefficients:

$$= -3 \cdot \frac{a^1}{a^5}$$

Second, divide the a's by subtracting exponents:

$$= -3 \cdot a^{-4}$$

Next, apply the definition of negative exponents to eliminate them:

$$= \frac{-3}{1} \cdot \frac{1}{a^{+4}} \quad \Longleftarrow \boxed{a^{-4} = \frac{1}{a^4}}$$

Finally, Combine the results into a single fraction:

$$= -\frac{3}{a^4}$$

(b) To divide $\dfrac{2m^6n^2}{8m^3n^3}$

Think of this as the product of three separate quotients:

$$\frac{2m^6n^2}{8m^3n^3} = \frac{2}{8} \cdot \frac{m^6}{m^3} \cdot \frac{n^2}{n^3}$$

Simplify each quotient separately:

$$= \frac{1}{4} \cdot m^3 \cdot n^{-1}$$

Eliminate negative exponents:

$$= \frac{1}{4} \cdot m^3 \cdot \frac{1}{n} \quad \Longleftarrow \quad \boxed{\begin{array}{c} n^{-1} = \frac{1}{n^{+1}} \\ = \frac{1}{n} \end{array}}$$

Combine the three factors into a single fraction:

$$= \frac{1}{4} \cdot \frac{m^3}{1} \cdot \frac{1}{n} = \frac{m^3}{4n}$$

→ **Your Turn**

Practice using the definition of negative exponents by giving two answers for each of the following problems—one with negative exponents, and one using fractions to eliminate the negative exponents.

(a) $\dfrac{a^2}{a^6}$ (b) $\dfrac{8x^2}{-2x^3}$ (c) $\dfrac{6^3}{6^5}$ (d) $\dfrac{5x^4y^2}{15xy^7}$ (e) $\dfrac{12x^6 - 9x^3 + 6x}{3x^2}$

→ **Solutions**

(a) $\dfrac{a^2}{a^6} = a^{2-6} = a^{-4} = \dfrac{1}{a^4}$ (b) $\dfrac{8x^2}{-2x^3} = \dfrac{8}{-2} \cdot \dfrac{x^2}{x^3} = -4x^{-1} = \dfrac{-4}{x}$

(c) $\dfrac{6^3}{6^5} = 6^{3-5} = 6^{-2} = \dfrac{1}{6^2} = \dfrac{1}{36}$

(d) $\dfrac{5x^4y^2}{15xy^7} = \dfrac{5}{15} \cdot \dfrac{x^4}{x} \cdot \dfrac{y^2}{y^7} = \dfrac{1}{3}x^3y^{-5} = \dfrac{1}{3} \cdot \dfrac{x^3}{1} \cdot \dfrac{1}{y^5} = \dfrac{x^3}{3y^5}$

(e) $\dfrac{12x^6 - 9x^3 + 6x}{3x^2} = \dfrac{12x^6}{3x^2} - \dfrac{9x^3}{3x^2} + \dfrac{6x}{3x^2} = 4x^4 - 3x + 2x^{-1} = 4x^4 - 3x + \dfrac{2}{x}$

Now turn to Exercises 7-7 for a set of problems on multiplication and division of algebraic expressions.

Exercises 7-7 **Multiplying and Dividing Algebraic Expressions**

A. Simplify by multiplying.

1. $5 \cdot 4x$ 2. $(a^2)(a^3)$

3. $-3R(-2R)$ 4. $3x \cdot 3x$

5. $(4x^2y)(-2xy^3)$

6. $2x \cdot 2x \cdot 2x$

7. $0.4a \cdot 1.5a$

8. $x \cdot x \cdot A \cdot x \cdot 2x \cdot A^2$

9. $\frac{1}{2}Q \cdot \frac{1}{2}Q \cdot \frac{1}{2}Q \cdot \frac{1}{2}Q$

10. $(pq^2)(\frac{1}{2}pq)(2.4p^2q)$

11. $(-2M)(3M^2)(-4M^3)$

12. $2x(3x - 1)$

13. $-2(1 - 2y)$

14. $ab(a^2 - b^2)$

15. $p^2 \cdot p^4$

16. $3 \cdot 2x^2$

17. $5x^3 \cdot x^4$

18. $4x \cdot 2x^2 \cdot x^3$

19. $2y^5 \cdot 5y^2$

20. $2a^2b \cdot 2ab^2 \cdot a$

21. $xy \cdot x^2 \cdot xy^3$

22. $x(x + 2)$

23. $p^2(p + 2p)$

24. $xy(x + y)$

25. $5^6 \cdot 5^4$

26. $3^2 \cdot 3^7$

27. $10^2 \cdot 10^{-5}$

28. $2^{-6} \cdot 2^9$

29. $-3x^2(2x^2 - 5x^3)$

30. $6x^3(4x^4 + 3x^2)$

31. $2ab^2(3a^2 - 5ab + 7b^2)$

32. $-5xy^3(4x^2 - 6xy + 8y^2)$

B. Divide as indicated. Express all answers using positive exponents.

1. $\dfrac{4^5}{4^3}$

2. $\dfrac{10^9}{10^3}$

3. $\dfrac{x^4}{x}$

4. $\dfrac{y^6}{y^2}$

5. $\dfrac{10^4}{10^6}$

6. $\dfrac{m^5}{m^{10}}$

7. $\dfrac{8a^8}{4a^4}$

8. $\dfrac{-15m^{12}}{5m^6}$

9. $\dfrac{16y^2}{-24y^3}$

10. $\dfrac{-20t}{-5t^3}$

11. $\dfrac{-6a^2b^4}{-2ab}$

12. $\dfrac{36x^4y^5}{-9x^2y}$

13. $\dfrac{48m^2n^3}{-16m^6n^4}$

14. $\dfrac{15c^6d}{20c^2d^3}$

15. $\dfrac{-12a^2b^2}{20a^2b^5}$

16. $\dfrac{100xy}{-10x^2y^4}$

17. $\dfrac{6x^4 - 8x^2}{2x^2}$

18. $\dfrac{9y^3 + 6y^2}{3y}$

19. $\dfrac{12a^7 - 6a^5 + 18a^3}{-6a^2}$

20. $\dfrac{-15m^5 + 10m^2 + 5m}{5m}$

When you have completed these exercises, check your answers to the odd-numbered problems in the Appendix, then turn to Exercises 7-8 to learn about scientific notation.

In technical work you will often deal with very small and very large numbers which may require a lot of space to write. For example, a certain computer function may take 5 nanoseconds (0.000000005 s), and in electricity, 1 kilowatt-hour (kWh) is equal to 3,600,000 joules (J) of work. Rather than take the time and space to write all the zeros, we use scientific notation to express such numbers.

Definition of Scientific Notation

In scientific notation, 5 nanoseconds would be written as 5×10^{-9} s, and 1 kWh would be 3.6×10^6 J. As you can see, a number written in **scientific notation** is a product of a number between 1 and 10 and a power of 10.

SCIENTIFIC NOTATION

A number is expressed in scientific notation when it is in the form

$P \times 10^k$

where P is a number less than 10 and greater than or equal to 1, and k is an integer.

Numbers that are powers of ten are easy to write in scientific notation:

$$1000 = 1 \times 10^3$$
$$100 = 1 \times 10^2$$
$$10 = 1 \times 10^1 \quad \text{and so on}$$

In general, to write a positive decimal number in scientific notation, first write it as a number between 1 and 10 times a power of 10, then write the power using an exponent.

Example 1

$$3,600,000 = 3.6 \times 1,000,000$$

A multiple of 10

A number between 1 and 10

$$= 3.6 \times 10^6$$

$$0.0004 = 4 \times 0.0001 = 4 \times \frac{1}{10,000} = 4 \times \frac{1}{10^4} = 4 \times 10^{-4}$$

Note If you need to review negative exponents, see page 462. ◀

Converting to Scientific Notation

For a shorthand way of converting to scientific notation, use the following four-step procedure illustrated by Example 2 (a) and (b):

Example 2

Steps for Converting to Scientific Notation	**(a)**	**(b)**
Step 1 If the number is given without a decimal point, rewrite it with a decimal point.	57,400 = 57,400. Decimal point	0.0038 Decimal point already shown
Step 2 Place a mark ∧ after the first nonzero digit.	5 ∧ 7400.	0.003 ∧ 8
Step 3 Count the number of digits from the mark to the decimal point.	5 ∧ 7400. 4 digits	0.003 ∧ 8 3 digits
Step 4 Place the decimal point in the marked position. Use the resulting number as the multiplier P. Use the number of digits from the mark ∧ to the original decimal point as the exponent. If the shift is to the right, the exponent is positive; if the shift is to the left, the exponent is negative.	5.74×10^4 Exponent is +4 because decimal point shifts 4 digits *right.*	3.8×10^{-3} Exponent is −3 because decimal point shifts 3 digits *left.*

Here are several more examples.

Example 3

(a) $150,000 = 1 \wedge 50000. = 1.5 \times 10^5$
Shift 5 digits right. Exponent is +5.

(b) $0.00000205 = 0.000002 \wedge 05 = 2.05 \times 10^{-6}$
Shift 6 digits left. Exponent is −6.

(c) $47 = 4 \wedge 7. = 4.7 \times 10^1$
Shift 1 digit right.

→ **Your Turn**

For practice, write the following numbers in scientific notation.

(a) 2900 (b) 0.006 (c) 1,960,100

(d) 0.0000028 (e) 600 (f) 0.0001005

→ **Solutions**

(a) $2,900 = 2 \wedge 900. = 2.9 \times 10^3$
Shift 3 digits right.

(b) $0.006 = 0.006 \wedge = 6 \times 10^{-3}$
Shift 3 digits left.

(c) $1,960,100 = 1 \wedge 960100. = 1.9601 \times 10^6$
Shift 6 digits right.

(d) $0.0000028 = 0.000002 \wedge 8 = 2.8 \times 10^{-6}$

Shift 6 digits left.

(e) $600 = 6 \wedge 00. = 6 \times 10^{2}$

Shift 2 digits right.

(f) $0.0001005 = 0.0001 \wedge 005 = 1.005 \times 10^{-4}$

Shift 4 digits left.

Learning Help   When a positive number greater than or equal to 10 is written in scientific notation, the exponent is positive. When a positive number less than 1 is written in scientific notation, the exponent is negative. When a number between 1 and 10 is written in scientific notation, the exponent is 0. ◄

Converting from Scientific Notation to Decimal Form   To convert a number from scientific notation to decimal form, shift the decimal point as indicated by the power of 10—to the right for a positive exponent, to the left for a negative exponent. Attach additional zeros as needed.

Example 4

(a)   Attach additional zeros.

$$3.2 \times 10^{4} = 3 \, 2000. \;\; = \;\; 32{,}000$$

Shift the decimal point 4 places right.

For a positive exponent, shift the decimal point to the right.

(b)   Attach additional zeros.

$$2.7 \times 10^{-3} = 0.002 \, 7 = 0.0027$$

Shift the decimal point 3 places left.

For a negative exponent, shift the decimal point to the left.

→ Your Turn

Write each of the following numbers in decimal form.

(a) 8.2×10^{3} (b) 1.25×10^{-6}

(c) 2×10^{-4} (d) 5.301×10^{5}

→ Solutions

(a) $8.2 \times 10^{3} = 8 \, 200. = 8200$

(b) $1.25 \times 10^{-6} = 0.000001 \, 25 = 0.00000125$

(c) $2 \times 10^{-4} = 0.0002 = 0.0002$

(d) $5.301 \times 10^5 = 5\ 30100. = 530,100$

Multiplying and Dividing in Scientific Notation

Scientific notation is especially useful when we must multiply or divide very large or very small numbers. Although most calculators have a means of converting decimal numbers to scientific notation and can perform arithmetic with scientific notation, it is important that you be able to do simple calculations of this kind quickly and accurately without a calculator.

To multiply or divide numbers given in exponential form, use the rules given in Section 7-7. (See page 460.)

$$a^m \times a^n = a^{m+n}$$
$$a^m \div a^n = \frac{a^m}{a^n} = a^{m-n}$$

To multiply numbers written in scientific notation, work with the decimal and exponential parts separately.

$$(A \times 10^B) \times (C \times 10^D) = (A \times C) \times 10^{B+D}$$

Example 5

$26,000 \times 3,500,000 = ?$

Step 1 Rewrite each number in scientific notation.

$$= (2.6 \times 10^4) \times (3.5 \times 10^6)$$

Step 2 Regroup to work with the decimal and exponential parts separately.

$$= (2.6 \times 3.5) \times (10^4 \times 10^6)$$

Step 3 Multiply using the rule for multiplying exponential numbers.

$$= 9.1 \times 10^{4+6}$$
$$= 9.1 \times 10^{10}$$

When dividing numbers written in scientific notation, it may help to think of the division as a fraction.

$$(A \times 10^B) \div (C \times 10^D) = \frac{A \times 10^B}{C \times 10^D} = \frac{A}{C} \times \frac{10^B}{10^D}$$

Therefore,

$$(A \times 10^B) \div (C \times 10^D) = (A \div C) \times 10^{B-D}$$

Example 6

$45{,}000 \div 0.0018 = ?$

Step 1 Rewrite.
$$= \frac{4.5 \times 10^4}{1.8 \times 10^{-3}}$$

Step 2 Split up into two fractions. $= \frac{4.5}{1.8} \times \frac{10^4}{10^{-3}}$

Step 3 Divide using the rules for exponential division.

$$\boxed{4.5 \div 1.8} \quad = 2.5 \times 10^{4-(-3)}$$
$$= 2.5 \times 10^7 \qquad \boxed{4 - (-3) = 4 + 3 = 7}$$

If the result of the calculation is not in scientific notation, that is, if the decimal part is greater than 10 or less than 1, rewrite the decimal part so that it is in scientific notation.

Example 7

$$0.0000072 \div 0.0009 = \frac{7.2 \times 10^{-6}}{9 \times 10^{-4}}$$

$$= \frac{7.2}{9} \times \frac{10^{-6}}{10^{-4}}$$

$$= 0.8 \times 10^{-6-(-4)}$$

$$= 0.8 \times 10^{-2} \qquad \boxed{-6 - (-4) = -6 + 4 = -2}$$

But 0.8 is not a number between 1 and 10, so write it as
$0.8 = 8 \times 10^{-1}$.
Then, $0.8 \times 10^{-2} = 8 \times 10^{-1} \times 10^{-2}$
$$= 8 \times 10^{-3} \quad \text{in scientific notation}$$

→ **Your Turn**

Perform the following calculations using scientific notation, and write the answer in scientific notation.

(a) $1600 \times 350{,}000$ (b) $64{,}000 \times 250{,}000$

(c) 2700×0.0000045 (d) $15{,}600 \div 0.0013$

(e) $0.000348 \div 0.087$ (f) $0.00378 \div 540{,}000{,}000$

→ **Solutions**

(a) $(1.6 \times 10^3) \times (3.5 \times 10^5) = (1.6 \times 3.5) \times (10^3 \times 10^5)$
$$= 5.6 \times 10^{3+5}$$
$$= 5.6 \times 10^8$$

(b) $(6.4 \times 10^4) \times (2.5 \times 10^5) = (6.4 \times 2.5) \times (10^4 \times 10^5)$
$$= 16 \times 10^{4+5}$$
$$= 16 \times 10^9$$
$$= 1.6 \times 10^1 \times 10^9 \qquad \boxed{16 = 1.6 \times 10^1}$$
$$= 1.6 \times 10^{10}$$

(c) $(2.7 \times 10^3) \times (4.5 \times 10^{-6}) = (2.7 \times 4.5) \times (10^3 \times 10^{-6})$
$$= 12.15 \times 10^{3 + (-6)}$$
$$= 12.15 \times 10^{-3}$$
$$= 1.215 \times 10^1 \times 10^{-3}$$
$$= 1.215 \times 10^{-2}$$

> $3 + (-6) = -3$

(d) $(1.56 \times 10^4) \div (1.3 \times 10^{-3}) = \dfrac{1.56}{1.3} \times \dfrac{10^4}{10^{-3}}$
$$= 1.2 \times 10^{4-(-3)}$$
$$= 1.2 \times 10^7$$

> $4 - (-3) = 4 + 3 = 7$

(e) $(3.48 \times 10^{-4}) \div (8.7 \times 10^{-2}) = \dfrac{3.48}{8.7} \times \dfrac{10^{-4}}{10^{-2}}$
$$= 0.4 \times 10^{-4-(-2)}$$
$$= 0.4 \times 10^{-2}$$
$$= 4 \times 10^{-1} \times 10^{-2}$$
$$= 4 \times 10^{-3}$$

> $-4 - (-2) = -4 + 2 = -2$

> $0.4 = 4 \times 10^{-1}$

(f) $(3.78 \times 10^{-3}) \div (5.4 \times 10^8) = \dfrac{3.78}{5.4} \times \dfrac{10^{-3}}{10^8}$
$$= 0.7 \times 10^{-3-8}$$
$$= 0.7 \times 10^{-11}$$
$$= 7 \times 10^{-1} \times 10^{-11}$$
$$= 7 \times 10^{-12}$$

CALCULATORS AND SCIENTIFIC NOTATION

If a very large or very small number contains too many digits to be shown on the display of a scientific calculator, it will be converted to scientific notation. For example, if you enter the product

6480000 ⊠ **75000** ⊟

on a calculator, the answer will be displayed in one of the following ways:

> *4.86* × *10*^{11} or   *4.86* ^{11} or   *4.86 E 11*

Interpret all of these as

$6{,}480{,}000 \times 75{,}000 = 4.86 \times 10^{11}$

Similarly, the division

.000006 ÷ **48000000** ⊟

gives one of the following displays:

> *1.25* × *10*$^{-13}$ or   *1.25*$^{-13}$ or   *1.25 E −13*

Therefore, $0.000006 \div 48{,}000{,}000 = 1.25 \times 10^{-13}$
You may also enter numbers in scientific notation directly into your calculator using a key labeled ⒠⒠, ⒠ⓍⓅ, or ⓧ₁₀ˣ. For the calculation,

$0.000006 \div 48{,}000{,}000 \quad$ or $\quad (6 \times 10^{-6}) \div (4.8 \times 10^7)$

(continued)

enter

6 $\boxed{\text{EE}}$ $\boxed{(-)}$ **6** $\boxed{÷}$ **4.8** $\boxed{\text{EE}}$ **7** $\boxed{=}$ **(C7-3)**

and your calculator will again display $\quad$ 1.25×10^{-13}

If you wish to see all of your answers in scientific notation, look for a key marked $\boxed{\text{SCI/ENG}}$ or use the $\boxed{\text{SET UP}}$ or $\boxed{\text{MODE}}$ menus to switch to scientific mode.

Now turn to Exercises 7-8 for a set of problems on scientific notation.

Exercises 7-8 Scientific Notation

A. Rewrite each number in scientific notation.

1. 5000	2. 450	3. 90	4. 40,700
5. 0.003	6. 0.071	7. 0.0004	8. 0.0059
9. 6,770,000	10. 38,200	11. 0.0292	12. 0.009901
13. 1001	14. 0.0020	15. 0.000107	16. 810,000
17. 31.4	18. 0.6	19. 125	20. 0.74

21. Young's modulus for the elasticity of steel: 29,000,000 lb/in.^{2}

22. Thermal conductivity of wood: 0.00024 cal/(cm · s°C)

23. Power output: 95,500,000 watts

24. Speed of light: 670,600,000 mph

B. Rewrite each number in decimal form.

1. 2×10^5	2. 7×10^6	3. 9×10^{-5}	4. 3×10^{-4}
5. 1.7×10^{-3}	6. 3.7×10^{-2}	7. 5.1×10^4	8. 8.7×10^2
9. 4.05×10^4	10. 7.01×10^{-6}	11. 3.205×10^{-3}	12. 1.007×10^3
13. 2.45×10^6	14. 3.19×10^{-4}	15. 6.47×10^5	16. 8.26×10^{-7}

C. Rewrite each of the following in scientific notation and calculate the answer in scientific notation. If necessary, convert your answer to scientific notation. Round to one decimal place if necessary.

1. $2000 \times 40,000$	2. 0.0037×0.0000024
3. $460,000 \times 0.0017$	4. $0.0018 \times 550,000$
5. $0.0000089 ÷ 3200$	6. $0.000125 ÷ 5000$
7. $45,500 ÷ 0.0091$	8. $12,450 ÷ 0.0083$
9. $2,240,000 ÷ 16,000$	10. $25,500 ÷ 1,700,000$
11. $0.000045 ÷ 0.00071$	12. $0.000086 ÷ 0.000901$
13. $9,501,000 \times 2410$	14. 9800×0.000066
15. $0.0000064 ÷ 80,000$	16. 1070×0.0000055

17. $\dfrac{0.000056}{0.0020}$

18. $\dfrac{0.0507}{43{,}000}$

19. $\dfrac{0.00602 \times 0.000070}{72{,}000}$

20. $\dfrac{2{,}780{,}000 \times 512{,}000}{0.000721}$

21. $64{,}000 \times 2800 \times 370{,}000$

22. $0.00075 \times 0.000062 \times 0.014$

23. $\dfrac{0.0517}{0.0273 \times 0.00469}$

24. $\dfrac{893{,}000}{5620 \times 387{,}000}$

D. Solve.

1. **Allied Health** The levels of some blood components are usually expressed using scientific notation. For example, red blood cells are commonly expressed as 10^{12} cells/Liter (abbreviated as 10^{12}/L), and white blood cells are commonly expressed as 10^9 cells/Liter. Write the following amounts without using scientific notation:

   (a) A red blood cell value of 5.3×10^{12}/L.

   (b) A white blood cell value of 9.1×10^9/L.

   Write the following amounts using scientific notation:

   (c) A red blood cell value of 3,600,000,000,000/L.

   (d) A white blood cell value of 6,700,000,000/L.

2. **Marine Technology** Migrating gray whales swim 7000 miles north from the Sea of Cortez to the Alaskan coast in the spring and then return in the fall. Use scientific notation to calculate the total number of miles swum by a gray whale in an average 80-year life span. Give the answer both in scientific notation and as a whole number.

3. **HVAC** A brick wall 15 m by 25 m is 0.48 m thick. Under particular temperature conditions, the rate of heat flow through the wall, in calories per second, is given by the expression

   $$H = (1.7 \times 10^{-4}) \times \left(\frac{1500 \times 2500}{48} \right)$$ Calculate the value of this quantity to the nearest tenth.

4. **Physics** If an atom has a diameter of 4×10^{-8} cm and its nucleus has a diameter of 8×10^{-13} cm, find the ratio of the diameter of the nucleus to the diameter of the atom.

5. **Electrical Trades** The capacitance of a certain capacitor (in microfarads) can be found using the following expression. Calculate this capacitance.

   $$C = (5.75 \times 10^{-8}) \times \left(\frac{8.00}{3.00 \times 10^{-3}} \right)$$

6. **Physics** The energy (in joules) of a photon of visible light with a wavelength of 5.00×10^{-7} m is given by the following expression. Calculate this energy.

   $$E = \frac{(6.63 \times 10^{-34}) \times (3.00 \times 10^8)}{5.00 \times 10^{-7}}$$

Check your answers to the odd-numbered problems in the Appendix, then turn to Problem Set 7 on page 477 for some practice problems on the work of this chapter. If you need a quick review of the topics in this chapter, visit the chapter Summary first.

Objective

Review

Evaluate formulas and literal expressions. (p. 392)

Step 1 Place the numbers being substituted within parentheses, then substitute them in the formula being evaluated.

Step 2 After the numbers are substituted, do the arithmetic carefully. Be careful to use the proper order of operations.

> **Example:** If $x = 2$, $y = 3$, $a = 5$, and $b = 6$, then evaluate
>
> (a) $T = \dfrac{2(a + b + 1)}{3x} = \dfrac{2[(5) + (6) + 1]}{3(2)}$
>
> $= \dfrac{2(12)}{3(2)} = \dfrac{24}{6} = 4$
>
> (b) $A = x^2y = (2)^2(3) = 4(3) = 12$

Combine like terms. (p. 403)

Only like terms, those with the same literal parts, may be added and subtracted.

> **Example:** $5x - 3y - 8x + 2y = (5x - 8x) + (-3y + 2y)$
>
> $= -3x + (-1y)$
>
> $= -3x - y$

Remove parentheses properly. (p. 405)

Follow these rules for removing parentheses:

Rule 1 If the left parenthesis immediately follows a plus sign, simply remove the parentheses.

Rule 2 If the left parenthesis immediately follows a negative sign, change the sign of each term inside the parentheses, then remove the parentheses.

Rule 3 If the left parenthesis immediately follows a multiplier, multiply each term inside the parentheses by the multiplier.

> **Example:** (a) $4 + (x - 2) = 4 + x - 2 = 2 + x$
>
> (b) $5x - (x + 2) = 5x - x - 2 = 4x - 2$
>
> (c) $3(x^2 + 5x) - 4(2x - 3) = 3x^2 + 15x - 8x + 12$
>
> $= 3x^2 + 7x + 12$

Solve linear equations in one unknown, and solve formulas. (p. 409–443)

First, remove parentheses as shown in the preceding objective, then combine like terms that are on the same side of the equation. Finally, use the two kinds of balancing operations (add/subtract and multiply/divide) to isolate the desired variable on one side of the equation. The solution will appear on the other side.

> **Example:** (a) To solve the equation, $\qquad$ $7 + 3(5x + 2) = 38$
>
> Remove parentheses (Rule 3). $\qquad$ $7 + 15x + 6 = 38$
>
> Combine like terms. $\qquad$ $13 + 15x = 38$
>
> Subtract 13 from each side. $\qquad$ $15x = 25$
>
> Divide each side by 15. $\qquad$ $x = 1\frac{2}{3}$

Objective	Review

(b) Solve for N: $\qquad S = \dfrac{N}{2} + 26$

Subtract 26 from each side. $\qquad S - 26 = \dfrac{N}{2}$

Multiply each side by 2. $\qquad 2(S - 26) = N$

Or $\qquad N = 2S - 52$

Translate English phrases and sentences into algebraic expressions and equations, and solve word problems. (p. 444)

To solve any word problem:

1. **Read** the problem carefully, noting what is given and what is unknown.

2. **Assign a variable** to represent the unknown. Use an expression involving the same variable to represent additional unknowns.

3. **Write an equation** using the unknown(s) and some given fact about the unknown(s). Look for signal words to help you translate words to symbols.

4. **Solve** the equation.

5. **Answer** the original question.

6. **Check** your solution.

Example: An electrician collected $1265.23 for a job that included $840 for labor, which was not taxed, and the remainder for parts, which were taxed at 6%. Determine how much of the total was tax.

Solution: Let x = cost of parts before tax. Then $0.06x$ is the amount of tax on the parts. The cost of the parts plus the tax on the parts is equal to the total cost minus the labor.

$x + 0.06x = \$1265.23 - \$840$

$1.06x = \$425.23$

$x = \$401.16$, rounded. This is the cost of the parts.

The amount of tax is: $0.06x = 0.06(401.16) = \$24.07$

Or $\$425.23 - \$401.16 = \$24.07$

The tax on the parts was $24.07.

Multiply algebraic expressions. (p. 458)

Group like factors together and use Rule 1:

$x^m \cdot x^n = x^{m+n}$

If there is a factor multiplying an expression in parentheses, be careful to multiply every term inside the parentheses by this factor. That is,

$a(b + c) = a \cdot b + a \cdot c$

Example: (a) $(6x^4y^2)(-2xy^2) = \underbrace{6(-2)} \cdot \underbrace{x^4 \cdot x} \cdot \underbrace{y^2 \cdot y^2}$

$= -12 \cdot x^5 \cdot y^4$

$= -12x^5y^4$

(b) $3x(y - 2x) = 3x \cdot y - 3x \cdot 2x$

$= 3xy - 6x^2$

Summary Basic Algebra

Objective

Divide algebraic expressions. (p. 460)

Review

Divide numerical factors and use Rule 2 to divide exponential expressions.

$$\frac{x^m}{x^n} = x^{m-n}$$

If a negative exponent results, use the following rule to convert to a positive exponent:

$$x^{-n} = \frac{1}{x^n}$$

Example: (a) $\dfrac{6a^2b^4}{18a^5b^2} = \dfrac{6}{18} \cdot \dfrac{a^2}{a^5} \cdot \dfrac{b^4}{b^2} = \dfrac{1}{3} \cdot a^{-3} \cdot b^2$

$$= \frac{1}{3} \cdot \frac{1}{a^3} \cdot \frac{b^2}{1}$$

$$= \frac{b^2}{3a^3}$$

(b) $\dfrac{12m^4 - 9m^3 + 15m^2}{3m^2} = \dfrac{12m^4}{3m^2} - \dfrac{9m^3}{3m^2} + \dfrac{15m^2}{3m^2}$

$$= 4m^2 - 3m + 5$$

Convert between decimal notation and scientific notation. (p. 465)

A number expressed in scientific notation has the form $P \times 10^k$, where P is less than 10 and greater than or equal to 1, and k is an integer. To convert between decimal notation and scientific notation, use the method explained on pages 465–468 and illustrated here.

Example: (a) $0.000184 = 0.0001{\scriptstyle\wedge}84 = 1.84 \times 10^{-4}$

4 digits left

(b) $213,000 = 2{\scriptstyle\wedge}13000. = 2.13 \times 10^5$

5 digits right

Multiply and divide in scientific notation. (p. 468)

$$(A \times 10^B) \times (C \times 10^D) = (A \times C) \times 10^{B+D}$$

$$(A \times 10^B) \div (C \times 10^D) = (A \div C) \times 10^{B-D}$$

Convert your final answer to proper scientific notation.

Example: (a) $(3.2 \times 10^{-6}) \times (4.5 \times 10^2) = (3.2 \times 4.5) \times (10^{-6+2})$

$$= 14.4 \times 10^{-4}$$

$$= 1.44 \times 10^1 \times 10^{-4}$$

$$= 1.44 \times 10^{-3}$$

(b) $(1.56 \times 10^{-4}) \div (2.4 \times 10^3) = \dfrac{1.56}{2.4} \times 10^{-4-3}$

$$= 0.65 \times 10^{-7}$$

$$= 6.5 \times 10^{-1} \times 10^{-7}$$

$$= 6.5 \times 10^{-8}$$

7 Problem Set

Basic Algebra

Answers to odd-numbered problems are given in the Appendix.

A. Simplify.

1. $4x + 6x$

2. $6y + y + 5y$

3. $3xy + 9xy$

4. $6xy^3 + 9xy^3$

5. $3\frac{1}{3}x - 1\frac{1}{4}x - \frac{5}{8}x$

6. $8v - 8v$

7. $0.27G + 0.78G - 0.65G$

8. $7y - 8y^2 + 9y$

9. $3x \cdot 7x$

10. $(4m)(2m^2)$

11. $(2xy)(-5xyz)$

12. $3x \cdot 3x \cdot 3x$

13. $3(4x - 7)$

14. $2ab(3a^2 - 5b^2)$

15. $(4x + 3) + (5x + 8)$

16. $(7y - 4) - (2y + 3)$

17. $3(x + y) + 6(x - y)$

18. $(x^2 - 6) - 3(2x^2 - 5)$

19. $\dfrac{6^8}{6^4}$

20. $\dfrac{x^3}{x^8}$

21. $\dfrac{-12m^{10}}{4m^2}$

22. $\dfrac{6y}{-10y^3}$

23. $\dfrac{-16a^2b^2}{-4ab^3}$

24. $\dfrac{8x^4 - 4x^3 + 12x^2}{4x^2}$

B. Find the value of each of the following. Round to two decimal places when necessary.

1. $L = 2W - 3$ for $W = 8$

2. $M = 3x - 5y + 4z$ for $x = 3, y = 5, z = 6$

3. $I = PRt$ for $P = 800, R = 0.06, t = 3$

4. $I = \dfrac{V}{R}$ for $V = 220, R = 0.0012$

5. $V = LWH$ for $L = 3\frac{1}{2}, W = 2\frac{1}{4}, H = 5\frac{3}{8}$

6. $N = (a + b)(a - b)$ for $a = 7, b = 12$

7. $L = \dfrac{s(P + p)}{2}$ for $s = 3.6, P = 38, p = 26$

8. $f = \dfrac{1}{8N}$ for $N = 6$

Name _____

Date _____

Course/Section _____

9. $t = \dfrac{D - d}{L}$ for $D = 12, d = 4, L = 2$

10. $V = \dfrac{gt^2}{2}$ for $g = 32.2, t = 4.1$

C. Solve the following equations. Round your answer to two decimal places if necessary.

1. $x + 5 = 17$
2. $m + 0.4 = 0.75$
3. $e - 12 = 32$
4. $a - 2.1 = -1.2$
5. $12 = 7 - x$
6. $4\frac{1}{2} - x = -6$
7. $5x = 20$
8. $-4m = 24$
9. $\frac{1}{2}y = 16$
10. $2x + 7 = 13$
11. $-4x + 11 = 35$
12. $0.75 - 5f = 6\frac{1}{2}$
13. $0.5x - 16 = -18$
14. $5y + 8 + 3y = 24$
15. $3g - 12 = g + 8$
16. $7m - 4 = 11 - 3m$
17. $3(x - 5) = 33$
18. $2 - (2x + 3) = 7$
19. $5x - 2(x - 1) = 11$
20. $2(x + 1) - x = 8$
21. $12 - 7y = y - 18$
22. $2m - 6 - 4m = 4m$
23. $9 + 5n + 11 = 10n - 25$
24. $6(x + 4) = 45$
25. $5y - (11 - 2y) = 3$
26. $4(3m - 5) = 8m + 20$

Solve the following formulas for the variable shown.

27. $A = bH$ for b
28. $R = S + P$ for P
29. $P = 2L + 2W$ for L
30. $P = \dfrac{w}{F}$ for F
31. $S = \frac{1}{2}gt - 4$ for g
32. $V = \pi R^2 H - AB$ for A

D. Rewrite in scientific notation.

1. 7500
2. 12,800
3. 0.041
4. 0.000236
5. 0.00572
6. 0.00000482
7. 447,000
8. 2,127,000
9. 80,200,000
10. 46,710
11. 0.00000705
12. 0.001006

Rewrite as a decimal.

13. 9.3×10^5
14. 6.02×10^3
15. 2.9×10^{-4}
16. 3.05×10^{-6}
17. 5.146×10^{-7}
18. 6.203×10^{-3}
19. 9.071×10^4
20. 4.006×10^6

Calculate using scientific notation, and write your answer in standard scientific notation. Round to one decimal place if necessary.

21. $45,000 \times 1,260,000$
22. $625,000 \times 12,000,000$
23. 0.0007×0.0043
24. 0.0000065×0.032
25. $56,000 \times 0.0000075$
26. 1020×0.00055
27. $0.0074 \div 0.00006$
28. $0.000063 \div 0.0078$
29. $96,000 \div 3,400,000$
30. $26,500,000 \div 12,000$
31. $0.00089 \div 37,000$
32. $123,500 \div 0.00077$

E. Practical Problems

1. **Aviation** In weight and balance calculations, airplane mechanics are concerned with the center of gravity (CG) of an airplane. The center of gravity may be calculated from the formula

$$CG = \frac{100(H - x)}{L}$$

where H is the distance from the datum to the empty CG, x is the distance from the datum to the leading edge of the mean aerodynamic chord (MAC), and L is the length of the MAC. CG is expressed as a percent of the MAC.

(All lengths are in inches.)
   (a) Find the center of gravity if $H = 180$, $x = 155$, and $L = 80$.
   (b) Solve the formula for L, and find L when $CG = 30\%$, $H = 200$, and $x = 150$.
   (c) Solve the formula for H, and find H when $CG = 25\%$, $x = 125$, and $L = 60$.
   (d) Solve the formula for x, and find x when $CG = 28\%$, $H = 170$, and $L = 50$.

2. **Electrical Trades** Electricians use a formula that states the level of light in a room, in foot-candles, is equal to the product of the fixture rating, in lumens, the coefficient of depreciation, and the coefficient of utilization, all divided by the area, in square feet.
   (a) State this as an algebraic equation.
   (b) Find the level of illumination for four fixtures rated at 2800 lumens each if the coefficient of depreciation is 0.75, the coefficient of utilization is 0.6, and the area of the room is 120 sq ft.
   (c) Solve for area.
   (d) Use your answer to (c) to determine the size of the room in which a level of 60 foot-candles can be achieved with 10,000 lumens, given that the coefficient of depreciation is 0.8 and the coefficient of utilization is 0.5.

3. **Sheet Metal Trades** The length of a piece of sheet metal is four times its width. The perimeter of the sheet is 80 cm. Set up an equation relating these measurements and solve the equation to find the dimensions of the sheet.

4. **Construction** Structural engineers have found that a good estimate of the crushing load for a square wooden pillar is given by the formula

$$L = \frac{25T^4}{H^2}$$

where L is the crushing load in tons, T is the thickness of the wood in inches, and H is the height of the post in feet. Find the crushing load for a 6-in.-thick post 12 ft high.

5. **Machine Trades** Two shims must have a combined thickness of 0.048 in. One shim must be twice as thick as the other. Set up an equation and solve for the thickness of each shim.

6. **Trades Management** Sal, Martin, and Gloria are partners in a welding business. Because of the differences in their contributions, it is decided that Gloria's share of the profits will be one-and-a-half times as much as Sal's share, and that Martin's share will be twice as much as Gloria's share. How should a profit of $48,200 be divided?

7. **Electrical Engineering** The calculation shown is used to find the velocity of an electron at an anode. Calculate this velocity (units are in meters per second).

$$V = \sqrt{\frac{2(1.6 \times 10^{-19})(2.7 \times 10^2)}{9.1 \times 10^{-31}}}$$

8. **Plumbing** The formula

$$N = \sqrt{\left(\frac{D}{d}\right)^5}$$

gives the approximate number N of smaller pipes of diameter d necessary to supply the same total flow as one larger pipe of diameter D. Unlike the formula in problem 24 on page 400, this formula takes into account the extra friction caused by the smaller pipes. Use this formula to determine the number of $\frac{1}{2}$-in. pipes that will provide the same flow as one $1\frac{1}{2}$-in. pipe.

9. **Sheet Metal Trades** The length L of the stretchout for a square vent pipe with a grooved lock-seam is given by

$$L = 4s + 3w$$

where s is the side length of the square end and w is the width of the lock. Find the length of stretchout for a square pipe $12\frac{3}{4}$ in. on a side with a lock width of $\frac{3}{16}$ in.

10. **HVAC** In planning a solar energy heating system for a house, a contractor uses the formula

$$Q = 8.33GDT$$

to determine the energy necessary for heating water. In this formula Q is the energy in Btu, G is the number of gallons heated per day, D is the number of days, and T is the temperature difference between tap water and the desired temperature of hot water. Find Q when G is 50 gallons, D is 30 days, and the water must be heated from $60°$ to $140°$.

11. **Sheet Metal Trades** The bend allowance for sheet metal is given by the formula

$$BA = N(0.01743R + 0.0078T)$$

where N is the angle of the bend in degrees, R is the inside radius of the bend in inches, and T is the thickness of the metal in inches. Find BA if N is $47°$, R is 0.725 in., and T is 0.0625 in. Round to three significant digits.

12. **Machine Trades** To find the taper per inch of a piece of work, a machinist uses the formula

$$T = \frac{D - d}{L}$$

where D is the diameter of the large end, d is the diameter of the small end, and L is the length. Find T if $D = 4.1625$ in., $d = 3.2513$ in., and $L = 8$ in.

13. **Automotive Trades** The fuel economy F of a car with a 3.5-liter diesel engine at speeds at or above 50 mph is given by the formula

$$F = -0.0007V^2 - 0.13V + 42$$

where F is in miles per gallon and V is the speed in miles per hour. What is the fuel economy at a speed of (a) 50 mph? (b) 80 mph? (Round to the nearest whole number.)

14. **Allied Health** Body surface area (BSA) is often used to calculate dosages of drugs for patients undergoing chemotherapy or those with severe burns. Body surface area can be estimated from the patient's height and weight.

$$\text{BSA} = \sqrt{\frac{\text{weight in kg} \times \text{height in cm}}{3600}}$$

The BSA will be in square meters.
(a) Calculate the BSA for a man who weighs 95 kg and is 180 cm tall.
(b) A chemotherapy drug is given at a dose of 20 mg/m^2. Calculate the proper dosage for the man in part (a).

15. **Electronics** Solve for the variable indicated in each of these electronics formulas.

(a) $R = \dfrac{KL}{A}$ Solve for L.   (b) $R_s = \dfrac{R_m \cdot I_m}{I_s}$ Solve for R_m.

(c) $X_c = \dfrac{1}{2\pi FC}$ Solve for C.   (d) $L_T = \dfrac{L_1 + L_2}{2L_m}$ Solve for L_m.

16. **Automotive Trades** The Static Stability Factor (SSF) is a measure of the relative rollover risk associated with an automobile.

$$\text{SSF} = \frac{1}{2} \times \frac{\text{tire-track width}}{\text{height of center of gravity}}$$

An SSF of 1.04 or less indicates a 40% or higher rollover risk, and an SSF of 1.45 or more indicates a 10% or less rollover risk. A model 2002 Avalunch SUV has a tire-track width of 56 in. and a center of gravity 26 in. above the road.

(a) Calculate the SSF for this SUV.
(b) By how much would auto designers need to lower the center of gravity to improve the SSF to 1.45? Round to the nearest inch.

17. **Agriculture** Agricultural researchers developed the following formula from a study on the effects of the salinity of irrigation water on the yield potential of rice fields:

$$Y = 100 - 9.1\,(S - 1.9)$$

In this formula, Y is the percent yield of the rice fields, and S is the average salinity of the irrigation water in decisiemens per meter (dS/m).

(a) If $S = 5.2$ dS/m, find the percent yield of the rice fields. Round to the nearest percent.
(b) To achieve a yield of 80%, what must be the salinity of the irrigation water? Round to the nearest 0.1 dS/m.

18. **Electronics** The inductance L in microhenrys of a coil constructed by a ham radio operator is given by the formula

$$L = \frac{R^2 N^2}{9R + 10D}$$

where R = radius of the coil
$\quad\quad\quad D$ = length of the coil
$\quad\quad\quad N$ = number of turns of wire in the coil
Find L if $R = 3$ in., $D = 6$ in., and $N = 200$.

19. **Construction** In the remodel of a two-story home, a stairwell opening, S, is determined by the formula,

$$S = U_{ru}\left[\frac{H + T}{U_{ri}} - 1\right]$$

where U_{ru} is the unit run, U_{ri} is the unit rise, H is the head clearance, and T is the floor thickness. The quantity in brackets must be rounded up to the next whole number before multiplying by U_{ru}. Find the stairwell opening if $U_{ru} = 9$ in., $U_{ri} = 7\frac{1}{2}$ in., $H = 84$ in., and $T = 10\frac{3}{8}$ in.

20. **Electronics** Assuming that there is no mutual coupling, the total inductance in this circuit can be calculated from the following formula.

$$L_T = \frac{L_1 L_2}{L_1 + L_2}$$

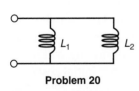

Problem 20

Calculate L_T for $L_1 = 200$ mH and $L_2 = 50$ mH. (See the figure.)

21. **Electronics** Calculate the total capacitance of the series-parallel circuit shown using the formula given. Round to the nearest $0.1\,\mu$F. (See the figure.)

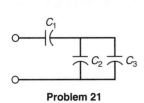

Problem 21

$C_1 = 14\mu$F
$C_2 = 4\mu$F
$C_3 = 6\mu$F

$$C_T = \frac{C_1(C_2 + C_3)}{C_1 + C_2 + C_3}$$

22. **Automotive Trades** The indicated horsepower (*ihp*) of an engine measures the power of the engine prior to friction loss. The *ihp* is given by

$$ihp = \frac{(MEP) \times D \times S}{792,000}$$

Where *MEP* represents the mean effective pressure of a typical cylinder in pounds per square inch (psi), *D* represents displacement in cubic inches (in.^{3}), and *S* represents engine speed in revolutions per minute (rpm).
   (a) Find the *ihp* of a 224-in.^{3} engine operating at 3800 rpm with an *MEP* of 170 psi. Round to three significant digits.
   (b) If the *ihp* of a 244-in.^{3} engine operating at 3250 rpm is 206, what is the *MEP* of a typical cylinder? Round to three significant digits.

23. **Electrical Trades** A certain electric utility charges residential customers $0.04 per kilowatt-hour (kWh) for the first 272 kWh, and $0.18 per kWh for the next 355 kWh. Therefore, when total usage *T* is between 272 kWh and 627 kWh, the monthly cost *C* for electricity can be expressed as

$$C = \$0.04(272) + \$0.18(T - 272)$$

How many kilowatt-hours would result in a monthly bill of $53?

24. **Automotive Trades** When balancing a crankshaft, bobweights are used in place of the piston and connecting rod. To choose the correct bobweight *B* for a cylinder, the following formula is used:

$$B = L + \frac{S + P}{2}$$

   where *L* = weight of the large end of the connecting rod
   S = weight of the small end of the connecting rod
   P = weight of the piston assembly

   (a) Find the bobweight if *L* = 518 grams, *S* = 212 grams, and *P* = 436 grams.
   (b) Solve the formula for *P*.

25. **Agriculture** Jean is mixing a batch of her CowVita supplement for cattle. She adds $3\frac{1}{2}$ times as much CowVita A as CowVita B, and the total amount of mixture is 120 lb. How much CowVita A is in the mix? Round to the nearest 0.1 lb.

26. **Trades Management** The Zeus Electronics repair shop charges $110 for the first hour of labor and $84 for each additional hour. The total labor cost for a lighting repair job was $656. Write an equation that allows you to solve for the total repair time, then solve this equation.

27. **Life Skills** Suppose you budgeted $1400 for annual fuel expenses and drove 12,000 miles per year. If gas cost $2.80 per gallon, what combined mileage would your vehicle need to meet your budget?

28. **Life Skills** Roxanne needs to buy a new car. She is debating between a hybrid and a nonhybrid model. The hybrid has an average fuel economy of 34 mi/gal, and it costs $32,000, but it will provide a tax credit of $1500. The nonhybrid has an average fuel economy of 22 mi/gal and costs $26,500. If she assumes an average gasoline cost of $2.60 per gallon, how many miles of driving will it take before the fuel savings of the hybrid compensate for its higher initial cost?

29. **Trades Management** Eric runs a small shop. He must set aside 22% of his revenue for taxes, and he spends another $3600 per month on other expenses. How much total revenue must he generate in order to clear a profit of $6000 per month? (Round to the nearest dollar.)

Practical Plane Geometry

Objective	Sample Problems		For help, go to

When you finish this chapter, you will be able to:

1. Measure angles with a protractor.

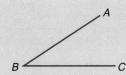

$\angle ABC =$ _____

Page 489

2. Classify angles.

(a)

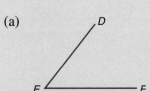

(a) $\angle DEF$ is _____

Page 490

(b)

(b) $\angle GHI$ is _____

(c)

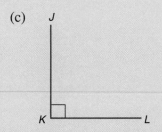

(c) $\angle JKL$ is _____

_____
Name

_____
Date

_____
Course/Section

Objective

Sample Problems

For help, go to

3. Use simple geometric relationships involving intersecting lines and triangles.

(a)

c/d
b
e/a
80°

Assume that the two horizontal lines are parallel.

∢a = _____

∢b = _____

∢c = _____

∢d = _____

∢e = _____

Page 492

(b)

x   70°

∢x = _____

Page 493

4. Identify polygons, including triangles, squares, rectangles, parallelograms, trapezoids, and hexagons.

(a)

6   12
117°
8

_____

Page 515

(b)

10
6   6
10

_____

Page 501

(c)

_____

(d)

4
4   4
4

_____

(e)

_____

Page 525

5. Use the Pythagorean theorem.

1.9"

x   1.2"

x = _____

Page 517

Round to one decimal place.

6. Find the area and perimeter of geometric figures.

(a)

7"
4"

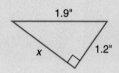

Area = _____

Page 520

(b)

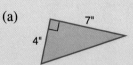

3 cm

8 cm

Area = _____

Page 507

Objective	Sample Problems	For help, go to

(c)

Area = _____
(Round to two decimal places.)

Page 526

Perimeter = _____

(d)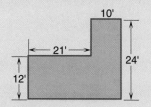

$r = 2.70"$

Area = _____

Page 542

Circumference = _____

Page 537

Use $\pi \approx 3.14$
and round both
answers to the
nearest tenth.

7. Solve practical prob-
lems involving area
and perimeter of
plane figures.

(a)

10'

21'

24'

12'

Flooring and Carpeting
If carpet costs $4.75 per
square foot, how much will
it cost to carpet the room in
the figure?

_____

Page 504

(b) **Landscaping** A circular flowerbed, with a
radius of 8.5 ft, is sur-
rounded by a circular
ring of lawn 12 ft
wide. At $2.10 per
square foot, how much
will it cost to sod the
lawn?

_____

Page 544

(Answers to these preview problems are given in the Appendix. Also, worked solutions
to many of these problems appear in the chapter Summary.)

If you are certain that you can work *all* these problems correctly, turn to page 553 for a
set of practice problems. If you cannot work one or more of the preview problems, turn
to the page indicated to the right of the problem. Those who wish to master this material
with the greatest success should turn to Section 8-1 and begin work there.

Chapter 8

Practical Plane Geometry

"NO! NO! I SAID BUILD AN *ARK!*"

© 1974 Reprinted by permission of *Saturda* and Orlando Busino and Omni International

Geometry is one of the oldest branches of mathematics. It involves study of the properties of points, lines, plane surfaces, and solid figures. Ancient Egyptian engineers used these properties when they built the Pyramids, and modern trades workers use them when cutting sheet metal, installing plumbing, building cabinets, and performing countless other tasks. When they study formal geometry, students concentrate on theory and actually prove the truth of various geometric theorems. In this chapter, we concentrate on the applications of the ideas of plane geometry to do technical work and the trades.

8-1 Angle Measurement

In geometry, the word *plane* is used to refer to a mathematical concept describing an infinite set of points in space. For practical purposes, a plane surface can be thought of as a perfectly flat, smooth surface such as a tabletop or a window pane.

Labeling Angles
An *angle* is a measure of the size of the opening between two intersecting lines. The point of intersection is called the *vertex,* and the lines forming the opening are called the *sides.*

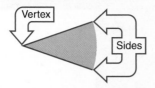

An angle may be identified in any one of the following ways:

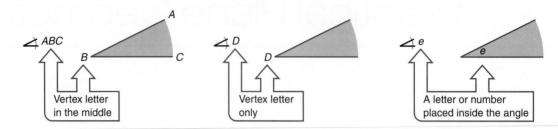

The angle symbol ∡ is simply a shorthand way to write the word "angle."

For the first angle, the middle letter is the vertex letter of the angle, while the other two letters represent points on each side. Notice that capital letters are used. The second angle is identified by the letter *D* near the vertex. The third angle is named by a small letter or number placed inside the angle.

The first and third methods of naming angles are most useful when several angles are drawn with the same vertex.

Example 1

The interior of angle *POR* (or *ROP*) is shaded.

The interiors of angles *a* and 1 are shaded.

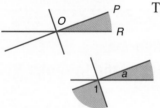

→ **Your Turn**

Name the angle shown in the margin using the three different methods.

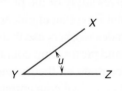

→ **Answers**

∡ *XYZ* or ∡ *ZYX* or ∡ *Y* or ∡ *u*

Measuring Angles

Naming or labeling an angle is the first step. Measuring it may be even more important. Angles are measured according to the size of their opening. The length of the sides of the angle is not related to the size of the angle.

and ⊿ are the same size angle.

Chapter 8 Practical Plane Geometry

The basic unit of measurement of angle size is the *degree*. Because of traditions going back thousands of years, we define a degree as an angle that is $\frac{1}{360}$ of a full circle, or $\frac{1}{90}$ of a quarter circle.

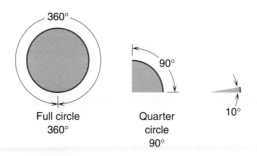

Full circle
360°

Quarter circle
90°

10°

The degree can be further subdivided into finer angle units known as *minutes* and *seconds* (which are not related to the time units).

1 degree, $1° = 60$ minutes $= 60'$

1 minute, $1' = 60$ seconds $= 60''$

An angle of 62 degrees, 12 minutes, 37 seconds would be written $62°12'37''$. Most trade work requires precision only to the nearest degree.

As you saw in Chapter 5, page 326, a protractor is a device used to measure angles. To learn to measure angles, first look at the protractor below.

A protractor that you may cut out and use is available in the Appendix on page 805.

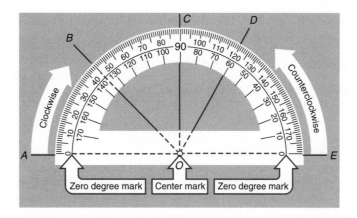

To measure an angle, place the protractor over it so that the zero-degree mark is lined up with one side of the angle, and the center mark is on the vertex. Read the measure in degrees where the other side of the angle intersects the scale of the protractor. When reading an angle clockwise, use the upper scale, and when reading counterclockwise, use the lower scale.

Example 2

(a) In the drawing, $\angle EOB$ should be read counterclockwise from the 0° mark on the right. $\angle EOB = 135°$.

(b) $\angle AOD$ should be read clockwise from the 0° mark on the left. $\angle AOD = 120°$.

→ **Your Turn**

Find (a) $\angle EOD$ (b) $\angle AOB$

(a) ∡*EOD* = 60°   Measure counterclockwise from the right 0° mark and read the lower scale.

(b) ∡*AOB* = 45°   Measure clockwise from the left 0° mark and read the upper scale.

Classifying Angles

An **acute angle** is an angle *less than* 90°. Angles *AOB* and *EOD* are both less than 90°, so both are acute angles.

Acute angles:

An **obtuse angle** is an angle *greater than* 90° and less than 180°. Angles *AOD* and *EOB* are both greater than 90°, so both are obtuse angles.

Obtuse angles:

C

X ———— *E*

Perpendicular lines

A **right angle** is an angle exactly equal to 90°. Angle *EXC* in the drawing at the left is a right angle.

Two straight lines that meet in a right or 90° angle are said to be **perpendicular.** In this drawing, *CX* is perpendicular to *XE*.

Notice that a small square is placed at the vertex of a right angle to show that the sides are perpendicular.

A **straight angle** is an angle equal to 180°.

Angle *BOA* shown here is a straight angle.

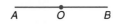

A   *O*   *B*

→ **Your Turn**

Use your protractor to measure each of the angles given and tell whether each is acute, obtuse, right, or straight. Estimate the size of the angle before you measure. (You will need to extend the sides of the angles to fit your protractor.)

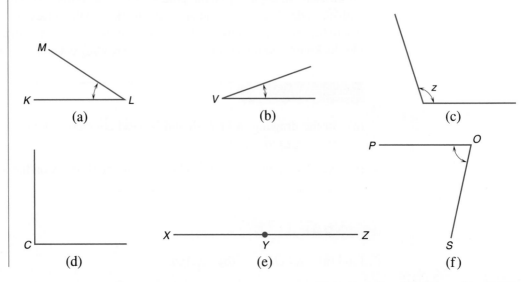

Drawing Angles

In addition to measuring angles that already exist, trades workers will also need to draw their own angles. To draw an angle of a given size follow these steps:

Step 1 Draw a line representing one side of the angle.

_____

Step 2 Place the protractor over the line so that the center mark is on the vertex, and the 0° mark coincides with the other end of the line. (*Note:* We could have chosen either end of the line for the vertex.)

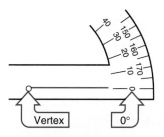

Step 3 Now place a small dot above the degree mark corresponding to the size of your angle. In this case the angle is 30°.

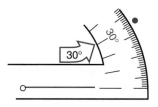

Step 4 Remove the protractor and connect the dot to the vertex.

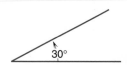

→ **Your Turn**

Now you try it. Use your protractor to draw the following angles.

$\angle A = 15°$ $\angle B = 112°$ $\angle C = 90°$

Here are our drawings:

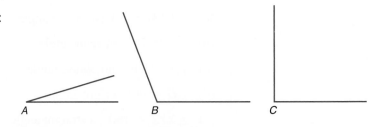

A   B   C

Angle Facts There are several important geometric relationships involving angles that you should know. For example, if we measure the four angles created by the intersecting lines shown,

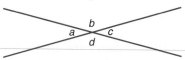

we find that

$\angle a = 30°$ $\angle b = 150°$ $\angle c = 30°$ and $\angle d = 150°$

Do you notice that opposite angles are equal?

For any pair of intersecting lines the opposite angles are called **vertical angles.** In the preceding drawing, angles a and c are a pair of vertical angles. Angles b and d are also a pair of vertical angles.

 are pairs of vertical angles.

An important geometric rule is that

> When two straight lines intersect, the vertical angles are always equal.

$\angle p = \angle r$
and $\angle q = \angle s$

If you remember that a straight angle = 180°, you can complete the following statements for the preceding drawing.

$\angle p + \angle q =$ _____ $\angle p + \angle s =$ _____

$\angle s + \angle r =$ _____ $\angle q + \angle r =$ _____

$\angle p + \angle q = 180°$ $\angle p + \angle s = 180°$

$\angle s + \angle r = 180°$ $\angle q + \angle r = 180°$

This leads to a second important geometry relation:

> When two lines intersect, each pair of adjacent angles sums to 180°.

→ **Your Turn**

Use these relationships to answer the following questions.

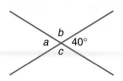

(a) $\sphericalangle a =$ _____

(b) $\sphericalangle b =$ _____

(c) $\sphericalangle c =$ _____

→ **Answers**

(a) $\sphericalangle a = 40°$ since   vertical angles are equal

(b) $\sphericalangle b = 140°$ since $\sphericalangle a + \sphericalangle b = 180°$

(c) $\sphericalangle c = 140°$ since   vertical angles are equal, $\sphericalangle b = \sphericalangle c$

A triangle is formed from three distinct line segments. The line segments are called the *sides* of the triangle, and the points where they meet are called its *vertices*. Every triangle forms three angles, and these are related by an important geometric rule.

If we use a protractor to measure each angle in the following triangle, *ABC*, we find

$\sphericalangle A = 69°$ $\sphericalangle B = 36°$ $\sphericalangle C = 75°$

Notice that the three angles of the triangle sum to 180°.

$69° + 36° + 75° = 180°$

Measuring angles on many triangles of every possible size and shape leads to the third geometry relationship:

> The interior angles of a triangle always add to 180°.

Example 3

We can use this fact to solve the following problems.

(a) If $\sphericalangle P = 50°$ and $\sphericalangle Q = 85°$, find $\sphericalangle R$.

By the relationship given, $\angle P + \angle Q + \angle R = 180°$, so

$\angle R = 180° - \angle P - \angle Q$
$\quad = 180° - 50° - 85°$
$\quad = 45°$

(b) If $\angle P = 27°\,42'$ and $\angle Q = 76°\,25'$, find $\angle R$.

Step 1 $\quad \angle R = 180° - 27°\,42' - 76°\,25'$

Step 2 Combine the two angles that are being subtracted.

$\quad 27°\,42' + 76°\,25' = (27 + 76)° + (42 + 25)'$
$\qquad\qquad\qquad\qquad = 103°\,67'$

Step 3 To simplify $103°\,67'$ use the fact that $60' = 1°$.

$\quad 103°\,67' = 103° + 60' + 7'$
$\qquad\qquad = 103° + 1° + 7' = 104°\,7'$

Step 4 Subtract this total from $180°$. First change $180°$ to $179°\,60'$.

$\quad 180° - 104°\,7' = 179°\,60' - 104°\,7'$
$\qquad\qquad\qquad = (179 - 104)° + (60 - 7)'$
$\qquad\qquad\qquad = 75°\,53'$

→ **Your Turn**

Find the missing angle in each problem.

(a) If $\angle X = 34°$ and $\angle Y = 132°$, find $\angle Z$

(b) If $\angle X = 29°\,51'$ and $\angle Z = 24°\,34'$, find $\angle Y$

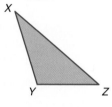

→ **Solutions**

(a) $\angle Z = 180° - \angle X - \angle Y$
$\quad = 180° - 34° - 132°$
$\quad = 180° - 166° = 14°$

(b) $\angle Y = 180° - \angle X - \angle Z$
$\quad = 180° - 29°\,51' - 24°\,34'$
$\quad = 180° - 53°\,85'$ ⟵ $\begin{array}{l} 53°85' = 53° + 60' + 25' \\ \qquad\quad = 53° + 1° + 25' \\ \qquad\quad = 54°25' \end{array}$
$\quad = 180° - 54°\,25'$
$\quad = 179°\,60' - 54°\,25'$
$\quad = 125°\,35'$

A fourth and final angle fact you will find useful involves parallel lines. Two straight lines are said to be **parallel** if they always are the same distance apart and never meet. If a pair of parallel lines is cut by a third line called a **transversal,** several important pairs of angles are formed.

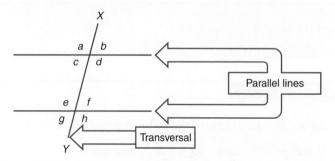

Angles *c* and *f* are called **alternate interior angles.** Angles *d* and *e* are also alternate interior angles.

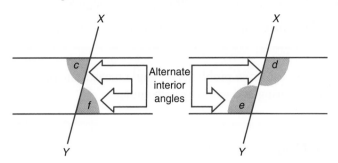

The angles are *interior* (or inside) the original parallel lines and they are on *alternate* sides of the transversal *XY.* The alternate interior angles are equal when the lines cut by the transversal are parallel.

The angles *a* and *e* form one of four pairs of **corresponding angles.** The others are *c* and *g, b* and *f,* and *d* and *h.*

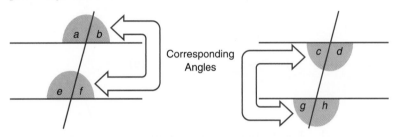

These angles are on the same side of the transversal, and they are on the same side of their respective parallel line. For example, *b* and *f* are both above their parallel lines while *c* and *g* are both below their lines. Corresponding angles are always equal when the lines cut by the transversal are parallel.

The geometric rule is:

> When parallel lines are cut by a third line, the corresponding angles are equal, and the alternate interior angles are equal.

→ **Your Turn**

Use this rule and the previous ones to answer the following questions. Assume that the two horizontal lines are parallel.

∡*p* = _____   ∡*q* = _____

∡*r* = _____   ∡*s* = _____

∡*t* = _____   ∡*u* = _____

∡*w* = _____

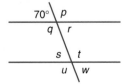

→ **Answers**

∡*p* = 110° since 70° + ∡*p* = 180°. Adjacent angles sum to 180°.

∡*q* = 110° since *p* and *q* are vertical angles.

∡*r* = 70° since it is a vertical angle to the 70° angle.

∡*s* = 70° since ∡*s* = ∡*r,* alternate interior angles.

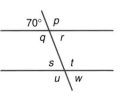

$\angle t = 110°$ since $\angle t = \angle q$, alternate interior angles, and $\angle p = \angle t$, corresponding angles.

$\angle u = 110°$ since $\angle u = \angle q$, corresponding angles.

$\angle w = 70°$ since $\angle u + \angle w = 180°$ and $\angle w = \angle r$, corresponding angles.

Now, ready for some problems on angle measurement? Turn to Exercises 8-1 for practice in measuring and working with angles.

Exercises 8-1 — Angle Measurement

A. Solve the following. If you need a protractor, cut out and use the one in the Appendix.

1. Name each of the following angles using the three kinds of notation shown in the text.

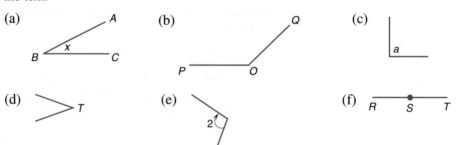

(a)   (b)   (c)

(d)   (e)   (f)

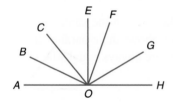

2. For the figure in the margin,

   (a) Name an acute angle that has AO as a side.

   (b) Name an acute angle that has HO as a side.

   (c) Name an obtuse angle that has AO as a side.

   (d) Name a right angle.

   (e) Use your protractor to measure $\angle AOB$, $\angle GOF$, $\angle HOC$, $\angle FOH$, $\angle BOF$, $\angle COG$.

3. Use your protractor to draw the following angles.

   (a) $\angle LMN = 29°$ (b) $\angle 3 = 100°$

   (c) $\angle Y = 152°$ (d) $\angle PQR = 137°$

   (e) $\angle t = 68°$ (f) $\angle F = 48°$

4. Measure the indicated angles on the following shapes. (You will need to extend the sides of the angles to improve the accuracy of your measurement.)

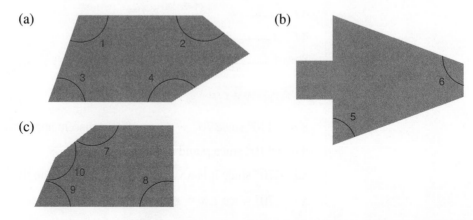

(a)   (b)

(c)

B. Use the geometry relationships to answer the following questions.

1. In each problem one angle measurement is given. Determine the others for each figure without using a protractor.

(a)
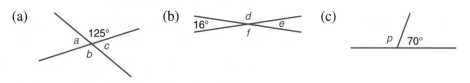
125°
a c
b

(b)
16° d
e
f

(c)
p 70°

(d) A ____ B
65° x
k t 75°
D C
Where *AB* is
parallel to *CD*

(e)
s
t
60°

(f)
q
150° w
p 60°

2. For each triangle shown, two angles are given. Find the third angle without using a protractor.

(a)
A
60° 52°
B C

(b) D
23°
20°
E
F

(c) G
90° 48°
H K

(d)
27°
20°

(e)
A
(The angles labeled *a* are equal.)
40°
a a

(f)
100°
27°

(g)
90°
27°

(h)
109°43'
25°32'

(i)
46°12'
73°46'

3. In each figure, two parallel lines are cut by a third line. Find the angles marked.

(a)
100°
w x
y
z

(b)
b c
f a
60° d
e g

(c)
n 80° p h
q m k

(d)
t x
s w
z u
y
40°

Problem 1

C. Applied Problems (*Note:* Figures accompany all problems.)

1. **Machine Trades** A machinist must punch holes *A* and *B* in the piece of steel shown by rotating the piece through a certain angle from vertex *C*. Use a protractor to find angle *ACB*.

2. **Plumbing** A plumber must connect the two pipes shown by first selecting the proper elbows. What angle elbows does he need?

3. **Carpentry** A carpenter needs to build a triangular hutch that will fit into the corner of a room as shown. Measure the three angles of the hutch.

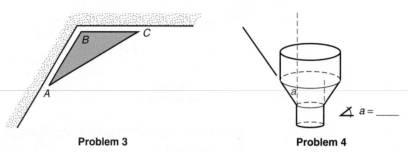

Problem 2

Problem 3   **Problem 4**

4. **Sheet Metal Trades** A sheet metal worker must connect the two vent openings shown. At what angle must the sides of the connecting pieces flare out?

5. **Electrical Trades** An electrician wants to connect the two conduits shown with a 45° elbow. How far up must she extend the lower conduit before they will connect at that angle? Give the answer in inches. (*Hint:* Draw the 45° connection on the upper conduit first.) Each dot represents a 1-in. extension.

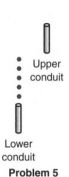

Upper
conduit

Lower
conduit

Problem 5

6. **Construction** At what angle must the rafters be set to create the roof gable shown in the drawing?

7. **Machine Trades** A machinist receives a sketch of a part he must make as shown. If the indicated angle *BAC* is precisely half of angle *BAD*, find ∡*BAD*.

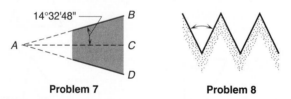

Problem 6

Problem 7   **Problem 8**

8. **Machine Trades** The drawing illustrates a cross section of a V-thread. Find the indicated angle.

9. **Carpentry** At what angles must drywall board be cut to create the wall shown in the figure?

Problem 9

10. **Carpentry** A carpenter wishes to make a semicircular deck by cutting eight equal angular pieces of wood as shown. Measure the angle of one of the pieces. Was there another way to determine the angle measure without using a protractor?

11. **Carpentry** A carpenter's square is placed over a board as shown in the figure. If ∡1 measures 74°, what is the size of ∡2?

Problem 10

Problem 11   **Problem 12**

12. **Carpentry** Board 1 must be joined to board 2 at a right angle. If ∡*x* measures 42°, what must ∡*y* measure?

Problem 13

13. **Welding**   In the pipe flange shown, the six bolt holes are equally spaced. What is the spacing of the holes, in degrees?

When you have finished these problems, check your answers to the odd-numbered problems in the Appendix, then continue in Section 8-2 with the study of plane figures.

8-2   Area and Perimeter of Polygons

Polygons   A *polygon* is a closed plane figure containing three or more angles and bounded by three or more straight sides. The word "polygon" itself means "many sides." The following figures are all polygons.

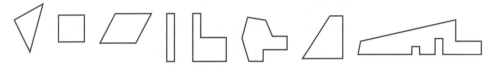

A figure with a curved side is *not* a polygon.

Every trades worker will find that an understanding of polygons is important. In this section you will learn how to identify the parts of a polygon, recognize some different kinds of polygons, and compute the perimeter and area of certain polygons.

In the general polygon shown, each *vertex* or corner is labeled with a letter: *A, B, C, D,* and *E*. The polygon is named simply "polygon *ABCDE*"—not a very fancy name, but it will do.

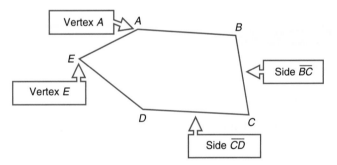

The *sides* of the polygon are named after the line segments that form each side: $\overline{AB}, \overline{BC}, \overline{CD}$, and so on. Notice that we place a bar over the letters to indicate that we are talking about a line segment. A *line segment* is a finite portion of a straight line; it has two endpoints.

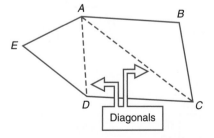

The *diagonals* of the polygon are the line segments connecting nonconsecutive vertices such as $\overline{AC}$ and $\overline{AD}$. These are shown as dashed lines in the figure. Other diagonals not shown are $\overline{EB}, \overline{EC}$, and $\overline{BD}$.

Example 1

Consider the following polygon, UVWXYZ:

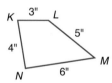

The sides are $\overline{UV}$, $\overline{VW}$, $\overline{WX}$, $\overline{XY}$, $\overline{YZ}$, and $\overline{ZU}$.

Note   The letters naming a side can be reversed: side $\overline{VU}$ is the same as side $\overline{UV}$. ◄

The vertices are U, V, W, X, Y, and Z.

There are nine diagonals; we need to be certain to find them all.

The diagonals are $\overline{UW}$, $\overline{UX}$, $\overline{UY}$, $\overline{VX}$, $\overline{VY}$, $\overline{VZ}$, $\overline{WY}$, $\overline{WZ}$, and $\overline{XZ}$.

Perimeter

For those who do practical work with polygons, the most important measurements are the lengths of the sides, the *perimeter*, and the *area*. The perimeter of any polygon is simply the distance around the outside of the polygon.

> To find the perimeter of a polygon, add the lengths of its sides.

Example 2

In the polygon *KLMN* the perimeter is

3 in. + 5 in. + 6 in. + 4 in. = 18 in.

The perimeter is a length; therefore, it has length units, in this case inches.

→ **Your Turn**

Find the perimeter of each of the following polygons.

(a)

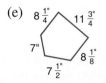

(b)

(c)

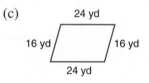

(d)

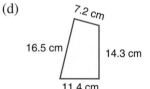

(e)

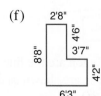

(f)

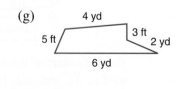

(g)

▶ **Careful**   As you may have discovered in problem (g) it is important that all sides be expressed in the same units before calculating the perimeter. ◀

Adding up the lengths of the sides will always give you the perimeter of a polygon, but handy formulas are needed to enable you to calculate the area of a polygon. Before you can use these formulas you must learn to identify the different types of polygons. Let's look at a few. In this section we examine the *quadrilaterals* or four-sided polygons.

Quadrilaterals   Figure *ABCD* is a *parallelogram.* In a parallelogram opposite sides are parallel and equal in length.

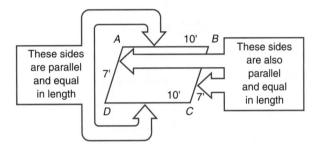

Figure *EFGH* is a *rectangle,* a parallelogram in which the four corner angles are right angles. The ∟ symbols at the vertices indicate right angles. Because a rectangle is a parallelogram, opposite sides are parallel and equal.

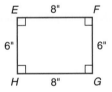

IJKL is a *square,* a rectangle in which *all* sides are the same length.

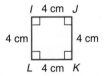

MNOP is called a *trapezoid.* A trapezoid contains two parallel sides and two nonparallel sides. $\overline{MN}$ and $\overline{OP}$ are parallel; $\overline{MP}$ and $\overline{NO}$ are not.

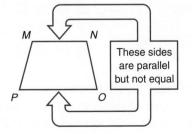

Here are more trapezoids:

If a four-sided polygon has none of these special features—no parallel sides—we call it a *quadrilateral*. *QRST* is a quadrilateral.

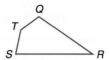

Name each of the following quadrilaterals. If more than one name fits, give the most specific name.

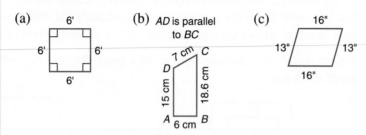

(a) 6'
     6' 6'
        6'

(b) *AD* is parallel to *BC*
    7 cm C
    D
    15 cm 18.6 cm
    A 6 cm B

(c) 16"
    13"   13"
        16"

(d) 11 yd
    16 yd   26 yd
         22 yd

(e) 24"
    18"   18"
        24"

→ Answers

(a) Square. (It is also a rectangle and a parallelogram, but "square" is a more specific name.)

(b) Trapezoid.

(c) Parallelogram.

(d) Quadrilateral.

(e) Rectangle. (It is also a parallelogram, but "rectangle" is a more specific name.)

Area and Perimeter of Rectangles

Once you can identify a polygon, you can use a formula to find its area. Next let's examine and use some area formulas for these quadrilaterals.

The area of any plane figure is the number of square units of surface within the figure.

In this rectangle,

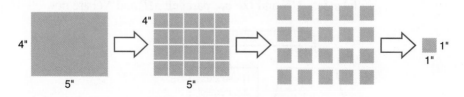

we can divide the surface into exactly 20 small squares each one inch on a side. By counting squares we can see that the area of this 4-in. by 5-in. rectangle contains 20 square inches. We abbreviate these area units as 20 square inches or 20 sq in.

Chapter 8 Practical Plane Geometry

As mentioned in Chapter 5, area units may be abbreviated using exponents. For example, the area unit square inches or sq in. results from multiplying inches × inches. Using the definition of exponents from Chapter 6, this can be expressed as inches^2 or in.^{2}. Similarly, square feet or sq ft can be written as ft^2, square centimeters as cm^2, and so on. Both types of units are used by trades workers, and we will use both in the remainder of this text. ◄

Of course there is no need to draw lines in the messy way shown above. We can find the area by multiplying the two dimensions of the rectangle.

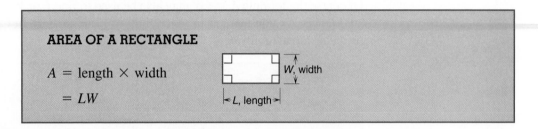

AREA OF A RECTANGLE

A = length × width

   = LW

The dimensions "length" and "width" are sometimes called "base" and "height." ◄

Example 3

Find the area of rectangle *EFGH*.

$L = 9$ in.

$W = 5$ in.

A, area $= LW = (9 \text{ in.})(5 \text{ in.})$

$A = 45$ in.^{2}

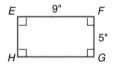

The formula is easy to use, but be careful. It is valid only for rectangles. If you use this formula with a different polygon, you will not get the correct answer. ◄

In technical work, it is important to both state measurements and round calculations to the proper precision and accuracy. However, in the trades and in many occupations, people do not normally state quantities with the exact accuracy with which they have been measured. For example, the length and width of a box might very well be measured accurately to two significant digits, but a carpenter would write it as 9 in. rather than 9.0 in. To avoid confusion in calculations, either state your unrounded answer or follow the rounding instructions that accompany the problems.

→ **Your Turn**

Find the area of rectangle *IJKL*.

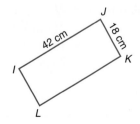

First, write the formula: $A = LW$

Second, substitute L and W: $= (42 \text{ cm})(18 \text{ cm})$

Third, calculate the answer = 756 cm^2
including the units.

 42 ⊗ 18 ⊜ → 756.

Of course the formula $A = LW$ may also be used to find L or W when the other quantities are known. As you learned in your study of basic algebra in Chapter 7, the following formulas are all equivalent.

$$A = LW \qquad L = \frac{A}{W} \quad \text{or} \quad W = \frac{A}{L}$$

Example 4

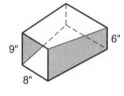

Sheet Metal Trades A sheet metal worker must build a heating duct in which a rectangular vent 6 in. wide must have the same area as a rectangular opening 8 in. by 9 in. Find the length of the vent.

The area of the opening is $A = LW$

$$= (8 \text{ in.})(9 \text{ in.}) = 72 \text{ in.}^2$$

The length of the vent is $L = \dfrac{A}{W}$

$$= \frac{72 \text{ in.}^2}{6 \text{ in.}} = \frac{72}{6} \frac{\text{in.} \times \cancel{\text{in.}}}{\cancel{\text{in.}}}$$

$$= 12 \text{ in.}$$

Another formula that you may find useful allows you to calculate the perimeter of a rectangle from its length and width.

PERIMETER OF A RECTANGLE

$P = 2L + 2W$

Example 5

We can use this formula to find the perimeter of the rectangle shown here.

Perimeter $= 2L + 2W$
$= 2(7 \text{ in.}) + 2(3 \text{ in.})$
$= 14 \text{ in.} + 6 \text{ in.}$
$= 20 \text{ in.}$

 2 ⊗ 7 ⊕ 2 ⊗ 3 ⊜ → 20.

→ **More Practice**

For practice on using the area and perimeter formulas for rectangles, work the following problems.

(a) Find the area and perimeter of each of the following rectangles.

(1) 8" 6"

(2) 12' 43'

(3) $9\frac{2}{3}$ yd $2\frac{2}{3}$ yd

(4) 1.80 m 1.80 m

(b) Find the area of a rectangle whose length is 4 ft 6 in. and whose width is 3 ft 3 in. (Give your answer in square feet in decimal form. Round to one decimal place.)

(c) **Sheet Metal Trades** A rectangular opening 12 in. long must have the same total area as two smaller rectangular vents 6 in. by 4 in. and 8 in. by 5 in. What must be the width of the opening?

(d) **Flooring and Carpeting** What is the cost of refinishing a wood floor in a room 35 ft by 20 ft at a cost of $2.25 per square foot?

(e) Find the area of a rectangular opening 20 in. wide and 3 ft long.

(f) **Landscaping** What is the cost of fencing a rectangular yard 45 ft by 58 ft at $30 per linear foot?

→ **Solutions**

(a) (1) $A = 48$ in.$^2, P = 28$ in. (2) $A = 516$ ft$^2, P = 110$ ft

   (3) $A = 25\frac{7}{9}$ yd$^2, P = 24\frac{2}{3}$ yd (4) $A = 3.24$ m$^2, P = 7.20$ m

(b) 4 ft 6 in. $= 4\frac{6}{12}$ ft $= 4.5$ ft
   3 ft 3 in. $= 3\frac{3}{12}$ ft $= 3.25$ ft
   4.5 ft $\times$ 3.25 ft ≈ 14.6 ft^2

(c) Total area of smaller vents $= (6 \times 4) + (8 \times 5) = 64$ in.^{2}
   Width of opening $= \dfrac{A}{L} = \dfrac{64 \text{ in.}^2}{12 \text{ in.}} = 5\frac{1}{3}$ in.

(d) Area $= 35$ ft $\times 20$ ft $= 700$ ft^2
   Cost $= 700$ ft$^2 \times \$2.25$ per ft$^2 = 700 \text{ ft}^2 \times \dfrac{\$2.25}{1 \text{ ft}^2} = \$1575$

(e) 720 in.^{2} or 5 ft^2

(f) $6180

 2 ⊗ 45 ⊞ 2 ⊗ 58 ⊜ ⊗ 30 ⊜ → 6180.

Area and Perimeter of Squares Did you notice that the rectangle in problem (a)(4) in the last set is actually a square? All sides are the same length. To save time when calculating the area or perimeter of a square, use the following formulas.

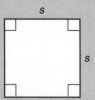

AREA OF A SQUARE

$A = s^2$

PERIMETER OF A SQUARE

$P = 4s$

Example 6

We can use these formulas to find the area and perimeter of the square shown.

Area, $A = s^2$
$$= (9\,\text{ft})^2$$
$$= (9\,\text{ft})(9\,\text{ft}) = 81\,\text{ft}^2$$

Remember s^2 means s times s and not $2s$.

Perimeter, $P = 4s$
$$= 4(9\,\text{ft})$$
$$= 36\,\text{ft}$$

A calculation that is very useful in carpentry, sheet metal work, and many other trades involves finding the length of one side of a square given its area. The following formula will help.

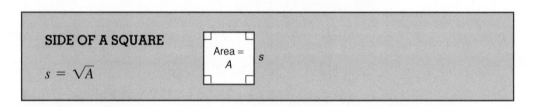

SIDE OF A SQUARE

Area = A

$s = \sqrt{A}$

Example 7

If a square opening is to have an area of 64 in.^{2} what must be its side length?

$s = \sqrt{A}$

$\quad = \sqrt{64\,\text{in.}^2}$

$\quad = 8\,\text{in.} \qquad (\sqrt{\text{in.}^2} = \text{in.})$

→ **More Practice**

Ready for a few problems involving squares? Try these.

(a) Find the area and perimeter of each of the following squares.

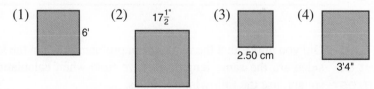

(1)   6'

(2) $17\frac{1}{2}$"

(3)   2.50 cm

(4)   3'4"

(b) Find the area of a square whose side length is $2\frac{1}{3}$ yd. Round to three significant digits.

(c) Find the length of the side of a square whose area is (1) 144 m², (2) 927 in.² Use a calculator and round to one decimal place.

(d) **Landscaping** At 29 cents per square foot, plus a $20 delivery charge, how much will it cost to purchase sod for a square lawn 24 ft 6 in. on a side? Round to the nearest cent.

(e) **Sheet Metal Trades** A square hot-air duct must have the same area as two rectangular ones 5 in. × 8 in. and 4 in. × 6 in. How long are the sides of the square duct?

→ **Answers**

(a) (1) $A = 36$ ft²; $P = 24$ ft

(2) $A = 306.25$ in.²; $P = 70$ in.

(3) $A = 6.25$ cm²; $P = 10$ cm

(4) $A \approx 11.11$ ft² (rounded) or $11\frac{1}{9}$ ft², using fractions; $P = 13$ ft 4 in.

(b) 5.44 yd²

(c) (1) 12 m   (2) 30.4 in.

(d) $194.07

(e) $(5 \times 8) + (4 \times 6) = 40$ in.² $+ 24$ in.² $= 64$ in.²
$\sqrt{64 \text{ in.}^2} = 8$ in.

For problem (c)(2),

C8-1

$\boxed{\vee}$ **927** $\boxed{=}$ → `30.4466747` or 30.4, rounded

For problem (d),

24.5 $\boxed{x^2}$ $\boxed{\times}$ **.29** $\boxed{+}$ **20** $\boxed{=}$ → `194.0725` or $194.07, rounded

---

Area and Perimeter of Parallelograms

You should recall that a parallelogram is a four-sided figure whose opposite pairs of sides are equal in length and parallel. Here are a few parallelograms:

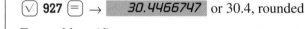

To find the area of a parallelogram use the following formula:

AREA OF A PARALLELOGRAM

$A = bh$

Notice that we do not use only the lengths of the sides to find the area of a parallelogram. The height h is the perpendicular distance between the base and the side opposite the base. The height h is perpendicular to the base b.

Example 8

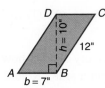

In parallelogram *ABCD,* the base is 7 in. and the height is 10 in. Find its area.

First, write the formula: $A = bh.$

Second, substitute the given values: $= (7 \text{ in.})(10 \text{ in.})$

Third, calculate the area including the units: $= 70 \text{ sq in.}$

Notice that we do not need to use the length of the slant side, 12 in., in this calculation.

→ **Your Turn**

Find the area of this parallelogram:

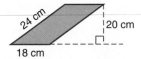

→ **Solution**

Area $= bh$

$= (18 \text{ cm})(20 \text{ cm})$

$= 360 \text{ cm}^2$

Be careful to use the correct dimensions.

This problem could also be worked using the 24-cm side as the base.

In this case the new height is 15 cm and the area is

$A = (24 \text{ cm})(15 \text{ cm})$

$= 360 \text{ cm}^2$

To find the perimeter of a parallelogram, simply add the lengths of the four sides or use the following formula:

PERIMETER OF A PARALLELOGRAM

$P = 2a + 2b$

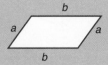

Example 9

Using the previous figure,

$P = 2a + 2b$

$= 2(24 \text{ cm}) + 2(18 \text{ cm})$

$= 48 \text{ cm} + 36 \text{ cm} = 84 \text{ cm}$

Find the perimeter and area for each parallelogram shown.

(a)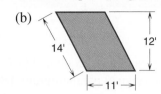

(b)

(c)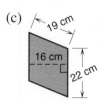

(d)

(e) A parallelogram-shaped opening has a base of 2 ft 9 in. and a height of 3 ft 3 in. Find its area. Your answer should be in square feet rounded to three significant digits.

→ **Answers**

(a) $P = 42$ in.; $A = 96$ in.^{2}

(b) $P = 50$ ft; $A = 132$ ft^2

(c) $P = 82$ cm; $A = 352$ cm^2

(d) $P = 58$ yd; $A = 125\frac{2}{3}$ yd^2

(e) $A = 8.94$ ft^2

  2.75 ☒ **3.25** ☐ → $\boxed{8.9375}$ ≈ 8.94

AREA OF A PARALLELOGRAM

You may be interested in where the formula for the area of a parallelogram came from. If so, follow this explanation.

Here is a typical parallelogram. As you can see, opposite sides $\overline{AB}$ and $\overline{DC}$ are parallel, and opposite sides $\overline{AD}$ and $\overline{BC}$ are also parallel.

Now let's cut off a small triangle from one side of the parallelogram by drawing the perpendicular line AE.

Next, reattach the triangular section to the other side of the figure, forming a rectangle. Notice that $EF = DC = b$.

(continued)

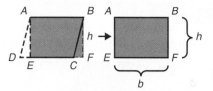

The area of the rectangle *ABFE* is $A = bh$. This is exactly the same as the area of the original parallelogram *ABCD*.

Area and Perimeter of Trapezoids

A trapezoid is a four-sided figure with only one pair of sides parallel. Here are a few trapezoids:

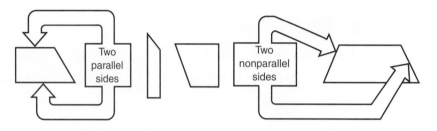

To find the area of a trapezoid use the following formula:

AREA OF A TRAPEZOID

$$A = \left(\frac{b_1 + b_2}{2}\right)h \qquad \text{or} \qquad A = \frac{h}{2}(b_1 + b_2)$$

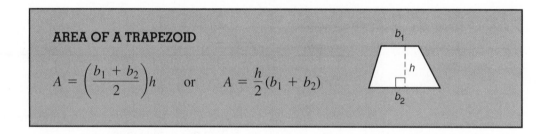

The factor $\left(\dfrac{b_1 + b_2}{2}\right)$ is the average length of the two parallel sides b_1 and b_2. The height h is the perpendicular distance between the two parallel sides.

The **perimeter** of a trapezoid is simply the sum of the lengths of all four sides.

Example 10

Find the area of this trapezoid-shaped metal plate.

The parallel sides, b_1 and b_2, have lengths 7 in. and 19 in. The height h is 6 in.

First, write the formula.
$$A = \left(\frac{b_1 + b_2}{2}\right)h$$

Next, substitute the given values.
$$= \left(\frac{7 \text{ in.} + 19 \text{ in.}}{2}\right)(6 \text{ in.})$$

Then, evaluate the quantity $\frac{7 + 19}{2}$.
$$= \left(\frac{26}{2}\text{ in.}\right)(6 \text{ in.})$$
$$= (13 \text{ in.})(6 \text{ in.})$$

Finally, multiply 13×6.
$$= 78 \text{ in.}^2$$

 7 ⊕ 19 ⊜ ⊙ 2 ⊗ 6 ⊜ →   *78.*

The perimeter of the trapezoid is $P = 8 \text{ in.} + 7 \text{ in.} + 9 \text{ in.} + 19 \text{ in.} = 43 \text{ in.}$

→ **Your Turn**

Find the area and perimeter of each of the following trapezoids.

(a)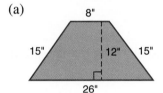

(b)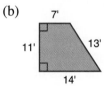

(c)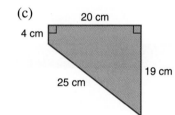

(d)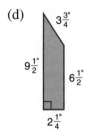

(e)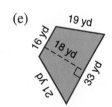

(f)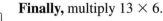

→ **Solutions**

(a) $A = \left(\dfrac{8 \text{ in.} + 26 \text{ in.}}{2}\right)(12 \text{ in.})$

    $= (17 \text{ in.})(12 \text{ in.})$

    $= 204 \text{ in.}^2$

$P = 8 \text{ in.} + 15 \text{ in.} + 26 \text{ in.} + 15 \text{ in.}$

    $= 64 \text{ in.}$

(b) $A = \left(\dfrac{7 \text{ ft} + 14 \text{ ft}}{2}\right)(11 \text{ ft})$

    $= (10.5 \text{ ft})(11 \text{ ft})$

    $= 115.5 \text{ ft}^2$

$P = 11 \text{ ft} + 7 \text{ ft} + 13 \text{ ft} + 14 \text{ ft}$

    $= 45 \text{ ft}$

(c) $A = \left(\dfrac{4 \text{ cm} + 19 \text{ cm}}{2}\right)(20 \text{ cm})$

    $= (11.5 \text{ cm})(20 \text{ cm})$

    $= 230 \text{ cm}^2$

$P = 4 \text{ cm} + 20 \text{ cm} + 19 \text{ cm} + 25 \text{ cm}$

    $= 68 \text{ cm}$

(d) $A = \left(\dfrac{9\frac{1}{2}\text{ in.} + 6\frac{1}{2}\text{ in.}}{2}\right)\left(2\frac{1}{4}\text{ in.}\right)$ $P = 3\frac{3}{4}\text{ in.} + 6\frac{1}{2}\text{ in.} + 2\frac{1}{4}\text{ in.} + 9\frac{1}{2}\text{ in.}$

$= (8\text{ in.})\left(2\frac{1}{4}\text{ in.}\right)$ $= 22\text{ in.}$

$= (8\text{ in.})(2.25\text{ in.})$

$= 18\text{ in.}^2$

(e) $A = \left(\dfrac{16\text{ yd} + 33\text{ yd}}{2}\right)(18\text{ yd})$ $P = 16\text{ yd} + 19\text{ yd} + 33\text{ yd} + 21\text{ yd}$

$= (24.5\text{ yd})(18\text{ yd})$ $= 89\text{ yd}$

$= 441\text{ yd}^2$

(f) $A = \left(\dfrac{16\text{ in.} + 6\text{ in.}}{2}\right)(12\text{ in.})$ $P = 13\text{ in.} + 16\text{ in.} + 13\text{ in.} + 6\text{ in.}$

$= (11\text{ in.})(12\text{ in.})$ $= 48\text{ in.}$

$= 132\text{ in.}^2$

> **Note** In a real job situation you will usually need to measure the height of the figure. Remember, the bases of a trapezoid are the parallel sides and h is the distance between them. ◄

Turn to Exercises 8-2 for practice in finding the area and perimeter of polygons.

Exercises 8-2 Area and Perimeter of Polygons

A. Find the perimeter of each polygon.

1.
5.7 m   6.2 m
8.1 m

2.
16 ft
87 ft

3.
31"
23"   19"
42"

4.
13'
14'
4'
11'

5.
10 in.   10 in.
6 in.   6 in.
16 in.

6.
8.1 m
27.2 m
25.2 m

7.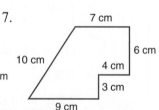
7 cm
10 cm   4 cm
18.7 m   3 cm
9 cm

8.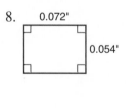
0.072"
0.054"

9.
6.15"
3.77"   4.82"
8.07"

10.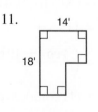
5.7'
7.2'   4.3'
0.9'
1.8'
3.9'

11.
14'
18'

12.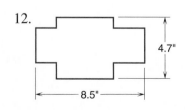
4.7"
8.5"

B. Find the perimeter and area of each figure. Round to the nearest tenth if necessary. (Assume right angles and parallel sides except where obviously otherwise.)

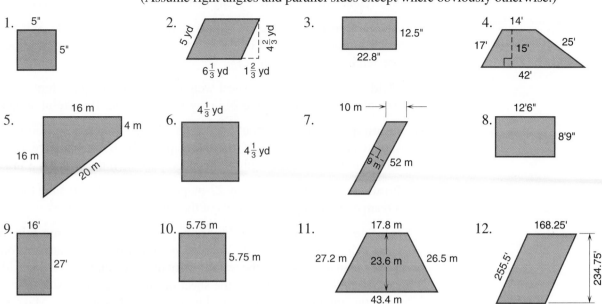

C. Practical Problems

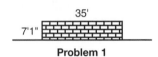

Problem 1

Problem 2

Problem 3

1. **Masonry**   How many bricks will it take to build the wall shown if each brick measures $4\frac{1}{4}$ in. $\times$ $8\frac{3}{4}$ in. including mortar? (*Hint:* Change the wall dimensions to inches.)

2. **Carpentry**   How many square feet of wood are needed to build the cabinet in the figure? (Assume that wood is needed for all six surfaces.)

3. **Flooring and Carpeting**   The floor plan of a room is shown in the figure. If the entire room is to be carpeted, how many square *yards* of carpet are needed? At $39.95 per square yard, how much will it cost?

4. **Construction**   The amount of horizontal wood siding (in square feet) needed to cover a given wall area can be calculated from the formula

    Area of siding (sq ft) = $K \times$ total area to be covered (sq ft)

    where K is a constant that depends on the type and size of the siding used. How much 1-by-6 rustic shiplapped drop siding is needed to cover a rectangular wall 60 ft by 14 ft? Use $K = 2.19$ and round to three significant digits.

5. **Flooring and Carpeting**   How many bundles of strip flooring are needed to cover a rectangular floor area 16 ft by 12 ft 6 in. in size if a bundle of strip flooring contains 24 sq ft, and if 30% extra must be allowed for side and end matching?

6. **Construction**   A large rectangular window opening measures $72\frac{5}{8}$ in. by $60\frac{3}{8}$ in. Calculate the area of this opening. Round to the nearest square inch.

7. **Masonry**   Find the area of one stretcher course of 20 concrete blocks if the blocks are $7\frac{5}{8}$ in. long and 4 in. high, and if the mortar joints are $\frac{3}{8}$ in. thick. Round to two significant digits. (*Hint:* Twenty blocks require 19 mortar joints.)

8. **Masonry**   To determine the approximate number of concrete blocks needed to construct a wall, the following formula is often used.

    Number of blocks per course = $N \times$ area of one course (sq ft)

    where N is a number that depends on the size of the block. How many 8-in. by 8-in. by 16-in. concrete blocks are needed to construct a foundation wall 8 in. thick with a total length of 120 ft, if it is laid five courses high? Use $N = \frac{9}{8}$ for this block size. (*Hint:* The blocks are set lengthwise. Translate all dimensions to inches.)

9. **Automotive Trades** A roll of gasket material is 18 in. wide. What length is needed to obtain 16 sq ft of the material? (*Careful:* The numbers are not expressed in compatible units.)

10. **Welding** A type of steel sheet weighs 3.125 lb per square foot. What is the weight of a piece measuring 24 in. by 108 in.? (*Hint:* Be careful of units.)

11. **Printing** A rule of thumb in printing says that the area of the typed page should be half the area of the paper page. If the paper page is 5 in. by 8 in., what should the length of the typed page be if the width is 3 in.?

12. **Printing** A ream of 17-in. by 22-in. paper weighs 16.0 lb. Find the weight of a ream of 19-in. by 24-in. sheets of the same density of paper.

13. **Printing** A sheet of 25-in. by 38-in. paper must be cut into 6-in. by 8-in. cards. The cuts can be made in either direction but they must be consistent. Which plan will result in the least amount of waste, cutting the 6-in. side of the card along the 25-in. side of the sheet or along the 38-in. side of the sheet? State the amount of waste created by each possibility. (*Hint:* Make a cutting diagram for each possibility.)

14. **HVAC** To determine the size of a heating system needed for a home, an installer needs to calculate the heat loss through surfaces such as walls, windows, and ceilings exposed to the outside temperatures. The formula

    $L = kDA$

    gives the heat loss L (in BTU per hour) if D is the temperature difference between the inside and outside, A is the area of the surface in sq ft, and k is the insulation rating of the surface. Find the heat loss per hour for a 7-ft by 16-ft single-pane glass window ($k = 1.13$) if the outside temperature is 40° and the desired inside temperature is 68°. Round to two significant digits.

15. **Construction** A playground basketball court 94 ft long and 46 ft wide is to be resurfaced at a cost of $3.25 per sq ft. What will the resurfacing cost?

16. **Construction** A rectangular structure 56 ft long and 38 ft wide is being built. The walls require studs 16 in. o.c., that is, 16 in. is the center-to-center distance separating the studs. Eight additional studs are needed for starters and corners. Determine the total number of studs required.

17. **Construction** A contractor needed a small workshop. He found a preengineered steel building advertised for $12,619. If the building is a 36-ft by 36-ft square, what is the cost of the building per square foot? (Round to the nearest cent.)

18. **Flooring and Carpeting** When expressed in English units, imported area rugs seem to come in odd sizes because they are converted from metric measurements to feet and inches. One such area rug measured 5 ft 3 in. by 7 ft 9 in. and cost $399.

    (a) What is the area of this rug in square feet? (*Hint:* Convert the inch measurements to fractions of a foot before calculating the area.)

    (b) What is the cost of the rug per square foot?

19. **Flooring and Carpeting** Because carpeting typically comes in 12-ft widths, the carpet for a 22-ft by 16-ft room must be laid out as shown in the diagram.

The four sub-regions shown have the following dimensions:

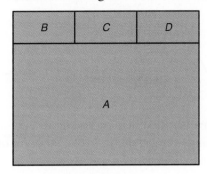

A 12 ft $\times$ 22$\frac{1}{4}$ ft	B 4 ft $\times$ 7$\frac{1}{2}$ ft
C 4 ft $\times$ 7$\frac{1}{2}$ ft	D 4 ft $\times$ 7$\frac{1}{2}$ ft

(*Note:* An extra $\frac{1}{4}$ ft has been added to each length to give the installer a margin of error.)

(a)   Determine the total area of carpet needed for this room.

(b)   At \$5.65 per square foot installed, how much will it cost to carpet the room?

When you have finished these problems, check your answers to the odd-numbered problems in the Appendix, then continue in Section 8-3 with the study of other polygons.

8-3 Triangles, Regular Hexagons, and Irregular Polygons

A *triangle* is a polygon having three sides and therefore three angles.

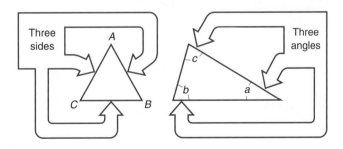

Just as there are several varieties of four-sided figures—squares, rectangles, parallelograms, and trapezoids—there are also several varieties of triangles. Fortunately, one area formula can be used with all triangles. First, you must learn to identify the many kinds of triangles that appear in practical work.

Triangles may be classified according to their sides.

An **equilateral triangle** is one in which all three sides have the same length. An equilateral triangle is also said to be *equiangular* because, if the three sides are equal, the three angles will also be equal. In fact, because the interior angles of any triangle always add up to 180°, each angle of an equilateral triangle must equal 60°.

An **isosceles triangle** is one in which at least two of the three sides are equal. It is always true that the two angles opposite the equal sides are also equal.

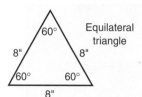

Two equal sides

Isosceles triangle

12"   12"

6"

Equal angles

A **scalene triangle** is one in which no sides are equal.

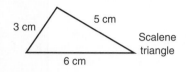

Equilateral triangle

60°

8"   8"

60°   60°

8"

Scalene triangle

3 cm   5 cm

6 cm

Triangles may be classified also according to their angles. An **acute triangle** is one in which all three angles are acute (less than 90°). An **obtuse triangle** contains one obtuse angle (greater than 90°).

A **right triangle** contains a 90° angle. Two of the sides, called the *legs,* are perpendicular to each other. The longest side of a right triangle is always the side opposite to the right angle. This side is called the *hypotenuse* of the triangle. An isosceles or scalene triangle can be a right triangle, but an equilateral triangle can never be a right triangle.

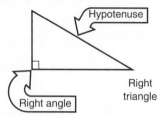

Acute triangle

Obtuse triangle

Hypotenuse

Right angle

Right triangle

→ **Your Turn**

Identify the following triangles as being equilateral, isosceles, or scalene. Also name the ones that are right triangles.

(a)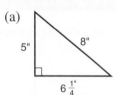

5"   8"

6 1/4"

(b)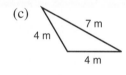

18'

18'

18'

(c)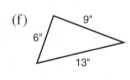

7 m

4 m

4 m

(d)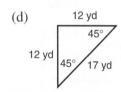

12 yd

45°

12 yd

45°   17 yd

(e)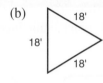

60°   60°

(f)

9"

6"

13"

→ **Answers**

(a) Scalene   (b) Equilateral   (c) Isosceles

(d) Isosceles   (e) Equilateral   (f) Scalene

(a) and (d) are also right triangles.

A Closer Look   Did you have trouble with (d) or (e)? The answer to each depends on the fact that the interior angles of a triangle add up to 180°. In (d) the two given angles add up to 90°. Therefore, the missing angle must be 90° if the three angles sum to 180°. This makes (d) a right triangle.

In (e) the two angles shown sum to 120°, so the third angle must be 180° − 120° or 60°. Because the three angles are equal we know the triangle is an equiangular triangle, and this means it is also equilateral. ◄

Pythagorean Theorem

The Pythagorean theorem is a rule or formula that allows us to calculate the length of one side of a right triangle when we are given the lengths of the other two sides. Although the formula is named after the ancient Greek mathematician Pythagoras, it was known to Babylonian engineers and surveyors more than a thousand years before Pythagoras lived.

PYTHAGOREAN THEOREM

For any right triangle, the square of the hypotenuse is equal to the sum of the squares of the other two sides.

$$c^2 = a^2 + b^2$$

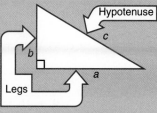

Geometrically, this means that if the squares are built on the sides of the triangle, the area of the larger square is equal to the sum of the areas of the two smaller squares. For the right triangle shown

$$3^2 + 4^2 = 5^2$$

$$9 + 16 = 25$$

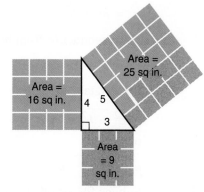

Notice that it does not matter whether leg a is 3 and leg b is 4 or vice versa. However, the longest side, opposite the right angle, must be the hypotenuse c.

Example 1

Let's use this formula to find the distance d between points A and B for this rectangular plot of land.

The Pythagorean theorem tells us that for the right triangle ABC,

$$d^2 = 36^2 + 20^2$$
$$= 1296 + 400$$
$$= 1696$$

Taking the square root of both sides of the equation, we have

$$d \approx 41 \text{ ft} \quad \text{rounded to the nearest foot}$$

Here are two different ways to find d using a calculator:

C8-2

$\boxed{\sqrt{}}$ **36** $\boxed{x^2}$ $\boxed{+}$ **20** $\boxed{x^2}$ $\boxed{=}$ → `41.18252056`

Or **36** $\boxed{x^2}$ $\boxed{+}$ **20** $\boxed{x^2}$ $\boxed{=}$ $\boxed{\sqrt{}}$ $\boxed{\text{ANS}}$ $\boxed{=}$ → `41.18252056`

With the first option, if your calculator does not automatically open parentheses with square root, then you must press $\boxed{(}$ after $\boxed{\sqrt{}}$.

This rule may be used to find any one of the three sides of a right triangle if the other two sides are given.

To solve for the hypotenuse c,

First, write the original formula

$$c^2 = a^2 + b^2$$

Then, take the square root of both sides of the formula.

$$\sqrt{c^2} = \sqrt{a^2 + b^2}$$

or

$$c = \sqrt{a^2 + b^2}$$

> **Note**   The square root of a sum is not equal to the sum of the square roots. For example $\sqrt{9 + 16} = \sqrt{25} = 5$, and this is not the same as $\sqrt{9} + \sqrt{16}$, which equals 7. The algebraic expression $\sqrt{a^2 + b^2}$ cannot be simplified further. ◄

To solve for one of the legs,

First, write the original formula,

$$c^2 = a^2 + b^2$$

Next, subtract b^2 from both sides,

$$c^2 - b^2 = a^2$$

Finally, take the square root of both sides,

$$\sqrt{c^2 - b^2} = \sqrt{a^2}$$

or

$$a = \sqrt{c^2 - b^2}$$

Solving for the other leg, b, yields a similar result.

$$b = \sqrt{c^2 - a^2}$$

These results are summarized in the following table.

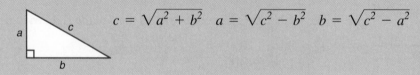

EQUIVALENT FORMS OF THE PYTHAGOREAN THEOREM

$$c = \sqrt{a^2 + b^2} \quad a = \sqrt{c^2 - b^2} \quad b = \sqrt{c^2 - a^2}$$

> **Careful**   These three formulas are valid only for a *right* triangle—a triangle with a right or 90° angle. ◄

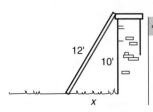

→ **Your Turn**

Use these formulas to solve the problem.

Masonry   A mason wants to use a 12-ft ladder to reach the top of a 10-ft wall. How far must the base of the ladder be from the base of the wall?

Use the formula $a = \sqrt{c^2 - b^2}$ to get

$$x = \sqrt{12^2 - 10^2}$$
$$= \sqrt{144 - 100}$$
$$= \sqrt{44}$$
$$\approx 6.63 \text{ ft or 6 ft 8 in., rounded to the nearest inch}$$

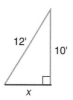

If you need help with square roots, pause here and return to page 373 for a review.

For some right triangles, the three side lengths are all whole numbers rather than fractions or decimals. Such special triangles have been used in technical work since Egyptian surveyors used them 2000 years ago to lay out rectangular fields for farming.

Here are a few "Pythagorean triple" right triangles:

    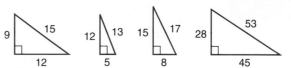

→ **More Practice**

Solve the following problems, using the Pythagorean theorem. (If your answer does not come out exactly, round to the nearest tenth.)

(a) Find the missing side for each of the following right triangles.

(1)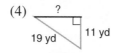
16'   ?   12'

(2) 48" ? 50"

(3) ? 8 cm 9 cm

(4) ? 19 yd 11 yd

(5) 24' ? 45'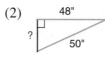

(6) ? $2\frac{1}{2}''$ $3\frac{1}{2}''$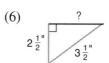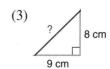

(7) Find the hypotenuse of a right triangle with legs measuring 4.6 m and 6.2 m.

(b) **Carpentry**   A rectangular table measures 4 ft by 6 ft. Find the length of a diagonal brace placed beneath the table top.

(c) **Manufacturing**   What is the distance between the centers of two pulleys if one is placed 9 in. to the left and 6 in. above the other?

(d) **Machine Trades**   Find the length of the missing dimension x in the part shown in the figure.

(e) **Carpentry**   Find the diagonal length of the stairway shown in the drawing.

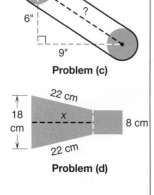

6"
?
9"

Problem (c)

22 cm
18 cm x 8 cm
22 cm

Problem (d)

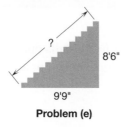

?
8'6"
9'9"

Problem (e)

→ **Answers**

(a) (1) 20 ft   (2) 14 in.   (3) 12.0 cm   (4) 15.5 yd

(5) 51 ft   (6) 2.4 in.   (7) 7.7 m

(b) 7.2 ft   or   7 ft $2\frac{1}{2}$ in.

(c) 10.8 in.

(d) 21.4 cm

(e) 12.9 ft or 12 ft 11 in., rounded to the nearest inch.

C8-3

In feet: $\boxed{\surd}$ **8.5** $\boxed{x^2}$ $\boxed{+}$ **9.75** $\boxed{x^2}$ $\boxed{=}$ → $\boxed{12.93493332}$ ≈ 12.9 ft

Converting the decimal part to inches:

$\boxed{-}$ **12** $\boxed{=}$ $\boxed{\times}$ **12** $\boxed{=}$ → $\boxed{11.21919984}$ ≈ 11 in.

A Closer Look   In problem (d), the first difficulty is to locate the correct triangle.

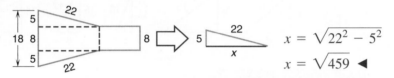

$$x = \sqrt{22^2 - 5^2}$$
$$x = \sqrt{459} \blacktriangleleft$$

Area of a Triangle   The area of any triangle, no matter what its shape or size, can be found using the same simple formula.

> **AREA OF A TRIANGLE**
>
> Area, $A = \frac{1}{2}bh$ or $A = \dfrac{bh}{2}$
>
> b = base
> h = height

Any side of a triangle can be used as the base. For a given base, the height is the perpendicular distance from the opposite vertex to the base. The following three identical triangles show the three different base-height combinations.

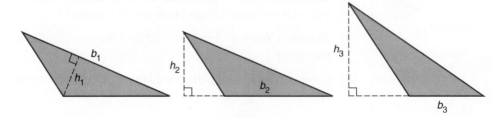

Notice that for the last two cases the height falls outside the triangle, and the base must be extended to meet the height at a right angle. For any such triangle, the three area calculations are equal:

$$A = \tfrac{1}{2}b_1h_1 = \tfrac{1}{2}b_2h_2 = \tfrac{1}{2}b_3h_3$$

Example 2

(a) In triangle *PQR* the base *b* is 13 in., and the height *h* is 8 in. Applying the formula, we have

Area, $A = \frac{1}{2}bh$

$= \frac{1}{2}(13 \text{ in.})(8 \text{ in.})$

$= 52 \text{ in.}^2$

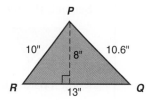

(b) In triangle *STU*,

$b = 25$ cm

$h = 16$ cm

Area, $A = \frac{1}{2}(25 \text{ cm})(16 \text{ cm})$

$= 200 \text{ cm}^2$

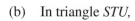

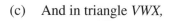

(c) And in triangle *VWX*,

$b = 20$ cm

$h = 12$ cm

Area, $A = \frac{1}{2}(20 \text{ cm})(12 \text{ cm})$

$= 120 \text{ cm}^2$

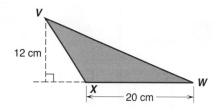

→ **Your Turn**

Find the area of triangle *ABC*.

→ **Solution**

Base, $b = 6.5$ ft

height, $h = 8$ ft

Area, $A = \frac{1}{2}bh$

$= \frac{1}{2}(6.5 \text{ ft})(8 \text{ ft})$

$= 26 \text{ ft}^2$

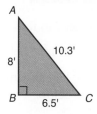

Notice that for a right triangle, the two sides meeting at the right angle can be used as the base and height.

AREA OF A TRIANGLE

The area of any triangle is equal to one-half of its base times its height, $A = \frac{1}{2}bh$.

To learn where this formula comes from, follow this explanation:

First, for any triangle *ABC*,

draw a second identical triangle *DEF*.

(continued)

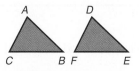

Second, flip the second triangle over and attach it to the first,

to get a parallelogram.

Finally, if the base and height of the original triangle are b and h, then these are also the base and height for the parallelogram.

The area of the parallelogram is

$A = bh$

so the area of the original triangle must be

$A = \frac{1}{2}bh$

Example 3

Suppose we need to find the area of the isosceles triangle shown.

We must first find the height.

If we draw the height perpendicular to the 6-in. base, it will divide the base in half. Then in the triangle below

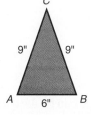

$h = \sqrt{9^2 - 3^2}$

$\quad = \sqrt{81 - 9}$

$\quad = \sqrt{72} \approx 8.485$ in.

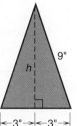

Now we can find the area:

$A = \frac{1}{2}bh$

$\quad \approx \frac{1}{2}(6 \text{ in.})(8.485 \text{ in.}) \approx 25.455 \text{ in.}^2$

$\quad \approx 25 \text{ in.}^2$ rounded

Using a calculator the entire calculation looks like this:

 C8-4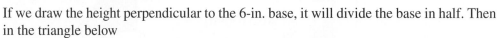

In general, to find the area of an isosceles triangle you may use the following formula.

AREA OF AN ISOSCELES TRIANGLE

Area, $A = \frac{1}{2}b\sqrt{a^2 - \left(\frac{b}{2}\right)^2}$

→ **Your Turn**

Construction An attic wall has the shape shown. How many square feet of insulation are needed to cover the wall? Round to the nearest square foot.

→ **Solution**

The triangle is isosceles since two sides are equal. Substituting $a = 7$ ft and $b = 12$ ft into the formula, we have

$$\text{Area}, A = \frac{1}{2}(12\text{ ft})\sqrt{(7\text{ ft})^2 - \left(\frac{12\text{ ft}}{2}\right)^2}$$

$$= 6\sqrt{49 - 36}$$

$$= 6\sqrt{13} \approx 21.63\text{ ft}^2, \text{ or } 22\text{ ft}^2 \quad \text{rounded}$$

C8-5 .5 ⊗ 12 ⊗ √ 7 x^2 ⊖ 6 x^2 ⊜ → **21.63330765**

For an equilateral triangle an even simpler formula may be used.

AREA OF AN EQUILATERAL TRIANGLE

Area, $A = \dfrac{a^2\sqrt{3}}{4} \approx 0.433a^2,$ approximately

---

→ **Your Turn**

Sheet Metal Trades What area of sheet steel is needed to make a triangular pattern where each side is 3 ft long?

Try it. Use the given formula and round to one decimal place.

→ **Solution**

Area, $A \approx 0.433a^2$, where $a = 3$ ft
$\approx 0.433(3\text{ ft})^2$
$\approx 0.433(9\text{ ft}^2)$
$\approx 3.9\text{ ft}^2$ rounded

→ **More Practice**

Work these problems for practice.

(a) In the following problems find the area of each triangle. (Round your answer to one decimal place if it does not come out exactly.)

(1)

15"

18"

(2)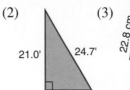

21.0' 24.7'

13.0'

(3)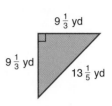

22.8 cm 32.1 cm 31.5 cm

35.6 cm

(4)

$9\frac{1}{3}$ yd

$9\frac{1}{3}$ yd

$13\frac{1}{5}$ yd

(5)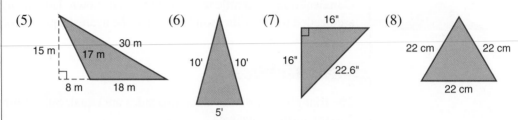

15 m 30 m 17 m

8 m 18 m

(6)

10' 10'

5'

(7)

16"

16" 22.6"

(8)

22 cm 22 cm

22 cm

13'3"

9'6"

Problem (b)

10' 8'

16'

Problem (c)

(b) **Carpentry**   Find the area, to the nearest square foot, of the triangular patio deck shown.

(c) **Flooring and Carpeting**   At a cost of 97 cents per square foot, what would it cost to tile the floor of a triangular work space shown?

→ **Answers**

(a) (1) 135 in.²   (2) 136.5 ft²   (3) 359.1 cm²

     (4) 43.6 yd²   (5) 135 m²   (6) 24.2 ft²

     (7) 128 in.²   (8) 209.6 cm²

For (a)(8): Using the formula for an equilateral triangle, we enter

.433 ⊠ 22 [x²] [=] →   *209.572*

(b) 63 sq ft   (c) $62.08

A Closer Look   In problem (a)(5) you should have used 18 m as the base and *not* 26 m. $b = 18$ m, $h = 15$ m.

In problem (c) you should have noticed that the height was given as 8 ft. ◀

HERO'S FORMULA

It is possible to find the area of any triangle from the lengths of its sides. The formula that enables us to do this was first devised almost 2000 years ago by Hero or Heron, a Greek mathematician. It is a complicated formula, and you may feel like a hero yourself if you learn how to use it.

(continued)

Area, $A = \sqrt{s(s - a)(s - b)(s - c)}$ For any triangle

where $s = \dfrac{a + b + c}{2}$

(s stands for "semiperimeter," or half of the perimeter.)

Example:

In the figure shown,

$a = 6$ in., $b = 5$ in., $c = 7$ in.

To find the area,

First, find s.

$$s = \frac{6 + 5 + 7}{2} = \frac{18}{2} = 9$$

Next, substitute to find the area:

$$A = \sqrt{9(9 - 6)(9 - 5)(9 - 7)}$$
$$= \sqrt{9 \cdot 3 \cdot 4 \cdot 2}$$
$$= \sqrt{216} \approx 14.7 \text{ in.}^2 \quad \text{rounded}$$

Example:

Find the area of this triangle.

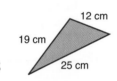

Step 1 $s = \dfrac{19 + 12 + 25}{2} = \dfrac{56}{2} = 28$

Step 2 $A = \sqrt{28(28 - 19)(28 - 12)(28 - 25)}$
$$= \sqrt{28 \cdot 9 \cdot 16 \cdot 3}$$
$$= \sqrt{12096} = 109.98\ldots \approx 110 \text{ cm}^2, \text{ rounded}$$

On a calculator, Step 2 looks like this:

C8-6

$\boxed{\surd}$ **28** $\boxed{\times}$ $\boxed{(}$ **28** $\boxed{-}$ **19** $\boxed{)}$ $\boxed{\times}$ $\boxed{(}$ **28** $\boxed{-}$ **12** $\boxed{)}$ $\boxed{\times}$ $\boxed{(}$ **28** $\boxed{-}$ **25** $\boxed{)}$ $\boxed{=}$

$\rightarrow$ *109.9818167*

Regular Hexagons

A polygon is a plane geometric figure with three or more sides. A *regular polygon* is one in which all sides are the same length and all angles are equal. The equilateral triangle and the square are both regular polygons.

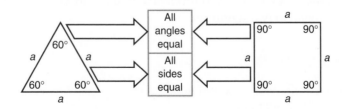

A *pentagon* is a five-sided polygon. A *regular pentagon* is one whose sides are all the same length and whose angles are equal.

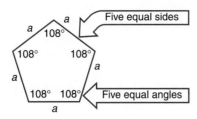

The regular *hexagon* is a six-sided polygon in which each interior angle is 120° and all sides are the same length.

Regular hexagon:
All sides equal.
All angles 120°.

If we draw the diagonals, six equilateral triangles are formed.

Because you already know that the area of one equilateral triangle is approximately $A \approx 0.433a^2$, the area of the complete hexagon will be 6 times $0.433a^2$ or $2.598a^2$.

AREA OF A REGULAR HEXAGON

Area, $A = \dfrac{3a^2\sqrt{3}}{2} \approx 2.598a^2$

Example 4

If $a = 6.00$ cm,
First, square the number in parentheses.
Then multiply.

 2.598 ⊠ **6** $\boxed{x^2}$ ⊟ → ▮ *93.528*

$A \approx 2.598(6\text{ cm})^2$
$\approx 2.598(36\text{ cm}^2)$
$\approx 93.5\text{ cm}^2$, rounded

→ Your Turn

Machine Trades Use this formula to find the area of a hexagonal plate 4.0 in. on each side. Round to two significant digits.

→ Solution

Area, $A \approx 2.598a^2$ $a = 4$ in.
$\approx 2.598(4\text{ in.})^2$
$\approx 2.598(16\text{ in.}^2)$
$\approx 42\text{ in.}^2$, rounded

Because the hexagon is used so often in practical and technical work, several other formulas may be helpful to you.

A hexagon is often measured by specifying the distance across the corners or the distance across the flats.

Distance across
the corners

Distance across
the flats

DIMENSIONS OF A REGULAR HEXAGON

Distance across the corners, $d = 2a$ or $a = 0.5d$

Distance across the flats, $f \approx 1.732a$ or $a \approx 0.577f$

Example 5

Machine Trades   For a piece of hexagonal bar stock where $a = \frac{1}{2}$ in., the distance across the corners would be

$$d = 2a \qquad \text{or} \qquad d = 1 \text{ in.}$$

The distance across the flats would be

$$f \approx 1.732a \qquad \text{or} \qquad f \approx 0.866 \text{ in.}$$

→ **Your Turn**

Machine Trades   If the cross-section of a hexagonal nut measures $\frac{3}{4}$ in. across the flats, find

(a) The side length, a.

(b) The distance across the corners, d.

(c) The cross-sectional area.

Round to three significant digits.

→ **Solutions**

$f = \frac{3}{4}$ in.   or   0.75 in.

(a) $a \approx 0.577f$
$\approx 0.577(0.75 \text{ in.})$
$\approx 0.433 \text{ in.}$

(b) $d = 2a$
$\approx 2(0.433 \text{ in.})$
$\approx 0.866 \text{ in.}$

(c) Area, $A \approx 2.598a^2$
$\approx 2.598(0.433 \text{ in.})^2$
$\approx 2.598(0.1875 \text{ in.}^2)$
$\approx 0.487 \text{ in.}^2$

→ **More Practice**

Work these problems for practice.

(a) Fill in the blanks with the missing dimensions of each regular hexagon. Round to two decimal places if necessary.

	a	d	f	A
(1)	2 in.			
(2)		$\frac{3}{4}$ in.		
(3)			6 mm	

(b) **Carpentry** A hex nut has a side length of $\frac{1}{4}$ in. From the following list, pick the smallest size wrench that will fit it.

(1) $\frac{1}{4}$ in. (2) $\frac{3}{8}$ in. (3) $\frac{7}{16}$ in. (4) $\frac{1}{2}$ in. (5) $\frac{5}{8}$ in.

→ **Answers**

		a	d	f	A
(a)	(1)	2 in.	4 in.	3.46 in.	10.39 in.^{2}
	(2)	$\frac{3}{8}$ in.	$\frac{3}{4}$ in.	0.65 in.	0.37 in.^{2}
	(3)	3.46 mm	6.92 mm	6 mm	31.14 mm^2

For problem (a)(3),

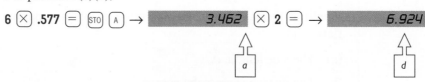

6 ⊗ .577 ⊟ STO A → ▮ 3.462 ▮ ⊗ 2 ⊟ → ▮ 6.924 ▮
⬆ a ⬆ d

2.598 ⊗ RCL A x^2 ⊟ ▮ 31.13818351 ▮ Area

(b) $\frac{7}{16}$ in.

A Closer Look

Let's take a closer look at problem (b).

By drawing a picture we can see that we must find f. Because $a = \frac{1}{4}$ in., we have

$f \approx 1.732a$
$\approx 1.732(0.25)$
$\approx 0.433 \text{ in.}$

By checking a list of wrench sizes in their decimal equivalents or converting the fractions given, we find that $\frac{7}{16}$ in. (0.4375 in.) is the smallest size that will fit the nut. ◄

CLASSIFICATION OF POLYGONS

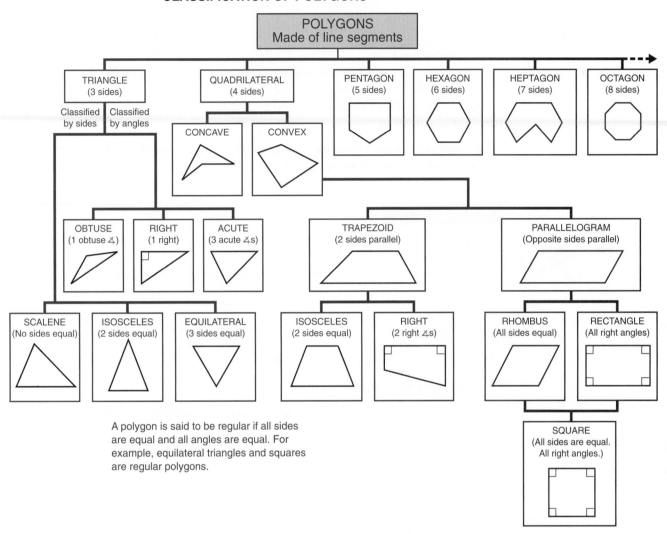

A polygon is said to be regular if all sides are equal and all angles are equal. For example, equilateral triangles and squares are regular polygons.

Irregular Polygons

Often, the shapes of polygons that appear in practical work are not the simple geometric figures that you have seen so far. The easiest way to work with irregular polygon shapes is to divide them into simpler, more familiar figures.

For example, look at this L-shaped figure:

There are two ways to find the area of this shape.

Method 1: Addition

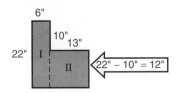

By drawing the dashed line, we have divided this figure into two rectangles, I and II. I is 22 in. by 6 in. for an area of 132 in.^{2}. II is 13 in. by 12 in. for an area of 156 in.^{2}. The total area is

$$132 + 156 = 288 \text{ in.}^2$$

Method 2: Subtraction

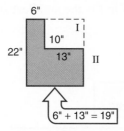

In this method we subtract the area of the small rectangle (I) from the area of the large rectangle (II). The area of I is

$$10 \times 13 = 130 \text{ in.}^2$$

The area of II is

$$22 \text{ in.} \times 19 \text{ in.} = 418 \text{ in.}^2$$
$$418 - 130 = 288 \text{ in.}^2$$

Notice that in the addition method we had to calculate the height of rectangle II as 22 in. − 10 in. or 12 in. In the subtraction method we had to calculate the width of rectangle II as 6 in. + 13 in. or 19 in.

Which method you use to find the area of an irregular polygon depends on your ingenuity and on which dimensions are given.

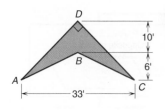

Example 6

Find the area of the shape *ABCD*.

You may at first wish to divide this into two triangles and add their areas.

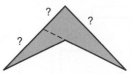 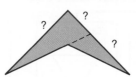

However, the dimensions provided do not allow you to do this.

The only option here is to subtract the area of triangle I from the area of triangle II.

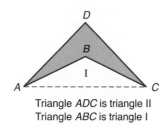

Triangle *ADC* is triangle II
Triangle *ABC* is triangle I

	Base	Height	Area
Triangle II	33 ft	16 ft	264 ft^2
Triangle I	33 ft	6 ft	− 99 ft^2
		Area =	165 ft^2

If both methods will work, pick the one that looks easier and requires the simpler arithmetic.

→ Your Turn

Find the area of this shape. Round to the nearest square inch.

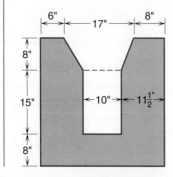

→ **Solution**

You might have been tempted to split the shape up into the five regions indicated in the figure. This will work, but there is an easier way.

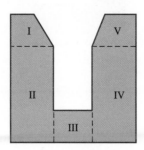

For the alternative shown in the next figure we need to compute only three areas, then we subtract II and III from I.

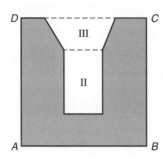

The outside rectangle *ABCD* is area I.

	Base	*Height*	*Area*	
Figure I	6 in. + 17 in. + 8 in. = 31 in.	8 in. + 15 in. + 8 in. = 31 in.	31 in. × 31 in.	= 961 in.^{2}
Figure II	10 in.	15 in.	10 in. × 15 in.	= 150 in.^{2}
Figure III (trapezoid)	17 in. and 10 in.	8 in.	$\left(\dfrac{17\text{ in.} + 10\text{ in.}}{2}\right)(8\text{ in.})$	= 108 in.^{2}

The total area is 961 in.^{2} − 150 in.^{2} − 108 in.^{2} = 703 in.^{2}

▶ **Learning Help**   When working with complex figures, organize your work carefully. Neatness will help eliminate careless mistakes. ◀

→ **More Practice**

Now, for practice, find the areas of the following shapes. Round to the nearest tenth.

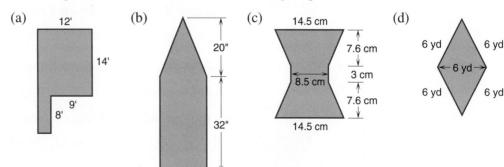

(a) 12' 14' 9' 8'

(b) 20" 32" 16"

(c) 14.5 cm 7.6 cm 3 cm 8.5 cm 7.6 cm 14.5 cm

(d) 6 yd 6 yd 6 yd 6 yd 6 yd

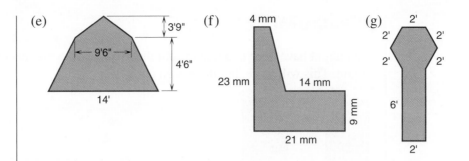

→ Solutions

(a) The figure shows how to set up the shape to use the subtraction method.

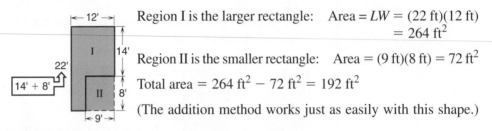

Region I is the larger rectangle: Area $= LW = (22\text{ ft})(12\text{ ft})$
$$= 264\text{ ft}^2$$

Region II is the smaller rectangle: Area $= (9\text{ ft})(8\text{ ft}) = 72\text{ ft}^2$

Total area $= 264\text{ ft}^2 - 72\text{ ft}^2 = 192\text{ ft}^2$

(The addition method works just as easily with this shape.)

(b) Divide this shape into a triangle (I) and a rectangle (II). Find the sum of the two areas.

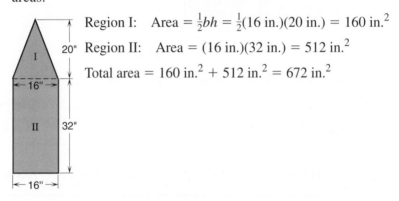

Region I: Area $= \frac{1}{2}bh = \frac{1}{2}(16\text{ in.})(20\text{ in.}) = 160\text{ in.}^2$

Region II: Area $= (16\text{ in.})(32\text{ in.}) = 512\text{ in.}^2$

Total area $= 160\text{ in.}^2 + 512\text{ in.}^2 = 672\text{ in.}^2$

(c) Divide this shape into two identical trapezoids (I and III) and a rectangle (II).

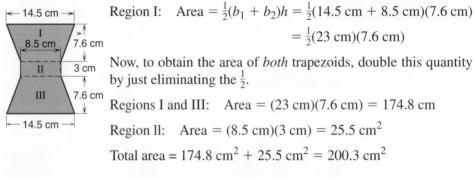

Region I: Area $= \frac{1}{2}(b_1 + b_2)h = \frac{1}{2}(14.5\text{ cm} + 8.5\text{ cm})(7.6\text{ cm})$
$$= \frac{1}{2}(23\text{ cm})(7.6\text{ cm})$$

Now, to obtain the area of *both* trapezoids, double this quantity by just eliminating the $\frac{1}{2}$.

Regions I and III: Area $= (23\text{ cm})(7.6\text{ cm}) = 174.8\text{ cm}$

Region II: Area $= (8.5\text{ cm})(3\text{ cm}) = 25.5\text{ cm}^2$

Total area $= 174.8\text{ cm}^2 + 25.5\text{ cm}^2 = 200.3\text{ cm}^2$

(d) Divide this shape into two identical equilateral triangles as shown.

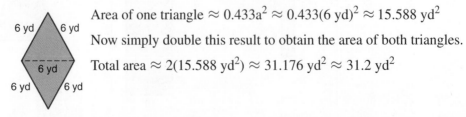

Area of one triangle $\approx 0.433a^2 \approx 0.433(6\text{ yd})^2 \approx 15.588\text{ yd}^2$

Now simply double this result to obtain the area of both triangles.

Total area $\approx 2(15.588\text{ yd}^2) \approx 31.176\text{ yd}^2 \approx 31.2\text{ yd}^2$

(e) Divide this shape into a triangle (I) and a trapezoid (II).

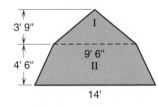

First, convert feet and inch measurements to decimal feet:

$$9'6'' = 9\frac{6}{12} \text{ ft} = 9.5 \text{ ft}$$

$$3'9'' = 3\frac{9}{12} \text{ ft} = 3.75 \text{ ft}$$

$$4'6'' = 4.5 \text{ ft}$$

Then, compute the two areas and find their sum.

Region I: Area $= \frac{1}{2}(9.5 \text{ ft})(3.75 \text{ ft}) = 17.8125 \text{ ft}^2$

Region II: Area $= \frac{1}{2}(9.5 \text{ ft} + 14 \text{ ft})(4.5 \text{ ft}) = 52.875 \text{ ft}^2$

Total area $= 17.8125 \text{ ft}^2 + 52.875 \text{ ft}^2 = 70.6875 \text{ ft}^2 \approx 70.7 \text{ ft}^2$, rounded

(f) Divide this shape into a trapezoid (I) and, a rectangle (II) as shown. Notice that the bottom base of the trapezoid is 21 mm − 14 mm = 7 mm, and that the height of the trapezoid is 23 mm − 9 mm = 14 mm.

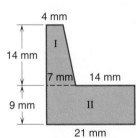

Region I: Area $= \frac{1}{2}(4 \text{ mm} + 7 \text{ mm})(14 \text{ mm}) = 77 \text{ mm}^2$

Region II: Area $= (21 \text{ mm})(9 \text{ mm}) = 189 \text{ mm}^2$

Total area $= 77 \text{ mm}^2 + 189 \text{ mm}^2 = 266 \text{ mm}^2$

(g) Divide this shape into a regular hexagon (I) and a rectangle (II) as shown.

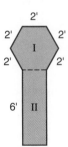

Region I: Area $\approx 2.598a^2 \approx 2.598(2 \text{ ft})^2 \approx 10.392 \text{ ft}^2$

Region II: Area $= (2 \text{ ft})(6 \text{ ft}) = 12 \text{ ft}^2$

Total area $\approx 10.392 \text{ ft}^2 + 12 \text{ ft}^2 \approx 22.392 \text{ ft}^2 \approx 22.4 \text{ ft}^2$, rounded

Now turn to Exercises 8-3 for a set of problems on the work of this section.

Exercises 8-3   Triangles, Regular Hexagons, and Irregular Polygons

A. Find the missing dimensions of each figure shown. All hexagons are regular. Round to three significant digits if necessary.

1.
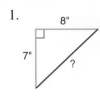

2.
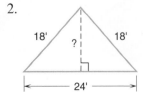

3.
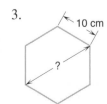

4.
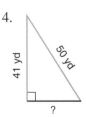

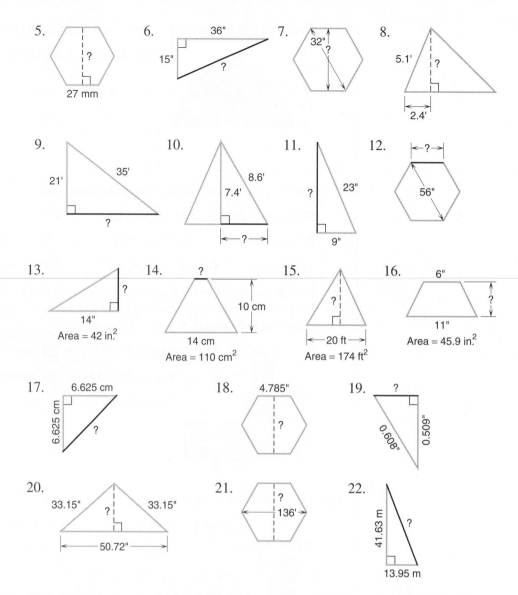

5. 27 mm ?

6. 36" 15" ?

7. 32" ?

8. 5.1' ? 2.4'

9. 35' 21' ?

10. 8.6' 7.4' ?

11. ? 23" 9"

12. ? 56"

13. ? 14" Area = 42 in.²

14. ? 10 cm 14 cm Area = 110 cm²

15. ? 20 ft Area = 174 ft²

16. 6" ? 11" Area = 45.9 in.²

17. 6.625 cm 6.625 cm ?

18. 4.785" ?

19. ? 0.608" 0.509"

20. 33.15" 33.15" ? 50.72"

21. ? 136'

22. 41.63 m ? 13.95 m

B. Find the area of each figure. All hexagons are regular. Round to three significant digits if necessary.

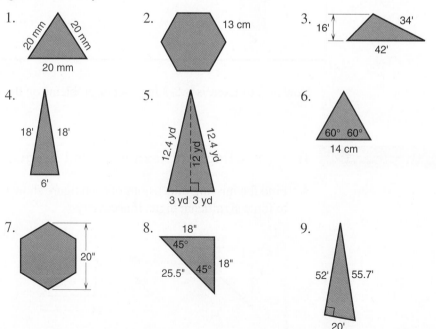

1. 20 mm 20 mm 20 mm

2. 13 cm

3. 16' 34' 42'

4. 18' 18' 6'

5. 12.4 yd 12.4 yd 12 yd 3 yd 3 yd

6. 60° 60° 14 cm

7. 20"

8. 18" 45° 25.5" 45° 18"

9. 52' 55.7' 20'

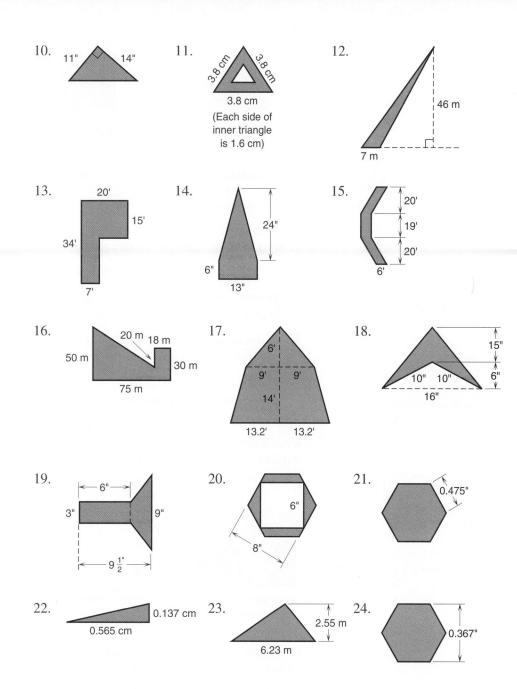

C. Practical Problems

1. **Painting** At 460 sq ft per gallon, how many gallons of paint are needed to cover the outside walls of a house as pictured below? Do not count windows and doors and round up to the nearest gallon.

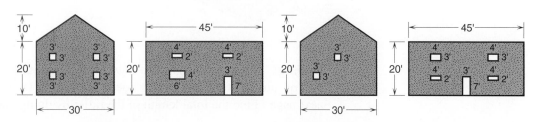

2. **Sheet Metal Trades** A four-sided vent connection must be made out of sheet metal. If each side of the vent is like the one pictured, how many total square inches of sheet metal will be used?

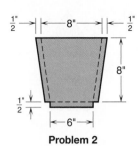

Problem 2

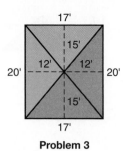

Problem 3

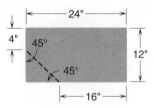

Problem 5

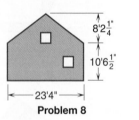

Problem 6

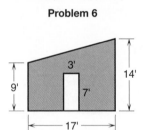

Problem 7

3. **Roofing** The aerial view of a roof is shown in the figure. How many squares (100 sq ft) of shingles are needed for the roof?

4. **Roofing** In problem 3, how many feet of gutters are needed?

5. **Carpentry** Allowing for a 3-ft overhang, how long a rafter is needed for the gable of the house in the figure? Round to the nearest tenth.

6. **Metalworking** A cut is to be made in a piece of metal as indicated by the dashed line in the figure. Find the length of the cut to the nearest tenth of an inch.

7. **Construction** How many square feet of drywall are needed for the wall shown in the figure?

8. **Construction** Find the area of the gable end of the house shown, not counting windows. Each window opening is 4 ft $3\frac{1}{4}$ in. by 2 ft $9\frac{1}{4}$ in. (Round to the nearest square foot.)

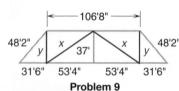

Problem 8   Problem 9

9. **Construction** Find the missing dimensions x and y on the bridge truss shown.

10. **Welding** The steel gusset shown is made in the shape of a right triangle. It will be welded in place along all three sides. Find the length of the weld.

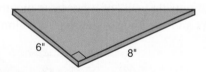

11. **Roofing** A roof is designed to have an $\frac{8}{12}$ slope, that is, 8 inches of rise per foot of run. The total run of one side of the roof is 16 ft. (See the figure on the next page.) Find the total length of the rafter by doing parts (a) through (d):

   (a) Use the Pythagorean theorem to find the *ULL*, unit line length.
   (b) Multiply the *ULL* by the run to find *R* in inches.
   (c) Add the length of the overhang as shown.
   (d) Give the answer in feet and inches to the nearest inch.

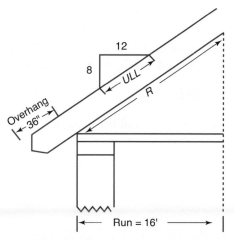

12. **Construction** A builder needs to pour a 4-in.-thick concrete slab that is shaped as shown in the figure.

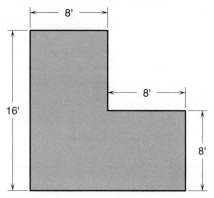

According to the manufacturer's guidelines for this thickness, approximately $\frac{3}{4}$ of a bag of concrete mix is required per square foot.

(a) Find the area of the slab.

(b) How many bags of the mix are needed for this job?

(c) At \$4.68 per bag, what will be the cost for purchasing the required amount of mix?

When you have finished these problems, check your answers to the odd-numbered problems in the Appendix, then continue in Section 8-4 with the study of circles.

8-4 Circles

A circle is probably the most familiar and the simplest plane figure. Certainly, it is the geometric figure that is most often used in practical and technical work. Mathematically, a circle is a closed curve representing the set of points some fixed distance from a given point called the *center.*

In this circle the center point is labeled O. The *radius r* is the distance from the center of the circle to the circle itself. The *diameter d* is the straight-line distance across the circle through the center point. It should be obvious from the drawing that the diameter is twice the radius.

Circumference

The *circumference* of a circle is the distance around it. It is a distance similar to the perimeter measure for a polygon. For all circles, the ratio of the circumference to the diameter, the distance around the circle to the distance across, is the same number.

$$\frac{\text{circumference}}{\text{diameter}} \approx 3.14159\ldots = \pi$$

The value of this ratio has been given the label π, the Greek letter "pi." The value of π is approximately 3.14, but the number has no simple decimal form—the digits continue without ending or repeating. Pressing the $\boxed{\pi}$ button on a scientific calculator will display the rounded ten-digit approximation of π:

C8-7 $\boxed{\pi}\boxed{=} \rightarrow$ `3.141592654`

The relationships between radius, diameter, and circumference can be summarized as follows:

CIRCLE FORMULAS

Diameter of a circle, $d = 2r$ or $r = \dfrac{d}{2}$

Circumference of a circle, $C = \pi d$ or $C = 2\pi r$

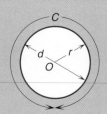

> **Note** If you are performing mental math with π, it is important to remember that π is slightly larger than 3. If you are doing calculations by hand or with a simple calculator, it is important to know that $\pi \approx 3.14$ or 3.1416. To reinforce this knowledge, we will sometimes instruct you to use one of these decimal equivalents for π. Otherwise, you may use the $\boxed{\pi}$ button on your scientific calculator. ◀

Example 1

We can use these formulas to find the radius and circumference of a hole with a diameter of 4.00 in.

Radius, $r = \dfrac{d}{2}$

$$= \frac{4 \text{ in.}}{2} = 2 \text{ in.}$$

Circumference, $C = \pi d$

$$= (\pi)(4 \text{ in.})$$

$$\approx 12.5664 \text{ in.} \approx 12.6 \text{ in.} \quad \text{rounded}$$

 $\boxed{\pi}\boxed{\times}$ 4 $\boxed{=} \rightarrow$ `12.56637061`

→ Your Turn

Use these formulas to solve the following practical problems. Round to two decimal places.

(a) **Carpentry** If a hole is cut using a $\frac{3}{4}$-in.-diameter wood bit, what will be the radius and circumference of the hole? Use 3.14 for π.

(b) **Sheet Metal Trades** An 8-ft strip of sheet metal is bent into a circular tube. What will be the diameter of this tube?

(c) **Carpentry** A circular redwood hot tub has a diameter of 5 ft. What length of steel band is needed to encircle the tub to hold the individual boards in position? Allow 4 in. for fastening. Use 3.1416 for π.

(d) **Machine Trades**   If eight bolts are to be equally spaced around a circular steel plate at a radius of 9 in., what must be the spacing between the centers of the bolts along the curve?

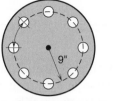

Problem (d)

Problem (e)

(e) **Construction**   At \$18.50 per foot, what will be the cost of the molding around the semicircular window shown? (Compute for the curved portion only.)

→ **Solutions**

(a) $r = \dfrac{d}{2}$

$= \dfrac{\frac{3}{4}\text{ in.}}{2} = \dfrac{3}{8}\text{ in.} \qquad \dfrac{\frac{3}{4}}{2} = \dfrac{3}{4} \cdot \dfrac{1}{2} = \dfrac{3}{8}$

$C = \pi d$
$\approx (3.14)\left(\frac{3}{4}\text{ in.}\right)$ ⟶
$\approx (3.14)(0.75\text{ in.})$ ⟵ | Change the fraction to a decimal. |
$\approx 2.36\text{ in.}$

(b)   If $C = \pi d$ then $d = \dfrac{C}{\pi}$

and $d \approx \dfrac{8\text{ ft}}{\pi}$

$\approx 2.55\text{ ft}$ rounded

(c) $C = \pi d$
$\approx (3.1416)(5\text{ ft})$
$\approx 15.708\text{ ft} \approx 15\text{ ft }8\frac{1}{2}\text{ in.}$

Adding 4 in., the length needed is $16\text{ ft }\frac{1}{2}\text{ in.}$

Notice that 0.708 ft is $(0.708)(12\text{ in.}) = 8.496$ in. or about $8\frac{1}{2}$ in.

3.1416 ⊗ **5** ⊜ → *15.708* ⊖ **15** ⊜ ⊗ **12** ⊜ → *8.496*

| Decimal feet |

| 0.708′ changed to inches |

(d) $C = 2\pi r$
$= 2(\pi)(9\text{ in.})$
$\approx 56.55\text{ in.}$

Because eight bolts must be spaced evenly around the 9-in. circle, divide by 8. Spacing $= 56.55 \div 8 \approx 7.07$ in.

(e) $C = \pi d$ for a complete circle
$= (\pi)(6\text{ ft})$
$\approx 18.8496\text{ ft}$

The length of molding is half of this or 9.4248 ft. The cost is

$(\$18.50)(9.4248\text{ ft}) \approx \$174.36$, rounded to the nearest cent.

Parts of Circles

Carpenters, plumbers, sheet metal workers, and other trades people often work with parts of circles. A common problem is to determine the length of a piece of material that contains both curved and straight segments.

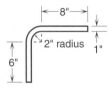

Example 2

Metalworking Suppose that we need to determine the total length of steel rod needed to form the curved piece shown. The two straight segments are no problem: we will need a total of 8 in. + 6 in. or 14 in. of rod for these. When a rod is bent, the material on the outside of the curve is stretched, and the material on the inside is compressed. We must calculate the circumference for the curved section from a *neutral line* midway between the inside and outside radius.

The inside radius is given as 2 in. The stock is 1 in. thick, so the midline or neutral line is at a radius of $2\frac{1}{2}$ in. The curved section is one-quarter of a full circle, so the length of bar needed for the curved arc is

$$C = \frac{2\pi r}{4}$$

Divide the circumference by 4 to find the length of the quarter-circle arc.

$$= \frac{\pi r}{2}$$

$$= \frac{(\pi)(2.5 \text{ in.})}{2}$$

$$\approx 3.93 \text{ in.} \quad \text{rounded to two decimal places.}$$

$2\frac{1}{2}$" radius

The total length of bar needed for the piece is

$$L \approx 3.93 \text{ in.} + 14 \text{ in.}$$
$$\approx 17.93 \text{ in.}$$

→ Your Turn

Metalworking Find the total length of ornamental iron $\frac{1}{2}$ in. thick needed to bend into this shape. Round to two decimal places.

→ Solution

Measuring to the midline we find that the upper curve (A) has a radius of $4\frac{1}{4}$ in. and the lower curve (C) has a radius of $3\frac{3}{4}$ in. Calculate the lengths of the two curved pieces this way:

$4\frac{1}{4}$ in. to midline

A

$3\frac{3}{4}$ in. to midline

C

Upper Curve (A)

(Half circle)

$$C = \frac{1}{2} \cdot 2\pi r = \pi r \quad \leftarrow \text{Multiply by } \frac{1}{2} \text{ for half circle.}$$

$$= (\pi)\left(4\frac{1}{4} \text{ in.}\right)$$

$$\approx 13.35 \text{ in.}$$

Lower Curve (C)

(Three-fourths circle)

$$C = \frac{3}{4} \cdot 2\pi r = \frac{3\pi r}{2} \quad \leftarrow \text{Multiply by } \frac{3}{4} \text{ for three fourths of a circle.}$$

$$= \frac{(3)(\pi)\left(3\frac{3}{4} \text{ in.}\right)}{2}$$

$$\approx 17.67 \text{ in.}$$

Chapter 8 Practical Plane Geometry

6 in.

B

Adding the length of the straight piece (B), the total length of straight stock needed to create this shape is

13.35 in. + 17.67 in. + 6 in. ≈ 37.02 in.

Once you have your work organized, you can perform the entire calculation on a calculator as follows:

$\boxed{\pi}$ $\boxed{\times}$ **4.25** $\boxed{+}$ **3** $\boxed{\times}$ $\boxed{\pi}$ $\boxed{\times}$ **3.75** $\boxed{\div}$ **2** $\boxed{+}$ **6** $\boxed{=}$ → $\boxed{37.02322745}$

→ **More Practice**

Find the length of stock needed to create each of the following shapes. Round to one decimal place.

(a)

(b)

(c)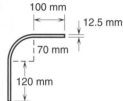

(d)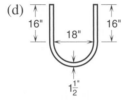

(e)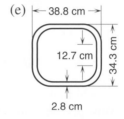

→ **Answers**

(a) 31.4 in.	(b) 31.8 in.	(c) 339.8 mm
(d) 62.6 in.	(e) 118.9 cm	

▶ **A Closer Look** On problem (e) a bit of careful reasoning will convince you that the dimensions are as follows:

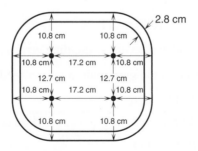

Each of the four corner arcs has radius 9.4 cm to the midline, and each is a quarter circle.

Radius = 10.8 cm − 1.4 cm
     = 9.4 cm ◀

Area of a Circle   To find the *area* of a circle use one of the following formulas.

AREA OF A CIRCLE

$$A = \pi r^2 \quad \text{or} \quad A = \frac{\pi d^2}{4} \approx 0.7854 d^2$$

If you are curious about how these formulas are obtained, don't miss the box on page 544.

Example 3

Let's use the formula to find the area of a circle with a diameter of 8 in.

$$A = \frac{\pi d^2}{4}$$

$$= \frac{(\pi)(8 \text{ in.})^2}{4}$$

$$\approx 50.2655 \text{ in.}^2 \quad \text{or} \quad 50.3 \text{ in.}^2 \quad \text{rounded}$$

8"

  $\boxed{\pi} \ \boxed{\times} \ \mathbf{8} \ \boxed{x^2} \ \boxed{\div} \ \mathbf{4} \ \boxed{=} \ \rightarrow \ \boxed{50.26548246}$

Of course you would find exactly the same area if you used the radius $r = 4$ in. in the first formula. Most people who need the area formula in their work memorize the formula $A = \pi r^2$ and calculate the radius r if they are given the diameter d.

Example 4

Forestry   After wildfires destroyed a portion of Glacier National Park, forest service biologists needed to determine the germination rate of new seedlings. Using a rope as a radius, they marked off a circular area of forest with a 40-ft radius and then counted 450 new seedlings that had germinated in that circle. What was the germination rate in seedlings per acre? Round to the nearest whole number.

Step 1   Find the area of the circle:

$$A = \pi r^2 = \pi(40)^2 = 1600 \ \pi \ \text{sq ft}$$

Keep the area as $1600 \ \pi$ to avoid a rounding error in Step 2.

Step 2   Use a proportion to find the germination rate in seedlings per acre.

$$\frac{450 \text{ seedlings}}{x \text{ seedlings}} = \frac{1600 \ \pi \text{ sq ft}}{43,560 \text{ sq ft}} \quad \Longleftarrow \boxed{\text{1 acre} = 43,560 \text{ sq ft}}$$

$$1600 \ \pi \ x = 19,602,000$$

$$x \approx 3900$$

The germination rate is approximately 3900 new seedlings per acre.

Now for practice try these problems.

(a) Find the areas of the following circles. Round to three significant digits.

(1)

(2)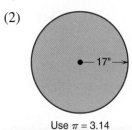

Use π = 3.14

(3) Find the area of the head of a piston with a radius of 6.5 cm.

(4)

19 yd

Use π = 3.14

(5)

2.2 mm

(b) **Painting**   Find the area of wood that must be stained to finish the top and bottom of a circular table 6 ft in diameter. Use π ≈ 3.14 and round to three significant digits.

(c) **Construction**   Find the area of stained glass used in a semicircular window with a radius of 9 in. Use π ≈ 3.14 and round to three significant digits.

(d) **Sheet Metal Trades**   What is the area of the largest circle that can be cut out of a square piece of sheet metal 4 ft 6 in. on a side? Round to the nearest hundredth.

(e) **Industrial Technology**   A pressure of 860 lb per square foot is exerted on the bottom of a cylindrical water tank. If the bottom has a diameter of 15 ft, what is the total force on the bottom? Round the final answer to three significant digits.

→ Solutions

(a) (1) $A = \pi r^2 = (\pi)(5 \text{ in.})^2 = (\pi)(25 \text{ in.}^2)$

   $\approx 78.5 \text{ in.}^2$ rounded

(2) $A = \pi r^2 \approx (3.14)(17 \text{ in.})^2$

   $\approx 907 \text{ in.}^2$

(3) $A = \pi r^2 = (\pi)(6.5 \text{ cm})^2$

   $\approx 133 \text{ cm}^2$ rounded

(4) $A = \frac{1}{2}(\pi r^2) \approx \frac{1}{2}(3.14)(9.5 \text{ yd})^2$

   $\approx 142 \text{ yd}^2$ rounded

(5) $A = \frac{1}{4}(\pi r^2) = \frac{1}{4}(\pi)(2.2 \text{ mm})^2$

   $\approx 3.80 \text{ mm}^2$ rounded

(b) $A = 2\pi r^2 \approx 2(3.14)(3 \text{ ft})^2$

   $\approx 56.5 \text{ ft}^2$

(c) $A = \frac{1}{2}(\pi r^2) \approx \frac{1}{2}(3.14)(9 \text{ in.})^2$

   $\approx 127 \text{ in.}^2$

(d) $A = \pi r^2 = (\pi)(2\text{ ft }3\text{ in.})^2$

$\quad\quad = (\pi)(2.25\text{ ft})^2 \approx 15.90\text{ ft}^2$

(e) $A = \pi r^2 = (\pi)(7.5\text{ ft})^2$

$\quad\quad \approx 176.715\text{ ft}^2$

$\quad\quad\text{Force} = 860\dfrac{\text{lb}}{\text{ft}^2} \times 176.715\ \cancel{\text{ft}^2}$

$\quad\quad\quad = 151{,}974.5 \text{ or } 152{,}000\text{ lb} \quad \text{rounded}$

 $\boxed{\pi}$ $\boxed{\times}$ **7.5** $\boxed{x^2}$ $\boxed{\times}$ **860** $\boxed{=}$ →  *151974.5446*

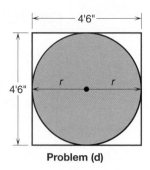

Problem (d)

AREA OF A CIRCLE

To find the formula for the area of a circle, first divide the circle into many pie-shaped sectors, then rearrange them like this:

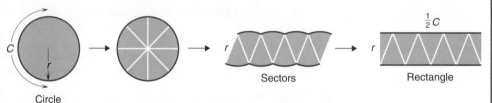

Circle · Sectors · Rectangle

When the sectors are rearranged, they fit into a rectangle whose height is r, the radius of the original circle, and whose base length is roughly half the circumference of the circle. The more sectors the circle is divided into, the better the fit. The area of the sectors is approximately equal to the area of the rectangle.

$$\text{Area of circle, } A = r \cdot \tfrac{1}{2}C$$
$$= r \cdot \tfrac{1}{2}(2\pi r)$$
$$= \pi r^2$$

Rings A circular *ring*, or annulus, is defined as the area between two concentric circles—two circles having the same center point. A washer is a ring; so is the cross section of a pipe, a collar, or a cylinder.

To find the area of a ring, subtract the area of the inner circle from the area of the outer circle. Use the following formula.

AREA OF A RING (ANNULUS)

Shaded area = area of outer circle − area of inner circle

$$A = \pi R^2 - \pi r^2$$
$$= \pi(R^2 - r^2) \quad \text{or} \quad A = \tfrac{1}{4}\pi(D^2 - d^2)$$
$$\approx 0.7854(D^2 - d^2)$$

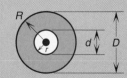

Example 5

To find the cross-sectional area of the wall of a ceramic pipe whose I.D. (inside diameter) is 8 in. and whose O.D. (outside diameter) is 10 in., substitute $d = 8$ in. and $D = 10$ in. into the second formula.

$$A = \tfrac{1}{4}(\pi)[(10 \text{ in.})^2 - (8 \text{ in.})^2]$$

$$\approx (0.7854)(100 \text{ in.}^2 - 64 \text{ in.}^2)$$

$$\approx (0.7854)(36 \text{ in.}^2)$$

$$\approx 28.3 \text{ in.}^2 \quad \text{rounded}$$

Note Notice that $\dfrac{\pi}{4}$ is equal to 0.7854 rounded to four decimal places. Also notice that the diameters are *squared first* and *then* the results are *subtracted*. ◄

→ **Your Turn**

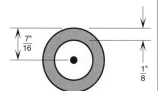

Try it. Find the area of a washer with an outside radius of $\frac{7}{16}$ in. and a thickness of $\frac{1}{8}$ in. Round to three significant digits.

→ **Solution**

$$R = \tfrac{7}{16} \text{ in.} \quad r = \tfrac{7}{16} \text{ in.} - \tfrac{1}{8} \text{ in.} \quad \text{or} \quad r = \tfrac{5}{16} \text{ in.}$$

then

$$A = \pi(R^2 - r^2)$$

$$= (\pi)\left[\left(\tfrac{7}{16}\text{ in.}\right)^2 - \left(\tfrac{5}{16}\text{ in.}\right)^2\right]$$

$$= (\pi)[(0.4375 \text{ in.})^2 - (0.3125 \text{ in.})^2]$$

$$\approx (\pi)[0.1914 \text{ in.}^2 - 0.0977 \text{ in.}^2]$$

$$\approx (\pi)[0.0937 \text{ in.}^2]$$

$$\approx 0.295 \text{ in.}^2 \quad \text{rounded}$$

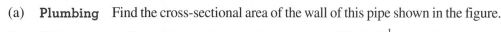
$\boxed{\pi}\ \boxed{\times}\ \boxed{(}\ \mathbf{7}\ \boxed{\text{A}^{\text{b}}_{\text{c}}}\ \mathbf{16}\ \boxed{x^2}\ \boxed{-}\ \mathbf{5}\ \boxed{\text{A}^{\text{b}}_{\text{c}}}\ \mathbf{16}\ \boxed{x^2}\ \boxed{=}\ \rightarrow\ \boxed{0.294524311}$

→ **More Practice**

Problem (a)

Solve the following problems involving rings. Round to two decimal places.

(a) **Plumbing**   Find the cross-sectional area of the wall of this pipe shown in the figure.

(b) Find the area of a washer with an inside diameter (I.D.) of $\frac{1}{2}$ in. and an outside diameter (O.D.) of $\frac{7}{8}$ in.

(c) **Machine Trades**   What is the cross-sectional area of a steel collar 0.6 cm thick with an inside diameter of 9.4 cm?

(d) **Landscaping**   A landscape company charges $2.10 per square foot to install sod. What would they charge for a circular ring of lawn, 6 ft wide, surrounding a flower garden with a radius of 20 ft?

(e) **Plumbing**   A pipe has an inside diameter of 8 in. How much larger is the cross-sectional area with a $\frac{1}{2}$-in.-thick wall than with a $\frac{3}{8}$-in.-thick wall?

→ **Answers**

(a) 351.86 in.^{2}

(b) 0.40 in.^{2}

(c) 18.85 cm^2

(d) $1820.87

(e) 3.49 in.^{2}

A Closer Look

In problem (c),

$$\text{Radius of inner hole} = \frac{9.4\,\text{cm}}{2} = 4.7\,\text{cm}$$

Radius of outer edge $= 4.7\,\text{cm} + 0.6\,\text{cm} = 5.3\,\text{cm}$

Then,

Area $= \pi \times (5.3^2 - 4.7^2) \approx 18.85\,\text{cm}^2$

In problem (e), organize your work like this:

	Thicker-Walled Pipe	*Thinner-Walled Pipe*
Inside diameter	8 in.	8 in.
Outside diameter	8 in. $+\frac{1}{2}$ in. $+\frac{1}{2}$ in. $= 9$	8 in. $+\frac{3}{8}$ in. $+\frac{3}{8}$ in. $= 8\frac{3}{4}$ in. $= 8.75$ in.
Cross-sectional area	$0.7854\,(9^2 - 8^2)$ $= 13.3518$ sq in.	$0.7854\,(8.75^2 - 8^2)$ ≈ 9.8666 sq in.
Difference	13.3518 sq in. $-$ 9.8666 sq in. ≈ 3.49 sq in.	

.7854 ⊗ ⦅ 9 x^2 ⊖ 8 x^2 ⦆ ⊖ .7854 ⊗ ⦅ 8.75 x^2 ⊖ 8 x^2 ⦆ ⊜

→ ▮ 3.4852125 ◀

Now turn to Exercises 8-4 for a set of review problems on circles.

Exercises 8-4 Circles

A. Find the circumference and area of each circle. Use the π key for π and round to three significant digits.

1.   14"

2.   21'

3.   6.1 m

4.   Radius = 7 cm

5. Circle of radius 12 ft.

6. Circle of diameter 0.50 m.

7. Circle of diameter 1.2 cm.

8. Circle of radius 0.4 in.

B. Find the area of each figure. Use the π key for π and round to the nearest tenth.

1. 8"; $6\frac{1}{2}$"

2. 18 cm; 2 cm

3. 0.8"; 2.7"

4. 1.5'; 3.2'

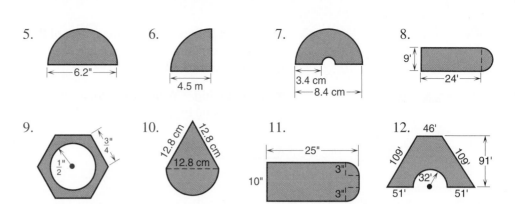

Find the length of stock needed to create each shape. Round to the nearest tenth.

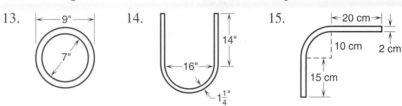

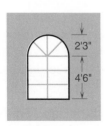

Problem 1

Problem 3

Problem 6

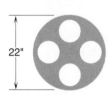

Problem 10

Problem 11

C. Practical Problems

1. **Metalworking**   What length of 1-in. stock is needed to bend a piece of steel into this shape? Use $\pi \approx 3.14$ and round to the nearest tenth.

2. **Landscaping**   (a) Calculate, to the nearest square foot, the cultivable area of a circular garden 38 ft in diameter if the outer 2 ft are to be used for a path. (b) Find the area of the path.

3. **Machine Trades**   What diameter must a circular piece of stock be to mill a hexagonal shape with a side length of 0.3 in.?

4. **Machine Trades**   In problem 3, how many square inches of stock will be wasted? (Round to the nearest hundredth.)

5. **Plumbing**   What is the cross-sectional area of a cement pipe 2 in. thick with an I.D. of 2 ft? Use $\pi \approx 3.14$ and round to the nearest whole number.

6. **Machine Trades**   In the piece of steel shown, $4\frac{1}{2}$-in.-diameter holes are drilled in a 22-in.-diameter circular plate. Find the area remaining. Use $\pi \approx 3.14$ and round to three significant digits.

7. **Plumbing**   How much additional cross-sectional area is there in a 30-in.-I.D. pipe 2 in. thick than one 1 in. thick? (Round to one decimal place.)

8. **Machine Trades**   What diameter round stock is needed to mill a hexagonal nut measuring $1\frac{1}{2}$ in. across the flats? (Round up to the nearest quarter of an inch.)

9. **Landscaping**   How many plants spaced every 6 in. are needed to surround a circular walkway with a 25-ft radius? Use $\pi \approx 3.14$.

10. **Machine Trades**   What diameter circular stock is needed to mill a square end 3 cm on a side? (Round to the nearest hundredth.)

11. **Carpentry**   At $11.25 per foot for the curved portion and $1.75 per foot for the straight portion, how much will it cost to put molding around the window pictured? Use $\pi \approx 3.14$.

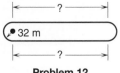

Problem 12

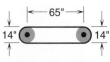

Problem 13

12. **Sports and Leisure** Given the radius of the semicircular ends of the track in the figure is 32 m, how long must each straightaway be to make a 400-m track? Use $\pi \approx 3.14$.

13. **Manufacturing** What is the total length of belting needed for the pulley shown? (Round to three significant digits.)

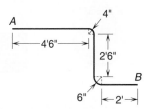

Problem 14

14. **Electrical Trades** An electrician bends a $\frac{1}{2}$-in. conduit as shown in the figure to follow the bend in a wall. Find the total length of conduit needed between A and B. Use $\pi \approx 3.14$.

Problem 15

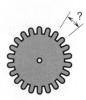

Problem 16

15. **Metalworking** How many inches apart will eight bolts be along the circumference of the metal plate shown in the figure? (Round to the nearest tenth of an inch.)

16. **Landscaping** How many square *yards* of concrete surface are there in a walkway 8 ft wide surrounding a tree if the inside diameter is 16 ft? (Round to one decimal place.) See the figure.

17. **Carpentry** The circular saw shown has a diameter of 18 cm and 22 teeth. What is the spacing between teeth? (Round to two significant digits.)

18. **Automotive Trades** A hose must fit over a cylindrical opening 57.2 mm in diameter. If the hose wall is 3.0 mm thick, what length of clamp strap is needed to go around the hose connection? (Round to two significant digits.)

19. **Automotive Trades** What length on the circumference would a 20° angle span on a 16-in. flywheel? (Round to the nearest hundredth.)

20. **Automotive Trades** In making a certain turn on a road, the outside wheels of a car must travel 8.75 ft farther than the inside wheels. If the tires are 23 in. in diameter, how many additional turns do the outside wheels make during this turn? (*Hint:* In one "turn," a tire travels a distance equal to its circumference. Convert 8.75 ft to inches.)

21. **Wastewater Technology** To find the velocity of flow (in feet per second, or fps) through a pipe, divide the rate of flow (in cubic feet per second, or cfs) by the cross-sectional area of the pipe (in square feet). If the rate of flow through an 8.0-in.-diameter pipe is 0.75 cfs, find the velocity of flow. (Round to the nearest tenth.)

22. **Plumbing** The pressure exerted by a column of water amounts to 0.434 psi of surface for every foot of height or *head*. What is the total force (pressure times area) exerted on the bottom surface of a cylindrical container 16 in. in diameter if the head is 4.5 ft? (Round to three significant digits.)

23. **Plumbing** To construct a bracket for a drain pipe, Jerry the plumber needs to know its circumference. Calculate the circumference of a pipe 42.3 cm in diameter. (Round to the nearest centimeter.)

24. **Culinary Arts**   A medium pizza at Papa's Pizza Place is 9 in. in diameter and a large pizza is 12 in. in diameter. If a medium pizza will serve four customers, how many servings will a large pizza provide?

25. **Sports and Leisure**   Two cars are racing around a circular track. The red car is traveling in a lane that is 82 ft from the center. The blue car is in a lane that is 90 ft from the center. If both cars are traveling at 120 mph, how much later, in seconds, will the blue car finish if they race one lap around the track. (Round to the nearest 0.1 sec.) (*Hint:* 60 mph ≈ 88 fps.)

26. **Forestry**   Forest service biologists roped off a circular area of recently burned forest to determine how well the forest was recovering. The circular sample area had a 32-ft radius and contained 220 new seedlings. Calculate the germination rate in seedlings per acre. Use π ≈ 3.14 and 1 acre = 43,560 sq ft. (Round to the nearest 10 seedlings.) See Example 4 on page 542.

Check your answers to the odd-numbered problems in the Appendix, then turn to Problem Set 8 on page 553 for practice over the work of Chapter 8. If you need a quick review of the topics in this chapter, visit the chapter Summary first.

Summary

Practical Plane Geometry

Objective	Review
Measure angles with a protractor. (p. 489)	Place the protractor over the angle so that the zero-degree mark is aligned with one side of the angle and the center mark is on the vertex. Read the measure, in degrees, where the other side of the angle intersects the scale of the protractor.

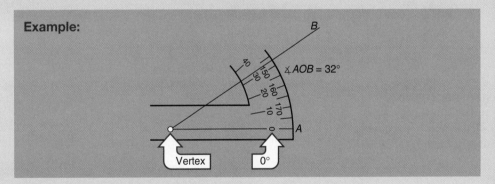

Example:

∡ AOB = 32°

Vertex   0°

Classify angles. (p. 490)	An acute angle measures less than 90°. A right angle measures exactly 90°. An obtuse angle measures greater than 90° and less than 180°. A straight angle measures exactly 180°.

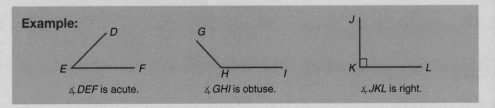

Example:

∡ DEF is acute.   ∡ GHI is obtuse.   ∡ JKL is right.

Objective

Use simple geometric relationships involving intersecting lines and triangles. (p. 492)

Review

When two straight lines intersect, the vertical angles are always equal.

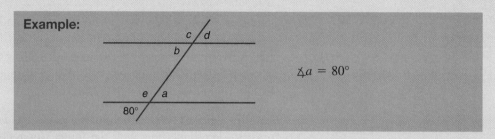

Example:

$\measuredangle a = 80°$

When parallel lines are cut by a third line, the corresponding angles are equal, and the alternate interior angles are equal.

Example: If the horizontal lines are parallel in the preceding figure, $\measuredangle a$ and $\measuredangle d$ are equal corresponding angles, therefore $\measuredangle d = 80°$. Also, $\measuredangle a$ and $\measuredangle b$ are equal alternate interior angles. Therefore $\measuredangle b = 80°$.

When two lines intersect, each pair of adjacent angles sums to 180°.

Example: In the preceding figure, $\measuredangle b$ and $\measuredangle c$ are adjacent angles, therefore:

$$\measuredangle b + \measuredangle c = 180° \quad \text{or} \quad 80° + \measuredangle c = 180°$$
$$\measuredangle c = 180° - 80° = 100°$$

The interior angles of a triangle always sum to 180°.

Example: $x + 70° + 90° = 180°$
$$x + 160° = 180°$$
$$x = 180° - 160° = 20°$$

Identify polygons, including triangles, squares, rectangles, parallelograms, trapezoids, and hexagons. (p. 501)

Triangles may be classified according to their sides or their angles as follows:

Equilateral: all sides equal
Isosceles: at least two sides equal
Scalene: no sides equal

Acute: all three angles acute
Obtuse: contains one obtuse angle
Right: contains a 90°angle

Example:

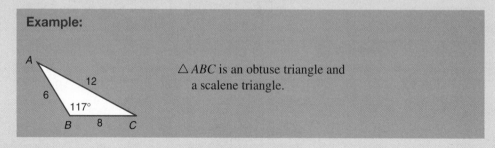

$\triangle ABC$ is an obtuse triangle and a scalene triangle.

The following quadrilaterals (four-sided polygons) are useful in the trades:

Parallelogram: opposite sides parallel and equal
Rectangle: parallelogram with four right angles

Objective	Review
	Square: rectangle with all sides equal
	Trapezoid: two parallel sides and two nonparallel sides

Example:

DEFG is a parallelogram.   *HIJK* is a trapezoid

A hexagon is a six-sided polygon. In a regular hexagon, all six sides are equal, and all six angles are equal.

Use the Pythagorean theorem. (p. 517)

In a right triangle, the square of the hypotenuse is equal to the sum of the squares of the other two sides (the legs).

Example:

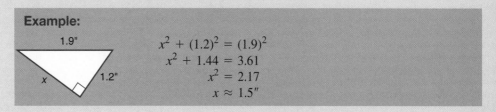

$$x^2 + (1.2)^2 = (1.9)^2$$
$$x^2 + 1.44 = 3.61$$
$$x^2 = 2.17$$
$$x \approx 1.5''$$

Find the area and perimeter of geometric figures.

Use the formulas listed in the following summary.

SUMMARY OF FORMULAS FOR PLANE FIGURES

Figure	Area	Perimeter
Rectangle	$A = LW$	$P = 2L + 2W$
Square	$A = s^2$	$P = 4s$
Parallelogram	$A = bh$	$P = 2a + 2b$
Triangle	$A = \frac{1}{2}bh$ or $\frac{bh}{2}$ $A = \sqrt{s(s-a)(s-b)(s-c)}$ where $s = \frac{a+b+c}{2}$	$P = a + b + c$
Equilateral triangle	$A = \frac{a^2\sqrt{3}}{4} \approx 0.433a^2$	$P = 3a$
Isosceles triangle	$A = \frac{b}{2}\sqrt{a^2 - \left(\frac{b}{2}\right)^2}$	$P = 2a + b$

(continued)

Objective

Review

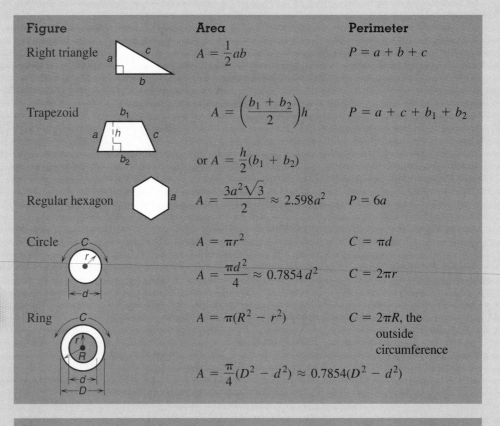

Figure	Area	Perimeter
Right triangle	$A = \dfrac{1}{2}ab$	$P = a + b + c$
Trapezoid	$A = \left(\dfrac{b_1 + b_2}{2}\right)h$ or $A = \dfrac{h}{2}(b_1 + b_2)$	$P = a + c + b_1 + b_2$
Regular hexagon	$A = \dfrac{3a^2\sqrt{3}}{2} \approx 2.598a^2$	$P = 6a$
Circle	$A = \pi r^2$ $A = \dfrac{\pi d^2}{4} \approx 0.7854\, d^2$	$C = \pi d$ $C = 2\pi r$
Ring	$A = \pi(R^2 - r^2)$ $A = \dfrac{\pi}{4}(D^2 - d^2) \approx 0.7854(D^2 - d^2)$	$C = 2\pi R$, the outside circumference

Example:

(a)

$$\text{Area} = \frac{1}{2}bh$$
$$= \frac{1}{2}(4)(7)$$
$$= 14 \text{ in.}^2$$

(b)

$$\text{Area} = \pi r^2$$
$$= 3.14(2.70)^2$$
$$\approx 22.9 \text{ in.}^2$$

$$\text{Circumference} = 2\pi r$$
$$= 2(3.14)(2.70)$$
$$\approx 17.0 \text{ in.}$$

Solve practical problems involving area and perimeter of plane figures. (p. 504)

Use the appropriate formula from the preceding summary to find the area or perimeter of the figure in the problem. Then perform any further calculations as needed.

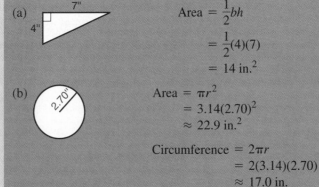

Example: A circular flowerbed with a radius of 8.5 ft is surrounded by a circular ring of lawn 12 ft wide. At $2.10 per square foot, how much will it cost to sod the lawn?

First, find the area of the ring of lawn:

Notice that the radius of the outer circle is the radius of the inner circle plus the width of the lawn, or $8.5 + 12 = 20.5$ ft.

$$A = \pi(R^2 - r^2) = \pi(20.5^2 - 8.5^2)$$
$$= 348\pi \text{ sq ft}$$

Then, multiply by $2.10 per sq ft:

$$(348\pi)(\$2.10) \approx \$2295.88$$

→ **Problem Set**

Practical Plane Geometry

Answers to the odd-numbered problems are given in the Appendix.

A. Solve the following problems involving angles.

Name each angle and tell whether it is acute, obtuse, or right.

1.

2.

3.

4.

5.

6.

Measure the indicated angles using a protractor.

7.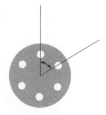

8.

9.

10.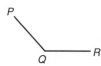

11.

12.

13. Use a protractor to draw angles of the following measures.

    (a) 65°   (b) 138°   (c) 12°   (d) 90°

14. Find the size of each indicated angle without using a protractor.

    (a)   (b)   (c)

The horizontal lines are parallel.

Name _____

Date _____

Course/Section _____

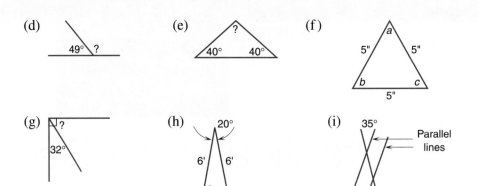

(d) 49° ?

(e) ? 40° 40°

(f) a 5" 5" b c 5"

(g) ? 32°

(h) 20° 6' 6' a

(i) 35° Parallel lines a

B. Solve the following problems involving plane figures.

Find the perimeter (or circumference) and area of each figure. Round to the nearest tenth.

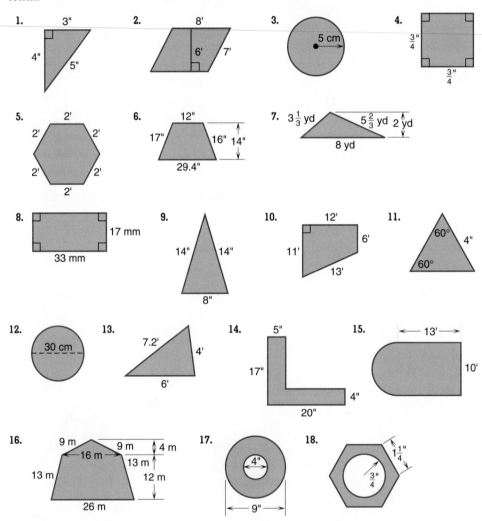

1. 3" 4" 5"

2. 8' 6' 7'

3. 5 cm

4. ¾" ¾"

5. 2' 2' 2' 2' 2' 2'

6. 12" 17" 16" 14" 29.4"

7. $3\frac{1}{3}$ yd $5\frac{2}{3}$ yd 2 yd 8 yd

8. 17 mm 33 mm

9. 14" 14" 8"

10. 12' 11' 6' 13'

11. 60° 4" 60°

12. 30 cm

13. 7.2' 4' 6'

14. 5" 17" 4" 20"

15. ← 13' → 10'

16. 9 m 9 m 4 m 16 m 13 m 13 m 12 m 26 m

17. 4" 9"

18. 1¼" ¾"

Find the missing dimensions of the following figures. Round to the nearest hundredth if necessary.

19. ? 5 mm

20. 18" 18" ? 9"

21. 2" ?

22.

23.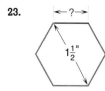

24.

C. Practical Problems

1. **Construction** What will it cost to pave a rectangular parking lot 220 ft by 85 ft at $3.50 per square foot?
2. **Metalworking** A steel brace is used to strengthen the table leg as shown. What is the length of the brace? Round to the nearest hundredth of an inch.

Problem 2

Problem 3

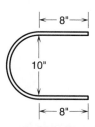

Problem 4

3. **Machine Trades** Holes are punched in a regular octagonal (8-sided) steel plate as indicated. Through what angle must the plate be rotated to locate the second hole?
4. **Masonry** How many square feet of brick are needed to lay a patio in the shape shown?
5. **Electrical Trades** A coil of wire has an average diameter of 8.0 in. How many feet of wire are there if it contains 120 turns? Round to two significant digits.
6. **Sheet Metal Trades** A sheet metal worker needs to make a vent connection in the shape of the trapezoidal prism shown in the drawing. Use a protractor to find the indicated angle.

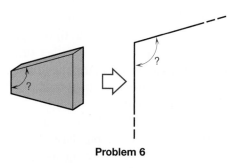

Problem 6

Problem 7

7. **Landscaping** At $16 per foot, how much does it cost to build a chain-link fence around the yard shown in the figure?
8. **Metalworking** Find the length of straight stock needed to bend $\frac{1}{2}$-in. steel into the shape indicated. Use $\pi \approx 3.14$.

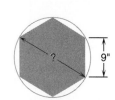

Problem 8

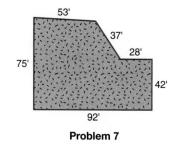

Problem 9

9. **Machine Trades** A hexagonal piece of steel 9 in. on a side must be milled by a machinist. What diameter round stock does he need? (See the figure.) Round to the nearest tenth of an inch.

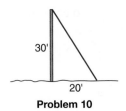

Problem 10

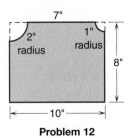

Problem 12

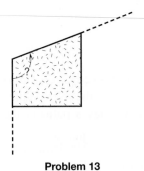

Problem 13

10. **Construction** How much guy wire is needed to anchor a 30-ft pole to a spot 20 ft from its base? Allow 4 in. for fastening and round to the nearest inch. (See the figure.)

11. **Carpentry** How many 4 ft by 8 ft sheets of exterior plywood must be ordered for the 4 walls of a building 20 ft long, 32 ft wide, and 12 ft high? Assume that there are 120 sq ft of window and door space. (*Note:* You can cut the sheets to fit, but you may not order a fraction of a sheet.)

12. **Metalworking** Find the perimeter of the steel plate shown in the figure. Round to the nearest tenth of an inch.

13. **Construction** At what angle must sheet rock be cut to conform to the shape of the wall shown? Measure with a protractor.

14. **Metalworking** A triangular shape with a base of 6 in. and a height of 9 in. is cut from a steel plate weighing 5.1 lb/sq ft. Find the weight of the shape (1 sq ft = 144 sq in.). (Round to the nearest ounce.)

15. **Construction** An air shaft is drilled from the indicated spot on the hill to the mine tunnel below. How long is the shaft to the nearest foot?

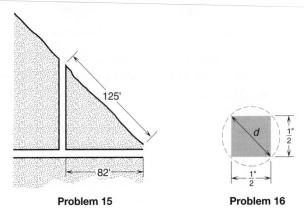

Problem 15

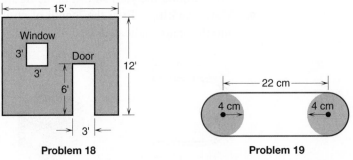

Problem 16

16. **Machine Trades** What diameter circular stock is needed to mill a square bolt $\frac{1}{2}$ in. on a side? (Round to three significant digits.) (See the figure.)

17. **Life Skills** A 17-ft roll of weather stripping costs $11.30, including tax. Suppose you need to weather strip the following: 6 windows measuring 4 ft by 6 ft; 8 windows measuring 3 ft by 2 ft; and all sides except the bottom of two doors 3 ft wide and 7 ft high.
    (a) How many rolls will you need?
    (b) What will it cost to purchase enough weather stripping?

18. **Construction** How many square feet must be plastered to surface the wall shown? (Round to three significant digits.)

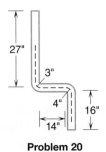

Problem 20

Problem 18

Problem 19

19. **Manufacturing** What length of belt is needed for the pulley shown? (Round to the nearest tenth.)

20. **Electrical Trades** Electrical conduit must conform to the shape and dimensions shown. Find the total length of conduit in inches. Use $\pi \approx 3.14$ and round to the nearest tenth.

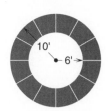

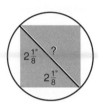

Problem 21

Problem 24

21. **Construction**   At $11 per square foot, how much will it cost to cover the circular pathway, shown in the figure, with pavers? Use $\pi \approx 3.14$.

22. **Carpentry**   Find the size of $\angle 2$ when the carpenter's square is placed over the board as shown.

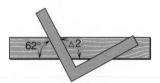

Problem 22

23. **Welding**   The top to a container is made of steel weighing 4.375 lb/sq ft. If the top is a 30.0 in. by 30.0 in. square, find its weight to three significant digits.

24. **Machine Trades**   A square bolt must be machined from round stock. If the bolt must be $2\frac{1}{8}$ in. on each side, what diameter stock is needed? (See the figure.) (Round to the nearest thousandth.)

25. **Automotive Trades**   A fan pulley has a diameter of $3\frac{7}{8}$ in. What is its circumference? (Round to four significant digits.)

26. **Fire Protection**   Firefighters report that a wildfire with a front about 10 miles long has consumed approximately 25,000 acres. What is the average width of the destruction in miles? (1 sq mi = 640 acres.) (Round to the nearest mile.)

27. **Automotive Trades**   If the tires of a car have a diameter of 24 in., how many miles will the car travel when the tires make 16,000 revolutions? (*Remember:* During one revolution of the tires the car travels a distance equal to the circumference of the tire.) (Round to the nearest mile.)

28. **Agriculture**   At the bean farm, the automatic seedling planter creates rows 30 in. apart and places seedlings 24 in. apart. If the bean field is a square 680 ft on a side, how many bean plants will the machine plant? (*Hint:* Add one row and one plant per row to account for rows and/or plants on all four sides of the square.)

Solid Figures

Objective	Sample Problems		For help, go to
When you finish this chapter, you will be able to:			
1. Identify solid figures, including prisms, cubes, cones, cylinders, pyramids, spheres, and frustums.	(a) 	_____	Page 561
	(b)	_____	
	(c)	_____	
2. Find the surface area and volume of solid objects (rounded to one decimal place).	(a) 2' / 4'	Volume = _____ Total surface area = _____	Page 581
	(b) Sphere r = 2.1 cm	Volume = _____ Surface area = _____	Page 585
	(c) 4" / 0.6"	Volume = _____	Page 563
	(d) 8" / 6"	Lateral surface area = _____ Volume = _____	Page 590

Name _____

Date _____

Course/Section _____

Objective	Sample Problems	For help, go to

(e)   Lateral surface area = _____ Page 576

$h = 4$ cm
$s = 5$ cm

Volume = _____

3. Solve practical problems involving solid figures.

(a) **Construction** Calculate the amount of water, in gallons, needed to fill the swimming pool shown. (1 cu ft ≈ 7.48 gal.) Round to the nearest 100 gallons. Page 564

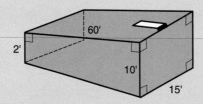

(b) **Metalworking** Find the weight, in ounces, of the steel pin shown if its density is 0.2963 lb/cu in. Round to three significant digits. Page 563

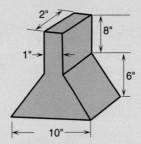

(c) **Roofing** The roof of a building is in the shape of a pyramid having a square base 84 ft on each side. If the slant height of the roof is 60 ft, what would be the cost of the roofing material needed, at $0.45 per sq ft? Page 573

(Answers to these preview problems are given in the Appendix. Also, worked solutions to many of these problems appear in the chapter Summary.)

If you are certain that you can work *all* these problems correctly, turn to page 601 for a set of practice problems. If you cannot work one or more of the preview problems, turn to the page indicated to the right of the problem. Those who wish to master this material with the greatest success should turn to Section 9-1 and begin work there.

Chapter 9

Solid Figures

By permission Jonny Hart and Creators Syndicate, Inc.

Plane figures have only two dimensions: length and width or base and height. A plane figure can be drawn exactly on a flat plane surface. Solid figures are three-dimensional: They have length, width, and height. Making an exact model of a solid figure requires shaping it in clay, wood, or paper. A square is a two-dimensional or plane figure; a cube is a three-dimensional or solid figure.

Trades workers encounter solid figures in the form of tanks, pipes, ducts, boxes, birthday cakes, and buildings. They must be able to identify the solid and its component parts and compute its surface area and volume.

9-1 Prisms

A *prism* is a solid figure having at least one pair of parallel surfaces that create a uniform cross-section. The figure shows a hexagonal prism.

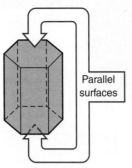

Hexagonal prism

Cutting the prism anywhere parallel to the hexagonal surfaces produces the same hexagonal cross-section.

All the polygons that form the prism are called *faces*. The faces that create the uniform cross-section are called the *bases*. They give the prism its name. The others are called lateral faces.

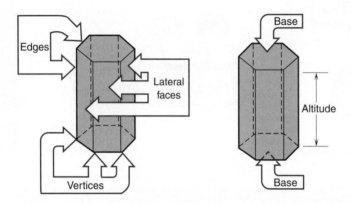

The sides of the polygons are the *edges* of the prism, and the corners are still referred to as *vertices*. The perpendicular distance between the bases is called the *altitude* of the prism.

In a *right prism* the lateral edges are perpendicular to the bases, and the lateral faces are rectangles. In this section you will learn about right prisms only.

Right prisms   Not right prisms

→ **Your Turn**

The components of a prism are its faces, vertices, and edges. Count the number of faces, vertices, and edges in this prism:

Number of faces = _____

Number of vertices = _____

Number of edges = _____

→ **Answers**

There are 7 faces (2 bases and 5 lateral faces), 15 edges, and 10 vertices on this prism.

Chapter 9 Solid Figures

Here are some examples of simple prisms.

Rectangular Prism   The most common solid in practical work is the *rectangular prism.* All opposite faces are parallel, so any pair can be chosen as bases. Every face is a rectangle. The angles at all vertices are right angles.

Cube   A *cube* is a rectangular prism in which all edges are the same length.

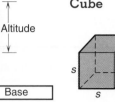

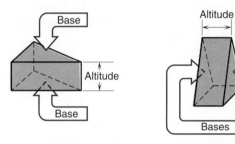

Triangular Prism   In a *triangular prism* the bases are triangular.

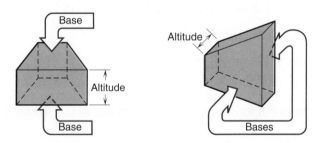

Trapezoidal Prism   In a *trapezoidal prism* the bases are identical trapezoids.

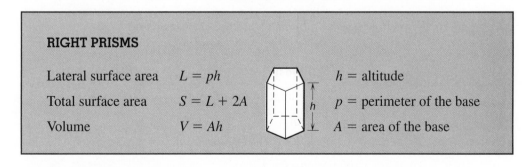

We know that the trapezoids and not the other faces are the bases because cutting the prism anywhere parallel to the trapezoid faces produces another trapezoidal prism.

Three important quantities may be calculated for any prism: the lateral surface area, the total surface area, and the volume. The following formulas apply to all right prisms.

RIGHT PRISMS

Lateral surface area	$L = ph$	h = altitude
Total surface area	$S = L + 2A$	p = perimeter of the base
Volume	$V = Ah$	A = area of the base

The lateral surface area L is the area of all surfaces *excluding* the two bases.

The total surface area S is the lateral surface area *plus* the area of the two bases.

The *volume* or capacity of a prism is the total amount of space inside it. Volume is measured in cubic units. (You may want to review these units in Chapter 5, pages 285 and 304.)

Note As with units of area, units of volume can be written using exponents. For example, cubic inches (cu in.) result from multiplying inches by inches by inches, which in exponent form is in.^{3}. Similarly, cubic yards (cu yd) can be expressed as yd^3, cubic meters (cu m) as m^3, and so on. Workers in the trades use both types of notation, and we shall use both in the remainder of this text. ◄

Example 1

Let's look at an example. Find the (a) lateral surface area, (b) total surface area, and (c) volume of the triangular cross-section duct shown.

(a) The perimeter of the base is

$$p = 6 \text{ in.} + 8 \text{ in.} + 10 \text{ in.} = 24 \text{ in. and } h = 15 \text{ in.}$$

Therefore, the lateral surface area is

$$L = ph = (24 \text{ in.})(15 \text{ in.})$$
$$= 360 \text{ in.}^2$$

(b) In this prism the bases are right triangles, so that the area of each base is

$$A = \frac{1}{2}bh$$
$$= \frac{1}{2}(6 \text{ in.})(8 \text{ in.})$$
$$= 24 \text{ in.}^2$$

Therefore, the total surface area is

$$S = L + 2A$$
$$= 360 \text{ in.}^2 + 2(24 \text{ in.}^2)$$
$$= 408 \text{ in.}^2$$

(c) The volume of the prism is

$$V = Ah$$
$$= (24 \text{ in.}^2)(15 \text{ in.})$$
$$= 360 \text{ in.}^3$$

→ **Your Turn**

Find L, S, and V for the trapezoidal right prism shown in the margin.

→ **Solution**

From the figure, $h = 10$ ft   and $p = 4$ ft $+ 15$ ft $+ 12$ ft $+ 17$ ft $= 48$ ft

The area of each trapezoidal base is

$$A = \left(\frac{b_1 + b_2}{2}\right)h$$

$$= \left(\frac{4 \text{ ft} + 12 \text{ ft}}{2}\right)(15 \text{ ft})$$

$$= 120 \text{ ft}^2$$

Therefore,

$L = ph$	$S = L + 2A$	$V = Ah$
$= (48 \text{ ft})(10 \text{ ft})$	$= 480 \text{ ft}^2 + 2(120 \text{ ft}^2)$	$= (120 \text{ ft}^2)(10 \text{ ft})$
$= 480 \text{ ft}^2$	$= 720 \text{ ft}^2$	$= 1200 \text{ ft}^3$

→ **More Practice**

Find the lateral surface area, total surface area, and volume for each of the following prisms. Round to the nearest whole number.

(a)   (b) 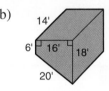   (c) 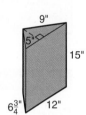   (d)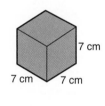

→ **Solutions**

(a) This is a rectangular prism—that is, the base is a rectangle.

First, find the altitude, h, of the prism and the perimeter, p, of the base:

$h = 4\frac{1}{2}$ in. $p = 3$ in. $+ 2$ in. $+ 3$ in. $+ 2$ in. $= 10$ in.

Then, find the area, A, of the base: $A = (3 \text{ in.})(2 \text{ in.}) = 6 \text{ in.}^2$

Now, we can proceed to calculate lateral area, L, total surface area, S, and volume, V:

$L = ph$	$S = L + 2A$	$V = Ah$
$= (10 \text{ in.})(4.5 \text{ in.})$	$= 45 \text{ in.}^2 + 2(6 \text{ in.}^2)$	$= (6 \text{ in.}^2)(4.5 \text{ in.})$
$= 45 \text{ in.}^2$	$= 57 \text{ in.}^2$	$= 27 \text{ in.}^3$

(b) This is a trapezoidal prism. The base is the trapezoid with $b_1 = 6$ ft, $b_2 = 18$ ft, and $h = 16$ ft.

The altitude of the prism is 14 ft and the perimeter of the trapezoidal base is 60 ft. The area of the trapezoidal base is 192 ft². Therefore,

$L = 840 \text{ ft}^2$ $S = 1224 \text{ ft}^2$ $V = 2688 \text{ ft}^3$

(c) This is a triangular prism. The base is a triangle with $b = 12$ in. and $h = 5$ in.

The altitude of the prism is 15 in. and the perimeter, p, of the triangular base is

$p = 9$ in. $+ 12$ in. $+ 6\frac{3}{4}$ in. $= 27\frac{3}{4}$ in.

The area of the triangular base is $A = \frac{1}{2}bh = \frac{1}{2}(12 \text{ in.})(5 \text{ in.}) = 30 \text{ in.}^2$

Therefore,

$L = 416.25 \text{ in.}^2 \qquad\qquad S = 476.25 \text{ in.}^2 \qquad\qquad V = 450 \text{ in.}^3$

$\approx 416 \text{ in.}^2$, rounded $\qquad \approx 476 \text{ in.}^2$

(d) For a cube, $L = 4s^2 = 196 \text{ cm}^2$

$S = 6s^2 = 294 \text{ cm}^2$

$V = s^3 \ = 343 \text{ cm}^3$

Converting Volume Units

In practical problems it is often necessary to convert volume units from cubic inches or cubic feet to gallons or similar units. Use the following conversion factors and set up unity fractions as shown in Chapter 5.

$1 \text{ cu ft (ft}^3) = 1728 \text{ cu in. (in.}^3)$ $1 \text{ cu yd (yd}^3) = 27 \text{ cu ft (ft}^3)$

$\approx 7.48 \text{ gallons}$

$\approx 28.32 \text{ liters}$ $1 \text{ cu in. (in}^3) \approx 16.39 \text{ cu cm (cm}^3)$

$\approx 0.0283 \text{ cu m (m}^3)$ $1 \text{ gal} = 231 \text{ cu in. (in.}^3)$

$\approx 0.806 \text{ bushel}$ $1 \text{ liter} = 1000 \text{ mL} = 1000 \text{ cu cm (cm}^3)$

Example 2

Construction Calculate the volume of water needed to fill this lap pool in gallons. Round to three significant digits.

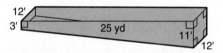

The pool is a prism with trapezoidal bases. The bases of the trapezoid are 3 ft and 11 ft, the height of the trapezoid is 25 yd or 75 ft, and the altitude of the prism is 12 ft.

First, find the area A of the base:

$$A = \left(\frac{b_1 + b_2}{2}\right)h = \left(\frac{3 \text{ ft} + 11 \text{ ft}}{2}\right)75 \text{ ft} = (7 \text{ ft})(75 \text{ ft}) = 525 \text{ ft}^2$$

Next, find the volume in cubic feet:

$$V = Ah = (525 \text{ ft}^2)(12 \text{ ft}) = 6300 \text{ ft}^3$$

Finally, use the conversion factor, $1 \text{ ft}^3 \approx 7.48 \text{ gal}$, to form a unity fraction that cancels cubic feet and replaces it with gallons:

$$V \approx 6300 \ \cancel{\text{ft}^3} \times \frac{7.48 \text{ gal}}{1 \ \cancel{\text{ft}^3}}$$

$V \approx 47,124 \text{ gal} \qquad$ or $\qquad 47,100 \text{ gal}$, rounded

 3 $\boxed{+}$ 11 $\boxed{=}$ $\boxed{\div}$ 2 $\boxed{\times}$ 75 $\boxed{\times}$ 12 $\boxed{\times}$ 7.48 $\boxed{=}$ →   47124.

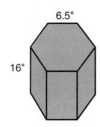

→ **Your Turn**

What volume of liquid, in liters, is needed to fill the container shown? Assume that the base is a regular hexagon and round your answer to two significant digits.

→ **Solution**

As shown in Chapter 8, the area of a hexagon can be calculated as

$A \approx 2.598a^2 \approx 2.598(6.5 \text{ in.})^2 \approx 109.7655 \text{ in.}^2$

And the volume of the prism is

$V = Ah \approx (109.7655 \text{ in.}^2)(16 \text{ in.}) \approx 1756.248 \text{ in.}^3$

To convert this to liters, first convert to either cubic feet or cubic centimeters, then use a second unity fraction to convert to liters. Using $1 \text{ ft}^3 = 1728 \text{ in.}^3$ and $1 \text{ ft}^3 \approx 28.32 \text{ L}$, the calculation looks like this:

$$V \approx 1756.248 \text{ in.}^3 \times \frac{1 \text{ ft}^3}{1728 \text{ in.}^3} \times \frac{28.32 \text{ liters}}{1 \text{ ft}^3}$$

$$\approx 28.78 \ldots \text{liters} \approx 29 \text{ liters, rounded.}$$

To convert to cubic centimeters first and then liters, use $1 \text{ in.}^3 \approx 16.39 \text{ cm}^3$ and $1 \text{ L} = 1000 \text{ cu cm}$ as follows:

$$V \approx 1756.248 \text{ in.}^3 \times \frac{16.39 \text{ cm}^3}{1 \text{ in.}^3} \times \frac{1 \text{ liter}}{1000 \text{ cm}^3}$$

$$\approx 28.78 \ldots \text{liters} \approx 29 \text{ liters, rounded}$$

In many practical problems the prism encountered has an irregular polygon for its base.

Example 3

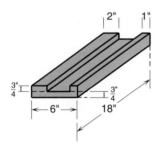

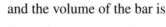

Metalworking   Steel weighs approximately 0.283 lb/in.^3 Find the weight of the steel bar shown. (Round to one decimal place.)

First, we must calculate the volume of this prism. To find this, we first calculate the cross-sectional area by subtracting.

$A = (6'' \times 1.5'') - (3'' \times \frac{3}{4}'')$

$\quad = 9 \text{ in.}^2 - 2.25 \text{ in.}^2$

$\quad = 6.75 \text{ in.}^2$

and the volume of the bar is

$V = Ah$

$\quad = (6.75 \text{ in.}^2)(18 \text{ in.})$

$\quad = 121.5 \text{ in.}^3$

Using $1 \text{ in.}^3 \approx 0.283 \text{ lb}$, we multiply by the unity fraction that will cancel cubic inches and replace it with pounds. The weight of the bar is

$$W = 121.5 \text{ in.}^3 \times \frac{0.283 \text{ lb}}{1 \text{ in.}^3}$$

$W = 34.3845 \text{ lb}$ or 34.4 lb rounded

The most common source of errors in most calculations of this kind is carelessness. Organize your work neatly. Work slowly and carefully. ◄

→ **More Practice**

Now try these practice problems.

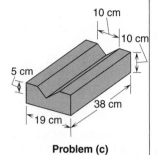

Problem (c)

(a) **Construction**   How many cubic *yards* of concrete does it take to pour a square pillar 2 ft on a side if it is 12 ft high? (Round to two significant digits.)

(b) **Wastewater Technology**   What is the capacity to the nearest gallon of a septic tank in the shape of a rectangular prism 12 ft by 16 ft by 6 ft?

(c) **Metalworking**   What is the weight of the steel V-block in the figure at 0.0173 lb/cm^3? Round to the nearest pound.

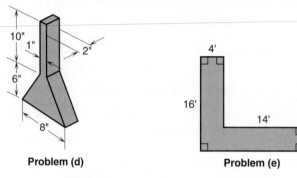

Problem (d)   **Problem (e)**

(d) **Metalworking**   Find the weight of the piece of brass pictured at 0.2963 lb/in.^{3} Round to the nearest tenth.

(e) **Landscaping**   The figure shows an overhead view of a raised bed built in a vegetable garden. A landscaper needs to fill the bed to a depth of 2 ft with soil. How many cubic *yards* of soil should she order? (Round to the nearest cubic yard.)

→ **Solutions**

(a) $V = Ah = (2 \text{ ft})(2 \text{ ft})(12 \text{ ft})$

$= 48 \text{ ft}^3$ but $1 \text{ yd}^3 = 27 \text{ ft}^3$

$= 48 \text{ ft}^3 \times \dfrac{1 \text{ yd}^3}{27 \text{ ft}^3}$

$= \dfrac{48}{27} \text{ yd}^3 \approx 1.8 \text{ yd}^3$

(b) $V = Ah = (12 \text{ ft})(16 \text{ ft})(6 \text{ ft})$

$= 1152 \text{ ft}^3$ but $1 \text{ ft}^3 \approx 7.48 \text{ gallons}$

$\approx 1152 \text{ ft}^3 \times \dfrac{7.48 \text{ gal}}{1 \text{ ft}^3}$

$\approx 8616.96 \text{ gal} \approx 8617 \text{ gal}$

(c)   **First,** find the area of the base. The cross-sectional view shown indicates that the base of the block is a rectangle with a triangle cut out. The area of the base is

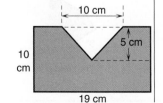

$A = \underbrace{(10 \text{ cm})(19 \text{ cm})}_{\text{rectangle}} - \underbrace{\tfrac{1}{2}(10 \text{ cm})(5 \text{ cm})}_{\text{triangle}}$

$= 190 \text{ cm}^2 - 25 \text{ cm}^2 = 165 \text{ cm}^2$

Next, find the volume.

$$V = Ah = (165 \text{ cm}^2)(38 \text{ cm})$$
$$= 6270 \text{ cm}^3$$

Finally, calculate the weight, W, by multiplying the volume by the density.

$$W \approx (6270 \text{ cm}^3)(0.0173 \text{ lb/cm}^3)$$
$$\approx 108.471 \text{ lb} \approx 108 \text{ lb}$$

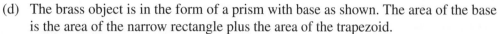

(d) The brass object is in the form of a prism with base as shown. The area of the base is the area of the narrow rectangle plus the area of the trapezoid.

$$A = \underbrace{(10 \text{ in.})(1 \text{ in.})}_{\text{rectangle}} + \underbrace{\tfrac{1}{2}(1 \text{ in.} + 8 \text{ in.})(6 \text{ in.})}_{\text{trapezoid}}$$

$$= 10 \text{ in.}^2 + 27 \text{ in.}^2 = 37 \text{ in.}^2$$

The altitude of the prism is the thickness of the brass or 2 in.
The volume of the object is

$$V = Ah = (37 \text{ in.}^2)(2 \text{ in.}) = 74 \text{ in.}^3$$

The weight of the object is

$$W \approx 74 \text{ in.}^3 \times 0.2963 \text{ lb/in.}^3 \approx 21.9262 \text{ lb} \approx 21.9 \text{ lb}$$

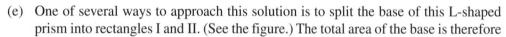

(e) One of several ways to approach this solution is to split the base of this L-shaped prism into rectangles I and II. (See the figure.) The total area of the base is therefore

$$A = \underbrace{(16 \text{ ft})(4 \text{ ft})}_{\text{Area I}} + \underbrace{(14 \text{ ft})(4 \text{ ft})}_{\text{Area II}} = 64 \text{ ft}^2 + 56 \text{ ft}^2 = 120 \text{ ft}^2$$

The altitude, h, of the prism is the depth of the soil, so that $h = 2$ ft. Therefore the volume, V, of soil in cubic feet is

$$V = Ah = (120 \text{ ft}^2)(2 \text{ ft}) = 240 \text{ ft}^3$$

Soil must be ordered by the cubic yard. Using $1 \text{ yd}^3 = 27 \text{ ft}^3$, we have

$$V = 240 \, \cancel{\text{ft}^3} \times \frac{1 \text{ yd}^3}{27 \, \cancel{\text{ft}^3}} = 8.888\ldots \text{ yd}^3 \approx 9 \text{ yd}^3$$

Now turn to Exercises 9-1 for more practice on calculating the surface area and volume of prisms.

Prisms

A. Find the volume of each of the following shapes. Round to the nearest tenth unless indicated otherwise.

1.

2.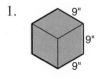

3.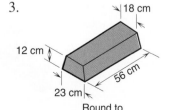

Round to nearest hundred

4.

Round to nearest cu in.

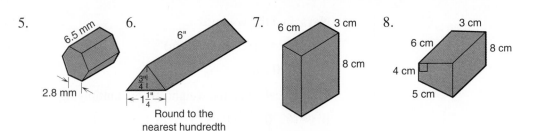

5. 6.5 mm 2.8 mm

6. 6" 3"/4 1¼"
Round to the
nearest hundredth

7. 6 cm 3 cm 8 cm

8. 3 cm 6 cm 8 cm 4 cm 5 cm

B. Find the lateral surface area and the volume of each of the following solids. (Round to the nearest tenth.)

1. 8' 10' 6'

2. 24 cm 8 cm 8 cm 8 cm

3. 14 mm 14 mm 14 mm

4. 10' 2' 14' 12' 16'

5. 7" 5" 4" 5" 12" 13"

6. 80 yd 60 yd 120 yd

C. Find the total outside surface area and the volume of each of the following solids. (Round to the nearest tenth.)

1. 5 yd 5 yd 5 yd

2. 9.3 m 7.4 m 5.2 m

3. 16' 16' 22'

4. ½" 3½" The base is a regular hexagon.

5. 13 cm 35 cm 25 cm 37 cm 36 cm

6. 6 yd 12 yd 2 yd 8 yd 15 yd

D. Practical Problems. Round to the nearest tenth.

1. **Manufacturing**   How many cubic *feet* of warehouse space are needed for 450 boxes 16 in. by 8 in. by 10 in.?

2. **Metalworking**   Find the weight of the piece of steel pictured. (Steel weighs 0.2833 lb/cu in.)

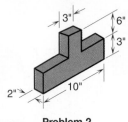

3" 6" 3" 2" 10"

Problem 2

Chapter 9 Solid Figures

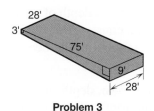

Problem 3

3. **Construction** How many gallons of water are needed to fill a swimming pool that approximates the shape in the figure? (Round to the nearest gallon.)

4. **Painting** How many gallons of paint are needed for the outside walls of a building 26 ft high by 42 ft by 28 ft if there are 480 sq ft of windows? One gallon covers 400 sq ft. (Assume that you cannot buy a fraction of a gallon.)

5. **Construction** Dirt must be excavated for the foundation of a building 30 yards by 15 yards to a depth of 3 yards. How many trips will it take to haul the dirt away if a truck with a capacity of 3 cu yd is used?

6. **Masonry** If brick has a density of 103 lb/cu ft, what will be the weight of 500 bricks, each $3\frac{3}{4}$ in. by $2\frac{1}{4}$ in. by 8 in.? (Round to the nearest pound.)

7. **Landscaping** Dirt cut 2 ft deep from a section of land 38 ft by 50 ft is used to fill a section 25 ft by 25 ft to a depth of 3 ft. How many cubic *yards* of dirt are left after the fill?

8. **Metalworking** Find the weight of the steel V-block in the figure if its density is 0.0173 lb/cm³.

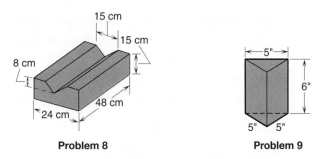

Problem 8   **Problem 9**

9. **Metalworking** Find the weight of the cast iron shape in the figure. (Cast iron weighs 0.2607 lb/cu in. The base is an equilateral triangle.)

10. **Metalworking** Find the weight of the piece of brass shown. Brass weighs 0.296 lb/cu in.

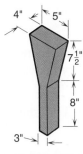

Problem 10

11. **Manufacturing** How high must a 400-gallon rectangular tank be if the base is a square 3 ft 9 in. on a side? (1 cu ft ≈ 7.48 gallons.)

12. **Landscaping** How many pounds of rock will you need to fill an area 25 ft by 35 ft to a depth of 2 in. if the rock weighs 1050 lb/cu yd? (Round to the nearest 10 lb.)

13. **Wastewater Technology** A rectangular sludge bed measuring 80 ft by 120 ft is filled to a depth of 2 ft. If a truck with a capacity of 3 cu yd is used to haul away the sludge, how many trips must the truck make to empty the sludge bed?

14. **Wastewater Technology** A reservoir with a surface area of 2.5 acres loses 0.3 in. per day to evaporation. How many gallons does it lose in a week? (1 acre = 43,560 sq ft, 1 sq ft = 144 sq in., 1 gal = 231 in.³) (Round to the nearest thousand.)

15. **Construction** A construction worker needs to pour a rectangular slab 28 ft long, 16 ft wide, and 6 in. thick. How many cubic yards of concrete does the worker need? (Round to the nearest tenth.)

16. **Construction** Concrete weighs 150 lb/cu ft. What is the weight of a concrete wall 18 ft long, 4 ft 6 in. high, and 8 in. thick?

17. **Masonry** The concrete footings for piers for a raised foundation are cubes measuring 18 in. by 18 in. by 18 in. How many cubic yards of concrete are needed for 28 footings?

18. **Plumbing**   A flush tank 21 in. by 6.5 in. contains water to a depth of 12 in. How many gallons of water will be saved per flush if a conservation device reduces the capacity to two-thirds of this amount? (Round to the nearest tenth of a gallon.)

19. **Roofing**   A flat roof 28 ft long and 16 ft wide has a 3-in. depth of water sitting on it. What is the weight of the water on the roof? (Water weighs 62.5 lb/cu ft.)

20. **Plumbing**   In a 4-hour percolation test, an 8-ft depth of water seeped out of a square pit 4 ft on a side. The plumbing code says that the soil must be able to absorb 5000 gallons in 24 hours. At this rate, will the soil be up to code?

21. **Plumbing**   The Waste-away Plumbing Company installs a rectangular septic tank 10 ft long by 5 ft wide and 6 ft deep. Calculate the capacity of the tank in gallons.

22. **Automotive Trades**   The fuel tank of a motor vehicle has the shape of a rectangular prism measuring 16 in. wide, 36 in. long, and 10 in. high. How many gallons will the tank hold? (1 gal $\approx$ 231 cu. in.) (Round to the nearest gallon.)

23. **Plumbing**   A pipe trench is roughly rectangular with dimensions 75 ft long by 2 ft wide and with an average depth of $3\frac{1}{2}$ ft. What volume of dirt was removed?

24. **Welding**   The parts bin shown is made of 16-gauge sheet metal. Calculate the volume in cubic feet and the total length of weld needed in feet.

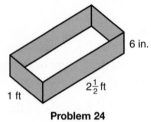

6 in.

$2\frac{1}{2}$ ft

1 ft

Problem 24

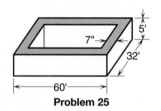

7"

5'

32'

60'

Problem 25

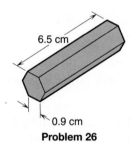

6.5 cm

0.9 cm

Problem 26

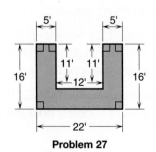

5'   5'

11' 11'

16'   16'

12'

22'

Problem 27

25. **Construction**   BiltWell Construction needs to create a concrete foundation 5 ft deep measuring 60 ft by 32 ft, outside dimensions, with walls 7 in. thick. How many cubic yards of concrete will they need?

26. **Metalworking**   Find the weight (in ounces) of the steel pin shown if its density is 0.0173 lb/cm^3. Round to two significant digits.

27. **Landscaping**   A landscaper has built a U-shaped raised bed in a vegetable garden as shown in the figure. How many cubic yards of soil should be ordered to fill the bed to a depth of 18 in.? (Round to the nearest cubic yard.)

28 **Landscaping**   A pool measuring 200 in. by 400 in. has a leak that causes the pool to lose $2\frac{3}{8}$ in. of water depth per day.

 (a)   How many cubic inches of water are lost per day?

 (b)   How many cubic feet of water are lost per day? (*Hint:* 1 cu ft = 1728 cu in.) (Round to the nearest cubic foot.)

 (c)   The local water district charges $3.75 per HCF (hundred cubic feet). If the leak is not fixed and the lost water is replaced, how much will this add to the resident's *monthly* water bill? (Assume a 31-day month.)

When you have completed these exercises, check your answers to the odd-numbered problems in the Appendix, then turn to Section 9-2 to study pyramids.

Pyramids A *pyramid* is a solid object with one base and three or more *lateral faces* that taper to a single point opposite the base. This single point is called the *apex*.

A *right pyramid* is one whose base is a regular polygon and whose apex is centered over the base. We will examine only right pyramids here.

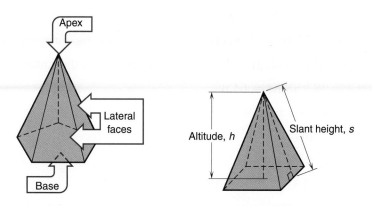

The *altitude* of a pyramid is the perpendicular distance from the apex to the base. The *slant height* of the pyramid is the height of one of the lateral faces.

Several kinds of right pyramids can be constructed:

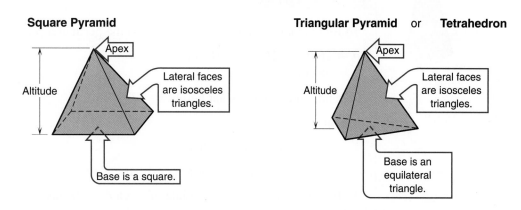

There are several very useful formulas relating to pyramids.

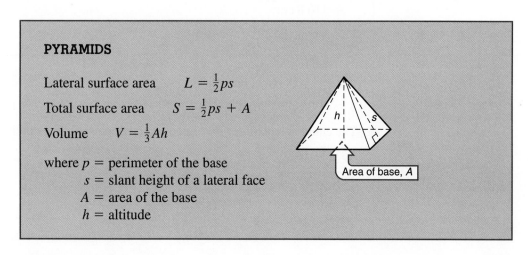

PYRAMIDS

Lateral surface area $\quad L = \frac{1}{2}ps$

Total surface area $\quad S = \frac{1}{2}ps + A$

Volume $\quad V = \frac{1}{3}Ah$

where p = perimeter of the base
$\quad s$ = slant height of a lateral face
$\quad A$ = area of the base
$\quad h$ = altitude

Example 1

The square pyramid shown has a slant height $s \approx 10.77$ ft, and an altitude $h = 10$ ft. To find the lateral surface area, first calculate the perimeter p of the base.

$p = 8 \text{ ft} + 8 \text{ ft} + 8 \text{ ft} + 8 \text{ ft} = 32 \text{ ft}$

The lateral surface area is

$L = \frac{1}{2}ps$

$\approx \frac{1}{2}(32 \text{ ft})(10.77 \text{ ft})$

$\approx 172.3 \text{ ft}^2$ This is the area of the four triangular lateral faces.

To find the volume, first calculate the area A of the square base.

$A = (8 \text{ ft})(8 \text{ ft}) = 64 \text{ ft}^2$

The volume is

$V = \frac{1}{3}Ah$

$= \frac{1}{3}(64 \text{ ft}^2)(10 \text{ ft})$

$\approx 213.3 \text{ ft}^3$ rounded

→ **Your Turn**

Find the lateral surface area and volume of this triangular pyramid. (*Remember:* The base is an equilateral triangle.) (Round to the nearest whole unit.)

→ **Solutions**

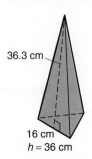

36.3 cm

16 cm
$h = 36$ cm

To find the lateral surface area, first find the perimeter p of the base.

$p = 16 \text{ cm} + 16 \text{ cm} + 16 \text{ cm}$
$\quad = 48 \text{ cm}$

Second, note that the slant height $s \approx 36.3$ cm.

Therefore, $L = \frac{1}{2}ps$

$\approx \frac{1}{2}(48 \text{ cm})(36.3 \text{ cm})$

$\approx 871.2 \text{ cm}^2$ or 871 cm^2 rounded

To find the volume, first calculate the area A of the base.

$A \approx 0.433s^2$ for an equilateral triangle
$\quad \approx 0.433(16 \text{ cm})^2$
$\quad \approx 110.8 \text{ cm}^2$

The volume is

$V = \frac{1}{3}Ah$

$\approx \frac{1}{3}(110.8 \text{ cm}^2)(36 \text{ cm})$

$\approx 1330 \text{ cm}^3$ rounded

The entire calculation for volume can be done as follows using a calculator:

.433 ⊠ 16 $\boxed{x^2}$ ⊠ 36 ⊟ 3 ⊟ →   *1330.176*

A

Multiplying by $\frac{1}{3}$ is the
same as dividing by 3

To calculate the **total** surface area, S, of a pyramid, add the area of the base to the lateral surface area.

Example 2

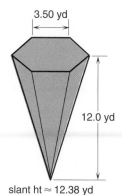

3.50 yd

12.0 yd

slant ht ≈ 12.38 yd

Find the total surface area of the hexagonal pyramid shown. (Round to three significant digits.)

First, find the lateral surface area, L.

$p = 6(3.5 \text{ yd}) = 21 \text{ yd}$

$s \approx 12.38 \text{ yd}$

$L = \frac{1}{2}ps \approx \frac{1}{2}(21 \text{ yd})(12.38 \text{ yd})$

$\approx 129.99 \text{ yd}^2$

Then find the area of the hexagonal base.

$A \approx 2.598 (3.5 \text{ yd})^2 \approx 31.8255 \text{ yd}^2$

Finally, add these areas and round.

$S = L + A \approx 129.99 \text{ yd}^2 + 31.8255 \text{ yd}^2$

$\approx 161.8155 \text{ yd}^2 \approx 162 \text{ yd}^2$, rounded.

The entire calculation can be performed on a calculator as follows:

 6 ⊗ 3.5 ⊗ 12.38 ⊗ .5 ⊕ 2.598 ⊗ 3.5 x^2 = → **161.8155**

$\underbrace{\qquad\qquad\qquad\qquad}_{\text{Lateral area}}$ $\underbrace{\qquad\qquad}_{\text{Base area}}$

→ More Practice

(a) Find the lateral surface area and volume of each pyramid. Round to three significant digits.

(1)

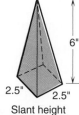

6"

2.5" 2.5"

Slant height
≈ 6.13"
Altitude = 6"

(2)

4'

3'

Slant height ≈ 4.58'
Altitude = 3'

(b) Find the total surface area and volume of this pyramid. Round to three significant digits.

13 cm

Altitude = 16 cm
Slant height ≈ 16.4 cm

(c) Find the volume of an octagonal pyramid if the area of the base is 245 in.^{2} and the altitude is 16 in. Round to the nearest cubic inch.

(a) (1) $p = 4(2.5 \text{ in.}) = 10 \text{ in.}$ $s \approx 6.13 \text{ in.}$

$L = \frac{1}{2}ps \approx \frac{1}{2}(10)(6.13)$

$\approx 30.65 \text{ in.}^2 \approx 30.7 \text{ in.}^2$, rounded

$V = \frac{1}{3}Ah = \frac{1}{3}(2.5 \text{ in.})^2(6 \text{ in.})$

$= 12.5 \text{ in.}^3$

(2) $p = 6(4 \text{ ft}) = 24 \text{ ft}$ $s \approx 4.58 \text{ ft}$

$L = \frac{1}{2}ps \approx \frac{1}{2}(24 \text{ ft})(4.58 \text{ ft})$

$\approx 54.96 \text{ ft}^2 \approx 55.0 \text{ ft}^2$, rounded

$\qquad\qquad\text{area of hexagonal base}$

$V = \frac{1}{3}Ah \approx \frac{1}{3}\overbrace{(2.598)(4 \text{ ft})^2}(3 \text{ ft})$

$\approx 41.568 \text{ ft}^3 \approx 41.6 \text{ ft}^3$, rounded

(b) **First,** find the lateral surface area, L.

$p = 3(13 \text{ cm}) = 39 \text{ cm}$ $s \approx 16.4 \text{ cm}$

$L = \frac{1}{2}ps \approx \frac{1}{2}(39 \text{ cm})(16.4 \text{ cm})$

$\approx 319.8 \text{ cm}^2$

Next, find A, the area of the base, an equilateral triangle.

$A \approx 0.433a^2$

$\approx 0.433(13 \text{ cm})^2 \approx 73.177 \text{ cm}^2$

Finally, add $L + A$ to obtain the total surface area, S.

$S \approx 319.8 \text{ cm}^2 + 73.177 \text{ cm}^2$

$\approx 392.977 \text{ cm}^2 \approx 393 \text{ cm}^2$, rounded

Having calculated the area of the base we can now find the volume.

$V = \frac{1}{3}Ah \approx \frac{1}{3}(73.177 \text{ cm}^2)(16 \text{ cm})$

$\approx 390.277\ldots \text{ cm}^3 \approx 390 \text{ cm}^3$, rounded

To use a calculator efficiently, find the area of the base first and store it in memory. You can then recall it when you calculate the volume.

Base area   Lateral area

Surface area: **.433** ⊠ **13** $\boxed{x^2}$ ⊟ ⓢⓉⓄ Ⓐ ⊞ **3** ⊠ **13** ⊠ **16.4** ⊠ **.5** ⊟

$\rightarrow$ ▨ **392.977**

Volume: ⓇⒸⓁ Ⓐ ⊠ **16** ⊟ **3** ⊟ $\rightarrow$ ▨ **390.2773333**

(c) $V = \frac{1}{3}Ah = \frac{1}{3}(245 \text{ in.}^2)(16 \text{ in.})$

$= 1306.666\ldots \text{ in.}^3 \approx 1307 \text{ in.}^3$

Frustum of a Pyramid   A *frustum* of a pyramid is the solid figure remaining after the top of the pyramid is cut off parallel to the base. Frustum shapes appear as containers, as building foundations, and as transition sections in ducts.

Most frustums used in practical work, such as sheet metal construction, are frustums of square pyramids or cones.

Every frustum has two bases, upper and lower, that are parallel and different in size. The altitude is the perpendicular distance between the bases.

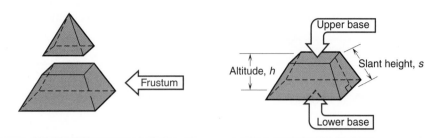

The following formulas enable you to find the lateral area and volume of any pyramid frustum.

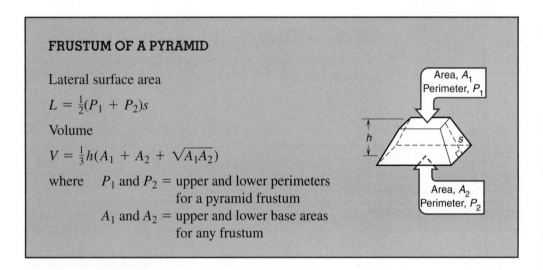

FRUSTUM OF A PYRAMID

Lateral surface area

$L = \frac{1}{2}(P_1 + P_2)s$

Volume

$V = \frac{1}{3}h(A_1 + A_2 + \sqrt{A_1A_2})$

where P_1 and P_2 = upper and lower perimeters
for a pyramid frustum
A_1 and A_2 = upper and lower base areas
for any frustum

Example 3

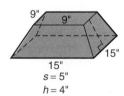

15"
$s = 5"$
$h = 4"$

To find the lateral surface area and volume of the pyramid frustum shown, first calculate the upper and lower perimeters.

Upper perimeter, $P_1 = 9$ in. $+ 9$ in. $+ 9$ in. $+ 9$ in. $= 36$ in.

Lower perimeter, $P_2 = 15$ in. $+ 15$ in. $+ 15$ in. $+ 15$ in. $= 60$ in.

Now substitute these values into the formula for lateral surface area.

$L = \frac{1}{2}(36\text{ in.} + 60\text{ in.})(5\text{ in.})$

   $= 240$ in.^{2}

To find the volume of this frustum, first calculate the upper and lower areas.

Upper area, $A_1 = (9\text{ in.})(9\text{ in.}) = 81$ in.^{2}

Lower area, $A_2 = (15\text{ in.})(15\text{ in.}) = 225$ in.^{2}

Now substitute these values into the formula for volume.

$V = \frac{1}{3}(4\text{ in.})\left(81\text{ in.}^2 + 225\text{ in.}^2 + \sqrt{81 \cdot 225}\right)$

   $= \frac{1}{3}(4\text{ in.})(81 + 225 + 135)\text{ in.}^2$

   $= 588$ in.^{3}

Use the following calculator sequence to find the volume:

C9-1

$9\,\boxed{x^2}\,\boxed{+}\,15\,\boxed{x^2}\,\boxed{+}\,\boxed{\sqrt{}}\,{}^{*}9\,\boxed{x^2}\,\boxed{\times}\,15\,\boxed{x^2}\,\boxed{)}\,\boxed{=}\,\boxed{\times}\,4\,\boxed{\div}\,3\,\boxed{=}\rightarrow$ 588.

*Parentheses must be opened here.

Notice that we multiply by the altitude 4 near the end to avoid needing a second set of parentheses. Notice also that dividing by 3 is easier than multiplying by $\frac{1}{3}$.

To find the total surface area of the frustum, add the areas A_1 and A_2 to the lateral surface area.

→ **Your Turn**

Find the total surface area and volume of the frustum shown.

→ **Solution**

$P_1 = 4(8\text{ ft}) = 32\text{ ft} \qquad P_2 = 4(14\text{ ft}) = 56\text{ ft}$

$L = \frac{1}{2}(32\text{ ft} + 56\text{ ft})(5\text{ ft}) = 220\text{ ft}^2$

$A_1 = (8\text{ ft})^2 = 64\text{ ft}^2 \qquad A_2 = (14\text{ ft})^2 = 196\text{ ft}^2$

Total surface area, $S = L + A_1 + A_2$

$S = 220\text{ ft}^2 + 64\text{ ft}^2 + 196\text{ ft}^2$

$\quad = 480\text{ ft}^2$

$V = \frac{1}{3}(4\text{ ft})(64\text{ ft}^2 + 196\text{ ft}^2 + \sqrt{64\cdot196})$

$\quad = 496\text{ ft}^3$

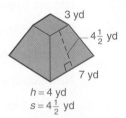

8'
5'
14'
$h = 4'$
$s = 5'$

→ **More Practice**

Practice with these problems.
Round to three significant digits.

(a) Find the lateral surface area and volume of this frustum.

3 yd
$4\frac{1}{2}$ yd
7 yd
$h = 4$ yd
$s = 4\frac{1}{2}$ yd

(b) Find the total surface area and volume of this frustum.

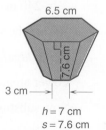

6.5 cm
7.6 cm
3 cm
$h = 7$ cm
$s = 7.6$ cm

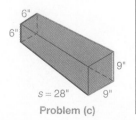

6"
6"
9"
9"
$s = 28"$
Problem (c)

(c) **HVAC** A connection must be made between two square vent openings in an air conditioning system. Find the total amount of sheet metal needed using the dimensions shown in the figure.

(d) **Construction** How many cubic yards of concrete are needed to pour the foundation shown in the figure? (Watch your units!) (Round to the nearest tenth.)

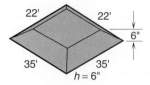

Problem (d)

→ **Solutions**

(a) $P_1 = 4(3 \text{ yd}) = 12 \text{ yd}$ $P_2 = 4(7 \text{ yd}) = 28 \text{ yd}$

$L = \frac{1}{2}(12 \text{ yd} + 28 \text{ yd})(4\frac{1}{2} \text{ yd}) = 90 \text{ yd}^2$

$A_1 = (3 \text{ yd})^2 = 9 \text{ yd}^2$ $A_2 = (7 \text{ yd})^2 = 49 \text{ yd}^2$

$V = \frac{1}{3}(4 \text{ yd})(9 \text{ yd}^2 + 49 \text{ yd}^2 + \sqrt{9 \cdot 49})$

  $= 105.33 \ldots \text{yd}^3 \approx 105 \text{ yd}^3$, rounded

(b) $P_1 = 6(6.5 \text{ cm}) = 39 \text{ cm}$ $P_2 = 6(3 \text{ cm}) = 18 \text{ cm}$

$L = \frac{1}{2}(39 \text{ cm} + 18 \text{ cm})(7.6 \text{ cm}) = 216.6 \text{ cm}^2$

$A_1 \approx 2.598 \,(6.5 \text{ cm})^2 \approx 109.7655 \text{ cm}^2$

$A_2 \approx 2.598 \,(3 \text{ cm})^2 \approx 23.382 \text{ cm}^2$

$S = L + A_1 + A_2$

$S \approx 216.6 \text{ cm}^2 + 109.7655 \text{ cm}^2 + 23.382 \text{ cm}^2$

  $\approx 349.7475 \text{ cm}^2 \approx 350 \text{ cm}^2$

$V \approx \frac{1}{3}(7 \text{ cm})(109.7655 + 23.382 + \sqrt{109.7655 \times 23.382})$

  $\approx 428.8865 \text{ cm}^3 \approx 429 \text{ cm}^3$, rounded

To make efficient use of a calculator, find A_1 and A_2 first and store them in different memory locations. You can then recall them when calculating both surface area and volume.

A_1: **2.598** $\boxed{\times}$ **6.5** $\boxed{x^2}$ $\boxed{=}$ $\boxed{\text{STO}}$ $\boxed{A}$ → $\boxed{\;109.7655\;}$

A_2: **2.598** $\boxed{\times}$ **3** $\boxed{x^2}$ $\boxed{=}$ $\boxed{\text{STO}}$ $\boxed{B}$ → $\boxed{\;23.382\;}$

S: $\underbrace{\textbf{6} \boxed{\times} \textbf{6.5} \boxed{+} \textbf{6} \boxed{\times} \textbf{3} \boxed{=} \boxed{\times} \textbf{7.6} \boxed{\times} \textbf{.5}}_{\text{Lateral area}} \boxed{+} \underbrace{\boxed{\text{RCL}} \boxed{A}}_{A_1} \boxed{+} \underbrace{\boxed{\text{RCL}} \boxed{B}}_{A_2} \boxed{=}$ → $\boxed{\;349.7475\;}$

C9-2 V: $\boxed{\text{RCL}} \boxed{A} \boxed{+} \boxed{\text{RCL}} \boxed{B} \boxed{+} \boxed{\sqrt{}} \,^* \boxed{\text{RCL}} \boxed{A} \boxed{\times} \boxed{\text{RCL}} \boxed{B} \boxed{)} \boxed{=} \boxed{\times} \textbf{7} \boxed{\div} \textbf{3} \boxed{=}$

  → $\boxed{\;428.8865\;}$

*Parentheses must be opened here.

(c) $P_1 = 24 \text{ in.}$ $P_2 = 36 \text{ in.}$

$L = \frac{1}{2}(24 \text{ in.} + 36 \text{ in.})(28 \text{ in.}) = 840 \text{ in.}^2$

(d) $A_1 = (22 \text{ ft})^2 = 484 \text{ ft}^2$ $A_2 = (35 \text{ ft})^2 = 1225 \text{ ft}^2$

$V = \frac{1}{3}(0.5 \text{ ft})(484 \text{ ft}^2 + 1225 \text{ ft}^2 + \sqrt{484 \cdot 1225})$

  $= 413.166 \ldots \text{ft}^3$

  $\approx 413.166 \, \cancel{\text{ft}^3} \times \dfrac{1 \text{ yd}^3}{27 \, \cancel{\text{ft}^3}} \approx 15.3 \text{ yd}^3$

Notice that the height, 6 in., was converted to 0.5 ft to agree with the other units. The answer was then converted from cu ft to cu yd.

Learning Help Always be certain that your units agree before entering them into a calculation. ◄

Now turn to Exercises 9-2 for more practice on finding the surface area and volume of pyramids and frustums of pyramids.

Exercises 9-2 Pyramids and Frustums of Pyramids

A. Find the lateral surface area and volume of the solid objects shown. Round to three significant digits, if necessary.

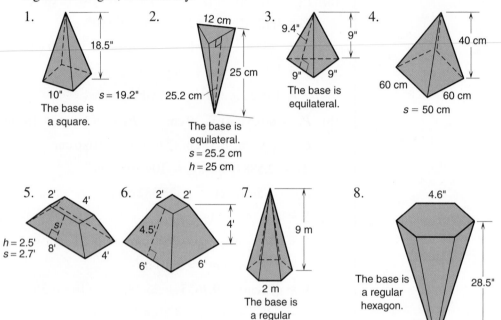

1.
18.5"
10" s = 19.2"
The base is a square.

2.
12 cm
25 cm
25.2 cm
The base is equilateral.
s = 25.2 cm
h = 25 cm

3.
9.4" 9"
9" 9"
The base is equilateral.

4.
40 cm
60 cm
60 cm
s = 50 cm

5.
2' 4'
s'
h = 2.5' 8'
s = 2.7' 4'

6.
2' 2'
4.5' 4'
6' 6'

7.
9 m
2 m
The base is a regular hexagon. s = 9.2 m

8.
4.6"
The base is a regular hexagon.
28.5"
s ≈ 28.8"

B. Find the total outside surface area and volume of the following solid objects. Round to three significant digits, if necessary. (Assume that all bases are regular.)

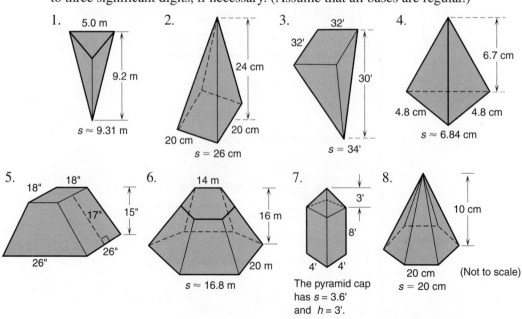

1.
5.0 m
9.2 m
s ≈ 9.31 m

2.
24 cm
20 cm
20 cm
s = 26 cm

3.
32'
32'
30'
s = 34'

4.
6.7 cm
4.8 cm 4.8 cm
s ≈ 6.84 cm

5.
18" 18"
17" 15"
26"
26"

6.
14 m
16 m
20 m
s ≈ 16.8 m

7.
3'
8'
4' 4'
The pyramid cap has s = 3.6' and h = 3'.

8.
10 cm
20 cm (Not to scale)
s = 20 cm

C. Practical Problems

1. **Roofing**   The roof of a building is in the shape of a square pyramid 25 m on each side. If the slant height of the pyramid is 18 m, how much will roofing material cost at $8.61 per sq m?

2. **Metalworking**   A pyramid-shaped piece of steel has a triangular base measuring 6.8 in. on each side. If the height of the piece is 12.4 in., find the weight of two dozen of these pyramids. Assume that the steel has a density of 0.283 lb/cu in. Round to the nearest pound.

3. **HVAC**   The Last Chance gambling casino is designed in the shape of a square pyramid with a side length of 75 ft. The height of the building is 32 ft. An air filtering system can circulate air at 20,000 cu ft/hr. How long will it take the system to filter all the air in the room?

4. **Agriculture**   How many bushels will the bin in the figure hold? (1 cu ft = 1.24 bushels.) (Round to the nearest bushel.)

5. **Sheet Metal Trades**   How many square inches of sheet metal are used to make the vent transition shown? (The ends are open.)

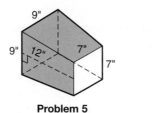

Problem 5

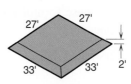

Problem 6

6. **Construction**   How many cubic yards of concrete are needed to pour the building foundation shown in the figure?

When you have completed these exercises, check your answers to the odd-numbered problems in the Appendix, then turn to Section 9-3 to study cylinders and spheres.

Problem 4 (figure: pyramid with dimensions 10', 10', 5', 4', 4')

9-3   Cylinders and Spheres

Cylinders

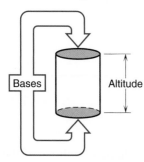

Bases   Altitude

A *cylinder* is a solid object with two identical circular bases. The *altitude* of a cylinder is the perpendicular distance between the bases.

A *right cylinder* is one whose curved side walls are perpendicular to its circular base. Whenever we mention the radius, diameter, or circumference of a cylinder, we are referring to those dimensions of its circular base.

Two important formulas enable us to find the lateral surface area and volume of a cylinder.

CYLINDERS

Lateral surface area

$L = Ch$ or $L = 2\pi rh = \pi dh$

Volume

$V = \pi r^2 h$ or $V = \frac{1}{4}\pi d^2 h \approx 0.7854 d^2 h$

where C is the circumference of the base, r is the radius of the base, d is the diameter of the base, and h is the altitude of the cylinder.

To find the total surface area of a cylinder, simply add the areas of the two circular bases to the lateral area. The formula is

Total surface area $= 2\pi r^2 + 2\pi rh = 2\pi r(r + h)$

Learning Help You can visualize the lateral surface area by imagining the cylinder wall "unrolled" into a rectangle as shown.

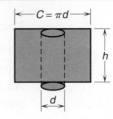

Lateral surface area $=$ area of rectangle

$$= L \times W$$

$$= \pi d \times h$$

$$= \pi dh \qquad \blacktriangleleft$$

Example 1

Use these formulas to find the lateral surface area and volume of the cylindrical container shown in the figure.

First, note that

$$d = 22 \text{ in.} \qquad r = \frac{d}{2} = 11 \text{ in.} \qquad h = 24 \text{ in.}$$

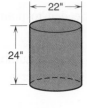

Then the lateral surface area is

$L = \pi dh$

$\quad = (\pi)(22 \text{ in.})(24 \text{ in.})$

$\quad \approx 1659 \text{ in.}^2 \quad$ rounded

and the volume is

$V = \pi r^2 h$

$\quad = (\pi)(11 \text{ in.})^2(24 \text{ in.})$

$\quad = (\pi)(121 \text{ in.}^2)(24 \text{ in.})$

$\quad \approx 9123 \text{ in.}^3 \quad$ rounded

→ Your Turn

Industrial Technology Find the lateral surface area and the volume in gallons of the cylindrical storage tank shown. Round answers to the nearest unit.

→ Solution

The diameter $d = 14$ ft, and the altitude $h = 30$ ft. Therefore,

$L = \pi dh$

$\quad = (\pi)(14 \text{ ft})(30 \text{ ft})$

$\quad \approx 1319 \text{ ft}^2 \quad$ rounded

$$V = \frac{1}{4}\pi d^2 h \approx 0.7854 d^2 h$$

$$\approx (0.7854)(14\text{ ft})^2(30\text{ ft})$$

$$\approx (0.7854)(196\text{ ft}^2)(30\text{ ft})$$

$$\approx 4618.152\text{ ft}^3 \quad \text{or} \quad 4618\text{ ft}^3 \quad \text{rounded}$$

To convert this to gallons, recall that $1\text{ ft}^3 \approx 7.48$ gal. Using a unity fraction,

$$V \approx 4618.152\text{ ft}^3 \times \frac{7.48\text{ gal}}{1\text{ ft}^3}$$

$$\approx 34{,}544\text{ gal} \quad \text{rounded}$$

After organizing your work for volume, you can key it into your calculator this way:

 $\boxed{\pi}$ $\boxed{\div}$ 4 $\boxed{\times}$ 14 $\boxed{x^2}$ $\boxed{\times}$ 30 $\boxed{\times}$ 7.48 $\boxed{=}$ → **34543.69618**

Notice that $\frac{1}{4}\pi$ is the same as $\frac{\pi}{4}$.

Careful When using an intermediate value in a later calculation, do not round it prematurely. For example, if you had rounded V to 4618 and used this value to calculate V in gallons, the result would be 34,543, which is not correct to the nearest gallon. ◄

→ More Practice

Try these problems for practice in working with cylinders.

(a) Find the lateral surface area and volume of each cylinder. Use 3.1416 for π and round to the nearest whole number.

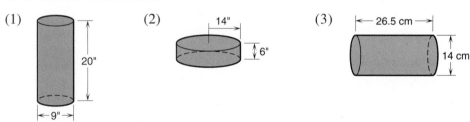

(1) 20" 9"

(2) 14" 6"

(3) 26.5 cm 14 cm

(b) Find the total surface area of a cylinder 9 yd high and 16 yd in diameter. Round to the nearest whole number.

(c) **Construction** How many cubic *yards* of concrete will it take to pour a cylindrical column 14 ft high with a diameter of 2 ft? Round to the nearest tenth.

(d) **Wastewater Technology** What is the capacity in gallons of a cylindrical tank with a radius of 8 ft and an altitude of 20 ft? Round to the nearest 100 gallons.

(e) **Industrial Technology** A pipe 3 in. in diameter and 40 ft high is filled to the top with water. If 1 cu ft of water weighs 62.4 lb, what is the weight at the base of the pipe? (Be careful of your units.) (Round to the nearest pound.)

(f) **Industrial Technology** The bottom and the inside walls of the cylindrical tank shown in the figure must be lined with sheet copper. How many square feet of sheet copper are needed? (Round to the nearest tenth.)

(g) **Painting** How many quarts of paint are needed to cover the outside wall (not including the top and bottom) of a cylindrical water tank 10 ft in diameter and 10 ft in height if one quart covers 100 sq ft? Use 3.14 for π. (Assume that you cannot buy a fraction of a quart.)

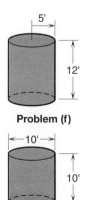

5' 12'

Problem (f)

10' 10'

Problem (g)

(a) (1) 565 in.^{2}, 1272 in.^{3} (2) 528 in.^{2}, 3695 in.^{3}

   (3) 1166 cm^2, 4079 cm^3

(b) $d = 16$ yd $r = 8$ yd $h = 9$ yd

   $S = 2\pi r(r + h)$

      $= 2\pi(8 \text{ yd})(8 \text{ yd} + 9 \text{ yd})$

      $\approx 854.513\ldots \text{ yd}^2 \approx 855 \text{ yd}^2$, rounded

(c) **First,** find the volume in ft^3.

   $V \approx 0.7854 d^2 h$

      $\approx 0.7854(2 \text{ ft})^2(14 \text{ ft})$

      $\approx 43.9824 \text{ ft}^3$

Then, convert this result to yd^3.

   $V \approx 43.9824 \, \cancel{\text{ft}^3} \times \dfrac{1 \text{ yd}^3}{27 \, \cancel{\text{ft}^3}}$

      $\approx 1.6289\ldots \text{ yd}^3 \approx 1.6 \text{ yd}^3$, rounded.

With a calculator, the entire operation may be keyed in as follows:

(d) **First,** find the volume in ft^3.

   $V = \pi r^2 h = \pi(8 \text{ ft})^2(20 \text{ ft})$

      $\approx 4021.24 \text{ ft}^3$

Then, convert this quantity to gallons.

   $V \approx 4021.24 \, \cancel{\text{ft}^3} \times \dfrac{7.48 \text{ gal}}{1 \, \cancel{\text{ft}^3}}$

      $\approx 30{,}078.8752 \text{ gal} \approx 30{,}100 \text{ gal}$, rounded

(e) **First,** convert the diameter, 3 in., to 0.25 ft to agree with the other units in the problem. **Next,** find the volume in ft^3.

   $V \approx 0.7854 \, d^2 h \approx 0.7854(0.25 \text{ ft})^2(40 \text{ ft})$

      $\approx 1.9635 \text{ ft}^3$

Finally, convert the volume to lb units.

   $V \approx 1.9635 \, \cancel{\text{ft}^3} \times \dfrac{62.4 \text{ lb}}{1 \, \cancel{\text{ft}^3}}$

      $\approx 123 \text{ lb}$, rounded

(f) The inside wall is the lateral surface area.

   $L = 2\pi r h = 2\pi(5 \text{ ft})(12 \text{ ft})$

      $= 120\pi \text{ ft}^2$

The bottom wall is a circle.

   $A = \pi r^2 = \pi(5 \text{ ft})^2$

      $= 25\pi \text{ ft}^2$

The total area of sheet copper is

$$L + A = 120\pi + 25\pi = 145\pi \text{ ft}^2$$

$$\approx 455.53 \ldots \text{ ft}^2 \approx 455.5 \text{ ft}^2, \text{ rounded}$$

> **Learning Help**   When you are adding two or more circular areas, it is usually simpler to express each in terms of π and substitute for π after finding their sum. ◀

(g) **First,** find the lateral surface area.

$$L = \pi dh \approx 3.14(10 \text{ ft})(10 \text{ ft})$$

$$\approx 314 \text{ ft}^2$$

Then, use a unity fraction to determine the amount of paint used.

$$L \approx 314 \, \cancel{\text{ft}^2} \times \frac{1 \text{ qt}}{100 \, \cancel{\text{ft}^2}} \approx 3.14 \text{ qt}$$

Because you cannot buy a fraction of a quart, you will need 4 qt of paint.

Spheres

The *sphere* is the simplest of all solid geometric figures. Geometrically, it is defined as the surface whose points are all equidistant from a given point called the *center*. The *radius* is the distance from the center to the surface. The *diameter* is the straight-line distance across the sphere on a line through its center.

The following formulas enable you to find the surface area and volume of any sphere.

SPHERE

Surface area

$$S = 4\pi r^2 \quad \text{or} \quad S = \pi d^2$$

Volume

$$V = \frac{4\pi r^3}{3} \quad \text{or} \quad V = \frac{\pi d^3}{6}$$

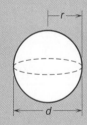

The volume formulas can be written approximately as

$$V \approx 4.1888 r^3 \qquad \text{and} \qquad V \approx 0.5236 d^3$$

Example 2

Sports and Leisure   A spherical basketball has a diameter of $9\frac{1}{2}$ in.

The surface area of the ball is

$$S = \pi d^2$$

$$= \pi(9.5 \text{ in.})^2$$

$$= \pi(90.25 \text{ in.}^2)$$

$$\approx 283.5 \text{ in.}^2$$

The volume of the ball is

$$V = \frac{\pi d^3}{6}$$

$$= \frac{\pi (9.5 \text{ in.})^3}{6}$$

$$\approx 448.9 \text{ in.}^3 \quad \text{rounded}$$

Using a calculator to find the volume, we have

C9-3 $\boxed{\pi}$ $\boxed{\times}$ **9.5** $\boxed{\wedge}$ **3** $\boxed{\div}$ **6** $\boxed{=}$ → `448.9205002`

───────────────────────────────

Note On some calculators, the power key, $\boxed{\wedge}$, may be labeled $\boxed{y^x}$ or $\boxed{x^y}$. ◀

→ **Your Turn**

Industrial Technology Find the surface area and the volume in gallons of a spherical tank 8 ft in radius. Round to the nearest whole unit.

→ **Solution**

The surface area is

$$S = 4\pi r^2$$

$$= 4\pi (8 \text{ ft})^2$$

$$= 4\pi (64 \text{ ft}^2) = 804.2477 \ldots \text{ ft}^2$$

$$\approx 804 \text{ ft}^2, \text{ rounded}$$

The volume is

$$V = \frac{4\pi r^3}{3}$$

$$= \frac{4\pi (8 \text{ ft})^3}{3} = 2144.66 \ldots \text{ ft}^3$$

$$\approx 2145 \text{ ft}^3, \text{ rounded}$$

$$\approx 2144.7 \text{ ft}^3 \times \frac{7.48 \text{ gallons}}{1 \text{ ft}^3} \quad \Longleftarrow \boxed{\text{Convert to gallons using a unity fraction}}$$

$$\approx 16{,}042 \text{ gallons} \quad \text{rounded to the nearest gallon}$$

The entire volume calculation can be keyed into a calculator like this:

4 $\boxed{\times}$ $\boxed{\pi}$ $\boxed{\times}$ **8** $\boxed{\wedge}$ **3** $\boxed{\div}$ **3** $\boxed{\times}$ **7.48** $\boxed{=}$ → `16042.06117`

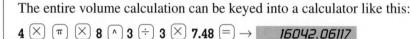

→ **More Practice**

For practice, try the following problems.

(a) Find the surface area and volume of each of the following spheres. Use $\pi = 3.14$ and round to the nearest whole unit.

    (1) radius = 15 in.   (2) diameter = 22 ft

    (3) radius = 6.5 cm   (4) diameter = 8 ft 6 in.

(b) **Wastewater Technology** How many gallons of water can be stored in a spherical tank 50 in. in diameter? (Use 1 gallon = 231 in.^{3}) (Round to the nearest gallon.)

(c) **Industrial Technology** What will be the weight of a spherical steel tank of radius 9 ft when it is full of water? The steel used weighs approximately 127 lb/ft^2 and water weighs 62.4 lb/ft^3. (Round to the nearest 100 lb.)

→ **Answers**

(a) (1) 2827 in.^{2}; 14,137 in.^{3}
    (2) 1521 ft^2; 5575 ft^3
    (3) 531 cm^2; 1150 cm^3
    (4) 227 ft^2; 322 ft^3

(b) 283 gallons

(c) 319,800 lb

A Closer Look In problem (b), the volume is approximately 65,450 in.^{3} Multiply by a unity fraction to convert to gallons.

$$65{,}450 \text{ in.}^3 \times \frac{1 \text{ gal}}{231 \text{ in.}^3} = \frac{65{,}450}{231} \text{ gal or about 283 gal}$$

In problem (c) the surface area is 1017.9 ft^2. To find the weight, multiply by a unity fraction:

$$1017.9 \text{ ft}^2 \times \frac{127 \text{ lb}}{1 \text{ ft}^2} \approx 129{,}300 \text{ lb} \quad \text{rounded}$$

The volume is 3053.6 ft^3 or the weight of water is

$$3053.6 \text{ ft}^3 \times \frac{62.4 \text{ lb}}{1 \text{ ft}^3} \approx 190{,}500 \text{ lb} \quad \text{rounded}$$

The total weight is approximately 129,300 lb + 190,500 lb = 319,800 lb.

We can perform the entire calculation on a calculator as follows.

First, calculate the weight of the tank.

C9-4 4 ⊗ π ⊗ 9 ⊗ ⊗ 127 ⊜ → ▨ *129270.2545*

Then, calculate the weight of the water and add it to the last answer, which is the weight of the tank.

4 ⊗ π ⊗ 9 ^ 3 ⊘ 3 ⊗ 62.4 ⊕ ANS ⊜ → ▨ *319816.6454* ◄

Now turn to Exercises 9-3 for a set of problems on cylinders and spheres.

Exercises 9-3 **Cylinders and Spheres**

A. Find the lateral surface area and volume of each of the following cylinders. Round to three significant digits and use $\pi \approx 3.14$.

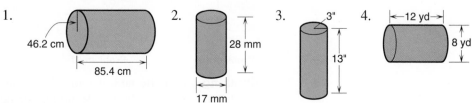

1. 46.2 cm; 85.4 cm
2. 28 mm; 17 mm
3. 3"; 13"
4. 12 yd; 8 yd

B. Find the total surface area and volume of each of the following solid objects. Round to three significant digits.

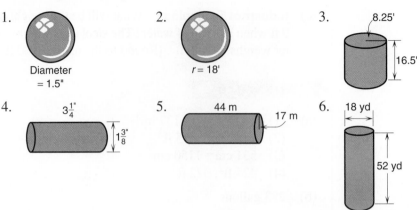

1. Diameter = 1.5"

2. r = 18'

3. 8.25' / 16.5'

4. 3¼" / 1⅜"

5. 44 m / 17 m

6. 18 yd / 52 yd

C. Practical Problems. Round to the nearest tenth unless otherwise directed.

1. **Painting**   A cylindrical tank has a radius of 6.5 ft and an altitude of 14 ft. If a gallon of paint will cover 120 ft² of surface, how much paint is needed to put two coats of paint on the entire surface of the tank? (Round to the nearest gallon.)

2. **Plumbing**   A 4-ft-high cylindrical pipe has a diameter of 3 in. If water weighs 62.4 lb/ft³, what is the weight of water in the pipe when filled to the top? (*Hint:* Change the diameter to feet.)

3. **Landscaping**   A layer of crushed rock must be spread over a circular area 23 ft in diameter. How deep a layer will be obtained using 100 ft³ of rock? (Round to the nearest hundredth of a foot.)

4. **Wastewater Technology**   How many liters of water can be stored in a cylindrical tank 8 m high with a radius of 0.5 m? (1 m³ contains 1000 liters.)

5. **Metalworking**   What is the weight of the piece of aluminum shown at 0.0975 lb/in.³? (*Hint:* Subtract the volume of the cylindrical hole.)

6. **Industrial Technology**   How high should a 50-gal cylindrical tank be if it must fit into an area allowing a 16-in. diameter? (1 gal = 231 in.³)

7. **Plumbing**   A marble-top bathroom sink has the shape of a hemispherical (a half-sphere) porcelain bowl with an inside diameter of 15 in. How much water, in gallons, will the bowl hold? (Round to the nearest gallon.)

8. **Sheet Metal Trades**   How many square inches of sheet copper are needed to cover the inside wall of a cylindrical tank 6 ft high with a 9-in. radius? (*Hint:* Change the altitude to inches.)

9. **Machine Trades**   What is the weight of the bushing shown if it is made of steel weighing 0.2833 lb/in.³? (The inner cylinder is hollow.)

10. **Manufacturing**   Find the capacity in gallons of the oil can in the figure. (1 gal = 231 in.³)

11. **Plumbing**   A septic tank has the shape shown. How many gallons does it hold? (1 ft³ ≈ 7.48 gallons.) (Round to the nearest gallon.)

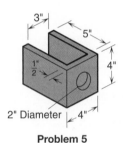

Problem 5

3" / 5" / ½" / 4" / 2" Diameter / 4"

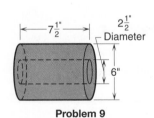

Problem 9

7½" / 2½" Diameter / 6"

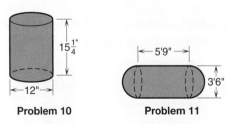

Problem 10   **Problem 11**

15¼" / 12"   5'9" / 3'6"

Chapter 9 Solid Figures

12. **Construction** How many cubic yards of concrete are needed to pour eight cylindrical pillars 12 ft 6 in. high, each with a diameter of 1 ft 9 in.?

13. **Machine Trades** At a density of 0.0925 lb/in.^{3}, calculate the weight of the aluminum piece shown. Round to the nearest hundredth.

Diameter of four outside holes: $\frac{1}{4}$ in.

Diameter of inside hole: $\frac{1}{3}$ in.

Thickness of piece: $\frac{1}{2}$ in.

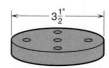

14. **Painting** A spherical tank has a diameter of 16.5 ft. If a gallon of paint will cover 80 ft^2, how many gallons are needed to cover the tank? Round to the nearest gallon.

15. **Industrial Technology** A cylindrical tank 28 in. in diameter must have a capacity of 75 gallons. How high must it be?

16. **Agriculture** The water tower shown consists of a 16-ft-high cylinder and a 10-ft-radius hemisphere. How many gallons of water does it contain when it is filled to the top of the cylinder? (Round to the nearest hundred.)

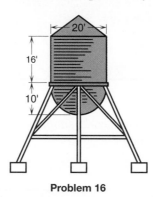

Problem 16

17. **Automotive Trades** In an automobile engine, the bore of a cylinder is its diameter, and the stroke of a cylinder is the height swept by the piston. Find the swept volume of a cylinder with a bore of 3.65 in. and a stroke of 4.00 in. (Round to three significant digits.)

18. **Automotive Trades** A cylindrical hose 24 in. long has an inside diameter of 1.4 in. How many quarts of coolant will it hold? (1 qt = 2 pt and 1 pt = 28.875 cu in.) (Round to two significant digits.)

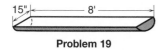

Problem 19

19. **Agriculture** The water trough shown in the figure is constructed with semicircular ends. Calculate its volume in gallons if the diameter of the end is 15 in. and the length of the trough is 8 ft. (Round to the nearest 0.1 gallon.) (*Hint:* Be careful of units.)

20. **Painting** The metal silo shown has a hemispherical top. It is to be painted with exterior weatherproofing paint. If a gallon of this paint covers 280 ft^2, how much paint is needed for two coats? (Do not include the bottom of the silo.) (Round your answer to the nearest gallon.)

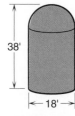

Problem 20

21. **Construction** An outdoor cylindrical fire pit has an interior radius of 30 in. Filling the fire pit with fireglass will require 91 lb of glass per cubic foot. How many pounds will be needed to fill the pit to a depth of 4 in.? (Round to the nearest pound.)

When you have completed these exercises, check your answers to the odd-numbered problems in the Appendix, then turn to Section 9-4 to learn about cones.

Cones and Frustums of Cones

Cones

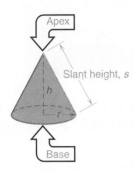

A *cone* is a pyramid-like solid figure with a circular base. The radius and diameter of a cone refer to its circular base. The *altitude* is the perpendicular distance from the apex to the base. The *slant height* is the apex-to-base distance along the surface of the cone.

The following formulas enable us to find the lateral surface area and volume of a cone.

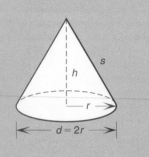

CONES

Lateral surface area

$$L = \pi rs \quad \text{or} \quad L = \tfrac{1}{2}\pi ds$$

Volume

$$V = \tfrac{1}{3}\pi r^2 h \quad \text{or} \quad V = \tfrac{1}{12}\pi d^2 h \approx 0.2618 d^2 h$$

where r = radius of the base
d = diameter of the base
h = altitude
s = slant height

To find the total surface area of a cone, simply add the area of the circular base to the lateral area. The formula would be

$$\text{Total surface area} = S = \pi r^2 + \pi rs = \pi r(r + s)$$

Example 1

To find the lateral surface area and volume of the cone shown, first note that $h = 15$ in., $s = 18$ in., $r = 10$ in.

Then the lateral surface area is

$$L = \pi rs$$
$$= \pi(10 \text{ in.})(18 \text{ in.})$$
$$= 565.486\ldots \text{ in.}^2 \quad \text{or} \quad 565 \text{ in.}^2 \quad \text{rounded}$$

and the volume is

$$V = \tfrac{1}{3}\pi r^2 h$$
$$= \tfrac{1}{3}\pi(10 \text{ in.})^2(15 \text{ in.})$$
$$= 1570.79\ldots \text{ in.}^3 \quad \text{or} \quad 1571 \text{ in.}^3 \quad \text{rounded}$$

Using a calculator to find the volume, we obtain

 $\pi \div 3 \times 10 \boxed{x^2} \times 15 = \rightarrow$ 『1570.796327』

13 cm · 12 cm · 10 cm

→ **Your Turn**

Find the lateral surface area, total surface area, and volume of this cone. (Round to the nearest whole unit.)

→ **Solution**

First, note that $h = 12$ cm $r = 5$ cm $s = 13$ cm

Lateral surface area $= L = \pi r s$
$$= \pi(5 \text{ cm})(13 \text{ cm})$$
$$\approx 204.2035\ldots \text{cm}^2 \quad \text{or} \quad 204 \text{ cm}^2 \quad \text{rounded}$$

Total surface area $= S = \pi r^2 + \pi r s$
$$\approx \pi(5 \text{ cm})^2 + 204.204 \text{ cm}^2$$
$$\approx 78.54 \text{ cm}^2 + 204.204 \text{ cm}^2$$
$$\approx 282.744 \quad \text{or} \quad 283 \text{ cm}^2 \quad \text{rounded}$$

 $\boxed{\pi} \boxed{\times} \, 5 \, \boxed{\times} \, 13 \, \boxed{=} \rightarrow$ **204.2035225** $\boxed{+} \boxed{\pi} \boxed{\times} \, 5 \, \boxed{x^2} \boxed{=} \rightarrow$ **282.743388**

                               ⬆   ⬆
                               L   S

Volume $= V = \frac{1}{3}\pi r^2 h$
$$= \frac{1}{3}\pi(5 \text{ cm})^2(12 \text{ cm})$$
$$= 314.159\ldots \text{cm}^3 \quad \text{or} \quad 314 \text{ cm}^3 \quad \text{rounded}$$

Learning Help You should realize that the altitude, radius, and slant height for any cone are always related by the Pythagorean theorem.

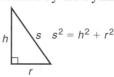

$$s^2 = h^2 + r^2$$

If you are given any two of these quantities, you can use this formula to find the third.

Notice that the volume of a cone is exactly one-third the volume of the cylinder that just encloses it. ◄

→ **More Practice**

Solve the following practice problems.

(a) Find the lateral surface area and volume of each of the following cones. (Use $\pi = 3.14$ and round to the nearest whole number.)

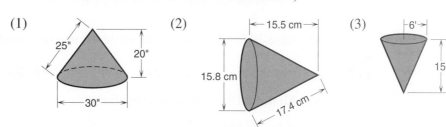

(1) 25" · 20" · 30"   (2) 15.5 cm · 15.8 cm · 17.4 cm   (3) 6' · 15'

(b) Find the total surface area of a cone with radius $\frac{3}{4}$ in., altitude 1 in., and slant height $1\frac{1}{4}$ in. (Round to two decimal places.)

(c) **Construction** A pile of sand dumped by a hopper is cone-shaped. If the diameter of the base is 18 ft 3 in. and the altitude is 8 ft 6 in., how many cubic feet of sand are in the pile? (Round to the nearest tenth.)

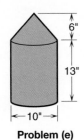

Problem (e)

(d) **Industrial Technology** What is the capacity in gallons of a conical drum 4 ft high with a radius of 3 ft? (1 cu ft ≈ 7.48 gal.) (Round to the nearest gallon.)

(e) **Metalworking** Find the weight of the cast iron shape shown at 0.26 lb/in.^{3} (Round to the nearest pound.)

(f) **Industrial Technology** Find the capacity in gallons (to the nearest tenth) of a conical oil container 15 in. high with a diameter of 16 in. (1 gal = 231 in.^{3})

→ **Solutions**

(a) (1) 1178 in.^{2}, 4712 in.^{3} (2) 432 cm^2, 1013 cm^3

(3) **First,** we must calculate the slant height using the Pythagorean theorem.

$$s^2 = h^2 + r^2$$
$$= 15^2 + 6^2 = 261$$
$$s = \sqrt{261} \approx 16.155 \text{ ft}$$

The lateral surface area, L, is

$$L = \pi rs \approx 3.14(6 \text{ ft})(16.155 \text{ ft})$$
$$\approx 304.3602 \text{ ft}^2 \approx 304 \text{ ft}^2$$

The volume, V, is

$$V = \tfrac{1}{3}\pi r^2 h \approx \tfrac{1}{3}(3.14)(6 \text{ ft})^2(15 \text{ ft})$$
$$\approx 565.2 \text{ ft}^3 \approx 565 \text{ ft}^3$$

(b) $S = \pi r(r + s) = \pi\left(\tfrac{3}{4}\text{ in.}\right)\left(\tfrac{3}{4}\text{ in.} + 1\tfrac{1}{4}\text{ in.}\right)$

$= \pi(0.75 \text{ in.})(2 \text{ in.})$

$\approx 4.71 \text{ in.}^2$

(c) $d = 18 \text{ ft } 3 \text{ in.} = 18.25 \text{ ft}$ $h = 8 \text{ ft } 6 \text{ in.} = 8.5 \text{ ft}$

$V = \tfrac{1}{12}\pi d^2 h = \tfrac{1}{12}\pi(18.25 \text{ ft})^2(8.5 \text{ ft})$

$= 741.16\ldots \text{ ft}^3 \approx 741.2 \text{ ft}^3$, rounded

(d) **First,** find the volume in cubic feet.

$V = \tfrac{1}{3}\pi r^2 h = \tfrac{1}{3}(\pi)(3 \text{ ft})^2(4 \text{ ft})$

$= 37.699\ldots \text{ ft}^3$

Then convert this to gallons.

$V \approx 37.699 \text{ ft}^3 \times \dfrac{7.48 \text{ gal}}{1 \text{ ft}^3} \approx 282 \text{ gal}$

Of course, with a calculator you would not need to round the intermediate result.

 $\boxed{\pi}$ $\boxed{\div}$ **3** $\boxed{\times}$ **3** $\boxed{x^2}$ $\boxed{\times}$ **4** $\boxed{\times}$ **7.48** $\boxed{=}$ → $\boxed{281.9893566}$

(e) **First,** find the volume in cubic inches.

$d = 10''$ so $r = 5''$

For the cylinder:

$V = \pi r^2 h = \pi(5 \text{ in.})^2(13 \text{ in.})$

$\approx 1021.02 \text{ in.}^3$

For the cone:

$V = \tfrac{1}{3}\pi r^2 h = \tfrac{1}{3}\pi(5 \text{ in.})^2(6 \text{ in.})$

$\approx 157.08 \text{ in.}^3$

Total volume, $V \approx 1178.1 \text{ in.}^3$

Then, convert to pounds.

$$\text{Weight} \approx 1178.1 \ \cancel{\text{in.}}^3 \times \frac{0.26 \ \text{lb}}{1 \ \cancel{\text{in.}}^3} \approx 306 \ \text{lb}$$

The entire calculation can be done as follows using a calculator:

 $\boxed{\pi}$ $\boxed{\times}$ **5** $\boxed{x^2}$ $\boxed{\times}$ **13** $\boxed{+}$ $\boxed{\pi}$ $\boxed{\div}$ **3** $\boxed{\times}$ **5** $\boxed{x^2}$ $\boxed{\times}$ **6** $\boxed{=}$ $\boxed{\times}$ **.26** $\boxed{=}$ → `306.3052837`

(f) The capacity, in cubic inches, is

$$V \approx 0.2618d^2h \approx 0.2618(16 \ \text{in.})^2(15 \ \text{in.}) \quad \left(\tfrac{\pi}{12} \approx 0.2618\right)$$
$$\approx 1005.312 \ \text{in.}^3$$

The capacity in gallons is

$$V \approx 1005.312 \ \cancel{\text{in.}}^3 \times \frac{1 \ \text{gal}}{231 \ \cancel{\text{in.}}^3}$$
$$\approx 4.352 \ \text{gal} \approx 4.4 \ \text{gal}, \ \text{rounded to the nearest tenth}$$

 $\boxed{\pi}$ $\boxed{\div}$ **12** $\boxed{\times}$ **16** $\boxed{x^2}$ $\boxed{\times}$ **15** $\boxed{\div}$ **231** $\boxed{=}$ → `4.351989898`

Learning Help | The three formulas for the volume of a cone were provided to give you a choice. You might want to memorize only the formula $V = \tfrac{1}{3}\pi r^2 h$, from which the others were derived. ◄

Frustum of a Cone

Frustum
of a cone

The frustum of a cone, like the frustum of a pyramid, is the figure remaining when the top of the cone is cut off parallel to its base.

Frustum shapes appear as containers, as funnels, and as transition sections in ducts.

Every frustum has two bases, upper and lower, that are parallel and different in size. The altitude is the perpendicular distance between bases.

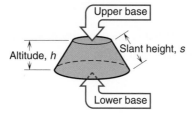

The following formulas enable you to find the lateral area and volume of any frustum of a cone.

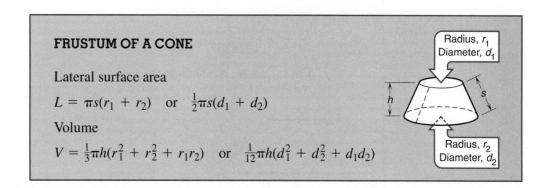

> **FRUSTUM OF A CONE**
>
> Lateral surface area
>
> $L = \pi s(r_1 + r_2)$ or $\tfrac{1}{2}\pi s(d_1 + d_2)$
>
> Volume
>
> $V = \tfrac{1}{3}\pi h(r_1^2 + r_2^2 + r_1 r_2)$ or $\tfrac{1}{12}\pi h(d_1^2 + d_2^2 + d_1 d_2)$

These formulas can be obtained from the equation for the lateral surface area and volume of the frustum of a pyramid, using πr^2 or $\tfrac{1}{4}\pi d^2$ for the area of a circle.

To find the *total surface area* of the frustum, add the areas of the circular ends to the lateral surface area.

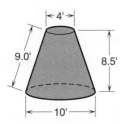

Example 2

To find the lateral surface area, total surface area, and volume of the frustum shown, use $s = 9.0$ ft, $r_1 = 2$ ft, $r_2 = 5$ ft, $h = 8.5$ ft, and $\pi \approx 3.1416$.

Lateral surface area, $\quad L \approx 3.1416(9)(2 + 5)$
$$\approx 197.9208 \text{ ft}^2 \quad \text{or} \quad 198 \text{ ft}^2 \quad \text{rounded}$$

Total surface area, $\quad A \approx L + \pi r_1^2 + \pi r_2^2$
$$\approx 197.9208 + 3.1416(2^2) + 3.1416(5^2)$$
$$\approx 289.0272 \text{ ft}^2 \quad \text{or} \quad 289 \text{ ft}^2 \quad \text{rounded}$$

Volume, $\quad V \approx \frac{1}{3}(3.1416)(8.5)(2^2 + 5^2 + 2 \cdot 5)$
$$\approx 347.1468 \text{ ft}^3 \quad \text{or} \quad 347 \text{ ft}^3 \quad \text{rounded}$$

→ More Practice

Practice with these problems.

(a) Find the lateral surface area and volume of each of the following frustums. (Round to the nearest whole unit and use $\pi = 3.14$.)

(1)
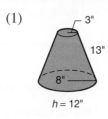
$h = 12"$

(2)
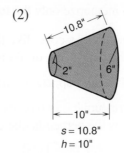
$s = 10.8"$
$h = 10"$

(3)
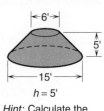
$h = 5'$

Hint: Calculate the slant height s using the Pythagorean theorem.

(b) **Industrial Technology**   What is the capacity in gallons of the oil can shown in the figure? (1 gal $= 231$ in.^{3}) (Round to the nearest tenth.)

$h = 12"$

→ Solutions

(a) (1)   449 in.^{2}; 1218 in.^{3} (2)   271 in.^{2}; 544 in.^{3}

(3)   **First,** find the slant height this way:

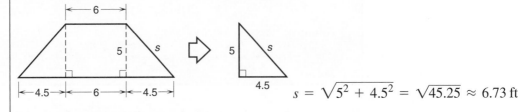

$$s = \sqrt{5^2 + 4.5^2} = \sqrt{45.25} \approx 6.73 \text{ ft}$$

Then, find the lateral surface area.

$$L = \tfrac{1}{2}\pi s(d_1 + d_2) \approx \tfrac{1}{2}(3.14)(6.73 \text{ ft})(6 \text{ ft} + 15 \text{ ft})$$
$$\approx 221.8881 \text{ ft}^2 \approx 222 \text{ ft}^2$$

The volume is

$$V = \tfrac{1}{12}\pi h(d_1^2 + d_2^2 + d_1 \cdot d_2)$$
$$\approx \tfrac{1}{12}(3.14)(5)(6^2 + 15^2 + 6\cdot 15)$$
$$\approx 459.225 \text{ ft}^3 \approx 459 \text{ ft}^3$$

(b) **First,** find the volume in cubic inches.

$$V = \tfrac{1}{12}\pi h(d_1^2 + d_2^2 + d_1 \cdot d_2)$$
$$= \tfrac{1}{12}(\pi)(12)(16^2 + 8^2 + 16\cdot 8)$$
$$\approx 1407.43 \text{ in.}^3$$

Then, translate this to gallons.

$$V \approx 1407.43 \text{ in.}^3 \times \frac{1 \text{ gal}}{231 \text{ in.}^3}$$
$$\approx 6.1 \text{ gal, rounded}$$

Using a calculator on this problem,

 $\boxed{\pi}\ \boxed{\div}\ 12\ \boxed{\times}\ 12\ \boxed{\times}\ \boxed{(}\ 16\ \boxed{x^2}\ \boxed{+}\ 8\ \boxed{x^2}\ \boxed{+}\ 16\ \boxed{\times}\ 8\ \boxed{)}\ \boxed{\div}\ 231\ \boxed{=} \rightarrow$ **6.092785752**

Now turn to Exercises 9-4 for more practice with cones and frustums of cones.

Exercises 9-4 Cones and Frustums of Cones

A. Find the lateral surface area and volume of each solid object. Round to three significant digits.

1.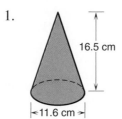
16.5 cm
11.6 cm

2.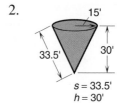
15'
33.5' 30'
$s = 33.5'$
$h = 30'$

3.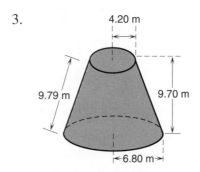
4.20 m
9.79 m   9.70 m
6.80 m

4.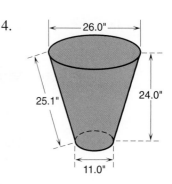
26.0"
25.1"   24.0"
11.0"

B. Find the total surface area and volume of each solid object. Round to three significant digits.

1.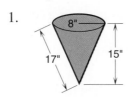
8"
17"   15"

2.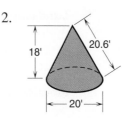
18'   20.6'
20'

3.

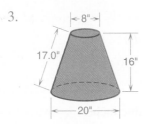

4.

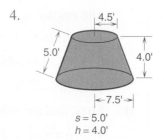

$s = 5.0'$
$h = 4.0'$

C. Find the volume of each figure. Use $\pi \approx 3.14$ and round to two significant digits.

1.

2.

3.

4.

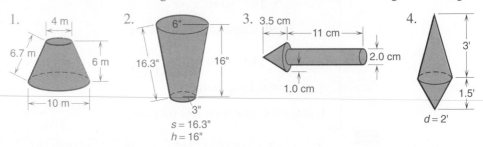

$s = 16.3"$
$h = 16"$

$d = 2'$

D. Practical Problems. Round to the nearest tenth if necessary.

Problem 2

Problem 3

1. **Landscaping** How many cubic meters of dirt are there in a pile, conical in shape, 10 m in diameter and 4 m high?

2. **Sheet Metal Trades** How many square centimeters of sheet metal are needed for the sides and bottom of the pail shown in the figure?

3. **Agriculture** How many bushels will the bin in the figure hold? (1 cu ft $\approx$ 1.24 bushels.)

4. **Landscaping** Decorative Moon Rocks cost $92 per cu yd. What will be the cost of a conical pile of this rock $3\frac{1}{3}$ yd in diameter and $2\frac{1}{4}$ yd high? (Round to the nearest dollar.)

5. **Metalworking** Calculate the weight of the piece of steel shown in the drawing. (This steel weighs 0.283 lb/cu in.)

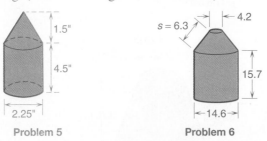

Problem 5

Problem 6

6. **Sheet Metal Trades** How many square centimeters of sheet metal will it take to make the open-top container shown in the figure? All measurements are in centimeters. (*Hint:* No metal is needed for the small opening at the top.)

7. **Manufacturing** A conical oil cup with a radius of 3.8 cm must be designed to hold 64 cu cm of oil. What should be the altitude of the cup?

8. **Landscaping** At an excavation site a conical pile of earth 9 ft high and 14 ft in diameter at the base must be hauled away by a small truck with a capacity of 2.6 cu yd. How many truckloads will it take to transport the soil? (Be careful with units!)

9. **Metalworking** Find the weight of the steel rivet shown in the figure. (Steel weighs 0.0173 lb/cu cm.)

2.0 cm

6.4 cm

3.2 cm

10.0 cm

Problem 9

10. **Welding** A cone-shaped hopper is constructed by first cutting a wedge from a circular piece of sheet metal [See figure (a).] The center of the circle is then hoisted up and the two edges of the missing wedge are welded together. [See figure (b).] Note that the radius of the original piece, 3.18 m, is now the slant height of the cone, and the radius of the cone is 2.56 m. Find the volume of the cone to the nearest tenth.

(a)

(b)

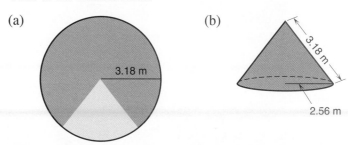

11. **Manufacturing** A yogurt carton is in the shape of the frustum of a cone. The upper diameter is 2 in., the lower diameter is $2\frac{5}{16}$ in., and the height is 3 in. How many ounces of yogurt does the carton hold? (*Hint:* 1 oz = 1.805 cu in.). (Round to the nearest tenth.)

When you have completed these exercises, go to Problem Set 9 on page 601 for a set of problems covering the work of this chapter. If you need a quick review of the topics in this chapter, visit the chapter Summary first.

Summary Solid Figures

Objective

Identify solid figures, including prisms, cubes, cones, cylinders, pyramids, spheres, and frustums. (pp. 561–593)

Review

A **prism** is a solid figure having at least one pair of parallel surfaces that create a uniform cross-section. A **cube** is a rectangular prism in which all edges are the same length. A **pyramid** is a solid object with one base that is a polygon and three or more lateral faces that taper to a single point, the apex. A **cylinder** is a solid object with two identical circular bases. A **cone** is a pyramid-like solid figure with a circular base. A **frustum** is the solid figure remaining after the top of a pyramid or cone is cut off by a plane parallel to the base. A **sphere** is the round surface created by all points that are equidistant from a given point, the center.

Example:

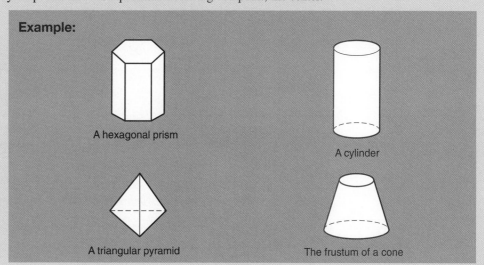

A hexagonal prism

A cylinder

A triangular pyramid

The frustum of a cone

Objective

Find the surface area and volume of solid objects.

Review

The following table provides a handy summary of the formulas for solid figures presented in this chapter.

SUMMARY OF FORMULAS FOR SOLID FIGURES

Figure	Lateral Surface Area	Volume
Prism	$L = ph$	$V = Ah$
Pyramid	$L = \frac{1}{2}ps$	$V = \frac{1}{3}Ah$
Cylinder	$L = \pi dh$	$V = \pi r^2 h$
	or $\quad L = 2\pi rh$	or $\quad V \approx 0.7854 d^2 h$
Cone	$L = \frac{1}{2}\pi ds$	$V = \frac{1}{3}\pi r^2 h$
	or $\quad L = \pi rs$	or $\quad V \approx 0.2618 d^2 h$
Frustum of pyramid	$L = \frac{1}{2}(P_1 + P_2)s$	$V = \frac{1}{3}h(A_1 + A_2 + \sqrt{A_1 A_2})$
Frustum of cone	$L = \frac{1}{2}\pi s(d_1 + d_2)$	$V = \frac{1}{12}\pi h(d_1^2 + d_2^2 + d_1 d_2)$
	or $\quad L = \pi s(r_1 + r_2)$	or $\quad V = \frac{1}{3}\pi h(r_1^2 + r_2^2 + r_1 r_2)$
Sphere	$S = 4\pi r^2$	$V = \frac{4}{3}\pi r^3$
	or $\quad S = \pi d^2$	or $\quad V \approx 4.1888 r^3$
		$V = \frac{1}{6}\pi d^3$
		or $\quad V \approx 0.5236 d^3$

Example:

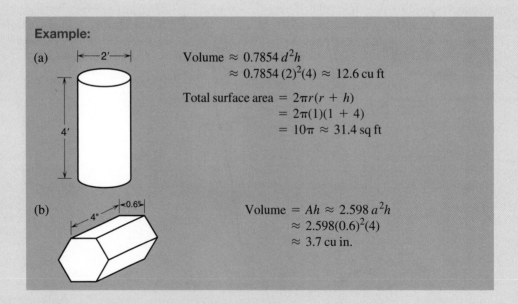

(a)

$$\text{Volume} \approx 0.7854\, d^2 h$$
$$\approx 0.7854\,(2)^2(4) \approx 12.6 \text{ cu ft}$$

$$\text{Total surface area} = 2\pi r(r + h)$$
$$= 2\pi(1)(1 + 4)$$
$$= 10\pi \approx 31.4 \text{ sq ft}$$

(b)

$$\text{Volume} = Ah \approx 2.598\, a^2 h$$
$$\approx 2.598(0.6)^2(4)$$
$$\approx 3.7 \text{ cu in.}$$

Objective

Solve practical problems involving solid figures.

Review

First, use the appropriate formula(s) to calculate the volume or surface area as needed. Then, use the information given in the problem to determine whether any additional calculation, such as a conversion to other units, is necessary.

Example: Calculate the amount of water, to the nearest 100 gal, needed to fill the swimming pool shown.

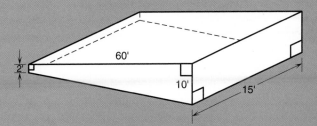

The pool is a trapezoidal prism.

$$V = Ah = \frac{h}{2}(b_1 + b_2) \cdot \text{height of prism}$$

$$= \frac{60}{2}(10 + 2) \cdot 15$$

$$= 5400 \text{ cu ft}$$

In gallons,

$$V \approx 5400 \text{ cu ft}\left(\frac{7.48 \text{ gal}}{1 \text{ cu ft}}\right)$$

$$\approx 40{,}392 \text{ gal} \approx 40{,}400 \text{ gal}$$

Solid Figures

Answers to odd-numbered problems are given in the Appendix.

A. Solve the following problems involving solid figures.

Find (a) the lateral surface area and (b) the volume of each of the following. (Round to the nearest tenth.)

1.

2.

3.

4.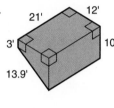

5.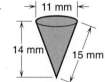

6.

7.

8.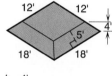

$h = 4'$
$s = 5'$

Find (a) the total surface area and (b) the volume of each of the following. (Round to the nearest tenth.)

9.

10.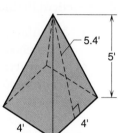

11.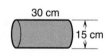

12.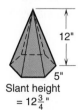

Slant height
$= 12\frac{3}{4}"$

13.

14.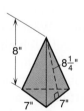

15. A sphere with a radius of 10 cm.

16.

Name

Date

Course/Section

17. A sphere with a diameter of 13 in.

18.

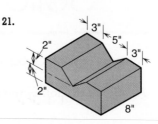

19.

h = 30"
s = 30.1"

20.

21.

B. Practical Problems

1. **Manufacturing**   How many boxes 16 in. by 12 in. by 10 in. will fit in 1250 ft^3 of warehouse space? (1 ft^3 = 1728 in.^{3})
2. **Construction**   How many cubic *yards* of concrete are needed to pour the highway support shown in the figure? (1 yd^3 = 27 ft^3.) (Round to nearest cubic yard.)

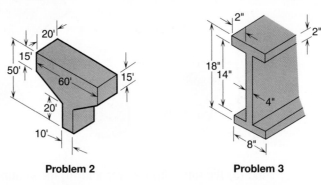

Problem 2   Problem 3

3. **Metalworking**   What is the weight of the 10-ft-long steel I-beam shown if the density of steel is 0.2833 lb/in.^{3}? (Round to the nearest pound.)
4. **Plumbing**   What is the capacity in gallons of a cylindrical water tank 3 ft in diameter and 8 ft high? (1 ft^3 ≈ 7.48 gal.) (Round to one decimal place.)
5. **Painting**   How many quarts of paint will it take to cover a spherical water tank 25 ft in diameter if one quart covers 50 ft^2?
6. **Construction**   A hole must be excavated for a swimming pool in the shape shown in the figure. How many trips will be needed to haul the dirt away if the truck has a capacity of 9 m^3? (*Remember:* A fraction of a load constitutes a trip.)

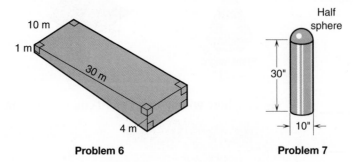

Problem 6   Problem 7

7. **Manufacturing**   How many cubic feet of propane will the tank in the figure contain? (1 ft^3 = 1728 in.^{3}) (Round to the nearest tenth of a cubic foot.)

8. **Industrial Technology** What must be the height of a cylindrical 750-gal tank if it is 4 ft in diameter? (1 ft^3 ≈ 7.48 gal.) (Round to the nearest inch.)

9. **Automotive Trades** At a density of 42 lb/ft^3, what is the weight of fuel in the rectangular gas tank pictured?

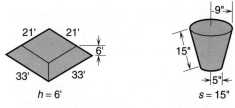

Problem 9	**Problem 10**

10. **Manufacturing** How many square inches of paper are needed to produce 2000 conical cups like the one in the figure? Round to the nearest 100 in.^{2}.

11. **Construction** How many cubic yards of concrete are needed to pour the monument foundation shown in the figure? Round to the nearest cubic yard.

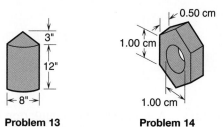

Problem 11	**Problem 12**

12. **Sheet Metal Trades** How many square inches of sheet metal are needed to make the bottom and sides of the pail in the figure? (Round to the nearest square inch and use π ≈ 3.14.)

13. **Industrial Technology** Find the capacity in gallons of the oil can in the figure. It is a cylinder with a cone on top. (1 gal = 231 in.^{3}) (Round to the nearest tenth of a gallon.)

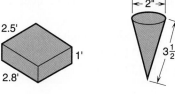

Problem 13	**Problem 14**

14. **Machine Trades** At a density of 7.7112 g/cm^3, find the weight of 1200 hex nuts, each 1.00 cm on a side and 0.50 cm thick with a hole 1.00 cm in diameter. (Round to two significant digits.)

15. **Industrial Technology** A cylindrical tank can be no more than 6 ft high. What diameter must it have in order to hold 1400 gal?

16. **Welding** A rectangular tank is made using steel strips 8 in. wide. The sides are 18 in. by 8 in., the ends are 8 in. by 8 in., and the top and bottom are 18 in. by 8 in. Calculate the volume of the tank and the length of the weld needed.

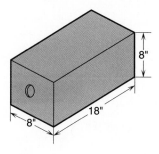

17. **Automotive Trades** An oil pan has the shape of a rectangular prism measuring 64 cm long by 32 cm wide by 16 cm deep. How many liters of oil will it hold? (1 liter = 1000 cu cm.) (Round to two significant digits.)

18. **Welding** Calculate the weight of a solid steel roller bar. It is 14.0 in. long and 3.0 in. in diameter. Use density of steel $\approx$ 0.2833 lb/cu in. (Round to the nearest pound.)

19. **Agriculture** A feeding trough is constructed from sheet metal as a half-cylinder, 26.0 ft long and 3.0 ft in diameter. When it is 80% filled with feed, what volume of feed does it hold? (Round to the nearest cubic foot.)

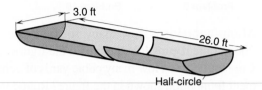

Half-circle

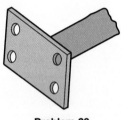

Problem 20

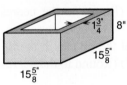

Problem 21

20. **Welding** Calculate the weight of the steel end-plate shown, if it is 20.0 cm on each side and 2.0 cm thick. The bolt holes are 1.0 cm in diameter. The density of steel is 0.7849 g/cu cm. (Round to two significant digits.)

21. **Masonry** A square pillar is built using $15\frac{5}{8}$ in.-square concrete blocks 8 in. high with $1\frac{3}{4}$-in.-thick walls. This creates a hollow square prism inside each block. (See the figure.) If the pillar is 10 blocks high, how many cubic *yards* of concrete will it require to fill the pillar. (Round to the nearest 0.1 cu yd.)

22. **Landscaping** In the figure, the solid slanted line represents the current grade of soil on a pathway between a building and a fence. The broken horizontal line represents the desired grade line. If the length of the pathway is 24 ft, what volume of soil (in cubic yards) must be brought in to achieve the desired grade? (*Hint:* The new soil forms a trapezoidal prism.) (Round to the nearest whole number.)

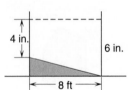

Careful: The figure is not drawn to scale, and the units are not all the same.

23. **Carpentry** A granite kitchen countertop is $1\frac{5}{16}$ in. thick. If the countertop is a rectangular shape measuring 28.8 in. by 82.5 in., and granite weighs 0.1 lb/cu in., how much does the countertop weigh? (Round to the nearest pound.)

24. **Landscaping** A landscaper installing a garden takes delivery of 5 cu yd of topsoil. If the topsoil is spread evenly over an area 18 ft by 30 ft, how many inches deep is the topsoil? (Be careful of units!)

Triangle Trigonometry

Objective	Sample Problems	For help, go to
When you finish this chapter, you will be able to:		
1. Convert angles between decimal degrees and degrees and minutes.	(a) Convert 43.4° to degrees and minutes. _____	Page 610
	(b) Convert 65°15′ to decimal degrees. _____	Page 611
2. Work with angles in radian measure.	(a) Express 46° in radians. _____	Page 613
	(b) Express 2.5 radians in degrees. _____	
	(c) Calculate the arc length and area of the circular sector shown. Round to one decimal place. _____	Page 614

$$40.1° \quad 6.0'' \quad 6.0''$$

| | (d) What is the angular speed in radians per second of a large flywheel that rotates through an angle of 200° in 2 sec? Round to the nearest tenth. _____ | Page 615 |

Name _____

Date _____

Course/Section _____

Objective **Sample Problems** **For help, go to**

3. Use the special right triangle relationships to find missing parts of these triangles. Round sides to the nearest tenth.

(a)

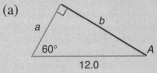

(b)

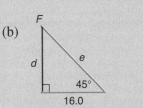

(a) $a =$ _____ Page 618

$b =$ _____

$A =$ _____

(b) $d =$ _____

$e =$ _____

$F =$ _____

4. Find the values of trig ratios.

(a) sin 26° _____ Page 627

(b) cos 84° _____ Page 629

(c) tan 43°20′ _____ Page 630

Round to three decimal places.

5. Find the angle when given the value of a trig ratio.

Find $\angle x$ to the nearest minute.

(a) $\sin x = 0.242$ $\angle x =$ _____ Page 633

(b) $\cos x = 0.549$ $\angle x =$ _____

Find $\angle x$ to the nearest tenth of a degree.

(c) $\tan x = 3.821$ $\angle x =$ _____

(d) $\sin x = 0.750$ $\angle x =$ _____

6. Solve problems involving right triangles. Round sides to the nearest tenth and angles to the nearest minute.

(a)

$X =$ _____ Page 635

$Y =$ _____

(b)

$\angle m =$ _____

$X =$ _____

(c) **Plumbing** In a pipe-fitting job, what is the run if the offset is $22\frac{1}{2}°$ and the length of set is $16\frac{1}{2}$ in.? run = _____ Page 642

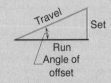

| Objective | Sample Problems | For help,
go to |

(d) **Machine Trades** Find the
angle of taper on the figure.   angle = _____

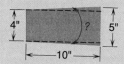 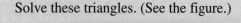 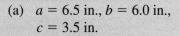

4"   ?   5"

|← 10" →|

7. Solve oblique trian-
gles. Round sides to
the nearest tenth and
angles to the nearest
degree.

Solve these triangles. (See the figure.)

(a) $a = 6.5$ in., $b = 6.0$ in.,
$c = 3.5$ in.

_____ Page 649

(b) $A = 68°$, $B = 79°$,
$b = 8.0$ ft

_____

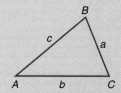

(Answers to these preview problems are given in the Appendix. Also, worked solutions
to many of these problems appear in the chapter Summary.)

If you are certain that you can work all of these problems correctly, turn to page 665 for
a set of practice problems. If you cannot work one or more of the preview problems, turn
to the page indicated to the right of the problem. Those who wish to master this material
with the greatest success should turn to Section 10-1 and begin work there.

Preview Chapter 10 Triangle Trigonometry

Triangle Trigonometry

"Not *the* Bermuda triangle?"
Drawing by Richter © 1977. The New Yorker Magazine, Inc

10-1 Angles and Triangles

The word *trigonometry* means simply "triangle measurement." We know that the ancient Egyptian engineers and architects of 4000 years ago used practical trigonometry in building the pyramids. By 140 B.C. the Greek mathematician Hipparchus had made trigonometry a part of formal mathematics and taught it as astronomy. In this chapter we look at only the simple practical trigonometry used in electronics, drafting, machine tool technology, aviation mechanics, and similar technical work.

In Chapter 8 you learned some of the vocabulary of angles and triangles. To be certain you remember this information, try these problems.

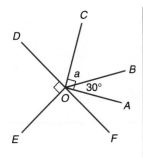

→ **Your Turn**

(a) Find each of the angles shown.

    (1) $\angle AOB$ = _____

    (2) $\angle DOF$ = _____

    (3) $\angle a$ = _____

    (4) $\angle DOE$ = _____

(b) Label each of these angles with the correct name: acute angle, obtuse angle, straight angle, or right angle.

    (1) 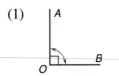   (2) 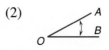   (3)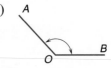

    (4) 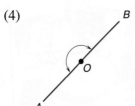   (5)   (6)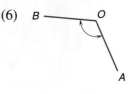

→ **Answers**

(a) (1) $\angle AOB = 30°$ (2) $\angle DOF = 180°$ (The symbol ∟ means "right angle" or 90°.)

    (3) $\angle a = 60°$ (4) $\angle DOE = 90°$

(b) (1) right angle   (2) acute angle   (3) obtuse angle

    (4) straight angle   (5) acute angle   (6) obtuse angle

If you missed any of these, you should return to Section 8-1 on page 487 for a quick review; otherwise, continue here.

Converting Angle Units

In most practical work angles are measured in degrees and fractions of a degree. Smaller units have been defined as follows:

> 60 minutes = 1 degree   abbreviated, $60' = 1°$
> 60 seconds = 1 minute   abbreviated, $60'' = 1'$

For most technical purposes angles can be rounded to the nearest minute. In the trades, angles are usually rounded to the nearest degree.

Sometimes in your work you will need to convert an angle measured in decimal degrees to its equivalent in degrees and minutes.

Example 1

To convert $17\frac{1}{2}°$ to degrees and minutes, follow these two steps:

Step 1 Write the angle as a sum of a whole number and a fraction.

$$17\frac{1}{2}° = 17° + \frac{1}{2}°$$

Step 2   Use a unity fraction to convert $\frac{1}{2}^\circ$ to minutes.

$$= 17^\circ + \left(\frac{1^\circ}{2} \times \frac{60'}{1^\circ} \right)$$

$$= 17^\circ 30'$$

→ **Your Turn**

Write this angle to the nearest minute:

$36.25^\circ = $ _____

→ **Solution**

$36.25^\circ = 36^\circ + 0.25^\circ$

$$= 36^\circ + \left(0.25^\circ \times \frac{60'}{1^\circ} \right) \quad \Longleftarrow \boxed{\text{Multiply the decimal part by a unity fraction.}}$$

$$= 36^\circ 15'$$

The reverse procedure is also useful.

Example 2

To convert the angle $72^\circ 6'$ to decimal form,

Step 1   Write the angle as a sum of degrees and minutes.

$$72^\circ 6' = 72^\circ + 6'$$

Step 2   Use a unity fraction to convert minutes to degrees.

$$= 72^\circ + \left(6' \times \frac{1^\circ}{60'} \right)$$

$$= 72^\circ + 0.1^\circ$$

$$= 72.1^\circ$$

CONVERTING ANGLE UNITS USING A CALCULATOR

Scientific calculators have special keys designed for entering angles in degrees and minutes and for performing angle conversions. There is normally a key marked $\boxed{\circ\,'\,''}$ that is used for entering and converting angles. You may instead see a key marked $\boxed{\blacktriangleright\text{DMS}}$ or $\boxed{\blacktriangleright\text{DD}}$ to be used for this purpose. Because the procedures for doing these conversions vary greatly from model to model, we leave it to each student to learn the key sequences from his or her calculator's instruction manual. You can always perform the conversions on your calculator using the methods of Examples 1 and 2, and we shall show them this way in the text. For example, to convert the angle 67.28° from decimal degrees to degrees and

(continued)

minutes, write down the degree portion, 67°, and then calculate the minutes portion as follows:

.28 ⊠ 60 ⊟ → [16.8]

Therefore, $67.28° = 67° 16.8'$ or $67° 17'$, rounded to the nearest minute.

To convert $49° 27'$ to decimal degrees, enter

49 ⊞ 27 ⊟ 60 ⊟ → [49.45]

Therefore, $49° 27' = 49.45°$.

→ More Practice

For practice in working with angles, rewrite each of the following angles as shown.

(a) Write in degrees and minutes.

    (1) $24\frac{1}{3}° = $ _____   (2) $64\frac{3}{4}° = $ _____

    (3) $15.3° = $ _____   (4) $38.6° = $ _____

    (5) $46\frac{3}{8}° = $ _____   (6) $124.8° = $ _____

(b) Write in decimal form.

    (1) $10°48' = $ _____   (2) $96°9' = $ _____

    (3) $57°36' = $ _____   (4) $168°54' = $ _____

    (5) $33' = $ _____   (6) $1°12' = $ _____

→ Answers

(a) (1) $24°20'$ (2) $64°45'$ (3) $15°18'$
     (4) $38°36'$ (5) $46°22\frac{1}{2}'$ (6) $124°48'$

(b) (1) $10.8°$ (2) $96.15°$ (3) $57.6°$
     (4) $168.9°$ (5) $0.55°$ (6) $1.2°$

Radian Measure In some scientific and technical work angles are measured or described in an angle unit called the *radian.* By definition,

$$1 \text{ radian} = \frac{180°}{\pi} \text{ or about } 57.296°$$

One radian is the angle at the center of a circle that corresponds to an arc exactly one radius in length.

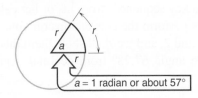

a = 1 radian or about 57°

To understand where the definition comes from, set up a proportion:

$$\frac{\text{angle } a}{\text{total angle in the circle}} = \frac{\text{arc length}}{\text{circumference of the circle}}$$

$$\frac{a}{360°} = \frac{r}{2\pi r}$$

$$a = \frac{360° \, \cancel{r}}{2\pi \cancel{r}}$$

$$a = \frac{180°}{\pi} \approx 57.295779°$$

Degrees to Radians To convert angle measurements from degrees to radians (rad), use the following formula:

$$\text{Radians} = \text{degrees} \cdot \frac{\pi}{180}$$

Use the $\boxed{\pi}$ button on your calculator unless you are told otherwise.

Example 3

(a) $43° = 43\left(\dfrac{\pi}{180}\right) = \dfrac{43 \cdot \pi}{180}$

  ≈ 0.75 rad   rounded

(b) $60° = 60\left(\dfrac{\pi}{180}\right) = \dfrac{60 \cdot \pi}{180}$

  $= \dfrac{\pi}{3}$ rad ≈ 1.0472 rad

Radians to Degrees To convert angles from radians to degrees, use the following formula:

$$\text{Degrees} = \text{radians} \cdot \frac{180}{\pi}$$

Example 4

(a) 1.3 rad $= 1.3\left(\dfrac{180}{\pi}\right) = \dfrac{1.3 \cdot 180}{\pi}$

  $\approx 74.5°$

(b) 0.5 rad $= 0.5\left(\dfrac{180}{\pi}\right) = \dfrac{0.5 \cdot 180}{\pi}$

  $= \dfrac{90}{\pi}$ degrees or about $28.6°$

→ **Your Turn**

Try these problems for practice in using radian measure.

(a) Convert to radians. (Round to two decimal places.)

(1) 10°   (2) 35°   (3) 90°   (4) 120°

(b) Convert to degrees. (Round to the nearest tenth.)

(1) 0.3 rad   (2) 1.5 rad   (3) 0.8 rad   (4) 0.05 rad

→ **Answers**

(a) (1) $\dfrac{10 \cdot \pi}{180} \approx 0.17$ rad   (2) 0.61 rad   (3) 1.57 rad   (4) 2.09 rad

(b) (1) $\dfrac{0.3 \cdot 180}{\pi} \approx 17.2°$ (2) 85.9°   (3) 45.8°   (4) 2.9°

Sectors The wedge-shaped portion of a circle shown is called a **sector.** Both its area and the length of arc S can be expressed most directly using radian measure.

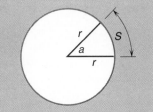

SECTORS

Arc length $S = ra$
Area $A = \frac{1}{2}r^2a$ where a is the central angle given in radians.

Example 5

Sheet Metal Trades A sheet metal worker wants to know the arc length and area of a sector with central angle 150° cut from a circular sheet of metal with radius 16.0 in.

First, calculate angle a in radians: $a = 150\left(\dfrac{\pi}{180}\right) \approx 2.618$ rad

Second, calculate the arc length S: $S \approx (16 \text{ in.})(2.618)$

$$\approx 41.9 \text{ in.}$$

Finally, calculate the area A: $A \approx \frac{1}{2}\left(16^2\right)(2.618)$

$$\approx 335 \text{ in.}^2$$

Using a calculator, we have

S: 16 ⊗ 150 ⊗ ⟨π⟩ ⟨÷⟩ 180 ⟨=⟩ → `41.88790205`

A: .5 ⊗ 16 ⟨x²⟩ ⊗ 150 ⊗ ⟨π⟩ ⟨÷⟩ 180 ⟨=⟩ → `335.1032164`

→ **Your Turn**

Calculate the arc length and area of a sector with central angle 80° and radius 25.0 ft. (Round to three significant digits.)

The central angle is $a = 80\left(\dfrac{\pi}{180}\right) \approx 1.396\,\text{rad}.$

The arc length is $S \approx 25 \times 1.396 \approx 34.9\,\text{ft}.$

The area is $A \approx \frac{1}{2} \times 25^2 \times 1.396 \approx 436\,\text{ft}^2.$

Linear and Angular Speed

Radian units are also used in science and technology to describe the rotation of an object. When an object moves along a straight line for a distance d in time t, its **average linear speed** v is defined as

$$v = \frac{d}{t} \quad \text{Linear speed}$$

Example 6

If a car travels 1800 ft in 24 seconds, its average linear speed for the trip is

$$v = \frac{d}{t} = \frac{1800\,\text{ft}}{24\,\text{sec}} = 75\,\text{ft/sec}$$

When an object moves along a circular arc for a distance S in time t, it goes through an angle a in radians, and its **average angular speed** w is defined as

$$w = \frac{a}{t} \quad \text{Angular speed}$$

Example 7

Manufacturing If a 24-in.-diameter flywheel rotates through an angle of 140° in 1.1 seconds, what is the average angular speed of the flywheel?

First, calculate the angle in radians. The angle traveled is $a = 140\left(\dfrac{\pi}{180}\right) \approx 2.44\,\text{rad}.$

Then, calculate the average angular speed, $w = \dfrac{a}{t} \approx \dfrac{2.44}{1.1} \approx 2.2\,\text{rad/sec}.$

140 ✕ π ÷ 180 ÷ 1.1 = → **2.221328139**

→ Your Turn

For practice in working with linear and angular speeds, try these problems.

(a) **Life Skills** If a bicycle travels 24 mi in 1.5 hr, find its average linear speed.

(b) **Manufacturing** The blade on a shop fan rotates at 8.0 rev/sec. Calculate its average angular speed. (Round to two significant digits.) (*Hint:* Each revolution is 360°.)

(c) **Manufacturing** A belt-driven drum of radius 28.4 cm makes one revolution every 0.250 sec. What is the linear speed of the belt driving the drum? (Round to three significant digits.) (*Hint:* If the belt doesn't slip, the distance traveled by the belt in one revolution is equal to the circumference of the drum.)

(d) **General Interest** A Ferris wheel in an amusement park takes an average of 40 sec for one nonstop revolution. Calculate its average angular speed. (Round to two significant digits.)

→ **Solutions**

(a) $v = \dfrac{d}{t} = \dfrac{24\,\text{mi}}{1.5\,\text{hr}} = 16\,\text{mi/hr}$

(b) Each revolution is 360°, therefore the fan blade travels through an angle of $8 \times 360° = 2880°$ in one second. Converting to radians,

$$2880\left(\frac{\pi}{180}\right) \approx 50.265\,\text{rad}$$

Using the formula for angular speed,

$$w = \frac{a}{t} \approx \frac{50.265\,\text{rad}}{1\,\text{sec}} \approx 50\,\text{rad/sec}$$

(c) The circumference of the drum is

$$C = 2\pi r = 2\pi(28.4\,\text{cm}) \approx 178.4\,\text{cm}$$

This is the distance traveled by any point on the belt in 0.250 sec. Substitute these values into the linear speed formula.

$$v = \frac{d}{t} \approx \frac{178.4\,\text{cm}}{0.25\,\text{sec}} \approx 714\,\text{cm/sec}$$

(d) One revolution is 360°. In radians this is

$$\frac{360 \times \pi}{180} \approx 6.28\,\text{rad}$$

Using the formula for angular speed,

$$w = \frac{a}{t} \approx \frac{6.28\,\text{rad}}{40\,\text{sec}} \approx 0.16\,\text{rad/sec}$$

Right Triangles Trigonometry is basically triangle measurement, and, to keep it as simple as possible, we can begin by studying only *right* triangles. You should remember from Chapter 8 that a right triangle is one that contains a right angle, a 90° angle. All of the following are right triangles. In each triangle the right angle is marked.

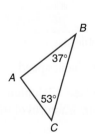

→ **Your Turn**

Is triangle *ABC* a right triangle?

→ **Solution**

You should remember that the sum of the three angles of any triangle is 180°. For triangle *ABC* two of the angles total 37° + 53° or 90°; therefore, angle *A* must equal 180° − 90° or 90°. The triangle *is* a right triangle.

Learning Help

To make your study of triangle trigonometry even easier, you may want to place a right triangle in *standard position*. For this collection of right triangles

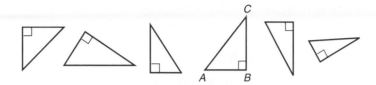

only triangle *ABC* is in standard position.

Place the triangle so that the right angle is on the right side, one leg (*AB*) is horizontal, and the other leg (*BC*) is vertical. The hypotenuse (*AC*) will slope up from left to right.

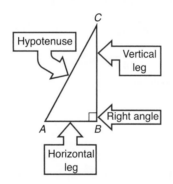

If you do not place a triangle in standard position, remember that the legs are the two sides forming the right angle, and the hypotenuse is the side opposite the right angle. The hypotenuse is always the longest side in a right triangle. ◄

Pythagorean Theorem

You should also recall from Chapter 8 that the *Pythagorean theorem* is an equation relating the lengths of the sides of any right triangle.

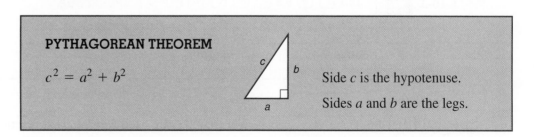

PYTHAGOREAN THEOREM

$$c^2 = a^2 + b^2$$

Side *c* is the hypotenuse.

Sides *a* and *b* are the legs.

This formula is true for every right triangle.

Careful

In the formula, the letters *a* and *b*, representing the legs, are interchangeable, but *c* must always represent the hypotenuse. ◄

Example 8

(a) In the triangle shown, the missing side is the hypotenuse c. To find c,

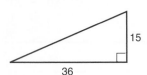

    first, substitute into the Pythagorean theorem $\quad c^2 = 15^2 + 36^2$

    and simplify. $\qquad\qquad\qquad\qquad\qquad\qquad c^2 = 225 + 1296$

$$c^2 = 1521$$

    Then, take the square root of both sides. $\qquad c = 39$

(b) In the triangle shown, one of the legs is missing. We can call it either a or b.

    Substitute into the Pythagorean theorem. $\qquad 19^2 = 9^2 + b^2$

$$361 = 81 + b^2$$

    Subtract 81 from both sides. $\qquad\qquad\qquad 280 = b^2$ or $b^2 = 280$

    Take the square root of both sides. $\qquad\qquad b = 16.733\ldots \approx 16.7$

→ Your Turn

For review, use this rule to find the missing side for each of the following right triangles.

(a)
    (b)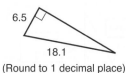
    (c)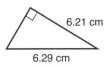

(Round to 1 decimal place)

→ Solutions

(a) $c^2 = 21^2 + 20^2$ (b) $18.1^2 = b^2 + 6.5^2$ (c) $6.29^2 = 6.21^2 + a^2$

    $c^2 = 441 + 400$ $327.61 = b^2 + 42.25$ $39.5641 = 38.5641 + a^2$

    $c^2 = 841$ $b^2 = 327.61 - 42.25$ $a^2 = 39.5641 - 38.5641$

    $c = 29$ in. $b^2 = 285.36$ $a^2 = 1$

                         $b \approx 16.9$ $a = 1$ cm

Here are two ways to work problem (b) on a calculator:

C10-3

18.1 $\boxed{x^2}$ $\boxed{-}$ **6.5** $\boxed{x^2}$ $\boxed{=}$ $\boxed{\surd}$ $\boxed{\text{ANS}}$ $\boxed{=}$ → $\boxed{\textit{16.89260193}}$

or $\boxed{\surd}$ * **18.1** $\boxed{x^2}$ $\boxed{-}$ **6.5** $\boxed{x^2}$ $\boxed{=}$ → $\boxed{\textit{16.89260193}}$

*Parentheses must be opened here.

Special Right Triangles

Three very special right triangles are used very often in practical work. Here they are:

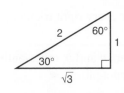
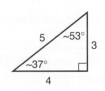

45°–45°–90°△   The first triangle is a *45° right triangle.* Its angles are 45°, 45°, and 90°, and it is formed by the diagonal and two sides of a square. Notice that it is an isosceles triangle—the two shorter sides or legs are the same length.

Our triangle has legs one unit long, but we can draw a 45° right triangle with any length sides. If we increase the length of one side, the others increase in proportion while the angles stay the same size.

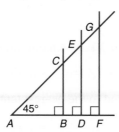

Triangle *ABC* is a 45° right triangle.
Triangle *ADE* is also a 45° right triangle.
Triangle *AFG* is also a 45° right triangle, . . . and so on.

No matter what the size of the triangle, if it is a right triangle and if one angle is 45°, the third angle is also 45°. The two legs are equal, and the hypotenuse is $\sqrt{2} \approx 1.4$ times the length of a leg.

Example 9

Refer to the previous figure for the following problems.

(a) If *AD* = 2 in., then

   DE = 2 in.   The legs are always equal in length. If *AD* is 2 in. long, *DE* is also 2 in. long.

   and

   $AE = 2\sqrt{2}$ in. or about 2.8 in.   The hypotenuse of a 45° right triangle is $\sqrt{2}$ times the length of each leg.

(b) If *AC* = 5 in., then

   $$AC = AB\sqrt{2}, \text{ so } AB = \frac{AC}{\sqrt{2}} = \frac{5}{\sqrt{2}}$$

Using a calculator gives

C10-4

$5\ \div\ \sqrt{}\ 2\ = \rightarrow$ **3.535533906** $\approx$ 3.5 in.

→ Your Turn

Plumbing   Parallel pipes are connected by 45° fittings as shown. Calculate the length of the connecting pipe from A to B. (Round to the nearest tenth.)

→ **Solution**

The connecting pipe AB is the hypotenuse of a 45°–45°–90° triangle. The length L is, therefore, $8.5\sqrt{2}$ in. ≈ 12.0 in.

→ **More Practice**

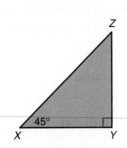

(a) If $XY = 4.6$ cm, find YZ and XZ.

(b) If $XZ = 15.8$ ft, find XY and YZ.

(Round to the nearest tenth.)

→ **Solutions**

(a) $YZ = 4.6$ cm

$XZ = 4.6\sqrt{2}$ cm ≈ 6.5 cm

(b) $XY = YZ = \dfrac{XZ}{\sqrt{2}} = \dfrac{15.8 \text{ ft}}{\sqrt{2}} \approx 11.2$ ft

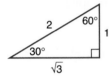

30°–60°–90° △ The second of our special triangles is the 30°–60° right triangle. In a 30°–60° right triangle the length of the side opposite the 30° angle, the smallest side, is exactly one-half the length of the hypotenuse. The length of the third side is $\sqrt{3}$ times the length of the smallest side.

To see where this relationship comes from, start with an equilateral triangle with sides two units long.

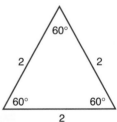

Draw a perpendicular from one vertex to the opposite side. This segment also bisects the opposite side.

Then in the 30°–60° right triangle the Pythagorean theorem gives

$$a^2 + 1^2 = 2^2$$
$$a^2 + 1 = 4$$
$$a^2 = 3$$
$$a = \sqrt{3}$$

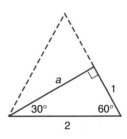

Again, we may form any number of 30°–60° right triangles, and in each triangle the hypotenuse is twice the length of the smallest side. The third side is $\sqrt{3}$, or about 1.7 times the length of the smallest side. When you find a right triangle with one angle equal to 30° or 60°, you automatically know a lot about it.

Example 10

(a) In this right triangle, if $A = 60°$ and $b = 8$ ft, then
$$c = 2b = 2 \cdot 8 \text{ ft} = 16 \text{ ft}$$
and $a = b\sqrt{3} = 8\sqrt{3} \approx 13.9$ ft

(b) If $A = 60°$ and $c = 24$ in., then
$$b = \frac{1}{2}c = \frac{1}{2}(24) = 12 \text{ in.}$$
and $a = b\sqrt{3} = 12\sqrt{3} \approx 20.8$ in.

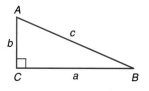

Note The triangle shown in Example 10 is labeled according to a standard convention. Angles are labeled with uppercase letters, and sides are labeled with lowercase letters. Furthermore, side a is opposite angle A, and b is opposite angle B, and side c, the hypotenuse, is opposite angle C, the right angle. ◄

Suppose you are given the longer leg a, opposite the 60° angle, and you need to find the shorter leg b, opposite the 30° angle. (See the figure for Example 10.)

We know that $\qquad\qquad\qquad\qquad a = b\sqrt{3}$

Therefore, dividing both sides by $\sqrt{3}$ $\qquad b = \dfrac{a}{\sqrt{3}}$

Example 11

If $A = 60°$ and $a = 32$ yd,

then $\qquad\qquad b = \dfrac{32}{\sqrt{3}} \approx 18.5$ yd

and $\qquad\qquad c = 2b \approx 37.0$ yd.

→ **Your Turn**

(a) In this right triangle, if $A = 30°$ and $a = 2$ in., calculate c, b, and B.

(b) If $A = 30°$, $c = 3$ in., calculate b and a.

(c) If $A = 30°$, $b = 18$ cm, calculate a and c.

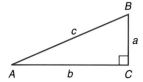

→ **Solutions**

(a) $c = 4$ in., $b = 2\sqrt{3}$ in. ≈ 3.5 in., and $B = 60°$

(b) In a 30°–60° right triangle, $a = \dfrac{1}{2}c = \dfrac{1}{2}(3) = 1.5$ in.
$$b = a\sqrt{3} = 1.5\left(\sqrt{3}\right) \approx 2.6 \text{ in.}$$

(c) $a = \dfrac{18}{\sqrt{3}} \approx 10.4$ cm

$\qquad c = 2a \approx 20.8$ cm

3–4–5 △ The third of our special triangles is the 3–4–5 triangle that was discussed in Chapter 8. If a triangle is found to have sides of length 3, 4, and 5 units (any units—inches, feet, centimeters), it must be a right triangle. The 3–4–5 triangle is the smallest right triangle with sides whose lengths are whole numbers.

Construction workers or surveyors can use this triangle to set up right angles.

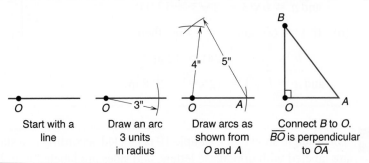

Start with a line

Draw an arc 3 units in radius

Draw arcs as shown from O and A

Connect B to O. $\overline{BO}$ is perpendicular to $\overline{OA}$

Notice that in any 3–4–5 triangle the acute angles are approximately 37° and 53°. The smallest angle is always opposite the smallest side.

You will find these three triangles appearing often in triangle trigonometry. Now, for some practice in working with angles and triangles, turn to Exercises 10-1.

Exercises 10-1 | Angles and Triangles

A. Label the shaded angles as acute, obtuse, right, or straight.

1.   2.   3.

4.   5. 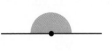   6.   7.

8.   9.   10.

B. Write the following angles in degrees and minutes. (Round to the nearest minute.)

1. $36\frac{1}{4}°$ 2. $73\frac{3}{5}°$ 3. 65.45°   4. 84.24°

5. $17\frac{1}{2}°$ 6. $47\frac{3}{8}°$ 7. 16.11°   8. 165.37°

Write the following angles in decimal degrees. (Round to the nearest hundredth if necessary.)

9. 27°30′   10. 80°15′   11. 154°39′   12. 131°6′

13. 57°3′   14. 44°20′   15. 16°53′   16. 16′

Write the following angles in radians. (Round to the nearest hundredth.)

17. 24.75°   18. 185.8°   19. 1.054°   20. $14\frac{3}{4}°$

21. $67\frac{3}{8}°$ 22. 9.65°   23. 80.60°   24. 216.425°

Write the following angles in degrees. (Round to the nearest hundredth.)

25. 0.10 rad   26. 0.84 rad   27. 2.1 rad   28. 3.5 rad

C. Solve the following problems involving triangles.

Find the lengths of the missing sides on the following triangles using the Pythagorean theorem. (Round to one decimal place.)

1.
24
10

2.
7
12

3.
7

4.
15
11

5.
15'
9'

6.
2"
3"

7.
11.3   11.2

8.
6.5 cm
5.6 cm

9.
4.8'
6.4'

10.
1.7"
2.4"

Each of the following right triangles is an example of one of the three "special triangles" described in this chapter. Find the quantities indicated without using the Pythagorean theorem. (Round the sides to the nearest tenth if necessary.)

11.
A
3'
c
3'

(a) A = _____

(b) c = _____

12.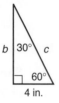
b   30° c
60°
4 in.

(a) b = _____

(b) c = _____

13.
B
a   5
3   A

(a) a = _____

(b) A = _____

(c) B = _____

14.
T   e
6 cm   30° t

(a) e = _____

(b) t = _____

(c) T = _____

15.
45°
A
7
k

(a) k = _____

(b) A = _____

16.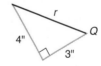
r
Q
4"
3"

(a) r = _____

(b) Q = _____

17.
a   10
60°
b   A

(a) b = _____

(b) a = _____

(c) A = _____

18.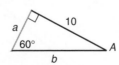
T   n
53°
m
37° 0.5"

(a) m = _____

(b) n = _____

(c) T = _____

19.

20.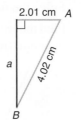

(a) $q =$ _____

(b) $U =$ _____

(a) $a =$ _____

(b) $A =$ _____

(c) $B =$ _____

Calculate the arc length S and area A for each of the following sectors. (Round to the nearest whole number if necessary.)

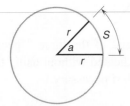

21. $a = 40°$

   $r = 30$ cm

22. $a = 120°$

   $r = 24$ ft

23. $a = 0.8$ rad

   $r = 110$ ft

24. $a = 2$ rad

   $r = 18$ m

25. $a = 72°$

   $r = 5$ in.

26. $a = 3$ rad

   $r = 80$ cm

27. $a = 0.5$ rad

   $r = 10$ m

28. $a = 180°$

   $r = 30$ in.

D. Practical Problems

1. **Sheet Metal Trades** A transition duct is constructed from a circular sector of radius 18 in. and central angle 115°. (a) What is the area of the metal used? (b) Calculate the arc length of the sector. (Round to the nearest tenth.)

2. **Landscaping** Calculate the area (in square feet) of a flower garden shaped like a circular sector with radius 60 ft and central angle 40°.

3. **Landscaping** In problem 2, if shrubs are planted every 2 ft along the outer border of the garden, how many shrubs are needed?

4. **Construction** The outfield fencing for a Little League field forms a circular sector with home plate as the center. The fence is placed at a uniform distance of 215 ft from home plate. The boundaries of the fence, which extends partway into foul territory, create an angle of 105° with home plate. At $32 per foot, how much will the fence cost? (Round to the nearest $10.)

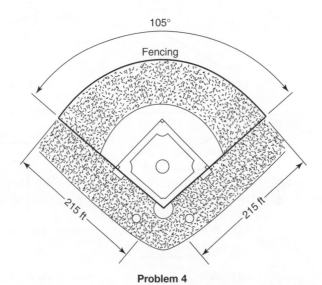

Problem 4

5. **Construction** In problem 4, topsoil must be bought to cover the sector to a depth of 4 in. How many cubic yards of topsoil are needed? (Be careful with the units.)

6. **Life Skills** If a car travels 725 miles in 12.5 hours, find its average linear speed.

7. **Manufacturing** The blades of a fan have a radius of 16.5 in. and they turn one revolution every 0.125 sec. Calculate the linear speed of the tip of a blade. (Round to three significant digits.)

8. **Manufacturing** A pulley with a radius of 36.2 cm rotates at 6.5 rev/sec. Find the linear speed of the belt driving the pulley. (Round to three significant digits.)

9. **Industrial Technology** If the indicator on a flow rate meter moves 75° in 16.4 sec, what is its average angular speed during this time? (Round to two decimal places.)

10. **Machine Trades** If a lathe makes 25 rev/sec, calculate its average angular speed. (Round to the nearest whole number.)

11. **General Interest** A Ferris wheel in an amusement park takes an average of 16 sec for each revolution. What is its average angular speed? (Round to two decimal places.)

12. **Plumbing** For the connection shown here, calculate the rise, the distance *H*, and the length *AB*. The elbows are 45° fittings. (Round to the nearest tenth.)

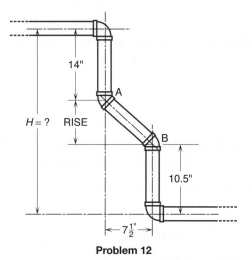

Problem 12

13. **Automotive Trades** During a wheel alignment, the camber on a certain car is measured to be 1°48′. If the maximum acceptable camber for this car is 1.750°, is the camber within this limit?

14. **Automotive Trades** The caster for a certain vehicle is measured to be 2.150°. The maximum allowable caster is specified to be 2°10′. Is the caster within the maximum limit?

Check your answers to the odd-numbered problems in the Appendix, then continue in Section 10-2.

10-2 Trigonometric Ratios

In Section 10-1 we showed that in *all* 45° right triangles, the adjacent side is equal in length to the opposite side, and the hypotenuse is roughly 1.4 times this length. Similarly, in *all* 30°–60° right triangles, the hypotenuse is twice the length of the side opposite the 30° angle, and the adjacent side is about 1.7 times this length.

The first skill needed for trigonometry is the ability to label the three sides of a right triangle according to convention: the hypotenuse, the side *adjacent* to a given acute angle, and the side *opposite* that angle.

In triangle ABC, side $\overline{AB}$ is the hypotenuse. It is the longest side and it is opposite the right angle.

For acute angle A:

$\overline{AC}$ is the adjacent side.

$\overline{BC}$ is the opposite side.

For acute angle B:

$\overline{BC}$ is the adjacent side.

$\overline{AC}$ is the opposite side.

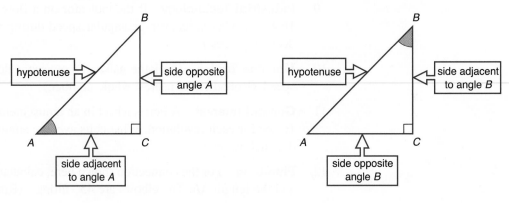

> **Careful** The hypotenuse is never considered to be either the adjacent side or the opposite side. It is not a leg of the triangle. ◄

→ **Your Turn**

Check your understanding of this vocabulary by putting triangle XYZ in standard position and by completing the following statements.

(a) The hypotenuse is _____.

(b) The opposite side for angle Y is _____.

(c) The adjacent side for angle Y is _____.

(d) The opposite side for angle Z is _____.

(e) The adjacent side for angle Z is _____.

→ **Answers**

(a) $\overline{YZ}$ (b) $\overline{XZ}$ (c) $\overline{XY}$

(d) $\overline{XY}$ (e) $\overline{XZ}$

Of course, the phrases "adjacent side," "opposite side," and "hypotenuse" are used only with right triangles.

→ **More Practice**

For each of the following right triangles, mark the hypotenuse with the letter c, mark the side adjacent to the shaded angle with the letter a, and mark the side opposite the shaded angle with the letter b.

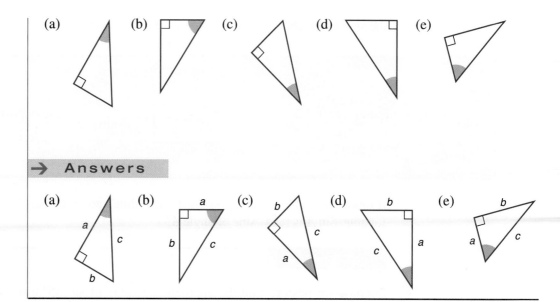

(a)   (b)   (c)   (d)   (e)

→ **Answers**

(a)   (b)   (c)   (d)   (e)

> **Learning Help**

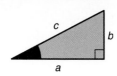

If you find it difficult to identify the adjacent side and the opposite side of a given angle in a right triangle, try placing the triangle in standard position. Position the triangle so that the given angle is at the lower left and the right angle is at the lower right. When the triangle is in this position, the adjacent side is always the horizontal leg, a, and the opposite side will always be the vertical leg, b. ◄

The key to understanding trigonometry is to realize that in *every* right triangle there is a fixed relationship connecting the size of either of the acute angles to the lengths of its adjacent side, its opposite side, and the hypotenuse.

For example, in all right triangles that contain the angle 20°,

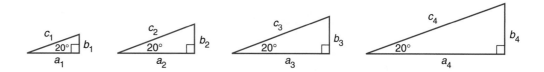

the ratio $\dfrac{\text{opposite side}}{\text{hypotenuse}}$ will always have the same value. Therefore,

$$\frac{b}{c} = \frac{b_1}{c_1} = \frac{b_2}{c_2} = \frac{b_3}{c_3} = \dots \text{ and so on}$$

This ratio will be approximately 0.342.

Sine Ratio

This kind of ratio of the side lengths of a right triangle is called a *trigonometric ratio*. It is possible to write down six of these ratios, but we will use only three here. Each ratio is given a special name.

For the triangle

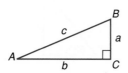

the **sine ratio,** abbreviated sin, is defined as follows:

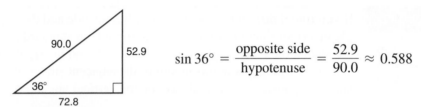

$$\sin A = \frac{\text{side opposite angle } A}{\text{hypotenuse}} = \frac{a}{c} \qquad\qquad \sin B = \frac{\text{side opposite angle } B}{\text{hypotenuse}} = \frac{b}{c}$$

(Pronounce "sin" as "sign" to rhyme with "dine" not with "sin." *Sin* is an abbreviation for *sine*. Sin A is read "sine of angle A.")

Example 1

Suppose we wish to find the approximate value of sin 36° using the triangle shown. Notice that the side opposite the 36° angle has length 52.9, and the hypotenuse has length 90.0.

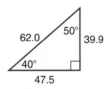

$$\sin 36° = \frac{\text{opposite side}}{\text{hypotenuse}} = \frac{52.9}{90.0} \approx 0.588$$

→ Your Turn

Use this triangle to find the approximate value of sin 50° to three decimal places.

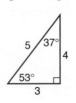

→ Solution

The side opposite the 50° angle has length 47.5, and the hypotenuse has length 62.0.

$$\sin 50° = \frac{\text{opposite side}}{\text{hypotenuse}} = \frac{47.5}{62.0} \approx 0.766$$

Every possible angle x will have some number sin x associated with it. If x is an acute angle, we can calculate the value of sin x from any right triangle that contains x.

Now, let's look again at those three special right triangles from Section 10-1.

 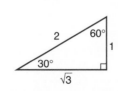 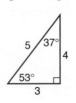

We can use these labeled figures to calculate the approximate values of sine for angles of 30°, 37°, 45°, 53°, and 60°.

Example 2

To find sin 60°, use the 30°–60° right triangle. Notice that the side opposite the 60° angle has length $\sqrt{3}$, and the hypotenuse has length 2. Therefore,

$$\sin 60° = \frac{\text{opposite side}}{\text{hypotenuse}} = \frac{\sqrt{3}}{2} = 0.866\ldots \approx 0.87$$

→ **Your Turn**

Use the three special right triangles to complete the table. Round to the nearest hundredth.

A	30°	37°	45°	53°	60°
$\sin A$					0.87

→ **Answers**

A	30°	37°	45°	53°	60°
$\sin A$	0.50	0.60	0.71	0.80	0.87

Note Sin A is a trigonometric *ratio* or trigonometric *function* associated with the angle A. The value of sin A can never be greater than 1, since the legs of any right triangle can never be longer than the hypotenuse. ◀

Cosine Ratio A second useful trigonometric ratio is the **cosine** ratio, abbreviated cos. This ratio is defined as follows:

$$\cos A = \frac{\text{side adjacent to angle } A}{\text{hypotenuse}} = \frac{b}{c} \qquad \cos B = \frac{\text{side adjacent to angle } B}{\text{hypotenuse}} = \frac{a}{c}$$

Example 3

In the triangle

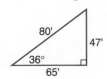

$$\cos 36° = \frac{65 \text{ ft}}{80 \text{ ft}} \quad \begin{array}{l} \longleftarrow \text{Adjacent side} \\ \longleftarrow \text{Hypotenuse} \end{array}$$

$$\cos 36° \approx 0.81 \quad \text{approximately}$$

Use the following special right triangle to find cos 45°. (Round to two decimal places.)

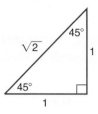

→ Solution

The hypotenuse is $\sqrt{2}$ and the adjacent side to the 45° angle is 1.

$$\cos 45° = \frac{\text{adjacent side}}{\text{hypotenuse}}$$

$$\cos 45° = \frac{1}{\sqrt{2}} = 0.707\ldots \approx 0.71$$

→ More Practice

Now use the drawing of the three special right triangles on page 628 to help you to complete this table. Round to two decimal places.

A	30°	37°	45°	53°	60°
cos A			0.71		

→ Answers

A	30°	37°	45°	53°	60°
cos A	0.87	0.80	0.71	0.60	0.50

Tangent Ratio

A third trigonometric ratio may also be defined. The ratio **tangent,** abbreviated tan, is defined as follows:

$$\tan A = \frac{\text{side opposite angle } A}{\text{side adjacent to angle } A} = \frac{a}{b} \qquad \tan B = \frac{\text{side opposite angle } B}{\text{side adjacent to angle } B} = \frac{b}{a}$$

Example 4

In the triangle

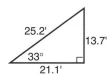

$$\tan 33° = \frac{13.7 \text{ ft}}{21.1 \text{ ft}}$$

Opposite side
Adjacent side

$$\tan 33° \approx 0.65$$

→ **Your Turn**

Use the following special right triangle to find tan 30°.
(Round to two decimal places.)

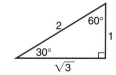

→ **Solution**

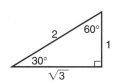

$$\tan 30° = \frac{\text{opposite side}}{\text{adjacent side}}$$

$$\tan 30° = \frac{1}{\sqrt{3}} \approx 0.58$$

→ **More Practice**

Complete this table for tan A using the three special right triangles. (Round to two decimal places.)

A	30°	37°	45°	53°	60°
tan A	0.58				

→ **Answers**

A	30°	37°	45°	53°	60°
tan A	0.58	0.75	1.00	1.33	1.73

Finding Values of Trigonometric Functions

You have been asked to calculate these "trig" ratios to help you get a feel for how the ratios are defined and to encourage you to memorize a few values. In actual practical applications of trigonometry, the values of the ratios are found by using a calculator with built-in trigonometric function keys.

To find the value of a trigonometric ratio on a multi-line calculator, first press the key of the given function, then enter the value of the angle, and finally press (=).

Your calculator may automatically open a parentheses when you press $\boxed{\sin}$, $\boxed{\cos}$, or $\boxed{\tan}$. However, unless there are additional calculations to perform before pressing $\boxed{=}$, it is not neccessary to close the parentheses in order to display the correct value. ◄

Example 5

To find sin 38°, enter this sequence:

C10-5 $\boxed{\sin}$ **38** $\boxed{=}$ → ▓▓ *0.615661475* ▓▓   sin 38° ≈ 0.616

Angles can be measured in units of radians and gradients as well as degrees, and most calculators are capable of accepting all three units of measure. Before entering an angle in degrees, make certain the calculator is in "degree mode" by either using the $\boxed{\text{DRG}}$ key or the $\boxed{\text{MODE}}$ key. Look for the "DEG" indication on the display. On some calculators absence of any mode indication means that it is in degree mode. ◄

→ **Your Turn**

Now you try it. Find the value of each of the following to two decimal places.

(a) cos 26°
(b) tan 67.8°
(c) sin 8°

(d) tan 14.25°
(e) cos $47\frac{1}{2}°$
(f) sin 39.46°

(g) sin 138°
(h) cos 151.6°
(i) tan 141.12°

→ **Solutions**

C10-6

(a) $\boxed{\cos}$ **26** $\boxed{=}$ → *0.898794046*   cos 26° ≈ 0.90

(b) $\boxed{\tan}$ **67.8** $\boxed{=}$ → *2.450425198*   tan 67.8° ≈ 2.45

(c) $\boxed{\sin}$ **8** $\boxed{=}$ → *0.139173101*   sin 8° ≈ 0.14

(d) $\boxed{\tan}$ **14.25** $\boxed{=}$ → *0.253967646*   tan 14.25° ≈ 0.25

(e) $\boxed{\cos}$ **47.5** $\boxed{=}$ → *0.675590208*   cos $47\frac{1}{2}°$ ≈ 0.68

(f) $\boxed{\sin}$ **39.46** $\boxed{=}$ → *0.63553937*   sin 39.46° ≈ 0.64

(g) $\boxed{\sin}$ **138** $\boxed{=}$ → *0.669130606*   sin 138° ≈ 0.67

(h) $\boxed{\cos}$ **151.6** $\boxed{=}$ → *-0.879648573*   cos 151.6° ≈ −0.88

(i) $\boxed{\tan}$ **141.12** $\boxed{=}$ → *-0.806322105*   tan 141.12° ≈ −0.81

- Notice in problems (g), (h), and (i) that the trigonometric functions exist for angles greater than 90°. The values of the cosine and tangent functions are negative for angles between 90° and 180°.

- Notice in problem (g) that

  sin 138° = 0.669130606 . . .

  and sin 42° = 0.669130606 . . . 42° = 180° − 138°

This suggests that the following formula is true:

  $\sin A = \sin(180° - A)$

This fact will be useful in Section 10-4. ◄

Angles in Degrees and Minutes

When finding the sine, cosine, or tangent of an angle given in degrees and minutes, either enter the angle as given using the ⌈∘ ′ ″⌉ key, or first convert the angle to decimal form as shown in the following example. See your calculator's instruction manual for the key sequence that will allow you to enter degrees and minutes directly.

Example 6

To find cos 54° 28′, the following sequence will work on all multi-line calculators:

C10-7

⌈cos⌉ * 54 ⌈+⌉ 28 ⌈÷⌉ 60 ⌈=⌉ → ▐ 0.581176491

*If your calculator does not automatically open parentheses with ⌈cos⌉, then you must press ⌈(⌉ here.

→ Your Turn

Try these problems for practice. Find each value to three decimal places.

(a) sin 71°26′   (b) tan 18°51′   (c) cos 42°16′   (d) cos 117°22′

→ Answers

(a) 0.948   (b) 0.341   (c) 0.740   (d) −0.460

Finding the Angle

In some applications of trigonometry it may happen that we know the value of the trigonometric ratio number and we need to find the angle associated with it. This is called finding the **inverse** of the trigonometric function.

To find the angle associated with any given value of a trigonometric function, press the ⌈2ⁿᵈ⌉ key (also labeled ⌈2ⁿᵈ F⌉ or ⌈SHIFT⌉), then press the key of the given function (⌈sin⌉, ⌈cos⌉, or ⌈tan⌉), and finally press ⌈=⌉. These inverse functions are labeled ⌈sin⁻¹⌉, ⌈cos⁻¹⌉, and ⌈tan⁻¹⌉ on your keyboard. Because these are second functions on all calculators, we will always include the ⌈2ⁿᵈ⌉ key in our key sequences as a reminder to press it first.

Note If your calculator automatically opens parentheses with the inverse trigonometric functions, it is not necessary to close them unless there are additional calculations to perform before pressing ⌈=⌉. ◄

Example 7

To find A if sin $A = 0.728$, enter the following:

C10-8

⌈2ⁿᵈ⌉ ⌈sin⁻¹⌉ .728 ⌈=⌉ → ▐ 46.718988

The angle is approximately 46.7°; therefore, sin 46.7° ≈ 0.728.* We can also express this result as $\sin^{-1} 0.728 \approx 46.7°$.

Careful Be certain your calculator is in the degree mode before you enter this kind of calculation. ◄

---

*We have found the angle between 0 and 90° that corresponds to the given trigonometric function value, but there are many angles greater than 90° whose sine ratio has this same value. Because such values of the inverse are generally not very important in the trades, we will not discuss them in detail here.

To find the angle in degrees and minutes, first follow the procedure described to find the angle in decimal degrees. Then convert to degrees and minutes using either the method of the following example, or the method described in your calculator's instruction manual.

Example 8

To find A in degrees and minutes for $\cos A = 0.296$, the following procedure will work on all multi-line calculators:

C10-9

Decimal degrees. Write 72°

Decimal portion converted to minutes.

Therefore $A \approx 72° \, 47'$.

Now turn to Exercises 10-2 for a set of practice problems on trigonometric ratios.

Exercises 10-2　Trigonometric Ratios

A. For each of the following triangles *calculate* the indicated trigonometric ratios of the given angle. Do not use the trigonometric function keys on your calculator. Round to nearest hundredth.

1.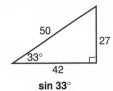

　　sin 33°

2.

　　cos 42°

3.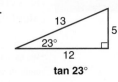

　　tan 23°

4.

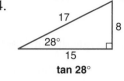

tan 28°

5.

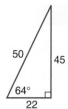

cos 64°

6.

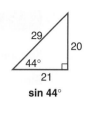

sin 44°

B. Find each of the following trig values. Round to three decimal places.

1. sin 27°   2. cos 38°   3. tan 12°   4. sin 86°

5. cos 79°   6. tan 6°   7. cos 87°   8. cos 6°30′

9. tan 50°20′   10. sin 75°40′   11. cos 41.25°   12. cos 81°25′

13. sin 50.4°   14. tan 74.15°   15. tan 81.06°   16. sin 12.6°

17. cos 98°   18. sin 106°   19. sin 144.2°   20. cos 134.5°

C. Find the acute angle *A*. Round to the nearest minute.

1. sin *A* = 0.974   2. cos *A* = 0.719

3. tan *A* = 2.05   4. sin *A* = 0.077

5. cos *A* = 0.262   6. tan *A* = 0.404

7. sin *A* = 0.168   8. cos *A* = 0.346

Find the acute angle *A*. Round to the nearest tenth of a degree.

9. tan *A* = 1.165   10. cos *A* = 0.662

11. cos *A* = 0.437   12. tan *A* = 0.225

13. tan *A* = 0.872   14. sin *A* = 0.472

15. sin *A* = 0.605   16. cos *A* = 0.154

Check your answers to the odd-numbered problems in the Appendix, then turn to Section 10-3 to learn how to use trigonometric ratios to solve problems.

10-3 Solving Right Triangles

In the first two sections of this chapter you learned about angles, triangles, and the trigonometric ratios that relate the angles to the side lengths in right triangles. Now it's time to look at a few of the possible applications of these trig ratios.

A right triangle has the following six parts:

*   Three angles: A right angle and two acute angles.
*   Three sides: The hypotenuse and two legs.

We can use trig ratios to find all unknown sides and angles of a right triangle if we know either of the following combinations:

1. The length of one side and the measure of one of the acute angles.

2. The lengths of any two sides.

Here is an example of the first combination.

Example 1

Suppose we need to find side *a* in the triangle shown. Use the following three-step approach:

Step 1 Decide which trig ratio is the appropriate one to use. In this case, we are given the hypotenuse (21 cm), and acute angle B (50°) and we need to find the side adjacent to the given angle *B*. The trig ratio that relates the adjacent side and the hypotenuse is the cosine.

$$\cos B = \frac{\text{adjacent side}}{\text{hypotenuse}}$$

Step 2 Write the cosine of the given angle in terms of the sides of the given right triangle.

$$\cos 50° = \frac{a}{21 \text{ cm}}$$

Step 3 Solve for the unknown quantity. In this case multiply both sides of the equation by 21.

$$a = 21 (\cos 50°)$$

$$\approx 21(0.64278 \ldots)$$

$$\approx 13.4985 \text{ cm} \quad \text{or} \quad 13 \text{ cm} \quad \text{rounded to 2 significant digits}$$

C10-10 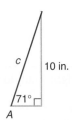 21 ⊗ (cos) 50 ⊜ → **13.4985398**

ROUNDING WHEN SOLVING RIGHT TRIANGLES

For practical applications involving triangle trigonometry, rounding will be dictated by specific situations or by job standards. To simplify rounding instructions in the remainder of this chapter, we shall adopt the following convention:

Sides rounded to:	correspond to	angles rounded to the nearest:
2 significant digits		degree
3 significant digits		0.1° or 10′
4 significant digits		0.01° or 1′

If side lengths are given as fractions, assume they are accurate to three significant digits.

→ **Your Turn**

Find the length of the hypotenuse *c* in the triangle at the left. (Round to the nearest tenth.) Work it out using the three steps shown in Example 1.

→ **Solution**

Step 1 We are given the side opposite to the angle 71°, and we need to find the hypotenuse. The trig ratio relating the opposite side and the hypotenuse is the sine.

$$\sin A = \frac{\text{opposite side}}{\text{hypotenuse}}$$

Step 2 $\sin 71° = \dfrac{10 \text{ in.}}{c}$

Step 3 Solve for c. $c \cdot \sin 71° = 10 \text{ in.}$

or $c = \dfrac{10 \text{ in.}}{\sin 71°} \approx \dfrac{10}{0.9455 \ldots}$

$\approx 10.6 \text{ in.}$

C10-11   10 ⊕ (sin) 71 (=) → `10.57620681`

Here is an example of the second combination, where we are given two sides.

Example 2

To find the measure of angle A in the triangle shown, follow these steps:

Step 1 We are given the opposite and adjacent sides to angle A. The tangent ratio involves these three parts.

$$\tan A = \frac{\text{opposite side}}{\text{adjacent side}}$$

Step 2 Substitute the given information.

$$\tan A = \frac{11.3}{16.5}$$

Step 3 To find the missing *angle*, use the *inverse* tangent.

$$A = \tan^{-1}\left(\frac{11.3}{16.5}\right) = 34.405 \ldots \approx 34.4°$$

C10-12 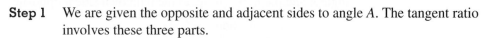 Here are two calculator options for this problem.

1. Calculate the trig ratio first, then find tangent inverse of "last answer":

   11.3 ⊕ **16.5** (=) (2nd) (tan⁻¹) (ANS) (=) → `34.40523429`

2. Perform the calculation exactly as it is written:

   (2nd) (tan⁻¹)* **11.3** ⊕ **16.5** (=) → `34.40523429`

*If your calculator does not automatically open parentheses here, you must press (.

There is no need to close parentheses in this calculation.

In the previous two examples, we were asked to find only one missing part of each right triangle. If we are asked to "solve the right triangle," then we must find *all* unknown sides and angles.

Example 3

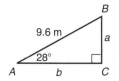

To *solve* the right triangle in the figure, we must find the lengths of the two unknown legs, a and b, and the unknown acute angle B.

First, find angle B. This is the easy part. Recall that the three angles of any triangle sum to 180°. Because angle C is a right angle, acute angles A and B will always sum to 90°.

Therefore,

$$B = 90° - A = 90° - 28° = 62°$$

Next, use a trig ratio to find one of the unknown legs. We are given the hypotenuse (9.6 m). If we use the given angle A to find its adjacent side b, we must use cosine.

$$\cos A = \frac{\text{adjacent side}}{\text{hypotenuse}}$$

$$\cos 28° = \frac{b}{9.6}$$

$$b = 9.6\,(\cos 28°)$$

$$b = 8.476\ldots \approx 8.5\text{ m} \quad \text{rounded}$$

Finally, find the length of the third side a. Because we now have two sides of the right triangle, we could use either the Pythagorean theorem or a trig equation. To avoid rounding errors, it is best to choose whichever method uses the original given information. The Pythagorean theorem would require the use of the calculated value from the previous step, so in this case we will use a trig equation. The sine ratio relates the given parts, angle A and the hypotenuse, to the unknown side a opposite angle A.

$$\sin A = \frac{\text{opposite side}}{\text{hypotenuse}}$$

$$\sin 28° = \frac{a}{9.6}$$

$$a = 9.6\,(\sin 28°)$$

$$a = 4.5069\ldots \approx 4.5\text{ m} \quad \text{rounded}$$

Therefore, the missing parts of the given triangle are: $B = 62°$, $b \approx 8.5$ m, $a \approx 4.5$ m.

→ Your Turn

Solve the right triangle shown. Express all angles in degrees and minutes.

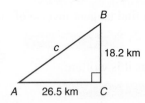

→ Solution

First, find one of the acute angles. To find angle A, notice that we are given the opposite and adjacent sides to angle A, so we must use the tangent ratio.

$$\tan A = \frac{\text{opposite side}}{\text{adjacent side}}$$

$$\tan A = \frac{18.2}{26.5}$$

$$A = \tan^{-1}\left(\frac{18.2}{26.5}\right)$$

$$A = 34.48\ldots° \approx 34° \, 30' \quad \text{rounded to the nearest } 10'$$

Next, find the other acute angle by subtraction.

$$B = 90° - A \approx 90° - 34° \, 30' \approx 55° \, 30'$$

Finally, determine the length of the third side. In this case, to use the given information, we choose the Pythagorean theorem.

$$c = \sqrt{a^2 + b^2}$$

$$c = \sqrt{(18.2)^2 + (26.5)^2}$$

$$c \approx 32.1 \text{ km}$$

Therefore, the missing parts of the given triangle are: $A \approx 34°30'$, $B \approx 55°30'$, $c \approx 32.1$ km.

→ **More Practice**

(a) Find c

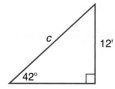

(b) Find angle B in decimal degrees.

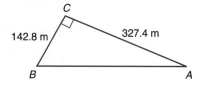

(c) Solve the triangle. Express angles in decimal degrees.

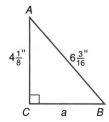

(d) Solve the triangle. Express angles in degrees and minutes.

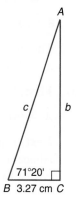

→ **Solutions**

(a) For the given angle, 12 ft is the length of the opposite side and c is the hypotenuse. Use the sine ratio.

$$\sin 42° = \frac{12}{c}$$

$$c = \frac{12}{\sin 42°} = 17.93 \ldots \approx 18 \text{ ft rounded}$$

(b) For the unknown angle B, 142.8 m is the length of the adjacent side and 327.4 m is the length of the opposite side. Use the tangent ratio.

$$\tan B = \frac{327.4}{142.8}$$

$$B = \tan^{-1}\left(\frac{327.4}{142.8}\right) = 66.434 \ldots° \approx 66.43° \text{ rounded}$$

(c) **First,** for angle A, we have the adjacent side and the hypotenuse. Use cosine:

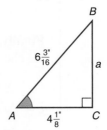

$$\cos A = \frac{4\frac{1}{8}}{6\frac{3}{16}} = \frac{2}{3}$$

$$A = \cos^{-1}\left(\frac{2}{3}\right) \approx 48.2°$$

C10-13

With a calculator, we recommend dividing the fractions first and then finding inverse cosine of "last answer."

4 $\boxed{A_c^b}$ 1 $\boxed{A_c^b}$ 8 $\boxed{\div}$ 6 $\boxed{A_c^b}$ 3 $\boxed{A_c^b}$ 16 $\boxed{=}$ $\rightarrow$ $\boxed{2/3}$

$\boxed{2^{nd}}$ $\boxed{\cos^{-1}}$ $\boxed{ANS}$ $\boxed{=}$ $\rightarrow$ $\boxed{48.1896851}$

Next, to find angle B subtract angle A from $90°$.

$$B = 90° - A \approx 90 - 48.2° \approx 41.8°$$

Finally, to find side a, use the Pythagorean theorem.

$$a = \sqrt{c^2 - b^2}$$

$$= \sqrt{\left(6\frac{3}{16}\right)^2 - \left(4\frac{1}{8}\right)^2}$$

$$\approx 4.61 \text{ in. rounded to three significant digits}$$

C10-14

6 $\boxed{A_c^b}$ 3 $\boxed{A_c^b}$ 16 $\boxed{x^2}$ $\boxed{-}$ 4 $\boxed{A_c^b}$ 1 $\boxed{A_c^b}$ 8 $\boxed{x^2}$ $\boxed{=}$ $\boxed{\sqrt{}}$ $\boxed{ANS}$ $\boxed{=}$ $\rightarrow$ $\boxed{4.611890204}$

Therefore, the missing parts of the given triangle are: $A \approx 48.2°$, $B \approx 41.8°$, $a \approx 4.61$ in.

(d) **First,** find the unknown acute angle A.

$$A = 90° - B = 90° - 71°20' = 18°40'$$

Next, find one of the missing sides. We have the acute angle B ($71° \, 20'$) and its adjacent side (3.27 cm). We could use tangent to find the opposite side AC or cosine to find the hypotenuse AB. Using tangent,

$$\tan 71°20' = \frac{b}{3.27}$$
$$b = 3.27 \, (\tan 71° \, 20')$$
$$b \approx 9.68 \text{ cm}$$

C10-15

3.27 $\boxed{\times}$ $\boxed{\tan}$ * 71 $\boxed{+}$ 20 $\boxed{\div}$ 60 $\boxed{=}$ $\rightarrow$ $\boxed{9.679337948}$

Finally, find the hypotenuse. Using the given angle B and its adjacent side, we must use the cosine ratio.

$$\cos 71° \, 20' = \frac{3.27}{c}$$
$$c = \frac{3.27}{\cos 71° \, 20'}$$
$$c \approx 10.2 \text{ cm}$$

Therefore, the missing parts of the given triangle are $A = 18° \, 40'$, $b \approx 9.68$ cm, $c \approx 10.2$ cm

---

* If your calculator does not automatically open parentheses with $\boxed{\tan}$, you must press $\boxed{(}$ here. You may also use your $\boxed{° \, ' \, ''}$ key to enter degrees and minutes.

Practical problems are usually a bit more difficult to set up, but they are solved with the same approach. Sometimes you must redraw a given figure in order to find a right triangle.

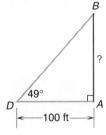

Example 4

Surveying Suppose you need to measure the distance between two points A and B on opposite sides of a river. You have a tape measure and a protractor but no way to stretch the tape measure across the river. How can you measure the distance AB?

Try it this way:

1. Choose a point C in line with points A and B. If you stand at C, A and B will appear to "line up."

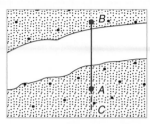

2. Construct a perpendicular to line BC at A and pick any point D on this perpendicular.

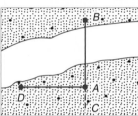

3. Find a point E where B and D appear to "line up." Draw DE and extend DE to F to form angle D.

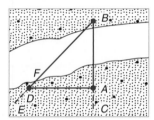

4. Triangle ABD is a right triangle. You can measure AD and angle D, and, using the trigonometric ratios, you can calculate AB.

   Suppose $AD = 100$ ft and angle $D = 49°$. To find AB, first draw the following right triangle.

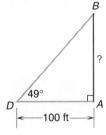

   AD is the adjacent side to angle D.

   AB is the side opposite angle D.

The trig ratio that relates the opposite to the adjacent side is the tangent.

$$\tan D = \frac{\text{opposite side}}{\text{adjacent side}}$$

$$\tan D = \frac{AB}{AD}$$

$$\tan 49° = \frac{AB}{100 \text{ ft}}$$

Multiply both sides of this equation by 100 to get

$$100(\tan 49°) = AB$$

Using a calculator, we have

$100 \times \tan 49 = \rightarrow$ `115.0368407`

Therefore, $AB \approx 115$ ft

Using trig ratios, you have managed to calculate the distance from A to B without ever coming near point B.

→ Your Turn

Metalworking   Find the angle of taper, x. (Round to the nearest degree.)

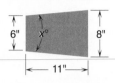

→ Solution

We draw a new figure and draw dashed line segment BC to create triangle ABC with angle m as indicated. Notice that angle x, the *taper angle,* is twice as large as angle m. We can find angle m from triangle ABC.

$$\tan m = \frac{\text{opposite side}}{\text{adjacent side}} = \frac{AC}{BC}$$

$$\tan m = \frac{1 \text{ in.}}{11 \text{ in.}} \approx 0.090909 \dots$$

$$m \approx \tan^{-1} 0.090909 \dots \approx 5.2°$$

$$x = 2m \approx 10.4° \approx 10°$$

Here are a few more applications for you to try.

→ More Practice

(a)   **Metalworking**   Find the angle a in the casting shown. (Round to the nearest 0.1°.)

(b)   **Plumbing**   In a pipe-fitting job, what is the length of set if the offset angle is $22\frac{1}{2}°$ and the travel is $32\frac{1}{8}$ in.?

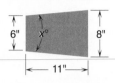

Problem (a)

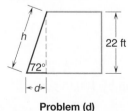

Problem (b)   **Problem (c)**

(c)   **Metalworking**   Find the width, w, of the V-slot shown.

(d)   **Machine Trades**   Find the dimensions d and h on the metal plate shown.

Problem (d)

(a) In this problem the angle a is unknown. The information given allows us to calculate tan a:

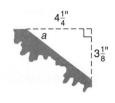

$$\tan a = \frac{\text{opposite side}}{\text{adjacent side}} = \frac{3\frac{1}{8} \text{ in.}}{4\frac{1}{4} \text{ in.}}$$

$$\tan a = \frac{25}{34}$$

$$a = \tan^{-1}\left(\frac{25}{34}\right) \approx 36.3°$$

(b) In this triangle the opposite side is unknown and the hypotenuse is given. Use the formula for sine:

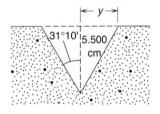

$$\sin 22\frac{1}{2}° = \frac{\text{opposite side}}{\text{hypotenuse}}$$

Then substitute, $\sin 22.5° = \dfrac{s}{32\frac{1}{8} \text{ in.}}$

And solve, $s = \left(32\frac{1}{8}\text{ in.}\right)(\sin 22.5°) \approx 12.3$ in.

(c) The V-slot creates an isosceles triangle whose height is 5.500 cm. The height divides the triangle into two identical right triangles as shown in the figure.

To find y,

$$\tan 31°10' = \frac{y}{5.500}$$

or

$$y = (5.500)(\tan 31°10')$$

$$y \approx 3.3265 \text{ cm}$$

$$w = 2y \approx 6.653 \text{ cm}$$

(d) $\sin 72° = \dfrac{22}{h}$

$$h = \frac{22}{\sin 72°} \approx 23 \text{ ft}$$

$$\tan 72° = \frac{22}{d}$$

$$d = \frac{22}{\tan 72°} \approx 7.1 \text{ ft}$$

Now turn to Exercises 10-3 for a set of problems on solving right triangles using trigonometric ratios.

Unless otherwise directed, round according to the rules in the box on page 636.

A. For each right triangle, find the missing quantity indicated below the figure. Express angles in decimal degrees.

1.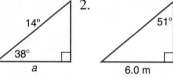

2.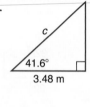

3.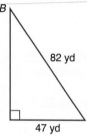

4.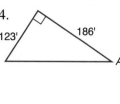

$a =$ _____ $b =$ _____ $c =$ _____ $A =$ _____

5.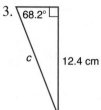

6.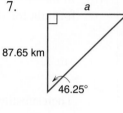

7.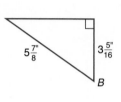

8.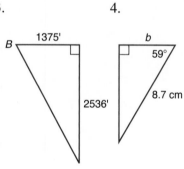

$c =$ _____ $B =$ _____ $a =$ _____ $B =$ _____

B. For each right triangle, find the missing quantity indicated below the figure. Express angles in degrees and minutes.

1. 2. 3. 4.

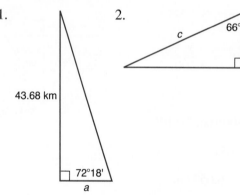

$a =$ _____ $c =$ _____ $B =$ _____ $b =$ _____

5. 6.

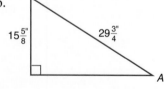

$a =$ _____ $A =$ _____

7. 8.

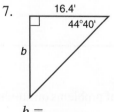

$b =$ _____ $B =$ _____

C. For each problem, use the given information to solve right triangle *ABC* for all missing parts. Express angles in decimal degrees.

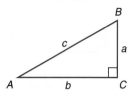

1. $a = 36$ ft, $B = 68°$
2. $b = 52.5$ m, $c = 81.6$ m
3. $A = 31.5°$, $a = 6\frac{3}{4}$ in.
4. $B = 58°$, $c = 5.4$ cm
5. $a = 306.5$ km, $c = 591.3$ km
6. $A = 47.4°$, $b = 158$ yd
7. $B = 74°$, $b = 16$ ft
8. $a = 12\frac{3}{8}$ in., $b = 16\frac{5}{16}$ in.
9. $A = 21° \; 15'$, $c = 24.75$ m

D. Practical Problems

Problem 1

1. **Manufacturing**   The most efficient operating angle for a certain conveyor belt is 31°. If the parts must be moved a vertical distance of 16 ft, what length of conveyor is needed?

2. **Metalworking**   Find the angle *m* in the casting shown. (Round to the nearest tenth of a degree.)

Problem 2

3. **Plumbing**   A pipe fitter must connect a pipeline to a tank as shown in the figure. The run from the pipeline to the tank is 62 ft 6 in., while the set (rise) is 38 ft 9 in.

    (a) How long is the connection? (Round to the nearest inch.)

    (b) Will the pipe fitter be able to use standard pipe fittings (i.e., $22\frac{1}{2}°$, 30°, 45°, 60°, or 90°)?

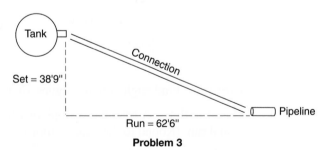
Problem 3

4. **Aviation**   A helicopter, flying directly over a fishing boat at an altitude of 1200 ft, spots an ocean liner at a 15° angle of depression. How far from the boat is the liner? (Round to the nearest hundred feet.)

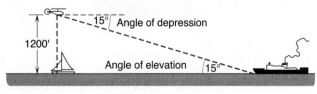
Problem 4

5. **Metalworking** Find the angle of taper on the steel bar shown if it is equal to twice *m*. (Round to the nearest minute.)

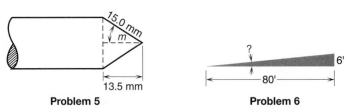

Problem 5   **Problem 6**

6. **Construction** A road has a rise of 6 ft in 80 ft. What is the gradient angle of the road? (Round to the nearest degree.)

7. **Carpentry** Find the length of the rafter shown. (Round to the nearest inch.)

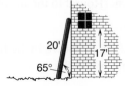

Problem 7

8. **Metalworking** Three holes are drilled into a steel plate. Find the distances *A* and *B* as shown in the figure. (Round to the nearest hundredth of an inch.)

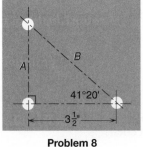

Problem 8   **Problem 9**

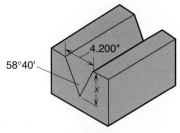

9. **Machine Trades** Find the depth of cut *x* needed for the V-slot shown. (Assume that the V-slot is symmetric.) (Round to the nearest hundredth.)

10. **General Trades** A 20-ft ladder leans against a building at a 65° angle with the ground. Will it reach a window 17 ft above the ground?

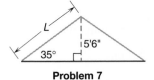

Problem 10

11. **Metalworking** Find the distance *B* in the special countersink shown. (Round to the nearest tenth.)

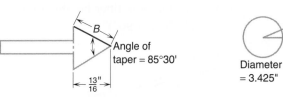

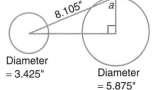

Problem 11   **Problem 12**

12. **Drafting** Find angle *a* in the figure. (Round to the nearest 0.01°.)

13. **Construction** A road has a slope of 2°25′. Find the rise in 6500 ft of horizontal run. (Round to the nearest foot.)

14. **Machine Trades** Find the missing dimension *d* in the $\frac{5}{16}$-in. flathead screw shown.

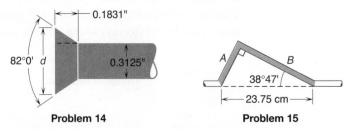

Problem 14   **Problem 15**

15. **Plumbing** What total length of conduit is needed for sections *A* and *B* of the pipe connection shown?

16. **Machine Trades**   Find the included angle *m* of the taper shown. (Round to the nearest minute.)

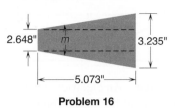

Problem 16

17. **Machine Trades**   Ten holes are spaced equally around a flange on a 5.0-in.-diameter circle. Find the center-to-center distance *x* in the figure.

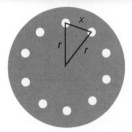

Problem 17

18. **Police Science**   Detectives investigating a crime find a bullet hole in a wall at a height of 7 ft 6 in. from the floor. The bullet passed through the wall at an angle of 34°. If they assume that the gun was fired from a height of 4 ft above the floor, how far away from the wall was the gun when it was fired? (Round to the nearest foot.)

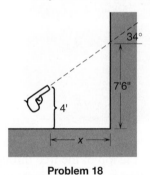

Problem 18

19. **Construction**   The Santa Teresa municipal building code specifies that public buildings have a handicapped access ramp with an incline of no more than 10°. What length of ramp is needed if the ramp climbs 3 ft 8 in.? Express the answer in feet and inches to the nearest inch. (*Hint:* Work the problem in inches only, then convert the answer to feet and inches.)

20. **Construction**   A gable roof is to be constructed with a slope of 28° and a run of 27 ft 6 in. Calculate the rise of the roof. Express the answer in feet and inches to the nearest inch.

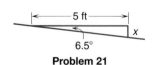

Problem 21

21. **Construction**   A crane is being set up on a slope of 6.5°. If the base of the crane is 5.0 ft wide, how many inches should the downhill side of the base be raised in order to level the crane? (Round to the nearest tenth.)

22. **General Interest**   The dispersion angle of a speaker indicates the maximum range of undistorted sound for listeners directly in front of the speaker. A dispersion angle of 90° means that the listener can be within 45 degrees on either side of the speaker and still receive good sound quality. The two speakers being used for a particular concert have a dispersion angle of 120°, are perpendicular to the stage, and are 10 ft from the audience. (See the figure.) What is

the maximum distance between the speakers so that no one in the first row of the audience is outside of the dispersion angle?

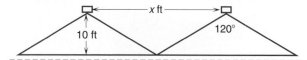

23. **Electrical Trades** An electrician is installing 1-m-square solar panels on the roof of an apartment building. The roof is flat and measures 20 m wide in the north–south direction. As shown in the figure, the panels are installed at an angle of 55° and must be separated by 0.2 m in the north to south direction to avoid shadowing. With these specifications, how many panels will fit along the 20-m side? (Round down to the next whole number.)

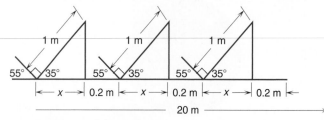

24. **Surveying** A surveyor wants to estimate the width *EF* of the ravine shown in the figure without crossing over to the other side. He walks 55 ft perpendicular to *EF* to point *G* and sights point *E*. He then measures angle *EGF* to be 36°. Calculate the width *EF*.

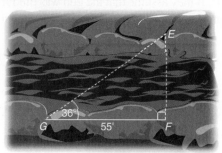

25. **Forestry** A forest ranger needs to estimate the height of the tree shown in the figure. She sights the top of the tree through a clinometer, a device that gives her angle of elevation as 68°. If she is standing 32 ft from the tree, and it is 5.5 ft from the ground to her eye level, how tall is the tree?

Turn to the Appendix to check your answers to the odd-numbered problems. Then go to Section 10-4 to study oblique triangles.

10-4 Oblique Triangles (Optional)

We have already seen how the trigonometric functions can be used to solve right triangles. To solve for the missing parts of a triangle that does not contain a right angle, we need two new formulas, the *law of sines* and the *law of cosines*. In this section we will derive these formulas and show how they may be used to solve triangles.

Oblique Triangles

Any triangle that is not a right triangle is called an **oblique triangle.** There are two types of oblique triangles: acute and obtuse. In an **acute triangle** all three angles are acute—each is less than 90°. In an **obtuse triangle** one angle is obtuse—that is, greater than 90°. Notice in these figures that the angles have been labeled A, B, and C, and the sides opposite these angles have been labeled a, b, and c, respectively.

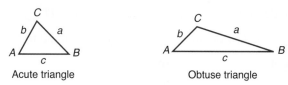

Acute triangle   Obtuse triangle

Law of Sines

The law of sines states that the lengths of the sides of a triangle are proportional to the sines of the corresponding angles. To show this for an acute triangle, first construct a perpendicular from C to side AB. Then in triangle ACD,

$$\sin A = \frac{h}{b} \quad \text{or} \quad h = b \sin A$$

and in triangle BCD,

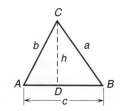

$$\sin B = \frac{h}{a} \quad \text{or} \quad h = a \sin B$$

Then $a \sin B = b \sin A$ or $\dfrac{\sin A}{a} = \dfrac{\sin B}{b}$

By repeating this process with a perpendicular from B to side AC, we can obtain a similar equation involving angles A and C and sides a and c. Combining these results gives the **law of sines.**

LAW OF SINES

$$\frac{\sin A}{a} = \frac{\sin B}{b} = \frac{\sin C}{c}$$

To *solve* a triangle means to calculate the values of all the unknown sides and angles from the information given. The law of sines enables us to solve any oblique triangle for two situations that we shall refer to as Case 1 and Case 2.

Case 1.   Two angles and a side are known.

Case 2.   Two sides and the angle opposite one of them are known.

Case 1

Example 1

For the triangle shown, $A = 47°$, $B = 38°$, and $a = 8.0$ in. Find all its unknown parts, in this case angle C and sides b and c.

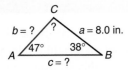

First, sketch the triangle as shown. Then since the angles of a triangle sum to 180°, find angle C by subtracting.

$$C = 180° - A - B = 180° - 47° - 38° = 95°$$

Next, apply the law of sines. Substitute A, B, and a to find b.

$$\frac{\sin B}{b} = \frac{\sin A}{a} \quad \text{gives} \quad b = \frac{a \sin B}{\sin A}$$

$$b = \frac{8.0 \sin 38°}{\sin 47°} \approx 6.7 \text{ in.}$$

C10-17 Using a calculator, we have **8** $\times$ $\boxed{\sin}$ **38** $\boxed{)}$ $\div$ $\boxed{\sin}$ **47** $\boxed{=}$ $\rightarrow$ `6.734486736`

Finally, apply the law of sines, again using A, C, and a to find c.

$$\frac{\sin C}{c} = \frac{\sin A}{a} \quad \text{gives} \quad c = \frac{a \sin C}{\sin A}$$

$$c = \frac{8.0 \sin 95°}{\sin 47°} \approx 11 \text{ in.}$$

In any triangle the largest angle is opposite the longest side, and the smallest angle is opposite the shortest side. Always check to make certain that this is true for your solution.

Careful Never use the Pythagorean theorem with an oblique triangle. It is valid only for the sides of a right triangle. ◀

→ Your Turn

Solve the triangle for which $c = 6.0$ ft, $A = 52°$, and $C = 98°$.

→ Solution

$B = 180° - 52° - 98° = 30°$

$$\frac{\sin A}{a} = \frac{\sin C}{c} \quad \text{or} \quad a = \frac{c \sin A}{\sin C}$$

$$a = \frac{6 \sin 52°}{\sin 98°} \approx 4.8 \text{ ft}$$

$$\frac{\sin B}{b} = \frac{\sin C}{c} \quad \text{or} \quad b = \frac{c \sin B}{\sin C}$$

$$b = \frac{6 \sin 30°}{\sin 98°} \approx 3.0 \text{ ft}$$

Case 2

In case 2, when two sides and the angle opposite one of these sides are known, the numbers may not always lead to a solution. For example, if $A = 50°$, $a = 6$, and $b = 10$, we may make a sketch of the triangle like this:

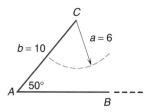

No matter how we draw side a it will not intersect line AB. There is no triangle with these measurements.

By looking at the given information before trying to solve the problem you can get an idea of what the solution triangle looks like. Let's examine all the possibilities.

- If $A = 90°$, then the triangle is a right triangle. Solve it using the methods of Section 10-3.
- If $A > 90°$ and $a > b$, then the triangle looks like this ⟶
- If $A > 90°$ and $a \leq b$, then there will be no solution, like this

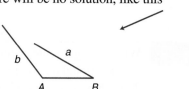

- If $A < 90°$ and $a = b \sin A$, then the triangle is a right triangle. You can solve it using the methods of Section 10-3. ⟶

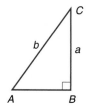

- If $A < 90°$ and $a \geq b$, then the triangle looks like this ⟶

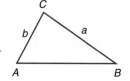

- If $A < 90°$ and $a < b$ and $a > b \sin A$, then the triangle looks either like this

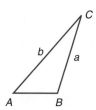   or like this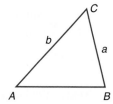

Because two different triangles may satisfy the given information, this is called the **ambiguous** case.

- If $A < 90°$ and $a < b \sin A$, there will be no solution.

Now, let's look at some examples.

Example 2

Given $A = 30.0°$ and $b = 10.0$ ft, solve the triangle where (a) $a = 12.0$ ft, (b) $a = 5.00$ ft, (c) $a = 4.50$ ft, (d) $a = 8.00$ ft.

(a) **First,** note that $a > b$ so the triangle will look like this:

Make a quick sketch of the triangle.

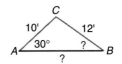

Second, solve for angle B using the law of sines.

$$\frac{\sin B}{b} = \frac{\sin A}{a}$$

$$\sin B = \frac{b \sin A}{a} = \frac{10 \sin 30°}{12}$$

$$\sin B = 0.4166\ldots$$

Find angle B using inverse sine.

$$B \approx 24.62° \approx 24.6°$$

C10-18 Using a calculator, we recommend the following sequence:

$10\,\boxed{\times}\,\boxed{\sin}\,30\,\boxed{)}\,\boxed{\div}\,12\,\boxed{=}\,\boxed{2^{nd}}\,\boxed{\sin^{-1}}\,\boxed{ANS}\,\boxed{=}\,\rightarrow\ \boxed{24.62431835}$

If your calculator does not automatically open parentheses with $\boxed{\sin}$, do not press $\boxed{)}$ after **30**.

▶ **Note** When using an intermediate answer in the next step of a problem, always carry at least one additional decimal place to avoid rounding errors. In this case the rounded answer is 24.6°, but we use 24.62° in performing the next calculation. ◀

Next, calculate angle C.　　$C = 180° - A - B \approx 180° - 30.00° - 24.62°$
$$C \approx 125.38° \approx 125.4°$$

Finally, use the law of sines to find side c.

$$\frac{\sin C}{c} = \frac{\sin A}{a} \quad \text{gives} \quad c = \frac{a \sin C}{\sin A} \approx \frac{12\,(\sin 125.38°)}{\sin 30°}$$

$$c \approx 19.6 \text{ ft}$$

(b) **First,** note that $a < b$. Then test to see if a is less than $b \sin A$. $b \sin A = 10 \sin 30° = 5$. Because $a = b \sin A$, the triangle is a right triangle and there is only one possible solution. We can show that the triangle is a right triangle by using the law of sines:

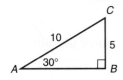

$$\frac{\sin B}{b} = \frac{\sin A}{a} \quad \text{gives} \quad \sin B = \frac{b \sin A}{a} = \frac{10 \sin 30°}{5}$$

$$\sin B = 1$$
$$B = 90°$$

Next, calculate angle C:　　$C = 180° - 30° - 90° = 60°$

Finally, calculate side c:　　$\cos A = \dfrac{c}{b}$

$$c = b \cos A = 10 \cos 30°$$
$$c \approx 8.66 \text{ ft}$$

(c) Note that $a < b$, then test to see if a is less than $b \sin A$. $b \sin A = 5$, and $4.5 < 5$; therefore, there is *no* triangle that satisfies the given conditions.

We can check this by using the law of sines:

$$\sin B = \frac{b \sin A}{a} = \frac{10 \sin 30°}{4.5} \approx 1.11$$

But the value of $\sin B$ can never be greater than 1; therefore, no such triangle exists. Using a calculator, we get

C10-19

10 $\boxed{\times}$ $\boxed{\sin}$ 30 $\boxed{)}$ $\boxed{\div}$ **4.5** $\boxed{=}$ $\boxed{2^{nd}}$ $\boxed{\sin^{-1}}$ $\boxed{ANS}$ $\boxed{=}$ → ▓▓▓▓▓▓▓▓ *Error*

The error message tells us that there is no solution.

(d) **First,** note that $a < b$. Because $8 > 5$, $a > b \sin A$, and we have the ambiguous case. There will be two possible triangles that satisfy these given conditions. Sketch them.

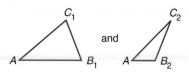

and

As given earlier (p. 632), $\sin A = \sin(180° - A)$. Therefore we know that $B_2 = 180° - B_1$.

Second, apply the law of sines to find angle B.

$$\frac{\sin A}{a} = \frac{\sin B}{b}$$

$$\sin B = \frac{b \sin A}{a} = \frac{10 \sin 30°}{8}$$

$$\sin B = 0.625$$

$$B \approx 38.68° \approx 38.7°$$

The two possible angles are

$$B_1 \approx 38.68° \qquad \text{and} \qquad B_2 \approx 180° - 38.68° \approx 141.32° \approx 141.3°$$

- For the acute value of angle B,

$$C_1 \approx 180° - 30.00° - 38.68° \approx 111.32° \approx 111.3°$$

and from the law of sines,

$$\frac{\sin 111.32°}{c_1} = \frac{\sin 30°}{8.00} \qquad \text{gives} \qquad c_1 \approx 14.9 \text{ ft}$$

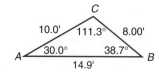

- For the obtuse value of angle B,

$$C_2 \approx 180° - 30.00° - 141.32° \approx 8.68° \approx 8.7°$$

and from the law of sines,

$$\frac{\sin 8.68°}{c_2} = \frac{\sin 30°}{8.00} \qquad \text{gives} \qquad c_2 \approx 2.41 \text{ ft}$$

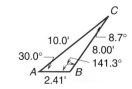

Summary of Procedures for Case 2

Step 1 Check the given angle. Is it $=$, $<$, or $> 90°$?

Step 2 If the angle is **equal to 90°,** the triangle is a right triangle.

Step 3 If the angle is **greater than 90°,** compare a and b. (a is the side opposite the given angle A.)

- If $a > b$,

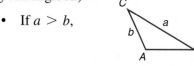

Use the law of sines to find this solution.

- If $a = b$, or $a < b$, there is no solution. Your calculator will display an error message when you try to calculate angle B.

Step 4 If the angle is **less than 90°,** compare a and b.

- If $a = b$ or $a > b$,

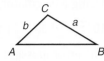

Use the law of sines to find this solution.

- If $a < b$, compare a and $b \sin A$.
- If $a = b \sin A$, the triangle is a right triangle. Solve using either the law of sines or right triangle trigonometry.

- If $a > b \sin A$, there are two solutions.

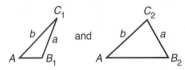

After finding B_1, subtract from 180° to find B_2, then complete both solutions.

- If $a < b \sin A$, there is no solution.

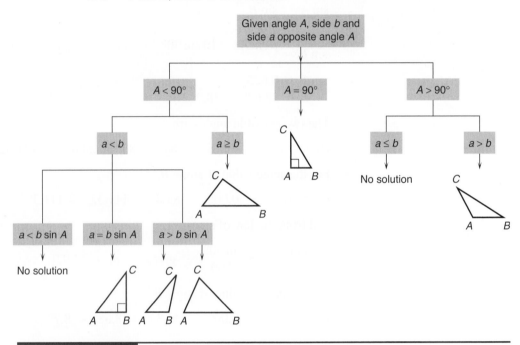

Given angle A, side b and side a opposite angle A

Example 3

If $A = 105.0°$ and $b = 125$ ft, solve the triangle where

    (a) $a = 95.0$ ft   (b) $a = 156$ ft

(a) The given angle is greater than 90°, and because $a < b$, from Step 3, we know there will be no solution. Check this using the law of sines:

$$\frac{\sin A}{a} = \frac{\sin B}{b} \quad \text{gives} \quad \sin B = \frac{b \sin A}{a} = \frac{125 \sin 105°}{95}$$

$$\sin B \approx 1.27$$

But this is not an acceptable value; $\sin B$ cannot be greater than 1.

(b) In this case $a > b$, and we expect from Step 3 to find a unique solution. Using the law of sines, we have

$$\sin B = \frac{b \sin A}{a} = \frac{125 \sin 105°}{156}$$

$$\sin B = 0.7739\ldots$$

$$B \approx 50.71° \approx 50.7°$$

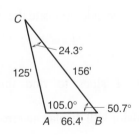

Then $C \approx 180° - 105.00° - 50.71° \approx 24.29° \approx 24.3°$

Use the law of sines to calculate side c:

$$\frac{\sin C}{c} = \frac{\sin A}{a} \qquad \text{gives} \qquad c = \frac{a \sin C}{\sin A} \approx \frac{156 \sin 24.29°}{\sin 105°}$$

$$c \approx 66.4 \text{ ft}$$

→ **Your Turn**

Use these steps and the law of sines to solve the triangle where $B = 36.0°$, $c = 16.4$ cm, and (a) $b = 20.5$ cm, (b) $b = 15.1$ cm. Round angles to the nearest $0.1°$ and sides to three significant digits.

→ **Solutions**

(a) The given angle is less than $90°$, so by Step 4 we need to compare the two sides. The side opposite the given angle is b, and b is greater than the other given side, c, so by Step 4 we know there is only one solution. Sketch it.

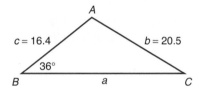

Now use the law of sines to solve for angle C.

$$\frac{\sin C}{c} = \frac{\sin B}{b} \qquad \text{gives} \qquad \sin C = \frac{c \sin B}{b} = \frac{16.4 \sin 36°}{20.5}$$

$$\sin C = 0.4702 \ldots$$

$$C \approx 28.049° \approx 28.0°$$

$$A \approx 180° - 28.05° - 36.00° \approx 115.95°$$

$$\frac{\sin A}{a} = \frac{\sin B}{b} \qquad \text{gives} \qquad a = \frac{b \sin A}{\sin B} \approx \frac{20.5 \sin 115.95°}{\sin 36°}$$

$$a \approx 31.4 \text{ cm}$$

(b) For this case $b < c$, so we need to compare b and $c \sin B$.

$$c \sin B = 16.4 \, (\sin 36°)$$

$$\approx 9.64$$

so $b > c \sin B$ and we know by Step 4 that there are two solutions.

$$\sin C = \frac{c \sin B}{b} = \frac{16.4 \sin 36°}{15.1}$$

$$\sin C = 0.6383 \ldots$$

$$C \approx 39.67° \approx 39.7°$$

The two possible solutions are

$$C_1 \approx 39.67° \qquad \text{and} \qquad C_2 \approx 180° - 39.67° \approx 140.33° \qquad \text{Sketch the triangles.}$$

$$\approx 140.3°$$

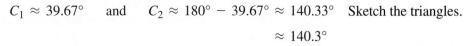

For the first value of angle C,

$$A_1 \approx 180° - 39.67° - 36.00° \approx 104.33° \approx 104.3°$$

and $\dfrac{\sin A_1}{a} = \dfrac{\sin B}{b}$ or $a \approx \dfrac{15.1 \sin 104.33°}{\sin 36°} \approx 24.9 \text{ cm}$

For the second value of C, $A_2 \approx 180° - 140.33° - 36.00° \approx 3.67°$ and

$$a \approx \frac{15.1 \sin 3.67°}{\sin 36°} \approx 1.64 \text{ cm}$$

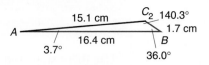

Law of Cosines

In addition to the two cases already mentioned, it is also possible to solve any oblique triangle if the following information is given.

Case 3. Two sides and the angle included by them.

Case 4. All three sides.

But the law of sines is not sufficient for solving triangles given this information. We need an additional formula called the **law of cosines.**

To obtain this law, in the triangle shown construct the perpendicular from C to side AB. Using the Pythagorean theorem, in triangle BCD,

$$a^2 = (c - x)^2 + h^2$$
$$= c^2 - 2cx + x^2 + h^2$$

and in triangle ACD,

$$b^2 = x^2 + h^2$$

Therefore,

$$a^2 = c^2 - 2cx + b^2$$
$$= b^2 + c^2 - 2cx$$

In triangle ACD, $\cos A = \dfrac{x}{b}$ or $x = b \cos A$; therefore,

$$a^2 = b^2 + c^2 - 2bc \cos A$$

By rotating the triangle ABC, we can derive similar equations for b^2 and c^2.

LAW OF COSINES

$$a^2 = b^2 + c^2 - 2bc \cos A$$
$$b^2 = c^2 + a^2 - 2ca \cos B$$
$$c^2 = a^2 + b^2 - 2ab \cos C$$

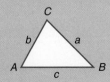

Learning Hint

Notice the similarity in these formulas. Rather than memorize all three formulas, you may find it easier to remember the following version of the law of cosines expressed in words.

The square of the length of any side of a triangle equals the sum of the squares of the lengths of the other two sides minus twice the product of the lengths of these two sides and the cosine of the angle between them. ◄

The law of cosines may be used to solve any oblique triangle for which the information in Case 3 and Case 4 is given.

Case 3

First, let's solve a triangle given two sides and the included angle.

Example 4

Solve the triangle with $a = 22.8$ cm, $b = 12.3$ cm, and $C = 42.0°$.

First, make a sketch and notice that the given angle C is the angle formed by the given sides a and b. Use the form of the law of cosines that involves the given angle to find the third side c.

$$c^2 = a^2 + b^2 - 2ab \cos C$$

$$c^2 = 22.8^2 + 12.3^2 - 2(22.8)(12.3) \cos 42°$$

Using a calculator, we get

C10-20 **22.8** $\boxed{x^2}$ $\boxed{+}$ **12.3** $\boxed{x^2}$ $\boxed{-}$ **2** $\boxed{\times}$ **22.8** $\boxed{\times}$ **12.3** $\boxed{\times}$ $\boxed{\cos}$ **42** $\boxed{=}$ $\boxed{\sqrt{}}$ $\boxed{ANS}$ $\boxed{=}$ $\rightarrow$ `15.94725463`

Therefore, $c \approx 15.9$ cm

Next, because we now know all three sides and one angle, we can use the law of cosines to find one of the remaining angles.

$$a^2 = b^2 + c^2 - 2bc \cos A$$

or $\quad \cos A = \dfrac{b^2 + c^2 - a^2}{2bc} \approx \dfrac{12.3^2 + 15.95^2 - 22.8^2}{2(12.3)(15.95)}$

Using a calculator gives us

12.3 $\boxed{x^2}$ $\boxed{+}$ **15.95** $\boxed{x^2}$ $\boxed{-}$ **22.8** $\boxed{x^2}$ $\boxed{=}$ $\boxed{\div}$ $\boxed{(}$ **2** $\boxed{\times}$ **12.3** $\boxed{\times}$ **15.95** $\boxed{)}$ $\boxed{=}$ $\boxed{2^{nd}}$ $\boxed{\cos^{-1}}$ $\boxed{ANS}$ $\boxed{=}$

C10-21

$\rightarrow$ `106.912924`

Therefore, $A \approx 106.9°$

Finally, find the remaining angle by subtracting.

$$B \approx 180° - 106.9° - 42.0° \approx 31.1°$$

→ Your Turn

Practice using the law of cosines by solving the triangle with $B = 34.4°$, $a = 145$ ft, and $c = 112$ ft.

→ Solution

Use the law of cosines to find the third side b.

$$b^2 = a^2 + c^2 - 2ac \cos B$$

$$b^2 = 145^2 + 112^2 - 2(145)(112) \cos 34.4°$$

$$b \approx 82.28 \text{ ft} \approx 82.3 \text{ ft}$$

Now find angle A. $\quad \cos A = \dfrac{b^2 + c^2 - a^2}{2bc} \approx \dfrac{82.28^2 + 112^2 - 145^2}{2(82.28)(112)}$

$$A \approx 95.3°$$
$$C \approx 180° - 34.4° - 95.3° \approx 50.3°$$

Case 4

Now we will solve a triangle given all three sides.

Example 5

Solve the triangle with sides $a = 106$ m, $b = 135$ m, and $c = 165$ m.

First, use the law of cosines to find the largest angle, the angle opposite the longest side. Side c is the longest side, so find C.

$$c^2 = a^2 + b^2 - 2ab \cos C$$

Solve for $\cos C$.

$$c^2 - a^2 - b^2 = -2ab \cos C$$

or

$$2ab \cos C = a^2 + b^2 - c^2$$

$$\cos C = \frac{a^2 + b^2 - c^2}{2ab} = \frac{106^2 + 135^2 - 165^2}{2(106)(135)}$$

$$\cos C = 0.078127\ldots$$

$$C \approx 85.5°$$

Next, use the law of sines to find one of the remaining angles. To find angle A,

$$\frac{\sin A}{a} = \frac{\sin C}{c} \qquad \text{or} \qquad \sin A = \frac{a \sin C}{c}$$

These are both known

$$\sin A \approx \frac{106 \sin 85.52°}{165} \approx 0.6404 \ldots$$

$$A \approx 39.8°$$

Finally, since the angles of a triangle must sum to 180°, we can calculate the remaining angle by subtracting.

$$B \approx 180° - 85.5° - 39.8° \approx 54.7°$$

| Careful | To avoid difficulties that can arise in the second step of the last example, you must find the largest angle first. Remember, the largest angle is always opposite the longest side. ◀ |

→ Your Turn

Now you try it. Solve the triangle with sides $a = 9.5$ in., $b = 4.2$ in., and $c = 6.4$ in.

→ Solution

First, find the largest angle A using the law of cosines.

$$a^2 = b^2 + c^2 - 2bc \cos A$$

$$\cos A = \frac{b^2 + c^2 - a^2}{2bc} = \frac{4.2^2 + 6.4^2 - 9.5^2}{2(4.2)(6.4)}$$

$$A \approx 126.07° \approx 126°$$

Next, use the law of sines to find angle B.

$$\frac{\sin B}{b} = \frac{\sin A}{a} \qquad \text{gives} \qquad \sin B = \frac{b \sin A}{a} \approx \frac{4.2 \sin 126.07°}{9.5}$$

$$B \approx 21°$$

Finally, subtract to find angle C.

$$C \approx 180° - 126° - 21° \approx 33°$$

The following table provides a summary showing when to use each law in solving oblique triangles. In each situation, remember that when two of the angles are known, the third may be found by subtracting from 180°.

When Given ...	Use ...	To Find ...
Any two angles and a side	Law of sines	A second side opposite one of the given angles. You may need to find the third angle first.
	Law of sines	The third side, opposite the third angle.
Two sides and the angle opposite one of them	Law of sines	A second angle, opposite one of the given sides. This is the ambiguous case.
	Law of sines	The third side, after finding the third angle by subtraction.
Two sides and the angle included by them	Law of cosines	The third side.
	Law of cosines	Any one of the remaining angles. (Then find the third angle by subtraction.)
All three sides	Law of cosines	The largest angle (opposite the longest side).
	Law of sines	One of the remaining angles. (Then find the third angle by subtraction.)

Now turn to Exercises 10-4 for a set of problems on solving oblique triangles.

Exercises 10-4 | Oblique Triangles

A. Solve each triangle.

1. $a = 6.5$ ft, $A = 43°$, $B = 62°$

2. $b = 17.2$ in., $C = 44.0°$, $B = 71.0°$

3. $b = 165$ m, $B = 31.0°$, $C = 110.0°$

4. $c = 2300$ yd, $C = 120°$, $B = 35°$

5. $a = 96.0$ in., $b = 58.0$ in., $B = 30.0°$

6. $b = 265$ ft, $c = 172$ ft, $C = 27.0°$

7. $a = 8.5$ m, $b = 6.2$ m, $C = 41°$

8. $b = 19.3$ m, $c = 28.7$ m, $A = 57.0°$

9. $a = 625$ ft, $c = 189$ ft, $B = 102.0°$

10. $b = 1150$ yd, $c = 3110$ yd, $A = 125.0°$

11. $a = 27.2$ in., $b = 33.4$ in., $c = 44.6$ in.

12. $a = 4.8$ cm, $b = 1.6$ cm, $c = 4.2$ cm

13. $a = 7.42$ m, $c = 5.96$ m, $B = 99.7°$

14. $b = 0.385$ in., $c = 0.612$ in., $A = 118.5°$

15. $a = 1.25$ cm, $b = 5.08$ cm, $c = 3.96$ cm

16. $a = 6.95$ ft, $b = 9.33$ ft, $c = 7.24$ ft

B. Practical Problems

1. **Landscaping** To measure the height of a tree, Steve measures the angle of elevation of the treetop as 46°. He then moves 15 ft closer to the tree and from this new point measures the angle of elevation to be 59°. How tall is the tree?

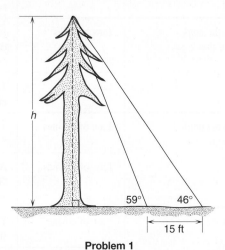

Problem 1

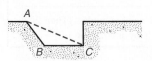

Problem 2

2. **Construction** In the channel shown, angle $B = 122.0°$, angle $A = 27.0°$, and side $BC = 32.0$ ft. Find the length of the slope AB.

3. **Construction** A triangular traffic island has sides 21.5, 46.2, and 37.1 ft. What are the angles at the corners?

4. **Construction** The lot shown is split along a diagonal as indicated. What length of fencing is needed for the boundary line?

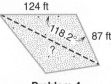

Problem 4

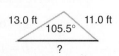

Problem 5

5. **Carpentry** Two sides of a sloped ceiling meet at an angle of 105.5°. If the distances along the sides to the opposite walls are 11.0 and 13.0 ft, what length of beam is needed to join the walls?

6. **Machine Trades** For the crankshaft shown, $AB = 4.2$ in., and the connecting rod $AC = 12.5$ in. Calculate the size of angle A when the angle at C is $12°$.

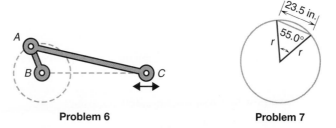

Problem 6   **Problem 7**

7. **Industrial Technology** If the chord of a circle is 23.5 in. long and subtends a central angle of $55.0°$, what is the radius of the circle?

Turn to the Appendix to check your answers to the odd-numbered problems. Then go to Problem Set 10 on page 665 for a set of problems covering the work of this chapter. If you need a quick review of the topics in this chapter, visit the chapter Summary first.

Summary Triangle Trigonometry

Objective	Review
Convert angles between decimal degrees and degrees and minutes. (p. 610)	Use the fact that $1° = 60'$ and set up unity fractions to convert.

Example: (a) $43.4° = 43° + 0.4°$

$$= 43° + \left(0.4° \times \frac{60'}{1°}\right)$$

$$= 43°24'$$

(b) $65°15' = 65° + 15'$

$$= 65° + \left(15' \times \frac{1°}{60'}\right)$$

$$= 65.25°$$

Convert angles between degree measure and radian measure. (p. 613)

To convert,

Degrees to radians → multiply by $\dfrac{\pi}{180}$

Radians to degrees → multiply by $\dfrac{180}{\pi}$

Example:

(a) $46° = 46\left(\dfrac{\pi}{180}\right) \approx 0.80$ radian

(b) 2.5 rad $= 2.5\left(\dfrac{180}{\pi}\right) \approx 143.2°$

Objective

Calculate the arc length and area of a sector.
(p. 614)

Review

Use the following formulas:

Arc length $S = ra$

Area $A = \dfrac{1}{2}r^2 a$

where r is the radius and a is the measure of the central angle in radians.

Example: Calculate the arc length and area of the circular sector shown.

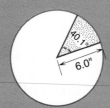

First, convert the central angle to radians.

$$40.1° = 40.1\left(\dfrac{\pi}{180}\right) \approx 0.700 \text{ rad}$$

Then, substitute into the formulas

$$S = ra = 6.0(0.7) \approx 4.2 \text{ in.}$$
$$A = \tfrac{1}{2}r^2 a = \tfrac{1}{2}(6.0)^2(0.7) \approx 12.6 \text{ in.}^2$$

Calculate both linear and angular speed.
(p. 615)

Average linear speed $v = \dfrac{d}{t}$, where d is the straight-line distance an object moves in time t. The average angular speed $w = \dfrac{a}{t}$, where a is the angle, in radians, through which an object moves in time t.

Example: Calculate the angular speed in radians per second of a large flywheel that rotates through an angle of 200° in 2 sec.

First, convert the angle to radians.

$$200° = 200\left(\dfrac{\pi}{180}\right) \approx 3.49 \text{ rad}$$

Then use the formula to calculate angular speed.

$$w = \dfrac{a}{t} = \dfrac{3.49 \text{ rad}}{2 \text{ sec}} \approx 1.7 \text{ rad/sec}$$

Use your knowledge of "special" right triangles to find missing parts.
(p. 618)

The following figures indicate the side ratios of the three special right triangles.

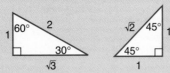

Example:

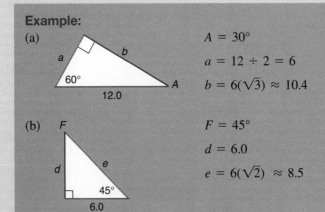

(a)

$A = 30°$

$a = 12 \div 2 = 6$

$b = 6(\sqrt{3}) \approx 10.4$

(b)

$F = 45°$

$d = 6.0$

$e = 6(\sqrt{2}) \approx 8.5$

Objective	Review
Find the values of trigonometric ratios. (p. 631)	Use the ⟨sin⟩ (sine), ⟨cos⟩ (cosine) and ⟨tan⟩ (tangent) buttons on your calculator.

Example: $\sin 26° \approx 0.438$ ⟨sin⟩ **26** ⟨=⟩ → *0.438371147*

$\tan 43° 20' \approx 0.943$ ⟨tan⟩ **43** ⟨+⟩ **20** ⟨÷⟩ **60** ⟨=⟩ → *0.943451341*

Find the angle when the value of its trigonometric ratio is given. (p. 633)	Use the ⟨2nd⟩ key on your calculator in combination with the ⟨sin⟩, ⟨cos⟩, and ⟨tan⟩ buttons to find $\sin^{-1}$, $\cos^{-1}$, and $\tan^{-1}$.

Example: If $\cos x = 0.549$

then $x = \cos^{-1} 0.549 \approx 56.7°$ or $56° 42'$

⟨2nd⟩ ⟨cos⁻¹⟩ **.549** ⟨=⟩ → *56.70156419*

Solve problems involving right triangles. (p. 635) 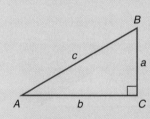	Use the following definitions to set up equations used to determine the missing quantities. For a right triangle,

$$\sin A = \frac{\text{side opposite angle } A}{\text{hypotenuse}} = \frac{a}{c} \qquad \sin B = \frac{\text{side opposite angle } B}{\text{hypotenuse}} = \frac{b}{c}$$

$$\cos A = \frac{\text{side adjacent to angle } A}{\text{hypotenuse}} = \frac{b}{c} \qquad \cos B = \frac{\text{side adjacent to angle } B}{\text{hypotenuse}} = \frac{a}{c}$$

$$\tan A = \frac{\text{side opposite angle } A}{\text{side adjacent to angle } A} = \frac{a}{b} \qquad \tan B = \frac{\text{side opposite angle } B}{\text{side adjacent to angle } B} = \frac{b}{a}$$

Example:

(a) To find X, $\sin 38° = \dfrac{X}{17}$ and $X = 17 \sin 38° \approx 10.5$

   To find Y, $\cos 38° = \dfrac{Y}{17}$ and $Y = 17 \cos 38° \approx 13.4$

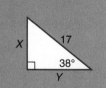

(b) In a pipe-fitting job, what is the run if the offset is $22\frac{1}{2}°$ and the length of set is $16\frac{1}{2}$ in.?

$$\tan 22\tfrac{1}{2}° = \frac{\text{set}}{\text{run}} = \frac{16\frac{1}{2}}{\text{run}}$$

$$\text{run} = \frac{16\frac{1}{2}}{\tan 22\frac{1}{2}°}$$

$$\approx 39.8 \text{ in.}$$

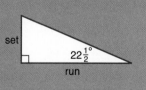

Solve oblique triangles. (p. 649)	When given two angles and a side, or two sides and the angle opposite one of them, use the law of sines to solve the triangle.

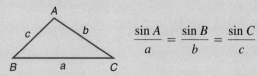

$$\frac{\sin A}{a} = \frac{\sin B}{b} = \frac{\sin C}{c}$$

Example: Solve the triangle where

$A = 68°$, $B = 79°$, and $b = 8.0$ ft.

Subtract from 180° to find the missing angle.

$C = 180° - 68° - 79° = 33°$

Use the law of sines to find both side a and side c.

$$\frac{\sin 68°}{a} = \frac{\sin 79°}{8.0}$$

$$a = \frac{8.0\,(\sin 68°)}{\sin 79°} \approx 7.6 \text{ ft}$$

$$\frac{\sin 33°}{c} = \frac{\sin 79°}{8.0}$$

$$c = \frac{8.0\,(\sin 33°)}{\sin 79°} \approx 4.4 \text{ ft}$$

When given three sides, or two sides and the included angle, use the law of cosines.

$$c^2 = a^2 + b^2 - 2ab \cos C \quad \text{or} \quad \cos C = \frac{a^2 + b^2 - c^2}{2ab}$$

Example: Solve the triangle where
$a = 6.5$ in., $b = 6.0$ in., and $c = 3.5$ in.

Use the second formula to find any two
of the angles.

$$\cos C = \frac{6.5^2 + 6.0^2 - 3.5^2}{2(6.5)(6.0)} \approx 0.846$$

$$C = \cos^{-1} 0.846 \approx 32°$$

$$\cos A = \frac{6.0^2 + 3.5^2 - 6.5^2}{2(6.0)(3.5)} \approx 0.143$$

$$A = \cos^{-1} 0.143 \approx 82°$$

Find the third angle by subtracting the other two from 180°.

$$B \approx 180° - 32° - 82° \approx 66°$$

Triangle Trigonometry

Answers to odd-numbered problems are given in the Appendix.

A. Convert the following angles as indicated.

Express in degrees and minutes. (Round to the nearest minute.)

1. 87.8°   **2.** 39.3°   **3.** 51.78°   **4.** 16.23°

Express in decimal degrees. (Round to the nearest tenth.)

5. 41°12′   **6.** 76°24′   **7.** 65°51′   **8.** 32°7′

Express in radians. (Round to three decimal places.)

9. 35°   **10.** 21.4°   **11.** 74°30′   **12.** 112.2°

Express in decimal degrees. (Round to the nearest tenth.)

13. 0.45 rad   **14.** 1.7 rad   **15.** 2.3 rad   **16.** 0.84 rad

B. Calculate the arc length S and area A for each of the following sectors.
(Round to nearest whole number if necessary.)

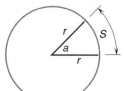

1. $a = 35°$ **2.** $a = 140°$ **3.** $a = 68°$ **4.** $a = 0.7$ rad
      $r = 26$ cm $r = 16$ ft $r = 11$ in. $r = 32$ m

C. Use a calculator to find the following trig values and angles.

Find each trig value to three decimal places.

1. cos 67°   **2.** tan 81°   **3.** sin 4°   **4.** cos 63°10′

5. tan 35.75°   **6.** sin 29.2°   **7.** sin 107°   **8.** cos 123°

Find the acute value of A to the nearest minute.

9. $\sin A = 0.242$ **10.** $\tan A = 1.54$ **11.** $\sin A = 0.927$

12. $\cos A = 0.309$ **13.** $\tan A = 0.194$ **14.** $\cos A = 0.549$

15. $\tan A = 0.823$ **16.** $\sin A = 0.672$ **17.** $\cos A = 0.118$

Name _____

Date _____

Course/Section _____

Find the acute value of *A* to the nearest tenth of a degree.

18. $\tan A = 0.506$ **19.** $\cos A = 0.723$ **20.** $\sin A = 0.488$

21. $\sin A = 0.154$ **22.** $\cos A = 0.273$ **23.** $\tan A = 2.041$

24. $\tan A = 0.338$ **25.** $\cos A = 0.608$ **26.** $\sin A = 0.772$

D. Find the missing dimensions in the right triangles as indicated.

Use the special right triangle relationships to solve problems 1–4. (Round all lengths to the nearest tenth and all angles to the nearest degree.)

1.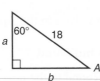

2.

3.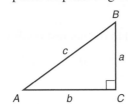

4.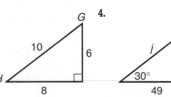

$a =$ _____	$d =$ _____	$\angle G \approx$ _____	$i =$ _____
$b =$ _____	$e =$ _____	$\angle H \approx$ _____	$j =$ _____
$\angle A =$ _____	$\angle D =$ _____		$\angle K =$ _____

For each problem, use the given information to solve right triangle *ABC* for all missing parts. Express angles in decimal degrees.

5. $a = 18$ mm, $A = 29°$ **6.** $B = 51°$, $c = 26$ ft

7. $b = 5.25$ m, $c = 7.35$ m   **8.** $a = 652$ yd, $B = 39.5°$

9. $a = 3\frac{1}{4}$ in., $b = 4\frac{1}{2}$ in.   **10.** $B = 65.5°$, $a = 32.4$ cm

11. $a = 12.0$ in., $c = 15.5$ in.   **12.** $A = 47°$, $c = 16$ ft

E. Solve the following oblique triangles.

1. $a = 17.9$ in., $A = 65.0°$, $B = 39.0°$

2. $a = 721$ ft, $b = 444$ ft, $c = 293$ ft

3. $a = 260$ yd, $c = 340$ yd, $A = 37°$

4. $b = 87.5$ in., $c = 23.4$ in., $A = 118.5°$

5. $a = 51.4$ m, $b = 43.1$ m, $A = 64.3°$

6. $a = 166$ ft, $c = 259$ ft, $B = 47.0°$

7. $a = 1160$ m, $c = 2470$ m, $C = 116.2°$

8. $a = 7.6$ in., $b = 4.8$ in., $B = 30°$

F. Practical Problems

1. **Machine Trades**   What height of gauge blocks is required to set an angle of 7°15′ for 10-in. plots? (Round to the nearest tenth.)

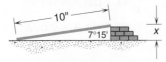

Problem 1

Chapter 10 Triangle Trigonometry

20 mph

220 mph v a

Problem 2

2. **Aviation** The destination of an airplane is due north, and the plane flies at 220 mph. There is a crosswind of 20.0 mph from the west. What heading angle a should the plane take? What is its relative ground speed v? (Round a to the nearest minute and v to three significant digits.)

3. **Carpentry** Six holes are spaced evenly around a circular piece of wood with a 4-in. diameter. Find the distance, d, between the holes. (*Hint:* The dashed line creates two identical right triangles.)

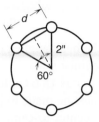

Problem 3

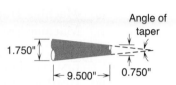

Angle of taper

1.750"   9.500"   0.750"

Problem 4

4. **Machine Trades** Find the angle of taper of the shaft in the figure. (Round to the nearest minute.)

5. **Electrical Trades** A TV technician installs an antenna 50 ft 6 in. tall on a flat roof. Safety regulations require a minimum angle of 30° between mast and guy wires.

   (a) Find the minimum value of X to the nearest inch.

   (b) Find the length of the guy wires to the nearest inch.

50'6" **30°** **X**

Problem 5

6. **Construction** A bridge approach shown below is 12 ft high. The maximum slope allowed is 15%—that is, $\frac{15}{100}$.

   (a) What is the length L of the approach?

   (b) What is the angle a of the approach?

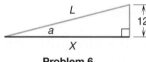

L a X 12'

$\left(Hint: \dfrac{15}{100} = \dfrac{12}{X} \right)$

Problem 6

Lead **Helix angle**

← Circumference →

Problem 7

7. **Machine Trades** The *helix angle* of a screw is the angle at which the thread is cut. The *lead* is the distance advanced by one turn of the screw. The circumference refers to the circumference of the screw as given by $C = \pi d$. The following formula applies:

$$\text{Tangent of helix angle} = \frac{\text{lead}}{\text{circumference}}$$

   (a) Find the helix angle for a 2-in.-diameter screw if the lead is $\frac{1}{8}$ in. (Round to the nearest minute.)

   (b) Find the lead of a 3-in.-diameter screw if the helix angle is $2°0'$.

8. **Construction** Find the lengths x and y of the beams shown in the bridge truss. (Round to the nearest tenth.)

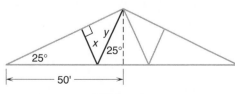

y x **25°** **25°**

← 50' →

Problem 8

9. **Machine Trades** Find the width *X* of the V-thread shown. (Round to the nearest thousandth.)

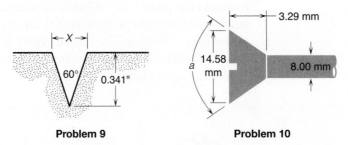

Problem 9   **Problem 10**

10. **Machine Trades** Find the head angle *a* of the screw shown.
11. **Machine Trades** A machinist makes a cut 13.8 cm long in a piece of metal. Then, another cut is made that is 18.6 cm long and at an angle of 62°0′ with the first cut. How long a cut must be made to join the two endpoints?
12. **Drafting** Determine the center-to-center measurement *x* in the adjustment bracket shown in the figure.

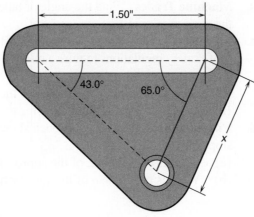

Problem 12

13. **Machine Trades** If a lathe makes 286 revolutions in 10 sec, calculate its average angular speed. (Round to the nearest whole number.)
14. **Automotive Trades** The wheels on a car have a radius of 12 in. and make 1 revolution every 0.12 sec. What is the linear speed of the tire in inches per second? (Round to two significant digits.)
15. **Construction** An amphitheater is in the shape of a circular sector with a smaller sector removed. (See the figure.) Find the area of the amphitheater shown as the shaded portion of the figure. (Round to the nearest thousand square feet.)
16. **Construction** The frontage of a lot at the end of a cul-de-sac is defined by an arc with central angle of 42°30′ and radius of 55.0 ft. Find the frontage distance in linear feet.

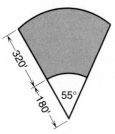

Problem 15

Objective	Sample Problems		For help, go to
When you finish this chapter, you will be able to:			
1. Solve a system of two linear equations in two variables.	(a) $5x + 2y = 20$	$x =$ _____	Page 673
	$3x - 2y = 12$	$y =$ _____	
	(b) $x = 3y$	$x =$ _____	
	$2y - 4x = 30$	$y =$ _____	
	(c) $3x - 5y = 7$	$x =$ _____	
	$6x - 10y = 14$	$y =$ _____	
	(d) $2y = 5x - 7$	$x =$ _____	
	$10x - 4y = 9$	$y =$ _____	
2. Solve word problems involving systems of equations in two variables.	(a) The sum of two numbers is 35. Twice the smaller decreased by the larger is 10. Find them.	_____	Page 688
	(b) **HVAC** The perimeter of a rectangular vent is 28 in. The length of the vent is 2 in. greater than its width. Find the dimensions.	_____	

_____
Name

_____
Date

_____
Course/Section

Objective	Sample Problems	For help, go to

(c) **Carpentry** A mixture of 800 drywall screws costs $42. If the mixture consists of one type costing 4 cents each and another type costing 6 cents each, how many of each kind are there? _____

3. Solve quadratic equations.

(a) $x^2 = 16$ $x =$ _____   Page 696

(b) $x^2 - 7x = 0$ $x =$ _____

(c) $x^2 - 5x = 14$ $x =$ _____

(d) $3x^2 + 2x - 16 = 0$ $x =$ _____

(e) $2x^2 + 3x + 11 = 0$ $x =$ _____

4. Solve word problems involving quadratic equations.

(a) **General Trades** Find the side length of a square opening whose area is 36 cm^2. _____   Page 703

(b) **Manufacturing** Find the radius of a circular pipe whose cross-sectional area is 220 sq in. _____

(c) **Electrical Trades** In the formula $P = RI^2$, find I (in amperes) if $P = 2500$ watts and $R = 15$ ohms. _____

(d) **HVAC** The length of a rectangular vent is 3 in. longer than its width. Find the dimensions if the cross-sectional area is 61.75 sq in. _____

(Answers to these preview problems are given in the Appendix. Also, worked solutions to many of these problems appear in the chapter Summary.)

If you are certain that you can work *all* these problems correctly, turn to page 713 for a set of practice problems. If you cannot work one or more of the preview problems, turn to the page indicated to the right of the problem. Those who wish to master this material with the greatest success should turn to Section 11-1 and begin work there.

Advanced Algebra

Source: © Dan Piraro King Features Syndicate

11-1 Systems of Equations

This chapter is designed for students who are interested in highly technical occupations. We explain how to solve systems of linear equations and how to solve quadratic equations. Before beginning, you may want to return to Chapter 7 to review the basic algebraic operations explained there. When you have had the review you need, return here and continue.

Before you can solve a system of two equations in two unknowns, you must be able to solve a single linear equation in one unknown. Let's review what you learned in Chapter 7 about solving linear equations.

Solution A *solution* to an equation such as $3x - 4 = 11$ is a number that we may use to replace x in order to make the equation a true statement. To find such a number, we *solve* the equation by changing it to an *equivalent* equation with only x on the left. The process goes like this:

Example 1

Solve: $3x - 4 = 11$

Step 1 Add 4 to both sides of the equation

$$3x \underbrace{- 4 + 4}_{0} = \underbrace{11 + 4}_{15}$$

$$3x = 15$$

Step 2 Divide both sides of the equation by 3.

$$\frac{3x}{3} = \frac{15}{3}$$

$x = 5$ This is the solution and you can check it by substituting 5 for x in the original equation.

 $3(5) - 4 = 11$

$15 - 4 = 11$ which is true.

→ Your Turn

Solve each of the following equations and check your answer.

(a) $\dfrac{3x}{2} = 9$ (b) $17 - x = 12$

(c) $2x + 7 = 3$ (d) $3(2x + 5) = 4x + 17$

→ Answers

(a) $x = 6$ (b) $x = 5$ (c) $x = -2$ (d) $x = 1$

A *system of equations* is a set of equations in two or more variables that may have a common solution.

Example 2

The pair of equations

$2x + y = 11$

$4y - x = 8$

has the common solution $x = 4, y = 3$. This pair of numbers will make each equation a true statement. If we substitute 4 for x and 3 for y, the left side of the first equation becomes

$2(4) + 3 = 8 + 3$ or 11, which equals the right side,

and the left side of the second equation becomes

$4(3) - 4 = 12 - 4$ or 8, which equals the right side.

The pair of values $x = 4$ and $y = 3$ satisfies *both* equations. This solution is often written $(4, 3)$, where the x-value is listed first and the y-value is listed second.

Note   Each of these equations by itself has an infinite number of solutions. For example, the pairs $(0, 11)$, $(1, 9)$, $(2, 7)$, and so on, all satisfy $2x + y = 11$, and the pairs $(0, 2)$, $\left(1, \frac{9}{4}\right), \left(2, \frac{5}{2}\right)$, and so on, all satisfy $4y - x = 8$. However, $(4, 3)$ is the only pair that satisfies both equations. ◄

→ **Your Turn**

By substituting, show that the numbers $x = 2$, $y = -5$ give the solution to the pair of equations

$$5x - y = 15$$
$$x + 2y = -8$$

→ **Solution**

The first equation is

$$5(2) - (-5) = 15$$
$$10 + 5 = 15, \text{ which is correct.}$$

The second equation is

$$(2) + 2(-5) = -8$$
$$2 - 10 = -8, \text{ which is also correct.}$$

Solution by Substitution   In this chapter we will show you two different methods of solving a system of two linear equations in two unknowns. The first method is called the method of *substitution.*

Example 3

To solve the pair of equations

$$y = 3 - x$$
$$3x + y = 11$$

follow these steps.

Step 1   *Solve* the first equation for x or y.

The first equation is already solved for y:

$$y = 3 - x$$

Step 2   *Substitute* this expression for y in the second equation.

$$3x + y = 11 \quad \text{becomes}$$
$$3x + \boxed{(3 - x)} = 11$$

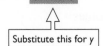
Substitute this for y

Step 3   *Solve* the resulting equation.

$$3x + (3 - x) = 11 \quad \text{becomes}$$
$$2x + 3 = 11 \quad \text{Subtract 3 from each side.}$$
$$2x = 8 \quad \text{Divide each side by 2.}$$
$$x = 4$$

Step 4   *Substitute* this value of x into the first equation and find a value for y.

$$y = 3 - x$$
$$y = 3 - (4)$$
$$y = -1$$

The solution is $x = 4, y = -1$ or $(4, -1)$.

Step 5   *Check* your solution by substituting it back into the second equation.

$$3x + y = 11 \quad \text{becomes}$$
$$3(4) + (-1) = 11$$
$$12 - 1 = 11 \quad \text{which is correct}$$

→ Your Turn

Try it. Use this substitution procedure to solve the system of equations

$$x - 2y = 3$$
$$2x - 3y = 7$$

→ Solution

Step 1   *Solve* the first equation for x by adding $2y$ to both sides of the equation.

$$x - 2y + \boxed{2y} = 3 + \boxed{2y}$$
$$x = 3 + 2y$$

Step 2   *Substitute* this expression for x in the second equation.

$$2x - 3y = 7 \quad \text{becomes}$$
$$2 \boxed{(3 + 2y)} - 3y = 7$$

Step 3   *Solve:*

$$6 + 4y - 3y = 7 \quad \text{Simplify by combining the } y\text{-terms.}$$
$$6 + y = 7 \quad \text{Subtract 6 from each side.}$$
$$y = 1$$

Step 4   *Substitute* this value of y in the first equation to find x.

$$x - 2y = 3 \quad \text{becomes}$$
$$x - 2(1) = 3$$

$$x - 2 = 3 \qquad \text{Add 2 to each side.}$$
$$x = 5$$

The solution is $x = 5$, $y = 1$ or $(5, 1)$.

Step 5 *Check* the solution in the second equation.

$$2x - 3y = 7 \qquad \text{becomes}$$
$$2(5) - 3(1) = 7$$
$$10 - 3 = 7 \qquad \text{which is correct.}$$

Of course, it does not matter which variable, x or y, we solve for in Step 1, or which equation we use in Step 4.

Example 4

In the pair of equations

$$2x + 3y = 22$$
$$x - y = 1$$

the simplest procedure is to solve the *second* equation for x to get $x = 1 + y$ and substitute this expression for x into the first equation.

The entire solution looks like this:

Step 1   In the second equation, add y to both sides to get $x = 1 + y$.

Step 2   When we substitute into the first equation,

$$2x + 3y = 22 \quad \text{becomes}$$
$$2\,(1 + y) + 3y = 22$$

Step 3   *Solve:*

Remove parentheses.	$2 + 2y + 3y = 22$
Combine terms.	$2 + 5y = 22$
Subtract 2 from each side.	$5y = 20$
Divide each side by 5.	$y = 4$

Step 4   *Substitute* 4 for y in the second equation.

$$x - y = 1 \qquad \text{becomes}$$
$$x - (4) = 1$$

or

$$x = 5 \qquad \text{The solution is } x = 5, y = 4.$$

 Step 5   *Check* the solution by substituting these values into the first equation.

$$2x + 3y = 22 \qquad \text{becomes}$$
$$2(5) + 3(4) = 22$$
$$10 + 12 = 22 \qquad \text{which is correct.}$$

Solve the following systems of equations by using the substitution method.

(a) $x = 1 + y$
$2y + x = 7$

(b) $3x + y = 1$
$y + 5x = 9$

(c) $x - 3y = 4$
$3y + 2x = -1$

(d) $y + 2x = 1$
$3y + 5x = 1$

(e) $x - y = 2$
$y + x = 1$

(f) $y = 4x$
$2y - 6x = 0$

→ Solutions

(a) **Step 1** The first equation is already solved for x.

Step 2 Substitute this expression for x into the second equation.

$$2y + \boxed{1 + y} = 7$$

Step 3 Solve for y.

$$3y + 1 = 7$$
$$3y = 6$$
$$y = 2$$

Step 4 Substitute 2 for y in the first equation.

$$x = 1 + (2)$$
$$\text{or } x = 3$$

Step 5 The solution is $x = 3$, $y = 2$.
Be sure to check it.

(b) **Step 1** You can easily solve either equation for y. Using the first equation, subtract $3x$ from both sides to obtain

$$y = 1 - 3x$$

Step 2 Substitute this expression for y into the second equation.

$$\boxed{1 - 3x} + 5x = 9$$

Step 3 Solve for x.

$$1 + 2x = 9$$
$$2x = 8$$
$$x = 4$$

Step 4 Substitute 4 for x in the equation from Step 1.

$$y = 1 - 3(4)$$
$$y = 1 - 12 = -11$$

Step 5 The solution is $x = 4$, $y = -11$. Check your answer.

(c) **Step 1** Use the first equation to solve for x.

$$x = 3y + 4$$

Step 2 Substitute this expression for x into the second equation.

$$3y + 2 \boxed{(3y + 4)} = -1$$

Step 3 Solve for y.

$$3y + 6y + 8 = -1$$
$$9y + 8 = -1$$
$$9y = -9$$
$$y = -1$$

Step 4 Substitute -1 for y in the equation from Step 1.

$$x = 3(-1) + 4$$
$$x = 1$$

Step 5 The solution is $x = 1, y = -1$.

(d) $x = 2, y = -3$ (e) $x = 1\frac{1}{2}, y = -\frac{1}{2}$ (f) $x = 0, y = 0$

Dependent and Inconsistent Systems

So far we have looked only at systems of equations with a single solution—one pair of numbers. Such a system of equations is called a *consistent* and *independent* system. However, it is possible for a system of equations to have no solution at all or to have very many solutions.

Example 5

The system of equations

$$y + 3x = 5$$
$$2y + 6x = 10$$

has *no unique* solution. If we solve for y in the first equation

$$y = 5 - 3x$$

and substitute this expression into the second equation

$$2y + 6x = 10$$

or $\quad 2 \boxed{(5 - 3x)} + 6x = 10$

This resulting equation simplifies to

$$10 = 10$$

There is no way of solving to get a unique value of x or y.

A system of equations that does not have a single unique number-pair solution but has an unlimited number of solutions is said to be *dependent*. The two equations are essentially the same. For the system shown in this example, the second equation is simply twice the first equation. There are infinitely many pairs of numbers that will satisfy this pair of equations. For example,

$$x = 0, y = 5$$
$$x = 1, y = 2$$
$$x = 2, y = -1$$
$$x = 3, y = -4$$

and so on.

If a system of equations is such that our efforts to solve it produce a false statement, the equations are said to be *inconsistent*. The system of equations has no solution.

Example 6

The pair of equations

$$y - 1 = 2x$$

$$2y - 4x = 7 \qquad \text{is inconsistent.}$$

If we solve the first equation for y

$$y = 2x + 1$$

and substitute this expression into the second equation,

$$2y - 4x = 7 \qquad \text{becomes}$$

$$2\,(2x + 1)\, - 4x = 7$$

$$4x + 2 - 4x = 7$$

$$\text{or} \qquad\qquad\qquad 2 = 7 \qquad \text{which is false.}$$

All the variables have dropped out of the equation, and we are left with an incorrect statement. The original pair of equations is said to be inconsistent, and it has no solution.

→ **Your Turn**

Try solving the following systems of equations.

(a) $2x - y = 5$ (b) $3x - y = 5$
     $2y - 4x = 3$ $6x - 10 = 2y$

→ **Solutions**

(a) Solve the first equation for y.

$$y = 2x - 5$$

Substitute this into the second equation.

$$2y - 4x = 3 \qquad \text{becomes}$$

$$2\,(2x - 5)\, - 4x = 3$$

$$4x - 10 - 4x = 3$$

$$-10 = 3 \qquad \text{This is impossible. There is no solution for this system of equations. The equations are inconsistent.}$$

(b) Solve the first equation for y.

$$y = 3x - 5$$

Substitute this into the second equation.

$$6x - 10 = 2y \qquad \text{becomes}$$

$$6x - 10 = 2\,(3x - 5)$$

$$6x - 10 = 6x - 10$$

$$0 = 0$$

This is true, but all of the variables have dropped out and we cannot get a single unique solution. The equations are dependent.

Solution by Elimination

The second method for solving a system of equations is called the *elimination method* or the *addition method*. When it is difficult or "messy" to solve one of the two equations for either x or y, the method of elimination may be the simplest way to solve the system of equations.

Example 7

In the system of equations

$$2x + 3y = 7$$

$$4x - 3y = 5$$

neither equation can be solved for x or y without introducing fractions that are difficult to work with. But we can simply add these equations, and the y-terms will be eliminated.

$$2x + 3y = 7$$

$$\underline{4x - 3y = 5}$$

$$6x + 0 = 12 \qquad \Longleftarrow \boxed{\text{Adding like terms eliminates } y.}$$

$$6x = 12$$

$$x = 2$$

Now substitute this value of x back into either one of the original equations to obtain a value for y.

The first equation	$2x + 3y = 7$
becomes	$2(2) + 3y = 7$
	$4 + 3y = 7$
Subtract 4 from each side.	$3y = 3$
Divide each side by 3.	$y = 1$

The solution is $x = 2$, $y = 1$, or $(2, 1)$.

 Check the solution by substituting it back into the second equation.

$$4x - 3y = 5$$

$$4(2) - 3(1) = 5$$

$$8 - 3 = 5 \qquad \text{which is correct.}$$

→ Your Turn

Try it. Solve the following system of equations by adding.

$$2x - y = 3$$

$$y + x = 9$$

First, rearrange the terms in the second equation
so that like terms are in the same column. ⟶

$$2x - y = 3$$
$$\underline{x + y = 9}$$

Then, add the columns of like terms to
eliminate y and solve for x.

$$3x + 0 = 12$$
$$x = 4$$

Finally, substitute 4 for x in the simpler
second equation to solve for y.

$$(4) + y = 9$$
$$y = 5$$

The solution is $x = 4$, $y = 5$, or $(4, 5)$. Check it.

Careful It is important to rearrange the terms in the equations so that the x and y terms appear in the same order in both equations.

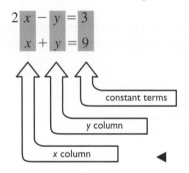

constant terms

y column

x column

◀

→ **More Practice**

Solve the following systems of equations by this process of elimination.

(a) $x + 5y = 17$ (b) $x + y = 16$ (c) $3x - y = -5$ (d) $\frac{1}{2}x + 2y = 10$

       $-x + 3y = 7$ $x - y = 4$ $y - 5x = 9$ $y + 1 = \frac{1}{2}x$

→ **Solutions**

(a)   **First,** add like terms to eliminate x and solve for y.

$$x + 5y = 17$$
$$\underline{-x + 3y = 7}$$
$$0 + 8y = 24$$
$$8y = 24$$
$$y = 3$$

Then, substitute 3 for y in the first equation, and solve for x.

$$x + 5(3) = 17$$
$$x + 15 = 17$$
$$x = 2$$ This solution is $x = 2$, $y = 3$, or $(2, 3)$.

$$-x + 3y = 7$$
$$-(2) + 3(3) = 7$$
$$-2 + 9 = 7 \qquad \text{which is correct.}$$

(b) **First,** add like terms to eliminate y and solve for x.

$$
\begin{array}{r}
x + y = 16 \\
\underline{x - y = 4} \\
2x + 0 = 20 \\
2x = 20 \\
x = 10
\end{array}
$$

Then, substitute 10 for x in the first equation, and solve for y.

$$(10) + y = 16$$
$$y = 6 \qquad \text{The solution is } x = 10, y = 6, \text{ or } (10, 6).$$

$$x - y = 4$$
$$(10) - (6) = 4$$
$$10 - 6 = 4 \qquad \text{which is correct.}$$

(c) **First,** note that x- and y-terms are not in the same order.

x-term is first ⇨	$3x - y = -5$
y-term is first ⇨	$y - 5x = 9$

Rearrange terms in the second equation. ⟶

$$
\begin{array}{r}
3x - y = -5 \\
\underline{-5x + y = 9} \\
-2x + 0 = 4 \\
-2x = 4
\end{array}
$$

Then, add like terms to eliminate y.

Next, solve for x. $\qquad\qquad x = -2$

Finally, substitute -2 for x in the
second equation and solve for y.

$$-5(-2) + y = 9$$
$$10 + y = 9$$
$$y = -1$$

The solution is $x = -2, y = -1$, or $(-2, -1)$.

Be sure to check your answer.

(d) **First,** note that like terms are not in the same order in the two equations.

x-term is first ⇨	$\frac{1}{2}x + 2y = 10$
y-term is first ⇨	$y + 1 = \frac{1}{2}x$

Therefore, we must rearrange terms in the second equation to match the order in the first equation.

Subtract 1 from both sides, and
subtract $\frac{1}{2}x$ from both sides. ⟶

Add like terms to eliminate x.

$$
\begin{array}{r}
\frac{1}{2}x + 2y = 10 \\
\underline{-\frac{1}{2}x + y = -1} \\
0 + 3y = 9 \\
3y = 9
\end{array}
$$

Solve for y. $\qquad\qquad\qquad y = 3$

Finally, substitute 3 for y in the first equation, and solve for x.

$$\frac{1}{2}x + 2(3) = 10$$

$$\frac{1}{2}x + 6 = 10$$

$$\frac{1}{2}x = 4$$

$$x = 8 \qquad \text{The solution is } x = 8, y = 3, \text{ or } (8, 3).$$

The check is left to you.

Multiplication with the Elimination Method

With some systems of equations neither x nor y can be eliminated by simply adding like terms.

Example 8

In the system

$$3x + y = 17$$

$$x + y = 7$$

adding like terms will not eliminate either variable. To solve this system of equations, multiply all terms of the second equation by -1 so that

$$\boxed{x + y = 7} \quad \boxed{\text{becomes}} \implies \boxed{-x - y = -7}$$

and the system of equations becomes

$$3x + y = 17$$

$$-x - y = -7$$

The system of equations may now be solved by adding the like terms as before.

$$3x + y = 17$$
$$\underline{-x - y = -7}$$
$$2x + 0 = 10 \qquad \Leftarrow \boxed{\text{Adding like terms eliminates } y.}$$
$$2x = 10$$
$$x = 5$$

Substitute 5 for x in the first equation, and solve for y.

$$3(5) + y = 17$$

$$15 + y = 17$$

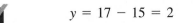 $$y = 17 - 15 = 2 \qquad \text{The solution is } x = 5, y = 2, \text{ or } (5, 2).$$

$$x + y = 7$$

$$(5) + (2) = 7$$

$$5 + 2 = 7 \qquad \text{which is correct.}$$

> **Careful** When you multiply an equation by some number, be careful to multiply *all* terms on *both* sides by that number. It is very easy to forget to multiply on the right side. ◀

Use this "multiply and add" procedure to solve the following system of equations.

$2x + 7y = 29$

$2x + y = 11$

First, multiply all terms in the
second equation by -1. ⟶

Then, add like terms to eliminate x.

Next, solve for y.

$$
\begin{array}{rcr}
2x + 7y = & 29 \\
\underline{-2x - y = } & -11 \\
0 + 6y = & 18 \\
6y = & 18 \\
y = & 3
\end{array}
$$

Finally, substitute 3 for y in the first equation, and solve for x.

$2x + 7(3) = 29$

$2x + 21 = 29$

$2x = 8$

$x = 4$ The solution is $x = 4$, $y = 3$, or $(4, 3)$.

Check the solution by substituting in the original equations.

$$
\begin{array}{ll}
2x + y = 11 & 2x + 7y = 29 \\
2(4) + (3) = 11 & 2(4) + 7(3) = 29 \\
8 + 3 = 11 & 8 + 21 = 29
\end{array}
$$

Solving by the multiply and add procedure may involve multiplying by constants other than -1, of course.

Example 9

Let's use this method to solve the following system of equations.

$7x + 4y = 25$

$3x - 2y = 7$

First, look at these equations carefully. Notice that the y-terms can be eliminated easily if we multiply all terms in the second equation by 2.

The y column $\boxed{\begin{array}{c} +4y \\ -2y \end{array}}$ becomes ⇒ $\boxed{\begin{array}{c} +4y \\ -4y \end{array}}$ when we multiply by 2.

$$\text{sum} = 0$$

The second equation becomes $\boxed{2}\,(3x) - \boxed{2}\,(2y) = \boxed{2}\,(7)$

or $\qquad\qquad\qquad\qquad\qquad 6x - 4y = 14$

and the system of equations is converted to the equivalent system

$$7x + 4y = 25$$
$$\underline{6x - 4y = 14}$$
$$13x + 0 = 39 \qquad \impliedby \boxed{\text{Adding like terms now eliminates } y.}$$
$$13x = 39$$
$$x = 3$$

Substitute 3 for x in the first equation, and solve for y.

$$7(3) + 4y = 25$$
$$21 + 4y = 25$$
$$4y = 4$$
$$y = 1 \qquad \text{The solution is } x = 3, y = 1, \text{ or } (3, 1).$$

Check the solution by substituting it back into the original equations.

→ **More Practice**

Try these problems to make certain you understand this procedure.

(a) $\quad 3x - 2y = 14$ $\qquad$ (b) $\quad 6x + 5y = 14$
$\qquad 5x - 2y = 22$ $\qquad\qquad\quad -2x + 3y = -14$

(c) $\quad 5y - x = 1$ $\qquad\quad$ (d) $\quad -x - 2y = 1$
$\qquad 2x + 3y = 11$ $\qquad\qquad\quad 19 - 2x = -3y$

→ **Solutions**

(a) Multiply each term in the first equation by -1.

$$\boxed{(-1)}\,(3x) - \boxed{(-1)}\,(2y) = \boxed{(-1)}\,(14)$$
$$-3x + 2y = -14$$

The system of equations is therefore

$$-3x + 2y = -14$$
$$\underline{5x - 2y = 22}$$
$$2x + 0 = 8 \qquad \impliedby \boxed{\text{Adding like terms now eliminates } y.}$$
$$x = 4$$

Substitute 4 for x in the first equation, and solve for y.

$$3(4) - 2y = 14$$
$$12 - 2y = 14$$
$$-2y = 2$$
$$-y = 1$$
$$y = -1 \qquad \text{The solution is } x = 4, y = -1, \text{ or } (4, -1).$$

$$3x - 2y = 14 \qquad\qquad 5x - 2y = 22$$
$$3(4) - 2(-1) = 14 \qquad 5(4) - 2(-1) = 22$$
$$12 + 2 = 14 \qquad\qquad 20 + 2 = 22$$
$$14 = 14 \qquad\qquad 22 = 22$$

(b) Multiply each term in the second equation by 3.

$$\boxed{(3)}(-2x) + \boxed{(3)}(3y) = \boxed{(3)}(-14)$$
$$-6x + 9y = -42$$

The system of equations is now

$$\begin{array}{rl} 6x + 5y = & 14 \\ -6x + 9y = & -42 \\ \hline 0 + 14y = & -28 \quad \longleftarrow \text{Adding like terms eliminates } x. \\ y = & -2 \end{array}$$

Substitute -2 for y in the first equation, and solve for x.

$$6x + 5(-2) = 14$$
$$6x - 10 = 14$$
$$6x = 24$$
$$x = 4 \qquad \text{The solution is } x = 4, y = -2, \text{ or } (4, -2).$$

Be certain to check your solution.

(c) Rearrange to put the terms in the first equation in the same order as they are in the second equation.

$$-x + 5y = 1$$
$$2x + 3y = 11$$

Multiply each term in the first equation by 2.

$$\begin{array}{rl} -2x + 10y = & 2 \\ 2x + 3y = & 11 \\ \hline 0 + 13y = & 13 \quad \longleftarrow \text{Adding like terms eliminates } x. \\ 13y = & 13 \\ y = & 1 \end{array}$$

Substitute 1 for y in the first equation, and solve for x.

$$5(1) - x = 1$$
$$5 - x = 1$$
$$-x = 1 - 5 = -4$$
$$x = 4 \qquad \text{The solution is } x = 4, y = 1, \text{ or } (4, 1).$$

Check it.

(d) Rearrange the terms in the second equation in the same order as they are in the first equation. (Subtract 19 from both sides and add $3y$ to both sides.)

$$-x - 2y = 1$$
$$-2x + 3y = -19$$

Multiply each term in the first equation by -2.

$$(-2)(-x) - (-2)(2y) = (-2)(1)$$
$$2x + 4y = -2$$

The system of equations is now

$$\begin{array}{r} 2x + 4y = -2 \\ -2x + 3y = -19 \\ \hline 0 + 7y = -21 \end{array}$$ ⟵ Adding like terms eliminates x.

$$7y = -21$$
$$y = -3$$

Substitute -3 for y in the first equation of the original problem, and solve for x.

$$-x - 2(-3) = 1$$
$$-x + 6 = 1$$
$$-x = 1 - 6 = -5$$
$$x = 5 \qquad \text{The solution is } x = 5, y = -3, \text{ or } (5, -3).$$

Check your solution.

Multiplying Both Equations

If you examine the system of equations

$$3x + 2y = 7$$
$$4x - 3y = -2$$

you will find that there is no single integer we can use as a multiplier that will allow us to eliminate one of the variables when the equations are added. Instead we must convert each equation separately to an equivalent equation, so that when the new equations are added one of the variables is eliminated.

Example 10

With the system of equations given, if we wish to eliminate the y variable, we must multiply the first equation by 3 and the second equation by 2.

First equation: $\boxed{3x + 2y = 7}$ $\boxed{\text{Multiply by 3}}$ ⟹ $\boxed{9x + 6y = 21}$

Second equation: $\boxed{4x - 3y = -2}$ $\boxed{\text{Multiply by 2}}$ ⟹ $\boxed{8x - 6y = -4}$

The new system of equations is

$$\begin{array}{r} 9x + 6y = 21 \\ 8x - 6y = -4 \\ \hline 17x + 0 = 17 \end{array}$$ ⟵ Adding like terms now eliminates y.

$$x = 1$$

Substitute 1 for x in the original first equation to solve for y.

$$3x + 2y = 7$$
$$3(1) + 2y = 7$$
$$3 + 2y = 7$$

$$2y = 4$$
$$y = 2 \qquad \text{The solution is } x = 1, y = 2, \text{ or } (1, 2).$$

Check this solution by substituting it back into the original pair of equations.

$3x + 2y = 7$	$4x - 3y = -2$
$3(1) + 2(2) = 7$	$4(1) - 3(2) = -2$
$3 + 4 = 7$	$4 - 6 = -2$
$7 = 7$	$-2 = -2$

→ **Your Turn**

Use this same procedure to solve the following system of equations:

$$2x - 5y = 9$$
$$3x + 4y = 2$$

→ **Solution**

We can eliminate x from the two equations as follows:

First equation: $\boxed{2x - 5y = 9}$ $\boxed{\text{Multiply by } -3}$ ⟹ $\boxed{-6x + 15y = -27}$

Second equation: $\boxed{3x + 4y = 2}$ $\boxed{\text{Multiply by } 2}$ ⟹ $\boxed{6x + 8y = 4}$

The new system of equations is

$$-6x + 15y = -27$$
$$\underline{6x + 8y = 4}$$
$$0 + 23y = -23 \qquad \Leftarrow \boxed{\text{Adding like terms now eliminates } x.}$$
$$y = -1$$

Substitute -1 for y in the first original equation to solve for x.

$$2x - 5y = 9$$
$$2x - 5(-1) = 9$$
$$2x + 5 = 9$$
$$2x = 4$$
$$x = 2 \qquad \text{The solution is } x = 2, y = -1, \text{ or } (2, -1).$$

▶ **A Closer Look** Of course, we could have chosen to eliminate the y-variable and we would have arrived at the same solution. To do this, we would multiply the top equation by 4 and the bottom equation by 5. Try it. ◀

→ **More Practice**

When you are ready to continue, practice your new skills by solving the following systems of equations.

(a) $2x + 2y = 4$ (b) $5x + 2y = 11$

      $5x + 7y = 18$ $6x - 3y = 24$

(c) $3x + 2y = 10$ (d) $-7x - 13 = 2y$

$\quad\quad 2x = 5y - 25$ $\quad\quad 3y + 4x = 0$

→ **Answers**

(a) $x = -2, y = 4$ (b) $x = 3, y = -2$

(c) $x = 0, y = 5$ (d) $x = -3, y = 4$

In (a) multiply the first equation by -5 and the second equation by 2.

In (b) multiply the first equation by 3 and the second equation by 2.

In (c) rearrange the terms of the second equation, then multiply the first equation by -2 and the second equation by 3.

In (d) rearrange the terms of the first equation to agree with the second equation, then multiply the first equation by 3 and the second equation by 2.

Word Problems

In many practical situations, not only must you be able to solve a system of equations, you must also be able to write the equations in the first place. You must be able to set up and solve word problems. In Chapter 7 we listed some "signal words" and showed how to translate English sentences and phrases to mathematical equations and expressions. If you need to review the material on word problems in Chapter 7, turn to page 445 now; otherwise, continue here.

Example 11

Translate the following sentence into *two* equations.

The sum of two numbers is 26 and their difference is 2. (Let x and y represent the two numbers.)

The <u>sum of two numbers</u> is 26   . . . <u>their difference</u> is 2.

$$x + y \qquad = 26 \qquad\qquad x - y \qquad = 2$$

The two equations are

$x + y = 26$

$x - y = 2$

We can solve this pair of equations using the elimination method.

The solution is $x = 14$, $y = 12$. Check the solution by seeing if it fits the original problem. The sum of these numbers is 26 ($14 + 12 = 26$) and their difference is 2 ($14 - 12 = 2$).

→ **Your Turn**

Ready for another word problem? Translate and solve this one:

The difference of two numbers is 14, and the larger number is three more than twice the smaller number.

→ **Solution**

The first phrase in the sentence should be translated as

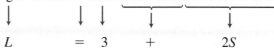

and the second phrase should be translated as

"... the larger number is three more than twice the smaller ..."

$$L = 3 + 2S$$

The system of equations is

$$L - S = 14$$
$$L = 3 + 2S$$

To solve this system of equations, substitute the value of L from the second equation into the first equation. Then the first equation becomes

$$(3 + 2S) - S = 14$$

or

$$3 + S = 14$$
$$S = 11$$

Now substitute this value of S into the first equation to find L.

$$L - (11) = 14$$
$$L = 25 \qquad \text{The solution is } L = 25, S = 11.$$

Check the solution by substituting it back into both of the original equations. Never neglect to check your answer.

___

→ **More Practice**

Translating word problems into systems of equations is a very valuable and very practical algebra skill. Translate each of the following problems into a system of equations, then solve.

(a) **Machine Trades** The total value of an order of nuts and bolts is $1.40. The nuts cost 5 cents each and the bolts cost 10 cents each. If the number of bolts is four more than twice the number of nuts, how many of each are there? (*Hint:* Keep all money values in cents to avoid decimals.)

(b) **Landscaping** The perimeter of a rectangular lot is 350 ft. The length of the lot is 10 ft more than twice the width. Find the dimensions of the lot.

(c) **Construction** A materials yard wishes to make a 900-cu ft mixture of two different types of rock. One type of rock costs $6.50 per cubic foot and the other costs $8.50 per cubic foot. If the cost of the mixture is to be $6500, how many cubic feet of each should go into the mixture?

(d) **Allied Health**   A lab technician wishes to mix a 5% salt solution and a 15% salt solution to obtain 4 liters of a 12% salt solution. How many liters of each solution must be added?

→ **Solutions**

(a)   In problems of this kind it is often very helpful to first set up a table:

Item	Number of Items	Cost per Item	Total Cost
Nuts	N	5	$5N$
Bolts	B	10	$10B$

We can write the first equation as

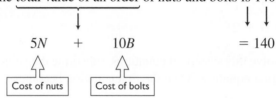

"The total value of an order of nuts and bolts is 140 cents."

$$5N \;+\; 10B \qquad\qquad = 140$$

Cost of nuts   Cost of bolts

The second equation would be

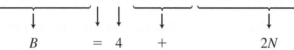

"... the number of bolts is four more than twice the number of nuts ..."

$$B \;=\; 4 \;+\; 2N$$

The system of equations to be solved is

$5N + 10B = 140$

$\qquad B = 4 + 2N$

Use substitution. Since B is equal to $4 + 2N$, replace B in the first equation with $4 + 2N$. Then solve for N.

$$5N + 10\,(4 + 2N) = 140$$

Remove parentheses. $5N + 40 + 20N = 140$

Add like terms. $25N + 40 = 140$

Subtract 40. $25N = 100$

Divide by 25. $N = 4$

When we replace N with 4 in the second equation,

$B = 4 + 2(4)$

$\quad = 4 + 8$

$\quad = 12$

There are 4 nuts and 12 bolts.

✔

$5N + 10B = 140 \qquad B = 4 + 2N$

$5(4) + 10(12) = 140 \qquad 12 = 4 + 2(4)$

$\quad 20 + 120 = 140 \qquad 12 = 4 + 8$

$\qquad\qquad 140 = 140 \qquad 12 = 12$

Chapter 11 Advanced Algebra

(b) Recalling the formula for the perimeter of a rectangle, we have

"The perimeter of a rectangular lot is 350 ft."

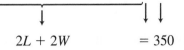

$$2L + 2W = 350$$

The second sentence gives us

"The length of the lot is 10 ft more than twice the width."

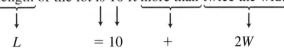

$$L = 10 + 2W$$

The system of equations to be solved is

$$2L + 2W = 350$$

$$L = 10 + 2W$$

Using substitution, we replace L with $\boxed{10 + 2W}$ in the first equation. Then solve for W.

$$
\begin{aligned}
2\,\boxed{(10 + 2W)} + 2W &= 350 \\
20 + 4W + 2W &= 350 \\
20 + 6W &= 350 \\
6W &= 330 \\
W &= 55 \text{ ft}
\end{aligned}
$$

Now substitute 55 for W in the second equation:

$$
\begin{aligned}
L &= 10 + 2(55) \\
&= 10 + 110 \\
&= 120 \text{ ft}
\end{aligned}
$$

The lot is 120 ft long and 55 ft wide.

✓
$$
\begin{array}{ll}
2L + 2W = 350 & L = 10 + 2W \\
2(120) + 2(55) = 350 & 120 = 10 + 2(55) \\
240 + 110 = 350 & 120 = 10 + 110 \\
350 = 350 & 120 = 120
\end{array}
$$

(c) First set up the following table:

Item	Amount (cu ft)	Cost per cu ft	Total Cost
Cheaper rock	x	$6.50	$6.50x$
More expensive rock	y	$8.50	$8.50y$

The first equation comes from the statement:

". . . a 900-cu ft mixture of two different types of rock."

$$900 = x + y$$

Consulting the table, we write the second equation as follows:

". . . the cost of the mixture is to be $6500."

$$6.50x + 8.50y = 6500$$

Multiply this last equation by 10 to get the system of equations

$$x + y = 900$$

$$65x + 85y = 65{,}000$$

Multiply the first equation by -65 and add it to the second equation:

$$-65x + (-65y) = -58{,}500$$

$$\underline{65x + 85y = 65{,}000}$$

$$20y = 6500$$

$$y = 325 \text{ cu ft}$$

Replacing y with 325 in the first equation, we have:

$$x + (325) = 900$$

$$x = 575 \text{ cu ft}$$

There should be 325 cu ft of the $8.50 per cu ft mixture, and 575 cu ft of the $6.50 per cu ft mixture.

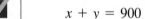

$x + y = 900$	$6.50x + 8.50y = 6500$
$575 + 325 = 900$	$6.50(575) + 8.50(325) = 6500$
$900 = 900$	$3737.5 + 2762.5 = 6500$
	$6500 = 6500$

(d)

Solution	Amount (liters)	Salt Fraction	Total Salt
5%	A	0.05	$0.05A$
15%	B	0.15	$0.15B$
12%	4	0.12	$0.12(4)$

The final solution is to contain 4 liters, so we have

$$A + B = 4$$

The second equation represents the total amount of salt:

$$0.05A + 0.15B = (0.12)4$$

Multiplying this last equation by 100 to eliminate the decimals, we have the system

$$A + B = 4$$

$$5A + 15B = 48$$

To solve this system, multiply each term in the first equation by -5 and add to get

$$-5A + (-5B) = -20$$

$$\underline{5A + 15B = 48}$$

$$10B = 28$$

$$B = 2.8 \text{ liters}$$

Substituting back into the first equation, we have

$$A + 2.8 = 4$$

$$A = 1.2 \text{ liters}$$

The technician must mix 1.2 liters of the 5% solution with 2.8 liters of the 15% solution.

$$A + B = 4 \qquad\qquad 5A + 15B = 48$$
$$1.2 + 2.8 = 4 \qquad 5(1.2) + 15(2.8) = 48$$
$$4 = 4 \qquad\qquad 6 + 42 = 48$$
$$48 = 48$$

Now turn to Exercises 11-1 for a set of problems on systems of equations.

Exercises 11-1 Systems of Equations

A. Solve each of the following systems of equations using the method of substitution. If the system is inconsistent or dependent, say so.

1. $y = 10 - x$
   $2x - y = -4$

2. $3x - y = 5$
   $2x + y = 15$

3. $2x - y = 3$
   $x - 2y = -6$

4. $2y - 4x = -3$
   $y = 2x + 4$

5. $3x + 5y = 26$
   $x + 2y = 10$

6. $x = 10y + 1$
   $y = 10x + 1$

7. $3x + 4y = 5$
   $x - 2y = -5$

8. $2x - y = 4$
   $4x - 3y = 11$

B. Solve each of the following systems of equations. If the system is inconsistent or dependent, say so.

1. $x + y = 5$
   $x - y = 13$

2. $2x + 2y = 10$
   $3x - 2y = 10$

3. $2y = 3x + 5$
   $2y = 3x - 7$

4. $2y = 2x + 2$
   $4y = 5 + 4x$

5. $x = 3y + 7$
   $x + y = -5$

6. $3x - 2y = -11$
   $x + y = -2$

7. $5x - 4y = 1$
   $3x - 6y = 6$

8. $y = 3x - 5$
   $6x - 3y = 3$

9. $y - 2x = -8$
   $x - \frac{1}{2}y = 4$

10. $x + y = a$
    $x - y = b$

11. $3x + 2y = 13$
    $5x + 3y = 20$

12. $3x + 4y = 11$
    $2x + 2y = 7$

C. Practical Problems

Translate each problem statement into a system of equations and solve.

1. The sum of two numbers is 39 and their difference is 7. What are the numbers?

2. The sum of two numbers is 14. The larger is two more than three times the smaller. What are the numbers?

3. Separate a collection of 20 objects into two parts so that twice the larger amount equals three times the smaller amount.

4. The average of two numbers is 25 and their difference is 8. What are the numbers?

5. **Automotive Trades** Four bottles of leather cleaner and three bottles of protectant cost $44. Three bottles of leather cleaner and four bottles of protectant cost $47. How much does a single bottle of each cost?

6. **Carpentry** The perimeter of a rectangular window is 14 ft, and its length is 2 ft less than twice the width. What are the dimensions of the window?

7. Harold exchanged a $1 bill for change and received his change in nickels and dimes, with seven more dimes than nickels. How many of each coin did he receive?

8. If four times the larger of two numbers is added to three times the smaller, the result is 26. If three times the larger number is decreased by twice the smaller, the result is 11. Find the numbers.

9. **Office Services** An office manager bought three cases of paper and five ink cartridges, but he forgot what each cost. He knows that the total cost was $167, and he recalls that a case of paper cost $13 more than an ink cartridge. What was the cost of a single unit of each?

10. **Sheet Metal Trades** The length of a piece of sheet metal is twice the width. The difference in length and width is 20 in. What are the dimensions?

11. **Electrical Trades** A 30-in. piece of wire is to be cut into two parts, one part being four times the length of the other. Find the length of each.

12. **Painting** A painter wishes to mix paint worth $30 per gallon with paint worth $36 per gallon to make a 12-gal mixture worth $33.50 per gallon. How many gallons of each should he mix?

13. **Allied Health** A lab technician wishes to mix a 10% salt solution with a 2% salt solution to obtain 6 liters of a 4% salt solution. How many liters of each should be added?

14. **Landscaping** The perimeter of a rectangular field is 520 ft. The length of the field is 20 ft more than three times the width. Find the dimensions.

15. **Landscaping** A 24-ton mixture of crushed rock is needed in a construction job; it will cost $800. If the mixture is composed of rock costing $30 per ton and $40 per ton, how many tons of each should be added?

When you have completed these problems check your answers to the odd-numbered problems in the Appendix. Then turn to Section 11-2 to learn about quadratic equations.

11-2 Quadratic Equations

Thus far in your study of algebra you have worked only with linear equations. In a linear equation, the variable appears only to the first power. For example, $3x + 5 = 2$ is a linear equation. The variable appears as x or x^1. No powers of x such as x^2, x^3, or x^4 appear in the equations.

An equation in which the variable appears to the second power, but to no higher power, is called a *quadratic equation*.

→ **Your Turn**

Which of the following are quadratic equations?

(a) $x^2 = 49$

(b) $2x - 1 = 4$

(c) $3x - 2y = 19$

(d) $5x^2 - 8x + 3 = 0$

(e) $x^3 + 3x^2 + 3x + 1 = 0$

Equations (a) and (d) are quadratic equations. Equation (b) is a linear equation in one variable. Equation (c) is a linear equation in two variables, x and y. Equation (e) is a cubic or third-order equation. Because (e) contains an x^3 term, it is not a quadratic.

Standard Form

Every quadratic equation can be put into a *standard quadratic form*.

$$ax^2 + bx + c = 0 \qquad \text{where } a \text{ cannot equal zero}$$

| x^2 term | x term | Constant term |

Every quadratic equation must have an x^2 term, but the x term or the constant term may be missing.

Example 1

$2x^2 + x - 5 = 0$ is a quadratic equation in standard form: the x^2 term is first, the x term second, and the constant term last on the left.

$x^2 + 4 = 0$ is also a quadratic equation in standard form. The x term is missing, but the other terms are in the proper order. We could rewrite this equation as

$$x^2 + 0 \cdot x + 4 = 0$$

→ **Your Turn**

Which of the following quadratic equations are written in standard form?

(a) $7x^2 - 3x + 6 = 0$ (b) $8x - 3x^2 - 2 = 0$ (c) $2x^2 - 5x = 0$
(d) $x^2 = 25$ (e) $x^2 - 5 = 0$ (f) $4x^2 - 5x = 6$

→ **Answers**

Equations (a), (c), and (e) are in standard form.

To solve a quadratic equation, it may be necessary to rewrite it in standard form.

Example 2

The equation

$x^2 = 25x$	becomes	$x^2 - 25x = 0$	in standard form
$8x - 3x^2 - 2 = 0$	becomes	$-3x^2 + 8x - 2 = 0$	in standard form
	or	$3x^2 - 8x + 2 = 0$	if we multiply all terms by -1
$4x^2 - 5x = 6$	becomes	$4x^2 - 5x - 6 = 0$	in standard form

In each case we add or subtract a term on both sides of the equation until all terms are on the left, then rearrange terms until the x^2 is first on the left, the x term next, and the constant term third.

Try it. Rearrange the following quadratic equations in standard form.

(a) $5x - 19 + 3x^2 = 0$ (b) $7x^2 = 12 - 6x$

(c) $9 = 3x - x^2$ (d) $2x - x^2 = 0$

(e) $5x = 7x^2 - 12$ (f) $x^2 - 6x + 9 = 49$

(g) $3x + 1 = x^2 - 5$ (h) $1 - x^2 + x = 3x + 4$

→ **Answers**

(a) $3x^2 + 5x - 19 = 0$ (b) $7x^2 + 6x - 12 = 0$

(c) $x^2 - 3x + 9 = 0$ (d) $-x^2 + 2x = 0$ or $x^2 - 2x = 0$

(e) $7x^2 - 5x - 12 = 0$ (f) $x^2 - 6x - 40 = 0$

(g) $x^2 - 3x - 6 = 0$ (h) $x^2 + 2x + 3 = 0$

Solutions to Quadratic Equations

The solution to a linear equation is a single number. The solution to a quadratic equation is usually a *pair* of numbers, each of which satisfies the equation.

Example 3

The quadratic equation

$$x^2 - 5x + 6 = 0$$

has the solutions

$$x = 2 \quad \text{or} \quad x = 3$$

To see that either 2 or 3 is a solution, substitute each into the equation.

For $x = 2$ For $x = 3$

$(2)^2 - 5(2) + 6 = 0$ $(3)^2 - 5(3) + 6 = 0$

$4 - 10 + 6 = 0$ $9 - 15 + 6 = 0$

$10 - 10 = 0$ $15 - 15 = 0$

→ **Your Turn**

Show that $x = 5$ or $x = 3$ gives a solution of the quadratic equation

$$x^2 - 8x + 15 = 0$$

→ **Solution**

$x^2 - 8x + 15 = 0$

For $x = 5$ For $x = 3$

$(5)^2 - 8(5) + 15 = 0$ $(3)^2 - 8(3) + 15 = 0$

$25 - 40 + 15 = 0$ $9 - 24 + 15 = 0$

$40 - 40 = 0$ $24 - 24 = 0$

Solving $x^2 - c = 0$ The easiest kind of quadratic equation to solve is one in which the x term is missing. Rather than write these in standard form, put the x^2-term on one side of the equation and the constant term c on the other side.

Example 4

To solve the quadratic equation

$$x^2 - 25 = 0$$

add 25 to both sides to rewrite it as

$$x^2 = 25$$

and take the square root of both sides of the equation.

$$\sqrt{x^2} = \sqrt{25}$$

or $x = \pm\sqrt{25}$

or $x = 5$ or $x = -5$

Both 5 and -5 satisfy the original equation.

For $x = 5$ | | For $x = -5$
$x^2 - 25 = 0$ | and | $x^2 - 25 = 0$
$(5)^2 - 25 = 0$ | | $(-5)^2 - 25 = 0$
$25 - 25 = 0$ | | $25 - 25 = 0$

► **Careful**   Every positive number has two square roots, one positive and the other negative. Both of them may be important in solving a quadratic equation. ◄

→ **Your Turn**

Solve each of the following quadratic equations and check *both* solutions. (Round to two decimal places if necessary.)

(a) $x^2 - 36 = 0$ (b) $x^2 = 8$ (c) $x^2 - 1 = 2$

(d) $4x^2 = 81$ (e) $3x^2 = 27$

(f) **Machine Trades**   The power P (in watts) required by a metal punch machine to punch six holes at a time is given by the formula

$$P = 3.17t^2$$

where t is the thickness of the metal being punched. What thickness can be punched using 0.750 watt of power?

→ **Solutions**

(a) $x^2 = 36$ (b) $x^2 = 8$

$x = \pm\sqrt{36}$ $x = \pm\sqrt{8}$

$x = 6$ or $x = -6$ $x \approx 2.83$ or $x \approx -2.83$ rounded

(c) $x^2 = 3$

$x = \pm\sqrt{3}$

$x \approx 1.73$ or $x \approx -1.73$

(d) $4x^2 = 81$

$x^2 = \dfrac{81}{4}$

$x = \pm\sqrt{\dfrac{81}{4}}$

$x = \dfrac{9}{2}$ or $x = -\dfrac{9}{2}$

(e) $3x^2 = 27$

$x^2 = 9$

$x = \pm\sqrt{9}$

$x = 3$ or $x = -3$

(f) First, substitute 0.750 for P:

$0.750 = 3.17t^2$

Then, solve the equation

$t^2 = \dfrac{0.750}{3.17}$

$t^2 \approx 0.2366$

$t \approx 0.486$

Because thickness cannot be negative, we need only consider the positive square root. The given power will allow the machine to punch holes in metal 0.486 in. thick.

Notice in each case that first we rewrite the equation so that x^2 appears alone on the left and a number appears alone on the right. Second, take the square root of both sides. The equation will have two solutions.

You should also notice that the equation

$x^2 = -4$

has no solution. There is no number x whose square is a negative number.

→ **More Practice**

Solve the following quadratic equations. (Round to two decimal places if necessary.)

(a) $x^2 - 3.5 = 0$ (b) $x^2 = 18$ (c) $6 - x^2 = 0$

(d) $9x^2 = 49$ (e) $7x^2 = 80$ (f) $\dfrac{3x^2}{5} = 33.3$

→ **Solutions**

(a) $x^2 = 3.5$

$x = \pm\sqrt{3.5}$

$x \approx 1.87$ or $x \approx -1.87$

(b) $x^2 = 18$

$x = \pm\sqrt{18}$

$x \approx 4.24$ or $x \approx -4.24$

(c) $6 - x^2 = 0$

$$x^2 = 6$$

$$x = \pm\sqrt{6}$$

$$x \approx 2.45 \quad \text{or} \quad x \approx -2.45$$

(d) $9x^2 = 49$

$$x^2 = \frac{49}{9}$$

$$x = \pm\sqrt{\frac{49}{9}}$$

$$x = \frac{7}{3} \quad \text{or} \quad x = -\frac{7}{3}$$

(e) $7x^2 = 80$

$$x^2 = \frac{80}{7}$$

$$x = \pm\sqrt{\frac{80}{7}}$$

$$x \approx \pm\sqrt{11.42857}$$

$$x \approx 3.38 \quad \text{or} \quad x \approx -3.38$$

(f) $\dfrac{3x^2}{5} = 33.3$

$$x^2 = \frac{(33.3)(5)}{3} = 55.5$$

$$x = \pm\sqrt{55.5}$$

$$x \approx 7.45 \quad \text{or} \quad x \approx -7.45$$

You can do problem (f) in one of the following two ways on a calculator:

C11-1

33.3 ⊠ **5** ⊡ **3** ⊟ ☑ [ANS] ⊟ → `7.449832213`

or ☑ * **33.3** ⊠ **5** ⊡ **3** ⊟ → `7.449832213`

———

*Parentheses must be opened here.

Quadratic Formula

The method just explained will work only for quadratic equations in which the x-term is missing. A more general method that will work for *all* quadratic equations is to use the *quadratic formula*. The solution of any quadratic equation

$$ax^2 + bx + c = 0$$

is

$$x = \frac{-b + \sqrt{b^2 - 4ac}}{2a} \quad \text{or} \quad x = \frac{-b - \sqrt{b^2 - 4ac}}{2a}$$

or

THE QUADRATIC FORMULA

$$x = \frac{-b \pm \sqrt{b^2 - 4ac}}{2a}$$

Careful Note that the entire expression in the numerator is divided by $2a$. ◄

Example 5

To solve the quadratic equation

$$2x^2 + 5x - 3 = 0$$

follow these steps.

Step 1 *Identify* the coefficients a, b, and c for the quadratic equation.

$$2x^2 + 5x - 3 = 0$$

$$\boxed{a = 2} \qquad \boxed{b = 5} \qquad \boxed{c = -3}$$

Step 2 *Substitute* these values of a, b, and c into the quadratic formula.

$$x = \frac{-(5) \pm \sqrt{(5)^2 - 4(2)(-3)}}{2(2)}$$

The $\pm$ sign means that there are two solutions, one to be calculated using the $+$ sign and the other solution calculated using the $-$ sign.

Step 3 *Simplify* these equations for x.

$$x = \frac{-5 \pm \sqrt{25 + 24}}{4}$$

$$= \frac{-5 \pm \sqrt{49}}{4}$$

$$= \frac{-5 \pm 7}{4}$$

$$x = \frac{-5 + 7}{4} \quad \text{or} \quad x = \frac{-5 - 7}{4}$$

$$x = \frac{2}{4} = \frac{1}{2} \quad \text{or} \quad x = \frac{-12}{4} = -3$$

The solution is $x = \frac{1}{2}$ or $x = -3$.

Step 4 *Check* the solution numbers by substituting them into the original equation.

 $2x^2 + 5x - 3 = 0$

For $x = \frac{1}{2}$ For $x = -3$

$$2\left(\frac{1}{2}\right)^2 + 5\left(\frac{1}{2}\right) - 3 = 0 \qquad\qquad 2(-3)^2 + 5(-3) - 3 = 0$$

$$2\left(\frac{1}{4}\right) + 5\left(\frac{1}{2}\right) - 3 = 0 \qquad\qquad 2(9) + 5(-3) - 3 = 0$$

$$\frac{1}{2} + 2\frac{1}{2} - 3 = 0 \qquad\qquad\qquad 18 - 15 - 3 = 0$$

$$3 - 3 = 0 \qquad\qquad\qquad\qquad 18 - 18 = 0$$

→ **Your Turn**

Use the quadratic formula to solve $x^2 + 4x - 5 = 0$.

→ **Solution**

Step 1 $x^2 + 4x - 5 = 0.$

$$\boxed{a = 1} \qquad \boxed{b = 4} \qquad \boxed{c = -5}$$

Step 2 $x = \dfrac{-(4) \pm \sqrt{(4)^2 - 4(1)(-5)}}{2(1)}$

Step 3   Simplify

$$x = \dfrac{-4 \pm \sqrt{36}}{2}$$

$$= \dfrac{-4 \pm 6}{2}$$

$x = \dfrac{-4 + 6}{2}$ or $x = \dfrac{-4 - 6}{2}$

$x = 1$ or $x = -5$ The solution is $x = 1$ or $x = -5$.

Step 4   ✓ $x^2 + 4x - 5 = 0$

For $x = 1$	For $x = -5$
$(1)^2 + 4(1) - 5 = 0$	$(-5)^2 + 4(-5) - 5 = 0$
$1 + 4 - 5 = 0$	$25 - 20 - 5 = 0$
$5 - 5 = 0$	$25 - 25 = 0$

Example 6

Here is one that is a bit tougher. To solve $3x^2 - 7x = 5$,

Step 1   Rewrite the equation in standard form.

$$3x^2 \ - \ 7x \ - \ 5 \ = 0$$

$a = 3$ $b = -7$ $c = -5$

$-(-7) = +7$

Step 2 $x = \dfrac{-(-7) \pm \sqrt{(-7)^2 - 4(3)(-5)}}{2(3)}$

$(-7)^2 = (-7) \cdot (-7) = 49$

Step 3   Simplify

$$x = \dfrac{7 \pm \sqrt{109}}{6}$$

$x = \dfrac{7 + \sqrt{109}}{6}$ or $x = \dfrac{7 - \sqrt{109}}{6}$

$\approx \dfrac{7 + 10.44}{6}$ or $\approx \dfrac{7 - 10.44}{6}$ rounded to two decimal places

$\approx \dfrac{17.44}{6}$ or $\approx \dfrac{-3.44}{6}$

≈ 2.91 or ≈ -0.57 The solution is

$x \approx 2.91$ or $x \approx -0.57$.

Step 4   Check both answers by substituting them back into the original quadratic equation. Use the following calculator sequence to find this solution.

To perform Step 3 efficiently using a calculator, compute $\sqrt{b^2 - 4ac}$ first and store it in memory. That way, you can recall it to more easily find the second answer. Also note that $(-7)^2$ is the same as $(+7)^2$, so enter the positive version to avoid having to square a negative. Here is the entire recommended sequence for finding both solutions:

C11-2

$\boxed{\sqrt{}}$ *$\mathbf{7}$ $\boxed{x^2}$ $\boxed{-}$ $\mathbf{4}$ $\boxed{\times}$ $\mathbf{3}$ $\boxed{\times}$ $\boxed{(-)}$ $\mathbf{5}$ $\boxed{=}$ $\boxed{\text{STO}}$ $\boxed{\text{A}}$ $\boxed{+}$ $\mathbf{7}$ $\boxed{=}$ $\boxed{\div}$ $\mathbf{6}$ $\boxed{=}$

$\rightarrow$ $\boxed{\textbf{2.906717751}}$ ≈ 2.91 (First answer)

*Parentheses must be opened here.

$\mathbf{7}$ $\boxed{-}$ $\boxed{\text{RCL}}$ $\boxed{\text{A}}$ $\boxed{=}$ $\boxed{\div}$ $\mathbf{6}$ $\boxed{=}$ $\rightarrow$ $\boxed{\textbf{-0.573384418}}$ ≈ -0.57 (Second answer)

$\underbrace{\qquad\qquad}_{\sqrt{b^2 - 4ac}}$

→ **More Practice**

Use the quadratic formula to solve each of the following equations. (Round to two decimal places if necessary.)

(a) $6x^2 - 13x + 2 = 0$ (b) $3x^2 - 13x = 0$

(c) $2x^2 - 5x + 17 = 0$ (d) $8x^2 = 19 - 5x$

→ **Solutions**

(a) $6x^2 - 13x + 2 = 0$

$a = 6$ $b = -13$ $c = 2$

$$x = \frac{-(-13) \pm \sqrt{(-13)^2 - 4(6)(2)}}{2(6)}$$

$$x = \frac{13 \pm \sqrt{121}}{12} = \frac{13 \pm 11}{12}$$

The solution is $x = \dfrac{13 + 11}{12}$ or $x = \dfrac{13 - 11}{12}$

$\qquad\qquad\qquad = 2$ or $= \dfrac{1}{6}$

The solution is $x = 2$ or $x = \dfrac{1}{6}$.

Check it.

(b) $3x^2 - 13x = 0$

$3x^2 - 13x + 0 = 0$

$a = 3$ $b = -13$ $c = 0$

$$x = \frac{-(-13) \pm \sqrt{(-13)^2 - 4(3)(0)}}{2(3)}$$

$$= \frac{13 \pm \sqrt{169}}{6}$$

The solution is $x = \dfrac{13 + 13}{6}$ or $x = \dfrac{13 - 13}{6}$

$\qquad\qquad\qquad = \dfrac{13}{3}$ or $= 0$

The solution is $x = 0$ or $x = \dfrac{13}{3}$.

(c) $2x^2 \quad - \quad 5x \quad + \quad 17 \quad = 0$

$\boxed{a = 2} \quad \boxed{b = -5} \quad \boxed{c = 17}$

$$x = \frac{-(-5) \pm \sqrt{(-5)^2 - 4(2)(17)}}{2(2)}$$

$$= \frac{5 \pm \sqrt{-111}}{4}$$

But the square root of a negative number is not acceptable if our answer must be a real number. This quadratic equation has no real number solution.

(d) $8x^2 = 19 - 5x$ or, in standard form

$8x^2 \quad + \quad 5x \quad - \quad 19 \quad = 0$

$\boxed{a = 8} \quad \boxed{b = 5} \quad \boxed{c = -19}$

$$x = \frac{-(5) \pm \sqrt{(5)^2 - 4(8)(-19)}}{2(8)}$$

$$= \frac{-5 \pm \sqrt{633}}{16}$$

$$\approx \frac{-5 \pm 25.159}{16} \qquad \text{Find the square root to three decimal places to ensure an accurate answer.}$$

The solution is $x \approx -1.88$ or $x \approx 1.26$ rounded

When you check a solution that includes a rounded value, the check may not give an exact fit. The differences should be very small if you have the correct solution.

Word Problems and Quadratic Equations

Example 7

Electrical Trades   In a dc circuit the power P dissipated in the circuit is given by the equation

$$P = RI^2$$

where R is the circuit resistance in ohms and I is the current in amperes. What current will produce 1440 watts of power in a 10-ohm resistor?

Substituting into the equation yields

$$1440 = 10I^2$$

or $I^2 = 144$

$$I = \pm\sqrt{144}$$

The solution is $I = 12$ amperes or $I = -12$ amperes.

Example 8

In many practical situations, it is necessary to determine what size circle has the same area as two or more smaller circles.

Plumbing Suppose we must determine the diameter of a water main with the same total area as two mains with diameters of 3 in. and 4 in. Because the problem involves diameters, we will use the formula,

Area of a circle $\approx 0.7854d^2$

We then write the following equation:

$$\underbrace{0.7854d^2}_{\substack{\text{Area of the} \\ \text{large main}}} = \underbrace{0.7854(3)^2 + 0.7854(4)^2}_{\substack{\text{Sum of the areas of the} \\ \text{two smaller mains}}}$$

Solving this quadratic equation for d,

$$0.7854d^2 = 7.0686 + 12.5664$$

$$0.7854d^2 = 19.635$$

$$d^2 = 25$$

$$d = 5 \text{ in.}$$

Only the positive square root is a valid solution.

Example 9

HVAC One side of a rectangular opening for a heating pipe is 3 in. longer than the other side. The total cross-sectional area is 70 sq in. To find the dimensions of the cross section

Let L = length, W = width.

Then $L = 3 + W$

and the area is

$$\text{area} = LW$$

or $\quad 70 = LW$

Substituting $L = \boxed{3 + W}$ into the area equation, we have

$$70 = \boxed{(3 + W)} \; W$$

or $\quad 70 = 3W + W^2$

or $\quad W^2 \; + \; 3W \; - \; 70 = 0 \qquad$ in standard quadratic form.

$$\boxed{a = 1} \quad \boxed{b = 3} \quad \boxed{c = -70}$$

To find the solution, substitute a, b, and c into the quadratic formula.

$$W = \frac{-(3) \pm \sqrt{(3)^2 - 4(1)(-70)}}{2(1)}$$

or $\quad W = \dfrac{-3 \pm \sqrt{289}}{2}$

$$= \frac{-3 \pm 17}{2}$$

The solution is

$$W = -10 \qquad \text{or} \qquad W = 7$$

Only the positive value makes sense. The answer is $W = 7$ in.

Substituting back into the equation $L = 3 + W$, we find that $L = 10$ in.

Check to see that the area is indeed 70 sq in.

→ Your Turn

In Chapter 9, the total surface area of a cylinder was found using the formula

$$S = 2\pi r^2 + 2\pi rh$$

Find the radius of a 15-in.-high cylinder with a total surface area of 339 sq in. Use $\pi \approx 3.14$ and round to the nearest inch.

→ Solution

First, substitute the given information into the formula:

$$S = 2\pi r^2 + 2\pi rh$$

$$339 = 2(3.14)r^2 + 2(3.14)r(15)$$

$$339 = 6.28r^2 + 94.2r$$

Then, subtract 339 from both sides and solve using the quadratic formula:

$$0 = 6.28r^2 + 94.2r - 339$$

Here $a = 6.28$, $b = 94.2$, and $c = -339$

$$r = \frac{-94.2 \pm \sqrt{(94.2)^2 - 4(6.28)(-339)}}{2(6.28)}$$

$$r = 2.999 \ldots \approx 3 \text{ in.} \quad \text{or} \quad r = -17.999 \ldots \approx -18 \text{ in.}$$

A negative radius cannot exist, so the radius of the cylinder is approximately 3 in.

→ More Practice

Solve the following problems. (Round each answer to two decimal places if necessary.)

(a) Find the side length of a square whose area is 200 sq m.

(b) **Sheet Metal Trades** The cross-sectional area of a rectangular duct must be 144 sq in. If one side must be twice as long as the other, find the length of each side.

(c) **Plumbing** Find the diameter of a circular pipe whose cross-sectional area is 3.00 sq in.

(d) **Automotive Trades** The SAE rating of an engine is given by $R = \dfrac{D^2 N}{2.5}$, where D is the bore of the cylinder in inches, and N is the number of cylinders. What must the bore be for an eight-cylinder engine to have an SAE rating of 33.8?

(e) **Metalworking** One side of a rectangular plate is 6 in. longer than the other. The total area is 216 sq in. How long is each side?

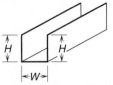

Problem (f)

(f) **Hydrology**   A field worker has a long strip of sheet steel 12 ft wide. She wishes to make an open-topped water channel with a rectangular cross section. If the cross-sectional area must be 16 ft², what should the dimensions of the channel be? (*Hint:* $H + H + W = 12$ ft.)

→ **Solutions**

(a) $A = s^2$

$200 = s^2$

$s = \pm\sqrt{200}$

≈ 14.14 m, rounded   The negative solution is not possible.

(b) $L \cdot W = A$

and $L = 2W$

Therefore, $(2W)\,W = A$

or $2W^2 = A$

$2W^2 = 144$

$W^2 = 72$

$W = +\sqrt{72}$

$\approx 8.485 \approx 8.49$ in.   rounded   for the positive root

Then $L = 2W$

or $L \approx 16.98$ in.   rounded

(c) $A = \dfrac{\pi d^2}{4}$

$3.00 = \dfrac{\pi d^2}{4}$

or $d^2 = \dfrac{4(3.00)}{\pi}$

$d^2 = 3.8197\dots$

$d \approx +\sqrt{3.8197}$

≈ 1.95 in.   rounded

C11-3 $\boxed{\checkmark}\ 4\ \boxed{\times}\ 3\ \boxed{\div}\ \boxed{\pi}\ \boxed{=} \rightarrow$ `1.954410048`

(d) $R = \dfrac{D^2 N}{2.5}$

$33.8 = \dfrac{D^2 \cdot 8}{2.5}$

or $D^2 = \dfrac{(33.8)(2.5)}{8}$

$D^2 = 10.5625$

$D = +\sqrt{10.5625}$

or $D = 3.25$ in.

(e) "One side . . . must be 6 in. longer than the other."

$$L = 6 + W$$

For a rectangle, area $= LW$.

Then,

$$216 = LW$$

or $\quad 216 = (6 + W)W$

$$216 = 6W + W^2$$

$$W^2 + 6W - 216 = 0 \qquad \text{in standard quadratic form}$$

$$\boxed{a = 1} \quad \boxed{b = 6} \quad \boxed{c = -216}$$

When we substitute into the quadratic formula,

$$W = \frac{-(6) \pm \sqrt{(6)^2 - 4(1)(-216)}}{2(1)}$$

$$= \frac{-6 \pm \sqrt{900}}{2}$$

$$= \frac{-6 \pm 30}{2}$$

The solution is

$$W = -18 \qquad \text{or} \qquad W = 12$$

Only the positive value is a reasonable solution. The answer is $W = 12$ in.
Substituting into the first equation yields

$$L = 6 + W$$

$$L = 6 + (12) = 18 \text{ in.}$$

(f)

$$H + H + W = 12$$

$$2H + W = 12$$

or $\qquad W = 12 - 2H$

Area $= HW$

or $\quad 16 = HW \qquad$ Substituting $W = \boxed{12 - 2H}$ into the area equation.

$$16 = H\,\boxed{(12 - 2H)}$$

$$16 = 12H - 2H^2 \quad \text{or} \quad 8 = 6H - H^2 \qquad \text{Divide each term by 2 to get a simpler equation.}$$

$$H^2 - 6H + 8 = 0 \qquad \text{in standard quadratic form}$$

$$\boxed{a = 1} \qquad \boxed{b = -6} \quad \boxed{c = 8}$$

Substituting a, b, and c into the quadratic formula yields

$$H = \frac{-(-6) \pm \sqrt{(-6)^2 - 4(1)(8)}}{2(1)}$$

$$H = \frac{6 \pm \sqrt{4}}{2}$$

$$H = \frac{6 \pm 2}{2}$$

The solution is $H = 4$ ft   or $H = 2$ ft

For $H = 4$ ft, $W = 4$ ft, since $16 = HW$
For $H = 2$ ft, $W = 8$ ft

The channel can be either 4 ft by 4 ft or 2 ft by 8 ft. Both dimensions give a cross-sectional area of 16 sq ft.

Now turn to Exercises 11-2 for more practice solving quadratic equations.

Exercises 11-2 Quadratic Equations

A. Which of the following are quadratic equations?

1. $5x - 13 = 23$
2. $2x + 5 = 3x^2$
3. $2x^3 - 6x^2 - 5x + 3 = 0$
4. $x^2 = 0$
5. $8x^2 - 9x = 0$

Which of the following quadratic equations are in standard form? For those that are not, rearrange them into standard form.

6. $7x^2 - 5 + 3x = 0$
7. $14 = 7x - 3x^2$
8. $13x^2 - 3x + 5 = 0$
9. $23x - x^2 = 5x$
10. $4x^2 - 7x + 3 = 0$

B. Solve each of these quadratic equations. (Round to two decimal places if necessary.)

1. $x^2 = 25$
2. $3x^2 - 27 = 0$
3. $5x^2 = 22x$
4. $2x^2 - 7x + 3 = 0$
5. $4x^2 = 81$
6. $6x^2 - 13x - 63 = 0$
7. $15x = 12 - x^2$
8. $4x^2 - 39x = -27$
9. $0.4x^2 + 0.6x - 0.8 = 0$
10. $0.001 = x^2 + 0.03x$
11. $x^2 - x - 13 = 0$
12. $2x - x^2 + 11 = 0$
13. $7x^2 - 2x = 1$
14. $3 + 4x - 5x^2 = 0$
15. $2x^2 + 6x + 3 = 0$
16. $5x^2 - 7x + 1 = 0$
17. $3x^2 = 8x - 2$
18. $x = 1 - 7x^2$
19. $0.2x^2 - 0.9x + 0.6 = 0$
20. $1.2x^2 + 2.5x = 1.8$

C. Practical Problems. (Round to the nearest hundredth.)

1. The area of a square is 625 mm^2. Find its side length.

2. **Manufacturing** The capacity in gallons of a cylindrical tank can be found using the formula

$$C = \frac{\pi D^2 L}{924}$$

where C = capacity in gallons
D = diameter of tank in inches
L = length of tank in inches

(a) Find the diameter of a 42-in.-long tank that has a capacity of 30 gal.

(b) Find the diameter of a 60-in.-long tank that has a capacity of 50 gal.

3. **Plumbing** The length of a rectangular pipe is three times longer than its width. Find the dimensions that give a cross-sectional area of 75 sq in.

4. **HVAC** Find the radius of a circular vent that has a cross-sectional area of 250 cm^2.

5. **Sheet Metal Trades** The length of a rectangular piece of sheet metal must be 5 in. longer than the width. Find the exact dimensions that will provide an area of 374 sq in.

6. **Construction** The building code in a certain county requires that public buildings be able to withstand a wind pressure of 25 lb/ft^2. The pressure P is related to the wind speed v in miles per hour by the formula $P = v^2/390$. What wind speed will produce the maximum pressure allowed? Round to two significant digits.

7. **Plumbing** Find the diameter of a circular pipe whose cross-sectional area is 40 sq in.

8. **Electrical Trades** If $P = RI^2$ for a direct current circuit, find I (current in amperes) if

(a) The power (P) is 405 watts and the resistance (R) is 5 ohms.

(b) The power is 800 watts and the resistance is 15 ohms.

9. **Automotive Trades** Total piston displacement is given by the following formula:

$$\text{P.D.} = \frac{\pi}{4}D^2LN$$

where P.D. = piston displacement, in cubic inches
$\quad\quad\quad D$ = diameter of bore of cylinder
$\quad\quad\quad L$ = length of stroke, in inches
$\quad\quad\quad N$ = number of cylinders

Find the diameter if

(a) P.D. = 400 cu in., $L = 4.5$ in., $N = 8$

(b) P.D. = 392.7 cu in., $L = 4$ in., $N = 6$

10. **Hydrology** An open-topped channel must be made out of a 20-ft-wide piece of sheet steel. What dimensions will result in a cross-sectional area of 48 sq ft?

11. **Sheet Metal Trades** A cylindrical water heater has a total surface area of 4170 in.^2. If the height of the water heater is 75 in., find the radius to the nearest inch.

12. **Construction** The area of an octagon can be approximated using the formula
$$A = 4.8275S^2$$

where S is the length of a side. How long should the length of a side be to construct an octagonal gazebo with an area of 400 ft^2? (Round to the nearest 0.01 ft.)

13. **Forestry** After wildfires destroyed a portion of Glacier National Park, forest service biologists calculated the germination rate of new seedlings. Using a rope as a radius, they marked off a circular area of forest equivalent to one-tenth of an acre. They then counted the number of new seedlings and multiplied by

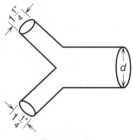

Problem 14

10 to obtain an estimate of the germination rate in seedlings per acre. What length of rope is needed to create the circle? (Round to the nearest 0.1 ft.)

14. **Landscaping**   Two small drain pipes merge into one larger pipe by means of a Y-connector. (See the figure.) The two smaller pipes have inside diameters of $1\frac{1}{4}$ in. each. The cross-sectional area of the large pipe must be at least as large as the sum of the areas of the two smaller pipes. What is the minimum diameter of the larger pipe? (Round to the nearest tenth of an inch.)

15. **Fire Protection**   The pressure loss F (in pounds per square inch) of a hose is given by

$$F = 2Q^2 + Q$$

where Q is the flow rate (in hundreds of gallons per minute) of water through the hose. If the pressure loss is 42.0 lb/sq in., what is the flow rate to the nearest gallon per minute?

16. **Automotive Trades**   The fuel economy E (in miles per gallon) of a certain vehicle is given by the formula,

$$E = -0.0088v^2 + 0.75v + 14$$

where v is the average speed of the vehicle (in miles per hour). What average speed must be maintained in order to have a fuel economy of 26 mi/gal? (Round to the nearest mile per hour.)

Check your answers to the odd-numbered problems in the Appendix. Then turn to Problem Set 11 on page 713 for a problem set covering both systems of equations and quadratic equations. If you need a quick review of the topics in this chapter, visit the chapter Summary first.

Summary   Advanced Algebra

Objective	Review
Solve a system of two linear equations in two variables. (p. 673)	Use either the substitution method or the elimination method to solve a system of equations. If both variables in the system drop out, the system has no unique solution.

Example:

(a) To solve the system of equations $5x + 2y = 20$
$$3x - 2y = 12$$

      Add the equations $\overline{8x + 0 \;\;= 32}$
      Solve for x $x = 4$

Substitute this back into the first equation: $5(4) + 2y = 20$
Solve for y: $20 + 2y = 20$
$$2y = 0$$
$$y = 0$$

The solution to the system of equations is $x = 4, y = 0$ or $(4,0)$

Objective	Review

(b) Another example. Solve the system
$$x = 3y$$
$$2y - 4x = 30$$

Substitute $3y$ for x in the second equation.

$$2y - 4(3y) = 30$$

Solve for y.
$$2y - 12y = 30$$
$$-10y = 30$$
$$y = -3$$

Substitute this back into the first equation and solve for x.

$$x = 3(-3) = -9$$

The solution to the system of equations is $x = -9$, $y = -3$.

Solve word problems involving systems of equations in two variables. (p. 688)

When there are two unknowns in a word problem, assign two different variables to the unknowns. Then look for two facts that can be translated into two equations. Solve this resulting system.

Example: A mixture of 800 drywall screws costs \$42. If the mixture consists of one type costing 4 cents each and another type costing 6 cents each, how many of each kind are there?

Let $x =$ the number of 4 cent screws and $y =$ the number of 6 cent screws. The total number of screws is

$$x + y = 800$$

The total cost in cents is

$$4x + 6y = 4200$$

Solve this system of equations by substitution or elimination.

The solution is $x = 300$, $y = 500$.

Solve quadratic equations. (p. 696)

If the x term is missing from a quadratic equation to be solved, isolate the x^2 term on one side, and take the square root of both sides. Remember that every positive number has two square roots.

Example: Solve $x^2 = 16$

Take the square root of both sides: $x = 4$ or $x = -4$

Solving quadratic equations using the quadratic formula. (p. 699)

If the x term is not missing, use the quadratic formula

$$x = \frac{-b \pm \sqrt{b^2 - 4ac}}{2a}$$

where $ax^2 + bx + c = 0$ is the equation being solved.

Objective	Review

Example: Solve $3x^2 + 2x - 16 = 0$

$a = 3, b = 2, c = -16$

Substitute into the quadratic formula

$$x = \frac{-2 \pm \sqrt{(2)^2 - 4(3)(-16)}}{2(3)}$$

$$= \frac{-2 \pm \sqrt{196}}{6}$$

$$x = \frac{-2 + 14}{6} \quad \text{or} \quad x = \frac{-2 - 14}{6}$$

$$x = 2 \quad \text{or} \quad x = -2\tfrac{2}{3}$$

Solve word problems involving quadratic equations. (p. 703)

To solve word problems involving quadratic equations, first assign a variable to the unknown quantity. If there is a second unknown, represent it with an expression involving the same variable. Finally, set up an equation based on the facts given in the problem statement.

Example: The length of a rectangular vent is 3 in. longer than its width. Find the dimensions of the vent if its cross-sectional area is 61.75 sq in.

Let $\quad\quad W =$ the width
then $\quad W + 3 =$ the length
and $W(W + 3) =$ the cross-sectional area

Therefore, the quadratic equation is $W(W + 3) = 61.75$ or $W^2 + 3W - 61.75 = 0$ and the quadratic formula becomes

$$W = \frac{-3 \pm \sqrt{256}}{2} = \frac{-3 \pm 16}{2}$$

$W = 6.5$ or $W = -9.5$

But the width cannot be a negative number, so the width is 6.5 in., and the length is $6.5 + 3 = 9.5$ in.

Advanced Algebra

Answers to odd-numbered problems are given in the Appendix.

A. **Solve and check each of the following systems of equations. If the system is inconsistent or dependent, say so.**

1.	$x + 4y = 27$ $x + 2y = 21$	**2.**	$3x + 2y = 17$ $x = 5 - 2y$	**3.**	$5x + 2y = 20$ $3x - 2y = 4$
4.	$x = 10 - y$ $2x + 3y = 23$	**5.**	$3x + 4y = 45$ $x - \frac{1}{3}y = 5$	**6.**	$2x - 3y = 11$ $4x - 6y = 22$
7.	$2x + 3y = 5$ $3x + 2y = 5$	**8.**	$5x = 1 - 3y$ $4x + 2y = -8$	**9.**	$3x - 2y = 10$ $4x + 5y = 12$

B. **Solve and check each of the following quadratic equations.**

1. $x^2 = 9$ **2.** $x^2 - 3x - 28 = 0$

3. $3x^2 + 5x + 1 = 0$ **4.** $4x^2 = 3x + 2$

5. $x^2 = 6x$ **6.** $2x^2 - 7x - 15 = 0$

7. $x^2 - 4x + 4 = 9$ **8.** $x^2 - x - 30 = 0$

9. $\dfrac{5x^2}{3} = 60$ **10.** $7x + 8 = 5x^2$

C. **Practical Problems**

For each of the following, set up either a system of equations in two variables or a quadratic equation and solve. (Round to two decimal places if necessary.)

1. The sum of two numbers is 38. Their difference is 14. Find them.
2. The area of a square is 196 sq in. Find its side length.
3. The difference of two numbers is 21. If twice the larger is subtracted from five times the smaller, the result is 33. Find the numbers.
4. **Carpentry** The perimeter of a rectangular door is 22 ft. Its length is 2 ft more than twice its width. Find the dimensions of the door.
5. **HVAC** One side of a rectangular heating pipe is four times as long as the other. The cross-sectional area is 125 cm^2. Find the dimensions of the pipe.
6. **Carpentry** A mixture of 650 nails costs $43.50. If some of the nails cost 5 cents apiece, and the rest cost 7 cents apiece, how many of each are there?
7. **Electrical Trades** In the formula $P = RI^2$ find the current I in amperes if
   (a) The power P is 1352 watts and the resistance R is 8 ohms.
   (b) The power P is 1500 watts and the resistance is 10 ohms.
8. **Sheet Metal Trades** The length of a rectangular piece of sheet steel is 2 in. longer than the width. Find the exact dimensions if the area of the sheet is 168 sq in.
9. **Electrical Trades** A 42-in. piece of wire is to be cut into two parts. If one part is 2 in. less than three times the length of the other, find the length of each piece.
10. **Landscaping** The perimeter of a rectangular field is 750 ft. If the length is four times the width, find the dimensions of the field.

Name _____

Date _____

Course/Section _____

11. **Sheet Metal Trades**   Find the diameter of a circular vent with a cross-sectional area of 200 in.^{2}

12. **Painting**   A painter mixes paint worth $28 per gallon with paint worth $34 per gallon. He wishes to make 15 gal of a mixture worth $32 per gallon. How many gallons of each kind of paint must be included in the mixture?

13. **Construction**   According to Doyle's log rule, the volume V (in board feet) produced from a log of length L (in feet) and diameter D (in inches) can be estimated from the formula

$$V = \frac{L(D^2 - 8D + 16)}{16}$$

What diameter log of length 20 ft will produce 90 board feet of lumber?

14. **Plumbing**   The length of a rectangular pipe is 3 in. less than twice the width. Find the dimensions if the cross-sectional area is 20 sq in.

15. **Landscaping**   Fifty yards of a mixture of decorative rock cost $6400. If the mixture consists of California Gold costing $120 per yard and a Palm Springs Gold costing $160 per yard, how many yards of each are used to make the mixture?

16. **Aviation**   The lift L (in pounds) on a certain type of airplane wing is given by the formula

$$L = 0.083v^2$$

where v is the speed of the airflow over the wing (in feet per second). What air speed is required to provide a lift of 2500 lb? (Round to two significant digits.)

17. **Plumbing**   Four small drain pipes flow into one large pipe as shown in the figure. Two of the pipes have 1-in. diameters and the other two have $1\frac{1}{2}$-in. diameters. The cross-sectional area of the large main pipe must be at least as large as the sum of the cross-sectional areas of the four smaller ones. What should be the minimum diameter of the large pipe? (Round to the nearest 0.01 in.)

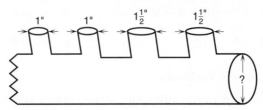

18. **Electronics**   For a certain temperature-sensitive electronic device, the voltage output V (in millivolts) is given by the formula

$$V = 3.1T - 0.014T^2$$

where T is the temperature (in degrees Celsius). What temperature is needed to produce a voltage of 150 mV?

19. **Transportation**   The stopping distance d (in feet) of a certain vehicle is approximated by the formula

$$d = 0.0611v^2 + 0.796v + 5.72$$

where v is the speed of the vehicle (in miles per hour) before the brakes are applied. If the vehicle must come to a stop within 140 ft, what is the maximum speed at which it can be traveling before the brakes are applied? (Round to the nearest mile per hour.)

20. **Automotive Trades**   The fuel economy F of a certain hybrid electric car at speeds between 50 mi/hr and 100 mi/hr is given by the formula

$$F = 0.005v^2 - 1.25v + 96$$

where F is in miles per gallon and v is the speed in miles per hour. What speed will result in a fuel economy of 39 mi/gal?

Preview

Statistics

Objective	Sample Problems	For help, go to

When you finish this chapter, you will be able to:

1. Read bar graphs, line graphs, and circle graphs.

Trades Management From the bar graph below,

(a) Determine the number of frames assembled by the Tuesday day shift. _____

(b) Calculate the percent decrease in output from the Monday day shift to the Monday night shift. _____

Page 717

Weekly Frame Assembly

Number of Frames

Day ■ Night

100
90
80
70
60
50
40
30
20
10
0

Monday Tuesday Wednesday Thursday Friday

Day of Week

Monthly Paint Jobs at Autobrite

Number of Jobs

70
60
50
40
30
20
10
0

January February March April May June

Month

Problems (c) and (d) refer to the line graph above,

Page 725

(c) Determine the maximum number of paint jobs and the month during which it occurred. _____

Name _____

(d) Calculate the percent increase in the number of paint jobs from January to February. _____

Date _____

(e) The average job for ABC Plumbing generates $227.50. Use the circle graph on the next page to calculate what portion of this amount is spent on advertising. _____

Course/Section _____

Page 729

Objective	Sample Problems	For help, go to

Percent of Business Expenditures

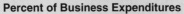
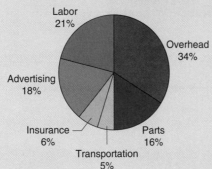

2. Draw bar graphs, line graphs, and circle graphs from tables of data.

Work Output	Day
150	Mon
360	Tues
435	Wed
375	Thurs
180	Fri

Trades Management For the data in the table, draw:

(a) A bar graph.   Page 720
(b) A line graph.   Page 727
(c) A circle graph.   Page 730

3. Calculate measures of central tendency: mean, median, and mode.

Find (a) the mean, (b) the median, and (c) the mode for the following set of lengths. All measurements are in meters.

12.7   16.2   15.5   13.9   13.2   17.1   15.5

(a) _____ Page 743
(b) _____ Page 744
(c) _____ Page 745

4. Calculate mean for data grouped in a frequency distribution.

Welding The regulator pressure settings on an oxygen cylinder used in oxyacetylene welding (OAW) must fall between a range of 110 and 160 pounds per square inch (psi) when welding material is 1 in. thick.

The following frequency distribution shows the pressures for 40 such welding jobs. Use these data to find the mean. _____ Page 747

Class Intervals	Frequency, F
110–119 psi	4
120–129 psi	8
130–139 psi	12
140–149 psi	10
150–159 psi	6

(Answers to these preview problems are given in the Appendix. Also, worked solutions to many of these problems appear in the chapter Summary.)

If you are certain you can work *all* these problems correctly, turn to page 759 for a set of practice problems. If you cannot work one or more of the preview problems, turn to the page indicated after the problem. Those who wish to master this material with the greatest success should turn to Section 12-1 and begin work there.

© Scott Adams/Dist. by United Features Syndicate, Inc.

12-1 Reading and Constructing Graphs

Working with the flood of information available to us today is a major problem for scientists and technicians, including workers in the trades. We need ways to organize and analyze numerical data in order to make it meaningful and useful. **Statistics** is a branch of mathematics that provides us with the tools we need to do this. In this chapter, you will learn the basics of how to prepare and read statistical graphs, calculate some statistical measures, and use these to help analyze data.

> **Note** In discussing statistics, we often use the word "data." Data refers to a collection of measurement numbers that describe some specific characteristic of an object or person or a group of objects or people. ◄

Reading Bar Graphs
Graphs allow us to transform a collection of measurement numbers into a visual form that is useful, simplified, and brief. Every graph tells a story that you need to be able to read.

A **bar graph** is used to display and compare the sizes of different but related quantities.

Example 1

General Trades   The following graph shows the average hourly wage for workers in six trades occupations. Hourly wages, in dollars per hour, are listed along the left side on the *vertical axis*. The six trade occupations being compared are listed, evenly spaced, along the *horizontal axis*.

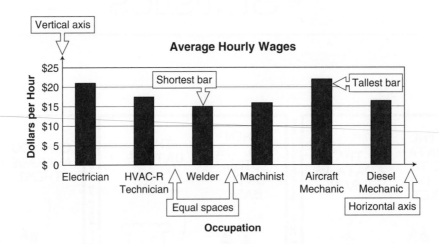

Some information is readily available from the graph: the highest hourly wage (tallest bar) is earned by the aircraft mechanic; the lowest average wage (shortest bar) is earned by the welder. When the top of a bar is directly opposite a mark on the vertical axis, it is easy to read its value. Welders earn an average wage of $15 per hour. When the top of a bar falls between two marks on the vertical axis, we must estimate its value. Aircraft mechanics earn an average wage of approximately $22 per hour.

Other calculations may be made from the information in the graph. Using the proportion technique from Chapter 4, we can calculate by what percent an electrician's hourly wage, $21, exceeds that of a welder, $15.

$$\frac{21 - 15}{15} = \frac{R}{100} \quad \text{or} \quad R = \frac{6 \times 100}{15} = 40\%$$

  ⎛ 21 ⊖ 15 ⎞ ⊗ 100 ⊙ 15 ⊜ → ▮ 40. ▮

On average, an electrician earns 40% more per hour than a welder.

Note   A bar graph like the one in the previous example, drawn with vertical bars, is called a *vertical bar graph*. It is sometimes helpful to construct a *horizontal bar graph* by simply switching the horizontal and vertical axes. ◀

→ **Your Turn**

General Trades   Using the graph in Example 1, calculate how much more the average aircraft machinic would earn in a 40-hr work week compared to the average machinist.

From the graph, we estimate the average aircraft mechanic's wage to be approximately $22 per hour. The average machinist's wage is approximately $16 per hour. Over a 40-hr work week, the difference would be

$$40(\$22 - \$16) = 40(\$6) = \$240$$

The average aircraft mechanic would earn approximately $240 more per week than the average machinist.

A *double bar graph* allows us to show side-by-side comparisons of related quantities on the same graph.

Example 2

The following horizontal double bar graph shows the effect of shielding on computer components exposed to radiation. The two bars compare the survival times of components that are shielded (dark blue) to those that are not shielded (light blue).

Notice that in drawing the graph, we arranged the computer components in order of survival time, labeled them from *A* to *F*, and marked them along the vertical axis. This is designed to make the graph easy to read.

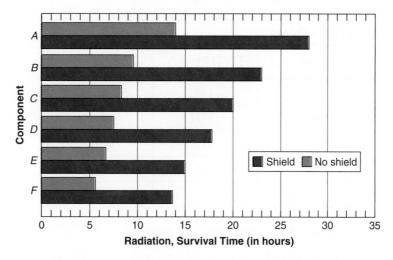

Computer Component vs. Radiation

The graph shows that component *C* survived about 8.5 hr without shielding and 20 hr with shielding.

→ **Your Turn**

Use the preceding graph to answer the following questions.

(a) Which computer component, with shielding, survived the longest?

(b) How much longer does component *E* survive when it is shielded compared to when it is unshielded.

(c) By what percent does the survival time of shielded component *B* exceed that of shielded component *D*?

(a)   Component *A* survived the radiation longest at 28 hr.

(b)   The difference is 15 hr − 7 hr = 8 hr.

(c)   Set up a proportion:

$$\frac{(\text{Time for } B) - (\text{Time for } D)}{\text{Time for } D} = \frac{R}{100}$$

$$\frac{23 - 17.5}{17.5} = \frac{R}{100}$$

Solve.
$$R = \frac{5.5 \times 100}{17.5} \approx 31.4\%$$

Component *B* survived the radiation about 31% longer than did component *D*.

Drawing Bar Graphs

Usually, a bar graph is created not from mathematical theory or abstractions, but is constructed from a set of measurement numbers.

Example 3

Suppose you want to display the following data in a bar graph showing sales of DVD players and televisions in five stores.

Quarterly Sales of DVD Players and TVs at Five Stores

Store	DVD Players	Televisions
Ace	140	65
Wilson's	172	130
Martin's	185	200
XXX	195	285
Shop-Rite	190	375

To draw a bar graph, follow these steps:

Step 1   Decide what type of bar graph to use. Because we are comparing two different items, we should use a double bar graph. The bars can be placed either horizontally or vertically. In this case let's make the bars horizontal.

Step 2   Choose a suitable spacing for the vertical (side) axis and a suitable scale for the horizontal (bottom) axis. Label each axis. When you label the vertical or side axis it should read in the normal way.

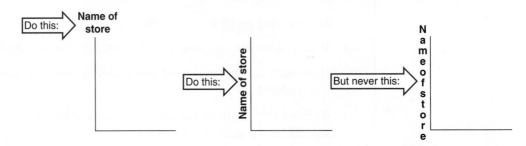

Chapter 12 Statistics

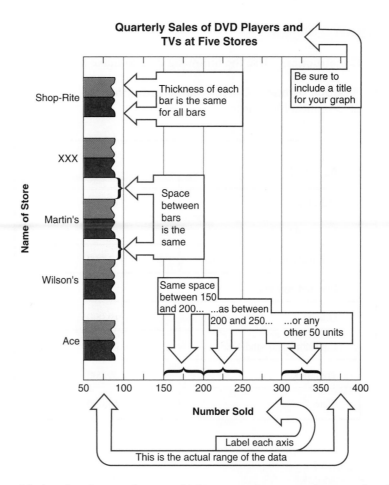

Notice that the numbers on the bottom axis are evenly spaced and cover the entire range of numbers in the data, in this case from 65 to 375.

Also notice that the stores were arranged in order of amount of sales. The biggest seller, Shop-Rite, is at the top, and the smallest seller, Ace, is at the bottom. This is not necessary, but it makes the graph easier to read.

Step 3 Use a straightedge to mark the length of each bar according to the data given. Round the numbers if necessary. Drawing your bar graph on graph paper will make the process easier.

The final bar graph will look like this:

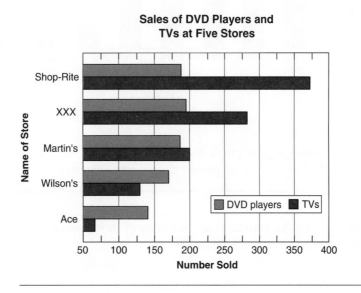

Construction   Use the following data to make a bar graph.

**Average Ultimate Compression Strength
of Common Materials**

Material	Compression Strength (psi)
Hard bricks	12,000
Light red bricks	1,000
Portland cement	3,000
Portland concrete	1,000
Granite	19,000
Limestone and sandstone	9,000
Trap rock	20,000
Slate	14,000

→ **Solution**

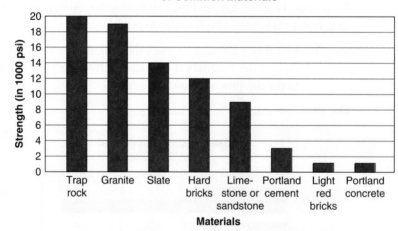

Don't worry if your graph does not look exactly like this one. The only requirement is that all of the data be displayed clearly and accurately.

A Closer Look   Notice that the vertical scale has units of *thousands* of pounds per square inch. It was drawn this way to save space. ◀

Energy Consumption of Gas Appliances

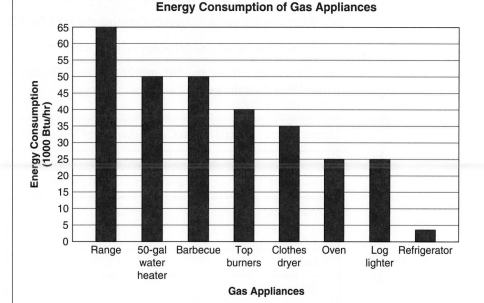

1. **Life Skills** Using the bar graph shown, answer the following questions.

   (a) Which appliance uses the most energy in an hour?

   (b) Which appliance uses the least energy in an hour?

   (c) How many Btu/hr does a gas barbecue use?

   (d) How many Btu/*day* (24 hr) would a 50-gal water heater use?

   (e) Is there any difference between the energy consumption of the range and that of the top burners plus oven?

   (f) How many Btus are used by a log lighter in 15 min?

   (g) What is the difference in energy consumption between a 50-gal water heater and a clothes dryer?

2. **Agriculture** The following double-bar graph compares the approximate costs per acre of growing corn in the United States in 2007 versus 2008 (projected).

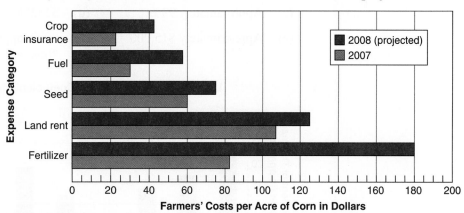

(Source: Wells Fargo Economics)

   (a) Which category contributed the most to cost per acre in 2007?

   (b) Which category was projected to increase the most in 2008?

(c) By what percent were fuel costs projected to increase in 2008?

(d) What was the total cost of seed for 250 acres of corn in 2007?

(e) What was the projected cost of renting the land for 120 acres of corn in 2008?

(f) What was the total cost per acre of all five categories in 2007? In 2008 (projected)?

3. **Metalworking** Draw the bar graph for the following information. (*Hint:* Use multiples of 5 on the vertical axis.)

Linear Thermal Expansion Coefficients for Materials

Material	Coefficient (in parts per million per °C)
Aluminum	23
Copper	17
Gold	14
Silicon	3
Concrete	12
Brass	19
Lead	29

→ **Solutions**

1. (a) Range   (b) Refrigerator
   (c) 50,000 Btu/hr   (d) 1,200,000 Btu/day
   (e) No   (f) 6250 Btu
   (g) 15,000 Btu/hr

2. (a) Land rent   (b) Fertilizer
   (c) Approximately 93%   (d) $15,000
   (e) Approximately $15,000   (f) 2007: approximately $300/acre; 2008: approximately $480/acre

3.

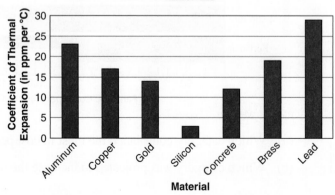

Linear Thermal Expansion Coefficients for Materials

Reading Line Graphs

A **line graph,** or **broken-line graph,** is a display that shows the change in a quantity, usually as it changes over a period of time.

The following broken-line graph shows monthly sales of tires at Treadwell Tire Company. The months of the year are indicated along the horizontal axis, while the numbers of tires sold are shown along the vertical axis. The actual data ranges from 550 to 950, so the vertical axis begins at 500 and ends at 1000, with each interval representing 50 tires. Each dot represents the number of tires sold in the month that is directly below the dot, and straight line segments connect the dots to show the monthly fluctuations in sales.

The following example illustrates how to read information from a broken-line graph.

Example 4

Trades Management   To find the number of tires sold in May, find May along the horizontal axis and follow the perpendicular line up to the graph. From this point, look directly across to the vertical axis and read or estimate the number sold. There were approximately 750 tires sold in May.

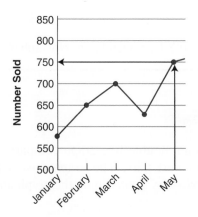

Other useful information can be gathered from the graph:

The highest monthly total was 950 in August.
The largest jump in sales occurred from July to August, with sales rising from 800 to 950 tires. To calculate the percent increase in August, set up the following proportion:

$$\frac{950 - 800}{800} = \frac{R}{100} \quad \text{or} \quad \frac{150}{800} = \frac{R}{100}$$

$$R = \frac{150 \cdot 100}{800}$$

$$= 18.75\%$$

The increase in sales from July to August was approximately 19%.

→ **Your Turn**

Trades Management Answer the following questions about the graph of monthly tire sales.

(a) What was the lowest monthly total, and when did this occur?

(b) How many tires were sold during the first three months of the year combined?

(c) During which two consecutive months did the largest drop in sales occur? By what percent did sales decrease?

→ **Solutions**

(a) Only 550 tires were sold in December.

(b) We estimate that 580 tires were sold in January, 650 in February, and 700 in March. Adding these, we get a total of 1930 for the three months.

(c) The largest drop in sales occurred from October to November, when sales decreased from 750 to 600. Calculating the percent decrease, we have

$$\frac{750 - 600}{750} = \frac{R}{100} \quad \text{or} \quad \frac{150}{750} = \frac{R}{100}$$

$$R = \frac{150 \cdot 100}{750}$$

$$= 20\%$$

Sales decreased by 20% from October to November.

→ **More Practice**

Life Skills The following double-line graph compares the median home prices in Duneville with those in Surf City. The graph covers the 13-month period from January 2007 through February 2008. Study the graph and answer the questions that follow.

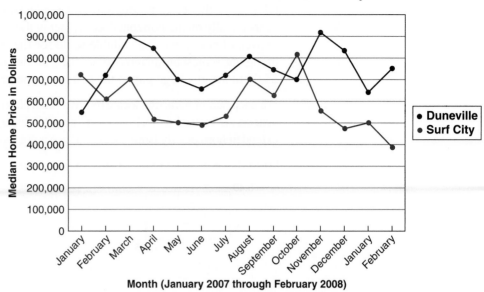

Median Home Prices in Duneville and Surf City

Month (January 2007 through February 2008)

(a) In which months did Duneville's median home price increase and Surf City's median home price decrease?

(b) In which month did the steepest decline in Surf City's median home price occur?

(c) What was the highest median home price in Surf City, and in which month did it occur?

(d) What was the lowest median home price in Duneville, and in which month did it occur?

(e) By what percent did Duneville's median price increase from January 2007 to January 2008?

→ **Answers**

(a) February 2007, November 2007, and February 2008

(b) November 2007

(c) $820,000 in October 2007

(d) $550,000 in January 2007

(e) approximately 18%

Drawing Line Graphs   Constructing a broken-line graph is similar to drawing a bar graph. If possible, begin with data arranged in a table of pairs. Then follow the steps shown in the example.

Example 5

Manufacturing   The given table shows the average unit production cost for an electronic component during the years 2003–2008.

Year	Production Costs (per unit)
2003	$5.16
2004	$5.33
2005	$5.04
2006	$5.57
2007	$6.55
2008	$6.94

Step 1 Draw and label the axes. According to convention time is plotted on the horizontal axis. In this case, production cost is placed on the vertical axis. Space the years equally on the horizontal axis, placing them directly on a graph line and not between lines. Choose a suitable scale for the vertical axis. In this case, each graph line represents an interval of $0.20. Notice that we begin the vertical axis at $5.00 to avoid a large gap at the bottom of the graph. Be sure to title the graph.

Step 2 For each pair of numbers, locate the year on the horizontal scale and the cost for that year on the vertical scale. Imagine a line extended up from the year and another line extended horizontally from the cost. Place a dot where these two lines intersect.

From the table, the cost in 2004 is $5.33.

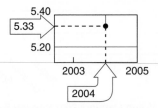

Step 3 After all the number pairs have been placed on the graph, connect adjacent points with straight-line segments.

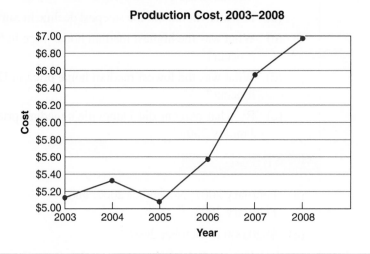

A double-line graph is one on which two separate sets of information are plotted. This kind of graph is very useful for comparing two quantities that vary over the same time period.

→ **Your Turn**

Manufacturing Plot the following data on the graph in the previous example to create a double-line graph.

Year	Shipping Cost (per unit)
2003	$5.30
2004	$5.61
2005	$6.05
2006	$6.20
2007	$6.40
2008	$6.50

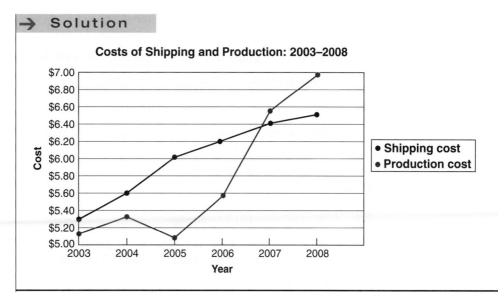

Costs of Shipping and Production: 2003–2008

The following circle graph gives the distribution of questions contained on a
comprehensive welding exam. The area of the circle represents the entire exam, and the
wedge-shaped sectors represent the percentage of questions in each part of the exam.
The percents on the sectors add up to 100%.

Reading Circle Graphs

A **circle graph,** or **pie chart,** is used to show what fraction of the whole of some quantity is represented by its separate parts.

Example 6

Welding   The following circle graph gives the distribution of questions contained on a
comprehensive welding exam. The area of the circle represents the entire exam, and the
wedge-shaped sectors represent the percentage of questions in each part of the exam.
The percents on the sectors add up to 100%.

**Number of Questions on a Comprehensive
Welding Exam**

The size of each sector is proportional to the percent it represents. If we wanted to know
which topic was asked about the most, we could visually pick out the largest sector,
Welding applications, or compare the percents shown to reach the same conclusion.

We can also use the percents on the graph to calculate additional information. For example, suppose we know that there are 150 questions on the exam, and we want to know
exactly how many questions deal with occupational skills. From the graph, we see that
24% of the questions relate to occupational skills, so we find 24% of 150 as follows:

$$\frac{S}{150} = \frac{24}{100}$$

$$S = \frac{24 \cdot 150}{100}$$

$$= 36$$

There are 36 questions relating to occupational skills.

→ **Your Turn**

Welding   Use the graph from Example 6 to answer the following questions.

(a)   What percent of the exam dealt with cutting?

(b)   Which question topic was asked about least?

(c)   If there were 125 questions on the exam, how many questions would deal with quality control?

→ **Solutions**

(a)   22%   (b)   Gouging

(c)   From the graph, we see that 16% of the questions deal with quality control. Using a proportion, we can calculate 16% of 125 as follows:

$$\frac{Q}{125} = \frac{16}{100}$$

$$Q = \frac{16 \cdot 125}{100}$$

$$= 20$$

There would be 20 questions on quality control.

Constructing Circle Graphs   If the data to be used for a circle graph is given in percent form, determine the number of degrees for each sector using the quick calculation method explained in Chapter 4. Convert each percent to a decimal and multiply by 360°.

Example 7

Automotive Trades   In a survey of automotive tasks, the following data were found to represent a typical work week.

Transmission	65%
Tune-up	15%
Front-end	10%
Diagnostics	5%
Exhaust	5%

Extend the table to show the calculations for the angles for each sector.

Transmission	65%	65% of 360° = 0.65 × 360° = 234°
Tune-up	15%	15% of 360° = 0.15 × 360° = 54°
Front-end	10%	10% of 360° = 0.10 × 360° = 36°
Diagnostic	5%	5% of 360° = 0.05 × 360° = 18°
Exhaust	5%	5% of 360° = 0.05 × 360° = 18°

Sum = 100%   Round to the nearest degree

Note   The sum of the angles should be 360° most of the time, but it may be a degree off due to rounding. ◄

Finally, using a protractor mark off the sectors and complete the circle graph. Notice that each sector is labeled with a category name and its percent.

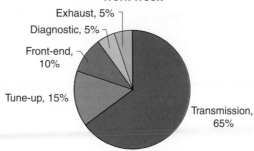

Automotive Technicians Work Week

Exhaust, 5%
Diagnostic, 5%
Front-end, 10%
Tune-up, 15%
Transmission, 65%

If the data for a circle graph are not given in percent form, first convert the data to percents and then convert the percents to degrees.

Example 8

Trades Management The Zapp Electric Company produces novelty electrical toys. The company has a very top-heavy compensation plan. The Chief Executive Officer is paid $175,000 yearly; the Vice President for finance, $70,000; a shop supervisor, $55,000; and 10 hourly workers, $25,000 each. Draw a circle graph of this situation.

It is helpful to organize our information in a table such as this:

Employee	Compensation	Percent	Angle
CEO	$175,000		
VP	70,000		
Supervisor	55,000		
10 hourly workers	250,000		

Sum = $550,000

First, find the sum of the compensations.

$175,000 + 70,000 + 55,000 + 250,000 = \$550,000$

Second, use this sum as the base to calculate the percents needed. Because all of the compensations end in three zeros, these may be dropped when setting up the proportions.

$$\frac{175}{550} = \frac{C}{100} \qquad \frac{70}{550} = \frac{V}{100} \qquad \frac{55}{550} = \frac{S}{100} \qquad \frac{250}{550} = \frac{E}{100}$$

$$C \approx 32\% \qquad V \approx 13\% \qquad S = 10\% \qquad E \approx 45\%$$

Third, use the percent values to calculate the angles in the last column.

$$(0.32)(360°) \approx 115° \qquad (0.13)(360°) \approx 47°$$

$$(0.10)(360°) = 36° \qquad (0.45)(360°) = 162°$$

The completed table:

Employee	Compensation	Percent	Angle
CEO	$175,000	32%	115°
VP	70,000	13%	47°
Supervisor	55,000	10%	36°
10 hourly workers	250,000	45%	162°

Sum = $550,000   Sum = 100%   Sum = 360°

Finally, use the angles in the last column to draw the circle graph. Label each sector as shown.

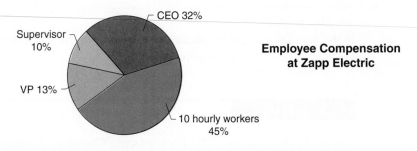

Employee Compensation at Zapp Electric

→ **Your Turn**

Plumbing   Jane the plumber keeps a record of the number of trips she makes to answer emergency calls. A summary of her records looks like this:

Trip Length	Number of Trips
Less than 5 miles	152
5–9 miles	25
10–19 miles	49
20–49 miles	18
50 or more miles	10

Calculate the percents and the angles for each category, and plot Jane's data in a circle graph.

→ **Solution**

The completed table shows the percents and angles for each category:

Trip Length	Number of Trips	Percent	Angle
Less than 5	152	60%	216°
5–9	25	10%	36°
10–19	49	19%	68°
20–49	18	7%	25°
50+	10	4%	14°
	254	100%	359°

Because of rounding, the percents will not always sum to exactly 100% and the angles will not always sum to exactly 360°. In this case, the angles added up to 359°. This will not noticeably affect the appearance of the graph.

The circle graph is shown here.

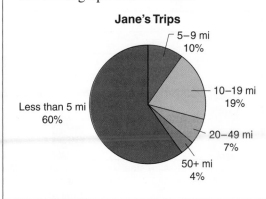

Jane's Trips

5–9 mi 10%

10–19 mi 19%

20–49 mi 7%

50+ mi 4%

Less than 5 mi 60%

Now turn to Exercises 12-1 for more practice in reading and constructing bar graphs, line graphs, and circle graphs.

Exercises 12-1 Reading and Constructing Graphs

A. Answer the questions following each graph.

1. **Automotive Trades**   The following bar graph shows the annual U.S. sales of hybrid vehicles from 2000–2008. Study the graph and answer the questions below.

U.S. Sales of Hybrid Vehicles: 2000–2008

(y-axis: Annual Sales in the U.S., 0 to 400,000 in increments of 50,000)
(x-axis: Year, 2000 to 2008)

(Source: Green Car Congress)

(a) During which year did the largest increase in hybrid sales occur?

(b) Approximately how many hybrid vehicles were sold in 2008?

(c) How many more hybrid vehicles were sold in 2006 than in 2003? (Estimate your answer to the nearest hundred thousand.)

(d) By what percent did sales increase from 2004 to 2007? (Round to the nearest hundred percent.)

(e) By what percent did sales decrease from 2007 to 2008? (Round to the nearest percent.)

2. **Machine Trades** The following bar graph shows the recommended cutting speeds when turning certain materials.

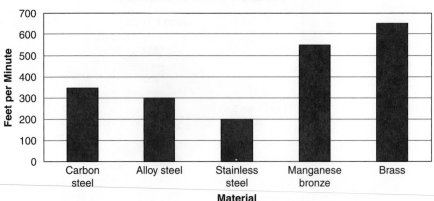

Recommended Cutting Speeds for Turning Certain Materials Using a Carbide Tool

(a) Which material requires the lowest cutting speed, and what is that speed?

(b) Which material requires the highest cutting speed, and what is that speed?

(c) What is the cutting speed for alloy steel?

(d) What is the difference in cutting speeds between manganese bronze and carbon steel?

3. **HVAC** The following double-bar graph represents the energy efficiency ratio (EER) and seasonal energy efficiency ratio (SEER) for different models of residential air conditioners. The EER is the measure of the instantaneous energy efficiency of the cooling equipment in an air conditioner. The SEER is the measure of the energy efficiency of the equipment over the entire cooling season. Both efficiency ratios are given in British thermal units per watt-hour (Btu/Wh).

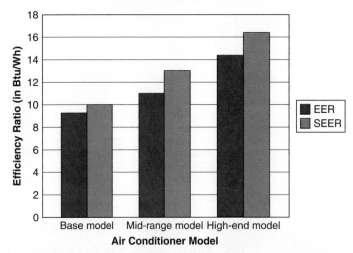

The EER and SEER for 3-ton Residential Central Air Conditioners

(a) What is the approximate EER for the base model?

(b) Which model has a SEER of about 13?

(c) Calculate the percent increase in SEER from the base model to the high-end model.

(d) Calculate the percent increase in the EER from the base model to the mid-range model.

4. **Police Science** The following bar graph represents crime statistics over a three-year period.

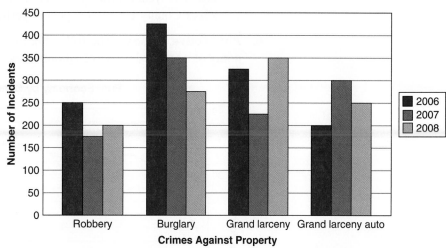

Yearly Incidence of Crimes for Gotham City from 2006–2008

(a) Over which two years did the number of incidents of grand larceny increase?

(b) Which two crimes had the highest number of occurrences in 2007?

(c) Which two crimes had the highest number of occurrences in 2006?

(d) In which category are crimes decreasing?

(e) Which crime had the lowest number of incidents in 2006? In 2008?

(f) What was the approximate total number of incidents in 2007?

(g) By what percent did grand larceny auto increase between 2006 and 2007?

5. **Allied Health** When Dr. Friedrich began working at the Zizyx County Hospital in 1998, his goal was to improve the quality and quantity of bone marrow transplants performed at the hospital. The number of successful bone marrow transplants performed at the Zizyx County Hospital from 1998–2007 is illustrated in the following broken-line graph.

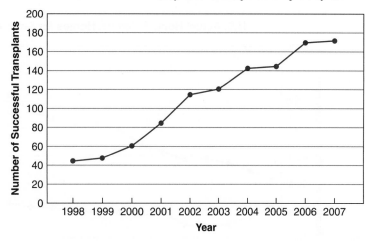

Bone Marrow Transplants at Zizyx County Hospital

(a) Approximately how many successful bone marrow transplants were performed in 1998?

(b) Approximately how many successful bone marrow transplants were performed in 2007?

(c) By about what percent did the number of successful bone marrow transplants increase in the five-year period from 2002–2007?

(d) If the bone marrow transplant program increases at the same rate over the next five years, how many successful bone marrow transplants can the Zizyx County Hospital expect to perform in 2012?

6. **Automotive Trades** The following line graph reflects the average fuel economy of a selected group of automobiles.

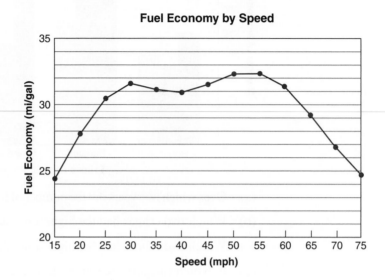

Fuel Economy by Speed

(a) What speed has the best fuel economy? The worst?

(b) What is the fuel economy at 65 mph? At 40 mph?

(c) At what speed is the fuel economy 29 mi/gal?

(d) After which two speeds does fuel economy begin to decrease?

7. **Construction** The following double-bar graph compares the number of actual home sales to the number of homes for sale in the United States from 1991–2007. Note that the vertical scale is in thousands—meaning that 5000, for example, actually means 5,000,000. Study the graph and answer the questions that follow.

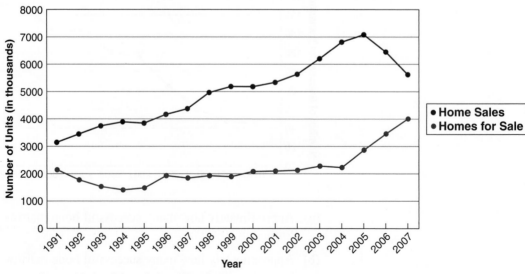

U.S. Actual Home Sales vs. Homes for Sale: 1991–2007

(Source: National Association of Realtors)

(a) In what year did the biggest drop in home sales occur? By approximately how many units did sales drop?

(b) Identify at least two years when the number of sales and the number of homes for sale both increased.

(c) In what year were home sales at their lowest level? In what year were home sales at their highest level?

(d) Based on your answer to part (c), how many more units were sold in the highest-selling year than in the lowest-selling year?

(e) By approximately what percent did the number of homes for sale increase between 2004 and 2007?

8. **General Interest** Study this circle graph and answer the questions that follow.

Land Area of the Earth by Continents

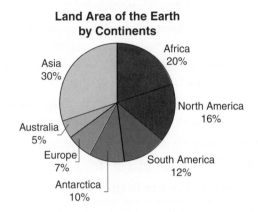

(a) Which are the largest two continents on the earth?

(b) Which two are the smallest?

(c) Which continent covers 12% of the earth?

9. **Life Skills** Study this circle graph and answer the questions that follow.

School Expenditures

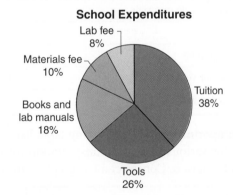

(a) Which category represents the largest expenditure?

(b) Which category represents the smallest expenditure?

(c) If a student spent a total of $2500, how much of this went toward tools and lab fees combined?

(d) If a student spent $300 on materials, how much would she spend on books and lab manuals?

10. **Allied Health** An assistant at a pharmaceutical company summarized the use of anti-obesity drugs in Zizyx County. Based on a survey of local pharmacies, the assistant estimated the percent of patients using each of the most common weight-loss medications, and presented the results in a circle graph.

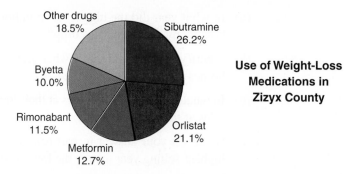

Use of Weight-Loss
Medications in
Zizyx County

(a) What is the most commonly used weight-loss medication Zizyx County?

(b) If a total of 2780 people in Zizyx County take weight-loss medication, how many of these are taking Sibutramine? How many more people take Orlistat than Metformin?

(c) After a Sports Center was opened in Zizyx County, the number of people taking weight-loss medications [see part (b)] decreased by 43%. How many people in Zizyx County were taking weight-loss medications after the opening of the Zizyx Sports Center?

(d) If the percents in the circle graph remained the same after the opening of the Zizyx Sports Center, how many patients were still taking Byetta?

B. From the following data, construct the type of graph indicated.

1. **Fire Protection** Plot the following data as a bar graph.

Causes of Fires in District 12

Cause of Fire	Number of Fires
Appliance	6
Arson	8
Electrical	18
Flammable materials	7
Gas	2
Lightning	2
Motor vehicle	10
Unknown	9

2. **Transportation** The following table lists the total carbon emissions per person for various forms of transportation for a round trip from Los Angeles to San Francisco. Construct a bar graph of these data.

Form of Transportation	Carbon Emissions (pounds)
Flying: smaller, newer plane	300
Flying: larger, older plane	406
Driving alone (car with 47 mi/gal)	273
Driving alone (car with 25 mi/gal)	527
Driving alone (car with 16 mi/gal)	793
Carpooling (2 people, car with 25 mi/gal)	263
Train	284

(Source: Terra Pass)

3. **General Interest** The following table shows the average cost of a gallon of regular gas in various countries during the summer of 2008. Construct a bar graph of these data.

Country	Price per Gallon of Regular Gasoline (converted to U.S. dollars)
United States	$4.11
Canada (Vancouver)	$5.50
Mexico (Mexico City)	$2.62
France (Paris)	$9.43
England (London)	$8.96
Norway (Oslo)	$9.85
Russia (Moscow)	$3.90
China (Beijing)	$3.40
Japan (Tokyo)	$6.30
Egypt (Cairo)	$1.24
South Korea (Seoul)	$7.33
Kenya (Nairobi)	$5.94
Australia (Sydney)	$5.57

(Source: AIRINC)

4. **Trades Management** Plot the following data as a bar graph.

Radish Tool and Dye Worker Experience

Length of Service	Number of Workers
20 years or more	9
15–19	7
10–14	14
5–9	20
1–4	33
Less than 1 year	5

5. **Construction** Plot the following data as a double-bar graph.

Sales of Construction Material: CASH IS US

Year	Wood (× $1000)	Masonry (× $1000)
2000	$289	$131
2001	$325	$33
2002	$296	$106
2003	$288	$92
2004	$307	$94
2005	$412	$89

6. **Automotive Trades**   The following table compares the torque in pound-feet (lb · ft) and the SAE horsepower ratings of six different models of 2008 Hondas. Construct a double-bar graph based on this information.

Model	Torque (lb · ft)	Horsepower Rating (hp)
Accord EX-L	248	268
Civic Hybrid	89	110
Civic Mugen Si	139	202
Civic Si Sedan	139	197
Fit Sport	105	109
S2000 CR	162	237

7. **Business and Finance**   Plot a broken-line graph for these data.

Earnings per Share ($) XYZ Co.

Year	Earnings per Share
1998	$0.7
1999	$0.5
2000	$1.5
2001	$3.0
2002	$3.2
2003	$1.8
2004	$2.5
2005	$2.4
2006	$3.5

8. **General Interest**   The following data show the total U.S. production of biofuels (in trillions of Btu) from 1998–2007. Use these data to construct a broken-line graph.

Year	Production (Btu, in trillions)	Year	Production (Btu, in trillions)
1998	206	2003	412
1999	215	2004	501
2000	238	2005	582
2001	260	2006	745
2002	315	2007	993

(Source: Energy Information Association)

9. **Fire Protection**   The following data show the number of total acres burned in wildfires in the United States from 1998–2007. The numbers have been rounded to the nearest hundred thousand. Plot a broken-line graph of these data.

Year	Acres Burned	Year	Acres Burned
1998	1,300,000	2003	4,000,000
1999	5,600,000	2004	8,100,000
2000	7,400,000	2005	8,700,000
2001	3,600,000	2006	9,900,000
2002	7,200,000	2007	9,300,000

(Source: National Interagency Fire Center)

10. **Allied Health** Plot a broken-line graph for these data.

Percentage of Male Nurses, 1890–2000

Year	Percent	Year	Percent
1890	13%	1950	2%
1900	10%	1960	2%
1910	7%	1970	3%
1920	4%	1980	4%
1930	2%	1990	6%
1940	1%	2000	7%

11. **Hydrology** The following table shows the daily evaporation and usage totals, in acre-feet, during a summer week at Bradbury Dam. Use these data to construct a double broken-line graph.

Day	Daily Evaporation (in acre-ft)	Daily Usage (in acre-ft)
Monday	62.3	108.7
Tuesday	58.2	130.2
Wednesday	51.5	117.9
Thursday	58.6	126.5
Friday	67.3	130.5
Saturday	67.3	130.2
Sunday	62.8	128.0

12. **Business and Finance** Plot a double broken-line graph for these data.

Actual and Projected Sales, IMD Corp. (× $1000)

Month	Actual	Projected
January	$10	$45
February	$18	$50
March	$12	$47
April	$22	$50

(continued)

Actual and Projected Sales, IMD Corp. (× $1000) (Continued)

Month	Actual	Projected
May	$40	$65
June	$39	$71
July	$50	$76
August	$42	$75
September	$35	$55
October	$37	$60
November	$41	$50
December	$44	$87

13. **Electrical Engineering** Plot the following data as a circle graph.

California's In-State Sources of Electricity

Source	Percent	Source	Percent
Renewable	11%	Natural gas	41%
Nuclear	13%	Large hydroelectric	19%
Coal	16%		

14. **Business and Finance** Plot a circle graph using these data.

Revenue for MEGA-CASH Construction Co., 2008 ($ billions)

First Quarter	Second Quarter	Third Quarter	Fourth Quarter
5.5	7.0	6.5	3.0

15. **Aviation** An aircraft mechanic spends 12.5% of a 40-hr week working on aircraft airframes, 37.5% of the week on landing gear, 43.75% of the week working on power plants and 6.25% of the week on avionics. Plot a circle graph using this information.

16. **Construction** The Minneapolis bridge disaster of 2007 has raised concern over the age and condition of other bridges in the country. The following table summarizes the age of U.S. bridges. Make a circle graph of these data.

Age of U.S. Bridges	Percentage
20 years or less	29%
21 to 30 years	14%
31 to 40 years	20%
41 years or older	37%

(Source: American Association of State Highway and Transportation Officials)

Check your answers to the odd-numbered problems in the Appendix, then turn to Section 12-2 to learn about measures of central tendency.

A **measure of central tendency** is a single number that summarizes an entire set of data. The phrase "central tendency" implies that these measures represent a central or middle value of the set. Those who work in the trades and other fields, as well as consumers, use these measures to help them make other important calculations, comparisons, and projections of future data. In this chapter, we shall study three measures of central tendency: the mean, the median, and the mode.

Mean   As we learned in Chapter 3, the **arithmetic mean,** also referred to as simply the **mean** or the **average,** of a set of data values is given by the following formula:

$$\text{Mean} = \frac{\text{sum of the data values}}{\text{the number of data values}}$$

Note   In this chapter, if the mean is not exact, we will round it to one decimal digit more than the least precise data value in the set. ◄

Example 1

Aviation   An airplane mechanic was asked to prepare an estimate of the cost for an annual inspection of a small Bonanza aircraft. The mechanic's records showed that eight prior inspections on the same type of plane had required 9.6, 10.8, 10.0, 8.5, 11.0, 10.8, 9.2, and 11.5 hr. To help prepare his estimate, the mechanic first found the mean inspection time of the past jobs as follows:

$$\text{Mean inspection time} = \frac{\text{sum of prior inspection times}}{\text{number of prior inspections}}$$

$$= \frac{9.6 + 10.8 + 10.0 + 8.5 + 11.0 + 10.8 + 9.2 + 11.5}{8}$$

$$= \frac{81.4}{8} = 10.175 \approx 10.18 \text{ hr}$$

To complete his cost estimate, the mechanic would then multiply this mean inspection time by his hourly labor charge.

→ Your Turn

Machine Trades   A machine shop produces seven copies of a steel disk. The thicknesses of the disks, as measured by a vernier caliper are, in inches, 1.738, 1.741, 1.738, 1.740, 1.739, 1.737, and 1.740. Find the mean thickness of the disks.

→ Solution

$$\text{Mean thickness} = \frac{1.738 + 1.741 + 1.738 + 1.740 + 1.739 + 1.737 + 1.740}{7}$$

$$= \frac{12.173}{7} = 1.739 \text{ in.}$$

 1.738 ⊕ 1.741 ⊕ 1.738 ⊕ 1.74 ⊕ 1.739 ⊕ 1.737 ⊕ 1.74 ⊜ ÷ 7 ⊜ →

| | *1.739* |

Most scientific calculators have special "STAT" and "DATA" functions that allow you to enter data and calculate statistical measures. However, before using any of these functions, you should learn to set up the calculations and compute these measures the "long way." ◀

Median

Another commonly used measure of central tendency is the median. The **median** of a set of data values is the middle value when all values are arranged in order from smallest to largest.

Example 2

Machine Trades To determine the median thickness of the disks in the previous Your Turn, first we arrange the seven measurements in order of magnitude:

1.737   1.738   1.738   1.739   1.740   1.740   1.741

The middle value is 1.739 because there are three values less than this measurement and three values greater than it. Therefore, the median thickness is 1.739 in. Notice that for these data, the median is equal to the mean. This will not ordinarily happen.

In the previous example, the median was easy to find because there was an odd number of data values. If a set contains an even number of data values, there is no single middle value, so the median is defined to be the mean of the two middle values.

→ Your Turn

Aviation Find the median inspection time for the data in the first example of this section.

→ Solution

First, arrange the eight prior times in order from smallest to largest as follows:

8.5   9.2   9.6   10.0   10.8   10.8   11.0   11.5

Next, because there is an even number of inspection times, there is no single middle value. We must locate the two middle times. To help us find these, we begin crossing out left-end and right-end values alternately until only the two middle values remain.

8̶.̶5̶   9̶.̶2̶   9̶.̶6̶ $\underbrace{10.0 \quad 10.8}$ 1̶0̶.̶8̶   1̶1̶.̶0̶   1̶1̶.̶5̶

The two middle times

Finally, calculate the mean of the two middle values. This is the median of the set.

Median inspection time $= \dfrac{10.0 + 10.8}{2} = 10.4$ hours

Notice that the median inspection time of 10.4 differs slightly from the mean time of 10.18.

The median and the mean are the two most commonly used "averages" in statistical analysis. The median tends to be a more representative measure when the set of data contains relatively few values that are either much larger or much smaller than the rest of the numbers. The mean of such a set will be skewed toward these extreme values and will therefore be a misleading representation of the data. ◀

Mode The final, and perhaps least used, measure of central tendency that we will consider is the mode. The **mode** of a set of data is the value that occurs most often in the set. If there is no value that occurs most often, there is no mode for the set of data. If there is more than one value that occurs with the greatest frequency, each of these is considered to be a mode.

Example 3

Aviation The mode inspection time for the Bonanza aircraft (see the previous example) is 10.8 hr. It is the only value that occurs more than once.

→ **Your Turn**

If possible, find the mode(s) of each set of numbers.

(a) 3, 7, 6, 4, 8, 6, 5, 6, 3   (b) 22, 26, 21, 30, 25, 28   (c) 9, 12, 13, 9, 11, 13, 8

→ **Solutions**

(a) The value 6 appears three times, the value 3 appears twice, and all others appear once. The mode is 6.

(b) No value appears more than once in the set. There is no mode.

(c) Both 9 and 13 appear twice each, and no other value repeats. There are two modes, 9 and 13. A set of data containing two modes is often referred to as a **bimodal** set.

→ **More Practice**

Find the mean, median, and mode of each set of numbers.

1. 22, 25, 23, 26, 23, 21, 24, 23

2. 6.5, 8.2, 7.7, 6.7, 8.9, 7.3, 6.9

3. 157, 153, 155, 157, 160, 153, 159, 158, 166, 154, 168

4. **Automotive Trades** A daily survey revealed the following prices for a gallon of regular unleaded gas at various stations in a certain metropolitan area:

    $2.39, $2.49, $2.39, $2.59, $2.45, $2.49, $2.35, $2.55, $2.39, $2.65

    Find the mean, median, and mode of these prices.

5. **Construction** For quality control purposes, 8-ft-long 2-by-4s are randomly selected to be carefully measured. Find the mean, median, and mode for the following measurements.

    $8'0000''$ $8'\dfrac{1''}{8}$ $8'\dfrac{1''}{16}$ $7'\dfrac{15''}{16}$ $7'\dfrac{7''}{8}$ $8'0000''$ $8'\dfrac{1''}{8}$ $8'\dfrac{3''}{8}$ $7'\dfrac{5''}{8}$ $8'0000''$

→ **Solutions**

1. Mean: 23.375 or 23.4, rounded; median: 23; mode: 23

2. Mean: 7.457 . . . or 7.46, rounded; median: 7.3; mode: none

3. Mean: 158.18 . . . or 158.2, rounded; median: 157; modes: 153 and 157

4. Mean: \$2.474; median: \$2.47; mode: \$2.39

5. Mean: $8\frac{1'}{80}$ or $8.0125'$; Median: $8'0000''$; Mode: $8'0000''$

Grouped Frequency Distributions

When a set of data contains a large number of values, it can be very cumbersome to deal with. In such cases, we often condense the data into a form known as a **frequency distribution.** In a frequency distribution, data values are often grouped into intervals of the same length, known as **classes,** and the number of values within each class is tallied. The result of each tally is called the **frequency** of that class.

Example 4

Drafting Consider the following data values, representing the number of hours during one week that a group of drafting students spent working on the conception, design, and modeling of a crankshaft.

```
19  27  56  37  61  42  39  53  73  46  30  48  59  26  45  39  21
33  30  37  16  25  15  24  13  23  41  34  27  56  32  45  17  31
46  16  30  32  63  24
```

To construct a grouped frequency distribution, we must first decide on an appropriate interval width for the classes. We shall use intervals of 9 hr beginning at 10 for this data.* We then construct a table with columns representing the "Class Intervals" and the "Frequency, F," or number of students whose hours on the project fell within each interval. To help determine the frequency within each class, use tally marks to count data points. Cross out each data point as it is counted. The final table looks like this:

Class Intervals	Tally	Frequency, F
10–19	⦀⦀	6
20–29	⦀⦀⦀	8
30–39	⦀⦀ ⦀⦀⦀	12
40–49	⦀⦀	7
50–59	⦀⦀	4
60–69	⦀	2
70–79	⦀	1

→ Your Turn

Welding Construct a frequency distribution for the following percent scores on a gas tungsten arc welding project. Use class intervals with a width of 14 percentage points beginning at 41%.

```
81  75  93  74  56  56  87  93  75  70  46  67  91  72  73  81  76
65  44  83  74  83  93  65  49  62  78  80  42  58  63  54  79  86
77  92  79  66  55  75
```

*Statisticians consider many factors when deciding on the width of the class intervals. This skill is beyond the scope of this text, and all problems will include an appropriate instruction.

Class Intervals	Tally	Frequency, F
41–55	卌	6
56–70	卌 卌	10
71–85	卌 卌 卌 ||	17
86–100	卌 ||	7

Mean of Grouped Data

To calculate the mean of data that are grouped in a frequency distribution, follow these steps:

Step 1 Find the midpoint, M, of each interval by calculating the mean of the endpoints of the interval. List these in a separate column.

Step 2 Multiply each frequency, F, by its corresponding midpoint, M, and list these products in a separate column headed "Product, FM."

Step 3 Find the sum of the frequencies and enter it at the bottom of the F column. We will refer to this sum as n, the total number of data values.

Step 4 Find the sum of the products from Step 2 and enter it at the bottom of the FM column.

Step 5 Divide the sum of the products from Step 4 by n. This is defined as the mean of the grouped data.

Example 5

Drafting To find the mean number of hours worked on the crankshaft design (see the previous example), extend the table as shown. The tally column has been omitted to save space. The results of each step in the process are indicated by arrow diagrams.

Class Intervals	Frequency, F	Midpoint, M	Product, FM
10–19	6	14.5	87
20–29	8	24.5	196
30–39	12	34.5	414
40–49	7	44.5	311.5
50–59	4	54.5	218
60–69	2	64.5	129
70–79	1	74.5	74.5
	$n = 40$		Sum = 1430

Step 5   Mean $= \dfrac{\text{sum of products, } FM}{n} = \dfrac{1430}{40} = 35.75$ hr

$= 35.8$ hr, rounded

Using a calculator for Steps 2–5, we enter:

6 $\times$ 14.5 $+$ 8 $\times$ 24.5 $+$ 12 $\times$ 34.5 $+$ 7 $\times$ 44.5 $+$ 4 $\times$ 54.5 $+$ 2 $\times$
64.5 $+$ 74.5 $=$ $\div$ 40 $=$ →   *35.75*

Note   The mean of grouped data is not equal to the mean of the ungrouped data unless the midpoint of every interval is the mean of all the data in that interval. Nevertheless, this method for finding the mean from a frequency distribution provides us with a useful approximation when there are a large number of data values. In the previous example, the mean of the ungrouped data is 35.775 hr, which rounds to the same result obtained from the grouped data. ◄

→ **Your Turn**

Welding   Find the mean score from the gas tungsten arc welding project (see the previous Your Turn).

→ **Solution**

Class Intervals	Frequency, F	Midpoint, M	Product, FM
		Step 1	Step 2
41–55	6	48	288
56–70	10	63	630
71–85	17	78	1326
86–100	7	93	651
	$n = 40$		Sum $= 2895$
	Step 3		Step 4

Step 5   Mean score $= \dfrac{\text{Sum of products, } FM}{n} = \dfrac{2895}{40} = 72.375\%$

$= 72.4\%$, rounded

A Closer Look   The mean of the ungrouped scores is 71.7%. There is a slight discrepancy with the mean of the grouped scores because the class intervals were relatively wide in this case. ◄

Now turn to Exercises 12-2 for more practice on measures of central tendency.

A. Find the mean, median, and mode for each set of numbers.

1. 9, 4, 4, 8, 7, 6, 2, 9, 4, 7

2. 37, 32, 31, 34, 36, 33, 35

3. 98, 79, 99, 79, 54, 52, 98, 58, 73, 62, 54, 54

4. 3.488, 3.358, 3.346, 3.203, 3.307

5. 123, 163, 149, 132, 183, 167, 105, 192

6. 7.9, 8.1, 9.5, 8.1, 9.6, 8.7, 7.7, 9.5, 8.1

7. 1188, 1176, 1128, 1126, 1356, 1151, 1313, 1344, 1367, 1396

8. 0.39, 0.39, 0.84, 0.11, 0.78, 0.18, 0.78

B. Construct an extended frequency distribution for each set of numbers and calculate the mean of the grouped data. Follow the instructions for interval width.

1. Use intervals of 0.09 beginning at 1.00.

1.52	1.68	1.46	1.13	1.89	1.44	1.56	1.09	1.81	1.79	1.23
1.46	1.91	1.40	1.30	1.35	1.79	1.61	1.80	1.26	1.29	1.40
1.57	1.53	1.28	1.45	1.72	1.35	1.93	1.86	1.95	1.70	1.88
1.76	1.05									

2. Use intervals of 49 beginning at 100.

245	106	112	321	235	209	263	215	168	350	340	401	179
433	286	141	358	468	166	498	341	171	119	362	264	325
225	391	133	127									

C. Applications

1. **Aviation** BF Goodrich produces brake pads for commercial airliners. Two production teams are working to produce 15-lb brake pads over a six-month period. Team A works on one furnace deck, while production team B works on a second furnace deck. A 15-lb carbon brake pad costs $1000 per pad to produce. Use the information in the table to answer the questions that follow.

Month	Team A	Team B
January	37,750 brake pads	40,000 brake pads
February	34,500	41,500
March	35,250	39,000
April	38,750	35,750
May	39,250	32,500
June	36,500	33,250

(a) Calculate the mean monthly production for team A.

(b) Calculate the mean monthly production for team B.

(c) Find the median monthly production for team A.

(d) Find the median monthly production for team B.

(e) Calculate the average cost per month for the two teams combined.

2. **General Interest** The U.S. Department of Labor reports the following average hourly earnings of workers in eight different sectors in 2008:

Sector	Average Hourly Earnings
Construction	$21.87
Leisure and hospitality	$10.84
Education and health services	$18.88
Professional and business services	$21.19
Trade, transportation, and utilities	$16.16
Financial	$20.27
Manufacturing	$16.97
Mining and logging	$22.50

(a) Find the mean hourly earnings for the eight sectors. (Round to the nearest cent.)

(b) Find the median hourly earnings for the eight sectors. (Round to the nearest cent.)

(c) In a 40-hr work week, how much more would an average construction worker earn compared to the mean earnings of the eight sectors?

(d) In a 40-hr work week, how much less would an average employee in the leisure and hospitality sector earn compared to the median earnings of the eight sectors?

(e) By what percent does the earnings of the average financial employee exceed the mean earnings of the eight sectors?

3. **Automotive Trades** The following table compares the curb weights of six different models of 2008 Hondas.

Model	Curb Weight (lb)
Accord EX-L	3545
Civic Hybrid	2900
Civic Mugen Si	2930
Civic Si Sedan	2945
Fit Sport	2475
S2000 CR	2790

(a) What is the mean of these curb weights?

(b) What is the median of these curb weights?

4. **Wastewater Technology**   The seven-day mean of settleable solids cannot exceed 0.15 milliliters per liter (mL/L). During the past seven days, the measurements of concentration have been 0.13, 0.18, 0.21, 0.14, 0.12, 0.11, and 0.15 mL/L. What was the mean concentration? Was it within the limit?

5. **Automotive Trades**   A mechanic has logged the following numbers of hours in maintaining and repairing a certain Detroit Series 6 diesel engine for ten different servicings.

   8.5   15.9   20.0   6.5   11.2   19.1   18.0   7.4   9.8   22.5

   (a)   Find the mean for the hours logged.

   (b)   Find the median for the hours logged.

   (c)   If the mechanic earns $17.95 per hour, how much does he earn doing the median servicing on this engine?

6. **Forestry**   A forest ranger wishes to determine the velocity of a stream. She throws an object into the stream and clocks the time it takes to travel a premeasured distance of 200 ft. She then divides the time into 200 to calculate the velocity in feet per second. For greater accuracy, she repeats this process four times and finds the mean of the four velocities. If the four trials result in times of 21, 23, 20, and 23 sec, calculate the mean velocity to the nearest 0.1 ft/sec.

7. **Hydrology**   The following table shows the monthly flow through the Darnville Dam from October through April.

Month	Volume (in Thousand Acre Feet, kAF)
October	92
November	133
December	239
January	727
February	348
March	499
April	1031

   (a)   Calculate the mean monthly water flow.

   (b)   Find the median monthly water flow.

   (c)   By what percent did the flow in March exceed the mean flow?

8. **Meteorology**   The National Weather Service provides data on the number of tornados in each state that occur throughout the year. Use the following statistics to answer questions (a)–(e).

Year	Kansas	Oklahoma
2000	70	46
2001	110	70
2002	104	23
2003	93	88
2004	125	65
2005	140	28

(a) Find the mean of the number of tornados in Kansas over the six-year period.

(b) Find the mean of the number of tornados in Oklahoma over the six-year period.

(c) Find the median of the number of tornados in Kansas over the six-year period.

(d) Find the median of the number of tornados in Oklahoma over the six-year period.

(e) Suppose a tornado causes an average $250,000 worth of damage. Use the mean in part (a) to calculate the average cost per year to repair tornado damage in Kansas.

9. **Automotive Trades** The following table compares the SAE horsepower ratings (hp) of six different models of 2008 Hondas.

Model	Horsepower Rating (hp)
Accord EX-L	268
Civic Hybrid	110
Civic Mugen Si	202
Civic Si Sedan	197
Fit Sport	109
S2000 CR	237

(a) What is the mean horsepower rating of these models? (Round your answer to one decimal digit.)

(b) What is the median horsepower rating of these models?

10. **Hydrology** The following table shows the daily evaporation and usage totals, in acre-feet, during a summer week at Bradbury Dam. Use these data to answer the questions.

Day	Daily Evaporation (in acre-ft)	Daily Usage (in acre-ft)
Monday	62.3	108.7
Tuesday	58.2	130.2
Wednesday	51.5	117.9
Thursday	58.6	126.5
Friday	67.3	130.5
Saturday	67.3	130.2
Sunday	62.8	128.0

(a) Find the mean daily evaporation.

(b) Find the median daily evaporation.

(c) Find the mean daily usage.

(d) Find the median daily usage.

11. **Allied Health** The Apgar score is widely used to assess the general health of infants shortly after birth. Scores range from 0 (very poor) to 10 (perfect). To evaluate the success of a new maternal health program, a hospital compared the Apgar scores of 20 infants whose mothers participated in the health program

(P = participant) with the scores of 20 infants whose mothers did not participate (NP = nonparticipant). The infants' scores for the two groups are shown in the following table.

P:	7	8	7	10	7	9	8	4	8	10	9	7	8	8	5	10	9	7	8	9
NP:	8	10	5	7	6	8	7	6	9	7	4	6	7	9	2	9	6	10	8	7

(a) Calculate the mean and median of the Apgar scores for the infants whose mothers participated in the program.

(b) Calculate the mean and median of the Apgar scores for the infants whose mothers did not participate in the program.

12. **Allied Health**   A pharmacist keeps careful track of the amount of medication sold each month in order to ensure that supplies are always available for her customers. However, she must avoid storing too much medication so that they do not expire before they are sold. For the anti-depressant medication Alljoy, the following amounts were sold in 2008:

Month	Packages Sold
January	242
February	273
March	187
April	163
May	155
June	135
July	93
August	148
September	159
October	174
November	238
December	256

(a) Calculate the mean monthly number of packages of Alljoy sold during the year.

(b) Find the median monthly number of packages of Alljoy sold during the year.

(c) Calculate the monthly mean for the winter months of January, February, and March.

(d) Calculate the monthly mean for the summer months of June, July, and August.

(e) What was the percent of decrease from winter [answer (c)] to summer [answer (d)]?

For Problems 13 and 14, construct an extended frequency distribution for each set of numbers and calculate the mean of the grouped data. Follow the instruction for interval width.

13. **Automotive Trades**   To best determine the fuel economy of a 2006 Honda Odyssey, an automotive technician calculates the mean gas mileage computed for 20 full tanks of gas.

(a) Use intervals of 0.9 beginning at 19.0 to compute the mean fuel economy (in miles per gallon) from the following data.

21.7 19.5 20.6 25.6 24.3 20.1 20.7 22.5 20.2 24.5
23.7 20.1 23.1 19.9 23.5 24.4 19.7 25.2 21.8 22.3

(b) At $3.00 per gallon, use the mean fuel economy to calculate the average cost per mile to operate this vehicle.

14. **Allied Health** A registered nurse has been carefully monitoring the blood work of a patient each day over a 24-day period. Each day a blood analysis was performed, and the patient's medication, in milligrams, was titrated (adjusted) accordingly. Use the values of the medication administered to determine the mean dose for the last 24 days. Use intervals of 0.49 beginning at 78.50.

Doses in mg

80.63 79.78 78.78 80.79 81.13 79.17 78.66 79.13
79.60 79.29 81.29 81.33 78.71 81.47 79.89 80.41
80.01 79.46 78.84 79.26 79.58 79.47 80.60 81.08

Check your answers to the odd-numbered problems in the Appendix, then turn to Problem Set 12 on page 759 for more practice on graphs and statistics. If you need a quick review of the topics in this chapter, visit the chapter summary first.

Summary **Statistics**

Objective

Read bar graphs, line graphs, and circle graphs. (pp. 717, 725, 729)

Review

For vertical bar graphs and all line graphs, locate numerical information along the vertical axis and categories or time periods along the horizontal axis. The numerical value associated with a bar or a point is the number on the vertical scale horizontally across from it. For horizontal bar graphs, the labeling of the axes is switched.

Example:

(a) Consider the double-bar graph below. To find the number of frames assembled by the Tuesday day shift, locate "Tuesday" on the horizontal axis. Then find the top of the day shift bar, and look horizontally across to the vertical axis. The day shift on Tuesday assembled approximately 70 frames.

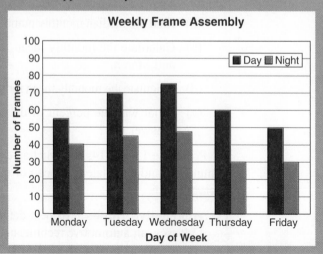

(b) Study the broken-line graph below. Suppose we wish to calculate the percent increase in the number of paint jobs from January to February. From the graph, we read the January job total to be 25, and the February total to be 35. The percent increase is:

$$\frac{35 - 25}{25} = \frac{R}{100}$$

$$\frac{10}{25} = \frac{R}{100}$$

$$R = \frac{10 \times 100}{25} = 40\%$$

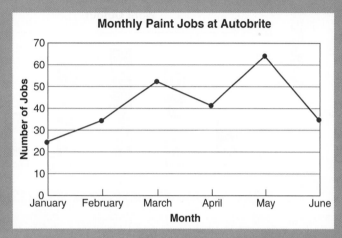

Monthly Paint Jobs at Autobrite

For circle graphs, or pie charts, each sector is usually labeled with a percent.

Example: Consider the circle graph below. Suppose the average plumbing job generates $227.50. To calculate the portion of this revenue spent on advertising, notice that the advertising sector is labeled 18%. Therefore, the portion of the average job spent on advertising would be

$$0.18 \times \$227.50 = \$40.95$$

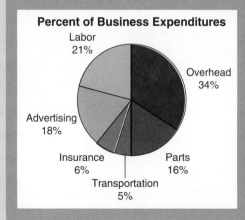

Percent of Business Expenditures

Draw bar graphs, line graphs, and circle graphs. (pp. 720, 727, 730)

For bar graphs and line graphs, construct a horizontal and vertical axis. Label the horizontal axis with categories or time periods and space these equally along the axis. Use the vertical axis to represent the numerical data. (The axes may be switched to make a horizontal bar graph.) Choose a convenient scale that will cover the range of numbers and provide reasonably precise readings for the graph. Be sure to title the graph.

Example: The data in the table were used to create the bar graph that follows.

Work Output	Day
150	Monday
360	Tuesday
435	Wednesday
375	Thursday
180	Friday

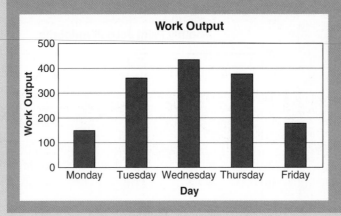

To create a circle graph, we must first convert each numerical value to its percent of the total. Then convert each percent to its corresponding portion of 360°. Finally, use a protractor to measure the sectors of the circle graph. Title the graph and label each sector with its category name and its percent of the total.

Example: For the work output data from the previous example, first we calculate that the total output for the week is 1500. The percent represented by Monday's output can then be determined as follows:

$$\text{Monday output} \rightarrow \frac{150}{1500} = \frac{R}{100} \qquad R = \frac{150 \times 100}{1500} = 10\%$$
$$\text{Weekly output} \rightarrow$$

Converting this percent to degrees, we have:

$$0.10 \times 360° = 36°$$

Continuing this process for the remaining days of the week, we obtain the results shown in the table. The completed circle graph is shown to the right of the table.

Day	Percent	Degrees
Monday	10%	36°
Tuesday	24%	86°
Wednesday	29%	104°
Thursday	25%	90°
Friday	12%	44°

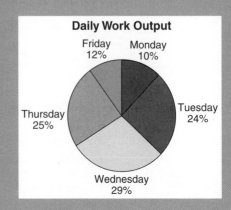

Objective

Calculate measures of central tendency: mean, median, and mode. (pp. 743–748)

Review

The following formula is used to find the mean:

$$\text{Mean} = \frac{\text{sum of the data values}}{\text{number of values}}$$

The median is the middle value when a set of data is arranged in order from smallest to largest. If there is an even number of data values, the median is the mean of the middle two values.

The mode is the data value that occurs most often in a set of data. There may be more than one mode or no mode at all.

> **Example:** Find (a) the mean, (b) the median, and (c) the mode for the following set of lengths. All measurements are in meters.
>
> 12.7 16.2 15.5 13.9 13.2 17.1 15.5
>
> (a) The sum of the lengths is 104.1, and there are 7 measurements. The mean is
>
> $$\frac{104.1}{7} = 14.871\ldots \approx 14.87\,\text{m}$$
>
> (b) Rearranging the lengths in numerical order, we have
>
> 12.7 13.2 13.9 15.5 15.5 16.2 17.1
>
> The median, or middle value, is 15.5 m.
>
> (c) There are two measurements of 15.5 m—this is the mode.

Statistics

Answers to odd-numbered problems are given in the Appendix.

A. Answer the questions based on the graphs pictured.

Questions 1 to 5 refer to Graph I.

Graph I   Electrical Trades

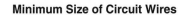

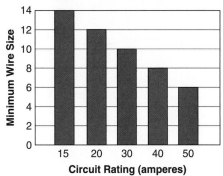

Minimum Size of Circuit Wires

1. In general, as amps increase, the necessary wire size _____.
   (increases, decreases)

2. What is the minimum size of wire needed for a circuit rating of 30 amps?

3. What is the minimum wire size needed for a circuit rating of 15 amps?

4. What circuit ratings could a wire size of 8 be used for?

5. What circuit ratings could a size 14 wire be used for?

Questions 6 to 15 refer to Graph II.

Graph II   Printing

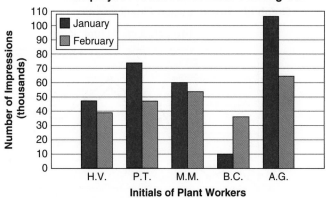

Employee Production at Miller Printing Co.

Name _____

Date _____

Course/Section _____

6. What was H.V.'s output in February?

7. What was P.T.'s output for January?

8. Who produced the most in January?

9. Who produced the least in January?

10. Who produced the most in February?

11. Who produced the least in February?

12. Which workers produced more in February than January?

13. Which workers produced less in February than January?

14. Determine the total production for January.

15. Determine the total production for February.

Problems 16 to 22 refer to Graph III.

Graph III   Automotive Trades

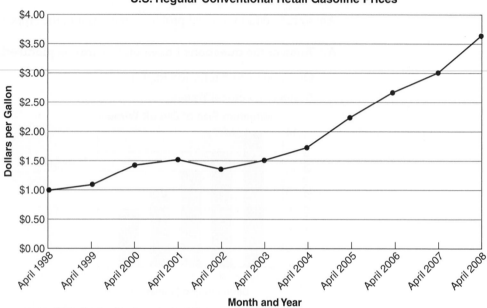

16. What was the cost per gallon of regular gas in April 2003?

17. At what other time was the cost per gallon the same as in April 2003?

18. During which time period did the cost of gas decrease?

19. During which two time periods did the largest price increases occur?

20. What was the overall increase in the cost per gallon over the time frame shown?

21. By what percent did the cost increase from April 1998 to April 2004?

22. How much more did it cost to fill up a 20-gal tank in 2007 than in 2001?

Problems 23 to 30 refer to Graph IV.

Graph IV   Business and Finance

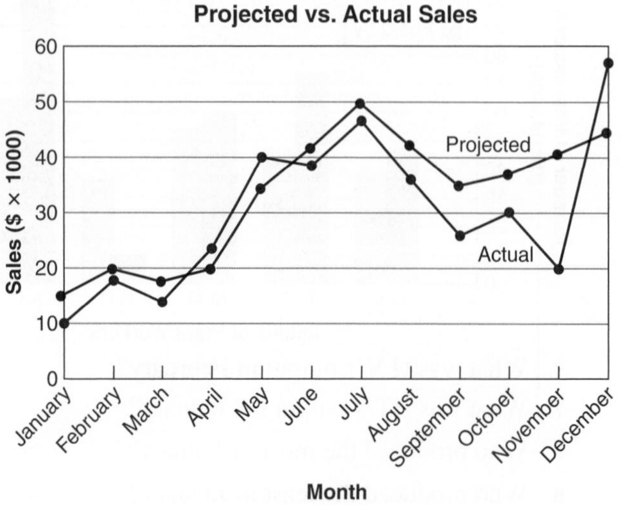

23. What was the actual sales total in October?

24. What was the projected sales total in May?

25. During which month were actual sales highest?

26. During which month were projected sales highest?

27. During which month were actual sales lowest?

28. During which month were projected sales lowest?

29. During which months were actual sales lower than projected sales?

30. During which month was the gap between actual and projected sales the largest? What was the difference?

Problems 31 to 35 refer to Graph V.

Graph V   Metalworking

Composition of Marine Bronze

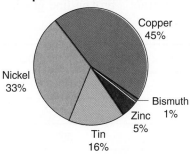

31. What percent of marine bronze is composed of tin?

32. What percent of this alloy is made up of quantities other than copper?

33. Without measuring, calculate how many degrees of the circle should have gone to nickel.

34. How many ounces of zinc are there in 50 lb of marine bronze?

35. How many grams of bismuth are there in 73 grams of this alloy?

B. Construct graphs from the data given.

1. **Construction**   Construct a bar graph based on these data.

Fire Resistance Ratings of Various Plywoods

Type of Wood	Fire Resistance Rating
A: $\frac{1}{4}$ in. plywood	10
B: $\frac{1}{2}$ in. plywood	25
C: $\frac{3}{4}$ in. tongue and groove	20
D: $\frac{1}{2}$ in. gypsum wallboard	40
E: $\frac{1}{2}$ in. gypsum—two layers	90
F: $\frac{5}{8}$ in. gypsum wallboard	60

2. **Masonry** Construct a double-bar graph based on these data.

Output of Bricklayers for the ABC Construction Co.

Initials of Employee	Bricks Laid	
	March	April
R.M.	1725	1550
H.H.	1350	1485
A.C.	890	1620
C.T.	1830	1950
W.F.	1175	1150
S.D.	2125	1875

3. **Meteorology** Construct a broken-line graph based on these monthly average high temperatures.

Month	Average High Temperature
January	42°
February	44°
March	50°
April	56°
May	65°
June	75°
July	81°
August	84°
September	77°
October	69°
November	57°
December	46°

4. **Allied Health** Construct a double broken-line graph based on these recorded blood pressures of a patient over a 12-hr period.

Time	Systolic	Diastolic	Time	Systolic	Diastolic
7 A.M.	124	85	1 P.M.	118	78
8	128	88	2	124	78
9	120	80	3	115	75
10	116	75	4	112	78
11	120	78	5	116	74
12 P.M.	114	80	6	120	80

5. **Metalworking** Construct a circle graph based on the alloy composition shown in the table.

Metal	Alloy Composition
Copper	43%
Aluminum	29%
Zinc	3%
Tin	2%
Lead	23%

6. **Fire Protection** Construct a circle graph based on the portion of calls represented by each category during a particular week at a fire station.

Type of Call	Number of Calls	Type of Call	Number of Calls
Fire	4	False alarm	10
Medical	9	Hazardous leak/spill	2
Traffic accident	12	Other	3

C. Find the mean, median, and mode for each set of numbers.

1. 89, 88, 84, 87, 89, 84, 84, 83, 85, 89, 84

2. 5.8, 3.2, 4.1, 3.5, 4.3, 5.2

3. 20.220, 20.434, 20.395, 20.324, 20.345, 20.356, 20.324, 20.258

4. 4287, 4384, 4036, 6699, 6491, 5460, 3182, 6321, 6642

5. 696, 658, 756, 727, 606, 607, 727, 635, 733, 637

6. 14.66, 17.70, 13.83, 15.53, 17.35, 12.18, 14.80

D. Calculate the mean of the grouped data in each problem.

1.

Class Intervals	Frequency, F
0–4.9 cu cm	12
5.0–9.9 cu cm	9
10.0–14.9 cu cm	3
15.0–19.9 cu cm	2
20.0–24.9 cu cm	6
25.0–29.9 cu cm	8
30.0–34.9 cu cm	10

2.

Class Intervals	Frequency, F
22–26 ft	4
27–31 ft	7
32–36 ft	11
37–41 ft	9
42–46 ft	5
47–51 ft	2

E. Applications

1. **Printing**   Technica Replica uses high-speed printers for large-volume copying. The following values represent the number of copies requested, followed by the number of impressions that were wasted during each run.
   (a) For each case, calculate the percent of waste and fill in the chart. (Round to the nearest 0.1%.)

Copies Requested	Wasted Impressions	Percent of Waste
15,500	409	
17,500	345	
18,000	432	
16,750	546	
14,000	316	
16,250	845	
18,500	538	

   (b) Find the median of the percent of waste.
   (c) Find the mean of the percent of waste.
   (d) There is an upcoming job that requires 17,000 usable copies. The supervisor decides to print 17,500. Based on the median percent of waste, should this provide enough of a cushion to complete the job?

2. **Meteorology**   The following table shows the average precipitation in Seattle, Washington, from March through August over the past 75 years.

Month	Inches	Month	Inches
March	3.75	June	1.50
April	2.50	July	0.75
May	1.75	August	1.25

   (a) Find the median monthly total for the six-month period.
   (b) Find the mean monthly total.

3. **Agriculture**   The following data from the United States Department of Agriculture represent the U.S. average price for corn in dollars per bushel.

Year	Price ($/Bushel)	Year	Price ($/Bushel)
1990–1991	$2.28	1999–2000	$1.82
1991–1992	$2.37	2000–2001	$1.85
1992–1993	$2.07	2001–2002	$1.97
1993–1994	$2.50	2002–2003	$2.32
1994–1995	$2.26	2003–2004	$2.42
1995–1996	$3.24	2004–2005	$2.06
1996–1997	$2.71	2005–2006	$2.00
1997–1998	$2.45	2006–2007	$3.04
1998–1999	$1.94	2007–2008	$4.00

(a) Determine the median yearly price from 1990–2008.

(b) Calculate the mean yearly price over this time period. (Round the nearest cent.)

(c) If a corn grower annually produces 3,000,000 bushels, use the mean price per bushel to calculate his average yearly earnings.

4. **Industrial Technology** The Energy Information Administration reports the following monthly prices for industrial customer natural gas in New York State (in dollars per thousand cubic feet) for the year 2008:

Month	Price	Month	Price
January	$12.36	July	$15.19
February	$11.97	August	$14.21
March	$12.61	September	$12.29
April	$12.77	October	$11.79
May	$13.52	November	$12.51
June	$14.73	December	$13.38

(a) What was the mean monthly price for the year?

(b) What was the median monthly price for the year?

(c) Use the median monthly price to compute the cost of 46,500 cu ft of natural gas.

5. **Police Science** To determine the speed that a vehicle was traveling at the time of an accident, police officers carefully measure the lengths of the skid marks left at the scene. There are many factors that determine the skid length, and a quick and simple formula for determining the speed is $S = \sqrt{30 \cdot L \cdot d \cdot e}$, where S is the speed in miles per hour, L is the skid length in feet, d is the drag factor, and e is the braking efficiency. The following table shows the skid lengths for 10 accidents that occurred on a particular stretch of an interstate during icy conditions this past winter.

Skid Length, L (ft)	Speed, S (mph)
514	
294	
726	
350	
216	
600	
476	
384	
486	
564	

(a) Use the given formula, with $d = 0.2$ (icy road) and $e = 1.0$, to calculate the speeds for the given skid lengths and fill in column 2. (Round to the nearest whole number.)

(b) Find the median speed.

(c) Find the mean speed.

6. **Automotive Trades** The following values from the National Highway Traffic Safety Administration give the average curb weights in pounds for domestic passenger cars in the United States from the years 1980 to 2003.

Curb Weights (lb)

3075	3101	3002	3153	3031	3033	2977	2982
3025	3062	3060	3071	3100	3046	3098	3146
3111	3143	3119	3124	3132	3168	3166	3161

(a) Construct a frequency distribution for the data using intervals of 49 beginning at 2976.

(b) Calculate the mean of the grouped data.

(c) The following chart represents the approximate fuel economy for certain curb weights.

Use the curb weight closest to the mean to determine the fuel economy for the average passenger car during this time period.

Fuel Economy	Curb Weight
30 mpg	2,500 lb
26 mpg	3,000 lb
23 mpg	3,500 lb
20 mpg	4,000 lb

7. **Sheet Metal Trades** The following data represent the gauges of the sheet metal used by a sheet metal worker during his last eight jobs.

28 GA   12 GA   16 GA   24 GA   22 GA   18 GA   30 GA   8 GA

(a) Use the chart to find the median thickness for these jobs in inches.

(b) Use the chart to find the mean thickness of the gauges used in millimeters.

Gauge	Thickness (in.)	Thickness (mm)
30	0.0157	0.3988
28	0.0187	0.4750
26	0.0217	0.5512
24	0.0276	0.7010
22	0.0336	0.8534
20	0.0396	1.0058
18	0.0516	1.3106
16	0.0635	1.6129
14	0.0785	1.9939
12	0.1084	2.7534
10	0.1382	3.5103
8	0.1681	4.2697

8. **Sheet Metal Trades** The following data represent the number of 48-in. by 96-in. (4-ft by 8-ft) sheets of aluminum used by the sheet metal worker for the 8 jobs in problem 7. Note that the number of sheets correspond to the gauges of sheet metal, respectively, as given in problem 7. (e.g., the first project used five 48-in. by 96-in. sheets of 28 gauge sheet metal).

5   7   10   16   9   12   3   4

(a) Calculate the total surface area, in square inches, of sheet metal used for each job. Then calculate the mean amount of surface area used per job.

(b) Find the total volume of aluminum, in cubic inches, used in each project by multiplying each surface area by the corresponding sheet thicknesses.

(c) Find the mean of the volumes from part (b).

(d) If aluminum costs approximately $0.236 per cubic inch, what is the cost of the average project?

Answers to Previews

Answers to Preview 1

1. (a) two hundred fifty thousand three hundred seventy-four
   (b) 1,065,008   (c) (1) 210,000   (2) 214,700

2. (a) 125   (b) 8607   (c) 37   (d) 2068   (e) 77

3. (a) 2368   (b) 74,115   (c) 640,140   (d) 334 remainder 2
   (e) 203

4. 445 lb

5. (a) 1, 2, 3, 4, 6, 12   (b) $2 \cdot 2 \cdot 3$

6. (a) 33   (b) 33   (c) 33   (d) 48

Answers to Preview 2

1. (a) $7\frac{3}{4}$ (b) $\frac{31}{8}$ (c) $\frac{20}{64}$ (d) $\frac{56}{32}$ (e) $\frac{5}{32}$ (f) $1\frac{7}{8}$

2. (a) $\frac{35}{256}$ (b) 3   (c) $\frac{9}{10}$ (d) $1\frac{1}{2}$ (e) $2\frac{3}{10}$ (f) 8

3. (a) $\frac{5}{8}$ (b) $1\frac{15}{16}$ (c) $\frac{11}{20}$ (d) $2\frac{11}{16}$

4. $1\frac{3}{8}$ in.

Answers to Preview 3

1. (a) 5.916   (b) 2.791   (c) 22.97   (d) 2.18225
   (e) 3256.25   (f) 225   (g) 3.045

2. 4.3

3. (a) 0.1875   (b) 106.0027   (c) twenty-six and thirty-five thousandths
   (d) 3.452   (e) 9.225   (f) 0.305

4. (a) $553.76   (b) 0.34 lb

Answers to Preview 4

1. (a) 5:2 or 2:5   (b) $8\frac{2}{3}$ to 1
2. (a) $x = 10$ (b) $y = 14.3$
3. (a) 64 ounces   (b) 18 ft   (c) 12 machines
4. (a) 25%   (b) 46%   (c) 500%   (d) 7.5%
5. (a) 0.35   (b) 0.0025   (c) $1\frac{3}{25}$
6. (a) 225   (b) 54   (c) 6.25% or $6\frac{1}{4}$%   (d) 75
       (e) 0.90%   (f) 10.4 lb   (g) $94.09   (h) 80%
       (i) 11.4%

Answers to Preview 5

1. (a) nearest hundredth   (b) nearest whole number   (c) nearest ten
   Measurement (a) is most precise.
       (d) 2 significant digits   (e) 3 significant digits   (f) 1 significant digit
2. (a) 44.2 sec   (b) 5.94 lb
3. (a) 94 sq ft   (b) 15 hr   (c) 13,000 lb
4. (a) 1040 oz   (b) 4.7 mi   (c) 504 sq in.
5. (a) 100 lb   (b) 85 qt
6. (a) 6.5 g   (b) 450 m   (c) 2850 sq m
7. (a) 12.7 gal   (b) 72 km/h   (c) 108°F

Answers to Preview 6

1. (a) -14 (b)
       (c) -250

2. (a) -8 (b) $11\frac{3}{4}$ (c) -7.5 (d) -3
3. (a) -3 (b) -8 (c) 10.9   (d) $-1\frac{7}{8}$
4. (a) -72 (b) 6   (c) 31   (d) $-\frac{3}{20}$ (e) 1500
5. (a) 64   (b) 4.2025
6. (a) 11   (b) 500   (c) 22
7. (a) 13   (b) 3.81

Answers to Preview 7

1. (a) 7   (b) 5   (c) $A = 12$ (d) $T = 4$ (e) $P = 120$
2. (a) $6ax^2$ (b) $-3x - y$ (c) $4x - 2$ (d) $3x^2 + 7x + 12$
3. (a) $x = 5$ (b) $x = 9$ (c) $x = 12$ (d) $N = 2S - 52$
       (e) $A = \dfrac{8M + BL}{L}$

4.   (a) $4A$ (b) i^2R (c) $E = \dfrac{100u}{I}$ (d) $R = \dfrac{12L}{D^2}$

5.   (a)   \$24.07   (b)   10.37 hr, or 11 hr, rounded up

6.   (a) $6y^2$ (b) $-12x^5y^4$ (c) $3xy - 6x^2$ (d) $-5x^5$ (e) $\dfrac{b^2}{3a^3}$

     (f) $4m^2 - 3m + 5$

7.   (a) 1.84×10^{-4} (b) 2.13×10^5 (c) 1.44×10^{-3}
     (d) 6.5×10^{-8}

Answers to Preview 8

1.   about $32°$

2. $\angle DEF$ is acute, $\angle GHI$ is obtuse, $\angle JKL$ is a right angle

3.   (a) $80°, 80°, 100°, 80°, 100°$ (b) $20°$

4.   (a)   obtuse and scalene triangle   (b)   parallelogram   (c)   trapezoid
     (d)   square   (e)   hexagon

5.   1.5 in.

6.   (a)   14 sq in.   (b) 24 cm^2 (c)   0.42 sq in., 2.4 in.
     (d)   22.9 sq in., 17.0 in.

7.   (a)   \$2337   (b)   \$2295.88

Answers to Preview 9

1.   (a)   cube   (b)   frustum of a cone   (c)   rectangular prism

2.   (a)   12.6 cu ft, 31.4 sq ft   (b) $38.8 \text{ cm}^3, 55.4 \text{ cm}^2$ (c)   3.7 cu in.
     (d)   80.5 sq in., 75.4 cu in.   (e) $180 \text{ cm}^2, 336 \text{ cm}^3$

3.   (a)   40,400 gal   (b)   24.3 lb   (c)   \$4536

Answers to Preview 10

1.   (a) $43°24'$ (b) $65.25°$

2.   (a)   0.80   (b) $143.2°$ (c)   4.2 in., 12.6 sq in.   (d)   1.7 rad/sec

3.   (a) $a = 6.0, b = 10.4, A = 30°$ (b) $d = 16.0, e = 22.6, F = 45°$

4.   (a)   0.438   (b)   0.105   (c)   0.943

5.   (a) $14°0'$ (b) $56°42'$ (c) $75.3°$ (d) $48.6°$

6.   (a) $X \approx 10.5, Y \approx 13.4$ (b) $\angle m = 44°25', X \approx 3.7$ in.
     (c) ≈ 39.8 in.   (d) $5°43'$

7.   (a) $A = 82°, B = 66°, C = 32°$ (b) $C = 33°, a = 7.6$ ft, $c = 4.4$ ft

Answers to Preview 11

1.   (a) $(4, 0)$ (b) $(-9, -3)$ (c)   dependent   (d)   inconsistent, no solution

2.   (a)   20 and 15   (b)   8 in. and 6 in.
     (c)   300 of the 4¢ screws, and 500 of the 6¢ screws

3. (a) $x = 4$ or $x = -4$ (b) $x = 0$ or $x = 7$
   (c) $x = 7$ or $x = -2$ (d) $x = -2\frac{2}{3}$ or $x = 2$ (e) no solution

4. (a) 6 cm (b) ≈ 8.4 in. (c) 12.9 amp (d) 6.5 in. by 9.5 in.

Answers to Preview 12

1. (a) 70 (b) 27% (c) 65 in May (d) 40% (e) $40.95

2. (a)

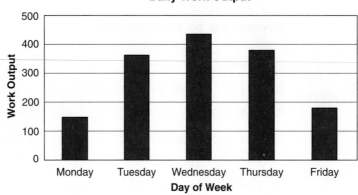

(b)

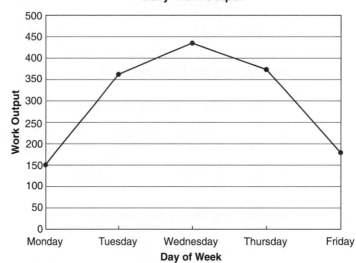

(c)

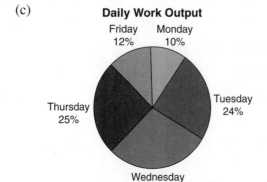

3. (a) 14.87 m (b) 15.5 m (c) 15.5 m

4. 136 psi

Answers to Odd-numbered Problems

Chapter 1 *Problems, page 7*

A.
10	11	11	12	16	7	9	15	12	13
10	13	15	8	14	9	13	18	9	11
16	8	10	17	9	14	12	7	10	11
13	13	12	16	9	11	10	10	14	17
11	14	11	10	11	14	13	12	13	15

B.
11	12	14	17	18	21	11	20	18	15
15	14	15	18	22	17	14	14	18	16
12	19	8	10	22	12	19	15	14	16

Exercises 1-1, page 14

A. 1. 70 3. 80 5. 123 7. 132 9. 393
11. 1390 13. 1009 15. 861 17. 9461 19. 11,428
21. 25,717 23. 11,071 25. 175,728 27. 663,264

B. 1. 1042 3. 2442 5. 7083 7. 6352 9. 6514
11. 64 13. 55 15. 357 17. 1,166,040

C. 1. three hundred fifty-seven
3. seventeen thousand ninety-two
5. two million thirty-four
7. seven hundred forty thousand one hundred six
9. one hundred eighteen million one hundred eighty thousand eighteen
11. 3006 13. 11,100 15. 4,040,006
17. 360 19. 4000 21. 230,000

D. 1. 4861 ft 3. 1636 screws 5. 1129 min
7. (a) 3607 watts (b) 1997 watts (c) 850 watts
9. 3114 11. $1196 13. 2929 ohms
15. 756 g 17. 3900 W
19. (a) 4264 (b) 3027 (c) 7291

E. 1. 97,001 kHz
   3. (a) $307,225   (b) $732,813   (c) $2,298,502   (d) $7156
   5. (a) #12 BHD: 11,453   (b) A3:3530
        #Tx: 258   A4:8412
        410 AAC: 12,715   B1:4294
        110 ACSR: 8792   B5:5482
        6B: 7425   B6:5073
                                C4:6073
                                C5:7779

Exercises 1-2, page 24

A. 1. 6   3. 2   5. 4   7. 3   9. 3   11. 8
   13. 9   15. 9   17. 3   19. 8   21. 7   23. 7
   25. 0   27. 8   29. 6   31. 6   33. 5   35. 4

B. 1. 13   3. 12   5. 15   7. 38   9. 46
   11. 25   13. 189   15. 281   17. 408   19. 273
   21. 574   23. 2809   25. 12,518 27. 4741   29. 47,593

C. 1. $459   3. 1758 ft   5. $330,535   7. 3 drums, by 44 liters
    9. 174 gal   11. 7750   13. 13,500 Ω   15. 500,000 hertz
   17. 122 CCF

D. 1.

Truck No.	1	2	3	4	5	6	7	8	9	10
Mileage	1675	1167	1737	1316	1360	299	1099	135	1461	2081

Total mileage 12,330

3. $24,431
5. (a) $2065
    (b)

Deposits	Withdrawals	Balance
		$6375
	$ 379	5996
$1683		7679
474		8153
487		8640
	2373	6267
	1990	4277
	308	3969
	1090	2879
	814	2065

Problems, page 30

A. 12   32   63   36   12   18   0   24   14   8
    48   16   45   30   10   9   72   35   18   4
    28   15   36   49   8   40   42   54   64   24
    20   0   25   27   81   6   1   48   16   63

B. 16   30   9   35   18   20   28   48   12   63
    32   0   18   24   9   25   24   45   10   72
    15   49   40   54   36   8   42   64   0   4
    25   27   7   56   36   12   81   0   2   56

Exercises 1-3, page 35

A. 1. 42   3. 48   5. 63   7. 54   9. 45
   11. 296   13. 576   15. 320   17. 290   19. 416
   21. 792   23. 1404   25. 282   27. 720   29. 5040
   31. 1938   33. 4484   35. 3822

B. 1. 37,515   3. 297,591   5. 378,012   7. 30,780
    9. 397,584   11. 7281   13. 25,000   15. 3,532,536
   17. 10,112   19. 89,577

C. 1. $3000   3. 1300 ft   5. 8000 envelopes
    7. 2430   9. 1200   11. 7650
   13. 23,040   15. 115 in. or   17. No
                            9 ft 7 in.
   19. 37,400 bu   21. 480 A   23. 88,000 mL

D. 1. $681,355 left
    3. (a) 111,111,111; 222,222,222; 333,333,333
        (b) 111,111; 222,222; 333,333
        (c) 1; 121; 12,321; 1,234,321; 123,454,321
        (d) 42; 4422; 444,222; 44,442,222; 4,444,422,222
    5. Alpha   Beta   Gamma   Delta   Tau
        $28,080 $45,560 $37,120 $72,520 $56,160

Exercises 1-4, page 46

A. 1. 9   3. Not defined   5. 10 rem. 1
    7. 8   9. 6   11. 23 rem. 6
   13. 51 rem. 4   15. 21   17. 37
   19. 23   21. 39   23. 9 rem. 1
   25. 22   27. 8 rem. 35

B. 1. 120   3. 56 rem. 8   5. 96
    7. 222 rem. 2   9. 305 rem. 5   11. 119
   13. 501   15. 604   17. 200
   19. 108 rem. 4   21. 600   23. 102 rem. 98
   25. 100 rem. 11   27. 17 rem. 123

C. 1. (a) 1, 2, 3, 6   (b) 2 × 3   3. (a) 1, 19   (b) Prime
    5. (a) 1, 2, 4, 5, 8, 10, 20, 40
        (b) 2 × 2 × 2 × 5

D. 1. 27 in.   3. 13 hr   5. 28   7. 7 in.
    9. $72   11. 48   13. 27 reams   15. 6

E. 1. (a) 21,021,731   (b) 449   (c) 93   (d) 27,270
    3. 19.148255 or 20 rivets to be sure
    5. 51 hours 40 minutes, or 52 hr, rounded
    7. 42 hours

Exercises 1-5, page 51

A. 1. 50 3. 36 5. 57 7. 4 9. 42 11. 6
13. 61 15. 112 17. 31 19. 17 21. 84 23. 4
25. 6 27. 36 29. 2 31. 8 33. 13 35. 7
37. 6 39. 5

B. 1. $3 \times \$34 + 5 \times \$39 = \$297$ 3. $12 \times \$25 - 3 \times \$6 = \$282$
5. $2 \times \$12 \times 40 + 3 \times \$20 \times 40 + \$3240 + \$500 = \$7100$
7. $(33 \times \$80) + (12 \times \$40) + (45 \times \$18) = \$3930$
9. China: 346, United States: 330; China "won"

C. 1. 8347 3. 7386 5. 5 7. 1359
9. 1691 11. 1458 13. 13,920 15. 63

Problem Set 1, page 57

A. 1. five hundred ninety-three
3. forty-five thousand two hundred six
5. two million four hundred three thousand five hundred sixty
7. ten thousand twenty
9. 408 11. 230,056 13. 64,700 15. 690
17. 18,000 19. 700,000

B. 1. 93 3. 528 5. 934 7. 15
9. 649 11. 195 13. 1504 15. 1407
17. 13,041 19. 230,384 21. 37 23. 57
25. 9 27. 18 29. 6 31. 115
33. 7 35. 1245

C. 1. (a) 1, 2, 4, 8 (b) $2 \times 2 \times 2$ 3. (a) 1, 31 (b) 31
5. (a) 1, 2, 3, 4, 6, 9, 12, 18, 36 (b) $2 \times 2 \times 3 \times 3$

D. 1. 43 ft 3. 1892 sq ft 5. 6 7. 207 lb
9. $1820 11. Yes 13. 650 gpm 15. 2839 lb
17. 24 hr 19. 87,780 cu in. 21. 193 rpm 23. 6 ft
25. $110 27. $365 29. (a) 17 gal (b) $34

Chapter 2 *Exercises 2-1, page 79*

A. 1. $\frac{7}{3}$ 3. $\frac{67}{8}$ 5. $\frac{23}{8}$ 7. $\frac{8}{3}$ 9. $\frac{29}{6}$

B. 1. $8\frac{1}{2}$ 3. $1\frac{3}{8}$ 5. $1\frac{1}{2}$ 7. $16\frac{4}{6}$ or $16\frac{2}{3}$ 9. $2\frac{16}{32}$ or $2\frac{1}{2}$

C. 1. $\frac{3}{4}$ 3. $\frac{3}{8}$ 5. $\frac{2}{5}$ 7. $\frac{4}{5}$
9. $4\frac{1}{4}$ 11. $\frac{21}{32}$ 13. $\frac{5}{12}$ 15. $\frac{19}{12}$

D. 1. 14 3. 8 5. 20 7. 36 9. 5 11. 42

E. 1. $\frac{3}{5}$ 3. $1\frac{1}{2}$ 5. $\frac{7}{8}$ 7. $\frac{6}{4}$ 9. $\frac{13}{5}$ 11. $\frac{5}{12}$

F. 1. $15\frac{3}{4}$ in. 3. $\frac{19}{6}, \frac{25}{8}$ 5. No. 7. $\frac{3}{5}$ 9. $\frac{1}{5}$ 11. $\frac{23}{32}$ in.

Exercises 2-2, page 85

A. 1. $\frac{1}{8}$ 3. $\frac{2}{15}$ 5. $2\frac{2}{3}$ 7. $1\frac{1}{9}$ 9. $2\frac{1}{2}$ 11. 3
   13. $3\frac{1}{4}$ 15. 69 17. 74 19. $10\frac{3}{8}$ 21. $\frac{1}{8}$ 23. $\frac{1}{15}$
   25. 2

B. 1. $\frac{1}{6}$ 3. $\frac{1}{2}$ 5. $\frac{3}{4}$ 7. $1\frac{5}{16}$ 9. 1
   11. $1\frac{1}{20}$ 13. $2\frac{5}{8}$ 15. 1

C. 1. $137\frac{3}{4}$ in. 3. $11\frac{2}{3}$ ft 5. 14 ft $1\frac{1}{2}$ in. 7. $318\frac{1}{2}$ mi
   9. $356\frac{1}{2}$ lb 11. $\frac{9}{10}$ in. 13. $9\frac{3}{4}$ in. 15. $348\frac{3}{4}$ min
   17. $10\frac{2}{3}$ hr 19. $5\frac{2}{5}$ in. 21. $22\frac{1}{2}$ picas 23. $31\frac{1}{8}$
   25. $3\frac{1}{4}$ in. 27. $\frac{15}{32}$ in. 29. $9\frac{1}{4}$ lb 31. $3\frac{7}{8}$ in.

Exercises 2-3, page 92

A. 1. $1\frac{2}{3}$ 3. $\frac{5}{16}$ 5. $\frac{1}{2}$ 7. $\frac{1}{4}$ 9. 9 11. $1\frac{1}{3}$
   13. $1\frac{1}{5}$ 15. 16 17. 18 19. $7\frac{1}{2}$ 21. $\frac{5}{6}$ 23. $\frac{1}{8}$

B. 1. 8 ft 3. 48 5. 84 7. 18
   9. 210 11. 7 sheets 13. 45 threads 15. 284 sq ft
   17. $\frac{1}{8}$ in. per yr 19. 6

Exercises 2-4, page 109

A. 1. $\frac{1}{4}$ 3. $\frac{3}{4}$ 5. $\frac{1}{2}$ 7. $\frac{2}{5}$ 9. $\frac{15}{16}$
   11. $1\frac{1}{2}$ 13. $\frac{3}{4}$ 15. $\frac{17}{24}$ 17. $\frac{1}{8}$ 19. $\frac{7}{16}$
   21. $\frac{29}{40}$ 23. $\frac{19}{40}$ 25. $\frac{5}{8}$ 27. $1\frac{3}{4}$ 29. $4\frac{1}{8}$
   31. $3\frac{8}{15}$ 33. $2\frac{3}{8}$ 35. $1\frac{7}{60}$

B. 1. $5\frac{1}{8}$ 3. $2\frac{13}{16}$ 5. $\frac{7}{8}$ 7. $2\frac{1}{8}$ 9. $2\frac{11}{16}$ 11. $3\frac{9}{10}$

C. 1. 9 in. 3. $11\frac{5}{8}$ in. 5. $1\frac{9}{16}$ in. 7. $25\frac{1}{4}$ c.i.
   9. $1\frac{1}{8}$ in. 11. $23\frac{5}{16}$ in. 13. $2\frac{13}{16}$ in. 15. $\frac{3}{8}$ in.
   17. $7\frac{3}{4}$ in. by $6\frac{1}{2}$ in. 19. $\frac{9}{16}$ in. 21. $4\frac{33}{64}$ in. 23. $\frac{3}{4}$ qt
   25. $4\frac{2}{3}$ cu yd

Problem Set 2, page 115

A. 1. $\frac{9}{8}$ 3. $\frac{5}{3}$ 5. $\frac{99}{32}$ 7. $\frac{13}{8}$ 9. $2\frac{1}{2}$ 11. $8\frac{1}{3}$
   13. $1\frac{9}{16}$ 15. $8\frac{3}{4}$ 17. $\frac{3}{16}$ 19. $\frac{3}{8}$ 21. $\frac{1}{6}$ 23. $1\frac{4}{5}$
   25. 9 27. 44 29. 68 31. 15 33. $\frac{7}{16}$ 35. $\frac{7}{8}$
   37. $\frac{3}{5}$ 39. $\frac{7}{4}$

B. 1. $\frac{3}{32}$ 3. $\frac{7}{12}$ 5. $1\frac{1}{4}$ 7. $\frac{5}{64}$ 9. $7\frac{1}{2}$ 11. 27
   13. $11\frac{2}{3}$ 15. 2 17. 32 19. $\frac{1}{6}$ 21. $\frac{7}{10}$ 23. $2\frac{4}{9}$

C.   1. $1\frac{1}{4}$ 3. $\frac{7}{32}$ 5. $1\frac{13}{30}$ 7. $\frac{3}{8}$ 9. $\frac{7}{16}$ 11. $\frac{23}{40}$

    13. $3\frac{3}{8}$ 15. $4\frac{1}{2}$ 17. $1\frac{19}{24}$ 19. $1\frac{1}{30}$ 21. $1\frac{1}{6}$ 23. $\frac{2}{5}$

D.   1. $37\frac{1}{8}$ in.   3. $22\frac{5}{16}$ in.; $22\frac{7}{16}$ in.   5. $\frac{25}{32}$ in.   7. $4\frac{3}{5}$ cu ft

    9. $92\frac{13}{16}$ in.   11. $\frac{15}{16}$ in.   13. $4\frac{3}{32}$ in.   15. $244\frac{1}{2}$ in.

    17. $1\frac{1}{16}$ in. and $\frac{11}{16}$ in.   19. $5\frac{3}{16}$ in.   21. $6\frac{2}{3}$ min   23. $2\frac{1}{2}$ in.

    25. $\frac{9}{32}$ in.   27. 225   29. 1 in.

    31. (a) $72\frac{1}{12}$ ft   (b) 865 sq ft   (c) $5190

Chapter 3   *Exercises 3-1, page 131*

A.   1. seventy-two hundredths
    3. twelve and thirty-six hundredths
    5. three and seventy-two thousandths
    7. three and twenty-four ten-thousandths
    9. 0.004   11. 6.7   13. 12.8
    15. 10.032   17. 0.0116   19. 2.0374

B.   1. 21.01   3. $15.02   5. 1.617   7. 828.6
    9. 63.7305   11. 6.97   13. $15.36   15. 42.33
    17. $22.02   19. 113.96   21. 45.195   23. $27.51
    25. 95.888   27. 15.16   29. 8.618   31. 31.23
    33. 17.608   35. 0.0776   37. 24.22   39. 1.748

C.   1. 0.473 in.   3. (a) 0.013 in.   (b) smaller; 0.021 in.   (c) #14
    5. *A:* 2.20 in.   *B:* 0.45 in.   *C:* 4.22 in.
    7. 2.267 in.   9. No   11. 3.4 hr   13. 3.37 in.
    15. 0.843 in.   17. 70.15 in.   19. 0.009 in.   21. 7.2 hr

D.   1. $308.24
    3. (a) 0.7399   (b) 4240.775   (c) 510.436   (d) 7.4262

Exercises 3-2, page 147

A.   1. 0.00001   3. 4   5. 0.84   7. 0.00003
    9. 0.07   11. 2.16   13. 6.03   15. 20
    17. 0.045   19. 60   21. 400   23. 6.6
    25. 605   27. 0.00378   29. 0.048   31. 45,000
    33. 0.00008   35. 0.000364 37. 0.000096 39. 1.705

B.   1. 42.88   3. 6.5   5. 79.14   7. 3.6494   9. 0.216

C.   1. 3.33   3. 10.53   5. 0.12   7. 33.86   9. 33.3
    11. 0.3   13. 0.2   15. 0.143   17. 65   19. 2.999

D.   1. $55   3. 18.8 lb
    5.

	W	C
A	16.91 lb	$16.57
B	23.18	20.63
C	8.64	9.07
D	3.04	6.54

      $T = 52.81

7. 48 lb 9. 153.37 ft 11. 0.075 volts 13. 336
15. 27 gal 17. $576 19. 2625 cycles 21. $899.57
23. 2660 lb 25. 8400 cu ft 27. $648 29. 5.7 in.
31. (a) $152.70 (b) $420.40 (c) $608.80

E. 1. 8.00000007 3. 4.2435 in. 5. $580 7. 26.2 psi
9. 11.67 mi 11. 23.1 mi/gal
13. (a) 0.255 mi/gal (b) 8.92 people · mi/gal
15. 130.052 therms 17. $143.54

Exercises 3-3, page 159

A. 1. 0.25 3. 0.75 5. 0.8 7. 0.29 9. 0.86
11. 0.75 13. 0.3 15. 0.42 17. 0.38 19. 0.81
21. 0.22 23. 0.46 25. 0.03 27. 0.019

B. 1. 4.385 3. 1.5 5. 7.88 7. 1.43 9. 3.64 11. 7.65

C. 1. 1.375 g; 5.2 tablets 3. 2.3 squares
5. (a) $64.94 (b) $385.94 (c) $316.74
(d) $295.83; Total: $1063.45
7. 0.3958 in., rounded 9. (a) 120.2 ft (b) 12 ft $6\frac{5}{8}$ in. 11. 4.5

D. 1.

Gauge No.	Thickness (in.)	Gauge No.	Thickness (in.)
7–0	0.5	14	0.078
6–0	0.469	15	0.07
5–0	0.438	16	0.063
4–0	0.406	17	0.056
3–0	0.375	18	0.05
2–0	0.344	19	0.044
0	0.313	20	0.038
1	0.281	21	0.034
2	0.266	22	0.031
3	0.25	23	0.028
4	0.234	24	0.025
5	0.219	25	0.022
6	0.203	26	0.019
7	0.188	27	0.017
8	0.172	28	0.016
9	0.156	29	0.014
10	0.141	30	0.013
11	0.125	31	0.011
12	0.109	32	0.01
13	0.094		

3. $699.48

Problem Set 3, page 165

A. 1. ninety-one hundredths
3. twenty-three and one hundred sixty-four thousandths
5. nine and three tenths
7. ten and six hundredths
9. 0.07 11. 200.8 13. 63.063 15. 5.0063

B. 1. 23.19 3. $19.29 5. 1.94 7. 88.26 9. 277.104
11. 239.01 13. 83.88 15. 33.672 17. 4.28 19. 1.92

C. 1. 0.00008   3. 0.84   5. 0.108   7. 19.866   9. 61.7
   11. 3.8556   13. 0.006   15. 18   17. 4.34   19. 0.23
   21. 2.78   23. 0.04   25. 0.526   27. 214.634 29. 27.007
   31. 1.5   33. 4.6   35. 47.2

D. 1. 0.0625   3. 0.15625   5. 1.375   7. 2.55
   9. $2.\overline{6}$ 11. 2.3125   13. 30.67   15. 78.57
   17. 0.13

E. 1. 291   3. 9.772 in.   5. 0.0021 in.   7. 489.4 lb
   9. (a) 0.1875 in.   (b) 0.15625 in.   (c) 0.375 in.   (d) 0.203125 in.
   11. (a) 0.0089 in.   (b) 0.0027 in.   (c) 0.55215 in. (d) 0.306175 in.
   13. 17.5 ft   15. Max: 2.525 in.; min: 2.475 in.   17. $229.90
   19. 56.7 cu ft   21. 18   23. 11.4 hr   25. 10 tablets
   27. $29.28   29. Last year was greater by $0.83   31. 3.375 V
   33. 25.8 mi/gal 35. 0.018 in.   37. $860.16   39. 100.3
   41. (a) $23.99   (b) $58.77

Chapter 4   *Exercises 4-1, page 188*

A. (In each case only the missing answer is given.)
   1. (a) 7:1   (b) 12:7   (c) 6   (d) 6
      (e) 45   (f) 9   (g) 16   (h) 50
      (i) 3:2   (j) 2:5
   3. (a) 16:12   (b) 8 ft   (c) 28 ft   (d) 6.88:12
      (e) 7.2:12   (f) 4 ft   (g) 20 ft   (h) 5 ft 1 in.

B. 1. $x = 12$ 3. $y = 100$ 5. $P = \frac{7}{6}$ 7. $A = 3.25$
   9. $T = 1.8$ 11. $x = 3$ 13. $x = 5.04$ 15. $L = 5$ ft
   17. $x = 4$ cm

C. 1. 5 cu in.   3. 21 pins   5. 5.25 cu yd   7. 288.75 gal
   9. 124.8 lb   11. $856   13. 375 ft   15. 37.9 lb
   17. 12.9 mL   19. 40,000 cal   21. 132 lb   23. 24 lb
   25. (a) 0.5 mL   (b) 0.33 mL   (c) 6.67 mL
   27. (a) 2   (b) $1\frac{1}{2}$ (c) $4\frac{1}{2}$ 29. $15,636   31. $3856

Exercises 4-2, page 202

A. 1. (a) $\frac{5}{8}$ in.   (b) 31.2 cm   (c) 8 ft   (d) 3 m
   3. (a) 100   (b) 9   (c) 25   (d) 1500

B. 1. 24 in.   3. 139 hp   5. 109.1 lb   7. 980 rpm
   9. (a) 5310 sq ft   (b) 8800 sq ft   11. 500 rpm   13. 6 hr
   15. 1470 ft   17. 0.96 sq in.   19. 4 in.   21. 22.4 psi
   23. $377.54   25. 14.4 hr, rounded
   27. (a) 6 in.   (b) 8 to 3   29. 458 lb

Exercises 4-3, page 214

A. 1. 32%   3. 50%   5. 25%   7. 4000%
   9. 200%   11. 50%   13. 33.5%   15. 0.5%
   17. 150%   19. 330%   21. 40%   23. 95%
   25. 30%   27. 60%   29. 120%   31. 604%
   33. 125%   35. 35%   37. $83\frac{1}{3}$%   39. 370%

B. 1. 0.06   3. 0.01   5. 0.71   7. 0.0025
9. 0.0625   11. 0.3   13. 8   15. 0.0025
17. 0.07   19. 0.56   21. 10   23. 0.90
25. 1.5   27. 0.0675   29. 0.1225   31. 0.012

C. 1. $\frac{1}{20}$ 3. $2\frac{1}{2}$ 5. $\frac{53}{100}$ 7. $\frac{23}{25}$ 9. $\frac{9}{20}$
11. $\frac{1}{12}$ 13. $\frac{6}{25}$ 15. $\frac{1}{2000}$ 17. $4\frac{4}{5}$ 19. $\frac{5}{8}$

Exercises 4-4, page 222

A. 1. 80%   3. 15   5. 54   7. 150
9. $66\frac{2}{3}\%$ 11. 100   13. $21.25   15. 1.5
17. 43.75%   19. 2.4   21. 40   23. 5000

B. 1. 225   3. 20¢   5. 160   7. 150%
9. 17.5   11. 427   13. 2%   15. 460
17. 50

C. 1. 88%   3. 5.5%   5. 2084 lb
7. $2.14   9. 9.6 in. by 12 in.   11. 1447 ft
13. 76%   15. 32%   17. 52,400 linear ft
19. 148–166 bpm   21. 304.3 cfm   23. Nadal (73% vs. 66%)
25. $750   27. 0.05 gal   29. $81,950

Exercises 4-5, page 245

1. 64%   3. 63.6%   5. 67.5%   7. 5%
9. 12.5%   11. 0.9%   13. 81.25 hp   15. 0.13%
17. (a) 0.03%   (b) 0.44%   (c) ±0.007 in.   19. 37.5%
21. $16.69   23. $261.75   25. $145   27. B, by $1.74
29. 25%   31. 15.6 MGD   33. 150 lb   35. 28%
37. $168.35   39. Approx. 2.8%   41. $36,315.14   43. 179 hp
45. (a) 8800 Btu/hr   (b) 10,080 Btu/hr
47. (a) 112 bpm   (b) 50%   49. 8.1%   51. 33.3%
53. (a) 27 lb   (b) 13%

Problem Set 4, page 253

A. 1. (a) 30 in.   (b) 5 to 2   (c) 25 in.

B. 1. $x = 35$ 3. $x = 20$ 5. $x = 14.3$

C. 1. 72%   3. 60%   5. 130%   7. 400%   9. $16\frac{2}{3}\%$

D. 1. 0.04   3. 0.11   5. 0.0125   7. 0.002   9. 0.03875
11. 1.15

E. 1. $\frac{7}{25}$ 3. $\frac{81}{100}$ 5. $\frac{1}{200}$ 7. $\frac{7}{50}$

F. 1. 60%   3. 5.6   5. 42   7. $28.97   9. 8000

G. 1. 29.92 in.   3. (a) $6977.25   (b) $598.50   (c) $1699
5. (a) $20.13   (b) $18.54   (c) 73¢   (d) $125.72   (e) $148.39
7. 1748 ft   9. (a) 0.06%   (b) 2.82%   (c) ±0.009 in.
(d) 0.03%   (e) ±0.62 mm

11. $7216.25   13. ±135 ohms; 4365 to 4635 ohms
15. 0.9%   17. 72%   19. 9.12 hp
21. 3.6%   23. (a) 66 fps   (b) 15 mph
25. $5\frac{1}{4}$ in.   27. 150 seconds or $2\frac{1}{2}$ minutes
29. 450 parts   31. $3\frac{3}{4}$ hr   33. 6 sacks
35. 960 lb   37. 1.47 amperes   39. $673.98
41. 35 employees   43. 83.96 psi   45. 158 hp
47. $368,600   49. Yes (13%)   51. 74
53. (a) 12 mL   (b) 6 mL   (c) 0.8 mL
55. (a) 17.65% (b) 460%

Chapter 5   *Exercises 5-1, page 275*

A.   *Precision to   Accuracy in
      nearest   significant digits*
1. tenth   2
3. whole number   3
5. thousandth   4
7. hundred   3
9. tenth   3

B. 1. 21 in.   3. 9.6 in.   5. 1.91 gal   7. 1.19 oz
    9. 50 psi

C. 1. 40 sq ft   3. 7.2 sq ft   5. 1.9 sq ft   7. 4.6 cu ft
    9. 63 mph   11. 1.6 hr

D. 1. (a) 1.88 in.   (b) 4.05 in.   (c) 3.13 sec   (d) 0.06 in.
        (e) 0.09 in.   (f) 1.19 lb   (g) 2.27 in.   (h) 0.59 in.
        (i) 0.38 lb

    3. (a) $1\frac{29}{32}$ in.   (b) $\frac{27}{32}$ in.   (c) $2\frac{11}{32}$ in.   (d) $\frac{21}{32}$ in.
        (e) $2\frac{3}{32}$ in.   (f) $\frac{9}{32}$ in.   (g) $\frac{19}{32}$ in.   (h) $\frac{22}{32}$ in.
        (i) $1\frac{17}{32}$ in.

E. 1. Yes   3. $\frac{30}{64}$ in.   5. 4.28 in.   7. $3\frac{5}{16}$ in.
    9. 354 mi   11. 0.013 sec

Exercises 5-2, page 290

A. 1. 51 in.   3. 18,000 ft   5. 504 qt   7. 3.1 atm
    9. 2.3 gal   11. 0.879   13. 8.7 yd   15. 1180 cu in.
    17. 0.80 mi   19. 83 in.   21. 20 lb 6 oz

B. 1. 860   3. 465.8   5. 0.563   7. 0.03
    9. $0.\overline{3}$ or $\frac{1}{3}$ 11. 12,800   13. 2700   15. 0.30

C. 1. 2.0 ft/sec   3. 28 in.   5. 32.6 ± 0.6 in.   7. 1265 sq ft
    9. 10.2 rps   11. 1.93 cfs   13. $466\frac{1}{2}$ in.   15. 8.5 gal
    17. 18,380 sq ft   19. 17 mi   21. The hybrid (by $1031.25)
    23. (a) 43,560   (b) 325,829   (c) 1613
    25. (a) Ford: 12.2 gal   (b) Ford: 12.9 gal
             Toyota: 13.2 gal   Toyota: 12.0 gal
    27. (a) 25 hands   (b) 14 rods   (c) 20 bones   (d) $6\frac{2}{3}$ hands
         (e) 64 ft

Exercises 5-3, page 309

A.   1. (c)   3. (a)   5. (b)   7. (a)   9. (a)
    11. (c)   13. (a)   15. (a)   17. (a)   19. (a)

B.   1. (b)   3. (a)   5. (b)   7. (b)   9. (b)   11. (a)

C.   1. 56 mm   3. 2500 sq m   5. 45 g   7. 1250 cu cm
    9. 420 cm   11. 16 cm/sec   13. 9620 m   15. 56.5 km
    17. 9.5 g   19. 0.58 liter   21. 1.4 mL   23. 0.65 L
    25. 4100 m/min

D.   1. 7.6   3. 3.2   5. 23   7. 68.9   9. 19
    11. 11.9   13. 6.1   15. 5.0   17. 50   19. 19
    21. 26.5 sq m   23. 3.59 cu yd   25. 427 mi   27. 220 sq m
    29. 6.35 lb/sq ft

E.   1.

in.	mm
0.030	0.762
0.035	0.889
0.040	1.016
0.045	1.143

in.	mm
$\frac{1}{16}$	1.588
$\frac{5}{64}$	1.984
$\frac{3}{32}$	2.381
$\frac{1}{8}$	3.175

in.	mm
$\frac{5}{32}$	3.969
$\frac{3}{16}$	4.763
$\frac{3}{8}$	9.525
$\frac{11}{64}$	4.366

    3. 12.4 mph; 20.0 km/h
    5. (a) 0.6093   (b) 564   (c) 0.89   (d) 0.057   (e) 1.02
    7. 12 km/liter   9. 9.3 sq m
    11. (a) 1.6093   (b) 21.59 cm × 27.94 cm   (c) 2.54   (d) 804.65
       (e) 0.0648   (f) 28.35; 453.6   (g) 7.57   (h) 37.85
    13. (a) 18 kV   (b) 0.435 V   15. 1.7 oz/qt   17. 338°F
    19. 560 mph   21. 62 lb   23. 30.3 sec

Exercises 5-4, page 333

A.   1. (a) $\frac{5}{8}$ in.   (b) $1\frac{7}{8}$ in.   (c) $2\frac{1}{2}$ in.   (d) 3 in.
       (e) $\frac{1}{4}$ in.   (f) $1\frac{1}{16}$ in.   (g) $2\frac{5}{8}$ in.   (h) $3\frac{1}{2}$ in.
    3. (a) $\frac{2}{10}$ in. = 0.2 in.   (b) $\frac{5}{10}$ in. = 0.5 in.   (c) $1\frac{3}{10}$ in. = 1.3 in.
       (d) $1\frac{6}{10}$ in. = 1.6 in.   (e) $\frac{25}{100}$ in. = 0.25 in.   (f) $\frac{72}{100}$ in. = 0.72 in.
       (g) $1\frac{49}{100}$ in. = 1.49 in.   (h) $1\frac{75}{100}$ in. = 1.75 in.

B.   1. 0.650 in.   3. 0.287 in.   5. 0.850 in.   7. 0.4068 in.
    9. 0.2581 in.   11. 0.0888 in.   13. 11.57 mm   15. 22.58 mm
    17. 18.94 mm

C.   1. 3.256 in.   3. 2.078 in.   5. 2.908 in.   7. 2.040 in.   9. 0.826 in.

D.   1. 62°21′   3. 34°56′   5. 35°34′   7. 20°26′

E.   1. 2.000 in. + 0.200 in. + 0.050 in. + 0.057 in. + 0.0503 in.
    3. 0.200 in. + 0.120 in. + 0.053 in. + 0.0502 in.
    5. 1.000 in. + 0.500 in. + 0.070 in. + 0.058 in. + 0.0509 in.
    7. 0.050 in. + 0.053 in. + 0.0509 in.
    9. 2.000 in. + 0.140 in. + 0.0505 in.

A. | Precision to nearest | Accuracy in significant digits |
|---|---|
| 1. tenth | 2 |
| 3. hundredth | 3 |
| 5. hundredth | 1 |
| 7. thousand | 1 |

B. 1. 3.5 sec   3. 0.75 in.   5. $16\frac{7}{8}$ in.   7. 21.1 mi/gal

9. 17.65 psi   11. 3.87 in.   13. 21 ft 3 in.

C. 1. 15 cm   3. ≈72°F   5. 820 g   7. 45,800 cu in.

9. 31.5 lb   11. 667 cm/sec   13. 23.3 km   15. 386 cu ft

17. 1.4 liters   19. 11.48 bu   21. 980 sq in.   23. 95 mg

25. 104 cm   27. 1850 cu cm   29. 53 mph

D. 1. (b)   3. (a)   5. (b)

E. 1. (a) $\frac{5}{10}$ in. = 0.5 in.   (b) $1\frac{3}{10}$ in. = 1.3 in.   (c) $\frac{17}{100}$ in. = 0.17 in.

(d) $1\frac{35}{100}$ in. = 1.35 in.   (e) $\frac{3}{4}$ in.   (f) $3\frac{3}{4}$ in.   (g) $\frac{5}{16}$ in.

(h) $1\frac{19}{32}$ in.   (i) $\frac{3}{8}$ in.   (j) $1\frac{27}{64}$ in.

3. (a) 2.156 in.   (b) 3.030 in.   (c) 0.612 in.   (d) 1.925 in.

(e) 1.706 in.   (f) 1.038 in.

F. 1. $22\frac{9}{16}$ in.   3. 6.1 mph   5. $\frac{15}{32}$ in.   7. 936 sq in.

9. 293°C   11. 7.623 in.   13. 0.28 lb/cu in.

15. 3.81 cm × 8.89 cm or 4 cm by 9 cm   17. 3.4 sq mi

19. 13.5 mi   21. 8 min   23. 864 bf   25. 36 mi/gal

27. (a) 6.60 m/sec   (b) 6.34 m/sec   (c) 1 hr 50 min 55 sec

Chapter 6   *Exercises 6-1, page 355*

A. 1. <   3. >   5. >   7. <

9. <   11. >   13. >   15. <

B. 1.
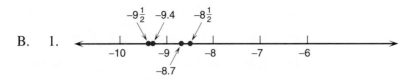

3.
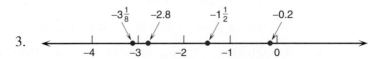

C. 1. −6   3. +12,000   5. −9   7. −80

9. −5   11. +$6000   13. −$17.50

D. 1. 5   3. −15   5. 17   7. 46

9. −16   11. 9   13. 29   15. −30

17. −0.9   19. −28.09   21. $-1\frac{1}{3}$ 23. $-1\frac{1}{4}$

25. −14   27. $-\frac{3}{4}$ 29. $1\frac{9}{16}$ 31. $-4\frac{5}{12}$

33. −10   35. −16   37. 2   39. 18

41. −230   43. 7   45. 3599   47. −26,620

E.   1.   +22 (22 above quota)   3.   −9° (9 degrees below zero)
     5.   −$136,000 (a loss of $136,000)   7.   0 mA

Exercises 6-2, page 361

A.   1.   −2   3.   −16   5.   9   7.   −6
     9.   8   11. 11   13. −18   15. −31
     17. 8   19. 26   21. 9   23. −19
     25. −17   27. 8   29. −87   31. −67
     33. −10   35. −42   37. 6   39. −1
     41. $-\frac{5}{9}$ 43. $-\frac{5}{8}$ 45. $-6\frac{1}{2}$ 47. $-8\frac{7}{12}$
     49. 9.2   51. −63.5   53. 41.02   55. −0.85

B.   1.   14,698 ft   3.   35°F   5. $1817
     7.   lost $5211   9.   420 V

Exercises 6-3, page 367

A.   1.   −63   3.   77   5.   −4   7.   12
     9.   −4   11. $12\frac{3}{8}$ 13. 5.1   15. $\frac{1}{5}$ or 0.2
     17. −90   19. 84   21. −28   23. 5
     25. −8   27. $\frac{1}{6}$ 29. $-2\frac{11}{20}$ 31. 0.4
     33. $-19\frac{1}{2}$ 35. −4.48   37. −90   39. −20
     41. −0.4

B.   1.   −19.3   3.   0.0019   5.   −3.16   7.   0.23
     9.   46   11. $-7\overline{0},000,000$

C.   1. $-8\frac{1}{3}$°C   3.   (a)   −120   (b)   40   5.   −2000 ft/min

Exercises 6-4, page 377

A.   1.   16   3.   64   5.   1000   7.   256
     9.   512   11. 625   13. −8   15. 729
     17. 5   19. 32   21. 196   23. 3375
     25. 108   27. 1125   29. 2700   31. 18
     33. 216   35. 19   37. 116   39. 2
     41. 71

B.   1.   9   3.   6   5.   5   7.   16
     9.   15   11. 18   13. 2.12   15. 3.52
     17. 14.49   19. 28.46   21. 31.62   23. 158.11
     25. 12.25   27. 54.77   29. 1.12   31. 13.42
     33. 51.26   35. 4.80   37. 375

C.   1.   13.6 ft   3.   89,000 psi   5.   96 fps   7.   26.6   9.   73 ft

Problem Set 6, page 381

A.   1.   −13, −8, −2, 4, 7
     3.   −180, −160, −150, −140, −120, 0

B.   1.   −10   3.   −6   5.   −15   7.   25
     9.   2.3   11. −$9.15   13. −16

C. 1. -60 3. 12   5. $7\frac{1}{32}$ 7. -0.4 9. -120 11. -16

D. 1. -43.75 3. -4 5. 64   7. 289
    9. 0.25   11. 0.0004   13. 0.000000001   15. 16.1604
    17. 16   19. 28   21. 0.4   23. 16.29
    25. 121.68   27. 432   29. 10   31. 0.5
    33. 8.94   35. 17.61   37. 2.05   39. 1.04
    41. 2.76   43. 0.78   45. 7.76

E. 1. $+7$ amp   3. $-22°F$ 5. 163 A   7. 700
    9. 3200 nanosec   11. (a) $-20°C$ (b) $-4°C$ 13. 33 ohms

Chapter 7 *Exercises 7-1, page 396*

A. 1. $6yx$ 3. $(a + c)(a - c)$ 5. $(n + 5)/4n$

B. 1. factor   3. factor   5. factor   7. factor

C. 1. $A = 6$ 3. $T = 13$ 5. $K = -16$ 7. $F = 19$
    9. $L = 1$ 11. $B = 13$

D. 1. $A = 11$ 3. $I = 15$ 5. $T = 23$ m^2
    7. $P = 40$ 9. $V = 121$ m^3

E. 1. $P = 39$ in.   3. $P = 96$ watts   5. $F = 104°$
    7. $A = \$12,500$ 9. $T = 19$ cm^2 11. $L = 18\frac{1}{4}$ in.
    13. $L = 5700$ in.   15. $X = 1000$ ft   17. $P = 2.64$ kW
    19. $L = 920$ cm   21. 36 lines   23. 0.525 in.
    25. $V = 4.2850$ cu in.   27. (a) 0.849 sq in.   (b) 0.601 sq in.
                                   (c) 13.4 sq in.   (d) 3.33 sq cm
    29. 208.9 cfm   31. 1.77 V   33. 79.84 ft
    35. (a) $-8.5°C$ (b) $-18.1°C$ or $-0.6°F$
    37. 2900 cc   39. (a) 46 mi/gal   (b) 28 mi/gal

Exercises 7-2, page 408

A. 1. $9y$ 3. $6E$ 5. $7B$
    7. $-2x^2$ 9. $2R^2 + 5R$ 11. $\frac{1}{8}x$
    13. $1\frac{1}{2} - 3.1W$ 15. $3x + 6xy$ 17. $5x^2 + x^2y + 3x$

B. 1. $3x^2 + 2x - 5$ 3. $4m^2 + 10m$
    5. $2 - x - 5y$ 7. $3a - 8 + 6b$
    9. $-10x^2 - 3x$ 11. $23 - 3x$
    13. $x + 8y$ 15. $-14 - 7w + 3z$
    17. $9x - 12y$ 19. $-56m - 48$
    21. $-9x - 15$ 23. $23m - 42$
    25. $3 - 8x - 12y$ 27. $28 - 6w$
    29. $6x + 8y - 24x^2 + 20y^2$ 31. $4x + 10y$
    33. $3x + 9y$ 35. $-2x^2 - 42x + 62$

Exercises 7-3, page 418

A. 1. $x = 9$ 3. $x = -24$ 5. $a = 8\frac{1}{2}$ 7. $y = 24.66$
    9. $x = -13$ 11. $a = 0.006$ 13. $z = 0.65$ 15. $m = -12$

17. $y = 6.5$ 19. $T = -18$ 21. $Q = 0.7$ 23. $x = 25$
25. $K = 4.24$

B. 1. 360 volts   3. 28.8 parts per   5. 1.715 m   7. 21 windings
                        million
    9. 32 ft   11. \$131,800   13. 1800 ft · lb

Exercises 7-4, page 427

A. 1. $x = 10$ 3. $x = 35$ 5. $m = 2.5$ 7. $n = 10$
    9. $z = -6$ 11. $x = 48$ 13. $n = -9$ 15. $x = \frac{1}{4}$
    17. $a = 7$ 19. $z = -5\frac{1}{2}$ 21. $x = \frac{2}{5}$ 23. $P = 3$
    25. $x = 10$ 27. $x = 5\frac{1}{3}$ 29. $x = -3$ 31. $x = -6$

B. 1. 5.5 hr   3. 32.5 years   5. 0.125 in.   7. 28.75 hr
    9. 7.5 yr

Exercises 7-5, page 440

A. 1. $x = 9$ 3. $n = 4\frac{5}{6}$ 5. $x = -7$ 7. $y = 7\frac{1}{2}$
    9. $x = 1$ 11. $c = 5$ 13. $x = -7$ 15. $t = \frac{1}{5}$
    17. $x = -2$ 19. $y = 2$ 21. $x = 2$ 23. $t = -3\frac{3}{4}$
    25. $x = 4\frac{2}{7}$ 27. $x = 3$

B. 1. $L = \dfrac{S}{W}$ 3. $I = \dfrac{V}{R}$ 5. $T = \dfrac{2S - WA}{W}$

    7. $B = \dfrac{P - 2A}{2}$ 9. $P = \dfrac{T(R + 2)}{R}$

C. 1. (a) $a = \dfrac{360L}{2\pi R}$ (b) $R = \dfrac{360L}{2\pi a}$ (c) $L = 5.23$ in.

    3. (a) $R_1 = \dfrac{V - R_2 i}{i}$ (b) $R_2 = 50$ ohms

    5. (a) $V_1 = \dfrac{V_2 P_2}{P_1}$ (b) $V_2 = \dfrac{V_1 P_1}{P_2}$ (c) $P_1 = \dfrac{V_2 P_2}{V_1}$

        (d) $P_2 = \dfrac{V_1 P_1}{V_2}$ (e) $P_1 = 300$ psi

    7. (a) $D = \dfrac{CA + 12C}{A}$ (b) 0.14 gram

    9. (a) $i_L = \dfrac{i_s T_p}{T_s}$ (b) $i_s = \dfrac{i_L T_s}{T_p}$ (c) $i_L = 22.5$ amp

    11. 2.5 ft   13. 7732 parts   15. 110.7°
    17. 3.74 in.

Exercises 7-6, page 455

A. 1. $H = 1.4W$ 3. $V = \frac{1}{4} h\pi d^2$ 5. $V = AL$

    7. $D = \dfrac{N}{P}$ 9. $W = 0.785\, hDd^2$

B. 1. $133\frac{1}{3}$ gal and $266\frac{2}{3}$ gal
    3. $13\frac{5}{7}$ ft, $6\frac{6}{7}$ ft, $3\frac{3}{7}$ ft

5. 22, 22, 22, 20, 18, 16, 14, 12, 10 blocks

7. Jo: $48,800; Ellen: $36,600

9. 21 (with some left over)

11. $6\frac{2}{3}$ hr   13. 5000 parts   15. $50.02   17. 8 hr

19. 36 hr   21. 22,400 mi   23. (a) 85,800 mi   (b) fewer

Exercises 7-7, page 463

A.   1. $20x$ 3. $6R^2$ 5. $-8x^3y^4$ 7. $0.6a^2$

   9. $\frac{1}{16}Q^4$ 11. $24M^6$ 13. $-2 + 4y$ 15. p^6

   17. $5x^7$ 19. $10y^7$ 21. x^4y^4 23. $3p^3$

   25. 5^{10} 27. 10^{-3} 29. $-6x^4 + 15x^5$

   31. $6a^3b^2 - 10a^2b^3 + 14ab^4$

B.   1. 4^2 3. x^3 5. $\dfrac{1}{10^2}$ 7. $2a^4$

   9. $-\dfrac{2}{3y}$ 11. $3ab^3$ 13. $-\dfrac{3}{m^4n}$ 15. $-\dfrac{3}{5b^3}$

   17. $3x^2 - 4$ 19. $-2a^5 + a^3 - 3a$

Exercises 7-8, page 471

A.   1. 5×10^3 3. 9×10^1 5. 3×10^{-3}

   7. 4×10^{-4} 9. 6.77×10^6 11. 2.92×10^{-2}

   13. 1.001×10^3 15. 1.07×10^{-4} 17. 3.14×10^1

   19. 1.25×10^2 21. 2.9×10^7 lb/sq in.   23. 9.55×10^7 watts

B.   1. 200,000   3. 0.00009   5. 0.0017   7. 51,000

   9. 40,500   11. 0.003205   13. 2,450,000   15. 647,000

C.   1. 8×10^7 3. 7.8×10^2 5. 2.8×10^{-9}

   7. 5×10^6 9. 1.4×10^2 11. 6.3×10^{-2}

   13. 2.3×10^{10} 15. 8×10^{-11} 17. 2.8×10^{-2}

   19. 5.9×10^{-12} 21. 6.6×10^{13} 23. 4.0×10^2

D.   1. (a) 5,300,000,000,000/L   (b) 9,100,000,000/L

      (c) 3.6×10^{12}/L   (d) 6.7×10^9/L

   3. 13.3 cal/sec   5. 1.53×10^{-4} μF

Problem Set 7, page 477

A.   1. $10x$ 3. $12xy$ 5. $1\frac{11}{24}x$ 7. $0.4G$

   9. $21x^2$ 11. $-10x^2y^2z$ 13. $12x - 21$ 15. $9x + 11$

   17. $9x - 3y$ 19. 6^4 21. $-3m^8$ 23. $\dfrac{4a}{b}$

B.   1. $L = 13$ 3. $I = 144$ 5. $V = 42\frac{21}{64}$ 7. $L = 115.2$

   9. $t = 4$

C.   1. $x = 12$ 3. $e = 44$ 5. $x = -5$ 7. $x = 4$

   9. $y = 32$ 11. $x = -6$ 13. $x = -4$ 15. $g = 10$

   17. $x = 16$ 19. $x = 3$ 21. $y = 3.75$ 23. $n = 9$

   25. $y = 2$ 27. $b = \dfrac{A}{H}$ 29. $L = \dfrac{P - 2W}{2}$ 31. $g = \dfrac{2S + 8}{t}$

D.   1. 7.5×10^3 3. 4.1×10^{-2} 5. 5.72×10^{-3}
    7. 4.47×10^5 9. 8.02×10^7 11. 7.05×10^{-6}
   13. 930,000   15. 0.00029   17. 0.0000005146
   19. 90,710   21. 5.7×10^{10} 23. 3.0×10^{-6}
   25. 4.2×10^{-1} 27. 1.2×10^2 29. 2.8×10^{-2}
   31. 2.4×10^{-8}

E.   1. (a) 31.25%   (b) $L = \dfrac{100(H - X)}{CG}; L = 166.7$ in.

      (c) $H = \dfrac{L \cdot CG + 100x}{100}; H = 140$ in.

      (d) $x = \dfrac{100H - L \cdot CG}{100}; x = 156$ in.

   3. $8W + 2W = 80; W = 8$ cm, $L = 32$ cm
   5. $2x + x = 0.048; x = 0.016$ in., $2x = 0.032$ in.
   7. 9.7×10^6 m/sec   9. $51\frac{9}{16}$ in.
  11. 0.617   13. (a) 34 mi/gal   (b) 27 mi/gal
  15. (a) $L = \dfrac{AR}{K}$ (b) $R_m = \dfrac{R_S \cdot I_S}{I_m}$ (c) $C = \dfrac{1}{2\pi FX_C}$

      (d) $L_m = \dfrac{L_1 + L_2}{2L_T}$

  17. (a) 70%   (b) 4.1 dS/m   19. 108 in.   21. 5.8 μF
  23. 506 kWh   25. 93.3 lb   27. 24 mi/gal   29. $12,308

Chapter 8   *Exercises 8-1, page 496*

A.   1. (a) $\angle B, \angle x, \angle ABC$ (b) $\angle O, \angle POQ$ (c) $\angle a$
      (d) $\angle T$ (e) $\angle 2$ (f) $\angle s, \angle RST$

B.   1. (a) $a = 55°, b = 125°, c = 55°$ (b) $d = 164°, e = 16°, f = 164°$
      (c) $p = 110°$ (d) $x = 75°, t = 105°, k = 115°$
      (e) $s = 60°, t = 120°$ (f) $p = 30°, q = 30°, w = 90°$
   3. (a) $w = 80°, x = 100°, y = 100°, z = 100°$
      (b) $a = 60°, b = 60°, c = 120°, d = 120°, e = 120°, f = 120°, g = 60°$
      (c) $n = 100°, q = 80°, m = 100°, p = 100°, h = 80°, k = 100°$
      (d) $z = 140°, s = 40°, y = 140°, t = 140°, u = 40°, w = 140°, x = 40°$

C.   1. $\angle ACB \approx 63°$ 3. $B \approx 122°, A \approx 28°, C \approx 30°$ 5. 2″
   7. 29°5′36″   9. $a = 112°, b = 68°$ 11. $\angle 2 = 106°$
  13. 60°

Exercises 8-2, page 512

A.   1. 20.0 m   3. 115 in.   5. 48 in.   7. 39 cm
   9. 22.81 in.   11. 64 ft

B.   *Perimeter*   *Area*   *Perimeter*   *Area*
   1. 20 in.   25 sq in.   3. 70.6 in.   285 sq in.
   5. 56 m   160 m²   7. 124 m   468 m²
   9. 86 ft   432 sq ft   11. 114.9 m   722.2 m²

C.   1. 960   3. $25\frac{1}{3}$ sq yd, $1012.07   5. 11 bundles
   7. 640 sq in.   9. 10 ft 8 in.   11. $6\frac{2}{3}$ in.
  13. Cut the 6-in. card along the 38-in. side; 18 cards, 86 sq in. of waste
  15. $14,053   17. $9.74   19. (a) 357 sq ft   (b) $2017.05

Exercises 8-3, page 533

A. 1. 10.6 in.   3. 20 cm   5. 46.8 mm   7. 27.7 in.
   9. 28 ft   11. 21.2 in.   13. 6 in.   15. 17.4 ft
   17. 9.37 cm   19. 0.333 in.   21. 118 ft

B. 1. 173 mm^2 3. 336 ft^2 5. 36 yd^2 7. 260 in.^2
   9. 520 ft^2 11. 5.14 cm^2 13. 433 ft^2 15. 354 ft^2
   17. 365 ft^2 19. 39 in.^2 21. 0.586 in.^2 23. 7.94 m^2

C. 1. 7 gal   3. 5 squares   5. 15.2 ft   7. 174.5 sq ft
   9. $y = 36$ ft 5 in., $x = 64$ ft 11 in.   11. 22 ft 3 in.

Exercises 8-4, page 546

A.   *Circumference*   *Area*
   1. 88.0 in.   616 sq in.
   3. 19.2 m 29.2 m^2
   5. 75.4 ft 452 ft^2
   7. 3.77 cm 1.13 cm^2

B. 1. 17.1 sq in.   3. 5.2 in.^2 5. 15.1 in.^2 7. 26.7 cm^2
   9. 0.7 sq in.   11. 246.1 in.^2 13. 25.1 in.   15. 52.3 cm

C. 1. 30.1 in.   3. 0.6 in.   5. 163 in.^2 7. 103.7 sq in.
   9. 314   11. $103.11   13. 174 in.   15. 2.4 in.
   17. 2.6 cm   19. 2.79 in.   21. 2.1 fps   23. 133 cm
   25. 0.3 sec

Problem Set 8, page 553

A. 1. $\angle 1$, acute   3. $\angle m$, acute   5. $\angle PQR$ or $\angle Q$, obtuse
   7. 63° approx.   9. 155° approx.   11. 70° approx.

B.   *Perimeter*   *Area*   *Perimeter*   *Area*
   1. 12 in. 6 in.^2 3. 31.4 cm 78.5 cm^2
   5. 12 ft 10.4 ft^2 7. 17 yd 8 yd^2
   9. 36 in.   53.7 sq in.   11. 12 in.   6.9 sq in.
   13. 17.2 ft   12.0 sq ft   15. 51.7 ft 169.3 ft^2
   17. 28.3 in.   51.1 sq in.   19. 5.77 mm
   21. 3.46 in.   23. $\frac{3}{4}$ in.

C. 1. $65,450   3. 45°   5. 250 ft   7. $5232
   9. 18 in.   11. 36   13. 110°   15. 94 ft
   17. (a) 14 rolls   (b) $158.20   19. 69.1 cm   21. $2210.56
   23. 27.3 lb   25. 12.17 in.   27. 19 mi

Chapter 9   *Exercises 9-1, page 569*

A. 1. 729 cu in.   3. $13,800 \text{ cm}^3$ 5. 132.4 mm^3 7. 144 cm^3

B.   *Lateral Surface*   *Volume*
      Area
   1. 280 sq ft   480 cu ft
   3. 784 mm^2 2744 mm^3
   5. 360 in.^2 480 in.^3

C.

	Total Surface Area	Volume
1.	150 sq yd	125 cu yd
3.	1457.8 ft^2	2816 ft^3
5.	5290 cm^2	23,940 cm^3

D.

1.	333.3 cu ft	3.	94,248 gal	5.	450 trips	7.	71.3 cu yd
9.	16.9 lb	11.	3.8 ft	13.	238 trips	15.	8.3 cu yd
17.	3.5 yd^3	19.	7000 lb	21.	2244 gal	23.	525 ft^3
25.	19.6 yd^3	27.	12 cu yd				

Exercises 9-2, page 580

A.

	Lateral Surface Area	Volume		Lateral Surface Area	Volume
1.	384 sq in.	617 cu in.	3.	127 sq in.	105 cu in.
5.	48.6 sq ft	46.7 cu ft	7.	55.2 sq m	31.2 cu m

B.

	Total Area	Volume		Total Area	Volume
1.	80.7 m^2	33.2 m^3	3.	3200 ft^2	10,200 ft^3
5.	2500 in.^{2}	7340 in.^{3}	7.	173 ft^2	144 ft^3

C.

1.	$7749	3.	3 hr	5.	384 sq in.

Exercises 9-3, page 587

A.

	Lateral Surface Area	Volume
1.	24,800 cm^2	572,000 cm^3
3.	245 sq in.	367 cu in.

B.

	Total Area	Volume
1.	7.07 in.^{2}	1.77 in.^{3}
3.	1280 ft^2	3530 ft^3
5.	6520 m^2	39,900 m^3

C.

1.	14 gal	3.	0.24 ft	5.	2.6 lb	7.	4 gal
9.	49.6 lb	11.	582 gal	13.	0.43 lb	15.	28.1 in.
17.	41.9 in.^{3}	19.	36.7 gal	21.	596 lb		

Exercises 9-4, page 595

A.

	Lateral Surface Area	Volume
1.	319 cm^2	581 cm^3
3.	338 m^2	939 m^3

B.

	Total Area	Volume
1.	628 in.^{2}	1010 in.^{3}
3.	1110 in.^{2}	2610 in.^{3}

C.

1.	240 m^3	3.	49 cm^3

D.

1.	104.7 m^3	3.	253.2 bu	5.	5.6 lb	7.	4.2 cm
9.	2.0 lb	11.	6.1 oz				

Problem Set 9, page 601

A. (a) *Lateral Surface Area*   (b) *Volume*   (a) *Lateral Surface Area*   (b) *Volume*

	(a) Lateral Surface Area	(b) Volume		(a) Lateral Surface Area	(b) Volume
1.	226.2 sq in.	452.4 cu in.	3.	43.2 sq ft	26.7 cu ft
5.	259.2 mm^2	443.5 mm^3	7.	377.0 ft^2	980.2 in.^{3}

	(a) Total Surface Area	(b) Volume		(a) Total Surface Area	(b) Volume
9.	384 ft^2	512 ft^3	11.	1767.2 cm^2	5301.5 cm^3
13.	450 sq ft	396 cu ft	15.	1256.6 cm^2	4188.8 cm^3
17.	530.9 sq in.	1150.3 cu in.	19.	1263.6 in.^{2}	2520 in.^{3}
21.	329.2 sq in.	312 cu in.			

B.
1. 1125 boxes   3. 2992 lb   5. 40 qt   7. 1.5 ft^3
9. 294 lb   11. 165 yd^3 13. 2.8 gal   15. 6.3 ft
17. 33 L   19. 74 ft^3 21. 0.3 yd^3 23. 312 lb

Chapter 10   *Exercises 10-1, page 622*

A.
1. acute   3. obtuse   5. straight   7. acute
9. right

B.
1. 36°15′   3. 65°27′   5. 17°30′   7. 16°7′
9. 27.5°   11. 154.65°   13. 57.05°   15. 16.88°
17. 0.43   19. 0.02   21. 1.18   23. 1.41
25. 5.73°   27. 120.32°

C.
1. 26   3. 24   5. 12 ft   7. 1.5
9. 8.0 ft   11. (a) 45°   (b) 4.2 ft
13. (a) 4   (b) 53°   (c) 37°
15. (a) 7   (b) 45°
17. (a) 11.5 ft   (b) 5.8 ft   (c) 30°
19. (a) 10 in.   (b) 45°   21. $S = 21$ cm, $A = 314$ cm^2
23. $S = 88$ ft, $A = 4840$ sq ft   25. $S = 6$ in., $A = 16$ sq in.
27. $S = 5$ m, $A = 25$ m^2

D.
1. (a) 325.2 in.^{2} (b) 36.1 in.   3. 21
5. 523 yd^3 7. 829 in./sec   9. 0.08 rad/sec
11. 0.39 rad/sec   13. No (1.8°)

Exercises 10-2, page 634

A.
1. 0.54   3. 0.42   5. 0.44

B.
1. 0.454   3. 0.213   5. 0.191   7. 0.052
9. 1.206   11. 0.752   13. 0.771   15. 6.357
17. −0.139   19. 0.585

C.
1. 76°54′   3. 64°0′   5. 74°49′   7. 9°40′
9. 49.4°   11. 64.1°   13. 41.1°   15. 37.2°

Exercises 10-3, page 644

A.
1. 11 in.   3. 13.4 cm   5. 4.65 m   7. 91.56 km

B.
1. 13.94 km   3. 61°32′   5. 4.65 m   7. 16.2 ft

C.   1. $A = 22°$, $b = 89$ ft, $c = 96$ ft
     3. $B = 58.5°$, $b = 11.0$ in., $c = 12.9$ in.
     5. $b = 505.7$ km, $B = 58.78°$, $A = 31.22°$
     7. $A = 16°$, $a = 4.6$ ft, $c = 17$ ft
     9. $B = 68.75°$ (or $68°45'$), $b = 23.07$ m, $a = 8.970$ m

D.   1.   31 ft   3. (a) 73 ft 6 in.   (b) No.   5. $51°41'$
     7.   115 in.   9. 3.74 in.   11. 1.1 in.   13. 274 ft
    15.   33.39 cm   17. 1.5 in.   19. 21 ft 1 in.   21. 6.8 in.
    23.   19   25. 85 ft

Exercises 10-4, page 659

A.   1. $C = 75°$, $c = 9.2$ ft, $b = 8.4$ ft   3. $A = 39.0°$, $a = 202$ m, $c = 301$ m
     5. $A = 55.9°$, $C = 94.1°$, $c = 116$ in. or $A = 124.1°$, $C = 25.9°$, $c = 50.6$ in.
     7. $c = 5.6$ m, $A = 92°$, $B = 47°$ 9. $b = 690$ ft, $A = 62.4°$, $C = 15.6°$
   11. $C = 94.2°$, $A = 37.5°$, $B = 48.3°$ 13. $b = 10.3$ m, $A = 45.4°$, $C = 34.9°$
   15. $B = 149.9°$, $A = 7.1°$, $C = 23.0°$

B.   1.   41 ft   3. $27.2°$, $52.1°$, $100.7°$
     5.   19.1 ft   7.   25.4 in.

Problem Set 10, page 665

A.   1. $87°48'$ 3. $51°47'$ 5. $41.2°$ 7. $65.9°$
     9.   0.611   11.   1.300   13. $25.8°$ 15. $131.8°$

B.   1. $S = 16$ cm, $A = 206$ cm^2 3. $S = 13$ in., $A = 72$ in.^{2}

C.   1.   0.391   3.   0.070   5.   0.720   7.   0.956
     9. $14°$ 11. $67°58'$ 13. $10°59'$ 15. $39°27'$
   17. $83°13'$ 19. $43.7°$ 21. $8.9°$ 23. $63.9°$
   25. $52.6°$

D.   1. $A = 9$, $B = 15.6$, $c = 30°$ 3. $g = 53°$, $h = 37°$
     5. $B = 61°$, $c = 37$ mm, $b = 32$ mm   7. $a = 5.14$ m, $A = 44.4°$, $B = 45.6°$
     9. $c = 5.55$ in., $A = 35.8°$, $B = 54.2°$ 11. $b = 9.81$ in., $A = 50.7°$, $B = 39.3°$

E.   1. $C = 76.0°$, $c = 19.2$ in., $b = 12.4$ in.
     3. $C = 52°$, $B = 91°$, $b = 430$ yd or $C = 128°$, $B = 15°$, $b = 110$ yd
     5. $B = 49.1°$, $C = 66.6°$, $c = 52.4$ m
     7. $A = 24.9°$, $B = 38.9°$, $b = 1730$ m

F.   1.   1.3 in.   3.   2 in.   5. (a) $X = 29$ ft 2 in.   (b) 58 ft 4 in.
     7. (a) $1°8'$ (b) 0.3 in.   9.   0.394 in.   11. 17.2 cm
   13.   180 rad/sec   15.   104,000 sq ft

Chapter 11   *Exercises 11-1, page 693*

A.   1. $x = 2$, $y = 8$ 3. $x = 4$, $y = 5$ 5. $x = 2$, $y = 4$
     7. $x = -1$, $y = 2$

B.   1. $x = 9$, $y = -4$ 3.   Inconsistent   5. $x = -2$, $y = -3$
     7. $x = -1$, $y = -1\frac{1}{2}$ 9.   Dependent   11. $x = 1$, $y = 5$

C. 1. $x + y = 39$ Solution: $x = 23$
     $x - y = 7$ $y = 16$

    3. $x + y = 20$ Solution: $x = 12$
        $2x = 3y$ $y = 8$

    5. $4L + 3P = 44$ Solution: $L = \$5$ (leather cleaner)
        $3L + 4P = 47$ $P = \$8$ (protectant)

    7. $10d + 5n = 100$ Solution: $n = 2$
        $d = n + 7$ $d = 9$

    9. $3P + 5C = 167$ Solution: $C = \$16$ (ink cartridge)
        $P = C + 13$ $P = \$29$ (case of paper)

   11. $x + y = 30$ Solution: $x = 24$ in.
        $x = 4y$ $y = 6$ in.

   13. $x + y = 6$ Solution: $x = 1.5$ liter (10%)
        $0.1x + 0.02y = 0.24$ $y = 4.5$ liter (2%)

   15. $x + y = 24$ Solution: $x = 16$ tons ($30 rock)
        $30x + 40y = 800$ $y = 8$ tons ($40 rock)

Exercises 11-2, page 708

A. 1. No   3. No   5. Yes
    7. $3x^2 - 7x + 14 = 0$ 9. $x^2 - 18x = 0$

B. 1. $x = 5$ or $x = -5$ 3. $x = 0$ or $x = 4\frac{2}{5}$
    5. $x = 4\frac{1}{2}$ or $x = -4\frac{1}{2}$ 7. $x \approx 0.76$ or $x \approx -15.76$
    9. $x \approx -2.35$ or $x \approx 0.85$ 11. $x \approx 4.14$ or $x \approx -3.14$
   13. $x \approx 0.55$ or $x \approx -0.26$ 15. $x \approx -2.37$ or $x \approx -0.63$
   17. $x \approx 2.39$ or $x \approx 0.28$ 19. $x \approx 3.69$ or $x \approx 0.81$

C. 1. 25 mm   3. 5 in. by 15 in.   5. 17 in. by 22 in.
    7. 7.14 in.   9. (a) 3.76 in.   (b) 4.56 in.
   11. 8 in.   13. 37.2 ft   15. 434 gal/min

Problem Set 11, page 713

A. 1. $x = 15, y = 3$ 3. $x = 3, y = 2\frac{1}{2}$ 5. $x = 7, y = 6$
    7. $x = 1, y = 1$ 9. $x = \frac{74}{23}, y = -\frac{4}{23}$

B. 1. $x = 3$ or $x = -3$ 3. $x \approx -1.43$ or $x \approx -0.23$
    5. $x = 0$ or $x = 6$ 7. $x = 5$ or $x = -1$
    9. $x = 6$ or $x = -6$

C. 1. $x + y = 38$ 3. $x - y = 21$ 5. $L = 4W$
       $x - y = 14$ $5y - 2x = 33$ $4W^2 = 125$
       $x = 26$ $x = 46$ $W \approx 5.59$ cm
       $y = 12$ $y = 25$ $L \approx 22.36$ cm
    7. (a) $I = 13$ amp   (b) $I = 12.25$ amp
    9. $x + y = 42$ 11. $d = 15.96$ in.   13. 12.5 in.
       $x = 3y - 2$
       $y = 11$ in.
       $x = 31$ in.
   15. $x + y = 50$ 17. 2.55 in.   19. 41 mph
       $120x + 160y = 6400$
       $x = 40$ yards (of California Gold)
       $y = 10$ yards (of Palm Springs Gold)

A. 1. (a) 2005   (b) approximately 330,000   (c) 200,000
        (d) 300%   (e) approximately 5%
    3. (a) 9 Btu/Wh   (b) the mid-range model
        (c) approximately 65%   (d) approximately 16% (from 9.5 to 11)
    5. (a) 45   (b) 172   (c) 50%   (d) 258
    7. (a) 2007; approximately 800,000   (b) any two of these: 1996, 1998,
                                                   2001, 2002, 2003, 2005
        (c) lowest: 1991, highest: 2005   (d) approximately 4,000,000
        (e) approximately 82%
    9. (a) tuition   (b) lab fee   (c) $850   (d) $540

B. 1.

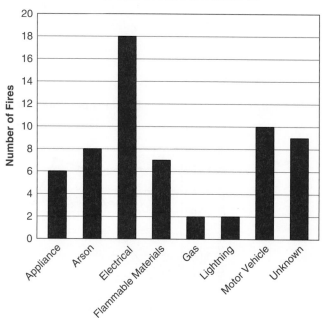

Causes of Fires in District 12

3.

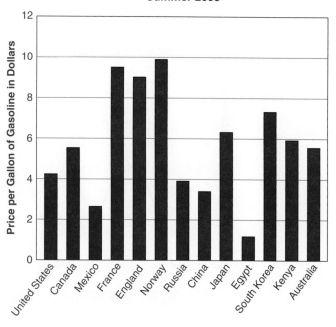

Worldwide Prices per Gallon of Regular Gas:
Summer 2008

5.

Yearly Sales of Construction Material for CASH IS US

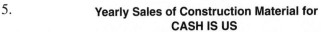
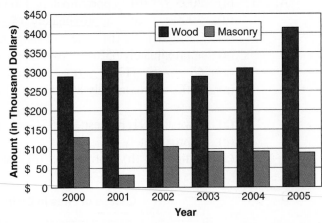

7.

Earnings per Share for XYZ Co.

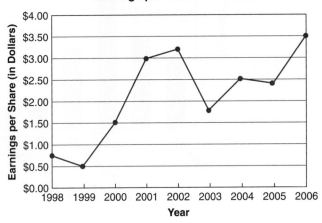

9.

Total Acres Burned in U.S. Wildfires: 1998–2007

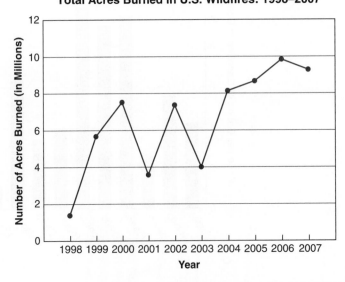

Answers to Odd-numbered Problems

11.

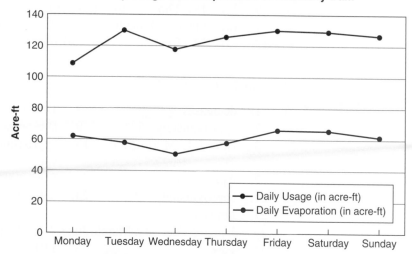

Daily Usage and Evaporation at Bradbury Dam

13.

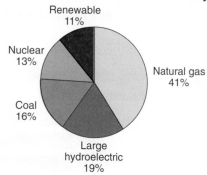

California's In-State Sources of Electricity

15.

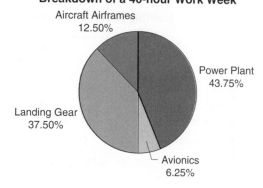

Aviation Mechanics Percent Breakdown of a 40-hour Work Week

Exercises 12-2, page 749

A.   *Mean*   *Median*   *Mode*

	Mean	Median	Mode
1.	6	6.5	4
3.	71.7	67.5	54
5.	151.8	156	none
7.	1254.5	1250.5	none

B.   1.   1.548

C.   1.   (a)   37,000   (b)   37,000   (c)   37,125
             (d)   37,375   (e) $74,000,000
    3.   (a)   2930.8 lb   (b)   2915 lb
    5.   (a)   13.89 hr   (b)   13.55 hr   (c) $243.22

7. (a) 438.4 kAF   (b) 348 kAF   (c) 13.8%
9. (a) 187.2 hp   (b) 199.5 hp
11. (a) Mean: 7.9, median: 8   (b) Mean: 7.1, median: 7
13. (a) 22.15 mi/gal   (b) $0.14

Problem Set 12, page 759

A. 1. decreases   3. 14   5. all circuits in the graph
     7. approximately 75,000   9. B.C.   11. B.C.
   13. H.V., P.T., M.M., A.G.   15. approximately 240,000 (Your answer may vary slightly.)
   17. April 2001   19. from 2004 to 2005 and from 2007 to 2008
   21. approximately 75%   23. $30,000   25. December
   27. January
   29. January, February, March, June, July, August, September, October, November
   31. 16%   33. 119   35. 0.73 g

B. 1.

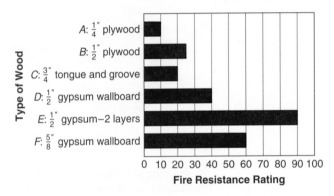

3.

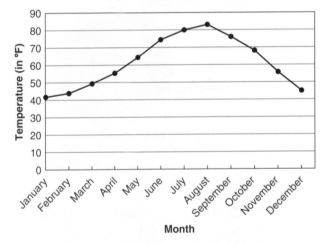

5.

C.

	Mean	Median	Mode
1.	86	85	84
3.	20.332	20.3345	20.324
5.	678.2	677	727

D.   1.   16.95 cu cm

E.   1.   (a)

Copies Requested	Wasted Impressions	Percent of Waste
15,500	409	2.6
17,500	345	2.0
18,000	432	2.4
16,750	546	3.3
14,000	316	2.3
16,250	845	5.2
18,500	538	2.9

   (b) 2.6%   (c) 2.96%   (d) Yes

3.   (a) $2.30   (b) $2.41   (c) $7,230,000

5.   (a)

Skid Length (in ft)	Speed (in mph)
514	56
294	42
726	66
350	46
216	36
600	60
476	53
384	48
486	54
564	58

   (b) 53.5 mph   (c) 51.9 mph

7.   (a) 0.0426   (b) 1.54685 mm

Calculator
Appendix

We have found that the majority of students using this text now use scientific calculators with two or more lines of display. Therefore, all calculator instructions and key sequences in the text reflect the way that these "multi-line" models operate. Generally speaking, on multi-line models, calculations can be entered exactly as they are written in textbooks or spoken aloud. Calculators with a one-line display window sometimes operate differently from multi-line models. For example, some calculations must be entered in the reverse order of the way they are written, and often times the $=$ key is not used to complete a calculation. For those students using these "single-line" models, we are providing this calculator appendix. Whenever the key sequences for your single-line model differ from the ones in the text, you will be alerted to consult this appendix by a reference number to the left of the calculator icon. For example, on page 355 in Chapter 6, you will see the following:

C6-1 $(-)$ **587** $+$ **368** $=$ → ▓▓▓▓▓▓ *−219.*

The "C6-1" indicates that an alternate key sequence for single-line calculators appears in this appendix as the first one in the Chapter 6 group.

Keep in mind that some of the key names on your calculator may be different from the ones that we use, and occasionally the operation may differ slightly. If any of our key sequences do not work on your calculator, consult your instruction manual for clarification.

Chapter 4

4-1 and **4-2**   Same sequences except do not press $=$ after $F \leftrightarrow D$.

4-3   **First,** find the discount and store it in memory.

   4 $\times$ **79.50** $\times$ **15** $\div$ **100** $=$ STO A → ▓▓▓▓▓▓ *47.7*

   Then, subtract it from the list price $318.

   318 $-$ RCL A $=$ → ▓▓▓▓▓▓ *270.3*

Chapter 5

5-1   Fraction: **.462** $\times$ **32** $=$ STO A → ▓▓▓▓▓▓ *14.784* Round to $\frac{15}{32}$.

   Error: **15** $-$ RCL A $=$ $\div$ **32** $=$ → ▓▓▓▓▓▓ *0.00675* or about 0.0068 in.

Chapter 6

6-1   587 [+/-] [+] 368 [=] → `-219.`

6-2 (a)   2675 [+/-] [+] 1437 [=] → `-1238.`

6-2 (b)   6.975 [+/-] [+] 5.2452 [+/-] [=] → `-12.2202`

6-2 (c)   .026 [+] .0045 [+/-] [=] → `0.0215`

6-2 (d)   683 [+/-] [+] 594 [+/-] [+] 438 [+] 862 [+/-] [=] → `-1701.`

6-3   26.75 [+/-] [−] 44.38 [+/-] [−] 16.96 [=] → `0.67`

6-4   3.25 [+/-] [×] 1.4 [+/-] [=] → `4.55`

6-5   3.75 [+/-] [÷] 6.4 [=] → `-0.5859375`

6-6   38 [x²] → `1444.`

6-7   25.4 [+/-] [x²] → `645.16`

6-8   .85 [+/-] [yˣ] 4 [=] → `0.52200625`

6-9 (a)   68 [x²] → `4624.`

6-9 (c)   38 [+/-] [x²] → `1444.`

6-9 (e)   .475 [+/-] [yˣ] 5 [=] → `-0.024180654` or −0.0242 rounded

6-10   237 [√] → `15.39480432` (The square root key may be labeled $\sqrt{x}$.)

6-11 (a)   760 [√] → `27.5680975` $\sqrt{760} \approx 27.6$

6-11 (b)   5.74 [√] → `2.39582971` $\sqrt{5.74} \approx 2.40$

6-12   38 [x²] [×] 3 [√] [÷] 4 [=] → `625.2703415`

6-13   17 [x²] [−] 15 [x²] [=] [√] → `8.`

6-14   12.4 [x²] [+] 21.6 [x²] [=] [√] → `24.90622412`

6-15   3 [×] 8.4 [x²] [×] 3 [√] [÷] 2 [=] → `183.3202575`

Chapter 7

7-1 (b)   9 [×] 160 [+/-] [÷] 5 [+] 32 [=] → `-256.`

7-2   2.75 [−] 14.25 [=] [÷] .2 [+/-] [=] → `57.5`

7-3   6 [EE] 6 [+/-] [÷] 4.8 [EE] 7 [=] → `1.25⁻¹³`

Chapter 8

8-1   927 [√] → `30.4466747`

8-2   36 [x²] [+] 20 [x²] [=] [√] → `41.18252056`

8-3   In feet: 8.5 [x²] [+] 9.75 [x²] [=] [√] → `12.93493332` ≈ 12.9 ft

In inches: [−] 12 [=] [×] 12 [=] → `11.21919984` ≈ 12 ft 11 in.

8-4 9 x^2 − 3 x^2 = √ × 6 ÷ 2 = → **25.45584412**

⬆ The Height ⬆ The Area

8-5 7 x^2 − 6 x^2 = √ × 12 ÷ 2 = → **21.63330765**

8-6 28 × (28 − 19) × (28 − 12) × (28 − 25) = √ =

→ **109.9818167**

8-7 π → **3.141592654**

Chapter 9

9-1 9 x^2 × 15 x^2 = √ + 9 x^2 + 15 x^2 = × 4 ÷ 3 =

→ **588.**

9-2 RCL A × RCL B = √ + RCL A + RCL B = × 7 ÷ 3 =

→ **428.8865**

9-3 π × 9.5 y^x 3 ÷ * 6 = → **448.9205002**

9-4 First, calculate the weight of the tank and store it in memory.

4 × π × 9 x^2 × 127 = STO → **129270.2545**

Then, calculate the weight of the water and add it to the tank weight in memory.

4 × π × 9 y^x 3 ÷ 3 × 62.4 + RCL = → **319816.6454**

Chapter 10

10-1 72.06 ▶DD → **72.1**

10-2 36.25 ▶DMS → **36°15′ 00″**

10-3 18.1 x^2 − 6.5 x^2 = √ → **16.89260193**

10-4 5 ÷ 2 √ = → **3.535533906**

10-5 38 sin → **0.615661475**

10-6 (a) 26 cos → **0.898794046**

10-6 (b) 67.8 tan → **2.450425198**

10-6 (c)–(i) Enter these in the same manner as problems (a) and (b): key in the angle measure first, then press the appropriate trig function key.

10-7 54 + 28 ÷ 60 = cos → **0.581176491**

10-8 .728 2nd sin^{-1} → **46.718988**

10-9 .296 2nd cos^{-1} → **72.78248857** ▶DMS → **72°46′ 56″9**

10-10 21 × 50 cos ** = → **13.4985398**

---

*After you press the ÷ key, there may be a slight delay while the calculator completes the y^x 3 calculation. If your calculator operates this way, do not continue until your screen shows a result.

**There may be a slight delay here while your calculator processes the cosine. Do not press = until your screen shows a result. This will happen whenever = or another operation follows a trig function key.

10-11 $10 \div 71$ [sin] [=] → `10.57620681`

10-12 $11.3 \div 16.5$ [=] [2nd] [tan⁻¹] → `34.40523429`

10-13   Same sequence as in text except eliminate [ANS] [=] at the end.

10-14   See 10-13.

10-15 $\underbrace{71 \; [+] \; 20 \; [\div] \; 60}$ [=] [tan] [×] 3.27 [=] → `9.679337948`

   or 71.20 [▸DD]

10-16 100 [×] 49 [tan] [=] → `115.0368407`

10-17 8 [×] 38 [sin] [÷] 47 [sin] [=] → `6.734486736`

10-18 10 [×] 30 [sin] [÷] 12 [=] [2nd] [sin⁻¹] → `24.62431835`

10-19 10 [×] 30 [sin] [÷] 4.5 [=] [2nd] [sin⁻¹] → `E.`

10-20 22.8 [x²] [+] 12.3 [x²] [−] 2 [×] 22.8 [×] 12.3 [×] 42 [cos] [=] [√]
   → `15.94725463`

10-21 12.3 [x²] [+] 15.95 [x²] [−] 22.8 [x²] [=] [÷] [(] 2 [×] 12.3 [×] 15.95 [)] [=] [2nd] [cos⁻¹]
   → `106.912924`

Chapter 11

11-1 33.3 [×] 5 [÷] 3 [=] [√] → `7.449832213`

11-2 7 [x²] [−] 4 [×] 3 [×] 5 [+/−] [=] [√] [STO] [A] [+] 7 [=] [÷] 6 [=] → `2.906717751`

   7 [−] [RCL] [A] [=] [÷] 6 [=] → `-0.573384418`

11-3 4 [×] 3 [÷] [π] [=] [√] → `1.954410048`

Protractor

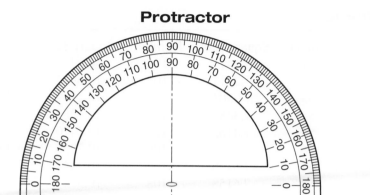

Metric Ruler

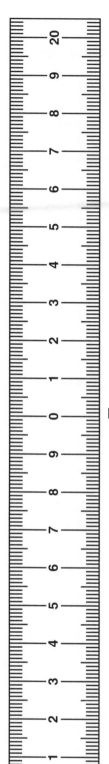

Multiplication Table

×	0	1	2	3	4	5	6	7	8	9	10
0	0	0	0	0	0	0	0	0	0	0	0
1	0	1	2	3	4	5	6	7	8	9	10
2	0	2	4	6	8	10	12	14	16	18	20
3	0	3	6	9	12	15	18	21	24	27	30
4	0	4	8	12	16	20	24	28	32	36	40
5	0	5	10	15	20	25	30	35	40	45	50
6	0	6	12	18	24	30	36	42	48	54	60
7	0	7	14	21	28	35	42	49	56	63	70
8	0	8	16	24	32	40	48	56	64	72	80
9	0	9	18	27	36	45	54	63	72	81	90
10	0	10	20	30	40	50	60	70	80	90	100

Table of Square Roots

Number	Square root	Number	Square root	Number	Square root	Number	Square root
1	1.0000	51	7.1414	101	10.0499	151	12.2882
2	1.4142	52	7.2111	102	10.0995	152	12.3288
3	1.7321	53	7.2801	103	10.1489	153	12.3693
4	2.0000	54	7.3485	104	10.1980	154	12.4097
5	2.2361	55	7.4162	105	10.2470	155	12.4499
6	2.4495	56	7.4833	106	10.2956	156	12.4900
7	2.6458	57	7.5498	107	10.3441	157	12.5300
8	2.8284	58	7.6158	108	10.3923	158	12.5698
9	3.0000	59	7.6811	109	10.4403	159	12.6095
10	3.1623	60	7.7460	110	10.4481	160	12.6491
11	3.3166	61	7.8102	111	10.5357	161	12.6886
12	3.4641	62	7.8740	112	10.5830	162	12.7279
13	3.6056	63	7.9373	113	10.6301	163	12.7671
14	3.7417	64	8.0000	114	10.6771	164	12.8062
15	3.8730	65	8.0623	115	10.7238	165	12.8452
16	4.0000	66	8.1240	116	10.7703	166	12.8841
17	4.1231	67	8.1854	117	10.8167	167	12.9228
18	4.2426	68	8.2462	118	10.8628	168	12.9615
19	4.3589	69	8.3066	119	10.9087	169	13.0000
20	4.4721	70	8.3666	120	10.9545	170	13.0384
21	4.5826	71	8.4261	121	11.0000	171	13.0767
22	4.6904	72	8.4853	122	11.0454	172	13.1149
23	4.7958	73	8.5440	123	11.0905	173	13.1529
24	4.8990	74	8.6023	124	11.1355	174	13.1909
25	5.0000	75	8.6603	125	11.1803	175	13.2288
26	5.0990	76	8.7178	126	11.2250	176	13.2665
27	5.1962	77	8.7750	127	11.2694	177	13.3041
28	5.2915	78	8.8318	128	11.3137	178	13.3417
29	5.3852	79	8.8882	129	11.3578	179	13.3791
30	5.4772	80	8.9443	130	11.4018	180	13.4164
31	5.5678	81	9.0000	131	11.4455	181	13.4536
32	5.6569	82	9.0554	132	11.4891	182	13.4907
33	5.7446	83	9.1104	133	11.5326	183	13.5277
34	5.8310	84	9.1652	134	11.5758	184	13.5647
35	5.9161	85	9.2195	135	11.6190	185	13.6015
36	6.0000	86	9.2736	136	11.6619	186	13.6382
37	6.0828	87	9.3274	137	11.7047	187	13.6748
38	6.1644	88	9.3808	138	11.7473	188	13.7113
39	6.2450	89	9.4340	139	11.7898	189	13.7477
40	6.3246	90	9.4868	140	11.8322	190	13.7840
41	6.4031	91	9.5394	141	11.8743	191	13.8203
42	6.4807	92	9.5917	142	11.9164	192	13.8564
43	6.5574	93	9.6437	143	11.9583	193	13.8924
44	6.6332	94	9.6954	144	12.0000	194	13.9284
45	6.7082	95	9.7468	145	12.0416	195	13.9642
46	6.7823	96	9.7980	146	12.0830	196	14.0000
47	6.8557	97	9.8489	147	12.1244	197	14.0357
48	6.9282	98	9.8995	148	12.1655	198	14.0712
49	7.0000	99	9.9499	149	12.2066	199	14.1067
50	7.0711	100	10.0000	150	12.2474	200	14.1421

Table of Trigonometric Functions

Angle	Sine	Cosine	Tangent	Angle	Sine	Cosine	Tangent
0°	0.0000	1.0000	0.0000	45°	0.7071	0.7071	1.000
1	0.0175	0.9998	0.0175	46	0.7193	0.6947	1.036
2	0.0349	0.9994	0.0349	47	0.7314	0.6820	1.072
3	0.0523	0.9986	0.0524	48	0.7431	0.6691	1.111
4	0.0698	0.9976	0.0699	49	0.7547	0.6561	1.150
5	0.0872	0.9962	0.0875	50	0.7660	0.6428	1.192
6	0.1045	0.9945	0.1051	51	0.7771	0.6293	1.235
7	0.1219	0.9925	0.1228	52	0.7880	0.6157	1.280
8	0.1392	0.9903	0.1405	53	0.7986	0.6018	1.327
9	0.1564	0.9877	0.1584	54	0.8090	0.5878	1.376
10	0.1736	0.9848	0.1763	55	0.8192	0.5736	1.428
11	0.1908	0.9816	0.1944	56	0.8290	0.5592	1.483
12	0.2079	0.9781	0.2126	57	0.8387	0.5446	1.540
13	0.2250	0.9744	0.2309	58	0.8480	0.5299	1.600
14	0.2419	0.9703	0.2493	59	0.8572	0.5150	1.664
15	0.2588	0.9659	0.2679	60	0.8660	0.5000	1.732
16	0.2756	0.9613	0.2867	61	0.8746	0.4848	1.804
17	0.2924	0.9563	0.3057	62	0.8829	0.4695	1.881
18	0.3090	0.9511	0.3249	63	0.8910	0.4540	1.963
19	0.3256	0.9455	0.3443	64	0.8988	0.4384	2.050
20	0.3420	0.9397	0.3640	65	0.9063	0.4226	2.145
21	0.3584	0.9336	0.3839	66	0.9135	0.4067	2.246
22	0.3746	0.9272	0.4040	67	0.9205	0.3907	2.356
23	0.3907	0.9205	0.4245	68	0.9272	0.3746	2.475
24	0.4067	0.9135	0.4452	69	0.9336	0.3584	2.605
25	0.4226	0.9063	0.4663	70	0.9397	0.3420	2.747
26	0.4384	0.8988	0.4877	71	0.9455	0.3256	2.904
27	0.4540	0.8910	0.5095	72	0.9511	0.3090	3.078
28	0.4695	0.8829	0.5317	73	0.9563	0.2924	3.271
29	0.4848	0.8746	0.5543	74	0.9613	0.2756	3.487
30	0.5000	0.8660	0.5774	75	0.9659	0.2588	3.732
31	0.5150	0.8572	0.6009	76	0.9703	0.2419	4.011
32	0.5299	0.8480	0.6249	77	0.9744	0.2250	4.331
33	0.5446	0.8387	0.6494	78	0.9781	0.2079	4.705
34	0.5592	0.8290	0.6745	79	0.9816	0.1908	5.145
35	0.5736	0.8192	0.7002	80	0.9848	0.1736	5.671
36	0.5878	0.8090	0.7265	81	0.9877	0.1564	6.314
37	0.6018	0.7986	0.7536	82	0.9903	0.1392	7.115
38	0.6157	0.7880	0.7813	83	0.9925	0.1219	8.144
39	0.6293	0.7771	0.8098	84	0.9945	0.1045	9.514
40	0.6428	0.7660	0.8391	85	0.9962	0.0872	11.43
41	0.6561	0.7547	0.8693	86	0.9976	0.0698	14.30
42	0.6691	0.7431	0.9004	87	0.9986	0.0523	19.08
43	0.6820	0.7314	0.9325	88	0.9994	0.0349	28.64
44	0.6947	0.7193	0.9657	89	0.9998	0.0175	57.29
45	0.7071	0.7071	1.0000	90	1.0000	0.0000	∞

Index

Index of Applications

Roofing

Chapter 1

Chapter 2

Chapter 3

Chapter 4

Chapter 5

Chapter 6

Chapter 7

Chapter 8

Chapter 9

Sheet Metal Trades

Chapter 2

Chapter 3

Chapter 4

Chapter 5

Chapter 6

Chapter 7

Chapter 8

Chapter 9

Chapter 10

Chapter 11

Chapter 12

Sports & Leisure

Chapter 1

Chapter 3

1